ANNALS OF THE NEW YORK ACADEMY OF SCIENCES

Volume 1003

EDITORIAL STAFF

Director, Publishing and New Media
SARAH GREENE

Managing Editor
JUSTINE CULLINAN

Associate Editor
STEVEN E. BOHALL

The New York Academy of Sciences
2 East 63rd Street
New York, New York 10021

THE NEW YORK ACADEMY OF SCIENCES
(Founded in 1817)

BOARD OF GOVERNORS, September 2003 – September 2004

TORSTEN N. WIESEL, *Chairman of the Board*
GERALD D. FISCHBACH, *Vice Chairman*
JOHN T. MORGAN, *Treasurer*
ELLIS RUBINSTEIN, *Chief Executive Officer* [ex officio]

Honorary Life Governors
WILLIAM T. GOLDEN   JOSHUA LEDERBERG

Governors

KAREN E. BURKE	PETER B. CORR	R. BRIAN FERGUSON
RONALD L. GRAHAM	MARNIE IMHOFF	WENDY EVANS JOSEPH
JACQUELINE LEO	RODERT W. LUCKY	PAUL MARKS
BRUCE McEWEN	RONAY MENSCHEL	JOHN F. NIBLACK
SANDRA PANEM	PETER RINGROSE	DAVID D. SABATINI
	LEE G. VANCE	DEBORAH WILEY

HELENE L. KAPLAN, *Counsel* [ex officio]   LARRY R. SMITH, *Secretary* [ex officio]

GLUTAMATE AND DISORDERS OF COGNITION AND MOTIVATION

ANNALS OF THE NEW YORK ACADEMY OF SCIENCES
Volume 1003

GLUTAMATE AND DISORDERS OF COGNITION AND MOTIVATION

Edited by Bita Moghaddam and Marina E. Wolf

The New York Academy of Sciences
New York, New York
2003

Copyright © 2003 by the New York Academy of Sciences. All rights reserved. Under the provisions of the United States Copyright Act of 1976, individual readers of the Annals are permitted to make fair use of the material in them for teaching or research. Permission is granted to quote from the Annals provided that the customary acknowledgment is made of the source. Material in the Annals may be republished only by permission of the Academy. Address inquiries to the Permissions Department (editorial@nyas.org) at the New York Academy of Sciences.

Copying fees: For each copy of an article made beyond the free copying permitted under Section 107 or 108 of the 1976 Copyright Act, a fee should be paid through the Copyright Clearance Center, Inc., 222 Rosewood Drive, Danvers, MA 01923 (www.copyright.com).

♾ The paper used in this publication meets the minimum requirements of the American National Standard for Information Sciences—Permanence of Paper for Printed Library Materials, ANSI Z39.48-1984.

Library of Congress Cataloging-in-Publication Data

Glutamate and disorders of cognition and motivation / edited by Bita Moghaddam and Marina E. Wolf.
    p. ; cm. — (Annals of the New York Academy of Sciences ; v. 1003)
Papers from a conference held on Apr. 13–15, 2003 in New Haven, Conn. by the New York Academy of Sciences and cosponsored by the National Institute of Mental Health and Eli Lilly and Company. Includes bibliographical references and index.
    ISBN 1-57331-476-5 (cloth : alk. paper) — ISBN 1-57331-477-3 (paper : alk. paper)
    1. Cognition-—Congresses. 2. Motivation (Psychology)—Congresses. 3. Glutamic acid—Physiological effect—Congresses.
    [DNLM: 1. Mental Disorders—etiology—Congresses. 2. Mental Disorders—physiopathology—Congresses. 3. Glutamic Acid—metabolism—Congresses. 4. Glutamic Acid—physiology—Congresses. 5. Receptors, Glutamate—physiology—Congresses. WM 140 G567 2003] I. Moghaddam, Bita. II. Wolf, Marina E. III. New York Academy of Sciences. IV. Title. V. Series. RC553.C64G588 2003 616.89'071—dc22

                                                                                                2003023023

GYAT/PCP
Printed in the United States of America
ISBN 1-57331-476-5 (cloth)
ISBN 1-57331-477-3 (paper)
ISSN 0077-8923

ANNALS OF THE NEW YORK ACADEMY OF SCIENCES

Volume 1003
November 2003

GLUTAMATE AND DISORDERS OF COGNITION AND MOTIVATION

Editors
BITA MOGHADDAM AND MARINA E. WOLF

This volume is the result of a conference entitled **Glutamate and Disorders of Cognition and Motivation** held on April 13–15, 2003 in New Haven, Connecticut by the New York Academy of Sciences and co-sponsored by the National Institute of Mental Health and Eli Lilly and Company.

CONTENTS

Dedication: Patricia S. Goldman-Rakic	xi
Preface. *By* BITA MOGHADDAM AND MARINA E. WOLF	xiii

Part I. Fundamental Aspects of Glutamate Neurotransmission

Synaptic Plasticity and AMPA Receptor Trafficking. *By* ROBERT C. MALENKA	1
Physiological Roles and Therapeutic Potential of Metabotropic Glutamate Receptors. *By* P. JEFFREY CONN	12
Genomics and Variation of Ionotropic Glutamate Receptors. *By* ROBERT H. LIPSKY AND DAVID GOLDMAN	22

Part II. Interactions between Glutamate and Monoamine Systems

Anatomical Substrates for Glutamate–Dopamine Interactions: Evidence for Specificity of Connections and Extrasynaptic Actions. *By* SUSAN R. SESACK, DAVID B. CARR, NATALIA OMELCHENKO, AND ALINE PINTO	36
Electrophysiological Interactions between Striatal Glutamatergic and Dopaminergic Systems. *By* ANTHONY R. WEST, STAN B. FLORESCO, ALI CHARARA, J. AMIEL ROSENKRANZ, AND ANTHONY A. GRACE	53

Part III. Glutamate and Schizophrenia

Molecular Abnormalities of the Glutamate Synapse in the Thalamus in Schizophrenia. *By* JAMES H. MEADOR-WOODRUFF, SARAH M. CLINTON, MONICA BENEYTO, AND ROBERT E. MCCULLUMSMITH	75

Glutamate Receptors and Transporters in the Hippocampus in Schizophrenia. *By* PAUL J. HARRISON, AMANDA J. LAW, AND SHARON L. EASTWOOD.... 94

Altered Cortical Glutamate Neurotransmission in Schizophrenia: Evidence from Morphological Studies of Pyramidal Neurons. *By* DAVID A. LEWIS, LEISA A. GLANTZ, JOSEPH N. PIERRI, AND ROBERT A. SWEET........... 102

Evaluating Glutamatergic Transmission in Schizophrenia. *By* CAROL A. TAMMINGA, ADRIENNE C. LAHTI, DEBORAH R. MEDOFF, XUE-MIN GAO, AND HENRY H. HOLCOMB....................................... 113

The NMDA Receptor Hypofunction Model of Psychosis. *By* NURI B. FARBER. 119

Glutamatergic Animal Models of Schizophrenia. *By* BITA MOGHADDAM AND MARK E. JACKSON... 131

Glutamate, Dopamine, and Schizophrenia: From Pathophysiology to Treatment. *By* MARC LARUELLE, LAWRENCE S. KEGELES, AND ANISSA ABI-DARGHAM... 138

Part IV. Glutamate and Addiction

Glutamate-Mediated Plasticity in Corticostriatal Networks: Role in Adaptive Motor Learning. *By* ANN E. KELLEY, MATTHEW E. ANDRZEJEWSKI, ANNE E. BALDWIN, PEPE J. HERNANDEZ, AND WAYNE E. PRATT........ 159

Glutamate Transmission and Addiction to Cocaine. *By* PETER W. KALIVAS, KRISTA MCFARLAND, SCOTT BOWERS, KAREN SZUMLINSKI, ZHENG-XIONG XI, AND DAVID BAKER............................. 169

NMDA Receptor Antagonism and the Ethanol Intoxication Signal: From Alcoholism Risk to Pharmacotherapy. *By* JOHN H. KRYSTAL, ISMENE L. PETRAKIS, EVGENY KRUPITSKY, CHRISTIAN SCHÜTZ, LOUIS TREVISAN, AND D. CYRIL D'SOUZA... 176

Short- and Long-Term Modulation of Synaptic Inputs to Brain Reward Areas by Nicotine. *By* ZARA M. FAGEN, HUIBERT D. MANSVELDER, J. RUSSEL KEATH, AND DANIEL S. MCGEHEE................................. 185

Glutamatergic Transmission in Opiate and Alcohol Dependence. *By* GEORGE ROBERT SIGGINS, GILLES MARTIN, MARISA ROBERTO, ZHIGUO NIE, SAMUEL MADAMBA, AND LUIS DE LECEA........................... 196

Exogenous and Endogenous Cannabinoids Control Synaptic Transmission in Mice Nucleus Accumbens. *By* DAVID ROBBE, GÉRARD ALONSO, AND OLIVIER J. MANZONI... 212

Plastic Control of Striatal Glutamatergic Transmission by Ensemble Actions of Several Neurotransmitters and Targets for Drugs of Abuse. *By* DAVID M. LOVINGER, JOHN G. PARTRIDGE, AND KA-CHOI TANG.............. 226

Mechanisms by which Dopamine Receptors May Influence Synaptic Plasticity. *By* MARINA E. WOLF, SIMONA MANGIAVACCHI, AND XIU SUN.......... 241

Part V. Glutamate and Mood Disorders

Glutamate and Depression: Clinical and Preclinical Studies. *By* IAN A. PAUL AND PHIL SKOLNICK.. 250

Regulation of Cellular Plasticity Cascades in the Pathophysiology and Treatment of Mood Disorders: Role of the Glutamatergic System. *By* CARLOS A. ZARATE JR., JING DU, JORGE QUIROZ, NEIL A. GRAY, KIRK D. DENICOFF, JASKARAN SINGH, DENNIS S. CHARNEY, AND HUSSEINI K. MANJI... 273

Clinical Studies Implementing Glutamate Neurotransmission in Mood Disorders. *By* GERARD SANACORA, DOUGLAS L. ROTHMAN, GRAEME MASON, AND JOHN H. KRYSTAL................................. 292

Part VI. Glutamate-Based Pharmacotherapies

A Role for Noradrenergic Transmission in the Actions of Phencyclidine and the Antipsychotic and Antistress Effects of mGlu2/3 Receptor Agonists. *By* CHAD J. SWANSON AND DARRYLE D. SCHOEPP................... 309

Converging Evidence of NMDA Receptor Hypofunction in the Pathophysiology of Schizophrenia. *By* JOSEPH T. COYLE, GUOCHUAN TSAI, AND DONALD GOFF.. 318

Glutamatergic Agents for Cocaine Dependence. *By* CHARLES DACKIS AND CHARLES O'BRIEN.. 328

Part VII. Poster Papers

Dopamine–Acetylcholine Interactions in the Modulation of Glutamate Release. *By* MARCO ATZORI, PATRICK KANOLD, JUAN CARLOS PINEDA, AND JORGE FLORES-HERNANDEZ................................. 346

N-Acetyl Cysteine–Induced Blockade of Cocaine-Induced Reinstatement. *By* DAVID A. BAKER, KRISTA MCFARLAND, RUSSELL W. LAKE, HUI SHEN, SHIGENOBU TODA, AND PETER W. KALIVAS....................... 349

AMPA- and NMDA-Associated Postsynaptic Protein Expression in the Human Dorsolateral Prefrontal Cortex. *By* MONICA BENEYTO AND JAMES H. MEADOR-WOODRUFF................................. 352

AGS3: A G-Protein Regulator of Addiction-Associated Behaviors. *By* M. S. BOWERS, R. W. LAKE, K. MCFARLAND, Y. K. PETERSON, S. M. LANIER, C. C. LAPISH, AND P. W. KALIVAS.................... 356

Changes in Electrophysiological Properties of Nucleus Accumbens Neurons Depend on the Extent of Behavioral Sensitization to Chronic Methamphetamine. *By* ANNE MARIE BRADY, STANLEY D. GLICK, AND PATRICIO O'DONNELL... 358

Evaluation of NMDA Receptors *in Vivo* in Schizophrenic Patients with [^{123}I]CNS 1261 and SPET: Preliminary Findings. *By* RODRIGO A. BRESSAN, KJELL ERLANDSSON, RACHEL S. MULLIGAN, ROGER N. GUNN, VINCENT J. CUNNINGHAM, JONATHAN OWENS, PETER J. ELL, AND LYN S. PILOWSKY.. 364

Effects of Naloxone-Precipitated Morphine Withdrawal on Glutamate-Mediated Signaling in Striatal Neurons *in Vitro*. *By* ELENA H. CHARTOFF, MARIA PAPADOPOULOU, CHRISTINE KONRADI, AND WILLIAM A. CARLEZON JR.. 368

Opposite Effects of GluR1 and PKA-Resistant GluR1 Overexpression in the Ventral Tegmental Area on Cocaine Reinforcement. *By* KWANG-HO CHOI, ZIA RAHMAN, SCOTT EDWARDS, STEPHANIE HALL, RACHAEL L. NEVE, AND DAVID W. SELF 372

Expression of ARHGEF11 mRNA in Schizophrenic Thalamus. *By* GENOVEVA DAVIDKOVA, ROBERT E. MCCULLUMSMITH, AND JAMES H. MEADOR-WOODRUFF 375

Structurally Dissimilar Antimanic Agents Modulate Synaptic Plasticity by Regulating AMPA Glutamate Receptor Subunit GluR1 Synaptic Expression. *By* JING DU, NEIL A. GRAY, CYNTHIA FALKE, PEIXIONG YUAN, STEVEN SZABO, AND HUSSEINI K. MANJI 378

Cocaine-Induced Expression Differences in Glutamate Receptor Subunits and Transporters in Amygdalae of Taste Aversion–Prone and Taste Aversion–Resistant Rats. *By* R. L. ELKINS, T. E. ORR, J. L. RAUSCH, Y. J. FEI, G. F. CARL, S. H. HOBBS, J. J. BUCCAFUSCO, AND G. L. EDWARDS 381

Cocaine-Induced Expression Differences in PSD-95/SAP-90–Associated Protein 4 and in Ca^{2+}/Calmodulin-Dependent Protein Kinase Subunits in Amygdalae of Taste Aversion–Prone and Taste Aversion–Resistant Rats. *By* R. L. ELKINS, T. E. ORR, J. L. RAUSCH, Y. J. FEI, G. F. CARL, S. H. HOBBS, J. J. BUCCAFUSCO, AND G. L. EDWARDS 386

Rapid AMPAR/NMDAR Response to Amphetamine: A Detectable Increase in AMPAR/NMDAR Ratios in the Ventral Tegmental Area Is Detectable after Amphetamine Injection. *By* L. J. FALEIRO, S. JONES, AND J. A. KAUER 391

Nucleus Accumbens Homer Proteins Regulate Behavioral Sensitization to Cocaine. *By* M. BEHNAM GHASEMZADEH, LINDSAY K. PERMENTER, RUSSELL W. LAKE, AND PETER W. KALIVAS 395

Altered Prefrontal Cortex–Nucleus Accumbens Information Processing in a Developmental Animal Model of Schizophrenia. *By* YUKIORI GOTO AND PATRICIO O'DONNELL 398

Lithium Regulates Total and Synaptic Expression of the AMPA Glutamate Receptor GluR2 *in Vitro* and *in Vivo*. *By* NEIL A. GRAY, JING DU, CYNTHIA S. FALKE, PEIXIONG YUAN, AND HUSSEINI K. MANJI 402

Prefrontal Group II Metabotropic Glutamate Receptor Activation Decreases Performance on a Working Memory Task. *By* MARY L. GREGORY, NICHOLAS E. STECH, RUSSELL W. OWENS, AND PETER W. KALIVAS 405

The Effects of Selective Orbitofrontal Cortex Lesions on the Acquisition and Performance of Cue-Controlled Cocaine Seeking in Rats. *By* DANIEL M. HUTCHESON AND BARRY J. EVERITT 410

In Vivo Characterization of Changes in Glycine Levels Induced by GlyT1 Inhibitors. *By* KIRK W. JOHNSON, AMY CLEMENS-SMITH, GEORGE NOMIKOS, RICHARD DAVIS, LEE PHEBUS, HARLAN SHANNON, PATRICK LOVE, KEN PERRY, JASON KATNER, FRANK BYMASTER, HONG YU, AND BETH J. HOFFMAN 412

Metabotropic Glutamate 5 Receptor Antagonist MPEP Decreased Nicotine and Cocaine Self-Administration but Not Nicotine and Cocaine-Induced Facilitation of Brain Reward Function in Rats. *By* P. J. KENNY, N. E. PATERSON, B. BOUTREL, S. SEMENOVA, A. A. HARRISON, F. GASPARINI, G. F. KOOB, P. D. SKOUBIS, AND A. MARKOU 415

Elucidation of Homer 1a Function in the Nucleus Accumbens Using Adenovirus Gene Transfer Technology. *By* M. S. BOWERS, R. W. LAKE, S. RUBINCHIK, J-Y. DONG, AND P. W. KALIVAS 419

Glutamate/Monoamine Interactions in the Limbic Thalamus. *By* ANTONIETA LAVIN 422

Changes in NMDA Receptor Subunit mRNAs and Cyclophilin mRNA during Development of the Human Hippocampus. *By* AMANDA J. LAW, CYNTHIA SHANNON WEICKERT, MAREE J. WEBSTER, MARY M. HERMAN, JOEL E. KLEINMAN, AND PAUL J. HARRISON 426

Blockade of the GlyT1 Glycine Transporter Prolongs Response to VTA Stimulation in Nucleus Accumbens Neurons. *By* BARBARA LEWIS AND PATRICIO O'DONNELL 431

Modulation of Inhibitory Transmission in the Rat Globus Pallidus by Activation of mGluR4. *By* MICHAEL J. MARINO, ORNELLA VALENTI, JULIE A. O'BRIEN, DAVID L. WILLIAMS JR., AND P. JEFFREY CONN 435

Expression of Transcripts for the Vesicular Glutamate Transporters in the Human Medial Temporal Lobe. *By* ROBERT E. MCCULLUMSMITH AND JAMES H. MEADOR-WOODRUFF 438

Metabotropic Glutamate Receptor Regulation of Extracellular Glutamate Levels in the Prefrontal Cortex. *By* ROBERTO I. MELENDEZ AND PETER W. KALIVAS 443

Cystine/Glutamate Antiporter Regulation of Vesicular Glutamate Release. *By* MEGAN M. MORAN, ROBERTO MELENDEZ, DAVID BAKER, PETER W. KALIVAS, AND JEREMY K. SEAMANS 445

Expression of the NR3A Subunit of the NMDA Receptor in Human Fetal Brain. *By* HELENA T. MUELLER AND JAMES H. MEADOR-WOODRUFF 448

Coupling of Glutamatergic Neurotransmission and Neuronal Glucose Oxidation over the Entire Range of Cerebral Cortex Activity. *By* ANANT B. PATEL, ROBIN A. DE GRAAF, GRAEME F. MASON, DOUGLAS L. ROTHMAN, ROBERT G. SHULMAN, AND KEVIN L. BEHAR 452

Real Time *in Vivo* Measures of L-Glutamate in the Rat Central Nervous System Using Ceramic-Based Multisite Microelectrode Arrays. *By* F. POMERLEAU, B. K. DAY, P. HUETTL, J. J. BURMEISTER, AND G. A. GERHARDT 454

TRH and Related Peptides: Homeostatic Regulators of Glutamate Transmission? *By* A. SATTIN A. E. PEKARY, R. L. LLOYD, M. PAULSON, J. A. MEYERHOFF, P. M. HINKLE, AND K. FAULL 458

L-Homocysteine Sulfinic Acid and L-Homocysteic Acid Stimulate Phosphoinositide Hydrolysis in Rat Cortical Neurons. *By* QI SHI, SANDRA J. HUFEISEN, JARDA T. WROBLEWSKI, JOSEPH H. NADEAU, AND BRYAN L. ROTH 461

Distinct Contributions of Glutamate Receptor Subtypes to Cognitive Set-Shifting Abilities in the Rat. *By* MARK R. STEFANI AND BITA MOGHADDAM 464

Evidence for a Relationship between Group 1 mGluR Hypofunction and Increased Cocaine and Ethanol Sensitivity in Homer2 Null Mutant Mice. *By* KAREN K. SZUMLINSKI, SHIGENOBU TODA, LAWRENCE D. MIDDAUGH, PAUL F. WORLEY, AND PETER W. KALIVAS 468

Bidirectional Modulation of Cystine/Glutamate Exchanger Activity in Cultured Cortical Astrocytes. *By* XING-CHUN TANG AND PETER W. KALIVAS 472

Dopamine–Glutamate Interactions in the Control of Cell Excitability in Medial Prefrontal Cortical Pyramidal Neurons from Adult Rats. *By* KUEI-YUAN TSENG AND PATRICIO O'DONNELL 476

Modulation of Excitatory Transmission onto Midbrain Dopaminergic Neurons of the Rat by Activation of Group III Metabotropic Glutamate Receptors. *By* ORNELLA VALENTI, MICHAEL J. MARINO, AND P. JEFFREY CONN 479

Difference in mGluR5 Interaction between Positive Allosteric Modulators from Two Structural Classes. *By* DAVID L. WILLIAMS JR., JULIE A. O'BRIEN, WEI LEMAIRE, TSING-BAU CHEN, RAYMOND S. L. CHANG, MARLENE A. JACOBSON, SOOKHEE N. HA, DAVID D. WISNOSKI, CRAIG W. LINDSLEY, CYRILLE SUR, MARK E. DUGGAN, DOUGLAS J. PETTIBONE, AND P. JEFFREY CONN 481

Index of Contributors 485

Financial assistance was received from:

Co-sponsors
- NATIONAL INSTITUTE OF MENTAL HEALTH—NATIONAL INSTITUTES OF HEALTH
- ELI LILLY AND COMPANY

Contributors
- MERCK RESEARCH LABORATORIES
- PFIZER GLOBAL RESEARCH AND DEVELOPMENT

The New York Academy of Sciences believes it has a responsibility to provide an open forum for discussion of scientific questions. The positions taken by the participants in the reported conferences are their own and not necessarily those of the Academy. The Academy has no intent to influence legislation by providing such forums.

Dedication

PATRICIA S. GOLDMAN-RAKIC

Patricia S. Goldman-Rakic (1937–2003) honored us with her presence at the conference in New Haven, Connecticut in April 2003 that provided the basis for this volume of the *Annals*. With a profound sense of loss, we dedicate this volume to her memory.

Preface

BITA MOGHADDAM[a] AND MARINA E. WOLF[b]

[a]Department of Neuroscience, University of Pittsburgh,
Pittsburgh, Pennsylvania 15260, USA

[b]Department of Neuroscience, Finch University of Health Sciences/The Chicago Medical School, North Chicago, Illinois 60064-3095, USA

For several decades, monoamine systems have been the major focus of research on cognitive, affective, and addictive disorders. In the late 1980s, a small number of investigators, including the organizers of this conference, began studying the possible role of glutamate in these disorders. This was a logical reflection of important developments in the understanding of glutamate neurotransmission. On the most basic level, the central importance of glutamate in cortical function forces consideration of glutamate transmission in any theory of pathology or therapeutic intervention for disorders with cognitive or affective components. Another important consideration is the primary role played by glutamate-containing projections, mainly originating from cortical and limbic regions, in determining the activity of monoaminergic neurons. Finally, the emergence of an understanding of the cellular basis of glutamate-dependent forms of plasticity over the past decade set the stage for evaluating glutamate's roles in disorders suspected to involve either abnormal plasticity during development (e.g., schizophrenia) or maladaptive plasticity during adulthood (e.g., addiction).

Despite considerable initial skepticism in the early days, the importance of glutamate transmission in cognitive and motivational disorders is now widely accepted. An exciting period is underway in which basic research is beginning to be translated into clinical developments. It therefore seemed an appropriate time to review the many important developments that have taken place in this field to date and consider the directions most likely to yield significant progress over the next decade.

The conference that forms the basis for the present volume took place in April 2003 in New Haven, Connecticut. Its objective was to bring together diverse groups of basic and clinical researchers who study the role of glutamate in cognitive disorders, such as schizophrenia, and disorders relating to motivation and affect, such as depression and addiction. By facilitating interactions between basic and clinical researchers, the conference aimed to accelerate the development of glutamate-based therapeutic approaches to these disorders and to enable basic scientists to design experiments with greater clinical relevance. An additional goal was to promote interactions between basic scientists with different research interests who have traditionally emphasized different aspects of glutamate transmission. For example, drug-abuse researchers have emphasized interactions between drugs and glutamate-dependent forms of neuronal plasticity, whereas schizophrenia researchers have focused more on neurochemical and anatomical measurements in animal models or alterations in glutamate receptor expression in postmortem human tissue.

The conference began with a series of plenary presentations, including one by Nobel Prize Laureate Paul Greengard. These sessions were designed to establish common ground by reviewing fundamental aspects of normal glutamate-mediated signaling and glutamate–monoamine interactions. They set the stage for the core of the meeting, which consisted of several sessions organized around glutamate's role in specific brain disorders: schizophrenia, drug addiction, and mood disorders. These presentations encompassed a wide range of methodological approaches, for example, behavioral, electrophysiological, neurochemical, and imaging studies. The meeting concluded with a session on glutamate-based therapeutic approaches for brain disorders. Because glutamate mediates fast synaptic neurotransmission throughout the central nervous system, targeting glutamate receptors has generally not been considered a feasible option for chronic treatment of psychiatric disorders. However, recent characterization of modulatory sites on ionotropic (ion channel–gating) glutamate receptors, as well as the discovery of at least eight subtypes of metabotropic glutamate receptors (which, similar to monoamine receptors, "modulate" rather than "mediate" neurotransmission), has suggested exciting potential approaches for altering glutamate neurotransmission in a functionally and regionally selective manner. Presentations in this final session highlighted progress toward the design of novel treatments based on our emerging understanding of the role of glutamate systems in schizophrenia, drug addiction, and mood disorders.

This volume retains the organization of the conference. The sections correspond to the conference sessions. Poster summaries are grouped at the end.

Synaptic Plasticity and AMPA Receptor Trafficking

ROBERT C. MALENKA

Nancy Friend Pritzker Laboratory, Department of Psychiatry and Behavioral Sciences, Stanford University School of Medicine, Palo Alto, California 94304-5485, USA

ABSTRACT: Alterations in neuronal activity can elicit long-lasting changes in the strength of synaptic transmission at excitatory synapses and, as a consequence, may underlie many forms of experience-dependent plasticity, including learning and memory. The best-characterized forms of such synaptic plasticity are the long-term depression (LTD) and long-term potentiation (LTP) observed at excitatory synapses in the CA1 region of the hippocampus. It is now well accepted that the trafficking of AMPA receptors to and away from the synaptic plasma membrane plays an essential role in both LTP and LTD, respectively. Here we review current models of AMPA receptor trafficking and how this trafficking may be regulated at the molecular level in order to produce the observed changes in synaptic strength. We also review recent work from our lab suggesting that synaptic plasticity in the mesolimbic dopamine system may contribute importantly to the neural adaptations elicited by drugs of abuse.

KEYWORDS: AMPA receptor; synaptic transmission; LTD; LTP; VTA; nucleus accumbens

INTRODUCTION

Long-lasting, activity-dependent changes in synaptic strength at excitatory synapses are thought to be critical for virtually all forms of experience-dependent plasticity, including learning and memory. Among the most widely studied and accepted models of synaptic plasticity in the mammalian brain are the long-term depression (LTD) and long-term potentiation (LTP) that are generated at excitatory synapses on hippocampal CA1 pyramidal cells. Both of these synaptic phenomena share the characteristic that they are triggered by a rise in postsynaptic calcium concentration due to activation of NMDA receptors.[1] Presumably different properties of the postsynaptic calcium signal, primarily its magnitude and perhaps its time course, activate different postsynaptic signaling cascades that lead to either LTP or LTD.[2]

Over the last 20 years, the specific cellular and molecular mechanisms responsible for LTP and LTD have been the object of intense study and debate. While much remains to be elucidated, recent studies have provided compelling evidence that

Address for correspondence: Robert C. Malenka, 1201 Welch Rd., Rm. P105, Stanford University School of Medicine, Palo Alto, CA 94304-5485. Voice: 650-724-2730; fax: 650-724-2753.
malenka@stanford.edu

alterations in synaptic activity lead to the regulated trafficking of AMPA receptors both to and from synapses and that this importantly contributes to the changes in synaptic strength during LTP and LTD. Here, we will briefly review this rapidly moving field. For more comprehensive reviews, see References 3–5.

AMPA RECEPTOR PROTEIN–PROTEIN INTERACTIONS

AMPA receptors are a subclass of ionotropic glutamate receptors found at virtually all excitatory synapses. They are multimeric protein assemblies likely consisting of combinations of four different subunits termed GluR1–4 or GluRA–D.[6–8] In

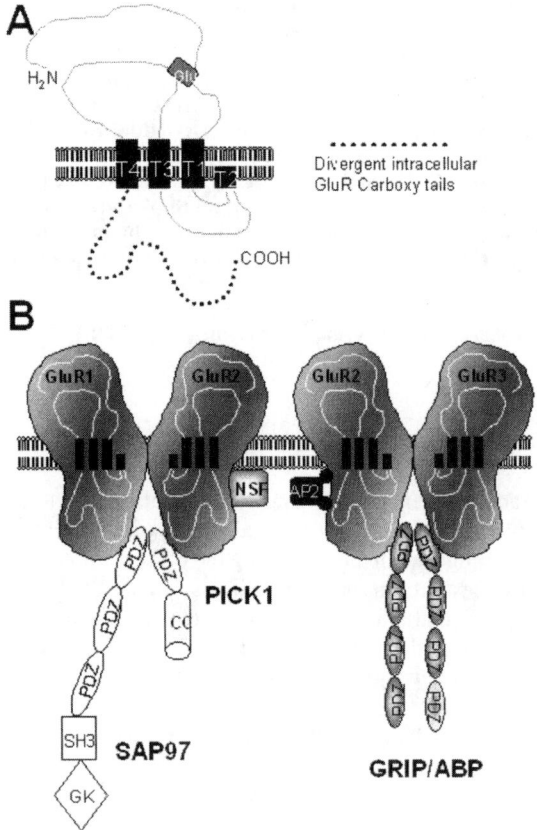

FIGURE 1. Diagrams showing the main protein–protein interactions of AMPA receptor subunits. (**A**) Schematic of membrane topology of AMPA receptor subunits. All subunits have a large extracellular N-terminus and an intracellular C-terminus of varying length. The ligand binding site is also shown. (**B**) The major forms of endogenous AMPA receptors (GluR1/2 and GluR2/3 heteromers) are shown with the domain structures of their main binding partners. GluR1 interacts with the PDZ protein SAP97. GluR2/3 interacts with the PDZ proteins GRIP/ABP and PICK1. GluR2 also interacts with NSF. The clathrin adaptor complex AP2 binds to GluR2 as well as to all other subunits.

the mature hippocampus, two types of AMPA receptors appear to dominate: heteromers consisting of GluR1 and 2, or GluR2 and 3.[9] The topology of each subunit is similar, consisting of four transmembrane domains (one of which does not completely transverse the membrane) and a large extracellular N-terminus (FIG. 1). Importantly, the individual subunits have intracellular C-termini that are unique in that they interact with different intracellular binding partners. As a consequence, the detailed subunit composition of AMPA receptors may greatly influence their trafficking and surface expression.

Many of the known protein–protein interactions of AMPA receptors involve intracellular proteins that contain so-called PDZ domains.[10] GluR1 appears to interact specifically with the PDZ containing protein SAP97, while GluR2 and GluR3 interact through a different class of PDZ domain with GRIP, ABP, and PICK1[4,5,11] (FIG. 1). Although many details remain to be worked out, the interactions of AMPA receptors with these various PDZ containing proteins appear to importantly influence both the targeting and clustering of AMPA receptors to specific subcellular regions as well as the stabilization of AMPA receptors both on the neuronal cell surface and in intracellular pools. Thus these interactions are thought to play key roles in determining how AMPA receptors are restricted from synaptic sites and how they may be delivered in an activity-dependent manner in order to respond to changes in synaptic inputs and elicit long-lasting changes in synaptic strength.

Another protein that plays an important role in AMPA receptor trafficking and hence synaptic plasticity is NSF (NEM-sensitive factor), which was originally identified as an ATPase that is required for membrane fusion processes during intracellular protein trafficking and presynaptic vesicle exocytosis.[12] NSF interacts with the intracellular C-terminus of GluR2 and is thought to be important for the delivery of AMPA receptors to the synaptic plasma membrane and/or their stabilization within the membrane.[4,5]

CONSTITUTIVE CYCLING OF AMPA RECEPTORS

An interesting aspect of the trafficking of membrane proteins is that they are often capable of undergoing constitutive as well as regulated insertion and removal. It appears that AMPA receptors also exhibit both forms of trafficking and that their basal rate of insertion and removal as well as the activity-dependent modulation of their trafficking are dependent upon their subunit composition (see below). The evidence for constitutive cycling derives from the observation that loading CA1 pyramidal cells with peptides that presumably inhibit specific protein–protein interactions involving AMPA receptors causes a "run-up" or "run-down" of synaptic responses. For example, a peptide that interferes with the NSF-GluR2 interaction causes a run-down of EPSCs as does NEM itself.[13,14] Conversely, peptides that interfere with the GluR2-GRIP/ABP interaction can cause a run-up of EPSCs[15,16] as do inhibitors of dynamin,[17] a GTPase that is involved in many forms of endocytosis (see below). Examination of the trafficking of recombinant receptor subunits in cultured hippocampal neurons has also yielded results that are consistent with the idea that AMPA receptors constitutively cycle into and out of the synaptic plasma membrane with a time course of minutes.[4,5] Thus it has become accepted that a significant proportion of synaptic AMPA receptors are continually being replaced via this mechanism.

The movement of AMPA receptors from the cytosol to the dendritic plasma membrane and subsequently to the synapse may involve an additional critical protein termed *stargazin*.[18,19] Stargazin was originally identified as the mutant protein that is responsible for the neurologic deficits in the stargazer mutant mouse. Cerebellar granule cells prepared from these mice lack functional synaptic and extrasynaptic AMPA receptors on their surface. Acute expression of stargazin in these mutant granule cells rescues both synaptic responses and the response to exogenous application of glutamate, indicating that stargazin is required for the normal surface expression of AMPA receptors. Biochemical and mutagenesis studies revealed that stargazin interacts with both AMPA receptor subunits and PDZ proteins, including PSD-95, which was originally identified based on its interaction with NMDA receptors.[20] Interestingly, a mutant form of stargazin lacking the PDZ-binding domain still is capable of delivering AMPA receptors to the cell surface but not to synapses. This and other observations have led to a model proposing that stargazin-AMPA receptor interactions are critical for the delivery of AMPA receptors to the cell surface and that a subsequent interaction between stargazin and PSD-95 (or other synaptic PDZ proteins) is required for their movement into synapses.[18,19]

AMPA RECEPTOR TRAFFICKING AND LTP

Almost two decades ago Lynch and Baudry proposed that LTP involved an increase in the number of synaptic glutamate receptors.[21] This idea was largely ignored over the next 10 years as a vigorous debate concerning whether LTP was primarily due to pre- or postsynaptic modifications took place. The idea was resurrected when electrophysiological evidence for the existence of "silent synapses" was presented.[22,23] Silent synapses are synapses that contain no or very small numbers of AMPA receptors but an easily detectable complement of NMDA receptors. Thus they are functionally silent at normal resting membrane potentials. Evidence was also presented that LTP involved the conversion of "silent" to "functional" synapses presumably due to the delivery of AMPA receptors to the silent synapses and their insertion into the synaptic plasma membrane. Consistent with this idea, loading cells with inhibitors of membrane fusion events was found to block LTP.[24]

The "silent synapse hypothesis" subsequently stimulated a major effort by a large number of groups to determine whether AMPA receptor trafficking contributed to LTP and LTD and the molecular mechanisms responsible for this activity-dependent trafficking. For LTP an important assumption of this model is that there are sufficient nonsynaptic stores of AMPA receptors that can be delivered rapidly to synapses. Indeed, both light and electron microscopic studies have provided evidence for intracellular stores of AMPA receptors as well as the existence of extrasynaptic receptors in the plasma membrane.[4,5]

There is also now reasonably strong evidence in support of the hypothesis that LTP involves the insertion of new AMPA receptors into the synaptic plasma membrane. In cultured hippocampal neurons, pharmacological manipulations that mimic LTP induction cause a rapid increase in the level of surface expression of AMPA receptors.[25,26] Similarly, in hippocampal slice cultures, a GFP-GluR1 fusion protein was "delivered" to dendritic spines in an activity- and NMDA receptor–dependent manner.[27] Most convincingly, when overexpressed in slice cultures, GluR1 forms homomeric channels that can be detected electrophysiologically when they are inserted into the synapse because, unlike normal endogenous synaptic AMPA receptors, they exhibit a profound inward rectification. Normally these overexpressed receptor subunits are not found at synapses, but, in response to LTP-inducing stimuli, they are rapidly incorporated into synapses such that they respond to synaptically released glutamate.[28] This delivery can also be triggered by overexpression of an active form of CaMKII,[28] a protein kinase known to be required for the triggering of LTP.[29] Consistent with the importance of GluR1 for the synaptic delivery of AMPA receptors during LTP, overexpression of a portion of the GluR1 intracellular tail, which presumably acts as a dominant negative inhibitor of GluR1 protein–protein interactions, blocked LTP.[30] Furthermore, knockout mice lacking GluR1 do not express LTP, at least in mature hippocampus.[31] Surprisingly, the critical substrate for the CaMKII-dependent delivery of AMPA receptors does not appear to be GluR1 itself but some other as yet unidentified protein.[28]

While GluR1 appears to be required for the activity-dependent delivery of AMPA receptors during LTP, GluR2/3 subunits may play a complementary role in the constitutive delivery pathway. Specifically, it has been proposed that the "rules" governing AMPA receptor trafficking are related to the length of the carboxyl tail. AMPA receptors with "long" cytoplasmic tails, such as GluR1 as well as GluR4, may normally be restricted from the synapse but can be delivered to the synapse during periods of enhanced synaptic activity (i.e., as would be seen during LTP). In contrast, GluR2/3 AMPA receptors may continually replace preexisting synaptic AMPA receptors in an activity-independent manner.[4,5] Indeed the molecular stoichiometry of AMPA receptors may influence whether they are inserted directly into the postsynaptic density (GluR2/3) or rather first get inserted into adjacent extrasynaptic plasma membrane and then diffuse laterally within the plasma membrane to synaptic sites (GluR1).[32,33]

A final additional complexity is the observation that early during postnatal development, LTP in the hippocampus does not require CaMKII or GluR1.[34,35] Instead LTP requires activation of PKA,[35] which appears to cause the delivery of GluR4, the expression of which is developmentally regulated with robust expression in early postnatal hippocampus and minimal expression in older hippocampus.[36]

LONG-TERM DEPRESSION AND AMPA RECPTOR ENDOCYTOSIS

An obvious corollary of the hypothesis that LTP involves synaptic delivery of AMPA receptors is that LTD involves the removal or loss of synaptic AMPA receptors. That this, in fact, can occur was first demonstrated using epitope-tagged recombinant AMPA receptor subunits and chronic manipulations of activity.[37] Subsequently, it was found that pharmacological activation of either AMPA receptors or NMDA receptors with agonist can elicit robust internalization/endocytosis of AMPA receptors.[38] This was shown to be a dynamin- and clathrin-dependent process[38] and likely involves direct binding of the AP2 clathrin adaptor complex to AMPA receptor subunits themselves.[39]

That AMPA receptor endocytosis does indeed play a critical role in LTD is supported by several lines of evidence. First, generation of LTD in cultured hippocampal neurons is accompanied by a decrease in the number of synaptic AMPA receptors.[40]

Second, manipulations that cause the loss of surface AMPA receptors, such as the injection of the NSF inhibitory peptide, prevent the subsequent expression of LTD.[13,41] Third, loading the cell with reagents that inhibit dynamin-dependent endocytosis blocks hippocampal LTD.[13] Interestingly, LTD at other synaptic connections also appears to involve endocytosis of AMPA receptors, specifically at parallel fiber-Purkinje cell synapses in the cerebellum[42] and excitatory synapses on dopamine cells in the ventral tegmental area.[43] These data suggest that AMPA receptor trafficking may be a fairly universal mechanism by which activity can modify synaptic strength.

In terms of intracellular signaling mechanisms, the regulated endocytosis of AMPA receptors in hippocampal neurons appears to share properties with synaptically evoked LTD. Both depend on rises in postsynaptic calcium concentration and are blocked by pharmacological inhibition of the calcium-dependent protein phosphatase calcineurin as well as protein phosphatase 1.[1,44,45] The critical required targets of these protein phosphatases are unknown. One hypothesis is that calcineurin acts on components of the endocytic machinery and enhances their functions.[3] Another obvious possibility is that dephosphorylation of specific AMPA receptor subunits is required for AMPA receptor endocytosis. Consistent with this idea, AMPA receptors appear to be dephosphorylated following LTD,[46] and a mouse expressing a mutant form of GluR1 with its major phosphorylation sites deleted does not express LTD.[47] However, the mechanisms controlling AMPA receptor endocytosis appear to be much more complicated, in that LTD in the cerebellum and in the ventral tegmental area both require AMPA receptor endocytosis but are due, at least in part, to activation of PKC and PKA, respectively.[42,43] Thus there may be multiple mechanisms by which AMPA receptor endocytosis can be enhanced, perhaps because of the existence of different subsets of binding partners in different cell types.

Because LTD is a long-lasting and presumably stable form of synaptic plasticity, an important question is whether the endocytosed AMPA receptors are targeted for degradation via the lysosome or are eventually recycled back to the cell surface. A study that measured the fate of surface biotinylated AMPA receptors in cultured hippocampal neurons found that whether internalized AMPA receptors were degraded or recycled back to the plasma membrane was influenced by the stimulus that triggered the endocytosis in the first place.[45] This observation lends additional complexity to the trafficking of AMPA receptors but also provides additional sites that may be subject to activity-dependent modulation.

SYNAPTIC PLASTICITY AND DRUGS OF ABUSE

Although AMPA receptor trafficking is clearly implicated in phenomena such as hippocampal LTP and LTD, which are primarily studied *in vitro*, a critical question is whether these mechanisms actually play a role *in vivo* in mediating various forms of experience-dependent plasticity. It has been shown that LTP and LTD generated *in vivo* in the hippocampus are accompanied by decreases and increases, respectively, in the level of AMPA receptors that were measured biochemically in fractions enriched in synaptic membranes.[48] More recently, using virally mediated overexpression of AMPA receptor subunits *in vivo* in somatosensory (barrel) cortex, it was found that alterations in sensory experience could stimulate the synaptic delivery and incorporation of the recombinant AMPA receptors.[49] This provides important evidence that the mechanisms of AMPA receptor trafficking elucidated from the study of reduced *in vitro* preparations likely apply to the intact, functioning brain.

We have attempted to examine whether synaptic plasticity mechanisms analogous to hipppocampal LTP and LTD occur *in vivo* by examining the synaptic changes elicited by drugs of abuse. A prominent hypothesis in the addiction field is that cellular mechanisms, such as synaptic plasticity, which are used during adaptive forms of experience-dependent plasticity, may also be important for the neural adaptations generated by acute and chronic exposure to drugs of abuse.[50,51] An important advantage of using administration of drugs of abuse as a model for experience-dependent plasticity is that the critical sites of action that are responsible for mediating the addictive properties of drugs of abuse have been extensively studied and clearly include the mesolimbic dopamine system, the chief components of which are the nucleus accumbens (NAc) and the ventral tegmental area (VTA). Thus we explored the hypothesis that *in vivo* exposure to drugs of abuse may elicit changes in synaptic strength at excitatory synapses in these brain regions due to synaptic plasticity mechanisms.

We first examined excitatory synaptic responses recorded from the VTA dopamine neurons in slices prepared from cocaine-treated animals and obtained evidence that a single *in vivo* exposure to cocaine caused a significant increase in synaptic strength using a mechanism that involved modification of AMPA receptors.[52] We also found evidence that this increase may share mechanisms with LTP. More recently we have found that not only cocaine but multiple different classes of drugs of abuse (i.e., amphetamine, morphine, nicotine, and ethanol) all cause an increase in strength at excitatory synapses on midbrain dopamine neurons when administered *in vivo*.[53] Importantly, nonabused psychoactive drugs, such as fluoxetine and carbamazepine, did not cause a change. Because stress has a profound facilitatory effect on the initiation and reinstatement of drug self-administration,[54] the effect of an acute stress was examined and, like drugs of abuse, was also found to cause an increase in synaptic strength in midbrain dopamine cells.[53]

These results suggest that plasticity at excitatory synapses on dopamine cells may be a key neural adaptation contributing to addiction and its interactions with stress. Since external stimuli that are associated with the firing of midbrain dopamine cells are granted high appetitive or motivational significance, we would suggest that by increasing synaptic drive onto these cells, drugs of abuse or stress enhance the motivational significance of drugs themselves as well as stimuli closely associated with drug seeking and self-administration. The detailed molecular mechanisms that are responsible for the observed synaptic changes are unknown. One intriguing hypothesis is that they are due to an LTP-like mechanism that involves the trafficking of GluR1-containing AMPA receptors to synapses. Indeed, overexpression of GluR1 in the VTA has been found to enhance the locomotor-stimulatory and rewarding properties of morphine.[55] This and several other observations have led to the specific hypothesis that elevated levels of GluR1 in the midbrain are an important trigger for the behavioral sensitization elicited by drugs of abuse.[56]

Because some of the long-term behavioral sequella of chronic administration of drugs of abuse clearly involve the NAc,[51] we also performed a similar study that involved preparing slices of this structure from animals who had been administered cocaine *in vivo*. In contrast to the changes observed in the VTA, chronic *in vivo* administration of cocaine was required to elicit detectable effects. Specifically, we examined synaptic strength at excitatory synapses in NAc slices that were prepared 10–14 days after repeated (5-day) *in vivo* administration of cocaine—a treatment that caused robust behavioral sensitization.[57] Neurons in the shell, but not the core region of NAc slices prepared from the cocaine-treated animals, showed a decrease in strength at excitatory synapses made by prelimbic cortical afferents. LTD was also diminished in these slices, suggesting that the decrease was due to mechanisms shared with LTD. As is the case for the synaptic changes in the VTA, the detailed mechanisms responsible for this drug-induced synaptic plasticity in the NAc are unclear. One intriguing hypothesis is suggested by the finding that persistent upregulation of the ΔFosB transcription factor, which is known to occur following repeated cocaine treatment, induces NAc expression of the AMPA receptor subunit GluR2.[58] Due to conductance differences in GluR2-containing versus non–GluR2-containing AMPA receptors, increases in GluR2 expression could potentially reduce AMPA receptor-mediated responses. This scenario, however, depends on the existence of a significant population of non–GluR2-containing synaptic AMPA receptors prior to cocaine exposure. If such a population existed, it should be identifiable due to the strong inward rectification of non-GluR2-containing receptors. Although this hypothesis remains intriguing, preliminary studies have failed to detect inward-rectifying AMPA receptor-mediated responses in NAc medium spiny neurons, suggesting that GluR2 incorporation does not explain the cocaine-induced depression. Another possibility is suggested by the fact that the acute administration of amphetamine to slices blocks LTP in the NAc.[59] This effect disappears in slices prepared from animals that have been repeatedly exposed to amphetamine. If this also occurs after *in vivo* cocaine exposure, such an action could initially enhance the likelihood of generating LTD.

CONCLUSIONS

We have briefly reviewed some of the evidence supporting the idea that activity-dependent modulation of the trafficking of AMPA receptors plays an important role in the expression of NMDA receptor–dependent LTP and LTD in the hippocampus (FIG. 2). On the basis of ultrastructural examination of synapses in the hippocampus (see, e.g., Ref. 60) and work in other systems, such as the neuromuscular junction,[61] it is attractive to speculate that the activity-dependent regulation of the number of synaptic AMPA receptors is an initial step in a more comprehensive and long-lasting reorganization of the entire ultrastructure of the synapse, processes that may involve the production of new synaptic connections and the pruning away of preexisting ones.[62]

We have also reviewed evidence that *in vivo* administration of drugs of abuse causes changes in excitatory synaptic strength in the NAc and VTA, two main components of the mesolimbic dopamine system, and that this may occur due to activation of the mechanisms that underlie LTP and LTD in these structures. While this latter work is still in its infancy, it, we hope, illustrates that the powerful *in vivo* effects of drugs of abuse may be a valuable model for studying the role of synaptic plasticity in mediating experience-dependent plasticity. Indeed, it is already apparent that, like other forms of experience-dependent plasticity, such as learning and

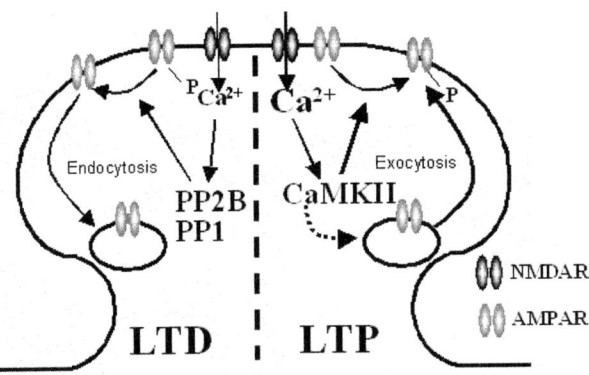

FIGURE 2. Simplified model of the intracellular pathways involved in LTD and LTP. LTD is triggered by a modest rise in calcium that activates protein phosphatase 2B (calcineurin) and protein phosphatase 1. This leads to the endocytosis of synaptic AMPA receptors as well as to their dephosphorylation. LTP is triggered by a large rise in calcium that activates CaMKII. This causes the delivery (exocytosis) of intracellular AMPA receptors to the synapse. CaMKII may also phosphorylate AMPA receptors directly, although this may not be required for their synaptic delivery.

memory, persistent drug-induced behavioral changes likely occur, in part, because of their ability to elicit long-lasting changes in synaptic weights in crucial brain circuits. Furthermore, it is important to note that the mesolimbic dopamine system did not evolve to respond to drugs of abuse but rather plays very important roles in adaptive behaviors, including various types of learning and memory. Thus examining the neural adaptations elicited by drugs of abuse will not only inform us about the pathophysiology of addiction but will also provide important information about how neural circuit modifications in the NAc and VTA contribute to normal, motivated behavior.

REFERENCES

1. MALENKA, R.C. 1994. Synaptic plasticity in the hippocampus: LTP and LTD. Cell **78:** 535–538.
2. MALENKA, R.C. & R.A. NICOLL. 1993. NMDA-receptor-dependent synaptic plasticity: multiple forms and mechanisms. Trends Neurosci. **16:** 521–527.
3. CARROLL, R.C. et al. 2001. Role of AMPA receptor endocytosis in synaptic plasticity. Nature Rev. Neurosci. **2:** 315–324.
4. SONG, I. & R.L. HUGANIR. 2002. Regulation of AMPA receptors during synaptic plasticity. Trends Neurosci **25:** 578–588.
5. MALINOW, R. & R.C. MALENKA. 2002. AMPA receptor trafficking and synaptic plasticity. Annu. Rev. Neurosci. **25:** 103–126.
6. WISDEN, W. & P.H. SEEBURG. 1993. Mammalian ionotropic glutamate receptors. Curr. Opin. Neurobiol. **3:** 291–298.
7. HOLLMANN, M. & S. HEINEMANN. 1994. Cloned glutamate receptors. Annu. Rev. Neurosci. **17:** 31–108.
8. ROSENMUND, C., Y. STERN-BACH & C.F. STEVENS. 1998. The tetrameric structure of a glutamate receptor channel. Science **280:** 1596–1599.

9. WENTHOLD, R.J. *et al.* 1996. Evidence for multiple AMPA receptor complexes in hippocampal CA1/CA2 neurons. J. Neurosci. **16:** 1982–1989.
10. SHENG, M. & C. SALA. 2001. PDZ domains and the organization of supramolecular complexes. Annu. Rev. Neurosci. **24:** 1–29.
11. HENLEY, J.M. 2003. Proteins interactions implicated in AMPA receptor trafficking: a clear destination and an improving route map. Neurosci. Res. **45:** 243–254.
12. ROTHMAN, J.E. 1994. Mechanisms of intracellular protein transport. Nature **372:** 55–63.
13. LÜSCHER, C. *et al.* 1999. Role of AMPA receptor cycling in synaptic transmission and plasticity. Neuron **24:** 649–658.
14. LUTHI, A. *et al.* 1999. Hippocampal LTD expression involves a pool of AMPARs regulated by the NSF-GluR2 interaction. Neuron **24:** 389–399.
15. DAW, M.I. *et al.* 2000. PDZ proteins interacting with C-terminal GluR2/3 are involved in a PKC-dependent regulation of AMPA receptors at hippocampal synapses. Neuron **28:** 873–886.
16. KIM, C.H. *et al.* 2001. Interaction of the AMPA receptor subunit GluR2/3 with PDZ domains regulates hippocampal long-term depression. Proc. Natl. Acad. Sci. USA **98:** 11725–11730.
17. LUSCHER, C. *et al.* 1999. Role of AMPA receptor cycling in synaptic transmission and plasticity. Neuron **24:** 649–658.
18. CHEN, L. *et al.* 2000. Stargazin regulates synaptic targeting of AMPA receptors by two distinct mechanisms. Nature **408:** 936–943.
19. SCHNELL, E. *et al.* 2002. Direct interactions between PSD-95 and stargazin control synaptic AMPA receptor number. Proc. Natl. Acad. Sci. USA **99:** 13902–13907.
20. KORNAU, H.C. *et al.* 1995. Domain interaction between NMDA receptor subunits and the postsynaptic density protein PSD-95. Science **269:** 1737–1740.
21. LYNCH, G. & M. BAUDRY. 1984. The biochemistry of memory: a new and specific hypothesis. Science **224:** 1057–1063.
22. ISAAC, J.T., R.A. NICOLL & R.C. MALENKA. 1995. Evidence for silent synapses: implications for the expression of LTP. Neuron **15:** 427–434.
23. LIAO, D., N.A. HESSLER & R. MALINOW. 1995. Activation of postsynaptically silent synapses during pairing-induced LTP in CA1 region of hippocampal slice. Nature **375:** 400–404.
24. LLEDO, P.M. *et al.* 1998. Postsynaptic membrane fusion and long-term potentiation. Science **279:** 399–403.
25. LU, W. *et al.* 2001. Activation of synaptic NMDA receptors induces membrane insertion of new AMPA receptors and LTP in cultured hippocampal neurons. Neuron **29:** 243–254.
26. PICKARD, L. *et al.* 2001. Transient synaptic activation of NMDA receptors leads to the insertion of native AMPA receptors at hippocampal neuronal plasma membranes. Neuropharmacology **41:** 700–713.
27. SHI, S.H. *et al.* 1999. Rapid spine delivery and redistribution of AMPA receptors after synaptic NMDA receptor activation. Science **284:** 1811–1816.
28. HAYASHI, Y. *et al.* 2000. Driving AMPA receptors into synapses by LTP and CaMKII: requirement for GluR1 and PDZ domain interaction. Science **287:** 2262–2267.
29. MALENKA, R.C. & R.A. Nicoll. 1999. Long-term potentiation—a decade of progress? Science **285:** 1870–1874.
30. SHI, S. *et al.* 2001. Subunit-specific rules governing AMPA receptor trafficking to synapses in hippocampal pyramidal neurons. Cell **105:** 331–343.
31. ZAMANILLO, D. *et al.* 1999. Importance of AMPA receptors for hippocampal synaptic plasticity but not for spatial learning. Science **284:** 1805–1811.
32. PASSAFARO, M., V. PIECH & M. SHENG. 2001. Subunit-specific temporal and spatial patterns of AMPA receptor exocytosis in hippocampal neurons. Nature Neurosci. **4:** 917–926.
33. BORGDORFF, A.J. & D. CHOQUET. 2002. Regulation of AMPA receptor lateral movements. Nature **417:** 649–653.
34. MACK, V. *et al.* 2001. Conditional restoration of hippocampal synaptic potentiation in GluR-A-deficient mice. Science **292:** 2501–2504.

35. YASUDA, H. et al. 2003. A developmental switch in the signaling cascades for LTP induction. Nature Neurosci. **6:** 15–16.
36. ESTEBAN, J.A. et al. 2003. PKA phosphorylation of AMPA receptor subunits controls synaptic trafficking underlying plasticity. Nature Neurosci. **6:** 136–143.
37. LISSIN, D.V. et al. 1998. Activity differentially regulates the surface expression of synaptic AMPA and NMDA glutamate receptors. Proc. Natl. Acad. Sci. USA **95:** 7097–7102.
38. CARROLL, R.C. et al. 1999. Dynamin-dependent endocytosis of ionotropic glutamate receptors. Proc. Natl. Acad. Sci. USA **96:** 14112–14117.
39. LEE, S.H. et al. 2002. Clathrin adaptor AP2 and NSF interact with overlapping sites of GluR2 and play distinct roles in AMPA receptor trafficking and hippocampal LTD. Neuron **36:** 661–674.
40. CARROLL, R.C. et al. 1999. Rapid redistribution of glutamate receptors contributes to long-term depression in hippocampal cultures. Nature Neurosci. **2:** 454–460.
41. LÜTHI, A. et al. 1999. Hippocampal LTD expression involves a pool of AMPARs regulated by the NSF-GluR2 interaction. Neuron **24:** 389–399.
42. WANG, Y.T. & D.J. LINDEN. 2000. Expression of cerebellar long-term depression requires postsynaptic clathrin-mediated endocytosis. Neuron **25:** 635–647.
43. GUTLERNER, J.L. et al. 2002. Novel protein kinase A-dependent long-term depression of excitatory synapses. Neuron **36:** 921–931.
44. BEATTIE, E.C. et al. 2000. Regulation of AMPA receptor endocytosis by a signaling mechanism shared with LTD. Nature Neurosci. **3:** 1291–1300.
45. EHLERS, M.D. 2000. Reinsertion or degradation of AMPA receptors determined by activity-dependent endocytic sorting. Neuron **28:** 511–525.
46. LEE, H.-K. et al. 2000. Regulation of distinct AMPA receptor phosphorylation sites during bidirectional synaptic plasticity. Nature **405:** 955–959.
47. LEE, H.K. et al. 2003. Phosphorylation of the AMPA receptor GluR1 subunit is required for synaptic plasticity and retention of spatial memory. Cell **112:** 631–643.
48. HEYNEN, A.J. et al. 2000. Bidirectional, activity-dependent regulation of glutamate receptors in the adult hippocampus in vivo. Neuron **28:** 527–536.
49. TAKAHASHI, T., K. SVOBODA & R. MALINOW. 2003. Experience strengthening transmission by driving AMPA receptors into synapses. Science **299:** 1585–1588.
50. HYMAN, S.E. & R.C. MALENKA. 2001. Addiction and the brain: the neurobiology of compulsion and its persistence. Nature Rev. Neurosci. **2:** 695–703.
51. EVERITT, B.J. & M.E. WOLF. 2002. Psychomotor stimulant addiction: a neural systems perspective. J. Neurosci. **22:** 3312–3320.
52. UNGLESS, M.A. et al. 2001. Single cocaine exposure in vivo induces long-term potentiation in dopamine neurons. Nature **411:** 583–587.
53. SAAL, D. et al. 2003. Drugs of abuse and stress trigger a common synaptic adaptation in dopamine neurons. Neuron **37:** 577–582.
54. PIAZZA, P.V. & M. LE MOAL. 1998. The role of stress in drug self-administration. Trends Pharmacol. Sci. **19:** 67–74.
55. CARLEZON, W.A., JR. et al. 1997. Sensitization to morphine induced by viral-mediated gene transfer. Science **277:** 812–814.
56. CARLEZON, W.A. & E.J. NESTLER. 2002. Elevated levels of GluR1 in the midbrain: a trigger for sensitization to drugs of abuse? Trends Neurosci. **25:** 610–615.
57. THOMAS, M.J. et al. 2001. Long-term depression in the nucleus accumbens: a neural correlate of behavioral sensitization to cocaine. Nature Neurosci. **4:** 1217–1223.
58. KELZ, M.B. et al. 1999. Expression of the transcription factor deltaFosB in the brain controls sensitivity to cocaine. Nature **401:** 272–276.
59. LI, Y. & J.A. KAUER. 2000. Amphetamine interferes with long-term potentiation in the nucleus accumbens. Soc. Neurosci. Abst. **26:** 1398.
60. TONI, N. et al. 1999. LTP promotes formation of multiple spine synapses between a single axon terminal and a dendrite. Nature **402:** 421–425.
61. SANES, J.R. & J.W. LICHTMAN. 2001. Induction, assembly, maturation and maintenance of a postsynaptic apparatus. Nature Rev. Neurosci. **2:** 791–805.
62. LÜSCHER, C. et al. 2000. Synaptic plasticity and dynamic modulation of the postsynaptic membrane. Nature Neurosci. **3:** 545–550.

Physiological Roles and Therapeutic Potential of Metabotropic Glutamate Receptors

P. JEFFREY CONN

Program in Translational Neuropharmacology, Department of Pharmacology, Vanderbilt University Medical Center, Nashville, Tennessee 37232-6600, USA

ABSTRACT: Discovery of mGlu receptors has dramatically influenced our understanding of glutamatergic neurotransmission in the central nervous system. This receptor family provides a mechanism by which activation by glutamate can regulate a number of important neuronal and glial functions that are not typically modulated by ligand-gated ion channels. This includes modulation of neuronal excitability, synaptic transmission, and various metabolic functions. Because of the ubiquitous distribution of glutamatergic synapses, discovery of the mGlu receptors immediately raised the likelihood that mGlu receptors would participate in most, if not all, major functions of the CNS. In addition, the wide diversity and heterogeneous distribution of mGlu receptor subtypes could provide an opportunity for development of pharmacological agents that selectively target specific CNS systems to achieve a therapeutic effect. Over the past decade, an increasing number of agonists and antagonists selective for specific mGlu receptor subtypes have been developed. Use of these pharmacological tools along with genetic approaches has led to major advances in our understanding of the roles of mGlu receptors in regulating CNS systems and animal behavior. These studies suggest that drugs active at mGlu receptors may be useful in treatment of a wide variety of neurological and psychiatric disorders.

KEYWORDS: metabotropic; glutamate receptor; mGluR; mGlu receptor; allosteric potentiator

The first evidence that the amino acid glutamate may serve as a neurotransmitter in the central nervous system (CNS) came in the late 1950s and early 1960s when glutamate and other acidic amino acids were found to excite a wide variety of central neurons. These findings spurred a massive research effort that eventually established glutamate as the primary excitatory neurotransmitter in the vertebrate CNS. Transmission at glutamatergic synapses was shown to involve the opening of ligand-gated cation channels (termed ionotropic glutamate or iGlu receptors) that were pharmacologically distinguished as NMDA, AMPA, and kainate receptor subtypes. It is now clear that glutamate plays a ubiquitous role as a fast excitatory transmitter and that

---

Address for correspondence: P. Jeffrey Conn, Program in Translational Neuropharmacology, Department of Pharmacology, Vanderbilt University Medical Center, 23rd Avenue South at Pierce, 452-B Preston Research Building, Nashville, Tennessee 37232-6600. Voice: 615-936-2478; fax: 615-343-6532.
    jeff.conn@vanderbilt.edu

most central neuronal circuits involve glutamatergic neurotransmission at some level.[1]

Until the mid-1980s, the actions of glutamate in the mammalian brain were thought to be mediated exclusively by activation of the glutamate-gated cation channels, which mediate the vast majority of fast excitatory synaptic transmissions in the mammalian brain. Fast synaptic transmission through networks of neurons can be modulated by activation of receptors coupled to second messenger systems through GTP-binding proteins. For instance, in a network of neurons connected by glutamatergic synapses, it was generally held that glutamate would elicit fast synaptic responses by activating members of the ionotropic glutamate receptor family. Neuromodulators from extrinsic afferents (e.g., acetylcholine, serotonin, and norepinephrine) could then modulate transmission through the network of glutamatergic neurons by activating GTP-binding protein-linked receptors and second messenger systems. Activation of G protein–coupled receptors (GPCRs) through extrinsic neu-

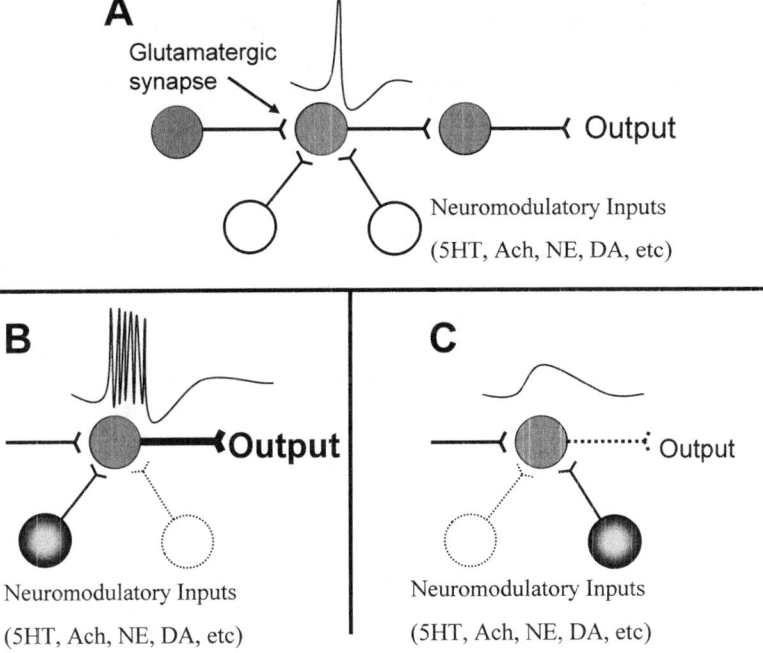

FIGURE 1. Traditional view of GPCR-mediated neuromodulation in the central nervous system. Before the discovery of the mGlu receptors, synaptic glutamate was thought to only elicit fast excitatory postsynaptic potentials (EPSPs). If EPSPs summated to reach action potential threshold, an action potential would be elicited (**A**). Neuromodulators could increase (**B**) or decrease (**C**) activity at a glutamatergic synapse by actions on neurotransmitter release or a variety of postsynaptic ion channels. These neuromodulators included serotonin (5HT), norepinephrine (NE), acetylcholine (ACh), dopamine (DA), as well as a number of other small molecules or neuroactive peptides. It is now known that mGlus can elicit the same neuromodulatory actions locally at the glutamatergic synapse.

romodulatory pathways can elicit a wide variety of effects in glutamatergic circuits, including increases or decreases in neurotransmitter release, changes in neuronal excitability, and changes in rate and patterns of action potential firing, among other changes. These modulatory effects can dramatically increase or decrease net transmission of information through a neuronal circuit (see FIG. 1).

Although there are a number of neurotransmitters that activate both ligand-gated ion channels and receptors coupled to second messenger systems, until recently, it was thought that all of the actions of glutamate were mediated by activation of ionotropic glutamate receptors and generation of fast synaptic responses. Because glutamate is the major excitatory neurotransmitter in the brain, separation of synapses involved in generation of fast synaptic responses and modulatory control of transmission through a neural circuit was thought to be the rule in the central nervous system, with relatively few examples of synapses in which a neurotransmitter elicits both fast and slow synaptic responses at a single synapse.

In the late 1970s and early 1980s, a handful of studies began to appear that now can be seen as providing evidence that glutamate may also have neuromodulatory effects that could be mediated by activation of GPCRs. For instance, Carl Cotman and colleagues first reported that L-2-amino-4-phosphonobutyric (L-AP4) inhibits transmission at excitatory synapses in the hippocampal formation.[2–4] Likewise, Watkins and coworkers reported that L-AP4 reduces evoked excitatory responses in the spinal cord.[5,6] These workers provided clear evidence that the effects of L-AP4 were not mediated by AMPA, kainate, or NMDA receptors and postulated mediation by a novel receptor subtype or other mechanism. However, the mechanism responsible for these effects of L-AP4 was not known, and these responses were generally thought to be mediated by antagonist actions of L-AP4 at iGlu receptors.

In the mid-1980s, direct evidence for the existence of glutamate receptors directly coupled to second messenger systems via G-proteins began to appear. Sladeczek *et al.*[7] were the first to report that glutamate analogues could activate receptors coupled to activation of phosphoinositide hydrolysis in striatal neurons. This was quickly followed by seminal papers by Nicoletti and coworkers demonstrating a similar response in hippocampal slices[8] and cerebellar granule cells,[9] and by a report by Sugiyama *et al.*[10] that expression of rat brain mRNA in *Xenopus* oocytes could lead to expression of a novel glutamate receptor that is coupled to activation of phosphoinositide hydrolysis.

One of the first major breakthroughs in clearly establishing the mGlu receptors as a unique pharmacological class came with the synthesis of *trans*-1-amino-1,3-cyclopentanedicarboxylate (*trans*-ACPD)[11] and the subsequent discovery that this compound selectively activates glutamate receptors coupled to activation of phosphoinositide hydrolysis while having no effect on the iGlu receptors.[12,13] These studies provided unequivocal evidence that activation of the phosphoinositide hydrolysis-coupled glutamate receptors did not require activation of any of the known iGlu receptor subtypes. This compound also provided a valuable tool to determine the physiological actions of mGlu receptor activation. Subsequent studies using *trans*-ACPD as a selective mGlu receptor agonist revealed that activation of these receptors induced modulatory effects in hippocampal neurons that were previously associated with activation of other families of GPCRs. Thus, *trans*-ACPD induced a slow depolarization, inhibited the AHP current, and blocked spike frequency adaptation in hippocampal pyramidal cells.[14] This was similar to an effect that was

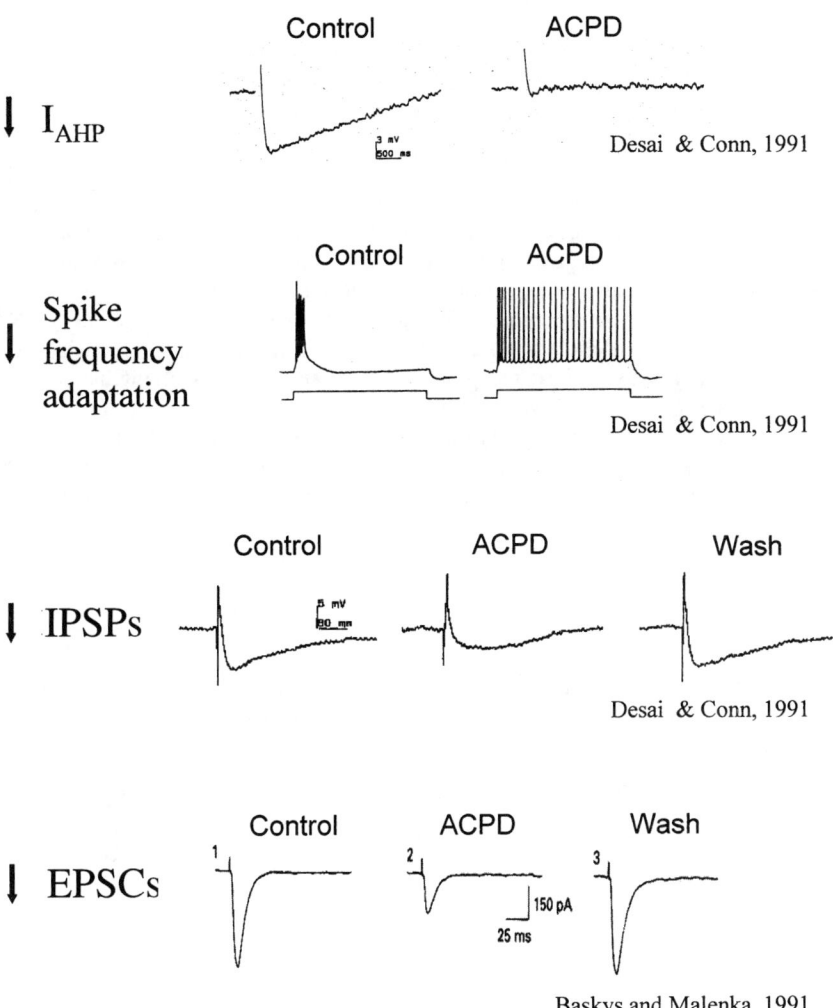

FIGURE 2. Neuromodulatory effects of *trans*-ACPD in hippocampal pyramidal cells. The selective mGlu receptor agonist *trans*-ACPD has a number of modulatory effects on hippocampal neurons that were previously associated with neuromodulators, such as acetylcholine and monoamines. These include a reduction of the slow afterhyperpolarization potassium current (I_{AHP}), blockade of spike frequency adaptation, and a reduction of both excitatory (EPSCs) and inhibitory (IPSCs) postsynaptic currents. (Modified from Desai & Conn[14] and Baskys & Malenka.[16])

observed with activation of these receptors by the less selective agonist quisqualate.[15] In addition, *trans*-ACPD reduced transmission at both excitatory[16] and inhibitory[14] synapses. These actions occurred in the absence of an increase in conductance, as occurs with activation of the iGlu receptors.

CLONING OF A FAMILY OF METABOTROPIC GLUTAMATE RECEPTORS

The first mGlu receptor cDNA was cloned independently by two groups using the same functional expression assay and is now generaly named mGlu1a.[17,18] Its deduced amino acid sequence revealed that this receptor shares no sequence homology with any other GPCR, suggesting that it could be a member of a new receptor gene family. Moreover, pharmacological studies suggested the existence of several G-protein–coupled glutamate receptors.[19] The search for mGlu receptor–related cDNA has now resulted in the isolation of seven other genes and several splice variants encoding mGlu receptors. These receptors are named mGlu1 through mGlu8.[19]

The 8 mGlu receptor subtypes are classified into three major groups on the basis of sequence homologies, coupling to second messenger systems, and selectivities for various agonists (FIG. 3).[19] mGlu receptors of the same group show about 70% sequence identity, whereas between groups this percentage falls to about 45%. Group I mGlu receptors, which include mGlu1 and mGlu5, couple primarily to Gq and increases in phosphoinositide hydrolysis. Group II mGlu receptors (mGlu2 and

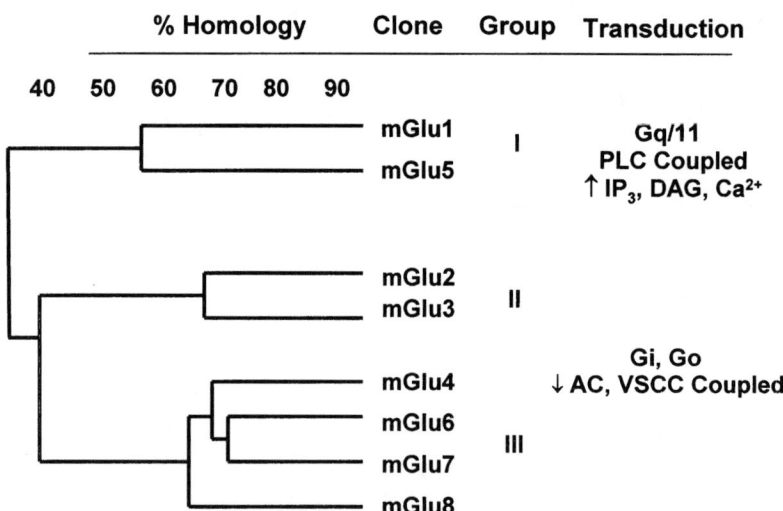

FIGURE 3. Dendrogram and classification of the members of the mGlu receptor family. The mGlu receptors are divided into three major groups, on the basis of sequence homology, pharmacological profile, and second messenger coupling.

mGlu3), and group III mGlu receptors (mGlu4, 6, 7, and 8) couple to Gi/o and associated effector pathways such as inhibition of adenylyl cyclase.

It is now clear that the mGlu receptors are widely distributed throughout the central nervous system and play important roles in regulating cell excitability and synaptic transmission.[19,20] One of the primary functions of the mGlu receptors is a role as presynaptic receptors involved in reducing transmission at glutamatergic synapses. The mGlu receptors also serve as heteroreceptors involved in reducing GABA release at inhibitory synapses. Finally, postsynaptically localized mGlu receptors often play an important role in regulating neuronal excitability and in regulating currents through ionotropic glutamate receptors. In general, the group I mGlu receptors most often serve as postsynaptic receptors involved in increasing neuronal excitability, whereas the group II and group III mGlu receptors often serve as presynaptic receptors involved in reducing neurotransmitter release. However, there are clear exceptions and examples of presynaptic group I mGlu receptors that can either increase or decrease transmitter release, and postsynaptic group I mGlu receptors that can hyperpolarize neurons. Also, there is clear evidence for postsynaptic group II and group III mGlu receptors.

The discovery of the mGlu receptors dramatically alters the traditional view of glutamatergic neurotransmission, since activation of mGlu receptors can modulate activity in glutamatergic circuits in a manner previously associated only with neuromodulators from nonglutamatergic afferents. However, unlike receptors for monoamines and other neuromodulators, the mGlu receptors provide a mechanism by which glutamate can modulate or fine tune activity at the same synapses at which it elicits fast synaptic responses. Because of the ubiquitous distribution of glutamatergic synapses, mGlu receptors have the potential of participating in a wide variety of functions of the CNS. In addition, the wide diversity and heterogeneous distribution of mGlu receptor subtypes provides an opportunity for developing pharmacological agents that selectively interact with mGlu receptors involved in only one or a limited number of CNS functions. Gaining a detailed understanding the specific roles of mGlu receptors could have a dramatic impact on development of novel treatment strategies for a variety of psychiatric and neurological disorders. Because of this, a great deal of effort has been focused on developing small molecules that selectively activate or inhibit specific mGlu receptor subtypes.

DEVELOPMENT OF SELECTIVE PHARMACOLOGICAL REAGENTS THAT INTERACT WITH THE mGlu RECEPTORS

In recent years, there have been tremendous advances in development of compounds that selectively activate or block specific mGlu receptor subtypes or act selectively on mGlu receptors belonging to only one subgroup. The development of selective pharmacological reagents for study of the mGlu receptors has been extensively reviewed.[21,22] One of the most important and earliest breakthroughs in mGlu receptor pharmacology was development of a highly selective agonist for the group II mGlu receptors by Jim Monn, Darryle Schoepp, and their coworkers at Eli Lilly.[21] This compound, termed LY354740, selectively activates mGlu2 and mGlu3 at nM concentrations that have no effect on any other known receptor subtypes. In addition, LY354740 is systemically active and penetrates the blood brain barrier. Since that time, workers at Lilly have developed a number of second and third generation agonists of group II mGlu receptors with higher affinity and improved pharmacokinetic properties. These compounds have provided extremely valuable tools for dissecting the physiological roles of group II mGlu receptors and the behavioral effects of group II mGlu receptor activation. Studies using these compounds suggest that group II mGlu receptor agonists may have utility for a number of therapeutic indications, including anxiety disorders, schizophrenia, Parkinson's disease, and addictive disorders.[23] Most recently, LY354740 has entered clinical development, and preliminary proof-of-concept studies have been presented that suggest this compound may have clinical efficacy for treatment of generalized anxiety disorder and panic attack. This represents a major breakthough in the mGlu receptor field. If LY354740 or other group II mGlu receptor agonists prove to have efficacy in treating anxiety disorders in phase III studies, this will provide the first clinically useful compound that was developed to act at glutamate receptors.

DEVELOPMENT OF ALLOSTERIC ANTAGONISTS OF mGlu RECEPTORS

LY354740 and related compounds are classical competitive agonists of mGlu receptors that act at the glutamate binding site. Unfortunately, except for the exciting success of workers at Lilly, it has been difficult to develop highly selective agonists and antagonists that act at the glutamate binding site of the mGlu receptors. This is due to the limited ability to modify glutamate-related compounds to obtain appropriate drug-like properties, while maintaining or improving affinity at the receptor. Also, the glutamate binding site is highly conserved, making it difficult to develop highly selective ligands.

Another major breakthrough in pharmacology of the mGlu receptors came with the discovery of CPCCOEt, a non–amino acid that selectively inhibits mGlu1 relative to other mGlu receptor subtypes.[22] Shortly after the discovery of CPCCOEt, Mark Varney and coworkers at Sibia (now Merck) and Novartis discovered highly selective antagonists of mGlu5.[22,24] The first of these compounds, SIB1757 and SIB1893, were identified using functional assays to screen small molecule libraries for non–amino acid compounds that inhibit mGlu5 responses. Unlike previous mGlu ligands, CPCCOEt, SIB1757, and SIB1893 do not bind to the glutamate binding site but act at an allosteric site to inhibit mGlu receptor activation of G proteins.

These initial Sibia leads were low-potency compounds that lacked appropriate pharmacokinetic properties for *in vivo* studies. However, SIB1757 and SIB1893 provided leads that eventually led to development of MPEP, which exhibited 100-fold greater potency and properties that allowed its use for *in vitro* and *in vivo* studies of mGlu5 function.[24] Most recently, the group at Merck Research Laboratories further modified the MPEP series of compounds to develop MTEP.[25] This compound has fewer off-target activities and more favorable pharmacokinetic and physical properties when compared to MTEP.

These compounds have been used to provide a wealth of information on mGlu5 function. *In vivo* studies with mGlu5 antagonists suggest that these compounds may have efficacy in a number of neurological and psychiatric disorders.[24] MPEP and MTEP have activity in a variety of rodent models of anxiety and also show potential efficacy as antinociceptive agents and for treatment of addictive disorders and Parkinson's disease. A number of noncompetitive antagonists for other mGlu receptor subtypes have been discovered.[22] Development of these noncompetitive antagonists of mGlu receptors has provided a major breakthrough in that it allowed development of non–amino acid antagonists with high selectivity for a single mGlu receptor subtype.

ALLOSTERIC POTENTIATORS OF mGlu RECEPTORS

The discovery of allosteric antagonists raises the exciting possibility that compounds could act at the same or related sites to potentiate rather than inhibit mGlu receptor function. Consistent with this, Knoflach et al.[26] recently reported discovery of potent, positive allosteric modulators of mGlu1a. The most potent of these compounds is Ro 67-4853. These compounds do not directly activate the receptor but shift the concentration response curve for agonist activation of mGlu1 to the left. As with the allosteric antagonists, the allosteric potentiators of mGlu1 were shown to act at a site in the transmembrane regions of the receptors. Furthermore, we have identified two distinct classes of selective allosteric potentiators of mGlu5. One of these is represented by DFB Pharmaceuticals (FIG. 4). Unlike previously identified allosteric regulators of GPCRs, the DFB family of compounds exhibits a range of

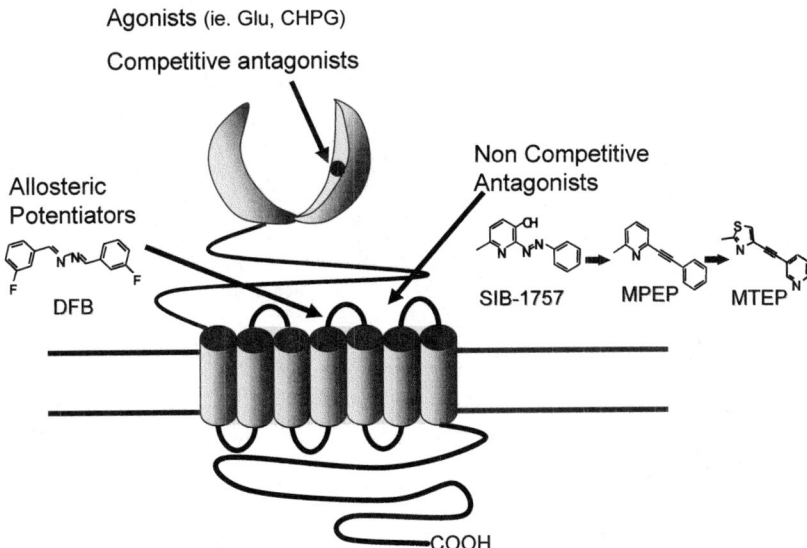

FIGURE 4. Schematic representation of an mGlu receptor and drug binding sites. Glutamate and competitive agonists and antagonists bind in the large extracellular domain, whereas allosteric antagonists and potentiators bind in the transmembrane spanning region. Also shown are the structures of the most selective allosteric antagonists and potentiators of mGlu5.

pharmacological activities, including potentiation of agonist responses, antagonism, and neutral competition of analogous potentiators or antagonists. Furthermore, DFB partially inhibits binding of a radioligand to the MPEP binding site. These findings expand our understanding of allosteric sites on GPCRs and suggest that small molecules can interact with allosteric sites with a range of intrinsic activities in a manner analogous to what is seen at traditional neurotransmitter binding sites. In addition, we have identified another class of allosteric potentiator, represented by N-{4-chloro-2-[(1,3-dioxo-1,3-dihydro-2H-isoindol-2-yl)methyl]phenyl}-2-hydroxybenzamide (CPPHA), that does not interact with the MPEP site. Thus, it is likely that multiple allosteric sites exist on mGlu5 that can respond to allosteric potentiators and inhibitors.

The discovery of allosteric potentiators of mGlu receptors raises the exciting possibility that this will provide a novel approach to development of selective agonists for different mGlu receptor subtypes. However, this class of compounds could offer several advantages over traditional neurotransmitter site agonists. Receptor modulation by binding to an allosteric site may afford very high subtype selectivity, assuming there would be less pressure to conserve such a site within a subfamily of receptors. Second, by selectively potentiating the response to the endogenous agonist, an allosteric potentiator should preserve activity-dependent activation of the receptor. This may provide a more physiologically relevant increase in receptor function. As such, it is conceivable that this could reduce adverse effects that may be seen with direct-acting agonists. Examples of this are clearly seen in the GABA-A receptor field where benzodiazepines offer clear advantages to direct activation of GABA-A receptors. Finally, it is conceivable that by maintaining activity-dependence of receptor activation, allosteric potentiators will induce less desensitization than traditional agonists. As additional allosteric potentiators of mGlu receptors are developed, each of these possibilities will be able to be tested in animal studies.

REFERENCES

1. WATKINS, J.C. 2000. L-Glutamate as a central neurotransmitter: Looking back. Biochem. Soc. Trans. **28:** 297–310.
2. WHITE, W.F, J.V. NADLER & C.W. COTMAN. 1979. The effect of acidic amino acid antagonists on synaptic transmission in the hippocampal formation in vitro. Brain Res. **164:** 177–194.
3. KOERNER, J.F. & C.W. COTMAN. 1981. Micromolar L-2-amino-4-phosphonobutyric acid selectively inhibits perforant path synapses from lateral entorhinal cortex. Brain Res. **216:** 192–298.
4. KOERNER, J.F. & C.W. COTMAN. 1982. Response of Schaffer collateral-CA1 cell synapses of the hippocampus to analogues of acidic amino acids. Brain Res. **251:** 105–115.
5. DAVIES, J. & J.C. WATKINS. 1982. Actions of D and L forms of 2-amino-5-phosphonovalerate and 2-amino-4-phosphonobutyrate in the cat spinal cord. Brain Res. **235:** 378–386.
6. EVANS, R.H., A.A. FRANCIS, A.W. JONES, *et al.* 1982. The effects of a series of ω-phosphonic ω-carboxylic amino acids of electrically evoked and excitant amino acid-induced responses in isolated spinal cord preparations. Br. J. Pharmacol. **75:** 65–75.
7. SLADECZEK, F., J.P. PIN, M. RECASENS, *et al.* 1985. Glutamate stimulates inositol phosphate formation in striatal neurones. Nature **317:** 717–719.
8. NICOLETTI, F., M.J. IADAROLA, J.T. WROBLEWSKI & E. COSTA. 1986. The activation of inositol phospholipid metabollism as a signal-transducing system for excitatory amino acids in primary cultures of cerebellar granule cells. J. Neurosci. **6:** 1905–1911.

9. NICOLETTI, F., J.L. MEEK, M.J. IADAROLA, et al. 1986. Coupling of inositol phospholipid metabolism with excitatory amino acid recognition sites in rat hippocampus. J. Neurochem. **46:** 40–46.
10. SUGIYAMA, H., I. ITO & C. HIRONO. 1987. A new type of glutamate receptor linked to inositol phospholipids metabolism. Nature (Lond.) **325:** 531–533.
11. CURRY, K., M.J. PEET, D.S. MAGNUSON & H. MCLENNAN. 1988. Synthesis, resolution, and absolute configuration of the isomers of the neuronal excitant 1-amino-1,3-cyclopentanedicarboxylic acid. J. Med. Chem. **31:** 864–867.
12. PALMER, E., D.T. MONAGHAN & C.W. COTMAN. 1988. Glutamate receptors and phosphoinositide metabolism: stimulation via quisqualate receptors is inhibited by N-methyl-D-aspartate receptor activation. Brain Res. **464:** 161–165.
13. DESAI, M.A. & P.J CONN. 1990. Selective activation of phosphoinositide hydrolysis by a rigid analogue of glutamate. Neurosci. Lett. **109:** 157–162.
14. DESAI, M.A. & P.J CONN. 1991. Excitatory effects of ACPD receptor activation in the hippocampus are mediated by direct effects on pyramidal cells and blockade of synaptic inhibition. J. Neurophysiol. **66:** 40–52.
15. CHARPAK, S., B.H. GAHWILER, K.Q. DO & T. KNOPFEL. 1990. Potassium conductances in hippocampal neurons blocked by excitatory amino-acid transmitters. Nature **347:** 765–767.
16. BASKYS, A. & R.C. MALENKA. 1991. Agonists at metabotropic glutamate receptors presynaptically inhibit EPSCs in neuronal rat hippocampus. J. Physiol. **444:** 687–701.
17. MASU, M., Y. TANABE, K. TSUCHIDA, et al. 1991. Sequence and expression of a metabotropic glutamate receptor. Nature **349:** 760–765.
18. HOUAMED, K.M., J.L. KUIJPER, T.L. GILBERT, et al. 1991. Cloning, expression, and gene structure of a G protein-coupled glutamate receptor from rat brain. Science **252:** 1318–1321.
19. CONN, P.J. & J.P. PIN. 1997. Pharmacology and functions of metabotropic glutamate receptors. Annu. Rev. Pharmacol. Toxicol. **37:** 205–237.
20. ANWYL, R. 1999. Metabotropic glutamate receptors: electrophysiological properties and role in plasticity. Brain Res. Brain Res. Rev. **29:** 83–120.
21. SCHOEPP, D.D., D.E. JANE & J.A. MONN. 1999. Pharmacological agents acting at subtypes of metabotropic glutamate receptors. Neuropharmacology **38:** 1431–1476.
22. GASPARINI, F., R. KUHN & J.P. PIN. 2002. Allosteric modulators of group I. Metabotropic glutamate receptors: novel subtype-selective ligands and therapeutic perspectives. Curr. Opin. Pharmacol. **2:** 43–49.
23. SCHOEPP, D.D. 2001. Unveiling the functions of presynaptic metabotropic glutamate receptors in the central nervous system. J. Pharmacol. Exp. Ther. **299:** 12–20.
24. SPOOREN, W.P., F. GASPARINI, T.E. SALT & R. KUHN. 2001. Novel allosteric antagonists shed light on mglu5 receptors and CNS disorders. Trends Pharmacol. Sci. **22:** 331–337.
25. COSFORD, N.D.P., L. TEHRANI, J. ROPPE, et al. 2003. 3-[2-Methyl-1,3-thiazol-4-ylethynyl]-pyridine: a potent and highly selective metabotropic glutamate subtype 5 receptor antagonist with anxiolytic activity. J. Med. Chem. **46:** 204–206.
26. KNOFLACH, F., V. MUTEL, S. JOLIDON, et al. 2001. Positive allosteric modulators of metabotropic glutamate 1 receptor: characterization, mechanism of action, and binding site. Proc. Natl. Acad. Sci. USA **98:** 13402–13407.

Genomics and Variation of Ionotropic Glutamate Receptors

ROBERT H. LIPSKY AND DAVID GOLDMAN

Laboratory of Neurogenetics, National Institute on Alcoholism and Alcohol Abuse, National Institutes of Health, Rockville, Maryland 20852, USA

> ABSTRACT: Sequencing of the human, mouse, and rat genomes has enabled a comprehensive informatics approach to gene families. This approach is informative for identification of new members of gene families, for cross-species sequence conservation related to functional conservation, for within-species diversity related to functional variation, and for historical effects of selection. This genome informatics approach also focuses our attention on genes whose genomic locations coincide with linkages to phenotypes. We are identifying ionotropic glutamate receptor (IGR) sequence variation by resequencing technologies, including denaturing high-performance liquid chromatography (dHPLC), for screening and direct sequencing, and by information mining of public (e.g., dbSNP and ENSEMBL) and private (i.e., Celera Discovery System) sequence databases. Each of the 16 known IGRs is represented in these databases, their positions on a canonical physical map (for example, the Celera map) are established, and comparison to mouse and rat sequences has been performed, revealing substantial conservation of these genes, which are located on different chromosomes but found within syntenic groups of genes. A collection of 38 missense variants were identified by the informatics and resequencing approaches in several of these receptor genes, including *GRIN2B*, *GRIN3B*, *GRIA2*, *GRIA3*, and *GRIK1*. This represents only a fraction of the sequence variation across these genes, but, in fact, these may constitute a large fraction of the common polymorphisms at these genes, and these polymorphisms are a starting point for understanding the role of these receptors in neurogenetic variation. Genetically influenced human neurobehavioral phenotypes that are likely to be linked to IGR genetic variants include addictions, anxiety/dysphoria disorders, post–brain injury behavioral disorders, schizophrenia, epilepsy, pain perception, learning, and cognition. Thus, the effects of glutamate receptor variation may be protean, and the task of relating variation to behavior difficult. However, functional variants of (1) catechol-*O*-methyltransferase, (2) serotonin transporter, and (3) brain-derived neurotrophic factor have recently been linked both to behavioral differences and to intermediate phenotypes, suggesting a pathway by which functional variation at IGRs can be tied to an etiologically complex phenotype.
>
> KEYWORDS: ionotropic glutamate receptor genes; missense variant; genomics; polymorphism

---

Address for correspondence: Robert H. Lipsky, Laboratory of Neurogenetics, NIAAA, NIH, 12420 Parklawn Drive, Suite 451, MSC 8110, Rockville, MD 20852. Voice: 301-402-5591; fax: 301-443-8579.
  rlipsky@mail.nih.gov

Ann. N.Y. Acad. Sci. 1003: 22–35 (2003). © 2003 New York Academy of Sciences.
doi: 10.1196/annals.1300.003

INTRODUCTION

Ionotropic glutamate receptors (IGRs) play the largest role in excitatory neurotransmission in the brain and play central roles in a variety of behaviors, such as learning, that are fundamental to survival. Glutamatergic pathways are directly responsible for the bulk of long-range interconnectivity in the human brain. Therefore, it is not surprising that this is a phylogenetically ancient family of receptors. Glutamate receptors are found in both animals and plants. The size of the glutamate receptor gene family is also consistent with the critical role of these receptors in connectivity—there are at least 16 IGRs in the human.

Glutamate receptor activation leads to calcium entry, which is fundamental in brain development and in forms of synaptic plasticity essential for learning and memory. In addition, earlier work has supported an essential role in neuronal survival, suggesting a role for glutamate receptors in neurological disorders, such as epilepsy, brain injury from trauma or focal cerebral ischemic stroke, and perhaps in disorders such as schizophrenia, Parkinson's disease, or Huntington's chorea, and amyotrophic lateral sclerosis (ALS). Although IGRs are largely expressed in the brain, they also have important roles in the peripheral nervous system, acting in other tissues to the modulate secretion of insulin by the pancreas, to regulate bone resorption, and to aid in the perception of pain. Originally defined on the basis of their agonist binding and electrophysiological properties, N-methyl-D-aspartate (NMDA), alpha-amino-3-hydroxy-5-methyl-4-isoxazole (AMPA), or kainate (KA), molecular cloning of cDNAs encoding IGR subunits during the late 1980s and early 1990s made it clear that these receptors were members of related gene families. The recent availability of largely complete genomic sequence data for rodent and human glutamate receptor genes has enabled a comprehensive informatics approach to understanding the functional diversity of this class of receptors.

RESULTS

Phylogenetically Ancient Gene Families

Ionotropic glutamate receptors are encoded by at least six gene families based on nucleotide and amino acid sequence identity (TABLE 1). There is a single family of AMPA receptors, two families of KA receptors, and three families of NMDA receptors. Sequence similarity, including partial conservation of exon-intron structure, suggests a common but ancient evolutionary origin for all IGR gene families. In humans, amino acid sequence similarities among the gene family members ranges from 16 to 62%. Thus, the individual genes within gene families are themselves ancient. The primate brain has undergone a large expansion in size and capacity. However, the IGR genes involved in long-range connectivity in primate brain are probably orthologous to genes found in other vertebrates.

The predicted topology of mammalian IGRs is shared across the receptor families and is illustrated for the NMDA receptor family in FIGURE 1. Each subunit has four predicted transmembrane segments. However, only three actually span the membrane (TM1, TM3, and TM4). The second segment (TM2) constitutes a cytoplasmic-facing membrane reentrant loop involved in forming the channel pore of the func-

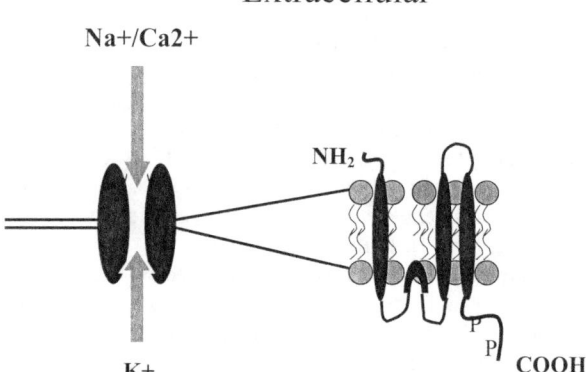

FIGURE 1. Ionotropic glutamate receptor topology.

TABLE 1. Human ionotropic glutamate receptors genes and their chromosomal locations

Receptor	Family	Subunit	Gene name	Chromosome
AMPA	1	GluR1	*GRIA1*	5q33
AMPA	1	GluR2	*GRIA2*	4q32-33
AMPA	1	GluR3	*GRIA3*	Xq25-26
AMPA	1	GluR4	*GRIA4*	11q22-23
KA	2	GluR5	*GRIK1*	21q21.1-22.1
KA	2	GluR6	*GRIK2*	6q16.3-q21
KA	2	GluR7	*GRIK3*	1p34-p33
KA	3	KA1	*GRIK4*	11q22.3
KA	3	KA-2	*GRIK5*	19q13.2
NMDA	4	NR1	*GRIN1*	9q34.3
NMDA	5	NR2A	*GRIN2A*	16p13.2
NMDA	5	NR2B	*GRIN2B*	12p12
NMDA	5	NR2C	*GRIN2C*	17q24-q25
NMDA	5	NR2D	*GRIN2D*	19q13.1qter
NMDA	6	NR3A	*GRIN3A*	9q34
NMDA	6	NR3B	*GRIN3B*	19p13.3

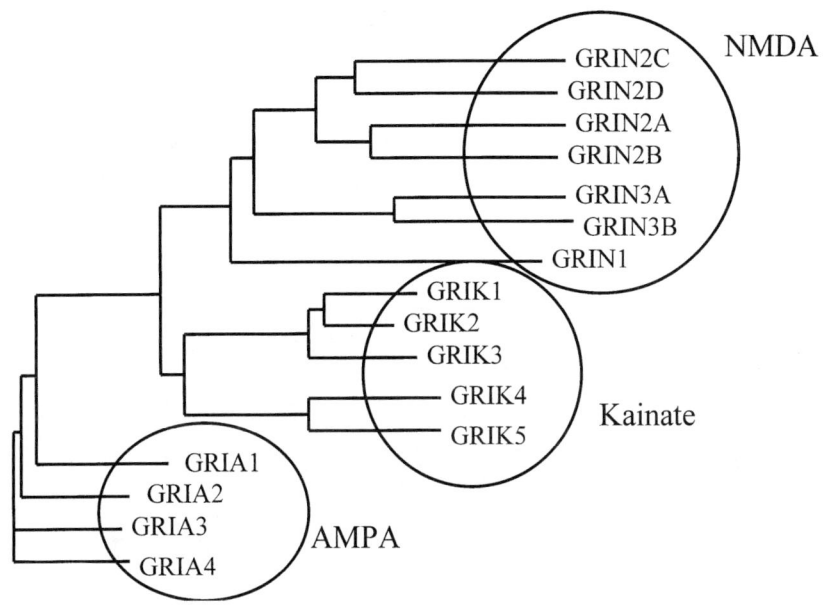

FIGURE 2. Human ionotropic glutamate receptor clades.

tional receptor. These segments are highly conserved in humans (particularly TM3) and play critical roles in channel function.

Parsimony analysis of amino acid sequences groups the IGRs into three distinct clades, which correspond to the three defining ligands: AMPA, KA, and NMDA (FIG. 2). Thus, ligand specificity parallels phylogeny.

Sequence conservation among the IGR genes has been used to discover new glutamate receptors in other eukaryotes, including plants. The availability of expressed sequence tag databases and the *Arabidopsis thaliana* genome project led to the discovery of three glutamate receptor-like gene families with 20 genes, as we found by BLAST searches using known *Arabidopsis* glutamate receptor-like receptor (GLR) cDNAs as the query sequence. Comparison of the human sequences to *Arabidopsis* glutamate receptor sequences representative of three known Arabidopsis clades (*At*GLR1.4, *At*GLR2.2, and *At*GLR 3.7) enables "rooting" of the ionotropic glutamate receptor cladogram. This comparison reveals moderate similarity between the mammalian and plant sequences. Focusing on proposed ligand binding and channel-pore and membrane-spanning domains, the sequence differences between receptors of the three functional classes are enhanced (FIG. 3). This focused comparison of functionally important domains reveals that the plant GLRs show highest similarity to human GRIN1 and GRIN3, suggesting that plant GLRs share some functional qualities with human NMDA receptors. Furthermore, NR1 and NR3 differ from the other NMDA receptor subunits (NR2) in that glycine is the primary ligand for NR1 and NR3, while NR2 receptors have glutamate-binding domains.

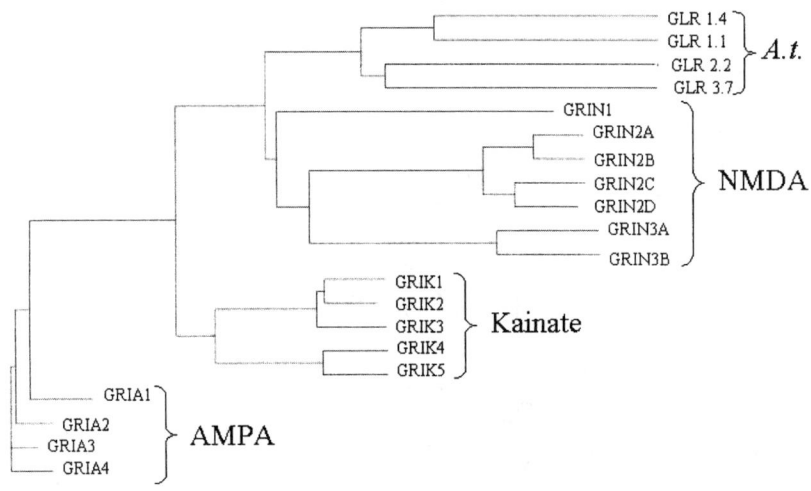

FIGURE 3. Human and *Arabidopsis thaliana* clades.

Chromosomal Localization and Conserved Synteny

The ionotropic glutamate receptor genes are dispersed over many chromosomes (TABLE 1). In humans, two pairs of genes are localized to the same chromosomal region. *GRIA4 and GRIK4* are located on the long arm of chromosome 11. However, they are separated by several million base pairs. Also, *GRIK5* and *GRIN2D* are located relatively close to each other on the long arm of chromosome 19. Syntenic regions have been identified in the mouse. However, the synteny of these two pairs of human IGR genes is not conserved in the mouse (TABLE 2). Therefore, although functional significance of these chromosomal colocalizations in the human cannot be ruled out, it is unlikely that these lead to coordinate regulation of gene expression. Because of conserved synteny (as shown in TABLE 2), genes in the immediate neighborhood of human IGRs are also in the immediate neighborhood of mouse IGRs. This would enable detection in the mouse of *cis*-effects on expression of IGRs and neighboring genes, if such effects are present. Also, linkages with converge on IGR genes can also implicate flanking genes in both species.

SNP Discovery Using dHPLC and Sequence Analysis

We have identified IGR sequence variation by resequencing technologies, including dHPLC for detection of new sequence variants and direct sequencing. We have also information mined publicly available (dbSNP, Ensembl) and private (Celera Discovery System) databases. We initially selected three genes for this approach. We chose the NMDA receptor subunit genes *GRIN1*, *GRIN2B*, and *GRIN3* because of the central role of NMDA receptors in neurotransmission, neuronal plasticity, and neuronal viability, and because of the potential for linkage to clinically important

TABLE 2. Comparison of human and murine glutamate receptor genes and their chromosomal locations

			Chromosome	
Receptor	Subunit[a]	Gene[a]	Human	Murine
AMPA	GluR4	GRIA4	11q22–23	3F3
KA	KA1	GRIK4	11q22.3	9A5.1
KA	KA-2	GRIK5	19q13.2	7A3
NMDA	NR2D	GRIN2D	19q13.1qter	7B27B3
NMDA	NR1	GRIN1	9q34.3	2A3
NMDA	Nr3B	GRIN3B	9q13.3	4B

[a]Human gene name and subunit classifications are used in describing species comparisons.

phenotypes, including addictions, anxiety/dysphoria, post–brain injury responses, pain perception, schizophrenia, and epilepsy. It has long been appreciated that alcohol acts directly on NMDA receptors and that the NMDA receptor antagonists, such as ketamine and MK801 produce schizophrenic symptoms and exacerbate symptoms in schizophrenic patients. Here we report on *GRIN1* and *GRIN2B* data based on genome informatics plus sequence variation detection across the entire coding sequences of these genes by dHPLC and DNA sequence analysis. *GRIN2B* expression is important developmentally, and *GRIN1* must be coexpressed to produce fully functional receptors. Sequence variants in other IGRs were obtained using the genome informatics approach.

A total of 480 unrelated individuals were resequenced across *GRIN1* and *GRIN2B* coding exons using dHPLC, as described previously.[4] The screening sample was ethnically diverse, representing Caucasians, Asians, African Americans, Native Americans, and a clinically diverse group, representing schizophrenia, OCD, alcoholism, depression, anorexia nervosa, and normal controls. A total of 2.9 Mb of genomic sequence was screened for novel sequence variants in the sample, which amounted to approximately 3 kb per individual per gene. Individual samples having variants detected by dHPLC were amplfied by PCR using the same primers and the products subjected to direct DNA sequence analysis. Both strands were sequenced, identifying six nonsynonymous *GRIN1* and *GRIN2B* substitutions (TABLE 3) plus one synonymous change. The allele frequencies provided in TABLE 3 are based on the deliberately ethnically diverse screening sample of 480 individuals. The exon 17 Cys744Tyr substitution was seen in only a single individual. All of the nonsynonomous variants discovered in *GRIN1* and *GRIN2B* were novel and not represented in public or commercially available databases. The number of variant sites normalized for the number of chromosomes and base pairs screened (θ) for the sequence changes across the coding exons of *GRIN1* and *GRIN2B*, and based on nonsynonymous substitutions, was 2.1×10^{-5}. The average heterozygosity per site (π) for nonsynonymous substitutions was 1.0×10^{-6}.

Also listed in TABLE 3 are other nonsynonymous substitutions at IGRs identified by a genome informatics approach. These data were compiled from translated exons from different IGR genes. Nucleotide sequence variants were detected by comparing available sequences via multiple sequence alignments and from entries in sequence variant databases. Of 16 IGRs examined using dbSNP BUILD 113 (indexed April 14, 2003) and Celera RefSNP Release R4.0 (indexed February 21, 2003), 12 contained at least one nonsynonymous SNP entry. There were also no reported nonsynonymous SNPs in *GRIA1*, *GRIA4*, *GRIK3*, and *GRIN2D*. As seen by the entries under the "minor allele frequency" column in TABLE 3, the vast majority of the SNPs reported in the databases have not been validated by genotyping multiple ethnically defined populations. Exceptions were *GRIK1* Leu902Ser (minor allele frequency 0.02), *GRIK5* Pro140His (minor allele frequency 0.2), *GRIN3B* Thr157Met (minor allele frequency 0.43), and *GRIN3B* Ala845Thr (minor allele frequency 0.19). Clearly, the validity of these computational sequence variants must be confirmed and their genotype frequencies determined in different human populations. Furthermore, it is clear from comparing our *GRIN1* and *GRIN2B* resequencing results to the genome informatics approach that many nonsynonymous substitution polymorphisms at other IGRs are not represented in the databases or in a publicly available sequence.

DISCUSSION

The IGR genes are well conserved phylogenetically. There is a high degree of amino sequence identity among IGR gene family members within humans. Three clades were determined that correspond to the three original groups defined by pharmacological and electrophysiological parameters. Conservation of ligand-receptor interactions and channel activity may be a major factor maintaining sequence conservation among the three clades.

Of the nonsynonymous SNPs discovered in IGR genes, the vast majority of these candidate functional variants were rare. Others have screened IGR genes for variants and also found a low rate of nonsynonymous substitutions.[5-8] Taken together, these data indicate that among the IGR genes, the *GRIN* family may be under the strongest selection for sequence conservation. This is also seen in the high degree of amino acid sequence similarity between human family members and those of other species, at least for the NR1 and NR2 subunits. Because other factors influence IGR activity, such as receptor subunit expression and subunit stoichiometry, it is likely that sequence variants influencing transcription and messenger RNA stability may profoundly alter receptor function. These sequence variants may be more common. Thus, the contribution of a particular sequence variant may have a variable influence, depending on the neuronal population and the environmental conditions influencing the expression of the gene. Under such conditions, the task of relating variation to behavior may be difficult. However, there are examples of functional variants that have recently been linked both to behavioral differences and to intermediate phenotypes in the brain, suggesting a pathway by which functional variants at IGRs can be linked to an etiologically complex phenotype.[1-3] Identifying such "intermediate phenotypes" in the central nervous system (CNS) can provide greater power to detect the effect of the candidate gene on phenotype. A comprehensive discussion of this issue is beyond the scope of this report. However, we have narrowed the perspec-

TABLE 3. Ionotropic glutamate receptor missense variants and their frequencies

Gene/variant	Minor allele frequency	Source
GRIN1		
Ala310Val	0.003	LNG[a]
Leu336Val	nd[b]	computational
Cys744Tyr	0.003	LNG
Asp765Tyr	0.003	LNG
GRIN2A		
Asn3Asp	nd	computational
Glu196Gln	nd	computational
Arg205Lys	nd	computational
Asn380Thr	nd	computational
GRIN2B		
Ser116Thr	0.01	LNG
Leu120Ile	0.03	LNG
Thr275Ala	0.02	LNG
Gly603Ser	nd	computational
Ile1167Val	nd	computational
Gly1168Ser	nd	computational
GRIN2C		
Trp1180Arg	nd	computational
Ala1266Thr	nd	computational
Cys1210Gly	nd	computational
GRIN3A		
Met362Val	nd	computational
Arg487Gly	nd	computational
Asp835Asn	nd	computational
Arg1041Gln	nd	computational
GRIN3B		
Thr157Met	0.43	computational
Arg404Trp	nd	computational
Thr577Met	nd	computational
Ala845Thr	0.19	computational
GRIA2		
Glu241Gly	nd	computational
GRIA3		
Leu525Phe	nd	computational

TABLE 3. Ionotropic glutamate receptor missense variants and their frequencies (*Continued*)

Gene/variant	Minor allele frequency	Source
Asn786Ser	nd	computational
Ile790Leu	nd	computational
GRIK1		
Val329Ile	nd	computational
Val442Ala	nd	computational
Ser474Leu	nd	computational
Val757Ile	nd	computational
Ala870Val	nd	computational
Leu902Ser	0.02	computational
GRIK2		
Ile867Met	nd	computational
GRIK4		
Glu700Gly	nd	computational
GRIK5		
Pro140His	0.2	computational

[a]LNG, this study.
[b]nd, not determined.

tive to those phenotypes that may influence vulnerability to addiction, schizophrenia, and pain perception in the context of different human studies and through the use of animal models. Where possible, we also discussed roles of functional variants and how they can provide approaches to determining an intermediate phenotype conferred by IGR variants that can be applied in human association and linkage studies.

IGR FUNCTION AND DRUG ADDICTION

In the medial prefrontal cortex, repeated cocaine produces tolerance of the extracellular dopamine response to subsequent cocaine injection. Previous studies have yielded pharmacologic and molecular evidence supporting the influence of AMPA/KA receptors in the medial prefrontal cortex dopamine response in rats exposed to acute cocaine and amphetamine and in mediating dopamine tolerance after chronic cocaine administration.[9–12] Collectively, these data support a role for these receptors on extracellular dopamine in the medial prefrontal cortex and in the midbrain, where these receptors may regulate dopamine release through a presynaptic mechanism. In other studies, overexpression of both GluR1 and GluR2 in nucleus accumbens shell neurons facilitates extinction of cocaine but not sucrose-seeking responses.[13]

To fully understand the molecular basis of addiction, we must determine the sequence of molecular and cellular events in the brain that occur during acute exposure to a drug—a sequence that ultimately leads to stable molecular and cellular changes after prolonged drug exposure. The pathway to addiction results from specific and nonspecific interactions with a variety of different proteins, including neurotransmitter receptors and their transporters, whose interaction is determined by each particular drug. For example, ethanol directly inhibits NMDA receptor function but helps facilitate gamma-amino butyric acid (GABA) activity. Both receptors activate ligand-gated channels. Opiates are ligands for a number of opioid receptors that signal through an inhibitory class of G-proteins. However, these interactions are likely to activate convergent intracellular signaling pathways that over time become altered to produce the phenotypes associated with addiction. These adaptations are "plastic" events that may become solidified as a molecular and cytoarchitectectural memory. A well-understood region thought to be critical in these initial adaptive changes following drug use is the mesolimbic dopamine system. For example, dopamine neurons of the midbrain extend axons into the nucleus accumbens, prefrontal cortex, and amygdala. These three regions are centrally involved in both acute and chronic responses to drugs of abuse, including alcohol. It should be noted that patients who have used psychostimulants may exhibit psychotic symptoms that resemble those of paranoid schizophrenia.[14] Patients with these symptoms are prone to relapse either following readministration of the drug or following stress. This phenomenon is thought to be identical to the stereotypical behavior seen in rodents after repeated administration of psychostimulants, termed behavioral sensitization. Behavioral sensitization must be the result of long-term neuronal plasticity, which may involve structural changes in the CNS. Glutamatergic neurons are also important in these responses, in particular, the medial prefrontal cortex, which plays an important role in working memory and "executive" functions. Specifically, dopamine D(1) receptors are thought to be central in enhancing NMDA-mediated activation through a PKA-dependent pathway.[15] These pathways may be critical to development of long-term potentiation (LTP) and long-term depression (LTD) of synaptic transmission, thought to be central in learned responses, which have also been observed *in vitro* and *in vivo* in the prefrontal cortex. Thus, it is plausible that dopamine and NMDA receptors interact in the prefrontal cortex. Ethanol and other drugs of abuse can modulate neuronal signaling systems. In particular, cAMP and phosphoinositide signaling pathways appear to be intracellular targets that mediate directly actions of ethanol (and other drugs). Central components of the post–receptor associated pathways of the cAMP signal transduction cascade include Gs protein, protein kinase A, and cAMP-response element binding protein (CREB). Thus, it is plausible that allelic variation in glutamate receptors (and second messenger pathways) may play a role in responses to early responses to alcohol and other drugs of abuse.

Identification of relevant phenotypes related to neural functioning is not yet fully appreciated, although behavioral phenotyping, through comparative studies in inbred strains of mice, including recombinant strains and gene knockouts have produced some major findings. For example, strain-specific, quantitative differences in responses of the prefrontal cortex in terms of dopamine metabolism and release were seen in restraint-induced stressed mice, comparing C57BL/6JIco mice with the DBA/2Jico strain.[16] It is also known that chronically stressed rats showed working memory impairment caused by reduced dopamine transmission.[17] Taken together, these results suggest a genetic control over the balance between mesocortical and mesoaccumbens dopamine responses to stress. It is possible to envisage similar comparisons in conditional knock-in mice for drugs of abuse. However, caution must be taken in selecting a phenotype that will not be confounded by environmental influences. An example of a confounding environmental influence is feeding, an example of an ecologically common experience, that was shown to abolish strain-related differences in certain behavior responses during amphetamine-induced place conditioning.[18]

ROLES OF IGR POLYMORPHISMS IN SCHIZOPHRENIA

Glutamate dysfunction is a central hypothesis, among a number of proposed hypotheses, to explain the origin of schizophrenia. This idea originated with the observation that phencyclidine (PCP) and other antagonists of NMDA receptors (ketamine and MK801) produce schizophrenia-like symptoms and exacerbate symptoms in schizophrenia patients. In addition, pharmacologically treated mice or transgenic mice with reduced expression of GluRζ (NR1 subunit in humans) displayed behavioral alterations similar to schizophrenia in humans.[19] Genetically deficient mice lacking GluRε1 subunit expression (NR2A in humans) also exhibited altered phenotypes associated with changes in monoaminergic neuronal activities seen in adulthood.[20]

Tests of linkage and association between *GRIN* loci and schizophrenia have been conducted. Recently, Ohtsuki *et al.*[21] detected eight *GRIN2B* SNPs in a Japanese population using single-stranded conformational polymorphism analysis (SSCP). Five sequence differences were within the coding region, but all of these variants encoded synonymous changes, and the remaining three variants were within the 3′ untranslated region (UTR). Five of the common polymorphisms were in strong linkage disequilibrium (LD), covering a genomic region of approximately 50 kb. Haplotype frequencies were estimated between the schizophrenic and control groups. On the basis of haplotype, a weak association with *GRIN2B* and susceptibility to schizophrenia was detected. Even more recently, a SNP in the *GRIN1* promoter was a candidate for susceptibility to schizophrenia, and a functional polymorphic (CT)n repeat in the *GRIN2A* promoter supported a role for reduced NR2A levels in brain with severe outcome in schizophrenia.[22] The influence of regulatory variants on NMDA receptor function may be large and are worthy of further study.

Kainate receptors have also been suggested to contribute to the pathogenesis of schizophrenia through excessive receptor activation. *GRIK2* is located within a region of chromosome 6 that was supported by linkage studies to contain a schizophrenia susceptibility gene. In a Japanese case-control population, haplotype analysis failed to support a role for *GRIK2* in schizophrenia.[23] However, a polymorphic missense variant in *GRIK3* (Ser310Ala) results in allele-specific differences in *GRIK3* mRNA levels that may be independent of RNA editing, possibly resulting in altered receptor protein expression.[24] Recently, an association study between the *GRIK3* Ser310Ala was performed in a European sample of 99 schizophrenic patients and 116 controls. A significant difference in the genotype frequency between the groups for the Ala allele, when considered dominant, was observed, suggesting a role for *GRIK3* as a susceptibility gene for schizophrenia.[25]

Finally, haplotypes of SNPs in LD have been tested for association of *GRIA4* with schizophrenia. A positive association was seen in a Japanese population.[26] However, functional variants affecting expression or receptor activity are not known.

GRIN2B AND PAIN PERCEPTION

Earlier evidence from animal models suggested the involvement of NMDA receptor subtypes in acute and chronic nociceptive responses. In rodents, injections of NMDA into a midbrain region associated with analgesia, the periaqueductal gray (PAG), increased the latency of nociceptive responses.[27,28] In contrast, injection of NMDA into the spinal cord decreased the latency of acute nociceptive responses.[29,30] These results suggested that there are differences between the neuroanatomical and neurochemical circuitry responsible for mediating analgesic responses measured in pain models. However, because antagonist specificity is a limitation of the pharmacological design, a molecular genetic approach may clarify apparent inconsistencies in pain response seen in different studies. It is known that four of the NMDA receptor subunit genes show differential mRNA expression patterns and differential functional properties in mice, rats, and humans.[31] In murine spinal cord, *GRIN2A* mRNA is widely expressed, while *GRIN2B* mRNA is localized to the lamina II of the dorsal horn. Expression of *GRIN2D* mRNA is low in spinal cord, while that of *GRIN2C* mRNA is absent. *GRIN1* mRNA is detected throughout the brain, while *GRIN2B* mRNA is localized to the forebrain. *GRIN2C* mRNA is largely restricted in expression to the molecular layer of the cerebellum and weakly to the thalamus and the olfactory bulb. Weak expression of *GRIN2D* mRNA is found in the diencephalon and the brain stem. With a pattern of *GRIN* expression established, pain responses of NMDA receptor subunit–deficient mice have been more consistent. Mice lacking either NR2A or NR2D subunits or both subunits showed no alterations in the acute nociceptive responses.[32] However, NR2B-deficient mice showed exaggerated responses in several acute nociceptive tests.[33] In addition, when NR2B was overexpressed in mouse forebrain by targeted methods, enhanced pain perception behavior was seen with prolonged exposure to noxious stimulation.[34] These results suggest that the NR2B subunit may play a key role in pain perception. Therefore, drugs targeting NR2B-containing NMDA receptors in regions related to nociceptive areas, such as the dorsal horn and the forebrain, could alleviate pain in humans. *GRIN2B* variants affecting either NR2B subunit levels or receptor function could be predictive of analgesic effects of drugs targeting this receptor.

PROSPECTS FOR THE FUTURE

By identifying susceptibility genes, we hope to gain a better understanding of the neurobiology of IGRs and of neurological disorders. Particularly in the case of the addictions, a greater knowledge of genetic risk factors may increase our ability to focus on important environmental determinants and to facilitate the development of effective prevention and treatment strategies.

REFERENCES

1. ZUBIETA, J.K., M.M. HEITZEG, Y.R. SMITH, *et al.* 2003. COMT val158met genotype affects mu-opioid neurotransmitter responses to a pain stressor. Science **299:** 1240–1243.
2. HARIRI, A.R., V.S. MATTAY, A. TESSITORE, *et al.* 2002. Serotonin transporter genetic variation and the response of the human amygdala. Science **297:** 400–4003.

3. EGAN, M.F., M. KOJIMA, J.H. CALLICOTT, *et al.* 2003. The BDNF val66met polymorphism affects activity-dependent secretion of BDNF and human memory and hippocampal function. Cell **112:** 257–269.
4. RUDOLPH, J.G., S. WHITE, C. SOKOLSKY, *et al.* 2002. Determination of melting temperature for variant detection using dHPLC: a comparison between an empirical approach and DNA melting prediction software. Genet Test. **6:** 169–176.
5. WILLIAMS, N.M., T. BOWEN, G. SPURLOCK, *et al.* 2002. Determination of the genomic structure and mutation screening in schizophrenic individuals for five subunits of the N-methyl-D-aspartate glutamate receptor. Mol. Psychiatry **7:** 508–514.
6. RICE, S.R., N. NIU, D.B. BERMAN, *et al.* 2001. Identification of single nucleotide polymorphisms (SNPs) and other sequence changes and estimation of nucleotide diversity in coding and flanking regions of the NMDAR1 receptor gene in schizophrenic patients. Mol. Psychiatry **6:** 274–284.
7. OHTSUKI, T., K. SAKURAI, H. DOU, *et al.* 2001. Mutation analysis of the NMDAR2B (GRIN2B) gene in schizophrenia. Mol. Psychiatry **6:** 211–216.
8. CARGILL, M., D. ALTSHULER, J. IRELAND, *et al.* 1999. Characterization of single-nucleotide polymorphisms in coding regions of human genes. Nat. Genet. **22:** 231–238.
9. WU, W.R., N. LI & B.A. SORG. 2002. Regulation of medial prefrontal cortex dopamine by alpha-amino-3-hydroxy-5-methylisoxazole-4-propionate/kainate receptors. Neuroscience **114:** 507–516.
10. THOMAS, M.J., C. BEURRIER, A. BONCI & R.C. MALENKA. 2001. Long-term depression in the nucleus accumbens: a neural correlate of behavioral sensitization to cocaine. Nat. Neurosci. **12:** 1217–1223.
11. GHASEMZADEH, M.B., L.C. NELSON, X.Y. LU & P.W. KALIVAS. 1999. Neuroadaptations in ionotropic and metabotropic glutamate receptor mRNA produced by cocaine treatment. J. Neurochem. **72:** 157–165.
12. MEAD, A.N., A. VASILAKI, C. SPYRAKI, *et al.* 1999. AMPA-receptor involvement in c-fos expression in the medial prefrontal cortex and amygdala dissociates neural substrates of conditioned activity and conditioned reward. Eur. J. Neurosci. **11:** 4089–4098.
13. SUTTON, M.A., E.F. SCHMIDT, K.H. CHOI, *et al.* 2003. Extinction-induced upregulation in AMPA receptors reduces cocaine-seeking behaviour. Nature **421:** 70–75.
14. SATO, M., C.C. CHEN, K. AKIYAMA & S. OTSUKI. 1983. Acute exacerbation of paranoid psychotic state after long-term abstinence in patients with previous methamphetamine psychosis. Biol. Psychiatry **18:** 429–440.
15. WANG, J. & P. O'DONNELL. 2001. D(1) dopamine receptors potentiate NMDA-mediated excitability increase in layer V prefrontal cortical pyramidal neurons. Cereb. Cortex **11:** 452–462.
16. VENTURA, R., S. CABIB & S. PUGLISI-ALLEGRA. 2001. Opposite genotype-dependent mesocorticolimbic dopamine response to stress. Neuroscience **104:** 627–631.
17. MIZOGUCHI, K., M. YUZURIHARA, A. ISHIGE, *et al.* 2000. Chronic stress induces impairment of spatial working memory because of prefrontal dopaminergic dysfunction. J. Neurosci. **20:** 1568–1574.
18. CABIB, S., C. ORSINI, M. LE MOAL & P.V. PIAZZA. 2000. Abolition and reversal of strain differences in behavioral responses to drug of abuse after a brief experience. Science **289:** 463–465.
19. MOHN, A.R., R.R. GAINETDINOV, M.G. CARON & B.H. KOLLER. 1999. Mice with reduced NMDA receptor expression display behaviors related to schizophrenia. Cell **98:** 427–436.
20. MIYAMOTO, Y., K. YAMADA, Y. NODA, *et al.* 2001. Hyperfunction of dopaminergic and serotonergic neuronal systems in mice lacking the NMDA receptor ε1 subunit. J. Neurosci. **21:** 750–757.
21. OHTSUKI, T., K. SAKURAI, H. DOU, M. TORU, *et al.* 2001. Mutation analysis of the NMDAR2B (GRIN2B) gene in schizophrenia. Mol. Psychiatry **6:** 211–216.
22. ITOKAWA, M., K. YAMADA, K. YOSHITSUGU, *et al.* 2003. A microsatellite repeat in the promoter of the N-methyl-D-aspartate receptor 2A subunit (GRIN2A) gene suppresses transcriptional activity and correlates with chronic outcome in schizophrenia. Pharmacogenetics **13:** 271–278.

23. SHIBATA, H., A. SHIBATA, H. NINOMIYA, et al. 2002. Association study of polymorphisms in the GluR6 kainate receptor gene (GRIK2) with schizophrenia. Psychiatry Res. **113:** 59–67.
24. SCHIFFER, H.H., G.T. SWANSON, E. MASLIAH & S.F. HEINEMANN. 2000. Unequal expression of allelic kainate receptor GluR7 mRNAs in human brains. J. Neurosci. **20:** 9025–9033.
25. BEGNI, S., M. POPOLI, S. MORASCHI, et al. 2002. Association between the ionotropic glutamate receptor kainte 3 (GRIK3) ser310ala polymorphism and schizophrenia. Mol. Psychiatry **7:** 416–418.
26. MAKINO, C., Y. FUJII, R. KIKUTA, et al. 2003. Positive association of the AMPA receptor subunit GluR4 gene (GRIA4) haplotype with schizophrenia: linkage disequilibrium mapping using SNPs evenly distributed across the gene region. Am. J. Med. Genet. **116B:** 16–22.
27. SIEGFRIED, B. & R.L. NUNES DE SOUZA. 1989. NMDA receptor blockade in the periaqueductal grey prevents stress-induced analgesia in attacked mice. Eur. J. Pharmacol. **168:** 239–242.
28. JACQUET, Y.F. 1988. The NMDA receptor: central role in pain inhibition in rat periaqueductal gray. Eur. J. Pharmacol. **154:** 271–276.
29. CHAPLAN, S.R., A.B. MALMBERG & T.L. YAKSH. 1997. Efficacy of spinal NMDA receptor antagonism in formalin hyperalgesia and nerve injury evoked allodynia in the rat. J. Pharmacol. Exp. Ther. **280:** 829–838.
30. AANONSEN, L.M. & G.L. WILCOX. 1987. Nociceptive action of excitatory amino acids in the mouse: effects of spinally administered opioids, phencyclidine and sigma agonists. J. Pharmacol. Exp. Ther. **243:** 9–19.
31. WATANABE, M. 1997. Developmental dynamics of gene expression for NMDA receptor channel. *In* The Ionotropic Glutamate Receptors. D.T. Mogaghan & R.J. Wenthold, Eds.: 189–218. Humana Press. Totowa, NJ.
32. MINAMI, T., J. SUGATANI, K. SAKIMURA, et al. 1997. Absence of prostaglandin E2-induced hyperalgesia in NMDA receptor epsilon subunit knockout mice. Br. J. Pharmacol. **120:** 1522–1526.
33. WAINAI, T., T. TAKEUCHI, N. SEO & M. MISHINA. 2001. Regulation of acute nociceptive responses by the NMDA receptor GluRepsilon2 subunit. Neuroreport **12:** 3169–3172.
34. WEI, F., G.D. WANG, G.A. KERCHNER, et al. 2001. Genetic enhancement of inflammatory pain by forebrain NR2B overexpression. Nat. Neurosci. **4:** 164–169.

Anatomical Substrates for Glutamate–Dopamine Interactions

Evidence for Specificity of Connections and Extrasynaptic Actions

SUSAN R. SESACK,[a] DAVID B. CARR,[b] NATALIA OMELCHENKO,[a] AND ALINE PINTO[c]

[a]*Departments of Neuroscience and Psychiatry, University of Pittsburgh, Pittsburgh, Pennsylvania 15260, USA*

[b]*Department of Physiology, Northwestern University, Chicago, Illinois 60611, USA*

[c]*Center for Molecular and Behavioral Neuroscience, Rutgers University, Newark, New Jersey 07102, USA*

ABSTRACT: For normal regulation of motor, affective, and cognitive functions, dopamine provides an essential modulation of glutamate transmission within multiple brain regions. This paper will review three principal anatomical substrates for such interactions. First, dopamine modulates the activity of glutamate neurons within the cerebral cortex. Evidence will be reviewed for dopamine regulation of pyramidal neurons in the prefrontal cortex via synaptic and extrasynaptic mechanisms and through indirect effects mediated by GABA cells. Second, glutamate neurons innervate dopamine cells within the ventral tegmental area. Evidence will be described for selective glutamate input from the prefrontal cortex or the brain stem tegmentum to different populations of dopamine cells. The third level of interaction occurs within target regions via convergent synaptic or extrasynaptic regulation of common neurons. Such convergence will be reviewed for the basal ganglia, prefrontal cortex, and amygdala. Together, these substrates for glutamate–dopamine interactions provide several mechanisms for normal regulation of brain function. Sites of modulatory interaction between dopamine and glutamate also suggest circuit alterations that might contribute to the pathophysiology of mental health disorders and provide potential sites for therapeutic intervention in these conditions.

KEYWORDS: amygdala; dopamine; GABA; glutamate; laterodorsal tegmentum; pedunculopontine tegmentum; prefrontal cortex; nucleus accumbens; striatum; ventral tegmental area

Address for correspondence: Susan R. Sesack, Department of Neuroscience, University of Pittsburgh, 446 Crawford Hall, Pittsburgh, PA 15260. Voice: 412-624-5158; fax: 412-624-9878.
sesack@bns.pitt.edu

Ann. N.Y. Acad. Sci. 1003: 36–52 (2003). © 2003 New York Academy of Sciences.
doi: 10.1196/annals.1300.066

INTRODUCTION

Midbrain dopamine (DA) neurons mediate crucial modulatory actions on their target neurons in the forebrain and, in so doing, provide an essential regulatory control over fine motor skills, motivated behaviors, and cognitive performance. A deficit in DA cell activity or transmission at DA receptors compromises normal function in target regions and contributes to the pathophysiology of neurological and mental health disorders, including Parkinson's disease, substance abuse, and schizophrenia. Moreover, the intricate physiology that is associated with neuromodulation, as opposed to typical excitatory or inhibitory transmission, make pharmaceutical manipulation of DA receptors a useful approach for treating mental disorders, even if there is no underlying defect in the DA system.

The functional role of DA is often inextricably linked with the actions of glutamate. This is not to say that DA has no effects independent of glutamate, but rather that DA's critical modulatory functions in the nervous system often involve regulation of or by glutamate. This chapter will focus on the anatomical substrates that facilitate this essential interrelationship between the two transmitters. The data is derived primarily from ultrastructural studies that combine tract-tracing with immunocytochemical methods. Interactions between DA and glutamate will be reviewed at three levels: DA innervation of glutamate neurons, glutamate innervation of DA cells, and the modulation of glutamate transmission by DA via convergent synaptic connections onto common dendritic targets. Interactions will be emphasized between DA neurons and glutamate pyramidal cells of the prefrontal cortex (PFC), a major synaptic glutamate target for DA, one of the principal sources of excitatory drive to DA cells, and a primary source of glutamate transmission that is modulated by DA via convergent synapses. At all three levels, evidence will be presented for specificity in the synaptic connections formed by DA neurons. Data will also be shown to support actions of DA that are mediated through extrasynaptic mechanisms, and a brief speculation will be offered to reconcile the seeming contradiction between DA acting via specific synaptic mechanisms as well as by extrasynaptic means.

REGULATION BY DA OF GLUTAMATE NEURONS IN TARGET AREAS

Direct Synapses onto Glutamate Cells in the PFC

Midbrain DA neurons project to a number of forebrain areas, where their targets primarily include GABA neurons, particularly within the basal ganglia.[1] However, DA projections to the cerebral cortex involve synaptic connections to both glutamate-containing pyramidal projection cells[2] and to GABA local circuit neurons.[3] Such connections have been established for both rat and monkey PFC,[2,4] as well as for primate motor and entorhinal cortices,[3,5] making it likely that this synaptic organization occurs for all cortical areas that receive DA input. DA axons examined by light microscopy in the primate PFC are observed in close proximity to multiple portions of the pyramidal cell dendritic tree but concentrate primarily at dendritic spines and distal dendritic shafts.[6] The proximal dendrites, main apical shaft, and soma generally do not receive such close contacts from DA axons. These findings replicate previous reports of DA synapses by electron microscopy[2,4] and indicate that DA axons are selective in the dendritic compartments of glutamate neurons that they synaptically regulate.

Our laboratory has been interested in whether there is specificity in the populations of pyramidal neurons targeted by DA afferents. Pyramidal cells exhibit a fairly uniform neurochemical phenotype, although they can be differentiated based on the principal target of their extrinsically directed axon. We therefore initiated a series of studies examining DA synaptic contacts onto different pyramidal neurons identified by their primary efferent target. The experiments involved injection of viral tract-tracing agents into the target area with the intention of infecting retrogradely labeled cells to the point where spines and distal dendrites contained extensive viral particles but prior to transneuronal infection of second order cells. Using this method, we demonstrated that DA axons in the PFC synapse onto pyramidal neurons that innervate the nucleus accumbens[7] (i.e., a major subcortical target) and onto pyramidal neurons that innervate the contralateral PFC[8] (i.e., a major cortical target). Unfortunately, further investigation of this question was prevented by the failure of certain pyramidal populations to transport the virus. Hence, the question of whether DA afferents to the PFC synaptically regulate different populations of pyramidal neurons remains to be more thoroughly addressed. However, on the basis of light microscopic studies of Krimer and Goldman-Rakic,[6] it appears that DA axons contact most classes of pyramidal cells.

Indirect Actions Mediated via Synapses onto GABA Cells

The specificity of DA synaptic organization in the PFC is clearly manifest at the level of GABA local circuit neurons. Following the demonstration that DA axons synapse onto GABA cells in multiple cortical areas,[3] our laboratory was interested in determining whether all classes of GABA neurons receive such input. Unlike pyramidal neurons, GABA cells can be differentiated by neurochemical phenotype, with each major class expressing primarily one calcium binding protein.[9] We therefore used immunocytochemistry for calcium binding proteins to determine which major cell populations were synaptically innervated by DA. These studies, which were performed in the monkey PFC, provided evidence that DA axons synapse onto GABA neurons that express parvalbumin[3] but not those expressing calretinin.[10] Examination of local circuit neurons containing calbindin was complicated by the presence of this protein in a substantial population of pyramidal neurons. However, to date, no significant DA innervation of this third major class of interneurons has been identified. The parvalbumin-containing population of local circuit neurons is known to include chandelier and wide-arbor cells that make proximal synaptic contacts onto the soma and axon initial segments of pyramidal neurons.[9,11] The fact that this cell class is synaptically regulated by DA indicates that DA exerts a specific and potent inhibitory indirect action on pyramidal neurons in addition to its direct innervation of selective pyramidal cell dendritic compartments.

Extrasynaptic Actions of DA

In contrast to the formation of synapses onto selective portions of pyramidal neurons and specific classes of local circuit neurons, several lines of evidence suggest that

DA fibers mediate actions via nonsynaptic mechanisms.[12] For example, when varicosities along DA axons are serially reconstructed, it is invariably observed that some do not form morphologically identified synapses.[4,13] Although such varicosities may represent storage sites for DA-containing vesicles, new evidence from other systems suggests that vesicle exocytosis can occur at ectopic locations.[14] In addition, it is likely that DA released from morphologically defined synapses can escape the synaptic cleft and diffuse in the extracellular space.[15] Consistent with this suggestion are observations regarding the distribution of the DA transporter, particularly within the PFC. In striatal axons, DA transporters are localized to nerve terminal membranes immediately adjacent to synaptic specializations,[16] a position that allows transporters to limit DA spillover from the synaptic cleft. However, DA axons in the PFC express considerably less DA transporter (with some axons appearing to express none), and transporter that is observed is located at a distance from sites of synaptic release.[17]

These observations suggest that DA has the capacity to act on receptors that are positioned at extrasynaptic locations, particularly within the PFC. Observations from electron microscopic and electrophysiological studies of glutamate neurons in the PFC are consistent with this suggestion. The main DA receptors localized by ultrastructural studies in the PFC are the D1 and D5 subtypes. Immunoperoxidase labeling of either receptor reveals a predominant localization to the spines and distal dendrites of pyramidal neurons,[18,19] that is, the principal compartments innervated by DA synapses. However, immunoreactivity for the D1 receptor within spines has been localized postsynaptic to excitatory, non-DA terminals,[18] and the D5 receptor is extensively localized to proximal dendritic shafts, including the main apical dendrites of pyramidal cells. The latter compartments rarely receive synaptic input from DA axons. Although these observations were based on immunoperoxidase methods that generally have inadequate subcellular resolution, they are consistent with a number of electrophysiological experiments that report D1 receptor-mediated actions of DA on voltage-gated ion channels localized to the soma and apical dendrites of pyramidal cells.[20–22] Such effects are generally produced by bath application of DA or receptor agonists and are unlikely to reflect the synaptic actions of DA mediated at distal sites.

Summary

Together, these observations suggest that DA signals target neurons via two modes, one that is synaptic and that demonstrates a considerable degree of specificity in the cell types and dendritic compartments innervated, and another that is extrasynaptic and therefore less specific. Such findings raise the question of why DA uses two modes of signaling. Although there may be several potential explanations for this tendency, it is possible that DA transmission conforms to the rules of stochastic resonance.[12] According to this view, extrasynaptic DA release and diffusion would produce a background tone of receptor activation that boosts rather than masks synaptic DA inputs that are likely to be weak by virtue of their small size and distal position. In this case, extrasynaptic DA would raise the probability of target neuron response to ensuing synaptic DA release. The boundaries of such a mode of communication would be constrained by the distribution of DA receptors and the distance between those receptors and DA release sites. Moreover, either raising or lowering extrasynaptic DA levels would be predicted to have a negative impact on cognitive function in the PFC, as such levels would either mask or insufficiently boost synaptic DA effects. This suggestion is consistent with neurochemical studies suggesting that some optimal level of DA D1 receptor stimulation is required for appropriate cognitive functions of the PFC.[23]

REGULATION BY GLUTAMATE OF MIDBRAIN DA NEURONS

Physiological Actions of Glutamate

In a reciprocal manner, DA neurons that modulate the activity of glutamate neurons in the cortex are themselves subject to afferent glutamate regulation within the ventral midbrain. Afferent control of DA neurons is critical for providing phasic disruptions in firing that signal behaviorally relevant events in an otherwise regular activity pattern. DA neurons are endowed with pacemaker potentials that allow them to fire in a tonic mode without excitatory drive from extrinsic sources.[24,25] However, both excitatory and inhibitory inputs[26] allow for variation in action potential frequency, including bursts and pauses,[24,27] that form an important component in the signaling capacity of DA cells.[28,29] For example, recordings of DA neurons in awake behaving monkeys demonstrate that DA cells fire a burst of action potentials to rewards that are unexpected.[29] When the animal learns to associate reward with a conditioned stimulus, DA neurons cease responding to the reward and respond instead to the conditioned stimulus. Similarly, DA cells suppress firing when anticipated rewards are not received. One interpretation of these findings is that DA neurons signal a "prediction error" that differentiates the presence of reward versus its expectation.[30] Alternatively, DA cell activity changes may signal attentional shifts associated with approach behaviors.[31] Regardless of interpretation, phasic changes in DA cell firing are likely to reflect the influence of synaptic inputs to these cells.

Our laboratory is interested in delineating the exact sources of excitatory and inhibitory inputs to different populations of DA cells. The consideration of DA neurons as comprising unique populations with different anatomy and physiology is based on long-standing observations regarding the unique neurochemical and electrophysiological properties of the DA neurons that project to the PFC versus striatal or limbic structures[32–34] and on observations that the ascending projections from the ventral tegmental area (VTA) arise from largely noncollateralized cells, at least in the rat.[35] Our initial studies on afferents to DA cells have focused on glutamate inputs in order to determine the extrinsic regions capable of driving burst firing in these neurons. The glutamate innervation of the VTA derives from only a limited number of sources, with the two major inputs coming from the PFC and the brain stem laterodorsal/pedunculopontine tegmentum (LDT/PPT).[36,37] Both structures activate bursts in DA cells[38,39] of the substantia nigra (SN) or VTA. The SN receives an additional glutamate input from the subthalamic nucleus, but a correlate of this structure for the VTA has yet to be identified.

Glutamate Afferents from the PFC

In the first investigation of its kind,[40] we examined whether two of the major DA cell populations, that is, those projecting to the PFC (mesoprefrontal) and those targeting the nucleus accumbens (NAc; mesoaccumbens), differed in their synaptic input from the PFC itself. The anatomical method for this study involved a combination of retrograde tract-tracing from the NAc or PFC, anterograde tract-tracing from the PFC, and immunocytochemical localization of phenotypic markers for DA or GABA cells. GABA neurons in the VTA project to the same targets as DA cells[41,42] and so provide an important comparison group. In the simple case of phenotypic targets, we observed that the PFC synapsed onto both DA and GABA neurons in the VTA. The former result replicated our prior findings,[36] whereas the latter result verified the expectation that some of the non–DA cells that receive synaptic input from the PFC are GABA neurons.

With regard to different cell populations, we then found that PFC afferents exhibited a considerable degree of specificity in their synaptic targets, such that they innervated mesoaccumbens GABA but not DA cells and mesoprefrontal DA but not GABA neurons.[40] These results are consistent with, and may potentially explain, some of the unique physiological properties of mesoprefrontal versus mesoaccumbens DA neurons. For example, DA cells that project to the PFC reportedly have a higher baseline firing rate, fire more action potentials in bursts, have a higher turnover and metabolism of DA, and are more sensitive to mild stressful stimuli.[32–34] The unique characteristics of mesoprefrontal DA neurons probably reflect a combination of differential pharmacology and afferent drive. With regard to synaptic input, heightened firing rate and burst firing may be due to an overall greater amount of excitatory drive from the PFC to the mesoprefrontal population. Differential PFC inputs may also underlie the greater sensitivity of mesoprefrontal DA neurons to stress, particularly as the PFC is itself highly stress responsive.[43]

Although differential PFC input is consistent with the known properties of mesoprefrontal DA neurons, our anatomical data seemed to be at odds with neurochemical observations for mesoaccumbens cells. Specifically, high-frequency electrical stimulation of the PFC had been reported to increase DA release within the NAc through a mechanism involving the VTA.[44] It is possible that some of this activation might be mediated via a disynaptic projection from the PFC to brain stem LDT neurons that project to the VTA (see below). However, subsequent analysis of the PFC to VTA pathway using low-frequency stimulation within the physiological range of PFC neurons indicated that such activation actually decreased the release of DA in the NAc.[45] This latter finding agrees with the anatomy described by our laboratory as well as with electrophysiological data suggesting that the more common response evoked in VTA DA cells by PFC stimulation is inhibition.[46] The inhibitory connection that appears to interface the PFC with mesoaccumbens DA neurons could derive either from the NAc or from the local connections of GABA mesoaccumbens neurons within the VTA.

Important implications may derive from our research when one considers the circuitry that might link hypofunction of the PFC in schizophrenia[47,48] to altered DA transmission. First, if the principal action of the PFC on mesoaccumbens DA neurons is indirect inhibition, then reduced PFC activity would remove this influence and promote excessive subcortical DA transmission. This suggestion is consistent with imaging studies that assess baseline and evoked DA release in the basal ganglia of schizophrenic patients.[49,50] Second, the direct excitatory connections of the PFC to mesoprefrontal DA neurons suggest that reduced transmission in this pathway would remove an important driving influence for DA afferents to the PFC, which would further degrade PFC function[51] in schizophrenia. In support of this idea, recent anatomical studies report a loss of deep-layer DA input to the PFC in this illness.[52] A more detailed version of the present model has been published,[53] and we acknowledge that it derives from the previous work of Carlsson.[54] According to Carlsson's model, the PFC provides both a braking and an accelerating influence on DA neurons as a whole. We suggest that the PFC selectively modulates different populations of DA cells in an opposing manner, slowing down the firing of mesoaccumbens neurons while accelerating activity of mesoprefrontal cells, and that this influence is lost in schizophrenia when the normal outflow of the PFC is reduced.

Glutamate Afferents from the Pontine Tegmentum

Our initial study of the PFC innervation to the VTA raised an obvious question as to the source of excitatory drive to mesoaccumbens DA neurons. We are currently engaged in an investigation of the other major source of glutamate input to the VTA from the brain stem LDT/PPT. In the monkey VTA, afferents from the PPT have been shown to synapse onto DA neurons,[37] although different populations of DA cells were not investigated in that study. In the rat, we chose to first examine inputs from the LDT, because this region appears to generate a more prominent projection to the VTA based on light microscopic anatomy and neurochemistry.[55,56] We have verified that the LDT in the rat forms synapses with excitatory morphology onto DA neurons with a strong preference for this cell type versus non–DA cells (i.e., approximately 80%). We have furthermore demonstrated that the LDT clearly innervates mesoaccumbens DA neurons,[57] a finding consistent with neurochemical studies reporting that the LDT regulates DA release in the NAc.[56,58] The LDT input to the VTA appears to be less specific than that from the PFC, in that the LDT also innervates mesoprefrontal DA cells, albeit at a lower frequency than the mesoaccumbens population. However, inputs from the LDT to GABA neurons do appear to be specific, preferring the mesoprefrontal and not mesoaccumbens GABA cells,[57] that is, the populations not targeted by PFC afferents to the VTA.[40]

These more preliminary observations with regard to LDT afferents suggest that mesoaccumbens and mesoprefrontal DA neurons are differentially regulated by different sources of excitatory drive. In this regard, it should be noted that the PFC input to the VTA consists of a single glutamate phenotype, whereas LDT afferents arise from cells with at least three different transmitter phenotypes: glutamate, acetylcholine, and GABA.[37,58–60] Our research has shown that mesoaccumbens DA neurons are synaptically innervated by cholinergic terminals (unpublished observations) that most likely originate from the LDT/PPT. This verifies and extends a previous observation that VTA DA neurons receive cholinergic synapses, although the cell populations innervated were not identified in that study.[61] Hence, some of the excitatory drive from the LDT to mesoaccumbens DA cells is likely to be mediated via acetylcholine as well as by glutamate,[58] possibly released from the same neurons.[62] We are currently investigating these phenotypes as well as presumed GABA inhibitory afferents from the LDT to the VTA with respect to their specific synaptic targets.

Summary

In summary, excitatory afferents from the PFC to the VTA synapse selectively onto DA neurons that project back to the PFC, whereas presumed excitatory inputs from the LDT innervate mesoaccumbens DA neurons with greater frequency than mesoprefrontal cells. These observations suggest that the PFC is responsible for regulating its own modulatory control by DA in a manner that may reflect a more general principle for DA afferent regulation. Specifically, we propose that different populations of DA neurons are innervated predominantly by the target areas to which they project, or by the regions that, in functional terms, are the most closely linked to the target area. In the case of the LDT, the above model predicts dual innervation of both mesoaccumbens and mesoprefrontal DA neurons, based on the connectivity of this structure. More specifically, the LDT receives afferent drive from both the PFC and the basal ganglia,[63] and so might be recruited to help regulate the modulatory DA tone within both cortical and basal ganglia structures. An important test of our hypothesis will be to examine VTA afferents from the PPT, as this region is similarly innervated by basal ganglia structures but receives little PFC input. We predict that PPT afferents to the VTA will selectively target mesoaccumbens DA cells, but this prediction requires experimental testing.

CONVERGENT REGULATION BY DA AND GLUTAMATE OF COMMON TARGET NEURONS

Dorsal and Ventral Striatum

Perhaps the level of modulatory interaction between DA and glutamate that is best known and most extensively researched is the ability of DA to regulate glutamate transmission at sites where these two inputs converge onto common target neurons. The first instance where this convergence was identified was in the dorsal striatum, where DA nerve terminals were found to synapse onto the heads or necks of spines emanating from medium spiny neurons that also received glutamate synaptic input from the sensorimotor cortex onto the same spines.[64,65] Since that time, this convergent arrangement, sometimes referred to as a triad, has been verified in the monkey dorsal striatum[66] and has been extended to include all sources of cortical input to different portions of the striatal complex that have been investigated to date: hippocampus, prefrontal cortex, and amygdala.[36,67–69]

Midline and intralaminar thalamic nuclei constitute another major source of excitatory, presumed glutamate inputs to the striatum. However, in the monkey dorsal striatum, DA axons were reported not to form convergent synaptic relationships with afferents from the centromedian thalamus,[66] suggesting that the convergence of DA and glutamate nerve terminals might be restricted to cortical afferents. Nevertheless, our laboratory has recently demonstrated in the rat NAc that DA axons synapse onto spines and dendrites that receive convergent synaptic input from the paraventricular nucleus of the thalamus,[70] suggesting that DA afferents are capable of modulating thalamostriatal glutamate drive. The incongruity with the prior publication may be due to species or regional differences, or both.

Prefrontal Cortex

Although there has been little difficulty in identifying the cortical and thalamic sources of glutamate that converge with DA axons in the striatum, the same has not held true for determining the sources of glutamate drive that are modulated by convergent DA synapses in the PFC. First, the incidence of spines that receive dual synaptic input from terminals with glutamate morphology and those with presumed inhibitory or modulatory function appears to be lower in cortical than in striatal structures.[71] Second, the incidence with which DA terminals synapse onto spines versus distal dendrites may be lower in the PFC than in the striatum,[12] suggesting that DA may mediate a more proximal, and therefore less specific (see below), modulation of glutamate drive in the cortex. Finally, we have examined several of the major glutamate afferents to the rat PFC, and although approximately 95% of their synapses are onto dendritic spines, we have failed to observe convergent synapses by DA axons onto the same spines. The inputs examined to date include the hippocampus, contralateral PFC, paraventricular nucleus of the thalamus, and basolateral amygdala.[70–73] The results are somewhat surprising, because these glutamate projections to the PFC also innervate the NAc, and each has been shown to converge with DA axons onto common spines in the latter structure. However, within the PFC, the spines that are innervated by these excitatory afferents either fail to exhibit synapses from secondary axon terminals or receive synaptic input from non–DA sources.[72]

Given that the PFC contains intrinsic glutamate neurons, in addition to receiving extrinsically derived glutamate, we have been exploring whether DA afferents might modulate glutamate synapses from the intrinsic collaterals of pyramidal neurons at discrete spines. Although some preliminary data is consistent with this suggestion,[74] the incidence of such interactions appears to be low. The current state of data in the rat PFC therefore suggests two possible, although not necessarily mutually exclusive, scenarios. First, multiple sources of glutamate afferents may synapse onto spines that receive convergent DA input, but each with a low incidence that has so far escaped detection. Second, DA axons may synapse onto spines that receive glutamate drive from a highly specific source that has yet to be identified. Additional studies are required to clarify these possibilities.

Amygdala

In addition to the PFC, DA axons innervate another forebrain target, the basolateral amygdala,[75] that both contains glutamate principal neurons and receives extrinsic glutamate drive that may be modulated by DA. The PFC is itself a substantial afferent to the basolateral complex, as well as to the medially adjacent intercalated cell masses of the amygdala.[76,77] Moreover, electrophysiological studies suggest that the response of amygdala neurons to PFC drive is regulated by DA receptors.[78] We recently examined whether PFC afferents to the amygdala target spines that receive convergent synaptic input from DA nerve terminals. Such triadic arrangements were observed, although the incidence of such relationships was infrequent,[79] suggesting that functional interactions between DA and PFC glutamate are likely to involve primarily extrasynaptic mechanisms in the amygdala. Consistent with this suggestion, we further observed that PFC axon terminals in the amygdala often synapsed onto spines that were immunoreactive for the DA D1 receptor.[80] Interestingly, in adjacent sections, PFC afferents were not found to contact spines or dendrites that were labeled for the D2 receptor, although PFC axons themselves contained D2 receptor immunolabeling.[80] We interpret these findings to indicate that DA modu-

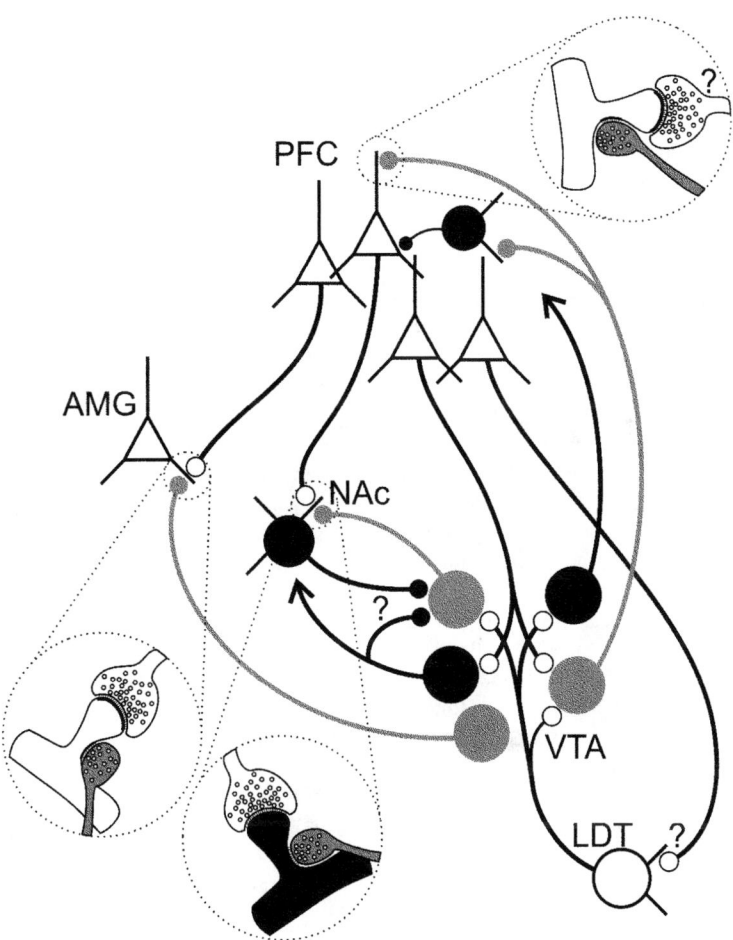

FIGURE 1. Schematic drawing of several major sites of synaptic interaction between dopamine and glutamate neurons in the rat brain, emphasizing connections that involve the prefrontal cortex and including those that involve indirect connections via GABA cells. Axonal projections ending in round varicosities denote identified synaptic connections; a ? at some of these sites indicates that ultrastructural studies have not yet verified connections that are presumed to exist based on light microscopic and/or physiological data. Axons ending in *open arrows* indicate projections whose synaptic targets remain to be identified. *Dashed areas* shown at higher magnification indicate sites at which dopamine and glutamate afferents converge synaptically onto the spines or distal dendrites of common target neurons. Within the PFC, the source of glutamate axons that synapse onto the same spines as dopamine inputs remains to be determined. Abbreviations: AMG, amygdala; LDT, laterodorsal tegmentum; NAc, nucleus accumbens; PFC, prefrontal cortex; VTA, ventral tegmental area. *White shapes,* glutamate; *black shapes,* GABA; *gray shapes,* dopamine. Adapted from a figure in a prior publication.[53]

lates PFC transmission in the amygdala via presynaptic D2 and postsynaptic D1 receptors and that both of these actions are likely to primarily involve extrasynaptic effects of DA.

Summary

The functional significance of the triadic relationship between DA and glutamate afferents targeting the same dendritic spines remains to be fully explained. Traditionally, DA has been thought to gate glutamate drive, either facilitating or attenuating transmission depending on various factors, including the types of DA and glutamate receptors expressed.[81] Such gating functions are consistent with the findings of *in vivo* electrophysiological studies reporting that the activity of target neurons evoked by stimulation of glutamate afferents is altered in the presence or absence of DA.[82–86]

The close proximity that is afforded by convergence of DA and glutamate inputs onto common target cells may also serve to facilitate presynaptic interactions between these transmitters. Clear axoaxonic synapses have not been reported in DA target regions, indicating that presynaptic interactions are likely to occur via extrasynaptic mechanisms. Hence, the distance over which DA must diffuse in order to reach receptors expressed by glutamate axons is minimized when these afferents synapse onto the same distal dendrites. The predominant DA receptor that has been localized to glutamate-type or identified cortical axons is the D2 subtype,[80,87,88] consistent with electrophysiological and neurochemical reports of presynaptic DA actions via this receptor.[89–92] Presynaptic glutamate receptors on DA nerve terminals have less supportive evidence in the anatomical literature, although at least one study reports NMDA type receptors on DA axons in the NAc,[93] and neurochemical studies indicate that glutamate can regulate DA release.[43,94,95]

As mentioned earlier, the results of *in vitro* recording studies indicate that DA has actions on the general excitability of target neurons through mechanisms involving voltage-gated ion channels on soma and proximal dendrites.[20–22,96,97] Although DA terminals rarely synapse at the latter sites,[4,13] functional D1 and/or D2 receptors have been localized to the extrasynaptic portions of the plasmalemmal surface of target neurons in the striatum and cortex.[18,19,98–100] Hence, DA may serve to modulate neuronal excitability generally, although such proximal extrasynaptic actions of DA would alter the efficacy with which glutamate depolarizes target neurons without regard to the specific sources of glutamate or the spines onto which glutamate axons synapse.

It should therefore be considered that the spatial and temporal coordination of DA and glutamate transmission that can occur via synapses onto discrete spines may serve a function that is more chronic than acute in nature. For example, the triadic arrangement may facilitate long-term DA modulation of glutamate transmission, depending on the frequency of synaptic activity at those particular sites. In this regard, the ability of DA to regulate glutamate synaptic plasticity in the striatum has been well documented.[101] Moreover, the loss of DA inputs to this region or blockade of DA receptors results in pruning of spine targets and morphological alterations in their cortical glutamate afferents.[102,103] Thus, DA axons may serve to stabilize the spine synapses of the particular glutamate inputs with which they share convergent targets.

By way of general conclusion, research from this and other laboratories has used ultrastructural methods to examine reciprocal anatomical connections between DA and glutamate neurons that also innervate common target regions. Cumulative evidence suggests that the sites at which DA and glutamate mediate synaptic interactions are fairly selective in terms of the neuron populations targeted either directly or indirectly, the dendritic compartments of target neurons that are specifically innervated, and the sources of glutamate that converge with DA axons at common sites. In addition, it seems clear that DA also mediates a number of important actions beyond the level of synapses, although these extrasynaptic effects are nevertheless constrained by the localization of DA receptors, the levels of extracellular DA achieved at these sites, and consequently the afferent drive of DA neuron firing. In future studies, it will be important to explore the detailed mechanisms whereby modulatory interactions between DA and glutamate cells regulate normal brain function and contribute to the pathophysiology of mental health disorders.

REFERENCES

1. PICKEL, V.M., A. TOWLE, T.H. JOH, et al. 1988. Gamma-aminobutyric acid in the medial rat nucleus accumbens: ultrastructural localization in neurons receiving monosynaptic input from catecholaminergic afferents. J. Comp. Neurol. **272:** 1–14.
2. GOLDMAN-RAKIC, P.S., C. LERANTH, S.M. WILLIAMS, et al. 1989. Dopamine synaptic complex with pyramidal neurons in primate cerebral cortex. Proc. Natl. Acad. Sci. USA **86:** 9015–9019.
3. SESACK, S.R., C.L. SNYDER & D.A. LEWIS. 1995. Axon terminals immunolabeled for dopamine or tyrosine hydroxylase synapse on GABA-immunoreactive dendrites in rat and monkey cortex. J. Comp. Neurol. **363:** 264–280.
4. SÉGUÉLA, P., K.C. WATKINS & L. DESCARRIES. 1988. Ultrastructural features of dopamine axon terminals in the anteromedial and the suprarhinal cortex of adult rat. Brain Res. **442:** 11–22.
5. ERICKSON, S.L., S.R. SESACK & D.A. LEWIS. 2000. The dopamine innervation of monkey entorhinal cortex: postsynaptic targets of tyrosine hydroxylase-immunoreactive terminals. Synapse **36:** 47–56.
6. KRIMER, L.S., R.L. JAKAB & P.S. GOLDMAN-RAKIC. 1997. Quantitative three-dimensional analysis of the catecholaminergic innervation of identified neurons in the macaque prefrontal cortex. J. Neurosci. **17:** 7450–7461.
7. CARR, D.B., P. O'DONNELL, J.P. CARD, et al. 1999. Dopamine terminals in the rat prefrontal cortex synapse on pyramidal cells that project to the nucleus accumbens. J. Neurosci. **19:** 11049–11060.
8. CARR, D.B. & S.R. SESACK. 2000. Dopamine terminals synapse on callosal projection neurons in the rat prefrontal cortex. J. Comp. Neurol. **425:** 275–283.
9. CONDÉ, F., J.S. LUND, D.M. JACOBOWITZ, et al. 1994. Local circuit neurons immunoreactive for calretinin, calbindin D-28k or parvalbumin in monkey prefrontal cortex: distribution and morphology. J. Comp. Neurol. **341:** 95–116.
10. SESACK, S.R., C.N. BRESSLER & D.A. LEWIS. 1995. Ultrastructural associations between dopamine terminals and local circuit neurons in the monkey prefrontal cortex: a study of calretinin-immunoreactive cells. Neurosci. Lett. **200:** 9–12.
11. WILLIAMS, S.M., P.S. GOLDMAN-RAKIC & C. LERANTH. 1992. The synaptology of parvalbumin-immunoreactive neurons in the primate prefrontal cortex. J. Comp. Neurol. **320:** 353–369.
12. SESACK, S.R. 2002. Synaptology of dopamine neurons. In Handbook of Experimental Pharmacology, Dopamine in the CNS I, Vol. 154/1. G. Di Chiara, Ed.: 63–119. Springer-Verlag. Berlin, Heidelberg.

13. DESCARRIES, L., K.C. WATKINS, S. GARCIA, et al. 1996. Dual character, asynaptic and synaptic, of the dopamine innervation in adult rat neostriatum: a quantitative autoradiographic and immunocytochemical analysis. J. Comp. Neurol. **375:** 167–186.
14. ZENISEK, D., J.A. STEYER & W. ALMERS. 2000. Transport, capture and exocytosis of single synaptic vesicles at active zones. Nature **406:** 849–854.
15. FUXE, K. & L.F. AGNATI. 1991. Two principal modes of electrochemical communication in the brain: volume versus wiring transmission. In Volume Transmission in the Brain: Novel Mechanisms for Neural Transmission. K. Fuxe & L.F. Agnati, Ed.: 1–9. Raven Press, Ltd. New York.
16. HERSCH, S.M., H. YI, C.J. HEILMAN, et al. 1997. Subcellular localization and molecular topology of the dopamine transporter in the striatum and substantia nigra. J. Comp. Neurol. **388:** 211–227.
17. SESACK, S.R., V.A. HAWRYLAK, M.A. GUIDO, et al. 1998. Dopamine axon varicosities in the rat prefrontal cortex exhibit sparse immunoreactivity for the dopamine transporter. J. Neurosci. **18:** 2697–2708.
18. SMILEY, J.F., A.I. LEVEY, B.J. CILIAX, et al. 1994. D_1 dopamine receptor immunoreactivity in human and monkey cerebral cortex: predominant and extrasynaptic localization in dendritic spines. Proc. Natl. Acad. Sci. **91:** 5720–5724.
19. BERGSON, C., L. MRZLJAK, J.F. SMILEY, et al. 1995. Regional, cellular, and subcellular variations in the distribution of D_1 and D_5 receptors in primate brain. J. Neurosci. **15:** 7821–7836.
20. GEIJO-BARRIENTOS, E. & C. PASTORE. 1995. The effects of dopamine on the subthreshold electrophysiological responses of rat prefrontal cortex neurons in vitro. Eur. J. Neurosci. **7:** 358–366.
21. YANG, C.R. & J.K. SEAMANS. 1996. Dopamine D1 receptor actions in layers V-VI rat prefrontal cortex neurons in vitro: modulation of dendritic-somatic signal integration. J. Neurosci. **16:** 1922–1935.
22. HENZE, D.A., G.R. GONZALEZ-BURGOS, N.N. URBAN, et al. 2000. Dopamine increases excitability of pyramidal neurons in primate prefrontal cortex. J. Neurophysiol. **84:** 2799–2809.
23. GOLDMAN-RAKIC, P.S., E.C. MULLY 3RD & G.V. WILLIAMS. 2000. D_1 receptors in prefrontal cells and circuits. Brain Res. Rev. **31:** 295–301.
24. GRACE, A.A. & B.S. BUNNEY. 1984. The control of firing pattern in nigral dopamine neurons: single spike firing. J. Neurosci. **4:** 2866–2876.
25. GRACE, A.A. & S. ONN. 1989. Morphology and electrophysiological properties of immunocytochemically identified rat dopamine neurons recorded in vitro. J. Neurosci. **9:** 3463–3481.
26. KITAI, S.T., P.D. SHEPARD, J.C. CALLAWAY, et al. 1999. Afferent modulation of dopamine neuron firing patterns. Curr. Opin. Neurobiol. **9:** 690–697.
27. GRACE, A.A. & B.S. BUNNEY. 1984. The control of firing pattern in nigral dopamine neurons: burst firing. J. Neurosci. **4:** 2877–2890.
28. GONON, F.G. 1988. Nonlinear relationship between impulse flow and dopamine released by rat midbrain dopaminergic neurons as studied by in vivo electrochemistry. Neuroscience **24:** 19–28.
29. SCHULTZ, W. 1998. Predictive reward signal of dopamine neurons. J. Neurophysiol. **80:** 1–27.
30. MONTAGUE, P.R., P. DAYAN & T.J. SEJNOWSKI. 1996. A framework for mesencephalic dopamine systems based on predictive Hebbian learning. J. Neurosci. **16:** 1936–1947.
31. REDGRAVE, P., T.J. PRESCOTT & K. GURNEY. 1999. Is the short-latency dopamine response too short to signal reward error? Tr. Neurosci. **22:** 146–151.
32. BANNON, M.J. & R.H. ROTH. 1983. Pharmacology of mesocortical dopamine neurons. Pharmacol. Rev. **35:** 53–68.
33. WHITE, F.J. 1996. Synaptic regulation of mesocorticolimbic dopamine neurons. Ann. Rev. Neurosci. **19:** 405–436.
34. TZSCHENTKE, T.M. 2001. Pharmacology and behavioral pharmacology of the mesocortical dopamine system. Prog. Neurobiol. **63:** 241–320.

35. SWANSON, L.W. 1982. The projections of the ventral tegmental area and adjacent regions: a combined fluorescent retrograde tracer and immunofluorescence study in the rat. Brain Res. Bull. **9:** 321–353.
36. SESACK, S.R. & V.M. PICKEL. 1992. Prefrontal cortical efferents in the rat synapse on unlabeled neuronal targets of catecholamine terminals in the nucleus accumbens septi and on dopamine neurons in the ventral tegmental area. J. Comp. Neurol. **320:** 145–160.
37. CHARARA, A., Y. SMITH & A. PARENT. 1996. Glutamatergic inputs from the pedunculopontine nucleus to midbrain dopaminergic neurons in primates: *Phaseolus vulgaris*-leucoagglutinin anterograde labeling combined with postembedding glutamate and GABA immunohistochemistry. J. Comp. Neurol. **364:** 254–266.
38. TONG, Z.-Y., P.G. OVERTON & D. CLARK. 1996. Stimulation of the prefrontal cortex in the rat induces patterns of activity in midbrain dopaminergic neurons which resemble natural burst events. Synapse **22:** 195–208.
39. LOKWAN, S.J., P.G. OVERTON, M.S. BERRY, et al. 1999. Stimulation of the pedunculopontine tegmental nucleus in the rat produces burst firing in A9 dopaminergic neurons. Neuroscience **92:** 245–254.
40. CARR, D.B. & S.R. SESACK. 2000. Projections from the rat prefrontal cortex to the ventral tegmental area: target specificity in the synaptic associations with mesoaccumbens and mesocortical neurons. J. Neurosci. **20:** 3864–3873.
41. VAN BOCKSTAELE, E.J. & V.M. PICKEL. 1995. GABA-containing neurons in the ventral tegmental area project to the nucleus accumbens in rat brain. Brain Res. **682:** 215–221.
42. CARR, D.B. & S.R. SESACK. 2000. GABA-containing neurons in the rat ventral tegmental area project to the prefrontal cortex. Synapse **38:** 114–123.
43. TAKAHATA, R. & B. MOGHADDAM. 1998. Glutamatergic regulation of basal and stimulus-activated dopamine release in the prefrontal cortex. J. Neurochem. **71:** 1443–1449.
44. TABER, M.T. & H.C. FIBIGER. 1993. Electrical stimulation of the medial prefrontal cortex increases dopamine release in the striatum. Neuropsychopharmacology **9:** 271–275.
45. JACKSON, M.E., A.S. FROST & B. MOGHADDAM. 2001. Stimulation of prefrontal cortex at physiologically relevant frequencies inhibits dopamine release in the nucleus accumbens. J. Neurochem. **78:** 920–923.
46. TONG, Z.-Y., P.G. OVERTON, C. MARTINEZCUE, et al. 1998. Do non-dopaminergic neurons in the ventral tegmental area play a role in the responses elicited in A10 dopaminergic neurons by electrical stimulation of the prefrontal cortex. Exp. Brain Res. **118:** 466–476.
47. LEWIS, D.A. 1995. Neural circuitry of the prefrontal cortex in schizophrenia. Arch. Gen. Psychiat. **52:** 269–273.
48. WEINBERGER, D.R. & K.F. BERMAN. 1996. Prefrontal function in schizophrenia: confounds and controversies. Phil. Trans. R. Soc. Lond. B **351:** 1495–1503.
49. BREIER, A., T.P. SU, R. SAUNDERS, et al. 1997. Schizophrenia is associated with elevated amphetamine-induced synaptic dopamine concentrations: evidence from a novel positron emission tomography method. Proc. Natl. Acad. Sci. USA **94:** 2569–2574.
50. LARUELLE, M. 2000. The role of endogenous sensitization in the pathophysiology of schizophrenia: implications from recent brain imaging studies. Brain Res. Rev. **31:** 371–384.
51. BROZOSKI, T.J., R.M. BROWN, H.E. ROSVOLD, et al. 1979. Cognitive deficit caused by regional depletion of dopamine in prefrontal cortex of rhesus monkey. Science **205:** 929–932.
52. AKIL, M., J.N. PIERRI, R.E. WHITEHEAD, et al. 1999. Lamina-specific alterations in the dopamine innervation of the prefrontal cortex in schizophrenic subjects. Am. J. Psychiat. **156:** 1580–1589.
53. SESACK, S.R. & D.B. CARR. 2002. Selective prefrontal cortex inputs to dopamine cells: implications for schizophrenia. Physiol. Behav. **77:** 513–517.

54. CARLSSON, A., N. WATERS, S. WATERS, *et al.* 2000. Network interactions in schizophrenia—therapeutic implications. Brain Res. Rev. **31:** 342–349.
55. OAKMAN, S.C., P.L. FARIS, P.E. KERR, *et al.* 1995. Distribution of pontomesencephalic cholinergic neurons projecting to substantia nigra differs significantly from those projecting to ventral tegmental area. J. Neurosci. **15:** 5859–5869.
56. BLAHA, C.D., L. F. ALLEN, S. DAS, *et al.* 1996. Modulation of dopamine efflux in the nucleus accumbens after cholinergic stimulation of the ventral tegmental area in intact, pedunculopontine tegmental nucleus-lesioned, and laterodorsal tegmental nucleus-lesioned rats. J. Neurosci. **16:** 714–722.
57. OMELCHENKO, N.V. & S.R. SESACK. 2002. Projections to the rat ventral tegmental area from the pedunculopontine-laterodorsal tegmentum synapse onto dopamine neurons that project to the nucleus accumbens. Soc. Neurosci. Abstr. **28:** 460.464.
58. FORSTER, G.L. & C.D. BLAHA. 2000. Laterodorsal tegmental stimulation elicits dopamine efflux in the rat nucleus accumbens by activation of acetylcholine and glutamate receptors in the ventral tegmental area. Eur. J. Neurosci. **12:** 3596–3604.
59. LAVOIE, B. & A. PARENT. 1994. Pedunculopontine nucleus in the squirrel monkey: cholinergic and glutamatergic projections to the substantia nigra. J. Comp. Neurol. **344:** 232–241.
60. FORD, B., C.J. HOLMES, L. MAINVILLE, *et al.* 1995. GABAergic neurons in the rat pontomesencephalic tegmentum: codistribution with cholinergic and other tegmental neurons projecting to the posterior lateral hypothalamus. J. Comp. Neurol. **363:** 177–196.
61. GARZÓN, M., R.A. VAUGHAN, G.R. UHL, *et al.* 1999. Cholinergic axon terminals in the ventral tegmental area target a subpopulation of neurons expressing low levels of the dopamine transporter. J. Comp. Neurol. **410:** 197–210.
62. LAVOIE, B. & A. PARENT. 1994. Pedunculopontine nucleus in the squirrel monkey: distribution of cholinergic and monoaminergic neurons in the mesopontine tegmentum with evidence for the presence of glutamate in cholinergic neurons. J. Comp. Neurol. **344:** 210–231.
63. SEMBA, K. & H.C. FIBIGER. 1992. Afferent connections of the laterodorsal and the pedunculopontine tegmental nuclei in the rat: a retro- and antero-grade transport and immunohistochemical study. J. Comp. Neurol. **323:** 387–410.
64. BOUYER, J.J., D.H. PARK, T.H. JOH, *et al.* 1984. Chemical and structural analysis of the relation between cortical inputs and tyrosine hydroxylase-containing terminals in rat neostriatum. Brain Res. **302:** 267–275.
65. FREUND, T.F., J.F. POWELL & A.D. SMITH. 1984. Tyrosine hydroxylase-immunoreactive boutons in synaptic contact with identified striatonigral neurons, with particular reference to dendritic spines. Neuroscience **13:** 1189–1215.
66. SMITH, Y., B.D. BENNETT, J.P. BOLAM, *et al.* 1994. Synaptic relationship between dopaminergic afferents and cortical or thalamic input in the sensorimotor territory of the striatum in monkey. J. Comp. Neurol. **344:** 1–19.
67. TOTTERDELL, S. & A.D. SMITH. 1989. Convergence of hippocampal and dopaminergic input onto identified neurons in the nucleus accumbens of the rat. J. Chem. Neuroanat. **2:** 285–298.
68. SESACK, S.R. & V.M. PICKEL. 1990. In the rat medial nucleus accumbens, hippocampal and catecholaminergic terminals converge on spiny neurons and are in apposition to each other. Brain Res. **527:** 266–279.
69. JOHNSON, L.R., R.L.M. AYLWARD, Z. HUSSAIN, *et al.* 1994. Input from the amygdala to the rat nucleus accumbens: its relationship with tyrosine hydroxylase immunoreactivity and identified neurons. Neuroscience **61:** 851–865.
70. PINTO, A., M. JANKOWSKI & S.R. SESACK. 2003. Projections from the paraventricular nucleus of the thalamus to the rat prefrontal cortex and nucleus accumbens shell: ultrastructural characteristics and spatial relationships with dopamine afferents. J. Comp. Neurol. **459:** 142–155.
71. CARR, D.B. & S.R. SESACK. 1996. Hippocampal afferents to the rat prefrontal cortex: synaptic targets and relation to dopamine terminals. J. Comp. Neurol. **369:** 1–15.
72. CARR, D.B. & S.R. SESACK. 1998. Callosal terminals in the rat prefrontal cortex: synaptic targets and association with GABA-immunoreactive structures. Synapse **29:** 193–205.

73. PINTO, A. & S.R. SESACK. 1999. Basolateral amygdala afferents to the rat prefrontal cortex: ultrastructure and relation to dopamine afferents. Soc. Neurosci. Abstr. **25:** 1216.
74. SESACK, S.R. & L.A.H. MINER. 1997. In the rat prefrontal cortex, dopamine terminals converge with local axon collaterals onto common dendritic processes. Soc. Neurosci. Abstr. **23:** 1213.
75. ASAN, E. 1997. Ultrastructural features of tyrosine-hydroxlyase-immunoreactive afferents and their targets in the rat amygdala. Cell Tissue Res. **288:** 449–469.
76. SESACK, S.R., A.Y. DEUTCH, R.H. ROTH, et al. 1989. Topographic organization of the efferent projections of the medial prefrontal cortex in the rat: an anterograde tract-tracing study using *Phaseolus vulgaris* leucoagglutinin. J. Comp. Neurol. **290:** 213–242.
77. BRINLEY-REED, M., F. MASCAGNI & A.J. MCDONALD. 1995. Synaptology of prefrontal cortical projections to the basolateral amygdala: an electron microscopic study in the rat. Neurosci. Lett. **202:** 45–48.
78. ROSENKRANZ, J.A. & A.A. GRACE. 2002. Cellular mechanisms of infralimbic and prelimbic prefrontal cortical inhibition and dopaminergic modulation of basolateral amygdala neurons in vivo. J. Neurosci. **22:** 324–327.
79. PINTO, A. & S.R. SESACK. 2001. Prefrontal cortex projections to the rat amygdala: ultrastructural relationship to dopamine and serotonin afferents. Soc. Neurosci. Abstr. **27:** 976.
80. PINTO, A. & S.R. SESACK. 2002. Prefrontal cortex projection to the rat amygdala: ultrastructural relationship to dopamine D1 and D2 receptors. Soc. Neurosci. Abstr. **28:** 587.586.
81. CEPEDA, C., N.A. BUCHWALD & M.S. LEVINE. 1993. Neuromodulatory actions of dopamine in the neostriatum are dependent on the excitatory amino acid receptor subtypes activated. Proc. Natl. Acad. Sci. USA **90:** 9576–9580.
82. BROWN, J.R. & G.W. ARBUTHNOTT. 1983. The electrophysiology of dopamine (D_2) receptors: a study of the actions of dopamine on corticostriatal transmission. Neuroscience **10:** 349–355.
83. YANG, C.R. & G.J. MOGENSON. 1984. Electrophysiological responses of neurones in the nucleus accumbens to hippocampal stimulation and the attenuation of the excitatory responses by the mesolimbic dopaminergic system. Brain Res. **324:** 69–84.
84. VIVES, F. & G.J. MOGENSON. 1986. Electrophysiological study of the effects of D_1 and D_2 dopamine antagonists on the interaction of converging inputs from the sensory-motor cortex and substantia nigra neurons in the rat. Neuroscience **17:** 349–359.
85. MANTZ, J., C. MILLLA, J. GLOWINSKI, et al. 1988. Differential effects of ascending neurons containing dopamine and noradrenaline in control of spontaneous activity and of evoked responses in the rat prefrontal cortex. Neuroscience **27:** 517–526.
86. KIYATKIN, E.A. & G.V. REBEC. 1996. Dopaminergic modulation of glutamate-induced excitations of neurons in the neostriatum and nucleus accumbens of awake, unrestrained rats. J. Neurophysiol. **75:** 142–153.
87. SESACK, S.R., C. AOKI & V.M. PICKEL. 1994. Ultrastructural localization of D2 receptor-like immunoreactivity in midbrain dopamine neurons and their striatal targets. J. Neurosci. **14:** 88–106.
88. WANG, H. & V.M. PICKEL. 2002. Dopamine D2 receptors are present in prefrontal cortical afferents and their targets in patches of the rat caudate-putamen nucleus. J. Comp. Neurol. **442:** 392–404.
89. YANG, C.R. & G.J. MOGENSON. 1986. Dopamine enhances terminal excitability of hippocampal-accumbens neurons via D2 receptor: role of dopamine in presynaptic inhibition. J. Neurosci. **6:** 2470–2478.
90. O'DONNELL, P. & A.A. GRACE. 1994. Tonic D_2-mediated attenuation of cortical excitation in nucleus accumbens neurons recorded in vitro. Brain Res. **634:** 105–112.
91. FLORES-HERNÁNDEZ, J., E. GALARRAGA & J. BARGAS. 1997. Dopamine selects glutamatergic inputs to neostriatal neurons. Synapse **25:** 185–195.
92. KALIVAS, P.W. & P. DUFFY. 1997. Dopamine regulation of extracellular glutamate in the nucleus accumbens. Brain Res. **761:** 173–177.

93. GRACY, K.N. & V.M. PICKEL. 1996. Ultrastructural immunocytochemical localization of the N-methyl-D-aspartate receptor and tyrosine hydroxylase in the shell of the rat nucleus accumbens. Brain Res. **739:** 169–181.
94. MORARI, M., M. MARTI, S. SBRENNA, *et al.* 1998. Reciprocal dopamine-glutamate modulation of release in the basal ganglia. Neurochem. Int. **33:** 383–397.
95. HOWLAND, J.G., P. TAEPAVARAPRUK & A.G. PHILLIPS. 2002. Glutamate receptor-dependent modulation of dopamine efflux in the nucleus accumbens by basolateral, but not central, nucleus of the amygdala in rats. J. Neurosci. **22:** 1137–1145.
96. SURMEIER, D.J. & S.T. KITAI. 1997. State-dependent regulation of neuronal excitability by dopamine. Shinkei Seishin Yakurigaku Zasshi **17:** 105–110.
97. WEST, A.R. & A.A. GRACE. 2002. Opposite influences of endogenous dopamine D1 and D2 receptor activation on activity states and electrophysiological properties of striatal neurons: studies combining in vivo intracellular recordings and reverse microdialysis. J. Neurosci. **22:** 294–304.
98. YUNG, K.K., J.P. BOLAM, A.D. SMITH, *et al.* 1995. Immunocytochemical localization of D1 and D2 dopamine receptors in the basal ganglia of the rat: light and electron microscopy. Neuroscience **65:** 709–730.
99. CAILLÉ, I., B. DUMARTIN & B. BLOCH. 1996. Ultrastructural localization of D1 dopamine receptor immunoreactivity in rat striatonigral neurons and its relation with dopaminergic innervation. Brain Res. **730:** 17–31.
100. DUMARTIN, B., I. CAILLÉ, F. GONON, *et al.* 1998. Internalization of D1 dopamine receptor in striatal neurons *in vivo* as evidence of activation by dopamine agonists. J. Neurosci. **18:** 1650–1661.
101. CALABRESI, P., A. PISANI, D. CENTONZE, *et al.* 1997. Synaptic plasticity and physiological interactions between dopamine and glutamate in the striatum. Neurosci. Biobehav. Rev. **21:** 519–523.
102. MESHUL, C.K., R.K. STALLBAUMER, B. TAYLOR, *et al.* 1994. Haloperidol-induced morphological changes in striatum are associated with glutamate synapses. Brain Res. **648:** 181–195.
103. ARBUTHNOTT, G.W., C.A. INGHAM & J.R. WICKENS. 2000. Dopamine and synaptic plasticity in the neostriatum. J. Anat. **196:** 587–596.

Electrophysiological Interactions between Striatal Glutamatergic and Dopaminergic Systems

ANTHONY R. WEST,[a] STAN B. FLORESCO,[b] ALI CHARARA,[c]
J. AMIEL ROSENKRANZ,[d] AND ANTHONY A. GRACE[c]

[a]*Department of Neuroscience, Finch University of Health Sciences/The Chicago Medical School, North Chicago, Illinois 60064, USA*

[b]*Department of Psychology, University of British Columbia, Vancouver British Columbia, V6T 1Z4, Canada*

[c]*Departments of Neuroscience and Psychiatry, University of Pittsburgh, Pittsburgh, Pennsylvania 15260, USA*

[d]*Department of Neuroscience, Baylor College of Medicine, Houston, Texas 77005, USA*

ABSTRACT: Glutamatergic and dopaminergic systems play a primary role in frontal-subcortical circuits involved in motor and cognitive functions. Considerable evidence has emerged indicating that the complex interaction between these neurotransmitter systems within the dorsal striatum and nucleus accumbens is critically involved in the gating of information flow in these highly integrative brain regions. As a result, disruptions of the interaction between glutamate and dopamine has been proposed as a pathological basis for a number of disorders, including the pathophysiology of schizophrenia. In this chapter, we discuss recent studies that have significantly advanced our understanding of the reciprocal interactions between glutamatergic and dopaminergic systems within the striatal complex in the normal brain and in pathological states.

KEYWORDS: glutamate; dopamine; striatum; nucleus accumbens; medium spiny neuron

INTRODUCTION

Distinct groups of neurons clustered within the mammalian midbrain use the catecholamine dopamine (DA) as their neurotransmitter.[1] These neurons are located primarily within the substantia nigra pars compacta and the ventral tegmental area and give rise to a dense efferent projection terminating in the dorsal (caudate-putamen) and ventral (nucleus accumbens, NAc) striatum, as well as in the limbic cortex and associated subcortical structures.[2–4] Within the striatal complex, these DAergic

---

Address for correspondence: Anthony R. West, Ph.D., Department of Neuroscience, Finch University of Health Sciences/The Chicago Medical School, 3333 Green Bay Road, North Chicago, IL 60064. Voice: 847-578-8658; fax: 847-578-8515.
westa@finchcms.edu

Ann. N.Y. Acad. Sci. 1003: 53–74 (2003). © 2003 New York Academy of Sciences.
doi: 10.1196/annals.1300.004

projections form synaptic contacts primarily on dendritic spines of medium-sized GABAergic projection neurons (MSNs) and the aspiny dendrites of striatal interneurons.[5,6] Ultrastructural studies have also shown that DAergic terminals appose both symmetrical and asymmetrical synapses,[7] and a significant proportion of DA inputs do not make classical synaptic contacts onto postsynaptic structures.[4,8] Thus, DAergic afferents to the dorsal striatum and NAc are in a position to modulate synaptic transmission by releasing DA in the vicinity of glutamatergic (GLUergic) synapses or by causing more diffuse increases in extracellular DA levels.[4,9–11]

In addition to the DA projection, the dorsal striatum receives a massive input of glutamate (GLU)-releasing afferents from the entire neocortical mantle and midline thalamic nuclei.[12] In contrast to the dorsal striatum, the core and shell regions of the NAc receive GLUergic afferents from multiple regions of the prefrontal cortex (PFC), hippocampal formation, basolateral amygdala, and thalamus.[13–16] Like the DA system, the primary targets of the GLUergic inputs to the striatal complex are the dendritic spines of MSNs.[5] Additional studies have shown that GLUergic inputs to the striatal complex also target the dendrites and soma of striatal interneurons receiving DAergic innervation.[6,17] The close proximity of GLUergic and DA inputs onto striatal target neurons indicates that GLU and DA transmission interact via both pre- and postsynaptic mechanisms within functionally related networks to control striatal output.[4,8–10] Behavioral studies have indicated that interactions between DAergic and GLUergic inputs in the dorsal striatum and NAc contribute to the expression of a variety of psychomotor behaviors.[18]

The complexity of striatal DA–GLU interactions is compounded by observations indicating that cortical and subcortical GLUergic pathways can influence the activity of midbrain DA cells via multiple direct and/or indirect pathways.[19] Moreover, GLUergic activation of striatonigral feedback pathways involved in regulating midbrain DA cell activity may be critically involved in directing information flow between the limbic and motor regions of the striatal complex.[20] Thus, GLU-DA interactions in ventral striatal subregions are also likely to influence the activity of more dorsal striatal subregions via nonreciprocal striatonigrostriatal feedback pathways.[20] Previous reports have thoroughly summarized the immense literature pertaining to reciprocal interactions between GLU–DA systems at the presynaptic[21,22] and postsynaptic[23–27] levels. In this chapter, we will focus primarily on recent studies that have significantly advanced our understanding of the nature of interactions between GLUergic and DAergic systems within the striatal complex and midbrain, and on the role of these interactions in regulating striatal function under normal and pathophysiological conditions associated with schizophrenia.

INFORMATION INTEGRATION IN THE STRIATAL COMPLEX

Intracellular recordings in the intact animal (FIG. 1) or in organotypic cocultures have shown that the membrane potential of MSNs often exhibits characteristic shifts (~1 Hz) between a hyperpolarized resting state termed the "down state" and a depolarized plateau potential or "up state."[28–31] The transition from the down state to the up state is driven by temporally synchronous GLUergic synaptic transmission. Thus, two-state membrane potential fluctuations are not observed in the dorsal striatum following decortication and thalamic transection[28] and in the NAc following fim-

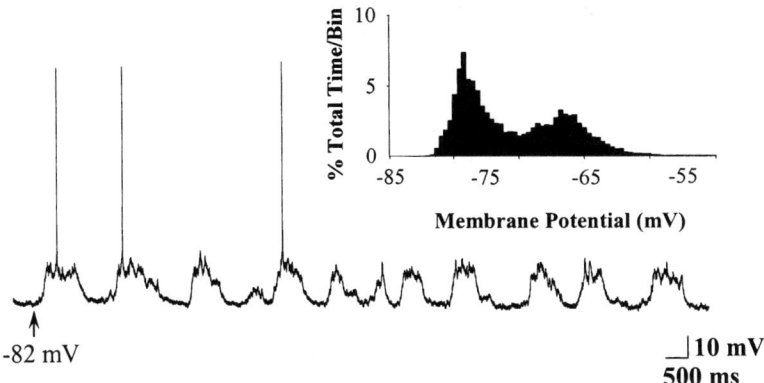

FIGURE 1. *In vivo* intracellular recordings from striatal MSNs. In vivo, MSNs exhibit a bistable membrane activity state characterized by rapid and spontaneous transitions from a hyperpolarized "down state" to a depolarized plateau "up state." *Inset:* Example of a time interval plot of membrane potential activity (30-s recordings sampled at 10 kHz) recorded from the same neuron. The bimodal distribution of the membrane potential indicates the neuron has bistable membrane properties characteristic of MSNs.

bria-fornix transection.[29] Additionally, neurons recorded in slice preparations typically do not exhibit two-state membrane potential fluctuations.[24] Thus, proper function of GLUergic afferents to the striatal complex is critical for enabling the membrane of MSNs to depolarize to the up state and fire action potentials.

Cortical and hippocampal GLUergic inputs to the striatal complex are also thought to play an important role in synchronizing the activity of functionally related striatal MSN subcircuits.[32–34] Thus, simultaneous intracellular recordings from pairs of MSNs have revealed that state transitions across cells are highly correlated.[35] Intense corticostriatal signaling has also been shown to increase electrotonic coupling between MSNs via an indirect mechanism involving nitric oxide (NO)–producing interneurons.[36] NO-producing interneurons are innervated by corticostriatal and nigrostriatal afferents[17,37] and make synaptic contacts on the spines of MSNs in close proximity to the DA and GLU inputs.[6] A NO-mediated synchronization of specific striatal output neuron clusters may be important for promoting the simultaneous transition of coupled bistable neurons to the depolarized up state.[36] This is consistent with our studies showing that intrastriatal infusion of a NO-generating compound increased the firing rate and burst length of striatal neurons under basal conditions and during electrical stimulation of the PFC.[38] Additionally, neurons responding to PFC stimulation exhibited decreased responsiveness to afferent excitation following local infusion of a NO scavenger.[38] Thus, these studies indicate that the activation of NO signaling by GLUergic inputs may play an important role in synchronizing the up state activity and, potentially, the spike discharge of functionally coupled striatal MSNs.

The excitability of MSNs is also highly dependent on the relative timing of different active synaptic inputs. Thus, the additive effect of temporally correlated synaptic inputs is thought to induce a short-term increase in the intrinsic excitability of

MSNs recorded in the dorsal striatum[39] and NAc.[40] Interestingly, when the MSN is in the up state, Ca^{2+} transients occurring in the distal dendrites are highly correlated with the length of the somatic spike burst, indicating that information regarding the action potentials generated at the cell body is backpropagated and integrated with ongoing synaptic activity.[41] Ultimately, the noisy subthreshold membrane potential activity occurring in the up state is thought to be determined by the balance between synaptic drive and voltage-dependent Na^+, K^+, and Ca^{2+} conductances.[26] Additionally, the interaction between afferent drive and intrinsic membrane properties of MSNs is likely to be potently modulated by DA and other neuromodulators known to regulate ion channel conductances and presynaptic GLU release (see below).

THE REGULATION OF DOPAMINE TRANSMISSION BY GLUTAMATERGIC SYSTEMS

In addition to their role in regulating MSN activity, GLUergic systems appear to be critically involved in regulating steady-state extracellular DA levels in the striatal complex via multiple direct and indirect pathways.[21,22,38,42,43] The net effect of endogenous striatal GLUergic tone on extracellular DA levels is likely to be facilitatory in nature, as several studies have shown that inactivation of corticostriatal pathways via excitotoxic lesions, ablations, or pharmacological methods results in decreases in extracellular GLU and DA levels in anesthetized and freely moving rats.[44–46]

Conversely, elevations in endogenous extracellular GLU levels, induced following intrastriatal infusions of GLU reuptake blockers, was observed to increase striatal DA release[47–49] via an ionotropic GLU receptor-dependent mechanism.[47] Furthermore, the facilitatory influence of GLU on striatal DA efflux appears to be behaviorally relevant. Thus, conditioned emotional responses (CER) to aversive stimuli have been shown to increase DA levels in a multiphasic manner in the NAc.[50] The delayed phase of the conditioned increase in DA efflux was shown to occur via a process that was abolished by intrastriatal infusion of the NMDA receptor antagonist MK-801.[50] Interestingly, multiple studies have recently reported that electrical stimulation of the basolateral amygdala (BLA) induces an increase in DA efflux in the NAc that occurs following cessation of the stimulation.[51,52] This effect is blocked by local infusions of ionotropic GLU receptor antagonists and persists after inactivation of the ventral tegmental area (VTA) or PFC, indicating that the BLA facilitates the local release of DA in the NAc via a GLUergic mechanism.[52,53] Consistent with the above studies and the vast majority of studies performed *in vitro*,[22] recent *in vivo* studies using sophisticated techniques with high-time resolution (< 1 min) and sensitivity[49,54] also indicate that GLU primarily facilitates both spontaneous and evoked DA release in the rat striatum.

In addition to the direct facilitatory effects of GLU on DA transmission, evidence has accumulated indicating that NMDA receptor activation indirectly augments extracellular DA levels via the production of NO.[38] The activation of endogenous NO signaling produces a concurrent increase in extracellular GLU and DA via a mechanism that is dependent on amplification of GLUergic transmission, and AMPA and NMDA receptor activation.[38] Robust activation of endogenous NO signaling inhibits NMDA receptor activation and blocks NMDA-mediated DA release.[38] These studies indicate that, in addition to their role in synchronizing MSN activity (see above), NO interneurons may function as an important local modulatory system involved in synchronizing GLU-DA transmission and regulating GLUergic control over extracellular DA levels.

DIRECT AND INDIRECT REGULATION OF MIDBRAIN DOPAMINE NEURON ACTIVITY BY GLUTAMATERGIC AFFERENTS

DA cells recorded in the intact animal fire in two distinct modes: tonic irregular single-spike firing occurring at rates of 2–8 spikes per second, or bursts of 3–8 spikes occurring at high frequencies of approximately 10–20 Hz with a pause between subsequent bursts.[55] GLUergic inputs from the PFC to the midbrain have been shown to synapse specifically on the mesocortical DA cells[57] and likely play a role in modulating the activity states of this subpopulation of DA neurons. The PFC is also likely to control the activity of mesolimbic and nigrostriatal DA cells indirectly via its influence on GABAergic cells in the striatum[58,59] and VTA/SN.[60] Prefrontal cortical or hippocampal activation most likely has its major influence in modulating the activity of mesolimbic DA neurons via excitation of NAc output neurons that affect DA neurons via both monosynaptic inputs[61] and through its influence on GABAergic neurons in the pallidum and VTA/SN.[62–64] Additionally, DA cell burst firing is regulated by the subthalamic nucleus (STN),[65] pedunculopontine tegmental nucleus (PPTg),[66] and amygdalar[67] projections. These subcortical regions are likely to be driven by excitatory input from the motor, prefrontal, and limbic cortices.[62,68] DA cell burst firing may also be under a suppressive influence by inhibitory inputs arising from the striatopallidal system or via local GABAergic neurons.[64,69] This is consistent with our studies showing that transection of GABAergic striatonigral inputs decreases burst-firing in nigrostriatal DA neurons.[70]

THE TONIC-PHASIC MODEL OF DOPAMINE SYSTEM REGULATION

Considerable evidence indicates that striatal DA neurotransmission occurs in two dissociable temporal modes, "tonic" and "phasic," both of which are regulated by GLUergic forebrain structures (FIG. 2).[10,25,71] Phasic DA transmission is proposed to be the DA signal that mediates rapid behaviorally relevant activation of the DA system and consists of DA released from the DA axonal varicosity as a result of DA cell-burst firing (FIG. 2B). Phasic DA transmission is a high-amplitude transient signal, in which intrasynaptic DA concentrations are estimated to reach into the low millimolar range (e.g., 1.6 mM in NAc),[9] where it is proposed to selectively act on DA receptors that are intra- or perisynaptically located. Additionally, phasic DA signaling is proposed to be terminated rapidly via reuptake mechanisms involved in eliminating DA from the synaptic cleft by the high-affinity DA transporter (DAT).

In contrast to the phasic DA signal, tonic DA transmission represents the extrasynaptic pool of DA that is present at steady-state concentrations within the extracellular space (FIG. 2A). Studies examining the impact of 6-OHDA lesions on extracellular DA levels[72] indicate that the tonic DA pool is tightly regulated and maintained within a narrow concentration range (e.g., approx. 4–20 nM in NAc)[73] even when the DA system is compromised. As indicated above, tonic extracellular

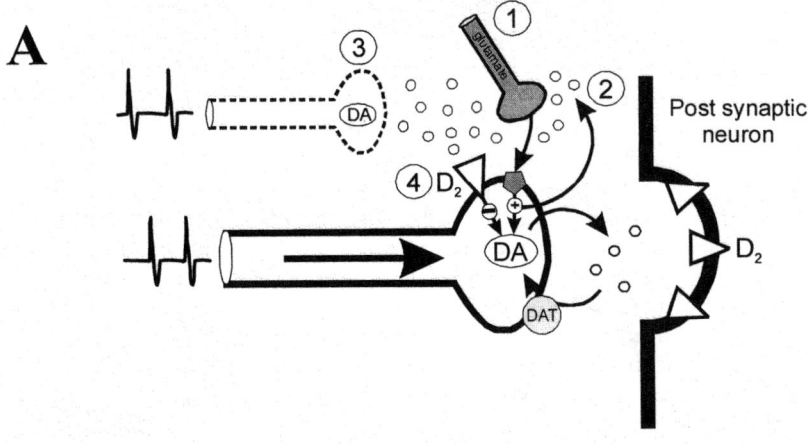

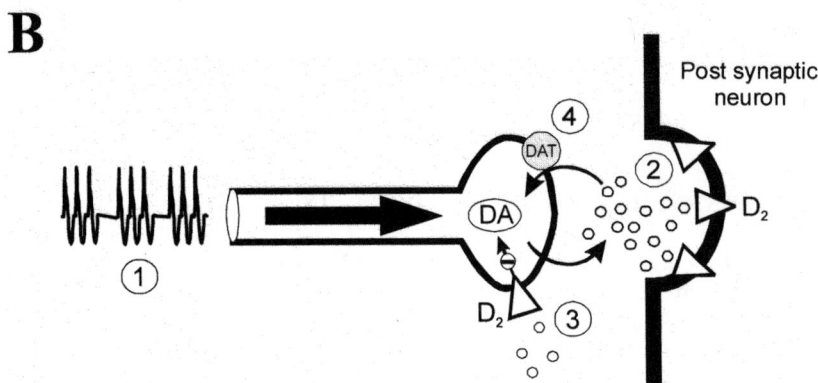

FIGURE 2. "Tonic" (**A**) and "phasic" (**B**) modes of DA neurotransmission. (**A**) As originally proposed by Grace,[10] local GLUergic inputs to the striatal complex (1) play a key role in maintaining a steady-state level of tonic extracellular DA. GLU is thought to increase DA efflux from nonsynaptic sites via direct and indirect mechanisms (2). Additionally, Floresco et al.[74] have recently shown that the extracellular concentration of tonic DA can be upregulated following the disinhibition of slow firing or quiescent DA neurons (3) via inactivation of the ventral pallidum. This pool of tonic DA controls the responsiveness of the DA terminal to action potentials via the activation of DA autoreceptors (4). (**B**) Phasic DA transmission is initiated by excitatory afferent drive at the level of the DA cell body involved in inducing burst firing (1). Phasic DA transmission plays a primary role in activating postsynaptic DA receptors (2) and is downregulated by tonic DA acting on DA terminal autoreceptors (3) and reuptake (4) via the DA transporter (DAT).

DA levels are maintained primarily by local GLUergic signaling pathways and DA released by overflow from the synapse and/or at nonsynaptic sites during tonic firing activity. Although the tonic and phasic modes of DA neurotransmission are maintained via different GLUergic and GABAergic mechanisms, it is postulated that tonic extracellular DA levels modulate the phasic (synaptic) component of DA signaling through the activation of synthesis- and release-regulating autoreceptors. These DA autoreceptors are believed to be located extrasynaptically on the DA terminal and play an important role in suppressing synaptic DA levels. Thus, tonic extracellular DA levels act to downregulate the responsivity of the DA system to bursts of action potentials thought to be generated during behavioral activation.[10]

Recent studies from our laboratory have focused on characterizing the subcortical circuits involved in regulating the tonic and phasic modes of DA neurotransmission in the NAc.[74] We have observed that inactivation of the ventral pallidum caused a selective increase in the overall number of spontaneously active DA neurons (i.e., population activity) in the VTA (FIG. 3A), presumably via a reduction in tonic GABAergic inhibition of subsets of VTA DA neurons. Although this manipulation increased the DA cell population activity, it did not affect the average firing rate or burst firing of recorded DA neurons. In contrast, activation of GLUergic/cholinergic inputs from the PPTg to DA neurons in the VTA had the opposite effect: a marked increase in burst firing of already active DA neurons, but no effect on the overall population activity or average firing rate (FIG. 3A). In studies using *in vivo* microdialysis, we observed that manipulations that increased the population activity of DA neurons resulted in consistent increases in extracellular DA levels in the NAc. Surprisingly, manipulations that selectively enhanced the burst firing of DA neurons caused no discernable increase in DA efflux (FIG. 3B). In light of these data, it is apparent that extracellular (tonic) levels of DA are not altered dramatically by increases in DA cell burst firing. However, the disinhibition of nonfiring DA cells can increase the overall population activity of midbrain DA neurons and result in an elevation of tonic DA transmission in the NAc.

The observation that manipulations that selectively increased burst firing of DA neurons produced no discernable increase in extracellular DA levels in the NAc was surprising, considering that bursting is generally thought to be the primary mechanism by which behaviorally relevant DA release is facilitated.[9,55] We posited that the lack of effect of enhanced bursting of DA neurons on mesoaccumbens DA efflux may be due to the actions of the DAT. The DAT is located at the borders of the synaptic junction and therefore plays an important role in limiting diffusion of DA from the synaptic space, while still allowing for large intrasynaptic DA transients. In a direct test of this hypothesis, we blocked DAT activity using continuous local infusions of nomifensine, which resulted in a ninefold increase in extracellular DA levels. When we then enhanced burst firing of DA neurons via activation of the pedunculopontine nucleus, we saw a further increase in DA release that was ~300% of baseline values (FIG. 3B). This finding indicates that a selective increase in burst firing of DA neurons induces a massive increase in DA release at the terminal level. However, under normal conditions, the DAT limits dramatically the amount of DA that escapes out of the synaptic cleft, thereby occluding detection of a sustained and measurable increase in extracellular DA.

When viewed collectively, these data provide strong support for the notion that there are at least two distinct components of DA neurotransmission at the terminal

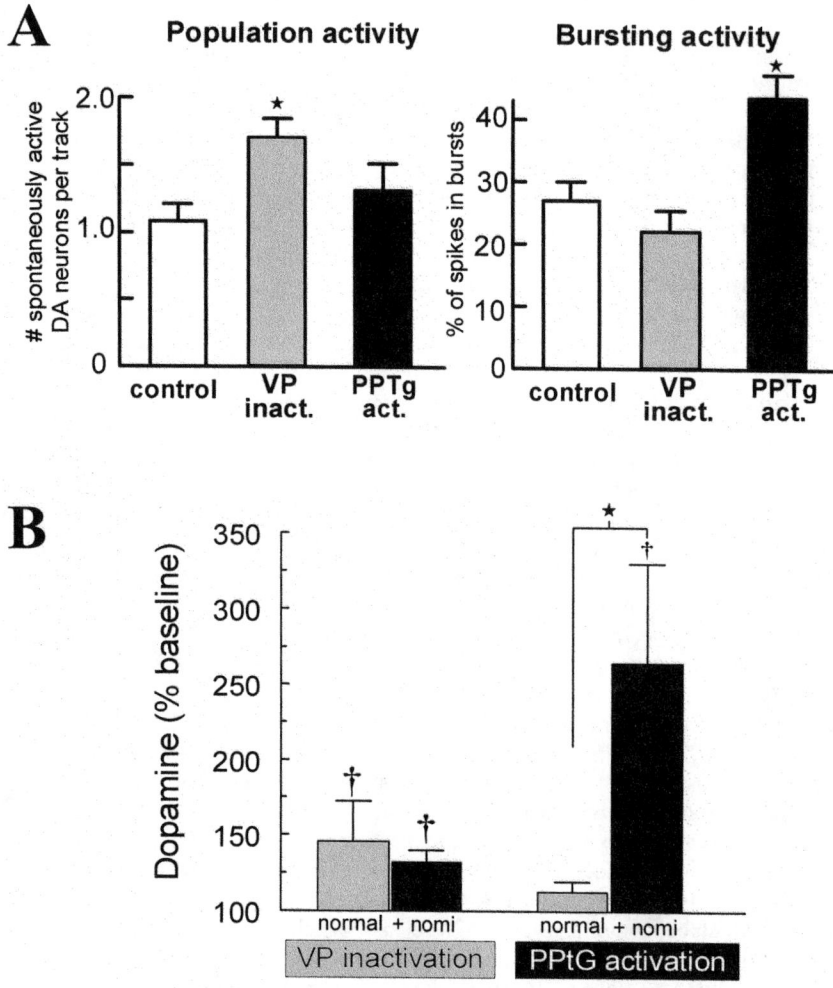

FIGURE 3. DA neuron activity states and NAc DA release are differentially modulated by subcortical nuclei. (**A**) Electrophysiological data. Inactivation of the VP via local muscimol and baclofen infusions selectively increased DA neuron population activity (i.e., number of spontaneously active DA neurons) without altering burst firing (*gray bars*). In contrast, activation of the PPTg via local bicuculline infusions had no effect on population activity but significantly increased burst firing of DA neurons (*black bars*). (**B**) Neurochemical data. Inactivation of the VP increases extracellular DA levels in the NAc, but activation of the PPTg, or control treatments, were without effect. Local application of the DA uptake blocker nomifensine (nomi) increased baseline DA levels by ~9 fold (data not shown). Inactivation of the VP during nomi infusion produced a comparable increase in DA release relative to controls. In contrast, activation of the PPTg during nomi infusion induced a 3-fold increase in DA release. *Daggers* denote significant increase from baseline at $P < .05$, and *star* denotes significant difference between PPTg activation groups with or without nomifensine (Dunnett's test).

level. The phasic component, driven primarily by bursting events at the level of the DA cell body, is highly compartmentalized and spatially limited by the DAT. Phasic DA transmission likely serves as an important teaching signal by altering synaptic strengths of selected inputs to a particular ensemble of MSNs in the striatum.[27,75] In contrast, slower changes in tonic levels of DA are not influenced by bursting of DA neurons; rather these levels are regulated by the overall levels of activity of the entire population of DA neurons, as well as by GLU-mediated mechanisms in the NAc. Tonic DA communication, acting via volume transmission, would have a much more spatially distributed influence over a large number of MSNs in the NAc, and likely serves a distinctly different function, both at the cellular and behavioral level, to the role played by intrasynaptic phasic DA transmission. An important goal for future studies on this topic is to elucidate the distinct actions that each of these dissociable components of the DA signal exert on the activity of MSNs in the striatal complex.

MODULATION OF STRIATAL GLUTAMATERGIC SYNAPTIC TRANSMISSION BY DOPAMINE

DA has been shown to have variable effects on cell excitability and synaptic transmission in MSNs of the dorsal striatum and NAc.[23–27] It is likely that the impact of DA receptor activation on MSN activity in the intact system is dependent on multiple factors, including the mode of DA transmission (i.e., tonic or phasic), relative contribution of D_1 and D_2 receptors involved in mediating the DAergic modulation, steady-state membrane potential of the neuron, and ongoing synaptic activity across excitatory and inhibitory synapses. Additionally, it should be kept in mind that the modulatory influence of DA on synaptic transmission in the striatum is likely to vary across neurochemically and anatomically distinct striatal subregions. The impact of DA signaling on MSN activity is also likely to depend on the experimental preparation (*in vitro* versus *in vivo*) and whether DA receptors are activated under physiologically relevant conditions. Although a clear picture of DA modulation in the striatal complex has not yet emerged, recent studies discussed below have advanced our understanding of how the physiological release of endogenous DA modulates activity states and responsiveness of MSNs to afferent inputs.

Considerable evidence now exists indicating that in the intact animal, the activation of D_1 receptors by *endogenous* DA appears to facilitate membrane depolarization in striatal MSNs and, in some cases, enhance spontaneous and evoked activity when the cell is strongly depolarized. This facilitatory D_1 receptor-mediated effect is particularly relevant when the neuron is in the depolarized up state. Thus, we have recently shown that intrastriatal infusion of the D_1 antagonist SCH 23390 depressed the amplitude of naturally occurring up states and decreased the maximal depolarized membrane potential measured in the up state (FIG. 4A).[76]

Additionally, local SCH 23390 infusion potently inhibited activity evoked by intracellular injection of depolarizing current delivered when the cell was in the down state, indicating that *in vivo* the physiological activation of D_1 receptors by endogenous DA enhances the effectiveness of persistent membrane depolarizations occurring at both hyperpolarized and depolarized membrane potentials (FIG. 5A and C).[76] These results are consistent with studies showing that DA depletion induces a depression of naturally occurring up states and corticostriatal transmission in striatal MSNs

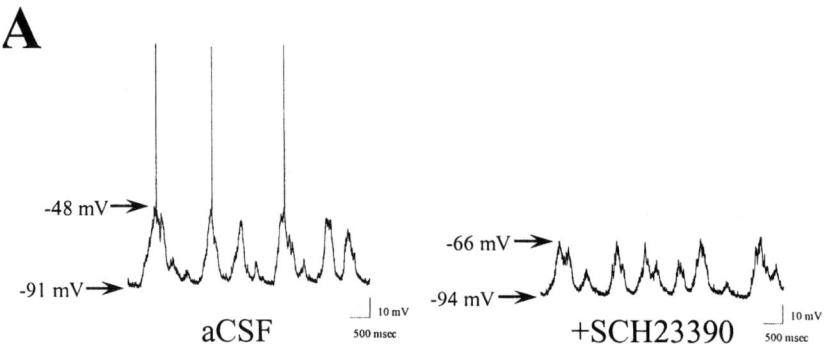

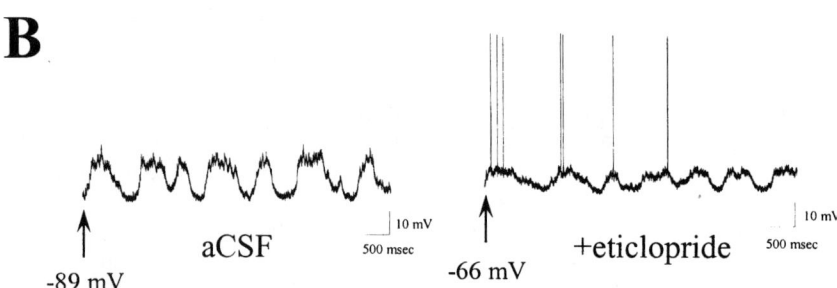

FIGURE 4. Intrastriatal infusion of DA antagonists modulate the membrane activity of striatal MSNs recorded *in vivo*. (**A**) *Left*: During aCSF (vehicle) infusion this striatal MSN exhibited rapid spontaneous shifts in steady state membrane potential and spontaneous spike discharge. *Right*: Following local infusion of the D_1 antagonist SCH 23390 (10 µM, 10 min), the amplitude of up events was depressed and a cessation of action potential discharge was observed. (**B**) *Left*: During aCSF (vehicle) infusion this striatal MSN exhibited bistable membrane activity but did not exhibit spontaneous spike discharge. *Right*: During local infusion of the D_2 antagonist eticlopride (20 µM, 4.5–5.5 min), the membrane potential of this same cell depolarized considerably, leading to the discharge of multiple action potentials. Adapted from West & Grace.[76]

recorded *in vivo*.[77] Additionally, augmentation of endogenous DA transmission following electrical or chemical stimulation of the VTA depolarized the membrane potential[78,79] and increased the spontaneous and evoked activity of the majority of tested neurons in a manner sensitive to D_1 receptor antagonist administration.[80]

There is evidence that the ability of D_1 receptor stimulation to potentiate up state transitions in striatal MSNs may shed light into the behavioral relevance of the up and down state transition and its modulation by the DA system, and how this may be functionally relevant to normal basal ganglia function. Thus, recent studies by Wickens and colleagues have shown that electrical stimulation of substantia nigra (using stimulation parameters found to be optimal for reward-related learning) potentiates corticostriatal synaptic activity via D_1 receptor activation.[81] Importantly, these investigators also demonstrated in parallel studies that the degree of D_1 receptor-mediated potentiation of corticostriatal signaling was negatively correlated with the amount of time needed for the rats to learn to lever press for reinforcing stimulation of the substantia nigra,[81] suggesting that the D_1 potentiation of corticostriatal function at the physiological level had a positive impact on reinforcement learning as well.

This is further substantiated by extracellular recording studies in the NAc that have shown that increases in endogenous DA levels evoked by tetanic stimulation of the fimbria potentiated hippocampal-evoked firing activity in a manner that was blocked by systemic administration of a D_1 or NMDA receptor antagonist.[82] Similarly, increases in endogenous DA efflux in the NAc evoked by tetanic stimulation of the BLA potentiated BLA-evoked firing activity via a D_1 and NMDA receptor–mediated mechanism.[83] These finding are also consistent with previous extracellular studies showing that electrical stimulation of DA pathways in the medial forebrain bundle increases the activity of a subpopulation of striatal neurons in a manner that is dependent on D_1 receptor activation.[11] Iontophoretic application of DA using low ejection currents has also been shown to facilitate the excitatory effects of GLU on striatal and accumbal MSNs via D_1 receptor activation.[84,85] Taken together, the above studies indicate that DA D_1 receptor activation acts to facilitate synaptic plasticity across active corticostriatal synapses involved in mediating the learning of behavioral responses.

Studies using voltage-clamp techniques have also provided important information regarding the ion channels that may be involved in the DA D_1 receptor-mediated facilitation of MSN activity.[26] These studies indicate that when the cell membrane potential is clamped at a depolarized level (i.e., −55 mV) thought to approximate the up state, DA D_1 receptor activation enhances L-type Ca^{2+} currents and evoked spike discharge.[86–88] Additionally, DA D_1 receptor activation reduces $GABA_A$ receptor currents in striatal MSNs[89] and facilitates NMDA receptor-mediated responses.[23] Furthermore, evidence is emerging from biochemical studies indicating that protein interactions between D_1 and NMDA and/or AMPA receptors are important for receptor expression and trafficking,[90,91] and immediate early gene expression in MSNs in the NAc.[92,93] Taken together, these studies indicate that activation of DA D_1 receptors on dendritic spines near GLUergic input enhances NMDA- and L-type calcium channel-dependent depolarizations and stimulates intracellular signaling cascades and the synthesis of structural proteins involved in use-dependent plasticity.

In contrast to the above reports, multiple studies have indicated that DA also modulates MSNs via inhibitory pathways involved in suppressing activity evoked by depolarizing current and synaptic activation, and that this occurs via a D_2-dependent mechanism. Although some studies have suggested that D_2 receptor-dependent attenuation of EPSPs may occur in the dorsal striatum only following DA depletions[24] or not at all,[94] this is not consistent with the majority of the literature. Thus, studies *in vivo* have reported that VTA stimulation, DA iontophoresis, or DAergic drug administration reduced stimulus-evoked excitatory responses from either hippocampal or amygdaloid afferents.[15,82,83,95] In the intact animal this inhibitory influence appears to be mediated primarily by D_2 receptors, as local D_2 antagonist infusions have been shown to depolarize both the up and down state membrane potentials of MSNs (FIG. 4B) and increase their responsiveness to depolarizing current (FIG. 5B and C) and electrical stimulation of the PFC.[76] These observations are in agreement with studies using striatal brain slices showing that bath-applied DA or

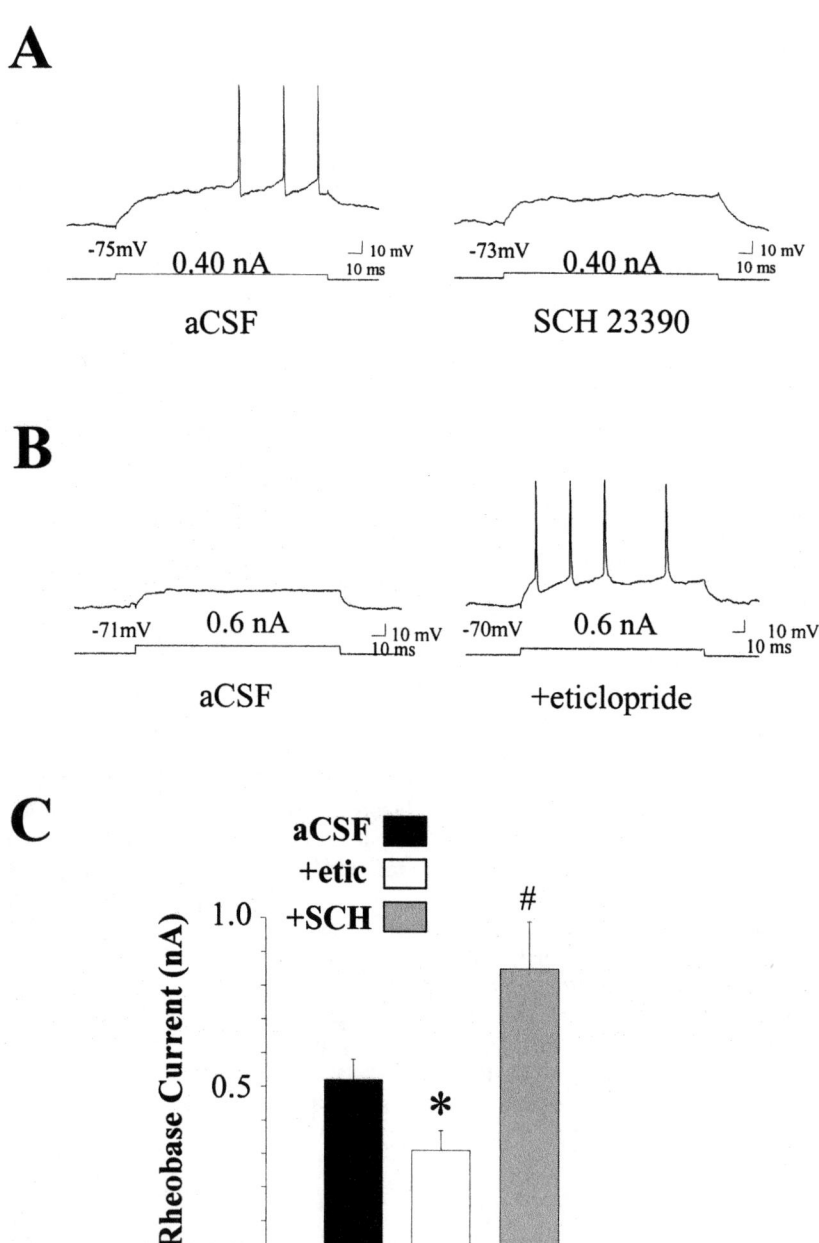

FIGURE 5. Intrastriatal infusion of DA antagonists modulate the membrane excitability of striatal MSNs recorded *in vivo*. (**A**) *Left*: Response of a single cell to a suprathreshold amplitude of depolarizing current injected during aCSF infusion. *Right*: Response of the same cell to depolarizing current injected during SCH 23390 infusion. (**B**) *Left*: Response

D_2 agonists decreased the amplitude of EPSPs evoked by electrical stimulation of corticostriatal pathways[96–98] and EPSCs evoked by local striatal stimulation.[99] Evidence also exists indicating that D_2 receptor-mediated inhibition of corticostriatal inputs is at least partially mediated via presynaptic receptors.[96,97] Recent neuroanatomical studies have provided evidence that D_2 receptors are present on prefrontal cortical afferents and their target neurons in the dorsal striatum.[100] Additional studies have shown that MSNs recorded in D_2 receptor-deficient mice exhibited an increase in the frequency of spontaneous synaptic activity and large amplitude depolarizations that were not observed in wild type mice.[101] Mice lacking the D_2 receptor also exhibited greater synaptic activity following application of potassium channel blockers and $GABA_A$ receptor antagonists.[101] Given the above studies, it is likely that presynaptic and postsynaptic D_2 receptors play a significant role in inhibiting corticostriatal signaling in striatal MSNs.

Additional studies using *in vitro* brain slice preparations of the NAc have demonstrated that DA and psychostimulants depress excitatory synaptic transmission via D_1- and D_2-mediated mechanisms.[94,102,103] Currently, little is known regarding the relative contribution of D1- and D2-like receptors to the modulation of excitatory in-

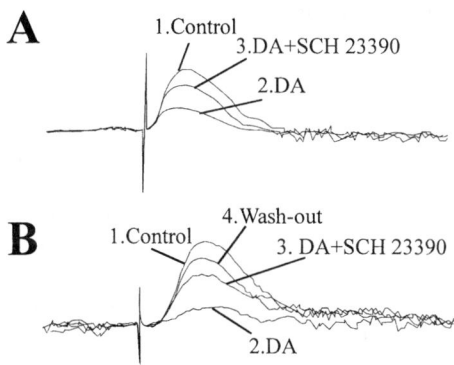

FIGURE 6. DA D_1 receptor-mediated modulation of amygdalar and hippocampal-evoked responses *in vitro*. (**A**) Stimulation of DA receptors attenuated the amplitude of the EPSP evoked by stimulation of the amygdala; this was reversed by administration of the D_1 antagonist SCH23390. (**B**) Stimulation of DA receptors attenuated the amplitude of the EPSP evoked by stimulation of the hippocampus; this was reversed by administration of the D_1 antagonist SCH23390. These responses could also be mimicked by administration of a D_1 agonist, but not a D_2 agonist (Charara & Grace[104]).

---

of a single cell to a suprathreshold amplitude of depolarizing current injected during aCSF infusion. *Right*: Response of the same cell to the same amplitude of depolarizing current injected during eticlopride infusion (~8–10 min). (**C**) Comparisons across striatal MSNs revealed that eticlopride infusion increased neuron excitability expressed as a decrease in the mean ± SEM minimal current amplitude (rheobase current) required to reach threshold (*$P < .05$, ANOVA, Dunn's test), whereas SCH 23390 was observed to decrease neuron excitability (#$P < .05$, ANOVA, Dunn's test). Adapted from West & Grace.[76]

puts from the hippocampus and amygdala to NAc MSNs. In order to clarify this issue, we have examined the effects of DA agonists on EPSPs evoked by selectively stimulating amygdalar and hippocampal inputs in an *in vitro* brain slice preparation in which each afferent system was preserved.[104] Our studies indicate that D_1 receptor activation primarily depresses the excitatory postsynaptic responses of NAc neurons recorded during stimulation of hippocampal and amygdalar afferents (FIG. 6A and B). D_2 receptor stimulation did not attenuate EPSPs evoked following the stimulation of either amygdalar or hippocampal afferent systems. In a small population of NAc MSNs (14%), however, D_1 receptor stimulation induced a facilitation of the hippocampus-evoked response.[104] Taken together, these observations provide evidence for a DA receptor subtype-specific modulation of GLUergic inputs to the NAc, with D_2 activation attenuating PFC drive,[96] and D_1 activation attenuating BLA drive, but with hippocampal-evoked responses showing a bimodal form of modulation by D_1 stimulation.[104]

Given the above, we propose that by exerting both pre- and postsynaptic actions that alter GLUergic effects on MSN neurons, phasic DA acts to gate specific functionally related striatal output circuits to selectively transfer psychomotor information through the striatum and NAc to basal ganglia output centers. Thus, DA may functionally depress GLUergic signaling via presynaptic D_2 receptors and postsynaptic D_1 receptors in MSNs that are not receiving strong convergent excitatory input. In support of this, D_1 receptors have been shown to reduce the excitability of MSNs at hyperpolarized membrane potentials[24] and decrease the efficacy of synaptic potentials in the striatum via indirect transynaptic or extrasynaptic mechanisms.[102] We have recently observed a similar DA D_1 receptor-dependant gating mechanism in the BLA.[105] Thus, in general, by increasing inhibition and suppressing background synaptic drive, DA appears to be able to filter weaker, spurious activity via a mechanism involving D_1 and D_2 receptor activation. However, in strongly depolarized cells the D_2 suppression of GLUergic inputs is overcome by the ability of DA D_1 receptor activation to facilitate the depolarized up state. In the depolarized up state, NMDA receptor activation may also enhance excitatory D_1 receptor signaling by recruiting D_1 receptors from the interior of the cell to the plasma membrane.[91] Thus, the net impact of phasic D_1 and D_2 neurotransmission is likely to be dependent on whether convergent excitatory inputs are powerful enough to overcome ongoing inhibitory processes (i.e., potassium, $GABA_A$ channels, and DA effects) and allow D_1 and NMDA-mediated interactions and increases in calcium-dependent plasticity to occur in the distal dendrites of the MSN.

ABNORMAL STRIATAL GLUTAMATE–DOPAMINE INTERACTIONS IN SCHIZOPHRENIA

It is becoming evident that a developmentally related dysfunction in cortical and/ or limbic GLUergic transmission may underlie the neuropathology of schizophrenia.[25,106–108] Support for aberrant GLUergic function in schizophrenia comes from studies indicating that the psychotomimetic drug phencyclidine (PCP) induces a schizophrenia-like condition in humans via its noncompetitive antagonistic effects on NMDA receptors.[110] PCP-induced psychoses resemble not only the active positive symptoms but also the deficit state of schizophrenia, as subjects exhibit a variety of negative symptoms and cognitive impairment. Chronic PCP administration in monkeys has been shown to produce enduring decreases in PFC DA transmission, an effect that is associated with frontostriatal cognitive dysfunction and is reversed by administration of the atypical antipsychotic drug, clozapine.[111] Similar to PCP, drugs such as ketamine and MK-801, which also block NMDA receptor activation, have been shown to produce symptoms reminiscent of schizophrenia.[112] Interestingly, recent postmortem studies have revealed that patients with schizophrenia have elevated cortical levels of the endogenous ionotropic GLU receptor antagonist, kynurenate.[113] This condition could potentially contribute to the prefrontal cortical hypofrontality and abnormal DA neurotransmission observed in patients with schizophrenia. In support of this, pharmacological manipulations that increase endogenous kynurenic acid levels in the rat brain have been shown to significantly increase the phasic firing activity of VTA DA neurons.[114] The discovery that a variety of NMDA receptor antagonists activate catecholaminergic systems and elicit behavioral arousal in rats and psychotogenesis in humans also indicates that the interaction between GLUergic and DAergic systems plays an important role in the pathophysiology of schizophrenia.[109]

Functional imaging studies have suggested that the severity of negative symptoms observed in patients with schizophrenia is related to abnormalities in GLUergic circuits involving the PFC, thalamus, amygdala, and anterior temporal lobe regions. Extensive evidence also indicates that frontal and temporal abnormalities are associated with deficits in sensorimotor gating in patients with schizophrenia.[116] Similar deficits have been reproduced in rats following neonatal excitotoxic lesions or pharmacological disruption of the ventral hippocampal inputs to the ventral striatum.[117] Additionally, developmental disruptions of cortical regions, such as the parahippocampal cortex and related hippocampus via prenatal exposure to the antimitotic agent methylazoxymethanol acetate (MAM), result in several anatomical abnormalities and behavioral deficits in rats that are similar to those observed in patients with schizophrenia.[71] Prenatal exposure to MAM also decreased prepulse inhibition and induced an abnormal reactivity to stress in adult animals.[118] Additional deficits in hippocampal gating of PFC-afferent drive were observed in MAM-treated animals as compared to controls.[71]

Because the above-mentioned regions implicated in the pathophysiology of schizophrenia are known to regulate subcortical DAergic function at the level of the limbic striatum and midbrain (see above) via GLUergic inputs, it is likely that the apparent dysfunction of subcortical processing is related to dysregulation of tonic and phasic DA transmission.[10,25] Given that tonic DA levels are believed to be maintained primarily by GLUergic circuits, chronic deficits in striatal GLUergic tone resulting from a pathological dysregulation of frontal and/or temporal cortical pathways is predicted to result in reductions in tonic extracellular DA levels (FIG. 7).[10,25,71] Furthermore, postmortem studies reporting changes in GLU receptor densities in the putamen of schizophrenia patients[119] suggest that a compensatory up-regulation of NMDA receptors may occur in attempt to restore pathophysiological deficits in striatal GLUergic signaling. Under these conditions, the decreased modulatory capacity of the tonic DA system would release the phasic DA signal from autoinhibition, leading to abnormally intense synaptic signaling (FIG. 7). This unregulated phasic transmission could potentially disrupt the temporal integration of the DA modulatory signal and result in an excessive phasic activation

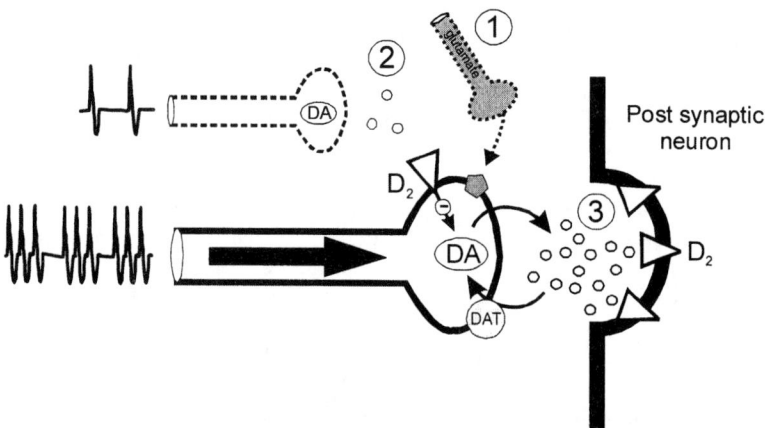

FIGURE 7. Dysregulation of "phasic" DA neurotransmission in schizophrenia. GLUergic inputs to the striatal complex are believed to be compromised in schizophrenia (1), resulting in a decreased pool of tonic extracellular DA (2) and a decreased capacity of the DA system to regulate phasic DA transmission via DA autoreceptors. This disruption in DA system homeostasis is thought to result in a supraphysiological increase in phasic DA transmission during behavioral activation and a pathological overactivation of postsynaptic DA receptors (3). Abnormally intense or sustained phasic DA neurotransmission may disrupt the balance between D_1 and D_2 receptor-mediated modulatory functions. For example, excessive D_2 receptor stimulation could result in decreased GLU neurotransmission, excessive inhibition of MSNs, and a desynchronization of striatal output pathways and information gating in the basal ganglia.

of postsynaptic DA receptors and an abnormal responsivity of the system to environmental stimuli.

In agreement with the above model, monkeys with excitotoxic lesions of the PFC exhibit enhanced behavioral responsiveness to drugs that potentiate presynaptic DA function.[120] The exaggerated increase in subcortical DA release in response to amphetamine observed in the neonatally lesioned monkeys is consistent with observations of functional imaging studies and amphetamine challenges in schizophrenia patients.[121,122] Furthermore, animals with neonatal lesions or pharmacological inactivations of the ventral hippocampus exhibit long-lasting behavioral changes indicative of a dysregulation of GLUergic and DAergic interactions.[117] Additionally, MSNs recorded in the NAc of animals with neonatal ventral hippocampal lesions exhibited abnormal increases in spike activity in response to VTA stimulation in a manner that was reversed by haloperidol.[79] MSNs recorded from these animals also exhibited abnormally regular up state transitions, indicating that the GLUergic signaling was altered by neonatal disruption of hippocampal development.[79]

When taken together, the above studies indicate that the underlying disturbance in schizophrenia may arise from abnormal interactions between subcortical DA systems and dysfunctional prefrontal and temporal cortical GLUergic afferents. Interestingly, aberrant GLU receptor activation in the NAc has also been shown to underlie the behavioral sensitization produced by repeated amphetamine administration.[123] Thus, abnormalities in the activity of the GLUergic inputs or NMDA receptor activation in the striatum and NAc that may occur in schizophrenia or through the use of psychotomimetics can result in a dysregulation of tonic DA release and abnormal phasic DA responses induced during behavioral activation or exposure to environmental stressors. Consistent with this tenant, schizophrenia patients have been shown to exhibit increased responsiveness to psychotomimetics and stressful stimuli.[121,122] It is likely that future studies integrating clinical observations of schizophrenia patients with preclinical studies of animals exhibiting developmentally compromised frontal and temporal cortical systems and dysfunctional subcortical GLU and DA interactions will improve our understanding of schizophrenia and our capacity to treat and prevent this devastating disease.

ACKNOWLEDGMENTS

The research contributing to the writing of this manuscript was supported by National Institutes of Health Grants MH 45156 and 57440 (A. Grace), the National Alliance for Research on Schizophrenia and Depression (A. West, A. Charara, and S. Floresco), and the Human Frontiers Science Program (S. Floresco).

REFERENCES

1. DAHLSTRÖM, A. & K. FUXE. 1964. Evidence for the existence of monoamine containing neurons in the central nervous system. I. Demonstration of monoamines in the cell bodies of brain stem neurons. Acta Physiol. Scan. **62** (suppl. 232): 1–55.
2. NAUTA, W.J.H. et al. 1978. Efferent connections and nigral afferents of the nucleus accumbens septi in the rat. Neuroscience **3**: 385–401.
3. VOORN, P. et al. 1986. The dopaminergic innervation of the ventral striatum in the rat: a light- and electron-microscopical study with antibodies against dopamine. J. Comp. Neurol. **251**: 84–99.
4. DESCARRIES, L. et al. 1996. Dual character, asynaptic and synaptic, of the dopamine innervation in adult rat neostriatum: a quantitative autoradiographic and immunocytochemical analysis. J. Comp. Neurol. **375**: 167–186.
5. SMITH, D. & P. BOLAM. 1990. The neural network of the basal ganglia as revealed by the study of synaptic connections of identified neurones. TINS **13**(7): 259–265.
6. HIDAKA, S. & S. TOTTERDELL. 2001. Ultrastructural features of the nitric oxide synthase-containing interneurons in the nucleus accumbens and their relationship with tyrosine hydroxylase containing terminals. J. Comp. Neurol. **431**: 139–154.
7. SESACK, S.R. & V.M. PICKEL. 1992. Prefrontal cortical efferents in the rat synapse on unlabeled neuronal targets of catecholamine terminals in the nucleus accumbens septi and on dopamine neurons in the ventral tegmental area. J. Comp. Neurol. **320**: 145–160.
8. BOUYER, J.J. et al. 1984. Chemical and structural analysis of the relation between cortical inputs and tyrosine hydroxylase-containing terminals in rat neostriatum. Brain Res. **302**: 267–275.
9. GARRIS, P.A. et al. 1994. Efflux of dopamine from the synaptic cleft in the nucleus accumbens of the rat brain. J. Neurosci. **14**: 6084–6093.
10. GRACE, A.A. 1991. Phasic versus tonic dopamine release and the modulation of dopamine system responsivity: a hypothesis for the etiology of schizophrenia. Neurosci. **41**:1–24.
11. GONON, F. 1997. Prolonged and extrasynaptic excitatory action of dopamine mediated by D1 receptors in the rat striatum in vivo. J. Neurosci. **17**: 5972–5978.

12. GERFEN, C.R. & C.J. WILSON. 1996. The basal ganglia. *In* Handbook of Chemical Neuroanatomy. A. Bjorklund & T. Hokfelt, Eds.: 371–468. Elsevier Science. London, UK.
13. SESACK, S.R. *et al.* 1989. Topographical organization of the efferent projections of the medial prefrontal cortex in the rat: an anterograde tract-tracing study with Phaseolus vulgaris leucoagglutinin. J. Comp. Neurol. **290**: 213–242.
14. MEREDITH, G.E. *et al.* 1990. Hippocampal fibers make synaptic contacts with glutamate decarboxylase-immunoreactive neurons in the rat nucleus accumbens. Brain Res. **513**: 329–334.
15. PENNARTZ, C.M.A. *et al.* 1994. The nucleus accumbens as a complex of functionally distinct neuronal ensembles: an integration of behavioral, electrophysiological and anatomical data. Prog. Neurobiol. **42**: 719–761.
16. FINCH, D.M. 1996. Neurophysiology of converging synaptic inputs from the rat prefrontal cortex, amygdala, midline thalamus, and hippocampal formation onto single neurons of the caudate/putamen and nucleus accumbens. Hippocampus **6**: 495–512.
17. VUILLET, J. *et al.* 1989. Ultrastructural correlates of functional relationships between nigral dopaminergic or cortical afferent fibers and neuropeptide Y-containing neurons in the rat striatum. Neurosci. Lett. **100**: 99–104.
18. KELLEY, A.M. & K.C. BERRIDGE. 2002. The neuroscience of natural rewards: relevance to addictive drugs. J. Neurosci. **22**(9): 3306–3311.
19. KITAI, S.T. *et al.* 1999. Afferent modulation of dopamine neuron firing patterns. Curr. Opin. Neurobiol. **9**: 690–697.
20. HABER, S.N. *et al.* 2000. Striatonigrostriatal pathways in primates form an ascending spiral from the shell to the dorsolateral striatum. J. Neurosci. **20**(6): 2369–2382.
21. CHÉRAMY, A. *et al.* 1998. Direct and indirect presynaptic control of dopamine release by excitatory amino acids. Amino Acids **14**: 63–68.
22. MORARI, M. *et al.* 1998. Reciprocal dopamine-glutamate modulation of release in the basal ganglia. Neurochem. Int. **33**: 383–397.
23. CEPEDA, C. & M.S. LEVINE. 1998. Dopamine and N-methyl-D-aspartate receptor interactions in the neostriatum. Dev. Neurosci. **20**: 1–18.
24. CALABRESI, P. *et al.* 2000. Synaptic transmission in the striatum: from plasticity to neurodegeneration. Prog. Neurobiol. **61**: 231–265.
25. GRACE, A.A. 2000. Gating of information flow within the limbic system and the pathophysiology of schizophrenia. Brain Res. Rev. **31**: 330–341.
26. NICOLA, S. *et al.* 2000. Dopaminergic modulation of neuronal excitability in the striatum and nucleus accumbens. Annu. Rev. Neurosci. **23**: 185–215.
27. O'DONNELL, P. 2003. Dopamine gating of forebrain neural ensembles. Eur. J. Neurosci. **17**: 429–435.
28. WILSON, C.J. 1993. The generation of natural firing patterns in neostriatal neurons. *In* Chemical Signaling in the Basal Ganglia, Progress in Brain Research. G.W. Arbuthnott & P.C. Emson, Eds.: **99**: 277–297. Elsevier. Amsterdam.
29. O'DONNELL, P. & A.A. GRACE. 1995. Synaptic interactions among excitatory afferents to nucleus accumbens neurons: hippocampal gating of prefrontal cortical input. J. Neurosci. **15**: 3622–3639.
30. WILSON, C.J. & Y. KAWAGUCHI. 1996. The origins of two-state spontaneous membrane potential fluctuations of neostriatal spiny neurons. J. Neurosci. **16**(7): 2397–2410.
31. PLENZ, D. & S.T. KITAI. 1998. Up and down states in striatal medium spiny neurons simultaneously recorded with spontaneous activity in fast-spiking interneurons studied in cortex-striatum-substantia nigra organotypic cultures. J. Neurosci. **18**(1): 266–283.
32. GOTO, Y. & P. O'DONNELL. 2001. Synchronous activity in the hippocampus and nucleus accumbens in vivo. J. Neurosci. **21**: RC131, 1–5.
33. GOTO, Y. & P. O'DONNELL. 2001. Network synchrony in the nucleus accumbens in vivo. J. Neurosci. **21**: 4498–4504.
34. MAHON, S. *et al.* 2001. Relationship between EEG potentials and intracellular activity of striatal and cortico-striatal neurons: an *in vivo* study under different anesthetics. Cereb. Cortex **11**: 360–373.
35. STERN, E.A. *et al.* 1998. Membrane potential synchrony of simultaneously recorded striatal spiny neurons in vivo. Nature **394**: 475–478.

36. O'DONNELL, P. & A.A. GRACE. 1997. Cortical afferents modulate striatal gap junction permeability via nitric oxide. Neuroscience **76:** 1–5.
37. FUJIYAMA, F. & S. MASUKO. 1996. Association of dopaminergic terminals and neurons releasing nitric oxide in the rat striatum: an electron microscopic study using NADPH-diaphorase histochemistry and tyrosine hydroxylase immunohistochemistry. Brain Res. Bull. **40**(2): 121–127.
38. WEST, A.R. *et al.* 2002. Regulation of striatal dopamine neurotransmission by nitric oxide: effector pathways and signaling mechanisms. Synapse **44:** 227–245.
39. MAHON, S. *et al.* 2000. Intrinsic properties of rat striatal output neurons and time-dependent facilitation of cortical inputs in vivo. J. Physiol. **527:** 345–354.
40. GOTO, Y. & P. O'DONNELL. 2002. Timing-dependent limbic-motor synaptic integration in the nucleus accumbens. Proc. Natl. Acad. Sci. USA **99**(20): 13189–13193.
41. KERR, J.N.D. & D. PLENTZ. 2002. Dendritic calcium encodes striatal neuron output during up-states. J. Neurosci. **22**(5): 1499–1512.
42. KREBS, M.O. *et al.* 1994. Role of dynorphin and GABA in the inhibitory regulation of NMDA-induced dopamine release in striosome- and matrix-enriched areas of the rat striatum. J. Neurosci. **14**(4): 2435–2443.
43. TABER, M.T. & H.C. FIBIGER. 1995. Electrical stimulation of the prefrontal cortex increases dopamine release in the nucleus accumbens of the rat: modulation by metabotropic glutamate receptors. J. Neurosci. **15:** 3896–3904.
44. MURASE, S. *et al.* 1993. Prefrontal cortex regulates burst firing and transmitter release in rat mesolimbic dopamine neurons studied in vivo. Neurosci. Lett. **157:** 53–56.
45. KARREMAN, M. & B. MOGHADDAM. 1996. The prefrontal cortex regulates the basal release of dopamine in the limbic striatum: an effect mediated by ventral tegmental area. J. Neurochem. **66:** 589–598.
46. SMOLDERS, I. *et al.* 1996. Extracellular striatal dopamine and glutamate after decortication and kainate receptor stimulation, as measured by microdialysis. J. Neuroschem. **66:** 2373–2380.
47. SEGOVIA, G. *et al.* 1997. Endogenous glutamate increases extracellular concentrations of dopamine, GABA, and taurine through NMDA and AMPA/kainate receptors in striatum of the freely moving rat: a microdialysis study. J. Neurochem. **69**(4): 1476–1483.
48. WEST, A.R. & M.P. GALLOWAY. 1997. Inhibition of glutamate reuptake potentiates endogenous nitric oxide facilitated dopamine efflux in the rat striatum: an in vivo microdialysis study. Neurosci. Lett. **230:** 21–24.
49. BERT, L. *et al.* 2002. In vivo temporal sequence of rat striatal glutamate, aspartate, and dopamine efflux during apomorphine, nomifensine, NMDA and PDC in situ administration. Neuropharmacology **43:** 825–835.
50. SAULSKAYA, N. & C.A. MARSDEN. 1995. Conditioned dopamine release: dependence upon N-methyl-D-aspartate receptors. Neuroscience **67:** 57–63.
51. JACKSON, M.E. & B. MOGHADDAM. 2001. Amygdala regulation of nucleus accumbens dopamine output is governed by the prefrontal cortex. J. Neurosci. **21**(2): 676–681.
52. HOWLAND, J.G. *et al.* 2002. Glutamate receptor-dependent modulation of dopamine efflux in the nucleus accumbens by basolateral, but not central, nucleus of the amygdala in rats. J. Neurosci. **22**(3): 1137–1145.
53. FLORESCO, S.B. *et al.* 1998. Basolateral amygdala stimulation evokes glutamate receptor-dependent dopamine efflux in the nucleus accumbens of the anaesthetized rat. Eur. J. Neurosci. **10:** 1241–1251.
54. KULAGINA, N.V. *et al.* 2001. Glutamate regulates the spontaneous and evoked release of dopamine in the rat striatum. Neurosci. **102**(1): 121–128.
55. GRACE, A.A. & B.S. BUNNEY. 1984. The control of firing pattern in nigral dopamine neurons: burst firing. J. Neurosci. **4:** 2877–2890.
56. OVERTON, P.G. & D. CLARK. 1997. Burst firing in midbrain dopaminergic neurons. Brain Res. Brain Res. Rev. **25:** 312–334.
57. CARR, D.B. & S.R. SESACK. 2000. Projections from the rat prefrontal cortex to the ventral tegmental area: target specificity in the synaptic associations with mesoaccumbens and mesocortical neurons. J. Neurosci. **20:** 3864–3873.
58. WEST, A.R. & A.A. GRACE. 2000. Striatal nitric oxide signaling regulates the neuronal activity of midbrain dopamine neurons *in vivo*. J. Neurophys. **83**(4): 1796–1808.

59. WEST, A.R. *et al.* 2002. Direct examination of local regulation of membrane activity in striatal and prefrontal cortical neurons *in vivo* using simultaneous intracellular recording and microdialysis. J. Pharmacol. Exp. Ther. **301:** 867–877.
60. SESACK, S.R. & D.B. CARR. 2002. Selective prefrontal cortex input to dopamine cells: implications for schizophrenia. Phys. Behav. **77:** 513–517.
61. GRACE, A.A. & B.S. BUNNEY. 1985. Opposing effects of striatonigral feedback pathways on midbrain dopamine cell activity. Brain Res. **333:** 271–284.
62. GROENEWEGEN, H.J. *et al.* 1996. The nucleus accumbens: gateway for limbic structures to reach the motor system? Prog. Brain Res. **107:** 485–511.
63. HABER, S.N. & J.L. FUDGE. 1997. The primate substantia nigra and VTA: integrative circuitry and function. Crit. Rev. Neurobiol. **11:** 323–342.
64. FLORESCO, S.B. *et al.* 2001. Glutamatergic afferents from the hippocampus to the nucleus accumbens regulate activity of ventral tegmental area dopamine neurons. J. Neurosci. **21**(13): 4915–4922.
65. SMITH, I.D. & A.A. GRACE. 1992. Role of the subthalamic nucleus in the regulation of nigral dopamine neuron activity. Synapse **12:** 287–303.
66. LOKWAN, S.J.A. *et al.* 1999. Stimulation of the pedunculopontine tegmental nucleus in the rat produces burst firing in A9 dopaminergic neurons. Neurosci. **92**(1): 245–254.
67. ROUILLARD, C. & A.S. FREEMAN. 1995. Effects of electrical stimulation of the central nucleus of the amygdala on the in vivo electrophysiological activity of rat nigral dopaminergic neurons. Synapse **21:** 348–356.
68. BEVAN, M.D. *et al.* 1995. The glutamate-enriched cortical and thalamic input to neurons in the subthalamic nucleus of the rat: convergence with GABA-positive terminals. J. Comp. Neurol. **361:** 491–511.
69. PALADINI, C.A. & J.M. TEPPER. 1999. $GABA_A$ and $GABA_B$ antagonists differentially affect the firing pattern of substantia nigra dopaminergic neurons in vivo. Synapse **32:** 165–176.
70. PUCAK, M.L. & A.A. GRACE. 1994. Regulation of substantia nigra dopamine neurons. Crit. Rev. Neurobiol. **9:** 67–89.
71. MOORE, H. *et al.* 1999. The regulation of forebrain dopamine transmission: relevance to the pathophysiology and psychopathology of schizophrenia. Biol. Psychiatry **46**(1): 40–55.
72. ROBINSON, T.E. & I.Q. WHISHAW. 1988. Normalization of extracellular dopamine in striatum following recovery of a partial unilateral 6-OHDA lesion of the substantia nigra: a microdialysis study in freely moving rats. Brain Res. **450:** 209–224.
73. PARSONS, L.H. & J.B. JUSTICE. 1992. Extracellular concentration and in vivo recovery of dopamine in the nucleus accumbens using microdialysis. J. Neurochem. **58:** 212–218.
74. FLORESCO, S.B. *et al.* 2003. Afferent modulation of dopamine neuron firing differentially regulates tonic and phasic dopamine transmission. Nat. Neurosci. **6**(9): 968–973.
75. SCHULTZ, W. *et al.* 1998. Reward prediction in primate basal ganglia and frontal cortex. Neuropharmacology **37:** 421–429.
76. WEST, A.R. & A.A. GRACE. 2002. Opposite influences of endogenous dopamine D_1 and D_2 receptor activation on activity states and electrophysiological properties of striatal neurons: studies combining *in vivo* intracellular recordings and reverse microdialysis. J. Neurosci. **22**(1): 294–304.
77. REYNOLDS, J.N.J. & J.R. WICKENS. 2000. Substantia nigra dopamine regulates synaptic plasticity and membrane potential fluctuations in the rat neostriatum, in vivo. Neuroscience **99**(2): 199–203.
78. YIM, C.Y. & G.J. MOGENSON. 1988. Neuromodulatory action of dopamine in the nucleus accumbens: an in vivo intracellular study. Neuroscience **26:** 403–415.
79. GOTO, Y. & P. O'DONNELL. 2002. Delayed mesolimbic system alteration in a developmental animal model of schizophrenia. J. Neurosci. **22**(20): 9070–9077.
80. GONON, F. & L. SUNDSTROM. 1996. Excitatory effects of dopamine released by impulse flow in the rat nucleus accumbens in vivo. Neurosci. **75:** 13–18.
81. REYNOLDS, J.N.J. *et al.* 2001. A cellular mechanism of reward-related learning. Nature **413:** 67–70.

82. FLORESCO, S.B. *et al.* 2001. Modulation of hippocampal and amygdalar-evoked activity of nucleus accumbens neurons by dopamine: cellular mechanisms of input selection. J. Neurosci. **21**(8): 2851–2860.
83. FLORESCO, S.B. *et al.* 2001. Dopamine D1, NMDA receptors mediate potentiation of basolateral amygdala-evoked firing of nucleus accumbens neurons. J. Neurosci. **21**: 6370–6376.
84. HU, X.T. & F.J. WHITE. 1997. Dopamine enhances glutamate-induced excitation of rat striatal neurons by cooperative activation of D_1 and D_2 class receptors. Neurosci. Lett. **224**: 61–65.
85. KIYATKIN, E.A. & G.V. REBEC. 1996. Dopaminergic modulation of glutamate-induced excitations of neurons in the neostriatum and nucleus accumbens of awake, unrestrained rats. J. Neurophys. **75**: 142–153.
86. CEPEDA, C. *et al.* 1998. Dopaminergic modulation of NMDA-induced whole cell currents in neostriatal neurons in slices: contribution of calcium conductances. J. Neurophys. **79**: 82–94.
87. SURMEIER, D.J. *et al.* 1995. Modulation of calcium currents by a D1 dopaminergic protein kinase/phosphatase cascade in rat neostriatal neurons. Neuron **14**(2): 385–397.
88. HERNANDEZ-LOPEZ, S. *et al.* 1997. D1 receptor activation enhances evoked discharge in neostriatal medium spiny neurons by modulating an L-type Ca^{2+} conductance. J. Neurosci. **17**: 3334–3342.
89. FLORES-HERNANDEZ, J. *et al.* 2000. D_1 receptor activation reduces $GABA_A$ receptor currents in neostriatal neurons through a PKA/DARPP-32/PPI signaling cascade. J. Neurophys. **83**: 2996–3004.
90. CHAO, S.Z. *et al.* 2002. D1 dopamine receptor stimulation increases GLuR1 phosphorylation in postnatal nucleus accumbens cultures. J. Neurochem. **81**: 984–992.
91. SCOTT, L. *et al.* 2002. Selective up-regulation of dopamine D1 receptors in dendritic spines by NMDA receptor activation. Proc. Natl. Acad. Sci. USA **99**(3): 1661–1664.
92. KONRADI, C. *et al.* 1996. Amphetamine and dopamine-induced immediate early gene expression in striatal neurons depends on post-synaptic NMDA receptors and calcium. J. Neurosci. **16**: 4231–4239.
93. KEEFE, K.A. *et al.* 1998. Effects of NMDA receptor antagonists on D1 dopamine receptor-mediated changes in striatal immediate early gene expression: evidence for involvement of pharmacologically distinct NMDA receptors? Dev. Neurosci. **20**(2-3): 216–2128.
94. NICOLA, S.M. *et al.* 1996. Psychostimulants depress excitatory synaptic transmission in the nucleus accumbens vai presynaptic D1-like dopamine receptors. J. Neurosci. **16**(5): 1591–1604.
95. YANG, C.R. & G.J. MOGENSON. 1984. Electrophysiological responses of neurones in the accumbens nucleus to hippocampal stimulation and the attenuation of the excitatory responses by the mesolimbic dopaminergic system. Brain Res. **324**: 69–84.
96. O'DONNELL, P. & A.A. GRACE. 1994. Tonic D2-mediated attenuation of cortical excitation in nucleus accumbens neurons recorded in vitro. Brain Res. **634**: 105–112.
97. HSU, K-S. *et al.* 1995. Presynaptic D_2 dopaminergic receptors mediate inhibition of excitatory synaptic transmission in rat neostriatum. Brain Res. **690**: 264–268.
98. LEVINE, M.S. *et al.* 1996. Neuromodulatory actions of dopamine on synaptically-evoked neostriatal responses in slices. Synapse **24**: 65–78.
99. UMEMIYA, M. & L.A. RAYMOND. 1997. Dopaminergic modulation of excitatory postsynaptic currents in rat neostriatal neurons. J. Neurophys. **78**: 1248–1255.
100. WANG, H. & V.M. PICKEL. 2002. Dopamine D2 receptors are present in prefrontal cortical afferents and their targets in patches of the rat caudate-putamen nucleus. J. Comp. Neurol. **442**: 392–404.
101. CEPEDA, C. *et al.* 2001. Facilitated glutamatergic transmission in the striatum of D2 dopamine receptor-deficient mice. J. Neurophys. **85**: 659–670.
102. HARVEY, J. & M.G. LACEY. 1996. Endogenous and exogenous dopamine depress EPSCs in rat nucleus accumbens in vitro via D1 receptors activation. J. Phys. **492**: 143–154.
103. ZHANG, X-F. *et al.* 2002. Repeated cocaine treatment decreases whole-cell calcium current in rat nucleus accumbens neurons. J. Pharmacol. Exp. Ther. **301**: 1119–1125.

104. CHARARA, A. & A.A. GRACE. 2003. Dopamine receptor subtypes selectively modulate excitatory afferents from the hippocampus and amygdala to rat nucleus accumbens neurons. Neuropsychopharmacology **28:** 1412–1421.
105. ROSENKRANZ, J.A. & A.A. GRACE. 2002. Cellular mechanisms of infralimbic and prelimbic prefrontal cotical inhibition and dopaminergic modulation of basolateral amygdala neurons in vivo. J. Neurosci. **22**(1): 324–337.
106. WEINBERGER, D.R. & B.K. LIPSKA. 1995. Cortical maldevelopment, anti-psychotic drugs, and schizophrenia: a search for common ground. Schizophrenia Res. **16:** 87–110.
107. TAMMINGA, C.A. 1998. Schizophrenia and glutamatergic transmission. Crit. Rev. Neurobiol. **12:** 21–36.
108. GRACE, A.A. 2003. Developmental dysregulation of the dopamine system and the pathophysiology of schizophrenia. *In* Neurodevelopment and Schizophrenia. M. Keshavan, Kennedy & R.M. Murray, Eds. Cambridge University Press. In press.
109. CARLSSON, A. *et al.* 2001. Interactions between monoamines, glutamate, and GABA in schizophrenia: new evidence. Annu. Rev. Pharmacol. Toxicol. **41:** 237–260.
110. JAVITT, D.C. & S.R. ZUKIN. 1991. Recent advances in the phencyclidine model of schizophrenia. Am. J. Psychiatry **148:** 1301–1308.
111. JENTSCH, J.D. *et al.* 1998. Enduring cognitive deficits and cortical dopamine dysfunction in monkeys after long-term administration of phencyclidine. Science **277:** 953–955.
112. KRYSTAL, J.H. *et al.* 1994. Subanesthetic effects of the noncompetitive NMDA antagonist, ketamine, in humans: psychotomimetic, perceptual, cognitive, and neuroendocrine responses. Arch. Gen. Psychiatry **51:** 199–214.
113. SCHWARCZ, R. *et al.* 2001. Increased cortical kynurenate content in schizophrenia. Biol. Psychiatry **50:** 521–530.
114. ERHARDT, S. & G. ENGBERG. 2002. Increased phasic activity of dopaminergic neurones in the rat ventral tegmental area following pharmacologically elevated levels of endogenous kynurenic acid. Acta Physiol. Scand. **175**(1): 45–53.
115. TAMMINGA, C.A. *et al.* 1992. Limbic system abnormalities identified in schizophrenia using positron emission tomography with fluorodeoxyglucose and neocortical alterations with deficit syndrome. Arch. Gen. Psychiatry **49:** 522–530.
116. GEYER, M.A. *et al.* 2001. Pharmacological studies of prepulse inhibition models of sensorimotor gating deficits in schizophrenia: a decade in review. Psychopharmacology **156:** 117–154.
117. LIPSKA, B.K. *et al.* 2002. Effects of reversible inactivation of the neonatal ventral hippocampus on behavior in the adult rat. J. Neurosci. **22**(7): 2835–2842.
118. GHAJARNIA, M. *et al.* 1998. Enhanced behavioral effects of phencyclidine in rats with developmental abnormalities of the temporal lobe. Soc. Neurosci. Abstr. **24:** 2177.
119. KORNHUBER, J. *et al.* 1989. [^{3}H]MK-801 binding studies in postmortem brain regions of schizophrenic patients. J. Neural Transm. **77:** 231–236.
120. WILKINSON, P.C. *et al.* 1997. Contrasting effects of excitotoxic lesions of the prefrontal cortex on the behavioral response to D-amphetamine and presynaptic and postsynaptic measures of striatal dopamine function in monkeys. Neurosci. **80**(3): 717–730.
121. BREIER, A. *et al.* 1997. Schizophrenia is associated with elevated amphetamine-induced synaptic dopamine concentrations: evidence from a novel positron emission tomography method. Proc. Natl. Acad. Sci. USA **94:** 2569–2574.
122. LARUELLE, M. *et al.* 1996. Single photon emission computerized tomography imaging of amphetamine-induced dopamine release in drug-free schizophrenic subjects. Proc. Natl. Acad. Sci. USA **93:** 9235–9240.
123. EVERITT, B.J. & M.E. WOLF. 2002. Psychomotor stimulant addiction: a neural systems perspective. J. Neurosci. **22**(9): 3312–3320.

Molecular Abnormalities of the Glutamate Synapse in the Thalamus in Schizophrenia

JAMES H. MEADOR-WOODRUFF, SARAH M. CLINTON, MONICA BENEYTO, AND ROBERT E. McCULLUMSMITH

Mental Health Research Institute and Department of Psychiatry, University of Michigan, Ann Arbor, Michigan 48109-0720, USA

ABSTRACT: Schizophrenia has been associated with dysfunction of glutamatergic neurotransmission. Synaptic glutamate activates pre- and postsynaptic ionotropic NMDA, AMPA, and kainate and metabotropic receptors, is removed from the synapse via five cell surface–expressed transporters, and is packaged for release by three vesicular transporters. In addition, there is a family of intracellular molecules enriched in the postsynaptic density (PSD) that target glutamate receptors to the synaptic membrane, modulate receptor activity, and coordinate glutamate receptor–related signal transduction. Each family of PSD proteins is selective for a given glutamate receptor subtype, the most well characterized being the NMDA receptor binding proteins PSD93, PSD95, NF-L, and SAP102. Besides binding glutamate receptors, many of these proteins also interact with cell surface proteins like cell adhesion molecules, ion channels, cytoskeletal elements, and signal transduction molecules. Given the complexity of the glutamate neurotransmitter system, there are many locations where disruption of normal signaling could occur and give rise to abnormal glutamatergic neurotransmission in schizophrenia. Using multiple cohorts of postmortem tissue, we have examined these synaptic molecules in schizophrenic thalamus. The expression of NR1 and NR2C subunit transcripts is decreased in the thalamus in schizophrenia. Interestingly, three intracellular PSD molecules that link the NMDA receptor to signal transduction pathways are also abnormally expressed. Additionally, several of the cell surface and vesicular transporters are abnormal in the schizophrenic thalamus. While occasional findings of abnormal receptor expression are made, the most dramatic and consistent alterations that we have found in the thalamus in schizophrenia involve the family of intracellular signaling/scaffolding molecules. We propose that schizophrenia has a glutamatergic component that involves alterations in the intracellular machinery that is coupled to glutamate receptors, in addition to abnormalities of the receptors themselves. Our data suggest that schizophrenia is associated with abnormal glutamate receptor–related intracellular signaling in the thalamus, and point to novel targets for innovative drug discovery.

KEYWORDS: glutamate synapse; schizophrenia; phencyclidine; excitatory amino acid transporters; thalamic anatomy

Address for correspondence: James H. Meador-Woodruff, M.D., Mental Health Research Institute, Department of Psychiatry, University of Michigan, 205 Zina Pitcher Place, Ann Arbor, MI 48109-0720. Voice: 734-936-2093; fax: 734-647-4130.
 jimmw@umich.edu

Ann. N.Y. Acad. Sci. 1003: 75–93 (2003). © 2003 New York Academy of Sciences.
doi: 10.1196/annals.1300.005

INTRODUCTION

Recently, increasing evidence has implicated glutamatergic abnormalities in the pathophysiology of schizophrenia. The glutamate hypothesis of schizophrenia was originally based on the observation that phencyclidine (PCP), an NMDA receptor antagonist, precipitates a schizophreniform psychosis in nonpsychiatrically ill persons.[1-7] Subsequent studies have noted that the psychomimetic effects of PCP and other NMDA receptor antagonists (such as ketamine) differ markedly in their effects compared to other psychotogenic substances. For example, effects of PCP intoxication may include both positive and negative psychotic symptoms, while effects of dopamine agonists are typically limited to positive symptoms. On the basis of these clinical observations, a hypothesis of NMDA receptor hypoactivity in schizophrenia was proposed. Supporting this initial hypothesis, administration of agonists of the glycine/D-serine coagonist site of the NMDA receptor modestly attenuates psychosis, especially negative symptoms, in persons afflicted with schizophrenia.[8-13] NMDA receptor activity, however, is just one component of glutamatergic neurotransmission, a tightly regulated neurotransmitter system that involves myriad other molecules. Recent efforts have focused not only on possible NMDA receptor abnormalities in schizophrenia, but also on the possibility of disturbances of other molecules associated with glutamatergic neurotransmission in this illness.

GLUTAMATERGIC NEUROTRANSMISSION: A BRIEF OVERVIEW

Glutamate neurotransmission requires three distinct cell types that comprise the typical glutamate synapse: an astrocyte, as well as both a presynaptic and a postsynaptic neuron (FIG. 1).[14-16] Glutamate is packaged into secretory vesicles in the presynaptic neuron by a family of at least three vesicular transporters (vGluT1–vGluT3) and released into the synapse.[17-21] Synaptic glutamate can stimulate both metabotropic and ionotropic glutamate receptors, located in receptor-specific distributions on pre- and postsynaptic neurons, as well as on astrocytes.[14-16,22] Glutamate receptor subtypes (FIG. 2) include a group of pharmacologically distinct ligand-gated ion channels (NMDA, AMPA, and kainate receptors) and the eight G-protein coupled metabotropic receptors (mGluR1–8).[23-25]

Glutamate is rapidly removed from the synapse by a family of at least five plasma membrane excitatory amino acid transporters (EAAT1–EAAT5), located on both synaptic neurons and astrocytes.[16,26-34] Recovered glutamate may enter the TCA cycle via conversion to α-ketoglutarate by glutamate dehydrogenase, and/or be converted to glutamine-by-glutamine synthetase and transported back into the synapse.[14] One particularly interesting pathway involves reuptake of glutamine into the presynaptic neuron. Presynaptic glutamine can be oxidized to glutamate by the enzyme glutaminase and repackaged into vesicles for release.[14] Glutaminase and the vesicular transporters have been proposed as markers for glutamatergic neurons. Glutaminase expression is enriched in glutamatergic neurons, and expression of the vesicular transporters appear to be exclusively in presynaptic glutamatergic terminals.[17-21,35,36] The packaging, release, and reuptake of glutamate are closely regulated, since excess glutamate may lead to excitotoxic cell death and/or seizures.[36]

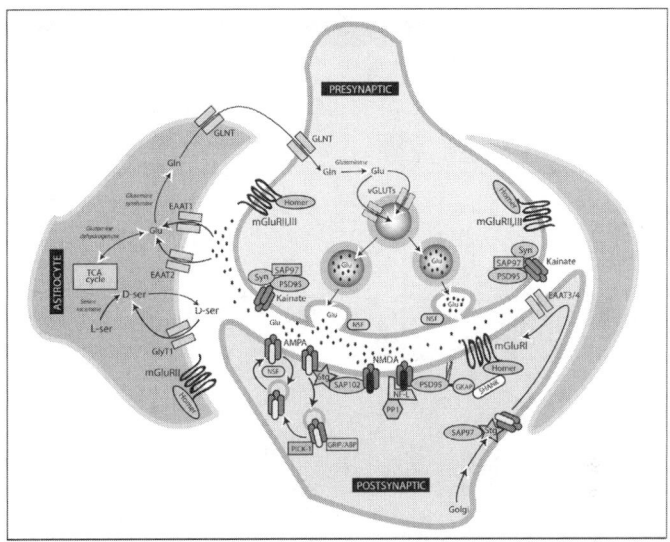

FIGURE 1. Diagram of the glutamate synapse. The glutamate synapse requires three discrete cell types: a presynaptic gluatamate-releasing neuron, a postsynaptic neuron, and an astrocyte. Glutamate is packaged into secretory vesicles by vesicular glutamate transporters. Once glutamate is released into the synapse, it can interact with a number of receptors that may exist on any of the three cell types. Each of these receptors, in turn, is associated with a series of intracellular signaling molecules that are specific for each receptor subtype. Most glutamate reuptake occurs via membrane-bound glutamate transporters located on the astrocyte, although these also exist on the postsynaptic neuron, for rapid inactivation of the actions of glutamate. Glutamate that is taken up by the astrocyte, in turn, is reduced to glutamine or alternatively enters intermediary metabolism. Glutamine is actively transported from the astrocyte back into the presynaptic neuron, where it is, in turn, oxidized to glutamate for repacking for release. In addition, astrocytes are responsible for the conversion of L-serine to D-serine, which is released by astrocytes and is an agonist of a modulatory site on the NMDA receptor complex.

Recent investigations have identified a novel group of molecules that functionally connect glutamate receptors and transporters with intracellular cytoskeletal and signaling elements. Using yeast two-hybrid techniques, families of proteins that have unique protein–protein interactions with NMDA, AMPA, kainate, and metabotropic receptors have been characterized. These intracellular proteins link these receptors to the cytoskeleton as well as to specific signal transduction pathways. Proteins that interact with the plasma membrane excitatory amino acid transporters (EAAT) have also been reported,[37–39] although their function is less well characterized.

The preceding description of glutamatergic neurotransmission is brief; further details about each class of molecules are provided below when discussed in the context of abnormalities of these molecules in the thalamus in schizophrenia.

	IONOTROPIC				METABOTROPIC	
	NMDA	AMPA	Kainate			
SUBUNITS	NR1 NR2A-D NR3A,B	GluR1 GluR2 GluR3 GluR4	GluR5 GluR6 GluR7 KA1 KA2	**RECEPTORS**	Group I	mGluR1 mGluR5
					Group II	mGluR2 mGluR3
Binding sites	Glutamate Glycine/D-serine Polyamines PCP/MK801 Proton Zinc ion	AMPA	Kainate		Group III	mGluR4 mGluR6 mGluR7 mGluR8
	Protein	subunit	Protein	Subunit	Protein	Subunit		Protein	Receptors
PSD Proteins	PSD95	NR2	ABP	GluR2/3	PSD95S	GluR5,6/KA2	**PSD proteins**	Homer 1	mGluR1/5
	PSD93	NR2	GRIP	GluR2/3	AP102	GluR6		Homer 2	mGluR1/5
	SAP102	NR2	NSF	GluR2/3	SAP97	GluR6		Homer 3	mGluR1/5
	CIPP	NR2	PICK-1	GluR2/3,14	PICK-1	GluR5,6		GRIP	mGluR3,4a,R6,R7a
	Densin-180	NR2	SAP97	GluR1	GRIP	GluR5,6		Syntenin	mGluR4a,R6,R7a,b
	NF-L	NR1-e21	Stargazin	GluR1-4	Syntenin	GluR5,6		PICK-1	mGluR3
	Yotiao	NR1-e21	Syntenin	GluR1-4					

FIGURE 2. Organization of the glutamate receptors. The glutamate receptors are both ligand-gated ion channels (ionotropic), as well as seven transmembrane domain G-protein–coupled receptors (metabotropic receptors). The ionotropic receptors cluster into three definable families, the NMDA, AMPA, and kainate types. These receptors are multimeric associations of specific subunits and have specific binding domains on the final receptor complexes. The metabotropic receptors are classed into three groups, based on similar pharmacological features. In the case of both the ionotropic and the metabotropic receptors, intracellular proteins associated with the postsynaptic density have been identified that have specific associations with both types of receptors. Many of these proteins are listed in this figure with specific protein:subunit associations indicated.

GLUTAMATE ABNORMALITIES IN SCHIZOPHRENIA

We recently reviewed a number of previous studies that have examined the expression of glutamate receptors in the brain in schizophrenia that have concentrated on cortical and medial temporal lobe structures.[40] These studies have generally revealed complex, region- and receptor-specific abnormalities. Several generalizations about glutamate receptor expression in the brain in schizophrenia emerge from this body of literature. First, most abnormalities have been reported in limbic cortical and hippocampal regions. Second, all three families of ionotropic glutamate receptors have been reported to be abnormal, with changes in binding sites as well as subunit changes suggestive of altered stoichiometry of subunit composition. Relatively few abnormalities have been reported for the metabotropic glutamate receptors. Finally, there are minimal to no meaningful changes in the expression of these receptors in striatal regions, indicating that abnormalities of these receptors in schizophrenia are likely region specific.

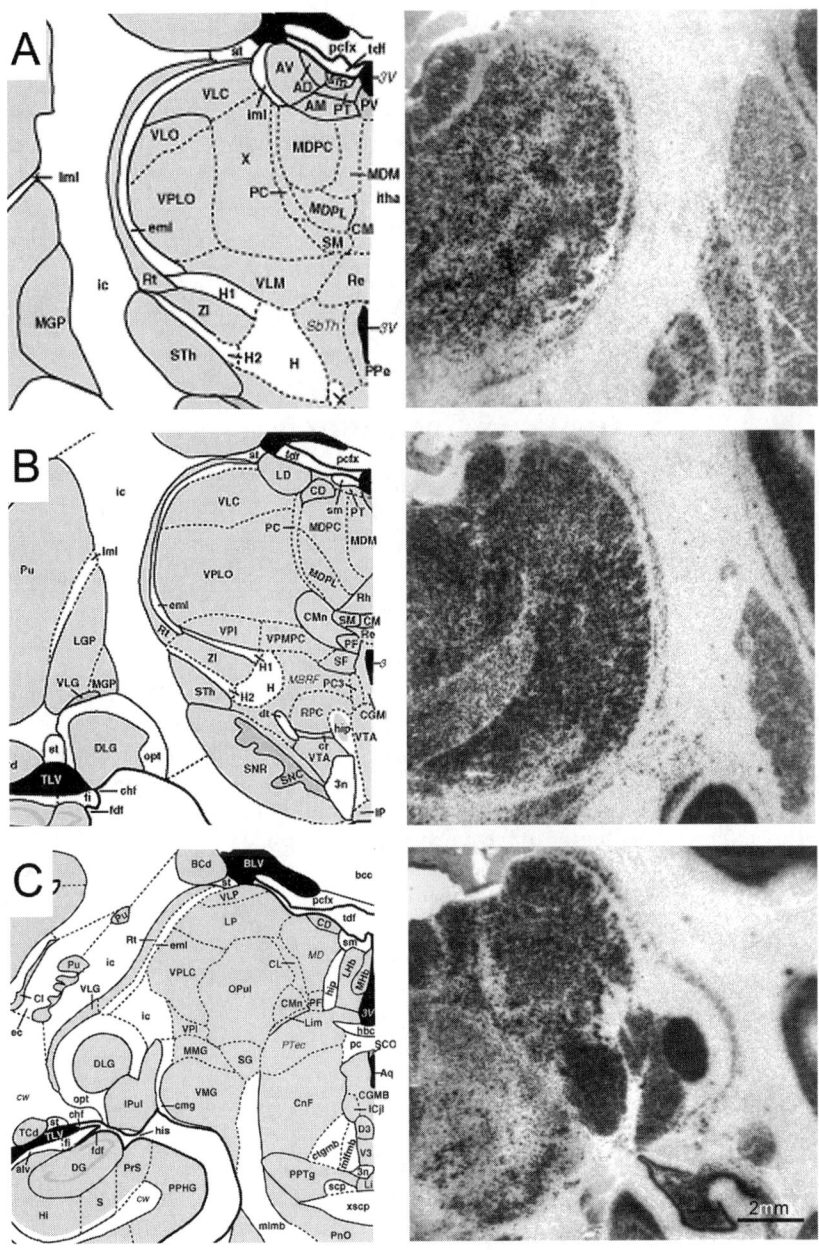

FIGURE 3. The thalamus is composed of multiple nuclei that have specific topographical projections to the neocortex, as well as to a number of subcortical structures. The *left column* demonstrates a schematic of these multiple nuclei at (**A**) a rostral level, (**B**) middle thalamus, and (**C**) more caudally. The *right column* shows representative *in situ* hybridiza-

THE THALAMUS IN SCHIZOPHRENIA

While prefrontal and temporal cortical dysfunction is often associated with the pathophysiology of schizophrenia, recent attention has focused on the role of the thalamus in this illness. Patients with schizophrenia experience a wide range of psychotic symptoms and cognitive deficits, including hallucinations and delusions; and negative symptoms, including a relative inability to pick up on social cues and focus their attention on a particular task.[41] Many of these symptoms may stem from a deficit in sensory processing.[41–44] The thalamus plays a central role in processing and integrating sensory information relevant to emotional and cognitive functions, and several lines of investigation now suggest the possibility of thalamic dysfunction in the pathophysiology of schizophrenia.[41–44]

THALAMIC ANATOMY AND CIRCUITRY

The thalamus is composed of numerous topographically organized nuclei (FIG. 3) that project to the cortex and several subcortical regions (FIG. 4).[45] Each relay nucleus receives modality-specific sensory input, such as somatosensory, visual, or auditory information, and is linked to a specific area of cerebral cortex, which processes this information and projects back to the thalamus. The thalamus is also critical for the regulation of states of consciousness, which, in turn, can influence the ability of the cortex to receive and process information.[42]

The thalamus contains three main cell types: excitatory relay cells, inhibitory interneurons, and the neurons of the reticular nucleus.[46] Large glutamatergic relay cells in the dorsal thalamus respond to sensory afferent input, and project to the cortex, which reciprocally projects back to the thalamus. Inhibitory interneurons also located in the dorsal thalamus impinge on the dendrites of nearby relay neurons and sensory afferents.[47] Neurons of the reticular nucleus reside in a thin sheet of γ-aminobutyric acid (GABA)-ergic neurons that encompass the dorsal thalamus. Reticular neurons receive excitatory input from collateral fibers of the thalamocortical and corticothalamic projections entering and leaving the dorsal thalamus. Reticular neurons, in turn, send axons into the dorsal thalamus to gate relay neuron activity.[46,47] Reticular neuron activity modulates the activity of dorsal thalamic relay cells and, consequently, is able to influence the ability of the dorsal thalamus to relay sensory information to the cortex.[42]

Thalamic afferents and efferents primarily use either glutamate or GABA as neurotransmitters. Thalamocortical projections, as well as corticothalamic and sensory afferents to the dorsal thalamus use glutamate, which acts on both ionotropic and metabotropic glutamate receptors expressed throughout the thalamus. The intrinsic interneurons and neurons of the reticular nucleus contain GABA, and certain thalamic nuclei (anterior ventral lateral and ventral medial) receive GABAergic input from

---

tion images of neurofilament-light (NF-L) mRNA, which labels the preponderance of cells in the subcortical structures that are observable in these three panels. The atlas diagrams and the *in situ* hybridization images are from a female macaque, although the appearance of these nuclei in the human thalamus is very similar.

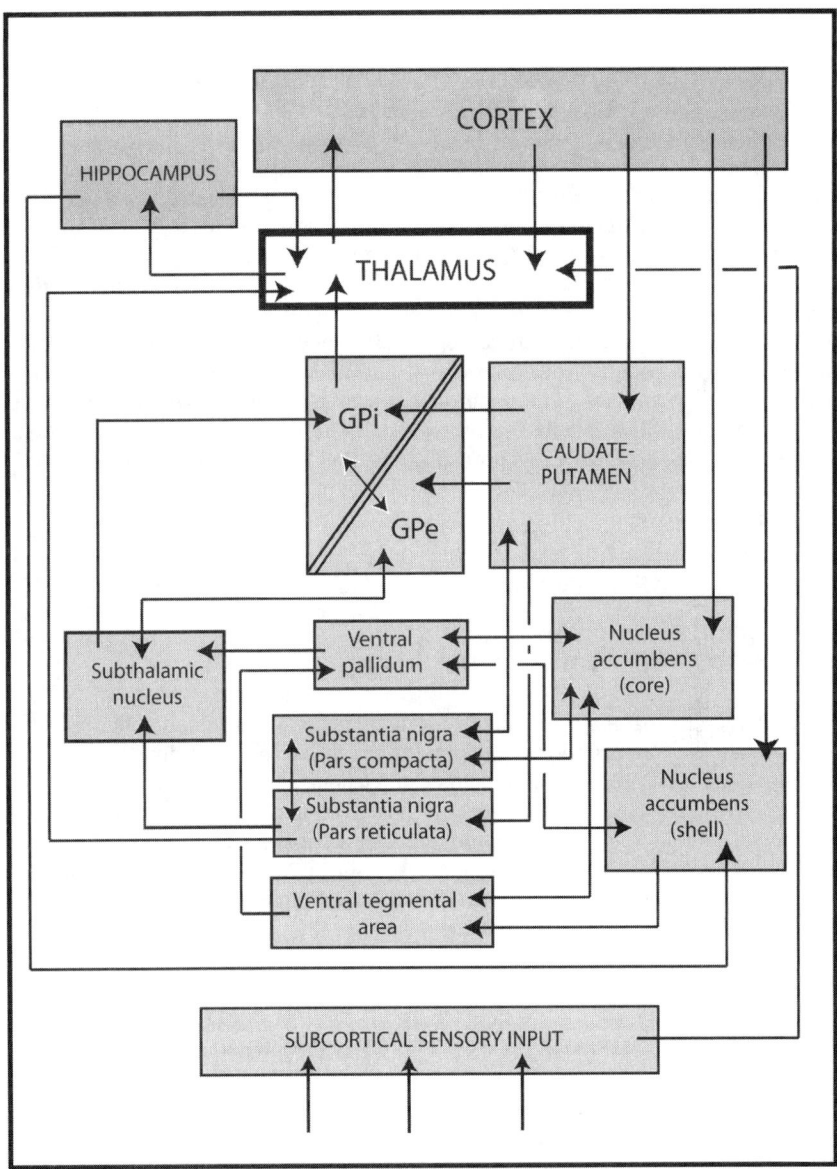

FIGURE 4. Diagram of thalamic efferents and afferents, demonstrating the central role of the thalamus in linking hippocampus, neocortex, and a number of subcortical structures.

the internal segment of the globus pallidus, ventral pallidum, and substantia nigra pars reticulata.[45] The thalamus also receives substantial noradrenergic, serotonergic, and cholinergic innervation, which likely modulates glutamatergic and GABAergic inputs.[46]

CELLULAR ABNORMALITIES OF THE THALAMUS IN SCHIZOPHRENIA

Postmortem morphometric studies have reported cellular abnormalities in the thalamus in schizophrenia, particularly in the dorsomedial (DM) nucleus (FIG. 3).[48–53] Using stereologic cell-counting procedures, Pakkenberg reported a 40% reduction of cell number and a 25% reduction in volume of the DM nucleus.[48] A number of other studies have confirmed these findings[50–53] and extended them to show that these changes are restricted to the densocellular division of the DM nucleus, which projects to the striatum and premotor cortical regions,[51] and the parvocellular subdivision of the DM nucleus, which projects to dorsal and lateral areas of the prefrontal cortex.[51,53] Further, another group reported decreased density of parvalbumin-immunoreactive varicosities (putative axon terminals) in middle layers of the prefrontal cortex, the primary target of DM projections.[54] Cell loss in the thalamus may not be restricted to the DM nucleus. Neuron number is also reportedly decreased in the pulvinar,[53] anteroventral/anteromedial,[52] and ventral lateral posterior nuclei,[55] but appears to be unchanged in the centromedian[53] and ventral posterior medial nuclei.[51]

VOLUMETRIC CHANGES IN THALAMUS: POSTMORTEM AND *IN VIVO* IMAGING STUDIES

In addition to postmortem findings of reduced volume and cell number in certain thalamic nuclei in schizophrenia, several groups have reported reduced overall thalamic volume in schizophrenia using magnetic resonance imaging (MRI).[56–65] More recent MRI studies have examined discrete thalamic nuclei and found decreased volume of the DM,[61,66] pulvinar,[66] central medial, anterior, and posteromedial thalamic nuclei.[61] Functional brain imaging studies, using positron emission tomography (PET) and single photon emission computed tomography (SPECT), have revealed decreased thalamic metabolism in schizophrenia.[67–74]

GLUTAMATE RECEPTOR ABNORMALITIES IN THE THALAMUS IN SCHIZOPHRENIA

The ionotropic receptors are composed of family-specific subunits (FIG. 2).[24] The AMPA receptor subunits are derived from four different genes, GluR1–GluR4 (FIG. 2). Kainate receptors are composed of subunits derived from genes for the low-affinity GluR5–GluR7 and high-affinity KA1-KA2 subunits (FIG. 2). Subunits associated with both the AMPA and kainate receptors exist in multiple forms due to alternative splicing and editing of their respective transcripts.[24] Accordingly, there is the potential for heterogeneity in both AMPA and kainate receptors, based on subunit composition and transcriptional modification of individual subunits. Subunit composition is physiologically relevant for both AMPA and kainate receptors, as unique pharmacological properties are associated with specific subunit complements in the final receptors. For example, GluR2-containing AMPA receptors have decreased calcium ion flux, which decreases the electrophysiological activity of these receptors.[75–79]

The NMDA receptor subunits are encoded by at least seven genes, NR1, NR2A–NR2D, and NR3A–NR3B (FIG. 2).[24,80] NR1 can be expressed as one of eight isoforms, due to the alternative splicing of exons 5, 21, and 22.[24,25,81] The pharmacological regulation of the NMDA receptor depends upon the unique combination of binding sites.[24] There is a primary site for the binding of glutamate. A separate glycine/D-serine binding site must also be occupied before glutamate can activate the ion channel. Modulatory binding sites for polyamines, pH, and zinc have also been identified. There is a site for ionic magnesium, which blocks the ion channel at physiological concentrations. This blockade is voltage dependent; partial depolarization of the cell membrane extrudes magnesium ions. Therefore, presynaptic glutamate release and postsynaptic predepolarization are both required for NMDA receptor activity. Finally, there is a site within the ion channel itself associated with the binding of uncompetitive antagonists of the NMDA receptor, such as PCP, ketamine, and MK-801. These antagonists are use dependent, that is, the ion channel must be opened for these compounds to bind to the receptor, so there must be cooperativity between multiple sites for occupancy of uncompetitive antagonists. Similar to the AMPA and kainate receptors, subunit composition confers physiological specificity: the NMDA binding sites are associated with different subunits, and their affinities can vary depending on subunit composition.[82–87]

Given that the thalamus has been shown to be abnormal in schizophrenia in earlier morphometric and *in vivo* imaging studies, combined with the hypothesis of glutamatergic abnormalities in schizophrenia, we and others have examined the expression of glutamate receptors in the thalamus in schizophrenia. We determined the expression of the transcripts encoding all of the subunits associated with the three families of ionotropic glutamate receptors, as well as multiple binding sites associated with the NMDA, AMPA, and kainate receptors.[88] In this study, we found marked decreases in the transcripts encoding both the NR1 and NR2C subunits of the NMDA receptor, particularly in limbic-associated nuclei (FIG. 5, TABLE 1).

Given that the NR1 transcript was decreased, we next determined if this decrease was associated with any specific subpopulation of NR1 isoforms. The NR1 gene contains 22 exons, and exons 5, 21, and 22 can be alternatively spliced, resulting in eight different isoforms. The 3′ exons 21 and 22 are particularly interesting, as it is this region of the NR1 subunit that interacts with various intracellular proteins of the postsynaptic density. In a recent study, we replicated the decreased expression of NR1 mRNA and, intriguingly, found that this reduction was exclusively associated with exon-22 containing isoforms.[89]

In addition, we observed decreased expression of some (but not all) of the binding sites associated with the NMDA complex: binding to both the polyamine site and the glycine/D-serine coagonist site was reduced in schizophrenia, in parallel with the decreased expression of the NR1 and NR2C subunits. We also found decreased expression of the GluR3 subunit of the AMPA receptor and the KA2 kainate receptor subunit, but both AMPA and kainate binding were preserved in schizophrenia (TABLE 1).[88]

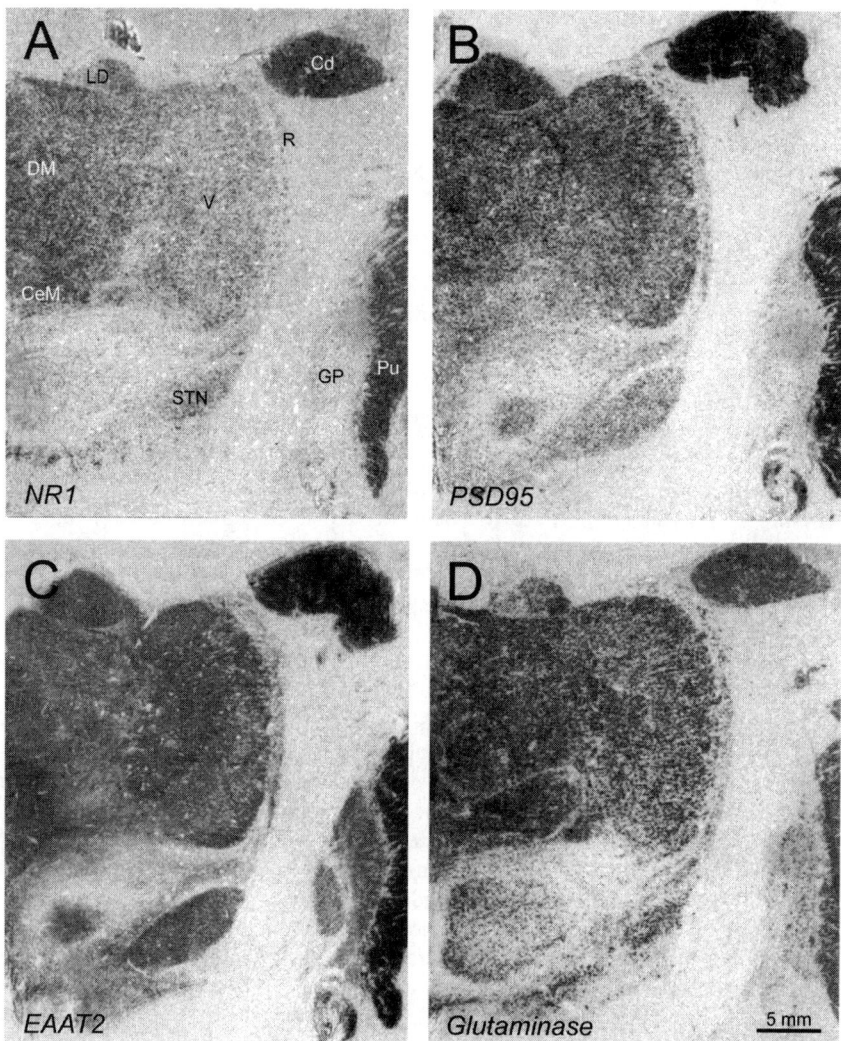

FIGURE 5. Representative *in situ* hybridization images of transcripts encoding key molecules associated with glutamatergic transmission. This is a section through the human thalamus at the level of the dorsomedial nucleus. (**A**) *In situ* hybridization of the NMDA subunit, NR1; (**B**) representative image of the intracellular signaling protein, PSD-95; (**C**) distribution of the transcript encoding the membrane-bound glutamate transporter, EAAT2; (**D**) expression of glutaminase mRNA, an enzyme localized to the presynaptic glutamate neuron responsible for the oxidation of glutamine to glutamate. Abbreviations of thalamic nuclei: DM, dorsomedial nucleus; LD, laterodorsal nucleus; V, nuclei of the ventral tier; R, reticular nucleus; CeM, central medial nucleus. In addition, several other structures are labeled for orienting purposes: Cd, tail of the caudate nucleus; Pu, putamen; GP, globus pallidus; STN, subthalamic nucleus.

TABLE 1. Glutamatergic abnormalities in the thalamus in schizophrenia

	Method	Findings	Reference
Glutamate receptors			
Glutamate binding (NMDAR)	Autoradiography	No change	88
Glycine site (NMDAR)		Decreased	
Polyamine site (NMDAR)		Decreased	
NMDA channel (MK-801)		No change	
AMPA receptor binding		No change	
Kainate receptor binding		No change	
NMDA NR1	In situ hybridization	Decreased	88
NMDA NR2A		No change	
NMDA NR2B		Decreased	
NMDA NR2C		Decreased	
NMDA NR2D		No change	
AMPA GluR1		Decreased	
AMPA GluR2		No change	
AMPA GluR3		Decreased	
AMPA GluR4		No change	
Kainate GluR5		No change	
Kainate GluR6		No change	
Kainate GluR7		No change	
Kainate KA1		No change	
Kainate KA2		Decreased	
mGluR1	In situ hybridization	No change	91
mGluR2		No change	
mGluR3		No change	
mGluR4		No change	
mGluR5		No change	
mGluR7		No change	
mGluR8		No change	
NMDA NR1	In situ hybridization	No change	90
NMDA NR2A		No change	
AMPA GluR2		No change	
AMPA GluR4		No change	
Kainate GluR5		No change	
Kainate GluR6		No change	
Glutamate transporters			
EAAT1	In situ hybridization	Increased	101
EAAT2		Increased	
EAAT3		No change	
VGLUT1		Not expressed	102
VGLUT2		Increased	
NMDAR-associated PSD proteins			
NF-L	In situ hybridization	Increased	89
PSD95		Increased	
SAP102		Increased	
PSD93		No change	

One other group has also examined the expression of a subset of the transcripts encoding the subunits of the ionotropic glutamate receptors in the thalamus in schizophrenia.[90] In this work, transcripts encoding the NR1 and NR2A subunits of the NMDA receptors, the GluR2 and GluR4 AMPA subunits, and the kainate receptor subunit, GluR6, were measured in six patients with schizophrenia and matched controls. No changes in any of these five subunits were found in schizophrenia (TABLE 1).

In another study, we examined the expression of the transcripts encoding seven of the eight cloned metabotropic glutamate receptors in the same subjects used in our other studies in the thalamus. There were no differences in any of these receptors between the subjects with schizophrenia and the comparison group (TABLE 1).[91] Together, these data indicate that there are abnormalities of glutamate receptor expression in the thalamus in schizophrenia. These abnormalities are restricted to the ionotropic receptors and particularly affect the NMDA subtype, which has changes of both subunit expression as well as binding domains for select sites on the NMDA receptor complex.

NMDA-ASSOCIATED INTRACELLULAR SIGNALING PROTEINS

When glutamate binds, the NMDA ion channel opens to permit sodium and calcium ions to enter, which, in turn, triggers multiple intracellular events. In recent years, it has become apparent that the NMDA receptor interacts with several intracellular proteins enriched in the postsynaptic density (PSD) (FIGS. 1 and 2). PSD95 and the related NR2-associated proteins, SAP102 (synapse-associated protein 102) and PSD93, contain several domains that bind the C-termini of NMDA receptor subunits, cytoskeletal proteins, and signal transduction enzymes. Through this array of protein–protein interactions, PSD95-like molecules promote NMDA receptor functions by clustering and anchoring the receptor at the PSD, modulating NMDA receptor sensitivity to glutamate and, perhaps most importantly, assembling a signaling complex to coordinate NMDA-mediated intracellular processes.[92]

Widespread interest in the NR2 subunit–associated PSD95 family of proteins has prompted the investigation of intracellular proteins that interact with the NR1 subunit.[93,94] Two proteins, NF-L (neurofilament-light chain)[93] and Yotiao,[94] have been identified as proteins that interact with exon 21–containing NR1 isoforms (FIGS. 1 and 2). NF-L, along with the two other neurofilament subunits, NF-heavy and medium chains, are among the most abundant cytoskeletal elements and play an important role in the maintenance of neuronal structure.[95] NF-L may also be involved in directing NMDA receptors to the PSD and/or linking it to the synaptic cytoskeleton.[93,96] NF-L interacts with protein phosphatase-1 (PP1), a major protein/serine/threonine phosphatase that is involved in numerous intracellular processes.[97] Although the functional significance of this NF-L:PP1 interaction is not fully understood, it has been suggested that NF-L may bind PP1 and position it to dephosphorylate other PSD proteins, such as NMDA receptor subunits, or CamKII.[98]

Although our earlier studies showed decreased expression of NR1 and NR2C subunit transcripts, and decreased binding at the polyamine and glycine/D-serine sites of the NMDA receptor,[88] we found a significant increase in the transcript expression of PSD95 and SAP102, but not PSD93 in the thalamus in schizophrenia (FIG. 5, TABLE 1).[89] The total pool of PSD proteins may be increased in schizophrenia in an attempt to compensate for decreased thalamic expression of NMDA receptor subunits. Enhanced expression of PSD molecules may lead to an enhanced association of the PSD proteins with the remaining NMDA subunits, an intracellular adaptation to attempt to maintain homeostasis of NMDA-related intracellular signaling in the face of decreased NMDA receptor expression, or in response to a general deficit in glutamate neurotransmission. Two other studies have examined the expression of PSD95 in schizophrenia and reported increased mRNA expression in the occipital cortex,[99] decreased expression in Brodmann area 9,[100] but no change in area 46 of the prefrontal cortex[99] or hippocampus.[100]

We found that NF-L mRNA expression was also elevated in the thalamus in schizophrenia (TABLE 1). Although a portion of NF-L in thalamic cells may associate with the NR1 subunit to participate in NMDA function, the majority of NF-L associates with the other neurofilament subunits to maintain the neuronal cytoskeleton. It is not possible to determine if the elevated NF-L mRNA we found in schizophrenia translates into greater amounts of NF-L protein interacting with NMDA receptors, which might serve to boost an impaired glutamatergic system, or if it is part of a more general cytoskeletal response to the illness.

ABNORMAL TRANSPORTER EXPRESSION

Multiple transporters for glutamate have been identified, including families of five membrane-bound and three vesicular transporters (FIG. 1). We have expanded our examination of molecules of the glutamate synapse in schizophrenia to include these proteins. Three of the five membrane transporters are expressed in the human thalamus, EAAT1, EAAT2, and EAAT3 (FIG. 5). EAAT1 and EAAT2 are both primarily associated with astrocytes, while EAAT3 is expressed on neurons. In the same subjects used in our earlier studies in the thalamus, we determined the expression of these three transporters.[101] Interestingly, both EAAT1 and EAAT2 were upregulated in schizophrenia, while EAAT3 was unchanged (TABLE 1). Both EAAT1 and EAAT2 are glial proteins, adding to a growing literature implicating glial, in addition to neuronal, abnormalities in schizophrenia.

We have also examined the expression of two of the vesicular transporters in those same subjects, vGluT1 and vGluT2 (previously identified as BNPi and DNPi, respectively). In the human thalamus, only DNPi/vGlut2 can be visualized. Similar to EAAT1 and EAAT2, the transcript encoding vGlut2 was upregulated in schizophrenia (TABLE 1).[102] These data indicate that glutamatergic abnormalities in the thalamus in schizophrenia occur not only at the receptor level, but also involve both membrane and vesicular transporters. Further, these data also suggest cellular abnormalities, not only of the postsynaptic neuron (receptor abnormalities), but also of the astrocyte (EAAT1 and EAAT2) and presynaptic neuron (vGluT2) associated with transmission at the glutamatergic synapse.

Several intracellular proteins have been identified that interact with EAATs. The protein Ajuba interacts with the amino terminus of EAAT2, the intracellular protein JWA interacts with EAAT3, and the proteins KIAA0302 and ARGHEF11 both interact with EAAT4.[37-39] The function of these EAAT-associated molecules has not been well characterized, but it has been proposed that they regulate the activity of the EAATs. We have recently found increased transcript expression of JWA and

ARGHEF11 in the thalamus in schizophrenia. The functional implications of such alterations are yet to be determined.

ENZYMATIC ABNORMALITIES

Few postmortem studies have examined expression or activity of glutamatergic enzymes in schizophrenia. We have reported increased glutaminase mRNA expression (FIG. 5) in the thalamus.[103] Interestingly, a fourfold increase in phosphate-activated glutaminase has been reported in the dorsolateral prefrontal cortex in schizophrenia.[104] Since dorsal thalamic nuclei send projections to the dorsolateral prefrontal cortex, the increase in thalamic glutaminase transcript expression that we noted could account for some of the increase in glutaminase enzymatic activity seen in the cortex.

CONCLUSIONS

These data suggest that abnormal glutamatergic neurotransmission occurs in the thalamus in schizophrenia. Abnormalities in glutamate receptor expression (especially the NMDA receptor), intracellular signaling proteins associated with the NMDA receptor, vesicular and cell-surface transporters, intracellular proteins associated with the membrane-bound transporters, and enzymes associated with the intracellular management of glutamate as a neurotransmitter have all been reported to be abnormal in the thalamus in schizophrenia. In addition, postmortem morphometric and stereological studies, as well as *in vivo* imaging data implicate thalamic dysfunction in schizophrenia. Multiple PET studies have reported reduced thalamic metabolism in patients with schizophrenia, particularly when they are engaged in complex cognitive tasks. Thalamic hypoactivity is associated with cell and volume loss, especially in the MD thalamus, but also in other limbic thalamic nuclei. Taken together, these data support thalamic dysfunction in schizophrenia. It is likely that these glutamatergic, structural, and metabolic thalamic abnormalities represent a primary thalamic deficit that underlies the pathophysiology of schizophrenia, although it is possible that these changes are secondary to cortical disturbances.

ACKNOWLEDGMENTS

This work was supported by NARSAD and MH53327.

REFERENCES

1. KRYSTAL, J.H. *et al.* 1994. Subanesthetic effects of the noncompetitive NMDA antagonist, ketamine, in humans. Psychotomimetic, perceptual, cognitive, and neuroendocrine responses. Arch. Gen. Psychiatry **51:** 199–214.
2. JAVITT, D.C. & S.R. ZUKIN. 1991. Recent advances in the phencyclidine model of schizophrenia. Am. J. Psychiatry **148:** 1301–1308.
3. LAHTI, A.C. *et al.* 1995. Ketamine activates psychosis and alters limbic blood flow in schizophrenia. Neuroreport **6:** 869–872.

4. LUBY, E. et al. 1962. Model psychoses and schizophrenia. Am. J. Psychiatry **119:** 61–67.
5. ITIL, T. et al. 1967. Effect of phencyclidine in chronic schizophrenics. Can. Psychiatr. Assoc. J. **12:** 209–212.
6. AANONSEN, L.M. & G.L. WILCOX. 1986. Phencyclidine selectively blocks a spinal action of N-methyl-D-aspartate in mice. Neurosci. Lett. **67:** 191–197.
7. JENTSCH, J.D. & R.H. ROTH. 1999. The neuropsychopharmacology of phencyclidine: from NMDA receptor hypofunction to the dopamine hypothesis of schizophrenia. Neuropsychopharmacology **20:** 201–225.
8. HERESCO-LEVY, U. et al. 2002. Placebo-controlled trial of D-cycloserine added to conventional neuroleptics, olanzapine, or risperidone in schizophrenia. Am. J. Psychiatry **159:** 480–482.
9. HERESCO-LEVY, U. et al. 1998. Double-blind, placebo-controlled, crossover trial of D-cycloserine adjuvant therapy for treatment-resistant schizophrenia. Int. J. Neuropsychopharmcol. **1:** 131–135.
10. HERESCO-LEVY, U. et al. 1999. Efficacy of high-dose glycine in the treatment of enduring negative symptoms of schizophrenia. Arch. Gen. Psychiatry **56:** 29–36.
11. EVINS, A.E. et al. 2002. D-Cycloserine added to risperidone in patients with primary negative symptoms of schizophrenia. Schizophr. Res. **56:** 19–23.
12. GREENE, R. et al. 2000. Short-term and long-term effects of N-methyl-D-aspartate receptor hypofunction. Arch. Gen. Psychiatry **57:** 1180–1181.
13. FARBER, N.B., J.W. NEWCOMER & J.W. OLNEY. 1999. Glycine agonists: what can they teach us about schizophrenia? Arch. Gen. Psychiatry **56:** 13–17.
14. DANBOLT, N.C. 2001. Glutamate uptake. Prog. Neurobiol. **65:** 1–105.
15. MASSON, J. et al. 1999. Neurotransmitter transporters in the central nervous system. Pharmacol. Rev. **51:** 439–464.
16. KANAI, Y., C.P. SMITH & M.A. HEDIGER. 1993. A new family of neurotransmitter transporters: the high-affinity glutamate transporters. FASEB J. **7:** 1450–1459.
17. TAKAMORI, S. et al. 2000. Identification of a vesicular glutamate transporter that defines a glutamatergic phenotype in neurons. Nature **407:** 189–194.
18. FREMEAU, R.T., JR. et al. 2001. The expression of vesicular glutamate transporters defines two classes of excitatory synapse. Neuron **31:** 247–260.
19. BELLOCCHIO, E.E. et al. 2000. Uptake of glutamate into synaptic vesicles by an inorganic phosphate transporter. Science **289:** 957–960.
20. AIHARA, Y. et al. 2000. Molecular cloning of a novel brain-type Na(+)-dependent inorganic phosphate cotransporter. J. Neurochem. **74:** 2622–2625.
21. BELLOCCHIO, E.E. et al. 1998. The localization of the brain-specific inorganic phosphate transporter suggests a specific presynaptic role in glutamatergic transmission. J. Neurosci. **18:** 8648–8659.
22. ZHOU, M. & H.K. KIMELBERG. 2001. Freshly isolated hippocampal CA1 astrocytes comprise two populations differing in glutamate transporter and AMPA receptor expression. J. Neurosci. **21:** 7901–7908.
23. BLEAKMAN, D. & D. LODGE. 1998. Neuropharmacology of AMPA and kainate receptors. Neuropharmacology **37:** 1187–1204.
24. HOLLMANN, M. & S. HEINEMANN. 1994. Cloned glutamate receptors. Annu. Rev. Neurosci. **17:** 31–108.
25. NAKANISHI, S. 1992. Molecular diversity of glutamate receptors and implications for brain function. Science **258:** 597–603.
26. NAGAO, S., S. KWAK & I. KANAZAWA. 1997. EAAT4, a glutamate transporter with properties of a chloride channel, is predominantly localized in Purkinje cell dendrites, and forms parasagittal compartments in rat cerebellum. Neuroscience **78:** 929–933.
27. BAR-PELED, O. et al. 1997. Distribution of glutamate transporter subtypes during human brain development. J. Neurochem. **69:** 2571–2580.
28. MILTON, I.D. et al. 1997. Expression of the glial glutamate transporter EAAT2 in the human CNS: an immunohistochemical study. Brain Res. Mol. Brain Res. **52:** 17–31.
29. LEHRE, K.P. et al. 1995. Differential expression of two glial glutamate transporters in the rat brain: quantitative and immunocytochemical observations. J. Neurosci. **15:** 1835–1853.

30. ROTHSTEIN, J.D. *et al.* 1994. Localization of neuronal and glial glutamate transporters. Neuron **13:** 713–725.
31. FURUTA, A., J.D. ROTHSTEIN & L.J. MARTIN. 1997. Glutamate transporter protein subtypes are expressed differentially during rat CNS development. J. Neurosci. **17:** 8363–8375.
32. FURUTA, A. *et al.* 1997. Cellular and synaptic localization of the neuronal glutamate transporters excitatory amino acid transporter 3 and 4. Neuroscience **81:** 1031–1042.
33. YAMADA, K. *et al.* 1996. EAAT4 is a post-synaptic glutamate transporter at Purkinje cell synapses. Neuroreport **7:** 2013–2017.
34. YAMADA, K. *et al.* 1997. Changes in expression and distribution of the glutamate transporter EAAT4 in developing mouse Purkinje cells. Neurosci. Res. **27:** 191–198.
35. LAAKE, J.H. *et al.* 1999. Postembedding immunogold labelling reveals subcellular localization and pathway-specific enrichment of phosphate activated glutaminase in rat cerebellum. Neuroscience **88:** 1137–1151.
36. MARAGAKIS, N.J. & J.D. ROTHSTEIN. 2001. Glutamate transporters in neurologic disease. Arch. Neurol. **58:** 365–370.
37. MARIE, H. *et al.* 2002. The amino terminus of the glial glutamate transporter GLT-1 interacts with the LIM protein Ajuba. Mol. Cell. Neurosci. **19:** 152–164.
38. LIN, C.I. *et al.* 2001. Modulation of the neuronal glutamate transporter EAAC1 by the interacting protein GTRAP3-18. Nature **410:** 84–88.
39. JACKSON, M. *et al.* 2001. Modulation of the neuronal glutamate transporter EAAT4 by two interacting proteins. Nature **410:** 89–93.
40. MEADOR-WOODRUFF, J.M., A.J. HOGG, *et al.* 2001. Striatal ionotropic glutamate receptor expression in schizophrenia, bipolar disorder, and major depressive disorder. Brain Res. Bull. **55**(5): 631–640.
41. ANDREASEN, N.C., S. PARADISO, *et al.* 1998. "Cognitive dysmetria" as an integrative theory of schizophrenia: a dysfunction in cortical-subcortical-cerebellar circuitry? Schizophr. Bull. **24**(2): 203–218.
42. JONES, E.G. 1997. Cortical development and thalamic pathology in schizophrenia. Schizophr. Bull. **23**(3): 483–501.
43. OKE, A.F. & R.N. ADAMS. 1987. Elevated thalamic dopamine: possible link to sensory dysfunctions in schizophrenia. Schizophr. Bull. **13**(4): 589–604.
44. SCHEIBEL, A.B. 1997. The thalamus and neuropsychiatric illness. J. Neuropsychiatry Clin. Neurosci. **9**(3): 342–353.
45. STERIADE, M., E.G. JONES, *et al.* 1997. Thalamus: Organization and Function. Elsevier. Amsterdam.
46. JONES, E.G. 1998. The thalamus of primates. *In* The Primate Nervous System, Part II. F.E. Bloom, A. Bjorklund & T. Hokfelt, Eds.: **14:** 1–246. Elsevier. New York.
47. SALT, T.E. & S.A. EATON. 1996. Functions of ionotropic and metabotropic glutamate receptors in sensory transmission in the mammalian thalamus. Prog. Neurobiol. **48**(1): 55–72.
48. PAKKENBERG, B. 1990. Pronounced reduction of total neuron number in mediodorsal thalamic nucleus and nucleus accumbens in schizophrenics. Arch. Gen. Psychiatry **47**(11): 1023–1028.
49. PAKKENBERG, B. 1992. The volume of the mediodorsal thalamic nucleus in treated and untreated schizophrenics. Schizophr. Res. **7**(2): 95–100.
50. PAKKENBERG, B. 1993. Leucotomized schizophrenics lose neurons in the mediodorsal thalamic nucleus. Neuropathol. Appl. Neurobiol. **19**(5): 373–380.
51. POPKEN, G.J., W.E. BUNNEY, JR., *et al.* 2000. Subnucleus-specific loss of neurons in medial thalamus of schizophrenics. Proc. Natl. Acad. Sci. USA **97**(16): 9276–9780.
52. YOUNG, K., K. MANAYE, *et al.* 2000. Reduced number of mediodorsal and anterior thalamic neurons in schizophrenia. Biol. Psychiatry **47**(11): 944–953.
53. BYNE, W., M.S. BUCHSBAUM, *et al.* 2002. Postmortem assessment of thalamic nuclear volumes in subjects with schizophrenia. Am. J. Psychiatry **159**(1): 59–65.
54. LEWIS, D.A., D.A. CRUZ, *et al.* 2001. Lamina-specific deficits in parvalbumin-immunoreactive varicosities in the prefrontal cortex of subjects with schizophrenia: evidence for fewer projections from the thalamus. Am. J. Psychiatry **158**(9): 1411–1422.
55. DANOS, P., B. BAUMANN, *et al.* 2002. The ventral lateral posterior nucleus of the thalamus in schizophrenia: a post-mortem study. Psychiatry Res. **114**(1): 1–9.

56. ANDREASEN, N.C., J.C. EHRHARDT, et al. 1990. Magnetic resonance imaging of the brain in schizophrenia. The pathophysiologic significance of structural abnormalities. Arch. Gen. Psychiatry 47(1): 35–44.
57. ANDREASEN, N.C., S. ARNDT, et al. 1994. Thalamic abnormalities in schizophrenia visualized through magnetic resonance image averaging. Science 266(5183): 294–298.
58. BUCHSBAUM, M.S. & E.A. HAZLETT. 1998. Positron emission tomography studies of abnormal glucose metabolism in schizophrenia. Schizophr. Bull. 24(3): 343–364.
59. DASARI, M., L. FRIEDMAN, et al. 1999. A magnetic resonance imaging study of thalamic area in adolescent patients with either schizophrenia or bipolar disorder as compared to healthy controls. Psychiatry Res. 91(3): 155–162.
60. FLAUM, M., V.W. SWAYZE, 2ND, et al. 1995. Effects of diagnosis, laterality, and gender on brain morphology in schizophrenia. Am. J. Psychiatry 152(5): 704–714.
61. GILBERT, A.R., D.R. ROSENBERG, et al. 2001. Thalamic volumes in patients with first-episode schizophrenia. Am. J. Psychiatry 158(4): 618–624.
62. GUR, R.E., V. MAANY, et al. 1998. Subcortical MRI volumes in neuroleptic-naive and treated patients with schizophrenia. Am. J. Psychiatry 155(12): 1711–1717.
63. STAAL, W.G., H.E. HULSHOFF POL, et al. 1998. Partial volume decrease of the thalamus in relatives of patients with schizophrenia. Am. J. Psychiatry 155(12): 1784–1786.
64. ETTINGER, U., X.A. CHITNIS, et al. 2001. Magnetic resonance imaging of the thalamus in first-episode psychosis. Am. J. Psychiatry 158(1): 116–118.
65. KONICK, L.C. & L. FRIEDMAN. 2001. Meta-analysis of thalamic size in schizophrenia. Biol. Psychiatry 49(1): 28–38.
66. BYNE, W., M.S. BUCHSBAUM, et al. 2001. Magnetic resonance imaging of the thalamic mediodorsal nucleus and pulvinar in schizophrenia and schizotypal personality disorder. Arch. Gen. Psychiatry 58(2): 133–140.
67. BUCHSBAUM, M.S., T. SOMEYA, et al. 1996. PET and MRI of the thalamus in never-medicated patients with schizophrenia. Am. J. Psychiatry 153(2): 191–199.
68. HAZLETT, E.A., M.S. BUCHSBAUM, et al. 1999. Three-dimensional analysis with MRI and PET of the size, shape, and function of the thalamus in the schizophrenia spectrum. Am. J. Psychiatry 156(8): 1190–1199.
69. RESNICK, S.M., R.E. GUR, et al. 1988. Positron emission tomography and subcortical glucose metabolism in schizophrenia. Psychiatry Res. 24(1): 1–11.
70. TAMMINGA, C.A., G.K. THAKER, et al. 1992. Limbic system abnormalities identified in schizophrenia using positron emission tomography with fluorodeoxyglucose and neocortical alterations with deficit syndrome. Arch. Gen. Psychiatry 49(7): 522–530.
71. VITA, A., S. BRESSI, et al. 1995. High-resolution SPECT study of regional cerebral blood flow in drug-free and drug-naive schizophrenic patients. Am. J. Psychiatry 152(6): 876–882.
72. ANDREASEN, N.C., D.S. O'LEARY, et al. 1996. Schizophrenia and cognitive dysmetria: a positron-emission tomography study of dysfunctional prefrontal-thalamic-cerebellar circuitry. Proc. Natl. Acad. Sci. USA 93(18): 9985–9990.
73. HECKERS, S., S.L. RAUCH, et al. 1998. Impaired recruitment of the hippocampus during conscious recollection in schizophrenia. Nat. Neurosci. 1(4): 318–323.
74. SILBERSWEIG, D.A., E. STERN, et al. 1995. A functional neuroanatomy of hallucinations in schizophrenia. Nature 378(6553): 176–179.
75. BURNASHEV, N., H. MONYER, et al. 1992. Divalent ion permeability of AMPA receptor channels is dominated by the edited form of a single subunit. Neuron 8:189–198.
76. GEIGER, J.R.P., T. MELCHER, et al. 1995. Relative abundance of subunit mRNAs determines gating and Ca^{2+} permeability of AMPA receptors in principal neurons and interneurons in rat CNS. Neuron 15:193–204.
77. HOLLMANN, M., M. HARTLEY, et al. 1991. Ca^{2+} permeability of KA-AMPA-gated glutamate receptor channels depends on subunit composition. Science 252: 851–853.
78. JONAS, P., C. RACCA, et al. 1994. Differences in Ca^{2+} permeability of AMPA-type glutamate receptor channels in neocortical neurons caused by differential GluR-B subunit expression. Neuron 12: 1281–1289.
79. SWANSON, G.T., S.K. KAMBOJ, et al. 1997. Single-channel properties of recombinant AMPA receptors depend on RNA editing, splice variation, and subunit composition. J. Neurosci. 17: 58–69.

80. ANDERSSON, O., A. STENQVIST, *et al.* 2001. Nucleotide sequence, genomic organization, and chromosomal localization of genes encoding the human NMDA receptor subunits NR3A and NR3B. Genomics **78**(3): 178–184.
81. DURAND, G.M., M.V.L. BENNETT, *et al.* 1993. Splice variants of the N-methyl-D-aspartate receptor NR1 identify domains involved in regulation of polyamines and protein kinase C. Proc. Natl. Acad. Sci. USA **90**: 6731–6735.
82. BOECKMAN, F.A. & E. AIZENMAN. 1994. Stable transfection of the NMDAR1 subunit in Chinese hamster ovary cells fails to produce a functional N-methyl-D-aspartate receptor. Neurosci. Lett. **173**: 189–192.
83. GALLAGHER, M.J., H. HUANG, *et al.* 1996. Interactions between ifenprodil and the NR2B subunit of the N-methyl-D-aspartate receptor. J. Biol. Chem. **271**: 9603–9611.
84. GRIMWOOD, S., B. LEBOURDELLES, *et al.* 1995. Recombinant human NMDA homomeric NMDAR1 receptors expressed in mammalian cells form a high-affinity glycine antagonist binding site. J. Neurochem. **64**: 525–530.
85. LYNCH, D.R., N.J. ANEGAWA, *et al.* 1994. N-methyl-D-Aspartate receptors: different subunit requirements for binding of glutamate antagonists, glycine antagonists, and channel-blocking agents. Mol. Pharmacol. **45**: 540–545.
86. MONYER, H.R. R. SPRENGEL, *et al.* 1992. Heteromeric NMDA receptors: molecular and functional distinction of subtypes. Science **256**: 1217–1220.
87. RODRIGUEZ-PAZ, J.M., Y. ANANTHARAM, *et al.* 1995. Block of the N-methyl-D-aspartate receptor by phencyclidine-like drugs is influenced by alternative splicing. Neurosci. Lett. **190**: 147–150.
88. IBRAHIM, H., A. HOGG, *et al.* 2000. Ionotropic glutamate receptor binding and subunit mRNA expression in thalamic nuclei in schizophrenia. Am. J. Psychiatry **157**: 1811–1823.
89. CLINTON, S.M., V. HAROUTUNIAN, *et al.* 2003. Altered transcript expression of NMDA receptor-associated postsynaptic proteins in the thalamus of subjects with schizophrenia. Am. J. Psychiatry **160**(6): 1100–1109.
90. POPKEN, G.J., M.G. LEGGIO, *et al.* 2002. Expression of mRNAs related to the GABAergic and glutamatergic neurotransmitter systems in the human thalamus: normal and schizophrenic. Thalamus & Related Systems **1**: 349–369.
91. RICHARDSON-BURNS, S.M., V. HAROUTUNIAN, *et al.* 2000. Metabotropic glutamate receptor mRNA expression in the schizophrenic thalamus. Biol. Psychiatry **47**: 22–28.
92. SHENG, M. & D. PAK. 2000. Ligand-gated ion channel interactions with cytoskeletal and signaling proteins. Annu. Rev. Physiol. **62**: 755–778.
93. EHLERS, M., E. FUNG, *et al.* 1998. Splice variant-specific interaction of the NMDA receptor subunit NR1 with neuronal intermediate filaments. J. Neurosci. **18**(2): 720–730.
94. LIN, J., M. WYSZYNSKI, *et al.* 1998. Yotiao, a novel protein of neuromuscular junction and brain that interacts with specific splice variants of NMDA receptor subunit NR1. J. Neurosci. **18**(6): 2017–2027.
95. XU, Z., D.L. DONG, *et al.* 1994. Neuronal intermediate filaments: new progress on an old subject. Curr. Opin. Neurobiol. **4**(5): 655–661.
96. EHLERS, M.D., W.G. TINGLEY, *et al.* 1995. Regulated subcellular distribution of the NR1 subunit of the NMDA receptor. Science **269**(5231): 1734–1737.
97. SHENOLIKAR, S. 1994. Protein serine/threonine phosphatases—new avenues for cell regulation. Annu. Rev. Cell Biol. **10**: 55–86.
98. TERRY-LORENZO, R., M. INOUE, *et al.* 2000. Neurofilament-L is a protein phosphatase-1-binding protein associated with neuronal plasma membrane and post-synaptic density. J. Biol. Chem. **275**(4): 2139–2446.
99. DRACHEVA, S., S.A. MARRAS, *et al.* 2001. N-methyl-D-aspartic acid receptor expression in the dorsolateral prefrontal cortex of elderly patients with schizophrenia. Am. J. Psychiatry **158**(9): 1400–1410.
100. OHNUMA, T., H. KATO, *et al.* 2000. Gene expression of PSD95 in prefrontal cortex and hippocampus in schizophrenia. NeuroReport **11**(14): 3133–3137.
101. SMITH, R.E., V. HAROUTUNIAN, *et al.* 2001. Expression of excitatory amino acid transporter transcripts in the thalamus of subjects with schizophrenia. Am. J. Psychiatry **158**(9): 1393–1399.
102. SMITH, R.E., V. HAROUTUNIAN, *et al.* 2001. Vesicular glutamate transporter transcript expression in the thalamus in schizophrenia. Neuroreport **12**(13): 1–3.

103. MCCULLUMSMITH, R.E., V. HAROUTUNIAN, et al. 2002. Expression of glutaminase transcripts in the thalamus of subjects with schizophrenia. Biol. Psychiatry **51:** 25.
104. GLUCK, M.R. et al. 2002. Implications for altered glutamate and GABA metabolism in the dorsolateral prefrontal cortex of aged schizophrenic patients. Am. J. Psychiatry **159:** 1165–1173.

Glutamate Receptors and Transporters in the Hippocampus in Schizophrenia

PAUL J. HARRISON, AMANDA J. LAW, AND SHARON L. EASTWOOD

Department of Psychiatry, Neurosciences Building, Warneford Hospital, University of Oxford, Oxford, OX3 7JX, United Kingdom

ABSTRACT: Postmortem studies, using various methods and directed at several molecular targets, have provided increasing evidence that glutamatergic neurotransmission is affected in schizophrenia. The bulk of the data are in the hippocampus, wherein there is reduced expression of one or more subunits for all three ionotropic receptors (NMDA, AMPA, and kainate). Presynaptic glutamatergic markers, notably the vesicular glutamate transporter VGLUT1, may also be decreased in schizophrenia, especially in older subjects. CA1 appears less affected than other subfields, and the decrements may be greater in the left than in the right hippocampus. The recently described susceptibility genes for schizophrenia all act upon glutamatergic synaptic transmission, which may, therefore, be part of the core pathophysiology of the disorder.

KEYWORDS: AMPA receptor; gene expression; kainate receptor; lateralization; medial temporal lobe; mRNA; NMDA receptor

There has been an increasing research focus on the glutamate system in schizophrenia, from both pathogenic and therapeutic perspectives.[1–3] Indeed, it now competes and interacts with dopamine as the neurotransmitter of greatest theoretical and experimental interest.[4] Unlike dopamine, however, *in vivo* imaging of the glutamate system, its receptors, and transporters remains problematic.[5] Hence empirical studies have continued to rely on postmortem brain tissue, for which, fortunately, an ever-expanding range of relevant methodologies is available.[6,7] The medial temporal lobe (dentate gyrus, hippocampus proper, subiculum, and parahippocampal cortex; abbreviated as *hippocampus* here) is the region with the most data and forms the focus of this overview of glutamate receptors and transporters in schizophrenia. Equivalent studies in other brain areas, and the conceptual framework for this research, are reviewed elsewhere.[2,8,9]

Address for correspondence: Prof. P.J. Harrison, Neurosciences Building, University Department of Psychiatry, Warneford Hospital, Oxford OX3 7JX, U.K. Voice: +44-1865-223730; fax: +44-1865-251076.

paul.harrison@psych.ox.ac.uk

HIPPOCAMPAL GLUTAMATE RECEPTORS AND TRANSPORTERS IN SCHIZOPHRENIA

Although the N-methyl-D-aspartate (NMDA) receptor is the subtype of most immediate relevance to glutamatergic theories of schizophrenia, there have been relatively few studies of NMDA receptor expression in the hippocampus in the disorder. Binding studies with ligands targeting the NMDA receptor complex do not show clear alterations.[10,11] *In situ* hybridization studies of NMDA receptor transcripts indicate decreased NR1 subunit mRNA,[11,12] together with increased NR2B mRNA and unchanged NR2A mRNA.[11] The altered proportion of NR2A and NR2B mRNAs recapitulates the situation in the neonatal human hippocampus (Law *et al.*, this meeting), of potential relevance to the various neurodevelopmental models of schizophrenia that involve the NMDA receptor. Relating the negative NMDA receptor ligand–binding data to the positive results for the NR subunit transcripts is difficult because of uncertainties regarding the quantitative relationship between mRNA and protein, and the relationship between receptor subunit composition and ligand-binding characteristics. Hence, detailed studies of NR subunit proteins using selective antibodies will be needed to clarify the status of hippocampal NMDA receptors in schizophrenia.

The other ionotropic glutamate receptors, α-amino-3-hydroxy-5-methyl-4-isoxazole propionic acid (AMPA) and kainic acid (KA) receptors, have been more extensively investigated in the hippocampus in schizophrenia. The results show a fairly consistent pattern, with most studies showing decreases in expression of subunit mRNAs, proteins, and ligand-binding sites.[8] In a series of investigations of AMPA receptor subunits, we found a marked reduction of GluR1 and GluR2 mRNAs,[13,14] as well as a lesser decrease of GluR1 and GluR2/3 proteins.[15] The splicing pattern for GluR2 mRNA also changed, with a preferential reduction of the "flop" isoform.[16] Of the KA receptor gene family, GluR6 and KA2 transcripts are lower,[17] consistent with reductions of GluR5/6/7 immunoreactivity in pyramidal neuron dendrites in schizophrenia.[18]

Hippocampal metabotropic glutamate receptors have yet to be investigated in any detail in schizophrenia. A small study of mGluR5 mRNA was negative.[19]

Hippocampal glutamate receptor expression has been measured after administration of antipsychotics to rodents. By and large, the results are negative and do not reproduce the findings in schizophrenia.[11,20-22] KA2 mRNA is increased by chronic haloperidol (i.e., the opposite result as in schizophrenia).[22] The only alteration seen in schizophrenia, which may, extrapolating from the rat data, be a medication effect, is the shift in GluR2 isoform ratio, which was reproduced by 2 weeks', though not 16 weeks', administration of haloperidol.[20,22]

The recently cloned vesicular glutamate transporters (VLGUT 1 and 2) provide a useful tool for the assessment of "presynaptic" glutamate neurons, since (unlike the reuptake transporters) they are highly selectively expressed in excitatory presynaptic terminals.[23,24] In schizophrenia, we have found a reduction of hippocampal VGLUT1 mRNA,[25] suggestive of a decline in the activity or number of glutamatergic synapses, and consistent with similar reductions seen earlier with the less selective marker, complexin II.[26,27] A recent immunohistochemical study has closely replicated these findings.[28] As in the rodent, VGLUT2 mRNA and protein are sparse in the human hippocampus, and neither gene product has yet proved reliably quantifiable (Ref. 25, and unpublished observations).

In summary, the hippocampus in schizophrenia is characterised by reductions in expression of AMPA and KA receptors, and less clearly also by decreases in NMDA receptors (NR1 subunit) and in presynaptic glutamate markers. As such, the postmortem findings are supportive of a glutamatergic involvement, affecting both pre- and postsynaptic elements in schizophrenia. The origin of the changes may be traced to schizophrenia susceptibility genes (see below). The direction of causality between the pre- and postsynaptic glutamatergic alterations remains unclear, although it might be postulated that the receptor alterations are a response to an altered innervation or activity of the synapses. The role of glia in altered glutamate signaling in schizophrenia must also be considered.[29] There is a large literature that attests to the diverse functional and pathophysiological correlates and consequences of altered abundance and proportions of glutamate receptor subtypes.[30,31] However, it is not known which of these is most relevant to schizophrenia, or even whether the changes in glutamate-receptor expression observed in the disorder are intrinsically pathological or a response (compensatory or maladaptive) to a primary abnormality elsewhere within or without the glutamate system. Postmortem studies cannot incisively address these questions; their primary role will continue to be to identify and characterize the glutamatergic abnormalities of schizophrenia, providing the data to justify and guide other approaches better suited to revealing the underlying molecular and cellular mechanisms.

GLUTAMATERGIC INVOLVEMENT MAY DIFFER BETWEEN SUBFIELDS AND HEMISPHERES

The above discussion has ignored the fact that the hippocampus is anatomically and functionally heterogeneous. Moreover, there is evidence that the hippocampal glutamate receptor changes in schizophrenia are not uniform; this information comes particularly from the *in situ* hybridization studies, and also from the receptor autoradiographic and immunohistochemical studies that have allowed analysis over the different subfields. As shown in TABLE 1, CA1 appears least affected, and CA4 most affected. The relative sparing of CA1 may contribute to the equivocal or negative results of some homogenate-based hippocampal studies.[33,34] Its basis is unknown, though a broadly similar pattern is seen for presynaptic proteins[35] and $GABA_A$ receptors,[36] suggesting that its explanation is not "glutamate-specific" (e.g., reflecting a unique profile or regulation of glutamate receptor expression by CA1 neurons[37]) but is perhaps related to more general features of the connectivity and characteristics of CA1 neurons.

The possibility that the left and right temporal lobes are differentially involved in schizophrenia has been advocated by several researchers, based on a variety of findings and theories (for a review, see Ref. 38). Asymmetrical alterations in hippocampal glutamate receptors may be one manifestation of this. The most striking evidence came from Kerwin and colleagues,[39] who found that the loss of [^{3}H]KA binding was entirely limited to the left hippocampus. Results of other studies have been less dramatic, but there is still sufficient suggestion of a greater left-sided involvement to indicate that laterality should continue to be investigated in glutamate receptor studies (TABLE 2).

TABLE 1. Alterations in glutamate receptors (mRNA, protein, and ligand binding) in schizophrenia in individual hippocampal subfields

Ref.	Parameter	Control/ Scz (n)	Schizophrenics vs. controls				(Percent difference)
			DG	CA4	CA3	CA1	Sub
	Binding sites						
10	[^{3}H]KA[a]	8/7	−53	−57	−17	−35	nd
10	[^{3}H]CNQX[a]	8/7	(−1)	−35	−36	(−18)	nd
32	[^{3}H]TCP[a]	15/15	(−15)	nd	−23	(−15)	(−14)
	mRNAs						
11	NR1 mRNA	26/27	−17	nd	−25	(−10)	(−15)
11	NR2B mRNA	8/9	nd	nd	+19	(+1)	(+10)
12	NR1 mRNA[a,b]	15/15	−43	nd	−26	(−12)	(−14)
13	GluR1 mRNA	8/6	(−58)	(−27)	−77	(−50)	−54
14	GluR1 mRNA	14/9	−47	−70	−62	(−19)	−38
14	GluR2 mRNA	14/9	−42	−50	−51	(−27)	−25
17	KA2 mRNA[a]	13/11	−30	(−37)	−34	(−4)	nd
25	VGLUT1 mRNA	12/13	−40	−50	−46	(−34)	(−48)
	Proteins						
15	GluR1	10/11	nd	(+4)	(−6)	(−8)	(−9)
15	GluR2/3	10/11	(−16)	−38	(−17)	(−8)	(−5)
	"Average"		−33	−40	−31	−18	−23

NOTE: Values in schizophrenia are expressed as a percent of the corresponding controls. Non-significant differences are shown in parentheses. Only studies that had significant differences in at least one subfield are included.
ABBREVIATIONS: nd, not determined; DG, dentate gyrus; Sub, subiculum; Scz, schizophrenics.
[a]Study examined both hemispheres; averaged data used here.
[b]Additional analyses of the published data.

ARE THE GLUTAMATERGIC CHANGES PROGRESSIVE?

Many structural and functional abnormalities first reported in chronic schizophrenia have since been shown to be present at, if not before, the onset of symptoms, increasing the likelihood that they are involved early in the disease process. This cannot yet be determined for glutamate receptors or transporters. However, some data suggest that the glutamate abnormalities may be progressive in the disease. For both VGLUT1 and complexin II, the reductions described above were in subjects with mean age at death in their seventh decade; reductions were less prominent[27] or absent[40] in a series of younger subjects. In the latter brain series, from the Stanley Neuropathology Consortium, hippocampal AMPA receptor binding is also unchanged.[41] Moreover, we have found significant inverse correlations between sever-

TABLE 2. Left–right differences in hippocampal glutamate receptors in schizophrenia

Ref.	Parameter	Control/Scz (n)	Schizophrenics vs. controls (percent difference)	
			Left	Right
	Binding sites			
10	[^{3}H]KA[a]	8/7	−56	−37
10	[^{3}H]CNQX[a]	8/7	−24	−7
39	[^{3}H]KA[a]	9/11	−43	+1
	mRNAs			
12	NR1 mRNA[b,c]	15/15	−33	−17
17	GluR6 mRNA[a]	13/11	−30	−23
17	KA2 mRNA[a]	13/11	−34	−26
	"Average"		−37	−18

NOTE: Only studies reporting both hemispheres, with at least one significant difference between cases and controls, are indicated. Results in schizophrenia are expressed as a percent of the corresponding controls. For autoradiographic studies, the value is the mean percentage difference averaged across all subfields.
[a]Left and right hippocampi studied in same brains.
[b]Left and right hippocampi studied in separate brains.
[c]Additional analyses of the published data.

al glutamatergic markers and increasing age in schizophrenics, but not in controls (Refs. 15 and 25, and unpublished observations). Clearly these observations are not conclusive, but taken together they do provide circumstantial evidence that the glutamatergic synaptic pathology of schizophrenia may be progressive. As such, it might contribute to the emergence and/or worsening of cognitive deficits that occurs in chronic schizophrenia and that is currently neuropathologically and neurochemically unexplained.[42]

GLUTAMATERGIC SYNAPSES AND THE ETIOLOGY OF SCHIZOPHRENIA

Given that (a) the majority of neurons use glutamate as a neurotransmitter, (b) most if not all neurons express glutamate receptors, and (c) all cells contain glutamate, the reported alterations in these indices in schizophrenia could be nonspecific. That is, they reflect a shift in the relative composition and/or activity of the tissue; while this explanation could still mean that the glutamatergic involvement has major functional and therapeutic implications, it would downplay its pathogenic significance. However, recent data suggest that glutamatergic neurotransmission may indeed have a more central pathophysiological role.

Moderate to strong evidence for allelic and haplotypic association with schizophrenia has recently been reported for several genes, including neuregulin-1, dysbindin, D-amino acid oxidase (DAAO), G72, and the regulator of G-protein signaling-4. Interestingly, all these genes play a role in synaptic plasticity and functioning, especially glutamatergic synapses; it may, therefore, be that an alteration or dysregulation of glutamate pathways is part of the core pathophysiology of the disorder.[43] For example, neuregulin is secreted with glutamate and regulates postsynaptic NMDA receptor pathways via ErbB receptors; dysbindin is present at hippocampal synapses, and DAAO encodes the enzyme that metabolizes the endogenous NMDA receptor modulator, D-serine.

Obviously, this scenario is highly speculative: the association of each gene with schizophrenia is still not proven beyond doubt, and it overlooks the many other roles that each gene might play. Nevertheless, it provides a rationale for further studies of components of the glutamate system, a framework within which to design the experiments, and a way to integrate genetic and neuropathological findings. It also leads to the parsimonious hypothesis that the other susceptibility genes yet to be discovered also act upon the same basic process. In this respect, the glutamate hypothesis of schizophrenia becomes homologous to the β-amyloid hypothesis of Alzheimer's disease, in which different genetic (and environmental) causes operate via a shared, critical effect on β-amyloid metabolism.

ACKNOWLEDGMENTS

Our work on schizophrenia is supported by the Stanley Medical Research Institute and the Wellcome Trust, U.K.

REFERENCES

1. OLNEY, J.W. & N.B. FARBER. 1995. Glutamate receptor dysfunction and schizophrenia. Arch. Gen. Psychiatry **52:** 998–1007.
2. TAMMINGA, C.A. 1998. Schizophrenia and glutamatergic transmission. Crit. Rev. Neurobiol. **12:** 21–36.
3. TSAI, G. & J.T. COYLE. 2002. Glutamatergic mechanisms in schizophrenia. Annu. Rev. Pharmacol. Toxicol. **42:** 165–179.
4. CARLSSON, A., N. WATERS, S. HOLM-WATERS, et al. 2001. Interactions between monoamines, glutamate, and GABA in schizophrenia: New evidence. Annu. Rev. Pharmacol. Toxicol. **41:** 237–260.
5. BRESSAN, R.A. & L.S. PILOWSKY. 2000. Imaging the glutamatergic system *in vivo*—relevance to schizophrenia. Eur. J. Nucl. Med. **27:** 1723–1731.
6. HARRISON, P.J. 1996. Advances in post mortem molecular neurochemistry and neuropathology: examples from schizophrenia research. Br. Med. Bull. **52:** 527–538.
7. BURNET, P.W.J., S.L. EASTWOOD & P.J. HARRISON. 2003. Laser-assisted microdissection: methods for the molecular analysis of psychiatric disorders at a cellular resolution. Biol. Psychiatry. In press.
8. MEADOR-WOODRUFF, J.H. & D.J. HEALY. 2000. Glutamate receptor expression in schizophrenic brain. Brain Res. Rev. **31:** 288–294.
9. KONRADI, C. & S. HECKERS. 2003. Molecular aspects of glutamate dysfunction: implications for schizophrenia and its treatment. Pharmacol. Therap. **97:** 153–179.
10. KERWIN, R.W., S. PATEL & B. MELDRUM. 1990. Quantitative autoradiographic analysis of glutamate binding sites in the hippocampal formation in normal and schizophrenic brain *post mortem*. Neuroscience **39:** 25–32.
11. GAO, X-M., K. SAKAI, R.C. ROBERTS, et al. 2000. Iontropic glutamate receptors and expression of *N*-methyl-D-aspartate receptor subunits in subregions of human hippocampus: effects of schizophrenia. Am J. Psychiatry **157:** 1141–1149.
12. LAW, A.J. & J.F.W. DEAKIN. 2001. Asymmetrical reductions of hippocampal NMDAR1 glutamate receptor mRNA in the psychoses. Neuroreport **12:** 2971–2974.

13. HARRISON, P.J., D. MCLAUGHLIN & R.W. KERWIN. 1991. Decreased hippocampal expression of a glutamate receptor gene in schizophrenia. Lancet **337:** 450–452.
14. EASTWOOD, S.L., B. MCDONALD, P.W.J. BURNET, *et al.* 1995. Decreased expression of mRNAs encoding non-NMDA glutamate receptors GluR1 and GluR2 in medial temporal lobe neurons in schizophrenia. Mol. Brain Res. **29:** 211–223.
15. EASTWOOD, S.L., R.W. KERWIN & P.J. HARRISON. 1997. Immuno-autoradiographic evidence for a loss of α-amino-3-hydroxy-5-methyl-4-isoxazole propionate-preferring non-N-methyl-D-aspartate glutamate receptors within the medial temporal lobe in schizophrenia. Biol. Psychiatry **41:** 636–643.
16. EASTWOOD, S.L., P.W.J. BURNET & P.J. HARRISON. 1997. GluR2 glutamate receptor subunit flip and flop isoforms are decreased in the hippocampal formation in schizophrenia: a reverse transcriptase-polymerase chain reaction (RT-PCR) study. Mol. Brain Res. **44:** 92–98.
17. PORTER, R.H.P., S.L. EASTWOOD & P.J. HARRISON. 1997. Distribution of kainate receptor subunit mRNAs in human hippocampus, neocortex and cerebellum, and bilateral reduction of hippocampal GluR6 and KA2 transcripts in schizophrenia. Brain Res. **751:** 217–231.
18. BENES, F.M., M.S. TODTENKOPF & P. KOSTOULAKOS. 2001. $GluR_{5,6,7}$ subunit immunoreactivity on apical pyramidal cell dendrites in hippocampus of schizophrenics and manic-depressives. Hippocampus **11:** 482–491.
19. OHNUMA, T., S. TESSLER, H. ARAI, *et al.* 2000. Gene expression of metabotropic glutamate receptor 5 and excitatory amino acid transporter 2 in the schizophrenic hippocampus. Mol. Brain Res. **85:** 24–31.
20. EASTWOOD, S.L., P. STORY, P.W.J. BURNET, *et al.* 1994. Differential changes in glutamate receptor subunit messenger RNAs in rat brain after haloperidol treatment. J. Psychopharmacol. **80:** 196–203.
21. FITZGERALD, L.W., A.Y. DEUTCH, G. GASIC, *et al.* 1995. Regulation of cortical and subcortical glutamate receptor subunit expression by antipsychotic drugs. J. Neurosci. **15:** 2453–2461.
22. EASTWOOD, S.L., R.H.P. PORTER & P.J. HARRISON. 1996. The effect of chronic haloperidol treatment on glutamate receptor subunit (GluR1, GluR2, KA1, KA2, NR1) mRNAs and glutamate binding protein mRNA in rat forebrain. Neurosci. Lett. **212:** 163–166.
23. FREMEAU, R.T., JR., M.D. TROYER, I. PAHNER, *et al.* 2001. The expression of vesicular glutamate transporters defines two classes of excitatory synapse. Neuron **31:** 247–260.
24. HERZOG, E., G.C. BELLENCHI, C. GRAS, *et al.* 2001. The existence of a second vesicular glutamate transporter specifies subpopulations of glutamatergic neurons. J. Neurosci. **21:** 1–6.
25. HARRISON, P.J. & S.L. EASTWOOD. 2003. Vesicular glutamate transporter (VGLUT) gene expression provides further evidence for glutamatergic synaptic pathology in the hippocampus in schizophrenia [Abstract]. Schizophr. Res. **60:** 62–63.
26. HARRISON, P.J. & S.L. EASTWOOD. 1998. Preferential involvement of excitatory neurons in medial temporal lobe in schizophrenia. Lancet **352:** 1669–1673.
27. EASTWOOD, S.L. & P.J. HARRISON. 2000. Hippocampal synaptic pathology in schizophrenia, bipolar disorder and major depression: a study of complexin mRNAs. Mol. Psychiatry **5:** 425–432.
28. SAWADA, K., A. BARR, S. TAKAHASHI, *et al.* 2003. Complexins I and II in hippocampus in schizophrenia [Abstract]. Schizophr. Res. **60:** 74–75.
29. NEDERGAARD, M., T. TAKOMA & A.J. HANSEN. 2002. Beyond the role of glutamate as a neurotransmitter. Nature Neurosci. Rev. **3:** 748–755.
30. LÜSCHER, C. & M. FRERKING. 2001. Restless AMPA receptors: implications for synaptic neurotransmission and plasticity. Trends Neurosci. **24:** 665–670.
31. WENTHOLD, R.J., K. PRYBYLOWSKI, S. STANDLEY, *et al.* 2003. Trafficking of NMDA receptors. Annu. Rev. Pharmacol. Toxicol. **43:** 335–358.
32. DEAN B., E. SCARR, R. BRADBURY & D. COPOLOV. 1999. Decreased hippocampal (CA3) NMDA receptors in schizophrenia. Synapse **32:** 67–69.
33. DEAKIN, J.F.W., P. SLATER, M.D.C. SIMPSON, *et al.* 1989. Frontal cortical and left temporal glutamatergic dysfunction in schizophrenia. J. Neurochem. **52:** 1781–1786.

34. BREESE, C.R., R. FREEDMAN & S. LEONARD. 1995. Glutamate receptor subtype expression in human postmortem brain tissue from schizophrenics and alcohol abusers. Brain Res. **674:** 82–90.
35. HARRISON, P.J. & S.L. EASTWOOD. 2001. Neuropathological studies of synaptic connectivity in the hippocampal formation in schizophrenia. Hippocampus **11:** 508–519.
36. BENES, F.M. 2000. Emerging principles of altered neural circuitry in schizophrenia. Brain Res. Rev. **31:** 251–269.
37. PELLEGRINI-GIAMPIETRO, D., R.S. ZUKIN, M.V.L. BENNETT, et al. 1992. Switch in glutamate receptor subunit gene expression in CA1 subfield of hippocampus following global ischemia in rats. Proc. Natl. Acad. Sci. USA **89:** 10499–10503.
38. HOLINGER, D., A. GALABURDA & P.J. HARRISON. 2000. Cerebral asymmetry. *In* The Neuropathology of Schizophrenia. Progress and Interpretation. P.J. Harrison & G.W. Roberts, Eds.: 151–171. Oxford University Press. Oxford.
39. KERWIN, R.W., S. PATEL, B.S. MELDRUM, et al. 1988. Asymmetrical loss of glutamate receptor subtype in left hippocampus in schizophrenia. Lancet **I:** 583–584.
40. MCCULLUM SMITH, R. & J. MEADOR-WOODRUFF. 2003. Expression of vesicular glutamate transporters one and two in medial temporal lobe structures in schizophrenia, bipolar disorder, and major depressive disorder [Abstract]. Schizophr. Res. **60** (suppl.): 65.
41. NOGA, T.J. & H. WANG. 2002. Further postmortem autoradiographic studies of AMPA receptor binding in schizophrenia. Synapse **45:** 250–258.
42. HARRISON, P.J. 1999. The neuropathology of schizophrenia: a critical review of the data and their interpretation. Brain **122:** 593–624.
43. HARRISON, P.J. & M.J. OWEN. 2003. Genes for schizophrenia? Recent findings and their pathophysiological implications. Lancet **361:** 417–419.

Altered Cortical Glutamate Neurotransmission in Schizophrenia

Evidence from Morphological Studies of Pyramidal Neurons

DAVID A. LEWIS,[a,b] LEISA A. GLANTZ,[c] JOSEPH N. PIERRI,[a] AND ROBERT A. SWEET[a]

Departments of [a]Psychiatry and [b]Neuroscience, University of Pittsburgh, Pennsylvania 15213, USA

[c]Department of Psychiatry, University of North Carolina-Chapel Hill, Chapel Hill, North Carolina 27599, USA

ABSTRACT: Multiple lines of evidence from pharmacological, neuroimaging, and postmortem studies implicate disturbances in cortical glutamate neurotransmission in the pathophysiology of schizophrenia. Given that pyramidal neurons are the principal source of cortical glutamate neurotransmission, as well as the targets of the majority of cortical glutamate–containing axon terminals, understanding the nature of altered glutamate neurotransmission in schizophrenia requires an appreciation of both the types of pyramidal cell abnormalities and the specific class(es) of pyramidal cells that are affected in the illness. In this chapter, we review evidence indicating that a subpopulation of pyramidal neurons in the dorsolateral prefrontal cortex exhibits reductions in dendritic spine density, a marker of the number of excitatory inputs, and in somal volume, a measure correlated with a neuron's dendritic and axonal architecture. Specifically, pyramidal neurons located in deep layer 3 of the dorsolateral prefrontal cortex and that lack immunoreactivity for nonphosphorylated neurofilament protein may be particularly involved in the pathophysiology of schizophrenia. The presence of similar changes in pyramidal neurons located in deep layer 3 of auditory association cortex suggests that a shared property, which remains to be determined, confers cell type–specific vulnerability to a subpopulation of cortical glutamatergic neurons in schizophrenia.

KEYWORDS: schizophrenia; pyramidal neurons; dorsolateral prefrontal cortex; deep layer 3

As described in other chapters in this volume, convergent lines of evidence from pharmacological, neuroimaging, and postmortem studies implicate disturbances in cortical glutamate neurotransmission in the pathophysiology of schizophrenia. For

Address for correspondence: David A. Lewis, M.D., Department of Psychiatry, University of Pittsburgh, 3811 O'Hara Street, W1651 BST, Pittsburgh, PA 15213. Voice: 412-624-3934; fax: 412-624-9910.

lewisda@msx.upmc.edu

example, cDNA microarray studies have demonstrated decreased expression of genes whose protein products are critical determinants of the efficacy of glutamate neurotransmission.[1] In addition, certain association areas of the cerebral cortex, such as those located in the superior and middle prefrontal and the superior temporal gyri, appear to be among the principal brain loci exhibiting both structural and functional disturbances in the illness.[2] In these regions, pyramidal neurons are the principal source of glutamate neurotransmission, as well as the targets of the majority of glutamate-containing axon terminals,[3] and thus are likely to be disturbed in subjects with schizophrenia.

Unlike cortical GABAergic neurons, which are readily divided into subclasses on the basis of their distinctive morphological, neurochemical, and electrophysiological features,[4] pyramidal cells have traditionally been considered to be more homogeneous in nature. However, it is well established that pyramidal neurons in different cortical layers tend to furnish projections to and receive inputs from different brain regions. For example, many pyramidal cells in layers 2–3 send axonal projections to other cortical regions; pyramidal neurons in layer 5 project to the striatum; and other subcortical structures and pyramidal neurons in layer 6 project to the thalamus.[5] In addition, more recent data indicate that, even within the same cortical layer, differences in the axonal projection targets of pyramidal neurons are associated with quantitative differences in dendritic morphology and qualitative differences in the gene products that they express. For example, pyramidal neurons in the supragranular layers of the monkey dorsolateral prefrontal cortex (DLPFC) with axons that project callosally have larger dendritic arbors and a greater density of dendritic spines than do neighboring pyramidal cells that furnish axons to the adjacent regions of the ipsilateral DLPFC.[6] Pyramidal neurons that furnish axonal projections to distant cortical regions also tend to have larger cell bodies and to express high levels of nonphosphorylated epitopes of neurofilament proteins compared to pyramidal cells in the same location that provide shorter corticocortical projections.[7] Thus, understanding the nature of altered glutamate neurotransmission in schizophrenia requires an appreciation of the types of pyramidal cell abnormalities and of the specific class(es) of pyramidal cells that are affected in the DLPFC and superior temporal cortex.

Among the different laminar groups of pyramidal neurons, those located in deep layer 3 appear to be of particular relevance to the pathophysiology of schizophrenia, given their role in both corticocortical and thalamocortical circuitry. In addition to sending principal axon projections to other cortical regions, these neurons furnish both local (within 300 µm of the cell body) and long-range axon collaterals that travel through the gray matter for up to several millimeters before arborizing in discrete stripe-like clusters.[8,9] Both the extrinsic and long-range intrinsic collaterals of these neurons target almost exclusively the dendritic spines of other pyramidal cells, whereas the synaptic targets of the local axon collaterals are equally divided between the dendritic spines of other pyramidal neurons and the dendritic shafts of the parvalbumin-containing class of GABA neurons.[10–12] In addition, pyramidal neurons in deep layer 3 receive "feed-forward" types of cortical projections[13] and are located in the termination zone of axon projections from the mediodorsal thalamus,[14] a nucleus that has been reported in a number of, but not all, studies to show reductions in neuronal number in schizophrenia.[15,16] Thus, pyramidal neurons in deep layer 3 are critically positioned to mediate the flow of excitatory transmission through both thalamocortical and corticocortical circuits.

In order to determine whether deep layer 3 pyramidal neurons exhibit morphological abnormalities that may be indicative of altered glutamate neurotransmission, we determined the density of dendritic spines, markers of excitatory inputs, on the basilar dendrites of Golgi-impregnated pyramidal neurons in the superficial and deep portions of layer 3 in DLPFC (area 46) from 15 schizophrenic, 15 normal control, and 15 nonschizophrenic psychiatric subjects.[17] We found a significant effect of diagnosis on spine density only for deep layer 3 pyramidal neurons (FIG. 1). In the subjects with schizophrenia, spine density on these neurons was decreased by 23% and 16% compared to the normal control ($P = .003$) and psychiatric subjects ($P = .08$), respectively. In contrast, spine density on neurons in superficial layer 3 of area 46 was decreased by only 15% and 13%, compared to the normal control and psychiatric subjects, respectively, differences that did not achieve statistical significance. Furthermore, spine density on deep layer 3 neurons did not significantly differ between psychiatric subjects treated with antipsychotic agents and normal controls.

Because dendritic spine density directly reflects the number of excitatory inputs to pyramidal neurons,[18] these findings, in concert with those of a pilot study that also reported decreased spine density on prefrontal layer 3 pyramidal neurons,[19] support the hypothesis that schizophrenia is associated with diminished excitatory connectivity in the DLPFC. This interpretation is consistent with reports of decreased synaptophysin protein, a marker of axon terminals, and of diminished neuropil measures in the DLPFC of subjects with schizophrenia.[20–23] Interestingly, dendritic spine density on DLPFC layer 3 pyramidal neurons undergoes a substantial decline during adolescence in primates.[24] In addition, the number of asymmetric (presumably excitatory) synapses changes in a similar age-related fashion in both monkey and human DLPFC.[25,26] These late developmental refinements in the excitatory circuitry of the DLPFC coincide with the age when the clinical manifestations of schizophrenia frequently first appear, suggesting that they may contribute to the pathophysiology of this disorder.[27] However, it remains unknown whether the presynaptic terminals to DLPFC layer 3 pyramidal neurons in subjects with schizophrenia never develop, are extensively pruned during adolescence, or are resorbed later in life. A testable hypothesis derived from gene expression profiling studies suggests that an interactive cascade of these events may be involved.[28]

The functional significance of a decrease in excitatory inputs depends on which populations of axon terminals are affected. In particular, the apparent intralaminar differences in pyramidal neuron spine density may be related to differences in the connectivity at different depths of layer 3. For example, as noted above, afferents from the thalamus terminate principally in deep layer 3 and layer 4,[14] and thus they are more likely to target the basilar dendrites of pyramidal neurons in deep layer 3 than of those located in superficial layer 3. The affected inputs may represent those from the mediodorsal thalamic nucleus since the number of neurons in this nucleus has been reported to be reduced in schizophrenia.[15] However, a reduction in thalamic inputs to the DLPFC cannot completely account for the decrease in dendritic spine density since thalamocortical terminals appear to comprise a small proportion (<10%) of the total excitatory inputs to the targeted cortical neurons, at least in cat visual cortex.[29] Assuming that the situation is similar in human DLPFC, then even a total absence of thalamocortical afferents would not be sufficient to account for the approximately 20% reduction in basilar dendritic spine density on deep layer 3 py-

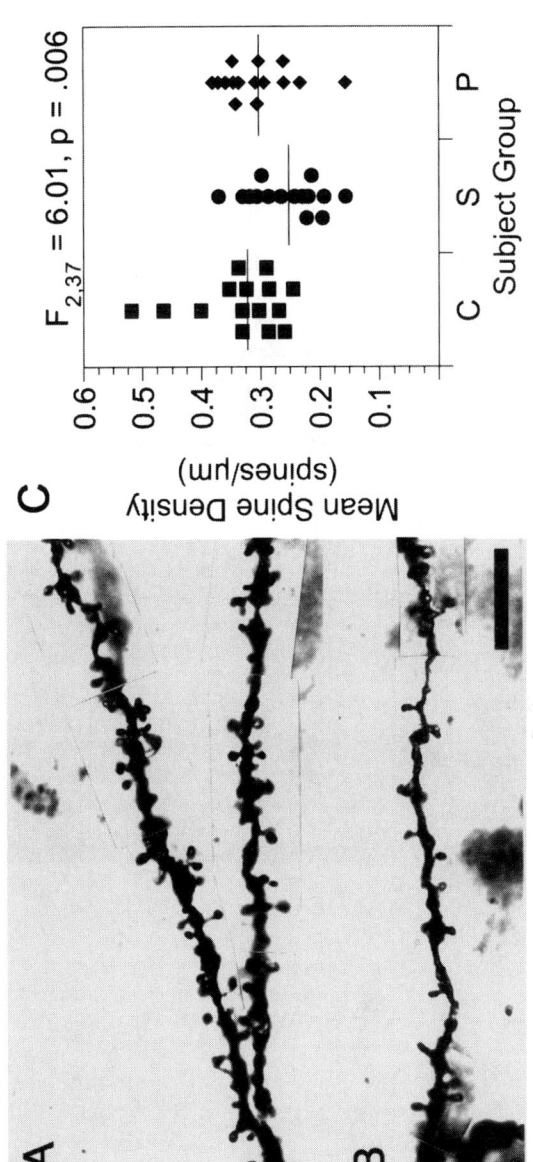

FIGURE 1. Brightfield photomicrograph illustrating Golgi-impregnated basilar dendrites and spines on PFC layer 3 pyramidal neurons from (**A**) a normal control subject and (**B**) a subject with schizophrenia. Calibration bar = 10 μm. Scatterplot (**C**) illustrating mean spine densities for 15 pyramidal neurons per subject in the deep portion of layer 3 in DLPFC area 46. Horizontal lines indicate group means for 15 control (C), 15 schizophrenic (S), and 15 nonschizophrenic psychiatric (P) subjects. Adapted from Lewis.[15]

ramidal cells observed in the subjects with schizophrenia. Two other major sources of excitatory inputs to both deep and superficial layer 3 DLPFC pyramidal neurons are intrinsic axon collaterals from other pyramidal neurons[8,9] and associational or callosal projections from other cortical regions.[9,30] Thus, the smaller decrease in spine density on superficial layer 3 pyramidal cells raises the possibility that abnormalities in thalamocortical afferents to deep layer 3 have an additive effect to a disturbance in cortical axon terminals that are distributed across layer 3. However, even if these interpretations are correct, they do not reveal the cause of the pathophysiological changes. For example, the inputs to DLPFC layer 3 pyramidal cells may not be reduced due to a more primary disturbance in the source of the inputs, but because an abnormality intrinsic to these pyramidal cells renders them unable to support a normal complement of excitatory inputs.

The functional integrity of these pyramidal neurons may be reflected in changes in their somal volume. For example, shifts in somal size may indicate disturbances in neuronal connectivity, given that somal size has been shown to be correlated with measures of a neuron's dendritic[31,32] and axonal architecture.[33,34] Indeed, the mean cross-sectional somal area of the Golgi-impregnated, deep layer 3 pyramidal neurons was decreased by 9.1% in the subjects with schizophrenia relative to normal control subjects, although this difference did not achieve statistical significance. Be-

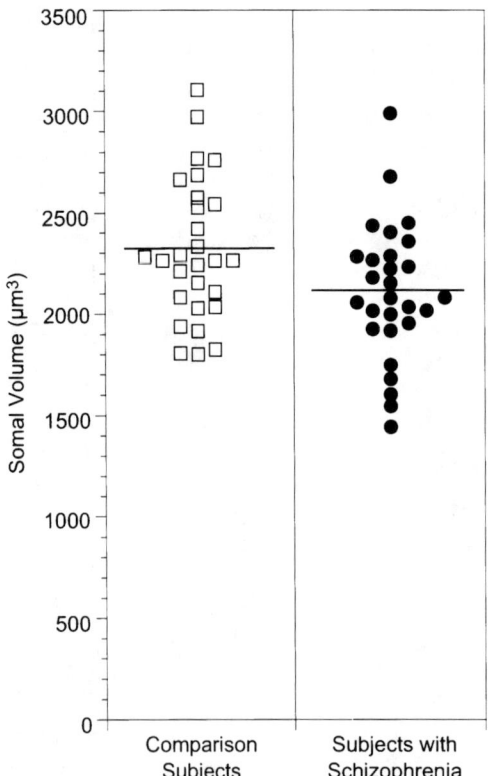

FIGURE 2. Geometric means of the somal volumes for Nissl-stained deep layer 3 pyramidal neurons in the DLPFC of comparison subjects and subjects with schizophrenia. *Cross-bars* indicate the geometric means for each diagnostic group. Reprinted with permission from Pierri *et al.*[35]

cause these findings were based on relatively small samples of neurons and subjects, we used stereological measures to estimate the somal volume of over 250 Nissl-stained pyramidal neurons per subject in deep layer 3 of the DLPFC area 9 in 28 subjects with schizophrenia, each of whom was matched to one normal comparison subject for sex, age, and postmortem interval.[35] The geometric mean of the somal volume estimates was significantly decreased by 9.2% in the subjects with schizophrenia (FIG. 2), a decrease that was not explained by either antipsychotic medication history or duration of illness. These findings are convergent with those of a previous study that found a decrease in the mean somal size of all layer 3 neurons in area 9 in subjects with schizophrenia, and a decrease in the density of the largest neurons in deep layer 3.[36] Together, these findings raised the hypothesis that the largest pyramidal neurons in deep layer 3 may be more affected than other DLPFC neurons in schizophrenia.

In order to test this hypothesis, we estimated the mean somal volume of deep layer 3 pyramidal neurons immunoreactive for nonphosphorylated neurofilament protein (NNFP) in 13 of the matched subject pairs used in our study of Nissl-stained neurons.[37] We included only subjects whose postmortem interval (PMI) was less than 18 hours because immunoreactivity for NNFP appears to decay with prolonged PMI.[38] In contrast to our prediction, the somal volume of NNFP-labeled pyramidal neurons was not decreased in the subjects with schizophrenia, whereas that of the Nissl-stained pyramidal neurons in the same subjects was significantly decreased by 14.2%. These comparisons suggest that NNFP-labeled deep layer 3 pyramidal neurons are less affected than other subpopulations of deep layer 3 pyramidal neurons in schizophrenia. It is noteworthy that the NNFP-positive population of deep layer 3 pyramidal neurons comprise only a subset of all deep layer 3 pyramidal neurons,[39] with the data from our study suggesting that NNFP-labeled neurons represent approximately 50% of deep layer 3 pyramidal neurons. Given that the NNFP-labeled neurons are less affected than the general population of deep layer 3 pyramidal neurons, the subtraction of the NNFP-labeled neurons would likely result in the remaining pyramidal neurons showing a mean decrease in somal size greater than the observed 14.2% decrease.

Because the total numbers of each type of neuron were not assessed in this study, the findings could reflect either a change in somal volume of certain pyramidal neurons or a change in the number of other classes of pyramidal neurons. For example, an increase or decrease, respectively, in the number of small or large pyramidal neurons not immunoreactive for NNFP, without a change in the number of NNFP-positive neurons, could produce the same estimates of somal volumes observed in this study. Although this question cannot be directly addressed from this study, estimates of cell density corrected for the differential z-axis shrinkage associated with the processing of each label do reveal the relative proportion of neurons of each type. For both the subjects with schizophrenia and the control subjects, approximately 50% of deep layer 3 pyramidal neurons were NNFP immunoreactive, and the relative densities of both the Nissl-stained and NNFP-labeled neurons did not differ across the two diagnostic groups.[37] These comparisons suggest that the estimates of mean somal size reflect a change in the size of the neurons that were measured, rather than changes in the numbers of neurons in different size classes. However, a definitive answer to this question requires unbiased estimates of the total number of each type of neuron in deep layer 3 of area 9.

What is the identity of the DLPFC deep layer 3 pyramidal neurons with reduced somal size and spine density? Connectivity studies in the nonhuman primate indicate that 70–100% of DLPFC pyramidal neurons furnishing long-range corticocortical projections are NNFP immunoreactive, whereas <20% of neurons participating in shorter-range corticocortical circuits within the DLPFC or across the corpus callosum are NNFP positive.[40] If these characteristics generalize to the human DLPFC, then pyramidal neurons that provide callosal and/or short-range corticocortical connections may be more affected in schizophrenia than are those that provide long-range ipsilateral connections. Support for this hypothesis may be found in functional neuroimaging studies, which show that abnormalities localized within the DLPFC, perhaps involving neurons that provide short-range and callosal connections, appear to account for deficits in working memory[41] and verbal fluency.[42] In fact, the latter study reports correlations suggesting that the connections between the DLPFC and the anterior cingulate cortex may be disrupted, whereas connections between the DLPFC and superior temporal gyrus are intact.

Understanding the characteristics of the affected class(es) of pyramidal neurons in schizophrenia may be informed by studies of other cortical regions that exhibit structural and functional deficits similar to those of the DLPFC in schizophrenia. For example, subjects with schizophrenia exhibit decreased gray matter volume of auditory association cortex of the superior temporal gyrus and deficits in the auditory sensory memory processes subserved by this region.[43,44] Since these findings parallel the *in vivo* observations of reduced gray matter volume and related working memory deficits associated with reduced somal volume of deep layer 3 pyramidal cells in the DLPFC, we hypothesized that deep layer 3 pyramidal cell somal volume would also be reduced in auditory association cortex in schizophrenia.[45] We used design-based stereology to estimate the somal volume of pyramidal neurons in deep layer 3 of auditory association cortex (area 42 in the superior temporal gyrus) in 18 of the subject pairs used in the Nissl-stain study of pyramidal neuron volume in the DLPFC described above.[35] Somal volume of deep layer 3 pyramidal cells in area 42 was significantly reduced by 13.1% in the subjects with schizophrenia. Reductions in somal volume were not associated with a history of antipsychotic use, alcohol use disorder, schizoaffective disorder, or death by suicide. The percent change in somal volume within pairs was highly correlated ($R = 0.67$, $P = .002$) between areas 42 and 9 (FIG. 3), suggesting that a common factor may affect deep layer 3 pyramidal cells in both regions.

As noted above, the reduced pyramidal cell somal volume, and associated reductions in dendritic spines, in the DLPFC in subjects with schizophrenia might be secondary to deafferentation, due to reduced numbers of neurons in the mediodorsal thalamic nucleus.[15] Though there is limited thalamic input from the mediodorsal thalamic nucleus to the auditory association cortex,[46] the pulvinar nucleus of thalamus, which does project to area 42, appears to be similarly affected in schizophrenia.[47] Alternatively, some other factor shared by populations of pyramidal neurons distributed across certain other regions may also need to be considered, since several studies have also observed reductions in somal volume in hippocampal pyramidal neurons.[48]

This review has emphasized the contribution that morphological abnormalities in a particular subset of cortical pyramidal neurons may make to disturbed cortical glutamate neurotransmission in schizophrenia. These observations appear to be

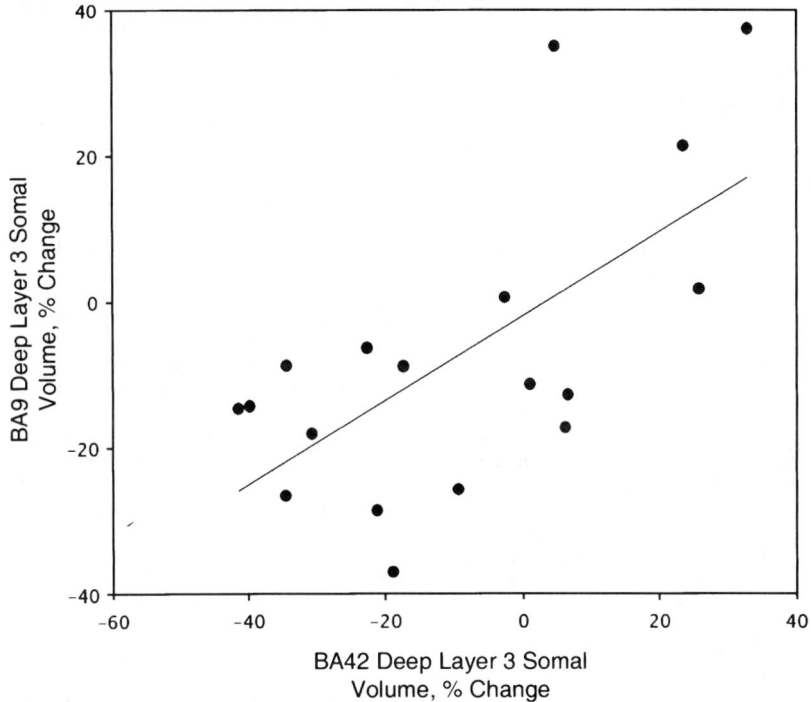

FIGURE 3. The relationship between differences in mean somal volumes of Nissl-stained deep layer 3 pyramidal cell within-subject pairs in auditory association cortex (area 42) and DLPFC (area 9). The changes were highly correlated between regions ($R = .67$; $P = .002$). Reprinted with permission from Sweet et al.[45]

robust, at least relatively selective for certain cell types, related to the disease process of schizophrenia rather than its treatment or comorbid factors, and replicated across laboratories. However, it is important to note that disturbances in other cell types may also contribute to alterations in cortical glutamate neurotransmission. For example, convergent lines of data indicate that the inhibitory control over pyramidal neuron output is altered in schizophrenia.[49] Thus, a comprehensive view of the nature of altered cortical glutamatergic transmission in schizophrenia requires not only discovering the identity of the affected classes of pyramidal neurons, but also understanding their role in cortical circuitry and the regulation of their activity by various inputs.

REFERENCES

1. MIRNICS, K., F.A. MIDDLETON, A. MARQUEZ, et al. 2000. Molecular characterization of schizophrenia viewed by microarray analysis of gene expression in prefrontal cortex. Neuron **28:** 53–67.

2. HARRISON, P.J. & D.A. LEWIS. 2003. Neuropathology in schizophrenia. *In* Schizophrenia. S. Hirsch & D.R. Weinberger, Eds.: Chapter 17: 310–325. Blackwell Science Ltd. Oxford.
3. WHITE, E.L. 1989. Cortical Circuits. Boston-Basel. Birkhauser.
4. DEFELIPE, J. 1997. Types of neurons, synaptic connections and chemical characteristics of cells immunoreactive for calbindin-D28K, parvalbumin and calretinin in the neocortex. J. Chem. Neuro. **14:** 1–19.
5. JONES, E.G. 1984. Laminar distribution of cortical efferent cells. *In* Cerebral Cortex, Vol. 1. A. Peters & E.G. Jones, Eds. 521–553. Plenum Press. New York.
6. SOLOWAY, A.S., M.L. PUCAK, D.S. MELCHITZKY & D.A. LEWIS. 2002. Dendritic morphology of callosal and ipsilateral projection neurons in monkey prefrontal cortex. Neuroscience **109:** 461–471.
7. HOF, P.R., L.G. UNGERLEIDER, M.J. WEBSTER, *et al.* 1996. Neurofilament protein is differentially distributed in subpopulations of corticocortical projection neurons in the macaque monkey visual pathways. J. Comp. Neurol. **376:** 112–127.
8. LEVITT, J.B., D.A. LEWIS, T. YOSHIOKA & J.S. LUND. 1993. Topography of pyramidal neuron intrinsic connections in macaque monkey prefrontal cortex (areas 9 & 46). J. Comp. Neurol. **338:** 360–376.
9. PUCAK, M.L., J.B. LEVITT, J.S. LUND & D.A. LEWIS. 1996. Patterns of intrinsic and associational circuitry in monkey prefrontal cortex. J. Comp. Neurol. **376:** 614–630.
10. MELCHITZKY, D.S., S.R. SESACK, M.L. PUCAK & D.A. LEWIS. 1998. Synaptic targets of pyramidal neurons providing intrinsic horizontal connections in monkey prefrontal cortex. J. Comp. Neurol. **390:** 211–224.
11. MELCHITZKY, D.S., G. GONZALEZ-BURGOS, G. BARRIONUEVO & D.A. LEWIS. 2001. Synaptic targets of the intrinsic axon collaterals of supragranular pyramidal neurons in monkey prefrontal cortex. J. Comp. Neurol. **430:** 209–221.
12. MELCHITZKY, D.S. & D.A. LEWIS. 2003. Preferential targeting of parvalbumin interneurons by local axon terminals of supragranular pyramidal neurons in monkey prefrontal cortex. Cereb. Cortex **13:** 452–460.
13. FELLEMAN, D.J. & D.C. VAN ESSEN. 1991. Distributed hierarchical processing in the primate cerebral cortex. Cereb. Cortex **1:** 1–47.
14. GIGUERE, M. & P.S. GOLDMAN-RAKIC. 1988. Mediodorsal nucleus: areal, laminar, and tangential distribution of afferents and efferents in the frontal lobe of rhesus monkeys. J. Comp. Neurol. **277:** 195–213.
15. LEWIS, D.A. 2000. Is there a neuropathology of schizophrenia? The Neuroscientist **6:** 208–218.
16. CULLEN, T.J., M.A. WALKER, N. PARKINSON, *et al.* 2003. A postmortem study of the mediodorsal nucleus of the thalamus in schizophrenia. Schizophr. Res. **60:** 157–166.
17. GLANTZ, L.A. & D.A. LEWIS. 2000. Decreased dendritic spine density on prefrontal cortical pyramidal neurons in schizophrenia. Arch. Gen. Psychiatry **57:** 65–73.
18. DEFELIPE, J. & I. FARINAS. 1992. The pyramidal neuron of the cerebral cortex: morphological and chemical characteristics of the synaptic inputs. Prog. Neurobiol. **39:** 563–607.
19. GAREY, L.J., W.Y. ONG, T.S. PATEL, *et al.* 1998. Reduced dendritic spine density on cerebral cortical pyramidal neurons in schizophrenia. J. Neurol. Neurosurg. Psychiatry **65:** 446–453.
20. GLANTZ, L.A. & D.A. LEWIS. 1997. Reduction of synaptophysin immunoreactivity in the prefrontal cortex of subjects with schizophrenia: regional and diagnostic specificity. Arch. Gen. Psychiatry **54:** 943–952.
21. KARSON, C.N., R.E. MRAK, K.O. SCHLUTERMAN, *et al.* 1999. Alterations in synaptic proteins and their encoding mRNAs in prefrontal cortex in schizophenia: a possible neurochemical basis for "hypofrontality." Mol. Psychiatry **4:** 39–45.
22. PERRONE-BIZZOZERO, N.I., A.C. SOWER, E.D. BIRD, *et al.* 1996. Levels of the growth-associated protein GAP-43 are selectively increased in association cortices in schizophrenia. Proc. Natl. Acad. Sci. USA **93:** 14182–14187.
23. SELEMON, L.D. & P.S. GOLDMAN-RAKIC. 1999. The reduced neuropil hypothesis: a circuit based model of schizophrenia. Biol. Psychiatry **45:** 17–25.

24. ANDERSON, S.A., J.D. CLASSEY, F. CONDÉ, et al. 1995. Synchronous development of pyramidal neuron dendritic spines and parvalbumin-immunoreactive chandelier neuron axon terminals in layer III of monkey prefrontal cortex. Neuroscience **67:** 7–22.
25. BOURGEOIS, J.-P., P.S. GOLDMAN-RAKIC & P. RAKIC. 1994. Synaptogenesis in the prefrontal cortex of rhesus monkeys. Cereb. Cortex **4:** 78–96.
26. HUTTENLOCHER, P.R. & A.S. DABHOLKAR. 1997. Regional differences in synaptogenesis in human cerebral cortex. J. Comp. Neurol. **387:** 167–178.
27. LEWIS, D.A. 1997. Development of the prefrontal cortex during adolescence: insights into vulnerable neural circuits in schizophrenia. Neuropsychopharmacology **16:** 385–398.
28. MIRNICS, K., F.A. MIDDLETON, D.A. LEWIS & P. LEVITT. 2001. Analysis of complex brain disorders with gene expression microarrays: schizophrenia as a disease of the synapse. Trends Neurosci. **24:** 479–486.
29. AHMED, B., J.C. ANDERSON, R.J. DOUGLAS, et al. 1994. Polyneuronal innervation of spiny stellate neurons in cat visual cortex. J. Comp. Neurol. **341:** 39–49.
30. BARBAS, H. 1992. Architecture and cortical connections of the prefrontal cortex in the Rhesus monkey. Adv. Neurol. **57:** 91–115.
31. HAYES, T.L. & D.A. LEWIS. 1996. Magnopyramidal neurons in the anterior motor speech region: dendritic features and interhemispheric comparisons. Arch. Neurol. **53:** 1277–1283.
32. JACOBS, B., L. DRISCOLL & M. SCHALL. 1997. Life-span dendritic and spine changes in areas 10 and 18 of human cortex: a quantitative Golgi study. J. Comp. Neurol. **386:** 661–680.
33. GILBERT, C.D. & J.P. KELLY. 1975. The projections of cells in different layers of the cat's visual cortex. J. Comp. Neurol. **63:** 81–106.
34. LUND, J.S., R.D. LUND, A.E. HENDRICKSON, et al. 1975. The origin of efferent pathways from the primary visual cortex, area 17, of the macaque monkey as shown by retrograde transport of horseradish peroxidase. J. Comp. Neurol. **164:** 287–304.
35. PIERRI, J.N., C.L.E. VOLK, S. AUH, et al. 2001. Decreased somal size of deep layer 3 pyramidal neurons in the prefrontal cortex in subjects with schizophrenia. Arch. Gen. Psychiatry **58:** 466–473.
36. RAJKOWSKA, G., L.D. SELEMON & P.S. GOLDMAN-RAKIC. 1998. Neuronal and glial somal size in the prefrontal cortex: a postmortem morphometric study of schizophrenia and Huntington disease. Arch. Gen. Psychiatry **55:** 215–224.
37. PIERRI, J.N., C.L.E. VOLK, S. AUH, et al. 2003. Somal size of prefrontal cortical pyramidal neurons in schizophrenia: differential effects across neuronal subpopulations. Biol. Psychiatry **54:** 111–120.
38. LEWIS, D.A. 2002. The human brain revisited: opportunities and challenges in postmortem studies of psychiatric disorders. Neuropsychopharmacology **26:** 143–154.
39. HOF, P.R., K. COX & J.H. MORRISON. 1990. Quantitative analysis of a vulnerable subset of pyramidal neurons in Alzheimer's disease: I. Superior frontal and inferior temporal cortex. J. Comp. Neurol. **301:** 44–54.
40. HOF, P.R. & J.H. MORRISON. 1995. Neurofilament protein defines regional patterns of cortical organization in the macaque monkey visual system: a quantitative immunohistochemical analysis. J. Comp. Neurol. **352:** 161–186.
41. BERTOLINO, A., G. ESPOSITO, J. H. CALLICOTT, et al. 2000. Specific relationship between prefrontal neuronal N-acetylaspartate and activation of the working memory cortical network in schizophrenia. Am. J. Psychiatry **157:** 26–33.
42. SPENCE, S.A., P.F. LIDDLE, M.D. STEFAN, et al. 2000. Functional anatomy of verbal fluency in people with schizophrenia and those at genetic risk: focal dysfunction and distributed disconnectivity reappraised. Br. J. Psychiatry **176:** 52–60.
43. MCCARLEY, R.W., C.G. WIBLE, M. FRUMIN, et al. 1999. MRI anatomy of schizophrenia. Biol. Psychiatry **45:** 1099–1119.
44. JAVITT, D.C., A.M. SHELLEY, G. SILIPO & J.A. LIEBERMAN. 2000. Deficits in auditory and visual context-dependent processing in schizophrenia: defining the pattern. Arch. Gen. Psychiatry **57:** 1131–1137.
45. SWEET, R.A., J.N. PIERRI, S. AUH, et al. 2003. Reduced pyramidal cell somal volume in auditory association cortex of subjects with schizophrenia. Neuropsychopharmacology **28:** 599–609.

46. MOLINARI, M., M.E. DELL'ANNA, E. RAUSELL, *et al.* 1995. Auditory thalamocortical pathways defined in monkeys by calcium-binding protein immunoreactivity. J. Comp. Neurol. **362:** 171–194.
47. BYNE, W., M.S. BUCHSBAUM, L.A. MATTIACE, *et al.* 2002. Postmortem assessment of thalamic nuclear volumes in subjects with schizophrenia. Am. J. Psychiatry **159:** 59–65.
48. Weinberger, D.R. 1999. Cell biology of the hippocampal formation in schizophrenia. Biol. Psychiatry **45:** 395–402.
49. VOLK, D.W. & D.A. LEWIS. 2002. Impaired prefrontal inhibition in schizophrenia: relevance for cognitive dysfunction. Physiol. Behav. **77:** 501–505.

Evaluating Glutamatergic Transmission in Schizophrenia

CAROL A. TAMMINGA, ADRIENNE C. LAHTI, DEBORAH R. MEDOFF, XUE-MIN GAO, AND HENRY H. HOLCOMB

University of Texas Southwestern Medical School, Dallas, Texas, 75390, USA

Maryland Psychiatric Research Center, University of Maryland, Baltimore, Maryland 21228, USA

ABSTRACT: Our findings with schizophrenia and the glutamate system have relied on the characterization of the clinical response of patients to ketamine and their functional brain imaging response (rCBF) to the drug. Prior to the human studies reported here, we had evaluated the region activation characteristics and pharmacology of PCP and its congener MK 801 in animals. What I will report in this paper has been individually reported elsewhere but brought together here in a new synthesis.

KEYWORDS: schizophrenia; psychosis; glutamatergic system; PCP; ketamine; MK-801; molecular targets

Schizophrenia, an illness of great medical and personal consequence, can be partially treated by existing treatments, but presently has no cure.[1] Molecular targets have yet to be defined. The pressing focus of discovery in the field is directed toward identifying one or more of the molecular targets that drive symptoms in schizophrenia.[2] Once new knowledge is in place, pharmaceutical research can devise new medications for the defined disease targets. So, considerable pressure is brought to bear on research initiatives to define disease pathophysiology. Once this is accomplished, the knowledge will enable development of new therapeutics that will alleviate mental and psychosocial suffering.[3]

When the first antipsychotic drugs were developed[4] (and their mechanism of action determined to be dopamine receptor blockade,[5] considerable research was directed toward defining dopaminergic dysfunction as pathophysiology in the illness.[5,6] This direction of research was largely negative, with the exception of the most recent studies on changes in dopamine release to pharmacologic stimulation in acute psychotic states.[7] Attention was also drawn to other likely potential causes of psychosis in the human brain. Other monoamines were explored.[8] Modulatory peptides were studied.[9] The GABA system, because of its ubiquitous inhibitory function, was studied and abnormalities noted.[10] The glutamate system was first

---

Address for correspondence: Carol A. Tamminga, M.D., Maryland Psychiatric Research Center, University of Maryland, P.O. Box 21247, Maple and Locust Streets, Baltimore, MD 21228. Voice: 410-402-6805; fax: 410-402-6882.

ctamming@mprc.umaryland.edu

identified as a potential "player" on the basis of a report citing reduced CSF glutamate levels in the illness,[11] even though this finding has not been replicated.[12] But interest in the glutamate system grew, largely based on the report that phencyclidine (PCP) is a ligand for a receptor associated with the NMDA ionophore. PCP is a known human psychotomimetic, an agonist at this receptor within the NMDA ionophore,[13] and indirectly antagonizes NMDA receptor function.[14] This connection between PCP action and the NMDA-sensitive glutamate receptor provided an organizing hypothesis. Thus, the idea developed that psychosis was due to reduced glutamatergic transmission at the NMDA-sensitive ionophore and several hypotheses were proposed to account for this.[15-17] While no mechanisms have yet been demonstrated, the idea that psychosis in schizophrenia may involve inadequate glutamatergic transmission at the NMDA receptor is widely acknowledged as plausible.

Now, clinical research tools have advanced to the point where new biological information can be discovered in human volunteers with schizophrenia. These tools include the use of drug probes, like ketamine (for the NMDA-sensitive glutamate system), and brain imaging routines that allow an assessment of *in vivo* brain chemistry and functional response to identify drug response and disease characteristics. These are especially important in brain diseases like schizophrenia where the symptoms are difficult to recreate in an animal and human information is often necessary to create an informative animal preparation.

Our findings with schizophrenia and the glutamate system have relied on the characterization of the clinical response of patients to ketamine and their functional brain imaging response (rCBF) to the drug. Prior to the human studies reported here, we had evaluated the region activation characteristics and pharmacology of PCP and its congener MK 801 in animals. What I report in this paper has been individually reported elsewhere but brought together here in a new synthesis.

Laboratory animals, given PCP, ketamine, or MK 801, show metabolic activation that is especially prominent in limbic cortex,[18] has a delayed component,[19] and can be blocked by D-cycloserine and, in some regions, by olanzapine (Gao, submitted for publication). The regions affected early after administration and most prominently in these experiments are the limbic cortex, both hippocampus and anterior cingulate. Then, at a delayed time point, well after the drug half-life, an inhibition of metabolism and of immediate early gene (IEG) activation occurs in neo- and limbic-cortex.[18,19] These data suggested to us that, despite the rather widespread distribution of the NMDA-sensitive glutamate receptor in mammalian brain, activation characteristics with PCP/ketamine appear to activate mainly a subset of the receptors, especially those in the limbic cortex. Whether this activation is due to pharmacologic differences, local microcircuitry characteristics, or an unknown reason is currently unclear, but of some interest.

With the idea of characterizing the human response of schizophrenia symptoms to NMDA-sensitive glutamate probes, ketamine was studied in schizophrenia. Other laboratories were also studying the actions of ketamine.[20,21] Because this is a drug that exacerbates symptoms, even though mildly, its use was initiated with considerable ethical review, safeguards, and clarity of informing of the patient populations. Its use was viewed in an analogous way to the use of a glucose tolerance test in a person with diabetes. Although in the studies done so far no personal gain has been claimed with the use of ketamine, either in diagnosis or in treatment, just a contribution to the potential understanding of the basis of the illness. Volunteers with schizophrenia have been willing to participate in order to increase basic knowledge in the area, and their family members have concurred.

Initially, volunteers with schizophrenia who were otherwise drug free, received ketamine in a placebo-controlled, rising, subanesthetic dose design. In a dose range of 0.1–0.5 mg/kg, ketamine caused a dose-related increase in psychotic symptoms in the volunteers, especially in positive symptoms like hallucinations, paranoid delusions and thought disorder.[22] The magnitude of the symptom increase in this dose range was approximately 25–35% of a person's drug-free symptom baseline. At least some of their symptoms were characteristic exacerbations of their baseline symptoms, not new drug-induced symptoms.[23] For example, if the person was experiencing hallucinations at baseline or a particular paranoid delusion, those hallucinations or that delusion increased, albeit mildly, in number and intensity, but new hallucinations or delusions were not necessarily induced. Ketamine did not create a new exogenous psychotic condition, rather it stimulated endogenous illness manifestations. Therefore, the symptom response seemed to be mediated by the illness condition, and was not just a stimulation of "normal" psychotomimetic circuits. This behavioral response of psychosis to ketamine is different than the response to amphetamine (for example), where there is no clear exacerbation of thought disorder or hallucinations in schizophrenia, but a classic paranoid response after repeated amphetamine administration.[6] This suggests that ketamine and its congeners may be unique psychotomimetic drugs in mimicking the psychosis of schizophrenia.

When these same volunteers were re-tested with ketamine while concurrently taking antipsychotic drugs, their response to ketamine was similar in type and magnitude, except that their psychosis score (BPRS psychosis subscale score) started from a lower (treated) baseline.[24] There was no evidence that first-generation antipsychotic treatment (haloperidol) could block the ketamine-induced psychotomimetic symptoms. Moreover, in an attempt to check whether this series of ketamine administrations had any long-term effects on the patient volunteers, their 6–12 month outcome was compared with a paired group of non-ketamine receiving research patients on the same inpatient research unit. The ketamine group did not have a worse outcome, indeed performed better than the group of non-ketamine volunteers.[25] Thus, there were no long-term detrimental effects in the patient volunteers from participating in the ketamine research study.

Because the behavioral symptoms of ketamine had been informative about symptoms in schizophrenia, it seemed indicated to (1) define the cerebral areas activated by the drug, (2) compare the activation of those cerebral areas between schizophrenia and control populations, and (3) identify those regions where the ketamine-induced symptoms correlated with rCBF changes. Ketamine was administered to volunteers with schizophrenia and a series of positron emission tomography (PET) scans with O-15 water were dynamically collected. First, the areas of activation were defined: ketamine activates the anterior cingulate cortex and the contiguous medial frontal cortex, and reduces rCBF activity in the parahippocampal cortex and in cerebellum.[22] With a bolus administration, the PET/O-15 H_2O scans were done at baseline (×3) and then repeated at 10-min intervals after drug administration, starting at 6 min (16, 26, 26, 36 min). In the areas of activation and inhibition, the peak rCBF response occurred within the first 16 min and regular change was not detectable after 36 min. This time course paralleled the symptom time course closely. In this analysis of dynamic activation patterns in the normal volunteers, there were three individual regions of activation and five of inhibition.[26]

Second, regions of activation/inhibition in the schizophrenia group were similar to the controls except for two areas: the anterior cingulate and the hippocampus. In the anterior cingulate, the activation in the patient group was greater in magnitude (area under the curve) than in the control group. In the hippocampus, the rCBF reduction was only apparent in the patient group and not in any region of hippocampus in the normal group (Holcomb, submitted for publication). Because in cerebral activation studies we have previously seen an inverse relationship between rCBF in hippocampus and rCBF in ACC (Medoff, submitted for publication), it seemed plausible that the ACC change in the schizophrenia group could be derived from the unusual ketamine-induced rCBF reduction in hippocampus that is unique to the schizophrenia group.

Third, we examined the behavioral correlates of ketamine-induced rCBF in both the control and the schizophrenia groups. Neither in cerebellum nor in the inferior frontal cortex, did correlations obtain between ketamine-induced behavioral effects (symptoms) and rCBF. But in the ACC and medial frontal area, correlations that were both significant and at a trend level developed between the ketamine-induced behaviors and rCBF in the normal and in the schizophrenia groups.[26] These data suggest that it is the ACC and medial frontal cortex that may be mediating the psychotomimetic properties of ketamine.

The cerebral activation pattern reported here is one that is established after a bolus intravenous administration of ketamine, not after an infusion. Because PCP and ketamine both establish a complex pattern of molecular changes in brain with respect to target and with respect to time, we thought it potentially confounding to evaluate an infusion where a possible delayed drug effect would be superimposed onto an ongoing acute administration drug action. The acute rCBF activation pattern seems to us to be the one that parallels the behavioral effects of ketamine with the most potent psychotomimetic actions apparent early (within 5–15 min) after drug administration.

What these human studies have detected is an abnormal response of the human hippocampus to ketamine in schizophrenia. Ketamine non-competitively blocks the NMDA-sensitive glutamate ionophore to block overall the signal at this excitatory ionophore. No response occurs in hippocampus to this probe in the normals, but an inhibition of rCBF does occur to the same stimulus in schizophrenia. An inhibition of the NMDA receptor does not produce an overall reduction in neuronal activity in the normal group but it does in the schizophrenia group. This result suggests that the schizophrenic hippocampus is more vulnerable to blockade of excitation than the normal organ, possibly because its glutamatergic tone is already reduced in this region. The hippocampus is a region where recurrent excitation, especially in CA_3, is routine and influential on transmission. Consequently, if there were a general defect in glutamatergic transmission in the entire brain, it might be only in hippocampus that early indication of the abnormality would be apparent. These data here are consistent with the idea that glutamatergic transmission is abnormally reduced in the hippocampus in schizophrenia,[27] either because this is the region where a subtle abnormality becomes manifest functionally or, perhaps, this is where a selective lesion exists. From a functional perspective, this is a localized lesion that in animal and humans could have great consequences for thought organization and for cognition.

The idea that excitatory glutamatergic transmission at the NMDA-sensitive receptor is diminished in schizophrenia, especially in hippocampus, is consistent with these data. This reduction could be caused by an exogenous substance,[28,29] by an irregularity in receptor composition,[30] or by a presynaptic abnormality coupled with the NMDA synapse and highly expressed in hippocampus. Perhaps the abnormality is highly expressed in hippocampus because of the unique recurrent excitatory feedback circuits characteristic of CA_3, not because of a molecular difference in schizophrenic hippocampus. It is interesting to speculate that the recently demonstrated therapeutic potential of the anticonvulsant drug divalproex sodium[31] is a pharmacologic agent through the action of valproic acid in the hippocampus to regulate excitatory transmission there. More knowledge about the illness determinants would allow the field to directly develop "silver bullet" treatments for the illness.

Glutamatergic transmission at the NMDA receptor may be involved in the generation of schizophrenia. If so, then ketamine (a pharmacologic probe that can be used in humans to modulate glutamatergic transmission) response data on symptom, cognitive, and rCBF parameters may be useful in finally understanding the biology of schizophrenia.

REFERENCES

1. TAMMINGA, C.A., G.K. THAKER & D.R. MEDOFF. 2002. Neuropsychiatric aspects of schizophrenia. *In* The American Psychiatric Publishing Textbook of Neuropsychiatry and Clinical Neurosciences. 4th edit. S.C. Yudofsky & R.E. Hales, Eds.: 989–1020. American Psychiatric Publishing, Inc. Washington, D.C.
2. HARRISON, P.J. 1999. The neuropathology of schizophrenia. A critical review of the data and their interpretation. Brain **122** (Pt 4): 593–624.
3. ANDREASEN, N.C. 1991. Schizophrenia: the characteristic symptoms [Review]. Schizophr. Bull. **17:** 27–49.
4. DELAY, J. & P. DENIKER. 1952. Le'traitement des psychoses par une methode neurolytique derivee de l'hibernotherapie. *In* Congres des Medecins Alienistes et Neurologistes de France: 497–502. Luxembourg.
5. CARLSSON, A. & M. LINDQUIST. 1963. Effect of chlorpromazine and haloperidol of formation of 3-methoxytyramine and normetanephrine in mouse brain. Acta Pharmacol. Toxicol.: 140–144.
6. SNYDER, S.H. 1972. Catecholamines in the brain as mediators of amphetamine psychosis. Arch. Gen. Psychiatry **27:** 169–179.
7. LARUELLE, M., A. ABI-DARGHAM, R. GIL, *et al.* 1999. Increased dopamine transmission in schizophrenia: relationship to illness phases. Biol. Psychiatry **46:** 56–72.
8. MELTZER, H.Y. 1995. Role of serotonin in the action of atypical antipsychotic drugs [Review]. Clin. Neurosci. **3:** 64–75.
9. NEMEROFF, C.B., W.W. YOUNGBLOOD, P.J. MANBERG, *et al.* 1983. Regional brain concentrations of neuropeptides in Huntington's chorea and schizophrenia. Science **221:** 972–975.
10. BENES, F.M., S.L. VINCENT, A. MARIE & Y. KHAN. 1996. Up-regulation of GABAA receptor binding on neurons of the prefrontal cortex in schizophrenic subjects. Neuroscience **75:** 1021–1031.
11. KIM, J.S., H.H. KORNHUBER, W. SCHMID-BURGK & B. HOLZMULLER. 1980. Low cerebrospinal fluid glutamate in schizophrenic patients and a new hypothesis on schizophrenia. Neurosci. Lett. **20:** 379–382.
12. ALTSHULER, L.L., M.F. CASANOVA, T.E. GOLDBERG & J.E. KLEINMAN. 1990. The hippocampus and para-hippocampus in schizophrenic, suicide, and control brains. Arch. Gen. Psychiatry **47:** 1029–1034.

13. ANIS, N.A., S.C. BERRY, N.R. BURTON & D. LODGE. 1983. The dissociative anesthetics, ketamine and phencyclidine, selectively reduce excitation of central mammalian neurones by N- methyl-D-aspartate. Br. J. Pharmacol. **79:** 565–575.
14. MONYER, H., N. BURNASHEV, D.J. LAURIE, *et al.* 1994. Developmental and regional expression in the rat brain and functional properties of four NMDA receptors. Neuron **12:** 529–540.
15. TAMMINGA, C.A. 1998. Schizophrenia and glutamatergic transmission. Crit. Rev. Neurobiol. **12:** 21-36.
16. TSAI, G. & J.T. COYLE. 2002. Glutamatergic mechanisms in schizophrenia. Annu. Rev. Pharmacol. Toxicol. **42:** 165–179.
17. OLNEY, J.W. & N.B. FARBER. 1995. Glutamate receptor dysfunction and schizophrenia. Arch. Gen. Psychiatry **52:** 998–1007.
18. TAMMINGA, C.A., K. TANIMOTO, S. KUO, *et al.* 1987. PCP-induced alterations in cerebral glucose utilization in rat brain: blockade by metaphit, a PCP-receptor-acylating agent. Synapse **1:** 497–504.
19. GAO, X.M. & C.A. TAMMINGA. 1996. Phencyclidine produces changes in NMDA and kainate receptor binding in rat hippocampus over a 48 hour time course. Synapse **23:** 274–279.
20. KRYSTAL, J.H, L.P. KARPER, J.P. SEIBYL, *et al.* 1994. Subanesthetic effects of the noncompetitive NMDA antagonist, ketamine, in humans: psychotomimetic, perceptual, cognitive, and neuroendocrine responses. Arch. Gen. Psychiatry **51:** 199–214.
21. MALHOTRA, A.K., D.A. PINALS, C.M. ADLER, *et al.* 1997. Ketamine-induced exacerbation of psychotic symptoms and cognitive impairment in neuroleptic-free schizophrenics. Neuropsychopharmacology **17:** 141–150.
22. LAHTI, A.C., H.H. HOLCOMB, D.R. MEDOFF & C.A. TAMMINGA. 1995. Ketamine activates psychosis and alters limbic blood flow in schizophrenia. NeuroReport **6:** 869–872.
23. LAHTI, A.C., B. KOFFEL, D. LAPORTE & C.A. TAMMINGA. 1995. Subanesthetic doses of ketamine stimulate psychosis in schizophrenia. Neuropsychopharmacology **13:** 9–19.
24. LAHTI, A.C., M.A. WEILER, T. MICHAELIDIS, *et al.* 2001. Effects of ketamine in normal and schizophrenic volunteers. Neuropsychopharmacology **25:** 455–467.
25. LAHTI, A.C., D. WARFEL, T. MICHAELIDIS, M.A. WEILER, *et al.* 2001. Long-term outcome of patients who receive ketamine during research. Biol. Psychiatry **49:** 869–875.
26. HOLCOMB, H.H., A.C. LAHTI, D.R. MEDOFF, *et al.* 2001. Sequential regional cerebral blood flow brain scans using pet with H(2)(15)O demonstrate ketamine actions in CNS dynamically. Neuropsychopharmacology **25:** 165–172.
27. MEDOFF, D.R., H.H. HOLCOMB, A.C. LAHTI & C.A. TAMMINGA. 2001. Probing the human hippocampus using rCBF: contrasts in schizophrenia. Hippocampus **11:** 543–550.
28. SCHWARCZ, R., A. RASSOULPOUR, H.Q. WU, *et al.* 2001. Increased cortical kynurenate content in schizophrenia. Biol. Psychiatry **50:** 521–530.
29. TSAI, G., D.C. GOFF, R.W. CHANG, *et al.* 1998. Markers of glutamatergic neurotransmission and oxidative stress associated with tardive dyskinesia. Am. J. Psychiatry **155:** 1207–1213.
30. HECKERS, S., D. GOFF, D.L. SCHACTER, *et al.* 1999. Functional imaging of memory retrieval in deficit vs nondeficit schizophrenia. Arch. Gen. Psychiatry **56:** 1117–1123.
31. CASEY, D.E. *et al.* 2003. Effect of divalproex combined with olanzapine or risperidone in patients with an acute exacerbation of schizophrenia. Neuropsychopharmacology **28:** 182–192.

The NMDA Receptor Hypofunction Model of Psychosis

NURI B. FARBER

Department of Psychiatry, Washington University, St. Louis, Missouri 63110-1093, USA

ABSTRACT: Antagonists of the NMDA glutamate receptor, including phencyclidine (PCP), ketamine, and CGS-19755, produce cognitive and behavioral changes in humans. In rodents these agents produce a myriad of histopathological and neurochemical changes. Several lines of evidence suggest that a large number of these drug-induced effects are dose-dependent manifestations of the same general disinhibition process in which NMDA antagonists abolish GABAergic inhibition, resulting in the simultaneous excessive release of acetylcholine and glutamate. Progressive increases in the severity of NMDA receptor hypofunction (NRHypo) within the brain produce an increasing range of effects on brain function. Underexcitation of NMDA receptors, induced by even relatively low doses of NMDA antagonist drugs, can produce specific forms of memory dysfunction without clinically evident psychosis. More severe NRHypo can produce a clinical syndrome very similar to a psychotic schizophrenic exacerbation. Finally, sustained and severe NRHypo in the adult brain is associated with a form of neurotoxicity with well-characterized neuropathological features. In this paper several of these effects of NMDA antagonists and a likely mechanism responsible for producing them will be reviewed. In addition the possible role of NRHypo in the pathophysiology of idiopathic psychotic disorders will be considered.

KEYWORDS: NMDA antagonists; dissociative anesthetics; dizocilpine maleate; phencyclidine; ketamine; psychosis; schizophrenia; bipolar disorder; psychoses

NMDA RECEPTOR FUNCTION/DYSFUNCTION IN THE ADULT HUMAN BRAIN

In the adult human brain the N-methyl-D-aspartate (NMDA) glutamate (Glu) transmitter system is thought to have an important role in memory and cognition, and in sensory information processing, to name but a few of its many functions. NMDA antagonists by blocking NMDA receptors produce an NMDA receptor hypofunction (NRHypo) state that is associated with psychiatric symptoms. PCP and ketamine, two NMDA antagonists, are classified as dissociative anesthetics. Early studies found that with this type of anesthesia adult patients exhibited psychiatric symptoms (referred to as "emergence reactions"), including maniacal excitation, catatonic

---

Address for correspondence: Nuri B. Farber, M.D., Washington University, Department of Psychiatry, Campus Box 8134, 660 S. Euclid Ave., St. Louis, MO 63110-1093. Voice: 314-362-2459; fax: 314-362-9902.

farbern@wustl.edu

signs, euphoria, hallucinations, delusions, and agitation.[1,2] More recent studies employing ketamine in subanesthetic doses have confirmed some of these initial findings, but the symptoms produced were milder and not as extensive.[3-5] In these carefully controlled later studies very low doses of ketamine produced selective impairments in explicit/declarative memory in the absence of psychosis. Higher subanesthetic doses of ketamine produced positive symptoms (delusions and hallucinations), and still higher subanesthetic doses produced thought disorder. Because these later studies were appropriately concerned with minimizing symptom severity and protecting human subjects from unpleasant experiences, they did not produce the full range of symptoms seen in the earlier studies using anesthetic doses of PCP and ketamine. However, the dose dependence of ketamine-induced core schizophrenia-like symptoms suggests that if higher doses of ketamine had been used in the later studies, the full variety and severity of effects seen in earlier studies probably would have been produced. These findings have stimulated interest in the possibility that an NRHypo mechanism could produce cognitive dysfunction and psychotic symptom formation in idiopathic psychotic disorders.

While schizophrenia has received the most attention as the disorder in which an NRHypo state might exist, the fact that higher doses of NMDA antagonists can produce maniacal excitation, catatonic signs, and euphoria suggests that such an NRHypo state also could be responsible for some of the signs and symptoms of bipolar and schizoaffective disorder. The possibility that NRHypo is involved in these disorders would be consistent with clinical experience that it is impossible to differentiate a manic psychosis from a schizophrenic one on cross-sectional examination.

CONSEQUENCES OF NRHypo IN THE ADULT RODENT BRAIN

To begin to understand the mechanism(s) underlying the consequences of the NRHypo state in humans, several research groups have begun examining the consequences of drug-induced NRHypo in adult rodents. One typical consequence is excessive release of Glu[6-8] and acetylcholine[9-11] (ACh) in the cerebral cortex. It has been proposed that this excessive release of excitatory transmitters and consequent overstimulation of postsynaptic neurons might explain the cognitive and behavioral disturbances associated with the NRHypo state.[7,12]

At doses approximately equal to or higher than those needed to produce measurable elevations in neurotransmitter release, NMDA antagonists begin to produce neurotoxicity in adult rodents. Initially, they induce reversible pathomorphological changes in pyramidal neurons in the retrosplenial cortex[13] (RSC). If NMDA receptor blockade is maintained for a prolonged interval, as occurs following a single high dose or repeated treatment with lower doses of an NMDA antagonist, neurons in the RSC and several other cerebrocortical and limbic regions of the adult rat brain undergo irreversible degeneration.[14-16]

The key feature of the mechanism (FIG. 1) that mediates the NRHypo neurotoxic process is that Glu, acting at NMDA receptors, functions in a multisynaptic circuit as a regulator of inhibitory tone. Glu accomplishes this regulatory function by tonically stimulating NMDA receptors on GABAergic interneurons, which, in turn, inhibit excitatory projections that convergently innervate vulnerable cerebrocortical neurons. NMDA receptor antagonists prevent Glu from driving GABAergic inhibi-

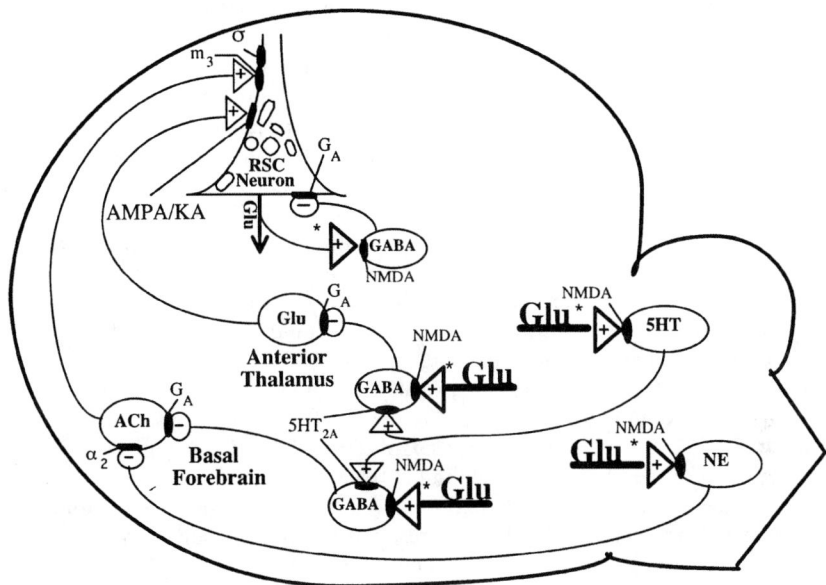

FIGURE 1. NRHypo disinhibition circuitry. Glu, acting through NMDA receptors on GABAergic, serotonergic, and noradrenergic neurons, maintains tonic inhibitory control over two major excitatory pathways that convergently innervate RSC neurons. Systemic administration of an NMDA antagonist blocks NMDA receptors, thereby abolishing inhibitory control over both of the excitatory inputs to the RSC neuron. The disinhibited excitatory pathways then simultaneously hyperactivate the RSC neuron, which would create chaotic disruption of multiple intracellular signaling systems, thereby causing immediate derangement of cognitive functions subserved by the afflicted neurons (psychotomimetic effects), and reversible or irreversible neuronal injury, depending on how long the disruption lasts. This circuit diagram focuses exclusively on RSC neurons. It is hypothesized that a similar disinhibition mechanism and similar, but not necessarily identical, neural circuits and receptor mechanisms mediate damage induced in other corticolimbic brain regions by sustained NRHypo. (+) = excitatory input; (−) = inhibitory input; ACh = acetylcholine; NE = norepinephrine; Glu = glutamate; GABA = γ-amino butyric acid; 5HT = serotonin; α_2 = α_2 subtype of adrenergic receptor; G_A = $GABA_A$ subtype of GABA receptor; m_3 = m_3 subtype of muscarinic cholinergic receptor; AMPA/KA = AMPA/KA subtype of Glu receptor; NMDA = NMDA subtype of Glu receptor; σ = sigma site; $5HT_{2A}$ = $5HT_{2A}$ subtype of serotonin receptor. Asterisks indicate the postulated sites where dopamine inputs may presynaptically regulate Glu release.

tory neurons, and this results in a loss of inhibitory control over two major excitatory projections to the cerebral cortex, one that is cholinergic and originates in the basal forebrain, and one that is glutamatergic and originates in the thalamus.[11,17,18] In addition to these basic features, the NRHypo circuitry includes noradrenergic[11,17,19] and serotonergic[20] neurons that presumably are driven by Glu through NMDA receptors and also perform an inhibitory function so that when NMDA receptors are hypofunctional the inhibitory restraint contributed by these elements is also lost.

One final aspect that may be quite important for understanding how disinhibition of this circuitry can trigger psychotic reactions is that the vulnerable cerebrocortical neurons are glutamatergic neurons that ordinarily control their own firing by activating an NMDA receptor on a GABAergic neuron in an inhibitory feedback loop.[21] When the NMDA receptor in this feedback loop is hypofunctional (e.g., blocked by NMDA antagonist drugs), GABAergic inhibition is lost, and the cerebrocortical neurons' control over their own firing is lost at the same time that these neurons are being hyperstimulated by disinhibited glutamatergic and cholinergic excitatory inputs. The expected result would be that the cerebrocortical neurons will bombard many other neurons in their projection fields with an abundance of unmodulated noise. This provides an explanation for the psychotomimetic reactions induced by NMDA antagonist drugs. A similar NRHypo mechanism could contribute to the psychotic process in idiopathic psychotic disorders.

NMDA antagonists (e.g., MK-801, ketamine, and PCP) also increase metabolism in certain corticolimbic regions.[22–25] While it is difficult to make specific comparisons across different research groups using different NMDA antagonists and different protocols, in general, the corticolimbic regions experiencing hypermetabolism tend to be the same corticolimbic regions that also develop NRHypo neurotoxicity. The increase in metabolism in these corresponding regions could be a reflection of a disinhibition syndrome in which Glu and ACh are excessively released at certain corticolimbic neurons that are injured in the NRHypo neurotoxic syndrome. Consitent with this proposal, clozapine and halothane reverse the hypermetabolism induced by NMDA antagonists,[24,26] just as they reverse NRHypo neurotoxicity.[27,28]

In addition to increasing the release of Glu and ACh, altering metabolism, and producing neurotoxicity, NRHypo also produces several other effects in the CNS. Soon after the original report of neuronal vacuolation, Dragunow and Faull[29] reported that MK-801 induced the production of c-Fos protein in these same neurons. Subsequent work has found that not only c-Fos but other immediate-early genes (IEG), including c-Jun, Jun-B, NGFI-A (a.k.a. zif268 and krox-24), NGFI-B, NGFI-C, and Nurr1, are activated by NRHypo.[30–34] In addition, the heat shock stress protein HSP70 (and its mRNA) is also induced in these vulnerable neurons by NRHypo.[35,36] Last, NRHypo also induces the expression of brain-derived growth factor (BDNF) mRNA.[31,37]

These additional changes in rodents have been suggested to underlie some of the cognitive and behavioral effects seen with NMDA antagonists in humans. The ability of some of the same pharmacological treatments, which have been shown to prevent NRHypo neurotoxicity, to prevent these other NMDA antagonist-induced responses (TABLE 1) suggests these other responses, when induced by NMDA antagonists, are also likely to be secondary to activation of the same NRHypo disinhibition mechanism. However, IEGs, heat shock proteins, and BDNF also can be induced after a wide variety of other stimuli, including trauma, injury, and neuronal depolarization. Thus, these additional changes, while particulary sensitive to the disruption induced by NRHypo, would not be specific for psychosis or the NRHypo disinhibition mechanism. Since the protein products of these IEGs recognize and bind certain DNA sequences and modify the transcription of target genes, NRHypo-induced production of IEGs might be a clue to understanding the more enduring intracellular and nuclear events that occur in response to the disinhibition syndrome induced by NRHypo.

TABLE 1. Agents that prevent NRHypo neurotoxicity prevent other NRHypo-induced phenomena

NRHypo-Induced Phenomenon	Protective Agent
Hypermetabolism	halothane[24] clozapine[26]
HSP70	antimuscarinics[35,59] GABAergics[60] non-NMDA Glu antagonists[61] sigma ligands[33,39] clozapine[60]
c-Fos, c-Jun, Jun-B, NGFI-A	GABAergics[32,62] antimuscarinics[31]
BDNF	GABAergics[37] antimuscarinics[31,37]

Thus, in the rodent, NRHypo produces a broad range of effects. Cognizant of the inherent difficulty in comparing results from different model systems that have different sensitivities of assessing change in brain function, it appears that the NRHypo disinhibition mechanism produces a range of effects, depending on its severity and duration. Mild versions of the NRHypo disinhibition syndrome produce mild elevations in neurotransmitter release, a loss of recurrent feedback inhibition, and elevated metabolism. More moderate degrees would probably begin to produce changes in IEGs and other cellular processes. Finally, even more severe degrees of NRHypo would result in neurotoxicity.

AGE DEPENDENCY OF NRHypo EFFECTS IN RODENTS

The ability of NRHypo to induce these changes in the brain is highly dependent on the age of the animal. The reversible vacuolar changes induced by MK-801 or PCP were not found in fetal (E17–19) or postnatal rats at day 15 or 30 of life. However, at 45 days of life (onset of puberty), partial susceptibility occurred, and this gradually increased to full susceptibility between 90 and 120 days of age.[38] PCP's induction of c-Fos and HSP70 has a similar age-dependency profile as does the formation of vacuoles.[39,40] Thus, in normal brain, these NRHypo-induced changes appear to occur only in the pubertal and postpubertal period.

THE NRHypo DISINHIBITED STATE MIGHT UNDERLIE NRHypo-INDUCED PSYCHOSIS IN HUMANS

Pharmacological Data

As noted above the main reason for studying the effects of NRHypo in rodents is to gain clues about the mechanism underlying the psychotomimetic effects of NMDA antagonists in humans. If the NRHypo disinhibited state does underlie the

the mental effects of NMDA antagonists, then the pharmacological treatments that ameliorate the NRHypo-induced effects in rodents should ameliorate NRHypo-induced psychosis. Information pertaining to this question, although incomplete, provides some interesting correlations. Agents that promote $GABA_A$ neurotransmission prevent the NRHypo state from releasing excessive acetylcholine[11] and prevent NRHypo neurotoxicity in the rat cerebral cortex,[18,35] and it is well recognized by anesthesiologists that these agents in sufficient dosage attenuate the psychotomimetic actions of ketamine.[1,2] The effect is dose dependent, with higher dosages more effective than lower dosages. The dose dependency of the effect might account for a reported negative finding of Krystal *et al.* with lorazepam.[41] α_2-Adrenergic agonists prevent the NRHypo state from releasing excessive acetylcholine[11] and prevent NRHypo neurotoxicity in the rat cerebral cortex[19] by acting in the basal forebrain where cholinergic cell bodies reside.[17] Furthermore, α_2-adrenergic agonists prevent ketamine from inducing positive schizophrenia-like symptoms.[42–44] Lamotrigine, an anticonvulsant, prevents NRHypo neurotoxicity in the rat cerebral cortex[45] and prevents ketamine-induced schizophrenia-like symptoms in human volunteers.[46] Clozapine is quite potent in blocking NRHypo neurotoxicity in the rat,[28] and clozapine has been reported to block ketamine-induced increases in positive symptoms in patients with schizophrenia.[47] Given the difficulties in making cross-species comparisons, it is rather remarkable that four classes of compounds (GABAergics, α_2-adrenergic agonists, clozapine, and lamotrigine) have been found to work in both models.

In addition, haloperidol, when used at its D_2 blocking dose, has been found to be minimally active acutely against ketamine-induced psychosis.[48] In rodents, while haloperidol can block NRHypo neurodegeneration, it does so only at a dose that is 10 times greater than its D_2 dose, whereas sulpiride, a D_2 selective antagonist, is ineffective.[28] These D_2 findings suggest that the D_2 receptor system is not an intrinsic component of the NRHypo disinhibition circuitry and is not directly involved in ketamine's production of psychosis (see below).

Age-Dependency Data

Thus, the pharmacological data reviewed above support the hypothesis that an NRHypo-induced disinhibited state may underlie both the neurotoxic and psychotomimetic effects of NMDA antagonists. Consistent with this hypothesis, prepubertal rats do not develop NRHypo-induced neurodegeneration,[38] just as prepubertal humans rarely develop psychosis after exposure to PCP or ketamine.[1,2]

Dose-Response Relationship among Different NRHypo Effects

The hypothesis that a similar NRHypo disinhibition syndrome underlies the psychotomimetic effect in humans and the neurotoxic effect in rodents does not mean that the neurotoxicity must exist in brains of people with NRHypo-induced psychosis. Instead, it is proposed that the severity of the NRHypo state is a critical variable in determinining the specific NRHypo effect seen. Specifically mild degrees of NRHypo would produce slight elevations in the release of ACh and Glu and slight decrements in local recurrent inhibition. These changes could produce mild derangements in memory function without significant behavioral changes. More moderate degrees of NRHypo would produce greater aberation in the circuit, resulting in more obvious cognitive and behavioral disturbances, but still not producing neurotoxicity. At this stage one might begin to see nuclear and cellular changes, like the production of certain IEGs, as a response to the aberrant functioning of the NRHypo circuit. Only in the case where NRHypo was severe would the NRHypo state produce enough of an increase in transmitter release and in the degree of postsynaptic m_3 muscarinic and non-NMDA Glu receptor overstimulation to produce first reversible and then irreversible neurotoxicity.

DOPAMINE, NRHypo DISINHIBITED STATE, AND PSYCHOSIS

The role of dopamine and D_2 receptors in the production and treatment of psychosis has been long established. What then is the relationship between the D_2 dopaminergic and NMDA glutamatergic systems? Partly on the basis of the well-described ability of NMDA antagonists to increase dopaminergic activity, many have assumed that NMDA antagonists cause excessive activity at D_2 receptors and that this excessive activity then directly produces psychosis. However, the lack of correlation in animals between the time course of NMDA antagonist-induced dopamine release and NMDA antagonist-induced behavioral changes suggests that the increase in dopamine release might not be responsible for the psychosis produced by NMDA antagonists.[49] The inability of D_2 antagonists to reverse ketamine-induced psychosis[48] is consistent with the proposal that changes in activation of D_2 receptors downstream from the NMDA receptor are not involved in the production of psychosis.

There is another role that D_2 receptors may play in the NRHypo model. As indicated above, D_2 receptors do not appear to function internally within the NRHypo disinhibtion network, but they may be exceedingly important extrinsic regulators of the network and modify the release of Glu at NMDA receptors. If this is the case in idiopathic psychotic disorders, a genetically determined aberration in the dopamine system, causing hyperinhibition of Glu release, would result in an NRHypo state that could explain psychotic symptom formation. Conversely, treatment with a D_2 antagonist could normalize Glu release, relieve the NRHypo state, and attenuate psychosis.

NRHypo DISINHIBITED STATE AND PSYCHIATRIC ILLNESS

It is proposed that in major psychotic disorders (e.g., schizophrenia, schizoaffective disorder, and bipolar disorder), the NRHypo state is instilled in the brain by a pathological event very early in life (e.g., prenatally). Because NRHypo does not produce the disinhibited mechanism or psychosis prior to puberty, the NRHypo state that was created prenatally can remain quiescent throughout childhood. Then, in adolescence, unknown maturational changes occur causing the NRHypo disinhibited mechanism to become active and symptoms to begin to appear. Since some brain regions and their associated mental functions may be more sensitive to disruption by NRHypo, different symptoms may present at different times even if brain maturation is relatively homogeneous. Alternatively, the maturation process could occur in particular brain regions at different times, so that mental functions subserved by different brain regions would become dysfunctional at different times. In either case, therefore, the full syndrome would unfold over time as each specific clinical sign appears in succession. For example, relatively subtle impairments in cognitive functions, such as attention and memory, might occur in early adolescence, while more obvious psychotic signs might not become present until later in adolescence.

How might an NRHypo lesion be instilled in the developing brain? Genetic abnormalities are known to be etiological factors in the major psychotic disorders. On the basis of the NRHypo circuitry, several likely candidates can be identified. Obviously abnormalities in NMDA receptor gene products would be candidates. In addition, various NRHypo "equivalent" conditions could similarly produce the relevant network disturbances. Recently several genetic abnormalities that would produce an NRHypo state have been linked to schizophrenia.[50–52] Primary disturbances in the GABAergic system or non-NMDA system would be other likely candidates. A wide variety of environmental factors could delete NMDA receptor-bearing neurons from the brain and render the immature brain into an NRHypo state (see below). Different genetic abnormalities are expected to be associated with different clinical syndromes (e.g., bipolar disorder vs. schizophrenia and schizophrenia spectrum disorders), which tend to aggregate together in families, and these different genetic abnormalities might affect the NRHypo circuit in different ways.

In addition, a wide range of nongenetic factors could interact differentially with these disease-related genetic abnormalities, modulating the NRHypo circuit and the clinical presentation of the disorder. Hypoxia/ischemia, which causes excessive amounts of Glu to be released at NMDA receptors, occurring at birth would be a prime candidate to produce a selective loss of NMDA receptor-bearing neurons and has long been recognized as a risk factor for idiopathic psychotic disorders. Similarly, *in utero* exposure to ethanol or a wide variety of other agents can also delete NMDA receptor-bearing neurons.[53] In a recent study,[54] a large cohort of human subjects with a history of *in utero* exposure to ethanol were studied as adults, and it was determined that a very high percentage (72%) of these individuals required psychiatric care for diverse adult-onset disorders, including a 40% incidence of psychosis and a 20% incidence of bipolar disorder. To establish the relevance of these observations to idiopathic psychotic disorders, it is not necessary to prove that hypoxia/ischemia, fetal ethanol exposure, or any other nongenetic agent by itself causes an adult psychotic disorder, mimicking all features of these idiopathic psychotic disorders. It is only necessary that these nongenetic factors act in concert with genetic factors to tip the balance and potentially cause a genetic predisposition for the disorder to be clinically expressed.[55]

Because of the key role played by NMDA receptor-bearing GABAergic neurons in the NRHypo disinhibition circuit, it will be important to determine whether the NMDA receptor-bearing neurons deleted by these nongenetic factors are GABAergic interneurons. The loss of NMDA receptor–bearing GABAergic neurons is of particular interest given that deficiencies in the GABAergic system exist in postmortem brain tissue of patients with schizophrenia, bipolar disorder, and schizoaffective disorder.[56–58] Because the GABAergic neuron in the feedback loop normally regulates the firing pattern of corticolimbic pyramidal neurons and aberrant firing of these neurons could cause considerable mental dysfunction, this population of GABAergic neurons might be differentially abnormal in patients with an idiopathic psychotic disorder. Indeed, an excellent formula for psychosis would be loss of GABAergic inhibition simultaneously in this feedback loop and in any one of the two main excitatory inputs to the pyramidal neuron, since this would cause the pyramidal neuron to be persistently hyperactived at the same time that it has lost feedback control over its firing on other neurons.

THE NRHypo HYPOTHESIS, IDIOPATHIC PSYCHOTIC DISORDERS, AND AGE OF ONSET

All three major idiopathic psychotic disorders typically have an age of onset in late adolescence or early adulthood. In order to account for growing evidence of an early developmental lesion in these disorders, investigators have postulated that maturational events must occur in the brain during adolescence that allow for the expression of symptoms. As discussed above, it has been shown that in normal rodents and humans the NRHypo disinhibited mechanism has a similar age of onset in adolescence, suggesting similar dependence on maturational events. The similarity in the age of onset between idiopathic psychotic illnesses and the drug-induced NRHypo model further supports the proposal that an NRHypo disinhibited mechanism underlies the signs and symptoms of these disorders. Information obtained about the specific changes underlying the age-dependency profile of the drug-induced NRHypo state should also provide clues to the specific events that must occur for the eventual expression of illness in humans.

SUMMARY

The NRHypo hypothesis can explain idiopathic psychotic disorders on the basis of genetic and/or nongenetic mechanisms instilling an NRHypo state with related circuit disinhibition in the developing brain. This state usually remains quiescent until early adulthood when maturational changes in brain circuitry make the brain vulnerable to the psychotogenic and neurotoxic potential of the NRHypo state. This hypothesis assumes that if the NRHypo state is mild or moderate only cognitive and behavioral signs will be expressed, but if the NRHypo state is particularly severe and of long duration, then cognitive, behavioral, and neurotoxic potential will be expressed. The severe NRHypo state could result in chronic severe symptoms complicated by ongoing structural brain changes and the potential for clinical deterioration. If this concept is correct, chronic treatment with certain protective drugs, including olanzapine, clozapine, lamotrigine, α_2 adrenergic agonists, and perhaps antimuscarinic agents, could be useful on the basis of studies that show that these drugs effectively arrest the acute neurotoxic process associated with the NRHypo state. In general, the animal model for NRHypo-associated neurotoxicity provides an opportunity to test pharmacological approaches for preventing the NRHypo state from hyperstimulating and injuring neurons. A careful analysis of the pharmacological interventions that are protective can also provide insights into the circuitry and receptor mechanisms that mediate this pathological process.

ACKNOWLEDGMENTS

This work was supported by AG11355 from the NIH.

REFERENCES

1. WHITE, P.F., W.L. WAY & A.J. TREVOR. 1982. Ketamine—its pharmacology and therapeutic uses. Anesthesiology **56:** 119–136.
2. REICH, D.L. & G. SILVAY. 1989. Ketamine: an update on the first twenty-five years of clinical experience. Can. J. Anaesth. **36:** 186–197.
3. KRYSTAL, J.H., L.P. KARPER, J.P. SEIBYL, et al. 1994. Subanesthetic effects of the noncompetitive NMDA antagonist, ketamine, in humans. Psychotomimetic, perceptual, cognitive, and neuroendocrine responses. Arch. Gen. Psychiatry **51:** 199–214.
4. MALHOTRA, A.K., D.A. PINALS, H. WEINGARTNER, et al. 1996. NMDA receptor function and human cognition: the effects of ketamine in healthy volunteers. Neuropsychopharmacology **14:** 301–307.
5. NEWCOMER, J.W., N.B. FARBER, V. JEVTOVIC-TODOROVIC, et al. 1999. Ketamine-induced NMDA receptor hypofunction as a model of memory impairment in schizophrenia. Neuropsychopharmacology **20:** 106–118.
6. MOGHADDAM, B. & B. ADAMS. 1998. Reversal of phencyclidine effects by a group II metabotropic glutamate receptor agonist in rats. Science **281:** 1349–1352.
7. MOGHADDAM, B., B. ADAMS, A. VERMA, et al. 1997. Activation of glutamatergic neurotransmission by ketamine: a novel step in the pathway from NMDA receptor blockade to dopaminergic and cognitive disruptions associated with the prefrontal cortex. J. Neurosci. **17:** 2921–2927.
8. NOGUCHI, K., R. JOHNSON & G. ELLISON. 1998. The effects of MK-801 on aspartate and glutamate levels in the anterior cingulate and retrosplenial cortices: an in vivo microdialysis study. Soc. Neurosci. Abstr. **24:** 233.
9. HASEGAWA, M., H. KINOSHITA, M. AMANO, et al. 1993. MK-801 increases endogenous acetylcholine release in the rat parietal cortex: a study using brain microdialysis. Neurosci. Lett. **150:** 53–56.
10. GIOVANNINI, M.G., D. MUTOLO, L. BIANCHI, et al. 1994. NMDA receptor antagonists decrease GABA outflow from the septum and increase acetylcholine outflow from the hippocampus: a microdialysis study. J. Neurosci. **14:** 1358–1365.
11. KIM, S.H., M.T. PRICE, J.W. OLNEY, et al. 1999. Excessive cerebrocortical release of acetylcholine induced by NMDA antagonists is reduced by GABAergic and α_2-adrenergic agonists. Mol. Psychiatry **4:** 344–352.
12. OLNEY, J.W. & N.B. FARBER. 1995. Glutamate receptor dysfunction and schizophrenia. Arch. Gen. Psychiatry **52:** 998–1007.
13. OLNEY, J.W., J. LABRUYERE & M.T. PRICE. 1989. Pathological changes induced in cerebrocortical neurons by phencyclidine and related drugs. Science **244:** 1360–1362.
14. ELLISON, G. 1994. Competitive and non-competitive NMDA antagonists induce similar limbic degeneration. Neuroreport **5:** 2688–2692.
15. CORSO, T.D., M.A. SESMA, T.I. TENKOVA, et al. 1997. Multifocal brain damage induced by phencyclidine is augmented by pilocarpine. Brain Res. **752:** 1–14.
16. HORVATH, Z.C., J. CZOPF & G. BUZSAKI. 1997. MK-801-induced neuronal damage in rats. Brain Res. **753:** 181–195.
17. FARBER, N.B., S.H. KIM, K. DIKRANIAN, et al. 2002. Receptor mechanisms and circuitry underlying NMDA antagonist neurotoxicity. Mol. Psychiatry **7:** 32–43.
18. FARBER, N.B., X.P. JIANG, K. DIKRANIAN, et al. Muscimol prevents NMDA antagonist neurotoxicity by activating GABAA receptors in several brain regions. Brain Res. In press.
19. FARBER, N.B., J. FOSTER, N.L. DUHAN, et al. 1995. α_2 Adrenergic agonists prevent MK-801 neurotoxicity. Neuropsychopharmacology **12:** 347–349.
20. FARBER, N.B., J. HANSLICK, C. KIRBY, et al. 1998. Serotonergic agents that activate $5HT_{2A}$ receptors prevent NMDA antagonist neurotoxicity. Neuropsychopharmacology **18:** 57–62.
21. GRUNZE, H.C., D.G. RAINNIE, M.E. HASSELMO, et al. 1996. NMDA-dependent modulation of CA1 local circuit inhibition. J. Neurosci. **16:** 2034–2043.
22. MEIBACH, R.C., D. GLICKS, R. COX, et al. 1979. Localisation of phencyclidine-induced changes in brain energy metabolism. Nature **282:** 625–626.

23. NELSOM, S.R., R.B. HOWARD, R.S. CROSS, et al. 1980. Ketamine-induced changes in regional glucose utilization in the rat brain. Anesthesiology **52:** 330–334.
24. KURUMAJI, A. & J. MCCULLOCH. 1989. Effects of MK-801 upon local cerebral glucose utilisation in conscious rats and in rats anaesthetised with halothane. J. Cereb. Blood Flow Metab. **9:** 786–794.
25. DUNCAN, G.E., S.S. MOY, D.J. KNAPP, et al. 1998. Metabolic mapping of the rat brain after subanesthetic doses of ketamine: potential relevance to schizophrenia. Brain Res. **787:** 181–190.
26. DUNCAN, G.E., J.N. LEIPZIG, R.B. MAILMAN, et al. 1998. Differential effects of clozapine and haloperidol on ketamine-induced brain metabolic activation. Brain Res. **812:** 65–75.
27. ISHIMARU, M., F. FUKAMAUCHI & J.W. OLNEY. 1995. Halothane prevents MK-801 neurotoxicity in the rat cingulate cortex. Neurosci. Lett. **193:** 1–4.
28. FARBER, N.B., M.T. PRICE, J. LABRUYERE, et al. 1993. Antipsychotic drugs block phencyclidine receptor-mediated neurotoxicity. Biol. Psychiatry **34:** 119–121.
29. DRAGUNOW, M. & R.L.M. FAULL. 1990. MK-801 induces c-fos protein in thalamic and neocortical neurons of rat brain. Neurosci. Lett. **111:** 39–45.
30. GASS, P., T. HERDEGEN, R. BRAVO, et al. 1993. Induction and suppression of immediate early genes in specific rat brain regions by the non-competitive N-methyl-D-aspartate antagonist, MK-801. Neuroscience **53:** 749–758.
31. HUGHES, P., M. DRAGUNOW, E. BEILHARZ, et al. 1993. MK801 induces immediate-early gene proteins and BDNF mRNA in rat cerebrocortical neurones. NeuroReport **4:** 183–186.
32. NAKAO, S.I., T. ADACHI, M. MURAKAWA, et al. 1996. Halothane and diazepam inhibit ketamine-induced c-fos expression in the rat cingulate cortex. Anesthesiology **85:** 874–882.
33. NAKKI, R., F.R. SHARP, S.M. SAGAR, et al. 1996. Effects of phencyclidine on immediate early gene expression in the brain. J. Neurosci. Res. **45:** 13–27.
34. GAO, X.M., T. HASHIMOTO & C.A. TAMMINGA. 1998. Phencyclidine (PCP) and dizocilpine (MK801) exert time-dependent effects on the expression of immediate early genes in rat brain. Synapse **29:** 14–28.
35. OLNEY, J.W., J. LABRUYERE, G. WANG, et al. 1991. NMDA antagonist neurotoxicity: mechanism and prevention. Science **254:** 1515–1518.
36. SHARP, F.R., P. JASPER, J. HALL, et al. 1991. MK-801 and ketamine induce heat shock protein HSP72 in injured neurons in posterior cingulate and retrosplenial cortex. Ann. Neurology **30:** 801–809.
37. CASTREN, E., M. DA PHENA BERZAGHI, D. LINDHOLM, et al. 1993. Differential effects of MK-801 on brain-derived neurotrophic factor mRNA levels in different regions of the rat brain. Exp. Neurology **122:** 244–252.
38. FARBER, N.B., D.F. WOZNIAK, M.T. PRICE, et al. 1995. Age specific neurotoxicity in the rat associated with NMDA receptor blockade: potential relevance to schizophrenia? Biol. Psychiatry **38:** 788–796.
39. SHARP, F.R., M. BUTMAN, S. WANG, et al. 1992. Haloperidol prevents induction of the hsp70 heat shock gene in neurons injured by phencyclidine (PCP), MK801, and ketamine. J. Neurosci. Res. **33:** 605–616.
40. SATO, D., A. UMINO, K. KANEDA, et al. 1997. Developmental changes in distribution patterns of phencyclidine-induced c-Fos in rat forebrain. Neurosci. Lett. **239:** 21–24.
41. KRYSTAL, J.H., L.P. KARPER, A. BENNETT, et al. 1998. Interactive effects of subanesthetic ketamine and subhypnotic lorazepam in humans. Psychopharmacology **135:** 213–229.
42. NEWCOMER, J.W., N.B. FARBER, G. SELKE, et al. 1998. Guanabenz effects on NMDA antagonist-induced mental symptoms in humans. Soc. Neurosci. Abstr. **24:** 525.
43. LEVANEN, J., M.L. MAKELA & H. SCHEININ. 1995. Dexmedetomidine premedication attenuates ketamine-induced cardiostimulatory effects and postanesthetic delirium. Anesthesiology **82:** 1117–1125.
44. HANDA, F., M. TANAKA, T. NISHIKAWA, et al. 2000. Effects of oral clonidine premedication on side effects of intravenous ketamine anesthesia: a randomized, double-blind, placebo-controlled study. J. Clin. Anaesth. **12:** 19–24.

45. FARBER, N.B., X.P. JIANG, C. HEINKEL, et al. 2002. Antiepileptic drugs and agents that inhibit voltage-gated sodium channels prevent NMDA antagonist neurotoxicity. Mol. Psychiatry **7:** 726–733.
46. ANAND, A., D.S. CHARNEY, D.A. OREN, et al. 2000. Attenuation of the neuropsychiatric effects of ketamine with lamotrigine. Arch. Gen. Psychiatry **57:** 270–276.
47. MALHOTRA, A.K., C.M. ADLER, S.D. KENNISON, et al. 1997. Clozapine blunts N-methyl-D-aspartate antagonist-induced psychosis: a study with ketamine. Biol. Psychiatry **42:** 664–668.
48. KRYSTAL, J.H., D.C. D'SOUZA, L.P. KARPER, et al. 1999. Interactive effects of subanesthetic ketamine and haloperidol in healthy humans. Psychopharmacology **145:** 193–204.
49. ADAMS, B. & B. MOGHADDAM. 1998. Corticolimbic dopamine neurotransmission is temporally dissociated from the cognitive and locomotor effects of phencyclidine. J. Neurosci. **18:** 5545–5554.
50. STRAUB, R.E., Y. JIANG, C.J. MACLEAN, et al. 2002. Genetic variation in the 6p22.3 gene DTNBP1, the human ortholog of the mouse dysbindin gene, is associated with schizophrenia. Am. J. Hum. Genet. **71:** 337–48.
51. CHUMAKOV, I., M. BLUMENFELD, O. GUERASSIMENKO, et al. 2002. Genetic and physiological data implicating the new human gene G72 and the gene for D-amino acid oxidase in schizophrenia. Proc. Natl. Acad. Sci. USA **99:** 13675–13680.
52. STEFANSSON, H., E. SIGURDSSON, V. STEINTHORSDOTTIR, et al. 2002. Neuregulin 1 and susceptibility to schizophrenia. Am. J. Hum. Genet. **71:** 877–892.
53. IKONOMIDOU, C., P. BITTIGAU, M.J. ISHIMARU, et al. 2000. Ethanol-induced apoptotic neurodegeneration and fetal alcohol syndrome. Science **287:** 1056–1060.
54. FAMY, C., A.P. STREISSGUTH & A.S. UNIS. 1998. Mental illness in adults with fetal alcohol syndrome or fetal alcohol effects. Am. J. Psychiatry **155:** 552–554.
55. LOHR, J.B. & H.S. BRACHA. 1989. Can schizophrenia be related to prenatal exposure to alcohol? Some speculations. Schizophr. Bull. **15:** 595–603.
56. BENES, F.M., J. MCSPARREN, E.D. BIRD, et al. 1991. Deficits in small interneurons in prefrontal and cingulate cortices of schizophrenic and schizoaffective patients. Arch. Gen. Psychiatry **48:** 996–1001.
57. WOO, T.U., R.E. WHITEHEAD, D.S. MELCHITZKY, et al. 1998. A subclass of prefrontal gamma-aminobutyric acid axon terminals are selectively altered in schizophrenia. Proc. Natl. Acad. Sci. USA **95:** 5341–5346.
58. GUIDOTTI, A., J. AUTA, J.M. DAVIS, et al. 2000. Decrease in reelin and glutamic acid decarboxylase$_{67}$ (GAD$_{67}$) expression in schizophrenia and bipolar disorder. Arch. Gen. Psychiatry **57:** 1061–1069.
59. TOMITAKA, S.I., K. HASHIMOTO, N. NARITA, et al. 1997. Regionally different effects of scopolamine on NMDA antagonist-induced heat shock protein HSP70. Brain Res. **736:** 255–258.
60. SHARP, F.R., M. BUTMAN, J. KOISTINAHO, et al. 1994. Phencyclidine induction of the hsp70 stress gene in injured pyramidal neurons is mediated via multiple receptors and voltage gated calcium channels. Neuroscience **62:** 1079–1092.
61. SHARP, J.W., D.L. PETERSEN & M.T. LANGFORD. 1995. DNQX inhibits phencyclidine (PCP) and ketamine induction of the hsp70 heat shock gene in the rat cingulate and retrosplenial cortex. Brain Res. **687:** 114–124.
62. NAGATA, A., S. NAKAO, E. MIYAMOTO, et al. 1998. Propofol inhibits ketamine-induced c-fos expression in the rat posterior cingulate cortex. Anesth. Analg. **87:** 1416–1420.

Glutamatergic Animal Models of Schizophrenia

BITA MOGHADDAM AND MARK E. JACKSON

Department of Neuroscience, University of Pittsburgh, Pittsburgh, Pennsylvania 15260, USA

ABSTRACT: Several lines of evidence, including recent genetic linkage studies implicating susceptibility genes for schizophrenia, make a strong case that abnormal NMDA receptor–mediated neurotransmission is a major locus for the pathophysiology of schizophrenia. Animal models that are relevant to putative NMDA dysfunction in schizophrenia have excellent face validity for several symptoms of schizophrenia and are important tools for the design of novel pharmacological intervention in schizophrenia. The present chapter includes a brief review of the utility of these models and the search for new medications that have the potential of normalizing glutamate neurotransmission in schizophrenia.

KEYWORDS: NMDA receptors; schizophrenia; cognitions; dopamine; prefrontal cortex

INTRODUCTION

Schizophrenia remains the most morbid and debilitating of all psychiatric disorders. Current modes of therapy, which are based on the accidental discovery of neuroleptics 50 years ago, do not treat the disease but merely ameliorate the psychotic symptoms in a subgroup of patients. Because all known antipsychotic drugs block at least one type of receptor for the neurotransmitter dopamine, the focus of preclinical and clinical research in this field has been primarily on the dopaminergic system. However, several decades of research focused on the dopaminergic system have made it evident that drugs that specifically target dopamine receptors are not sufficient for treatment of schizophrenia. Specifically, while dopamine receptor blockers, such as haloperidol, are generally effective in reducing the so-called positive symptoms of this disease (e.g., psychosis, hallucinations, and paranoia), these medications generally fail to treat the affective and cognitive symptoms of this disorder. These latter symptoms contribute greatly to the long-term morbidity of schizophrenia and the inability of patients to function in society. Thus, there is an acute need for novel approaches to the pharmacotherapy of this disorder that will, we hope, lead to the design of treatments that target the emotional symptoms and cognitive dysfunctions that are associated with schizophrenia.

Address for correspondence: Bita Moghaddam, Department of Neuroscience, University of Pittsburgh, 446 Crawford Hall, Pittsburgh, PA 15260. Voice: 412-624-2653; fax: 412-624-9198.
bita@pitt.edu

Undoubtedly, development of more effective treatments depends on a better understanding of the underlying functional pathology of schizophrenia and the development of appropriate animal models with similar functional abnormalities so that novel mechanistic ideas and pharmacotherapeutic approaches can be tested. Because of the accumulating evidence that glutamatergic neurotransmission may be abnormal in schizophrenics (chapters by Tamminga and Coyle in this volume), there is growing emphasis on glutamatergic animal models of schizophrenia and their utility for testing novel pharmacological strategies for treatment of schizophrenia. In this chapter, recent studies with these models that have had an impact in our understanding of the disease process and novel directions for pharmacological treatments are reviewed.

THE NMDA ANTAGONIST ANIMAL MODEL OF SCHIZOPHRENIA

Systemic injection of the dissociative anesthetic phencyclidine (PCP) and its analogue ketamine produce a behavioral syndrome in non-schizophrenics that closely resembles endogenous symptoms of schizophrenia and that is frequently misdiagnosed as acute schizophrenia.[1–5] These include positive symptoms such as paranoia, agitation, auditory hallucinations; negative symptoms such as apathy, poverty of thought, and social withdrawal; and cognitive deficits such as impaired working memory.

Despite their multifaceted actions at anesthetic doses, lower doses of PCP and ketamine are thought to selectively produce noncompetitive blockade of NMDA receptors by binding to a site located in the ion channel associated with this receptor.[5] This pharmacological characteristic of ketamine and PCP is generally considered the major contributing factor to the psychotomimetic properties of these drugs and is consistent with the "glutamate deficiency hypothesis of schizophrenia," which was first suggested by Kim *et al.*[6] and later elaborated on by other investigators (e.g., see Refs. 7–10).

In laboratory animals PCP, ketamine, and other NMDA antagonists, such as MK801, produce a complex behavioral profile that is related to the clinical effect of these drugs in humans. In particular, they produce impaired cognitive functions, such as impaired performance in working memory tasks, as well as altered social behavior, hyperactivity, stereotypy, and sensory gating deficits.[11–15] Hence, these drugs are routinely used as pharmacological animal models of schizophrenia.

In general, the NMDA antagonist model has several advantages over other pharmacological models of schizophrenia. The most important advantage is that this is the only animal model that has a clinical parallel. Using modern diagnostic criteria this model has been extensively characterized in humans. Healthy individuals treated with low doses of ketamine exhibit transient negative and positive symptoms[4] and cognitive deficits similar to those reported in patients with schizophrenia.[16] There are also ongoing clinical studies with ketamine in several clinical laboratories, which is especially useful for translating treatment strategies from animal experiments to the clinic.

Another strength of this model is that, unlike other pharmacological models, such as amphetamine and cannabinoid models, prolonged use of ketamine or PCP is not necessary for expression of psychosis. The first reports of PCP psychosis were during its initial use as an anesthetic, and case reports of PCP-precipitated psychosis after recreational use often include individuals who thought they were consuming other substances and had not experienced PCP before.[17] Furthermore, in clinical trials, a single exposure to ketamine or PCP produces schizophrenic-like symptomatology in non-schizophrenics, which may last for several hours or days.[18] Another important aspect of the NMDA antagonist model is that PCP psychosis is not responsive to conventional antipsychotic therapy.[2,17] This is supported by findings that only selective behavioral effects of PCP and ketamine in basic and clinical studies are ameliorated with dopamine antagonists.[19,20] This makes the PCP model especially useful for strategies aimed at developing novel approaches for treatment of schizophrenia.

The schizophrenia-like effects of NMDA antagonists in healthy volunteers suggest that the primary glutamatergic abnormality in schizophrenia may be reduced NMDA receptor function. Basic studies, therefore, have focused mostly on enhancing NMDA receptor function through mechanisms that produce subtle positive modulation of this receptor. The most common approach has been to activate the glycine/D-serine modulatory site on the NMDA receptor. In animal studies glycine, D-serine, or glycine uptake blockers, which increase intrasynaptic levels of glycine, ameliorate some of the behavioral effects of PCP.[21] Limited clinical studies have complemented basic studies by demonstrating that oral administration of D-serine or exogenous ligands, such as D-cycloserine, may be useful for treatment of the cognitive symptoms of schizophrenia.[22,23] One issue with these studies has been that the ligands used either have poor absorption or are not highly specific. However, more selective ligands are under development and are being characterized in basic studies using the NMDA antagonist model.[21]

In our laboratory, we have observed that selective NMDA receptor antagonists, such as AP5, increase the efflux of glutamate.[24] This finding suggested that although this class of drug reduces glutamatergic neurotransmission at the NMDA receptor, this class of drug may also work to stimulate glutamatergic neurotransmission at non-NMDA glutamate receptors. Similar to AP5, a dose-dependent increase in glutamate efflux in response to subanesthetic doses of ketamine and PCP was observed.[25] In addition, recent electrophysiological studies in awake animals suggest that systemic administration of NMDA antagonists increases the spontaneous firing rate of neurons in the prefrontal cortex (FIG. 1). Furthermore, some of the cognitive and dopaminergic effects of these drugs in the rodent are reduced by pretreatment with antagonists of non-NMDA glutamate receptors. Specifically, (1) activation of cortical dopamine release elicited by ketamine and PCP is reduced by local and systemic administration of non-NMDA antagonists.[25] This is important because the hyperdopaminergic state produced by ketamine and PCP is thought to mediate some of the behavioral effects of PCP and ketamine. (2) The performance decrement produced by ketamine and PCP in a delayed alternation task is ameliorated by non-NMDA antagonist pretreatment.[25] Consistent with a hyperglutamatergic mechanism accounting for some of the effects of ketamine, clinical trials also indicate a dramatic reduction in memory-related deficits caused by ketamine in healthy individuals following pretreatment with lamotrigine, a glutamate release inhibitor.[26]

Collectively, these findings suggested that NMDA antagonists may produce some of their adverse behavioral effects by increasing cortical glutamatergic neurotransmission at non-NMDA receptors. Thus, reduction of this excess glutamatergic out-

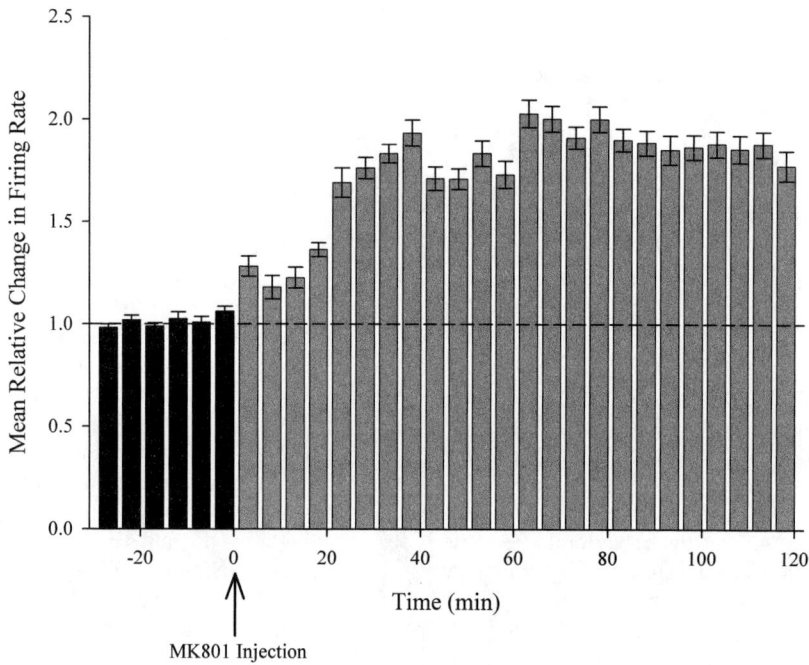

FIGURE 1. Mean firing rate increases of prefrontal cortex neurons after systemic injection of the NMDA antagonist MK801 (0.1 mg/kg). Data is the combined average of a total of 76 single units recorded from 6 different rats. Extracellular single unit recordings were obtained from awake animals using electrode arrays (eight microwires per array) chronically implanted into the prefrontal cortex. Changes in firing rate for individual units during a 2½ hour period were calculated by determining the average firing rate during 5-min bins and dividing by the average firing rate during the 30-min preinjection baseline period (*black bars*). Thus, a relative change of 1.0 (*dashed line*) indicates no change from baseline firing rates. Following systemic MK801 injection (*gray bars*), there is an immediate and sustained increase in PFC spontaneous firing rates, nearly doubling the average baseline firing rate. (Error bars = SEM).

put may have important clinical implications for the pharmacotherapy of cognitive deficits associated with schizophrenia. To test this mechanism, we pretreated animals with an agonist of group II metabotropic glutamate receptors (mGluR), which are localized presynaptically and have been shown to reduce the stimulated release of glutamate, presumably by presynaptic mechanisms (chapters by Conn and Schoepp in this volume). In these animals, we observed that activation of glutamate efflux by PCP was decreased. We also determined the effect of this drug in some of the behavioral disruptions produced by PCP, including activation of locomotor activity, stereotypical behavior, and disruption of working memory.[27] These basic studies have now been extended to clinical studies using the ketamine model of psychosis in healthy volunteers.[28] These studies so far suggest that pretreatment with a group II mGluR agonist diminishes some of the cognitive impairing effects of ketamine. Clinical trials in schizophrenia are forthcoming, but, regardless of whether this class of drugs is successful in treating symptoms of schizophrenia, the series of studies outlined above demonstrate the effectiveness of this model in providing novel therapeutic options for treatment of schizophrenia.

MUTANT MODELS

In addition to pharmacological models, recent mutant models have been introduced that may be relevant to putative genetic abnormalities of glutamate systems in patients with schizophrenia. Although only limited phenotypic characterization has been performed in these models, they are important models for understanding the processes that lead to the vulnerability to develop schizophrenia. One of the most interesting and clinically promising of these models are mice that are hypomorphic for neuregulin 1 (*NRG1*). Recent genome-wide scans of schizophrenia families in five distinct populations show that schizophrenia maps to chromosome 8p, and extensive fine-mapping of the 8p locus have identified *NRG1* as a candidate gene for schizophrenia.[29] *NRG1* is expressed at central nervous system synapses and has a clear role in the expression and activation of neurotransmitter receptors, in particular the NMDA receptors. Specifically, *NRG1* regulates expression of glutamate receptor subunits and directly activates ErbB4 receptors.[30] This member of the ErbB family of tyrosine kinases is colocalized with the NMDA receptor and is thought to regulate the kinetic properties of the NMDA receptor by phosphorylating the NR2 subunit of the NMDA receptor. Mutant mice heterozygous for either *NRG1* or its receptor, *ErbB4*, show a behavioral phenotype of impaired prepulse inhibition and hyperlocomotion that overlaps with the NMDA antagonist model of schizophrenia.[29] Furthermore, *NRG1* hypomorphs have fewer functional NMDA receptors than wild-type mice.

Another model with relevance to schizophrenia is the NMDAR1 (NR1) "knockdown" line of mutant mice.[31] These animals display exaggerated spontaneous locomotion and stereotypy, as well as deficits in social and sexual interactions. Although genetic or postmortem abnormalities in the NR1 subunit of NMDA receptors have not been found in schizophrenia, this is an interesting model that may add to our understanding of the long-term effects of congenital NMDA receptor hypofunction.

CONCLUSIONS

Recent lines of work suggest that abnormal NMDA receptor–mediated neurotransmission may contribute to the ontogeny and the pathophysiology of schizophrenia. Pharmacological and genetic animal models that are based on this mechanism appear to have excellent face validity for symptoms of schizophrenia and are likely to provide important tools for design of novel therapeutic options for schizophrenia.

REFERENCES

1. LUBY, E. *et al.* 1959. Study of a new schizophrenomimetic drug-sernyl. Am. Med. Assoc. Arch. Neurol. Psychiatry **81**: 363–369.

2. BURNS, R. & L.S. LERNER. 1976. Perspectives: acute phencyclidine intoxication. Clin. Toxicol. **9:** 477–501.
3. PEARLSON, G. 1981. Psychiatric and medical syndromes associated with phencyclidine (PCP) abuse. Johns Hopkins Med. J. **148:** 25–33.
4. KRYSTAL, J.H. *et al.* 1994. Subanesthetic effects of the noncompetitive NMDA antagonist, ketamine, in humans: psychotomimetic, perceptual, cognitive, and neuroendocrine responses. Arch. Gen. Psychiatry **51:** 199–214.
5. JAVITT, D.C. & S.R. ZUKIN. 1991. Recent advances in the phencyclidine model of schizophrenia. Am. J. Psychiatry **148:** 1301–1308.
6. KIM, J. *et al.* 1980. Low cerebrospinal fluid glutamate in schizophrenic patients and a new hypothesis on schizophrenia. Neurosci. Lett. **20:** 379–382.
7. ULAS J., C.W. COTMAN. 1993. Excitatory amino acid receptors in schizophrenia. Schizophr. Bull. **19:** 105–117.
8. OLNEY, J. & N. FARBER. 1995. Glutamate receptor dysfunction and schizophrenia. Arch. Gen. Psychiatry **52:** 998–1007.
9. COYLE, J. 1996. The glutamatergic dysfunction hypothesis for schizophrenia. Harvard Rev. Psychiatry **3:** 241–253.
10. TAMMINGA, C.A. 1998. Schizophrenia and glutamatergic transmission. Crit. Rev. Neurobiol. **12:** 21–36.
11. GREENBURG, B.D. & D.S. SEGAL. 1985. Acute and chronic behavioral interactions between phencyclidine (PCP) and amphetamine: evidence for a dopaminergic role in some PCP-induced behaviors. Pharmacol. Biochem. Behav. **23:** 99–105.
12. SCHMIDT, H.W. *et al.* 1989. Effects of intrastriatal blockade of glutamatergic transmission on the acquisition of T-maze and radial maze tasks. J. Neural Transm. **78:** 29–41.
13. STEINPREIS, R. *et al.* 1994. The effects of haloperidol and clozapine on PCP and amphetamine induced suppression of social behavior. Pharmacol. Biochem. Behav. **47:** 579–585.
14. VERMA, A. & B. MOGHADDAM. 1996. NMDA receptor antagonists impair prefrontal cortex function as assessed via spatial delayed alternation performance in rats: modulation by dopamine. J. Neurosci. **16:** 373–279.
15. MANSBACH, R.M. & M. GEYER. 1989. Effects of phencylidine and phecylidine biologs on sensorimotor gating in the rat. Neuropsychopharmacology **2:** 299–308.
16. ADLER, C.M. *et al.* 1999. Comparison of ketamine-induced thought disorder in healthy volunteers and thought disorder in schizophrenia. Am. J. Psychiatry **156:** 1646–1649.
17. RAINEY, J.J. & M. CROWDER. 1975. Prolonged psychosis attributed to phencyclidine: report of three cases. Am. J. Psychiatry **10:** 1076–1078.
18. BAKKER, C.B. & F.B. AMINI. 1961. Observations on the psychotomimetic effects of sernyl. Comp. Psychiatry **2:** 269–280.
19. KEITH, V.A., R.S. MANSBACH & M.A. GEYER. 1991. Failure of haloperidol to block the effects of phencyclidine and dizocilpine on prepulse inhibition of startle. Biol. Psychiatry **30:** 557–566.
20. KRYSTAL, J. *et al.* 1995. Modulating ketamine-induced thought disorder with lorazepam and haloperidol in humans. Schizophr. Res. **15:** 156a.
21. JAVITT, D.C. 2002. Glycine modulators in schizophrenia. Curr. Opin. Investig. Drugs **3:** 1067–1072.
22. GOFF, D.C. *et al.* 1999. A placebo-controlled trial of D-cycloserine added to conventional neuroleptics in patients with schizophrenia. [see comments]. Arch. Gen. Psychiatry **56:** 21–27.
23. HERESCO-LEVY, U. *et al.* 2002. Placebo-controlled trial of D-cycloserine added to conventional neuroleptics, olanzapine, or risperidone in schizophrenia. Am. J. Psychiatry **159:** 480–482.
24. LIU, J. & B. MOGHADDAM. 1995. Regulation of glutamate efflux by excitatory amino acid receptors: evidence for tonic inhibitory and phasic excitatory regulation. J. Pharmacol. Exp. Ther. **274:** 1209–1215.
25. MOGHADDAM, B. *et al.* 1997. Activation of glutamatergic neurotransmission by ketamine: a novel step in the pathway from NMDA receptor blockade to dopaminergic

and cognitive disruptions associated with the prefrontal cortex. J. Neurosci. **17:** 2921–2927.
26. ANAND, A. *et al.* 2000. Attenuation of the neuropsychiatric effects of ketamine with lamotrigine: support for hyperglutamatergic effects of N-methyl-D-aspartate receptor antagonists. Arch. Gen. Psychiatry **57:** 270–276.
27. MOGHADDAM, B. & B. ADAMS. 1998. Reversal of phencyclidine effects by a group II metabotropic glutamate receptor agonist in rats. Science **281:** 1349–1352.
28. Krystal, J.H., W. Abi-Saab, E. Perry, *et al.* 2003. Interaction of ketamine and an mGluR2/3 agonist in healthy volunteers. Biol. Psychiatry **53:** 312.
29. STEFANSSON, H. *et al.* 2002. Neuregulin 1 and susceptibility to schizophrenia. Am. J. Hum. Genet. **71:** 877–892.
30. OZAKI, M. *et al.* 1997. Neuregulin-beta induces expression of an NMDA-receptor subunit. Nature **390:** 691–694.
31. MOHN, A.R. *et al.* 1999. Mice with reduced NMDA receptor expression display behaviors related to schizophrenia. Cell **98:** 427–436.

Glutamate, Dopamine, and Schizophrenia

From Pathophysiology to Treatment

MARC LARUELLE, LAWRENCE S. KEGELES, AND ANISSA ABI-DARGHAM

Departments of Psychiatry and Radiology,
Columbia University College of Physicians and Surgeons,
New York, New York 10032, USA

ABSTRACT: The fundamental pathological process(es) associated with schizophrenia remain(s) uncertain, but multiple lines of evidence suggest that this condition is associated with (1) excessive stimulation of striatal dopamine (DA) D_2 receptors, (2) deficient stimulation of prefrontal DA D_1 receptors and, (3) alterations in prefrontal connectivity involving glutamate (GLU) transmission at N-methyl-D-aspartate (NMDA) receptors. This chapter first briefly discusses the current knowledge status for these abnormalities, with emphasis on results derived from clinical molecular imaging studies. The evidence for hyperstimulation of striatal D_2 receptors rests on strong pharmacological evidence and has recently received support from brain imaging studies. The hypothesis of deficient prefrontal cortex (PFC) D_1 receptor stimulation is almost entirely derived from preclinical studies. Preliminary imaging data compatible with this hypothesis have recently emerged. The NMDA hypofunction hypothesis originates mainly from indirect pharmacological data. The interactions between DA and GLU systems relevant to schizophrenia are then reviewed. Animal and imaging data supporting the general model that the putative DA imbalance in schizophrenia (striatal excess and cortical deficiency) might be secondary to NMDA hypofunction in the PFC and its connections are presented. Equally important are the potential consequences of this DA imbalance for NMDA function in the striatum and the cortex, which are subsequently discussed. In conclusion, it is proposed that schizophrenia is associated with strongly interconnected abnormalities of GLU and DA transmission: NMDA hypofunction in the PFC and its connections might generate a pattern of dysregulation of DA systems that, in turn, further weakens NMDA-mediated connectivity and plasticity.

KEYWORDS: dopamine; glutamate; schizophrenia; PET; SPECT

INTRODUCTION

Schizophrenia is a severe and chronic mental illness (or group of illnesses), associated with high prevalence (about 0.5–1% of the population suffers from this con-

---

Address for correspondence: Marc Laruelle, M.D., Associate Professor of Psychiatry and Radiology, Columbia University College of Physicians and Surgeons, New York State Psychiatric Institute, Unit 31, 1051 Riverside Drive, New York, NY 10032. Voice: 212-543-5388; fax: 212-568-6171.
ml393@columbia.edu

dition). Symptoms of schizophrenia typically emerge during adolescence or early adulthood. Psychotic symptoms include hallucinations, typically auditory, and delusions, which frequently involve persecution and/or megalomania. Psychotic symptoms and severe thought disorganization are often grouped under the term positive symptoms. Deficit symptoms, also commonly referred to as negative symptoms, manifest in many dimensions, such as affect (affect flattening), volition (apathy), speech (poverty of speech), pleasure (anhedonia), and social life (withdrawal). While the etiology and fundamental pathology of schizophrenia remain unclear, a large body of evidence suggests that alterations in several neurotransmitter systems are involved in the pathophysiological processes leading to the formation of these symptoms. Among these, the dopamine (DA) and glutamate (GLU) systems have received most attention, although other systems such as GABAergic, serotonergic, cholinergic, or opioid systems have also been implicated.

The putative role of DA systems in the pathophysiology and treatment of schizophrenia has been the subject of intense research efforts over the last fifty years. The first formulation of the DA hypothesis of schizophrenia proposed that hyperactivity of DA transmission was responsible for the positive symptoms observed in this disorder.[1] This hypothesis was based on the correlation between clinical doses of antipsychotic drugs and their potency to block DA D_2 receptors[2,3] and the psychotogenic effects of DA enhancing drugs.[4,5] Given the predominant localization of DA terminals and D_2 receptors in subcortical regions such as the striatum and the nucleus accumbens, the classical DA hypothesis of schizophrenia was concerned mostly with these subcortical regions.

More recently, increasing awareness of the importance of enduring negative and cognitive symptoms in this illness and their resistance to D_2 receptor antagonism has led to a reformulation of this classical DA hypothesis. Functional brain imaging studies have suggested that these symptoms might arise from altered prefrontal cortex (PFC) function.[6] A wealth of preclinical studies have emerged documenting the importance of prefrontal DA transmission at D_1 receptors (the main DA receptor in the neocortex) for optimal PFC performance.[7] These observations have led to the hypothesis that a deficit in DA transmission at D_1 receptors in the PFC might be implicated in the cognitive impairments and negative symptoms of schizophrenia.[8,9]

Besides DA, several lines of evidence support the hypothesis that schizophrenia might be associated with a persistent dysfunction of GLU transmission involving N-methyl-D-aspartate (NMDA) receptors.[10,11–14] Recently, preclinical data[10,15] and further receptor imaging studies in healthy humans[16] have converged to suggest that dysregulation of DA systems in schizophrenia may be secondary to a deficit in function of the GLU NMDA receptor. Furthermore, increased understanding of the modulatory role of DA on GLU transmission have started to unravel the mechanisms by which alterations of DA function in schizophrenia might in turn affect NMDA transmission.

In this paper, we briefly review recent imaging evidence suggesting that schizophrenia is associated with dysregulation of DA function, as well as imaging experiments supporting the hypothesis that this dysregulation might be secondary to NMDA dysfunction. Following presentation of the data, a general model accounting for these observations is presented. Finally, we speculate on the consequences of this DA dysregulation for NMDA-mediated transmission in corticostriatal circuits.

DOPAMINE AND GLUTAMATE FUNCTION IN SCHIZOPHRENIA: CURRENT STATUS

Striatal DA Function in Schizophrenia

Recent imaging studies documented the existence of dysregulation of striatal DA function in schizophrenia. Six studies reported rates of DOPA decarboxylase in patients with schizophrenia, using [^{18}F]DOPA[17–21] or [^{11}C]DOPA.[22] Five out of six studies reported increased accumulation of DOPA in the striatum of patients with schizophrenia. Several studies reported high DOPA accumulation in psychotic paranoid patients. These observations are compatible with higher DA synthesis activity in patients experiencing psychotic symptoms. However, since the relationship between DOPA decarboxylase and DA synthesis rate is unclear (DOPA decarboxylase is not the rate-limiting step of DA synthesis), a more direct measurement of DA synaptic output was required to further evaluate this issue.

Several groups recently demonstrated that, under specific conditions, *in vivo* neuroreceptor binding techniques can also be used to measure acute fluctuations in the concentration of endogenous transmitters in the vicinity of radiolabeled receptors.[23] Competition between radiotracers and transmitters for binding to neuroreceptors is the principle underlying this approach, although other mechanisms such as agonist-induced receptor internalization might also play a role.[23] So far, applications of this new paradigm have been developed mainly to study DA transmission at D_2 receptors. The amphetamine-induced reduction in [^{123}I]iodobenzamide (IBZM)- or

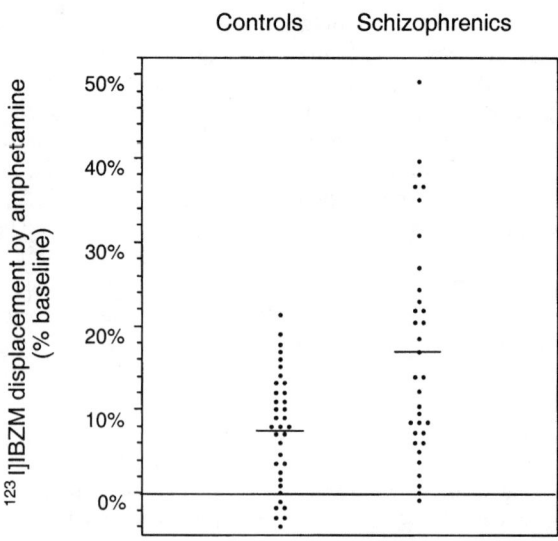

FIGURE 1. Effect of amphetamine (0.3 mg/kg) on [^{123}I]IBZM binding in healthy controls and untreated patients with schizophrenia. The *y* axis shows the percentage decrease in [^{123}I]IBZM binding potential induced by amphetamine, which is a measure of the increased occupancy of D_2 receptors by DA following the challenge.

[^{11}C]raclopride-binding potential (BP) has been well validated as an indirect measure of the changes in synaptic DA concentration induced by the challenge.[24–28]

Several studies reported that amphetamine-induced DA release is increased in patients with schizophrenia compared to matched healthy controls.[25,29–31] In our sample, the amphetamine-induced (0.3 mg/kg, i.v.) reduction in [^{123}I]IBZM BP was 7.5 ± 7.1% in control subjects ($N = 34$) and 17.1 ± 13.2% in patients with schizophrenia $N = 34$, $P < .001$, FIG. 1), a more than twofold greater response in the patients. A similar finding was reported by Breier and colleagues[25] using [^{11}C]raclopride, PET, and a smaller dose of amphetamine (0.2 mg/kg, i.v.). Providing that the affinity of D_2 receptors for DA is unchanged in this illness,[32] these data are consistent with increased amphetamine-induced DA release in schizophrenia.

The amphetamine effect on [^{123}I]IBZM BP was similar between chronic/previously treated patients and first episode/neuroleptic-naive patients, and both groups were significantly different from controls. In the previously treated group, no association was found between the duration of the neuroleptic-free period and the amphetamine-induced [^{123}I]IBZM displacement. Together, these data indicated that the exaggerated dopaminergic response to amphetamine exposure was not a prolonged side effect of previous neuroleptic exposure.

In patients with schizophrenia, the amphetamine challenge induced a significant increase in positive symptoms. The emergence or worsening of positive symptoms was transient, and patients returned to their baseline symptomatology within a few hours of the challenge. This observation provided evidence that exaggerated activation of DA transmission at D_2 receptors mediates the expression of psychotic symptoms following amphetamine challenge.

An important question raised by these studies is whether the stress associated with psychiatric hospitalization and/or the scanning procedure might account for the excess DA release measured in patients with schizophrenia, since stress activates DA release.[33,34] To investigate this issue, we studied amphetamine-induced DA release in a group of nonpsychotic unipolar depressed subjects.[35] Despite reporting preamphetamine anxiety levels higher than schizophrenic patients, patients with depression showed normal amphetamine-induced displacement of [^{123}I]IBZM. This finding supports the hypothesis that the increased amphetamine effect observed in patients with schizophrenia was not a nonspecific consequence of stressful conditions (although it could represent a specific interaction between stress and schizophrenia).

A major limitation of the amphetamine studies is that they measured changes in synaptic DA transmission following a nonphysiological challenge (i.e., amphetamine) and did not provide any information about "baseline" synaptic DA levels, that is, synaptic DA levels in the absence of pharmacological interventions. Using an acute DA depletion strategy, baseline occupancy of striatal D_2 receptors by DA was studied in acute patients with schizophrenia.[36] The results of this study suggested that DA occupies a greater proportion of striatal D_2 receptors in patients with schizophrenia compared to matched control subjects during first episode of illness and subsequent episodes of illness exacerbation. High synaptic level of DA at baseline was significantly associated with greater improvement of positive symptoms following six weeks of antipsychotic treatment.

In summary, imaging studies performed in the last decade have generated data consistent with the existence of dysregulated striatal DA function leading to hyperstimulation of D_2 receptors in schizophrenia. This dysregulation appears to be involved in both the pathogenesis and the treatment of positive symptoms.

Prefrontal DA Function in Schizophrenia

Indirect evidence supports the hypothesis that a deficit in prefrontal DA function might contribute to prefrontal impairment in schizophrenia. Clinical studies have suggested a relationship between low cerebro-spinal fluid homovanillic acid, a measure reflecting low DA activity in the prefrontal cortex, and poor performance at tasks involving working memory (WM) in schizophrenia.[37,38] Administration of DA agonists might have beneficial effects on the pattern of prefrontal activation measured with PET during these tasks.[39,40] More direct evidence for such a deficit was recently provided by one postmortem study suggesting a decrease in DA innervation in the dorsolateral prefrontal cortex (DLPFC),[41] but this result has not been replicated yet.

The only index of prefrontal DA transmission currently quantifiable with noninvasive *in vivo* imaging is D_1 receptor availability. The first PET radiotracer for the D_1 receptor to be introduced was the benzazepine [^{11}C]SCH 23390.[42] While SCH 23390 displayed relatively low selectivity toward 5-HT$_{2A}$ receptors *in vitro*,[43] studies in mice suggested that the *in vivo* binding of [^{11}C]SCH 23390 is selective for D_1 receptors, even in the PFC.[44] Yet, this selectivity has still to be demonstrated in primates. More recently, [^{11}C]NNC 112 was developed as a superior PET D_1 receptor radiotracer.[45,46] In humans, [^{11}C]NNC 112 provides higher specific to nonspecific ratios compared to [^{11}C]SCH 23390[46,47] and exquisite visualization of D_1 receptors, even in the neocortex. The *in vivo* selectivity of [^{11}C]NNC 112 has been demonstrated in monkeys,[46] and the reproducibility of measurement of [^{11}C]NNC 112 BP in the human PFC has been established.[48]

Three PET studies of prefrontal D_1 receptor availability in patients with schizophrenia have recently been published. Two studies were performed with [^{11}C]SCH 23390. The first reported decreased [^{11}C]SCH 23390 BP in the PFC,[49] and the other reported no change.[50] One study was performed with [^{11}C]NNC 112,[51] and reported increased [^{11}C]NNC 112 BP in the DLPFC and no change in other regions of the prefrontal cortex, such as the medial prefrontal cortex (MPFC) or the orbitofrontal cortex. In patients with schizophrenia, upregulated [^{11}C]NNC 112 BP was predictive of poor performance on a WM task (FIG. 2).

Many potential factors, including patient heterogeneity and differences in the boundaries of the sampled regions, might account for the discrepancies between the three imaging studies. Because of the prevalent view that schizophrenia is associated with a deficit in prefrontal DA activity, the impact of chronic DA depletion on the *in vivo* binding of [^{11}C]SCH 23390 and [^{11}C]NNC 112 was studied in rodents.[52] Chronic DA depletion was associated with increased *in vivo* PFC [^{11}C]NNC 112 binding, presumably reflecting a compensatory upregulation of D_1 receptors. Interestingly, chronic DA depletion did not result in enhanced *in vivo* binding of [^{3}H]SCH 23390 in the PFC, suggesting that this radiotracer is not sensitive to the effect of chronic DA depletion in the PFC. Thus, the increase in DLPFC [^{11}C]NNC 112 BP observed in schizophrenia might be related to a compensatory but inefficient upregulation of D_1 receptors following sustained DA deficit, and it is conceivable that

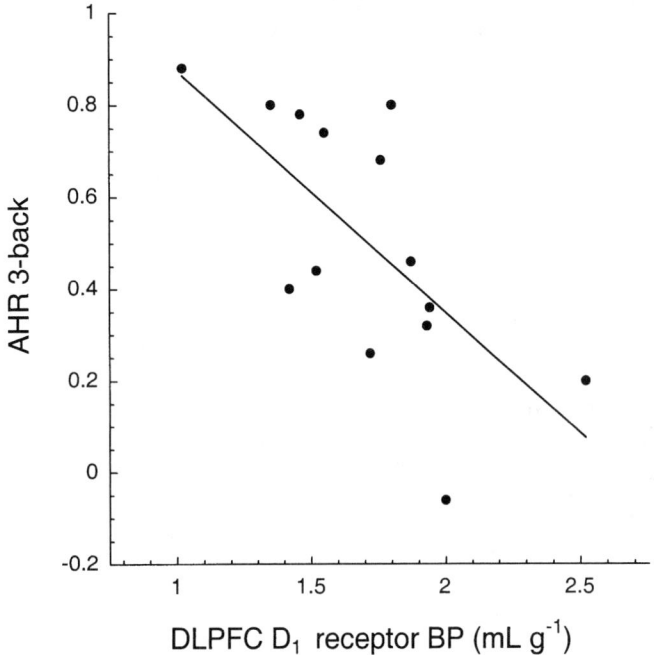

FIGURE 2. Relationship between upregulation of D_1 receptors in the DLPFC of untreated patients with schizophrenia and performance at WM task (3-back adjusted hit rate or AHR, lower values represent poorer performance).

such an upregulation might not be detectable with [^{11}C]SCH 23390. Studies with both radiotracers in the same patients are required to clarify this issue.

In conclusion, several indirect lines of evidence, including a recent PET study with [^{11}C]NNC 112,[51] are consistent with the hypothesis that, in schizophrenia, a deficit in prefrontal DA activity at D_1 receptors might contribute to the cognitive problems presented by these patients. Furthermore, a cortical DA deficit might contribute to the disinhibition of subcortical DA function, to the extent that the mesocortical DA system has an inhibitory effect on subcortical DA function.[9,21,33,53,54]

NMDA Hypofunction in Schizophrenia

Several lines of evidence suggest that schizophrenia might be associated with a persistent dysfunction of GLU transmission involving NMDA receptors.[10,11–14] Noncompetitive NMDA antagonists, such as phencyclidine (PCP) or ketamine, induce both positive and negative symptoms in healthy and schizophrenic subjects.[55,56] Unmedicated patients with schizophrenia are more sensitive than normal volunteers to the effects of NMDA antagonists.[57] In addition, adjunctive treatment with NMDA agonists, such as glycine,[58–60] D-cycloserine,[61–63] and D-serine,[64] might provide a modest symptom improvement in schizophrenia.[65]

However, direct evidence for NMDA dysfunction in schizophrenia is still lacking. Postmortem studies have detected a large number of abnormalities in the expression of GLU-related proteins, but few of these observations have been independently replicated, and a coherent picture has not yet emerged.[66–68] In addition, the lack of adequate radioligands to visualize the GLU system in the living brain is a major impediment to testing the NMDA hypofunction hypothesis of schizophrenia.

The fact that administrations of NMDA antagonists induce, in animals as well as in humans, a constellation of effects reminiscent of schizophrenia does not, per se, constitute evidence that NMDA function is impaired in schizophrenia. An alternate hypothesis is that schizophrenia might be associated with alterations in synaptic transmission not primarily due to NMDA hypofunction, but to alterations in other components of the synaptic machinery.[69] Due to the role of transmission in shaping synaptic contacts and organization (both during development and adulthood), synaptic dysfunction would result in abnormal connectivity. The consequences of this altered connectivity would be well modeled by blockade of NMDA transmission in an otherwise healthy brain. Under this proposition, administration of an NMDA antagonist, by inducing "dysconnectivity," might constitute an effective pharmacological intervention to mimic the illness process, and, as such, a very important research tool.

Another important point related to the NMDA hypofunction model is the difference between the effects of acute and chronic NMDA antagonist exposure, a point discussed in detail by Jentsch and Roth.[14] Schizophrenia being a chronic illness, alteration of NMDA transmission (or of synaptic connectivity modeled by NMDA hypofunction) is likely to be an enduring condition. This factor limits the relevance of observations made upon acute administration of NMDA antagonists. Thus, animal models and human models (such as chronic phencyclidine abusers) that involve sustained NMDA hypofunction might generate more informative data for schizophrenia compared to acute models.

In this context, the effects of sustained NMDA hypofunction on DA activity are particularly interesting. Long-term administration of NMDA antagonists in animal studies shows that sustained disruption of NMDA transmission induces alterations in DA transmission (reduction in mesocortical DA activity and excess subcortical reactivity) that are consistent with the abnormalities postulated by the DA imbalance hypothesis of schizophrenia.[14] Thus, both abnormalities in DA transmission in schizophrenia (cortical DA deficit and subcortical DA hyperactivity) might be related to a persistent alteration in synaptic connectivity, probably involving the PFC, and well modeled by a sustained dysfunction in NMDA transmission in the PFC.

GLUTAMATE–DOPAMINE INTERACTIONS

A Neuronal Circuitry Model of GLU–DA Interactions

The activity of DA neurons is modulated by projections involving GLU transmission from the PFC and other areas, such as the amygdala. A general model for GLU modulation of DA neurons in the substantia nigra (SN)/ventral tegmental area (VTA) has been introduced by Carlsson and collaborators.[16,70] This model provides an anatomical framework relating three fundamental putative neurochemical dysregula-

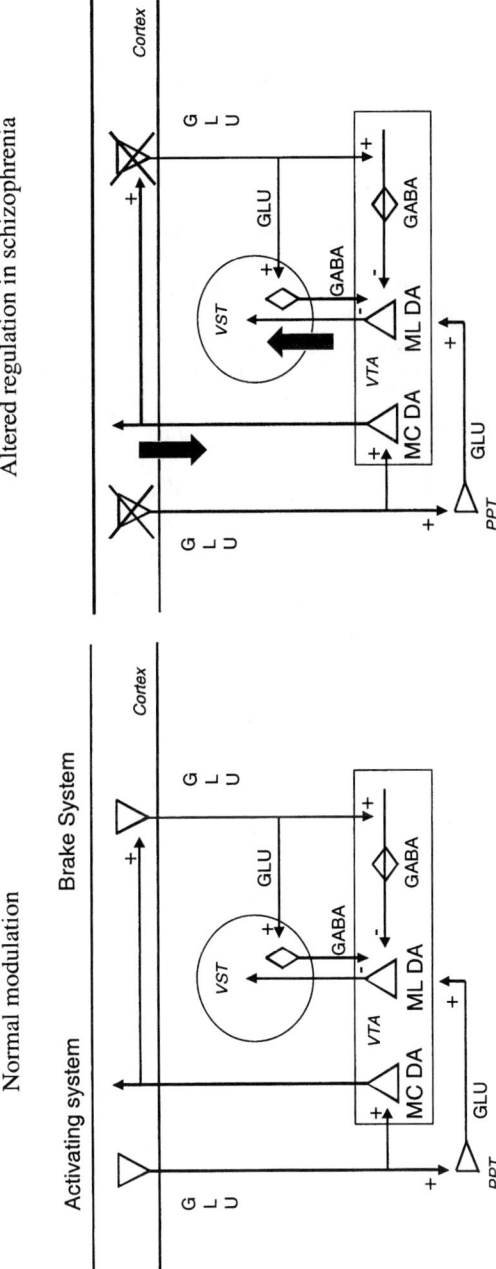

FIGURE 3. (A) Proposed model of modulation of DA cell activity by cortical projections. This model, adapted from Carlsson and colleagues,[70] proposes a bimodal modulation of DA activity in the ventral tegmental area (VTA) by glutamatergic (GLU) projections originating in the frontal cortex. Stimulation of VTA DA neurons by GLU projections is represented on the left ("activating system"). These neurons exert a tonic excitatory influence on DA activity. Evidence in rodents suggests that direct stimulation of DA neurons by GLU afferents from the PFC is restricted to DA neurons that project back to the cortex (mesocortical DA system, MC DA). Stimulation of mesolimbic (ML) DA neurons by GLU afferents from the cortex is probably polysynaptic, possibly involving a relay in the pedunculopontine tegmentum. The "brake system," represented on the right, exerts an inhibitory influence on DA activity via NMDA receptor-mediated stimulation of VTA GABAergic interneurons or striatotegmental GABA neurons, and comes predominantly into play when DA activity is increased (such as stress). In addition, the brake system regulating ML DA activity is activated by MC DA projections. **(B)** This model predicts that a deficiency in NMDA transmission in the cortex would result in decreased MC DA activity and would have unpredictable effects on ML DA activity under "baseline" conditions. Yet, it would result in an increase in stress (or amphetamine)-induced ML DA release. See text for references.

tions involved in the pathophysiology of schizophrenia, namely: (1) a deficit in GLU transmission; (2) a deficit in cortical DA transmission, and; (3) a dysregulation of striatal DA transmission.

According to this model, the PFC modulates the activity of midbrain DA neurons via both an activating pathway (the "accelerator") and an inhibitory pathway (the "brake"), allowing fine tuning of dopaminergic activity by the PFC (FIG. 3). The activating pathway is provided by direct and indirect glutamatergic projections onto the dopaminergic cells. A recent electron microscopy study in rodents suggests that direct stimulation of DA neurons by prefrontal afferents is restricted to DA neurons that project back to the cortex, while a polysynaptic mechanism, possibly involving the pedunculopontine tegmentum, mediates the prefrontal activation of mesolimbic DA neurons.[71] The inhibitory pathway is provided by PFC glutamatergic efferents to midbrain GABAergic interneurons and striatomesencephalic GABA neurons. The model of dual modulation of the mesolimbic DA system by the PFC has been recently confirmed by studies demonstrating that extracellular DA concentration in the accumbens is decreased or increased following low- or high-frequency PFC stimulation, respectively.[72] Furthermore, blockade of GLU transmission in the VTA increases DA release in the accumbens and decreases DA release in the PFC.[73] This observation demonstrates a GLU-mediated tonic inhibitory regulation of mesoaccumbens neurons and a tonic excitatory regulation of mesoprefrontal DA neurons.

In schizophrenia, a reduced prefrontal activity, possibly secondary to NMDA transmission deficiency, could result in a reduction of mesocortical DA activity (further worsening prefrontal related cognitive impairment), and, under conditions of stress (such as stimulation of the DA system by the amygdala), failure of the PFC to properly regulate DA activity in subcortical regions (FIG. 3). If sustained, this dysregulation of mesolimbic DA might precipitate positive symptoms. The scheme presented in FIGURE 3 encompasses only a limited aspect of GLU-DA interactions, leaving out interactions at the level of terminals and at the intracellular level. Nonetheless, it provides a general organizing principle and generates testable hypotheses. As this model is mainly derived from rodent studies, its relevance to humans remains to be ascertained. For example, evidence of glutamatergic projections from the PFC to midbrain DA cell bodies is still lacking in primates.

We recently undertook a number of imaging studies in humans to explore the validity of several predictions from this model. Specifically, we were interested in measuring the effects of NMDA blockade on subcortical DA release.

Imaging Studies of GLU–DA Interactions

Acute NMDA antagonist administration has little effect on striatal extracellular DA concentration. Preclinical measurements of striatal DA release with microdialysis under conditions of acute NMDA receptor blockade showed a small effect in rodents[15] and an absence of any detectable effect of PCP or ketamine on extracellular striatal DA levels in awake rhesus monkeys.[74]

We used PET and [^{11}C]raclopride to examine the effects of NMDA receptor blockade by ketamine[75] on striatal DA activity in healthy human subjects. No significant differences were observed between D_2 receptor availability measured with [^{11}C]raclopride prior to and during ketamine infusion (FIG. 4). Similarly, we failed to observe any effects of ketamine on the *in vivo* binding of the SPECT D_2 receptor

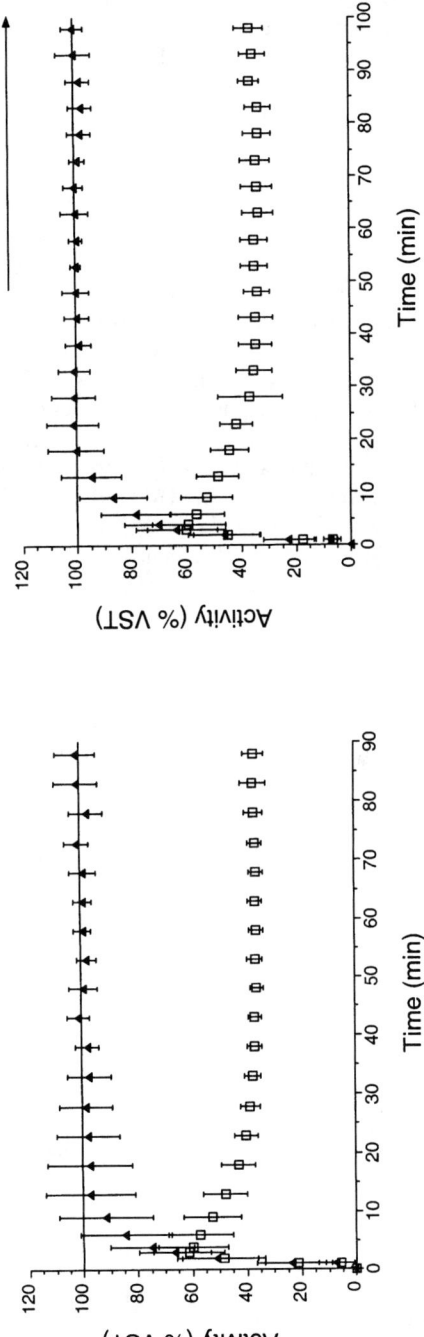

FIGURE 4. Activities in the ventral striatum (VST, *closed triangles*) and cerebellum (*open squares*) during control experiments (Mean ± SD, $N = 10$, *left panel*) and ketamine experiments ($N = 10$, *right panel*). For each subject, activities were expressed in percentage of the average VST activity during the equilibrium interval (30–90 min). Ketamine administration, initiated at 50 min (bolus of 0.12 mg/kg, followed by 0.65 mg/kg/h, *arrow*) did not induce detectable effects on regional [^{11}C]raclopride concentrations in the VST.

radioligand [^{123}I]IBZM in healthy subjects.[16] The results of these studies, which have recently been confirmed,[76] are consistent with the rodent and primate microdialysis studies cited above, and are compatible with the Carlsson model. However, these results contrast with the results of three earlier PET studies that found a significant reduction in D_2 receptor availability following ketamine.[77–79] Kegeles and colleagues[75] discuss these discrepant results.

In the rodent microdialysis study cited above,[15] the modulation of amphetamine-stimulated striatal DA release by acute NMDA receptor blockade was also examined. In contrast with NMDA-receptor blockade alone, these measurements showed not only a large effect of amphetamine alone, but a significant, roughly twofold further enhancement of this effect by acute NMDA receptor blockade.

The inhibition of dopaminergic cell firing following amphetamine is an important feedback mechanism by which the brain reduces the effect of amphetamine on DA release. The inhibition of dopaminergic cell firing induced by amphetamine is mediated both by stimulation of presynaptic D_2 autoreceptors, as well as by stimulation of the inhibitory pathway of the Carlsson model.[70,80] Following administration of amphetamine (i.e., under conditions in which the inhibitory pathway should be activated), NMDA receptor blockade is expected to result in an impairment of activation of the inhibitory pathway and therefore an exaggerated amphetamine-induced DA release, as seen in the rodent study.[15] We recently confirmed this mechanism in humans.[16] Amphetamine-induced decrease in [^{123}I]IBZM binding changed from 5.5 ±

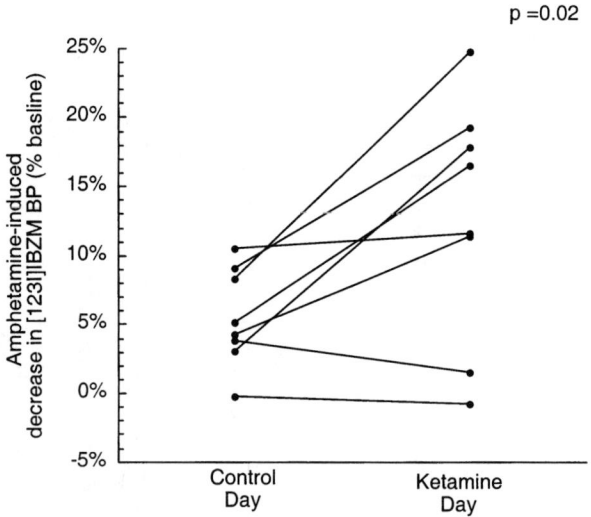

FIGURE 5. Ketamine modulation of striatal amphetamine-induced dopamine release in healthy volunteers, showing a significantly larger release in 8 subjects pretreated with intravenous ketamine compared to the same subjects under control conditions, that is, amphetamine alone (repeated measures ANOVA, $P = .023$). These data indicate that, in humans, amphetamine-induced DA release in the striatum is modulated by glutamatergic circuits involving NMDA transmission.

3.5% under control conditions (amphetamine alone, 0.25 mg/kg) to 12.8 ± 8.8% under conditions of NMDA blockade, induced by ketamine, $P = .023$ (FIG. 5).

The increase in amphetamine-induced DA release induced by ketamine (greater than twofold, FIG. 5) was comparable in magnitude to the exaggerated response to amphetamine alone seen in patients with schizophrenia (FIG. 1). These data are consistent with the hypothesis that the abnormally elevated DA release revealed by the amphetamine challenge in schizophrenia results from a disruption of glutamatergic neuronal systems regulating dopaminergic cell activity.

While our study showed an enhancement of amphetamine-induced striatal DA release by pretreatment with a noncompetitive NMDA receptor antagonist, an ambiguity persisted at the level of the interpretation of these data. The net effect of acute ketamine administration on GLU transmission is complex. Ketamine blocks NMDA receptors, but also induces GLU release, resulting in stimulation of other GLU receptors.[81,82] Therefore, it was unclear if the enhancement of amphetamine effects induced by ketamine was ultimately due to a deficit or an excess of GLU transmission.

Studies were performed in baboons with the metabotropic GLU (mGLU) receptor 2/3 agonist LY354740 to resolve this issue. To date, eight subtypes of mGLU receptor have been identified and classified into three subgroups.[83–85] Group II includes two subtypes, mGlu2 and mGlu3 receptors, which are located mainly perisynaptically. Activation of mGlu2/3 receptors reduces GLU release,[86,87] whereas mGlu2/3 receptor antagonists amplify the elicited release of GLU.[88] LY354740 is a highly selective agonist at mGlu2/3 receptors.[89] We investigated the effects of LY354740 on amphetamine-induced DA release using PET and the [^{11}C]raclopride displacement paradigm in baboons.[90] The amphetamine-induced decrease in [^{11}C]raclopride binding was measured under control conditions and following pretreatment with the mGlu2/3 receptor agonist LY354740 (20 mg/kg i.v.) in four baboons. Amphetamine (0.5 mg/kg i.v.) reduced [^{11}C]raclopride BP by 28 ± 7% under control conditions. Following LY354740 pretreatment, amphetamine-induced reduction in [^{11}C]raclopride BP was significantly enhanced (35 ± 7%, $P = .002$). The enhancement of the amphetamine-induced reduction in [^{11}C]raclopride BP by LY354740 was not a simple additive effect, as LY354740 alone did not reduce [^{11}C]raclopride BP. In contrast to ketamine, which blocks NMDA receptors but indirectly activates AMPA and kainate receptors, the effects of LY354740 on GLU transmission are not ambiguous. Therefore, this study clarified that inhibition of GLU transmission increases amphetamine-induced DA release, and provided additional support to the hypothesis that the dysregulation of DA function revealed by the amphetamine challenge in schizophrenia might stem from a deficit in GLU transmission.

Since the pharmacological model used in these studies involved acute administration of NMDA receptor antagonists, it is limited to an acute deficit of NMDA function, in contrast with the chronic abnormalities of neurotransmitter function in schizophrenia itself. This issue should be further addressed by studying human subjects who on a chronic or habitual basis self-administer agents such as phencyclidine or ketamine.

In summary, recent imaging data support the hypothesis that a deficit in GLU transmission might be involved in the alterations of DA regulation observed in patients with schizophrenia. Thus, alteration of DA function in schizophrenia might be secondary to NMDA dysfunction. However, one must also consider the impact of DA function on NMDA transmission to appreciate the potential complexity of these interactions in schizophrenia.

DOPAMINE–GLUTAMATE INTERACTIONS

DA–GLU Interaction in the Striatum

Cortical glutamatergic afferents and DA projections converge on GABAergic medium spiny neurons in the striatum, usually on dendritic shafts and spines.[91–93] At this convergence point, DA has potent modulatory effects on GLU transmission.[68,94,95] Overall, D_2 receptor stimulation inhibits NMDA-mediated GLU transmission and long-term potentiation (LTP), and D_1 receptor stimulation facilitates GLU transmission and LTP (FIG. 6).[96,97] The effect of D_2 receptor stimulation on GLU transmission involves both pre- and postsynaptic effects: D_2 stimulation inhibits GLU release and reduces the excitability of medium spiny neurons.[94,95,98–102] In contrast, D_1 receptor stimulation generally promotes NMDA function and medium spiny neuron excitability, more specifically when the cells are in a depolarized "upstate," due to the convergence of excitatory inputs.[102–108]

These opposite effects of D_1 and D_2 receptor stimulation on NMDA transmission in the striatum might be relevant to both the pathophysiology and the treatment of schizophrenia. From a pathophysiology standpoint, excess D_2 receptor stimulation in schizophrenia, as documented by the imaging studies reviewed above, would inhibit GLU-mediated information flow into cortico-striato-thalamic-cortical loops,

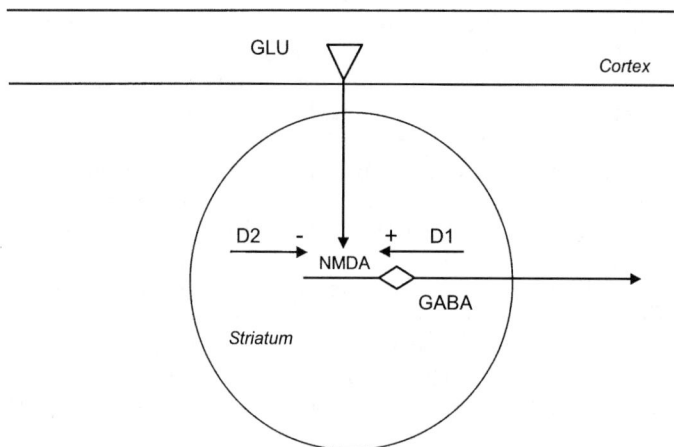

FIGURE 6. Opposite modulations of NMDA transmission by DA D_2 and D_1 receptors in GABAergic medium spiny neurons in the striatum. D_2 and D_1 receptors inhibit and facilitate, respectively, GLU transmission. Thus, an excess of D_2 receptor stimulation in schizophrenia would further impair NMDA-mediated information flow from the cortex into the striatum. By blocking D_2 receptors, antipsychotic drugs promote NMDA transmission. Conversely, D_1 receptor antagonists weaken NMDA transmission and are not antipsychotic drugs. DA release in response to D_2 receptor blockade might contribute to the therapeutic action of antipsychotic drugs by stimulating D_1 receptors. This model suggests that D_1 receptor agonists might have antipsychotic properties. Clozapine, by combining D_2 receptor antagonism and D_1 receptor agonism might provide a more comprehensive rescue of NMDA function compared to selective D_2 receptor antagonists or mixed D_2/D_1 antagonists.

and might worsen an already deficient NMDA transmission. As a result, the ability of the cortex to successfully send information to be processed in these loops would be impaired. Impairment of NMDA transmission would also inhibit plasticity in these loops. By blocking D_2 receptors (and, at least acutely, stimulating D_1 receptors by augmenting DA release), antipsychotic drugs restore GLU transmission in the striatum, the ability of the striatum to receive and process cortical information, and the plasticity required for the shaping of cognitive processes by experience.

Supporting the relevance of these D_1-D_2-GLU interactions in the striatum is the observation that D_1 receptor antagonists are not antipsychotic drugs. According to the data reviewed above, blockade of D_1 receptors is expected to weaken NMDA transmission. In clinical trials of D_1 receptor antagonists in schizophrenia, worsening of positive symptoms has been reported,[109–112] which might result from NMDA transmission impairment by D_1 receptor antagonists. An implication of this model is that augmentation of D_2 receptor antagonists by D_1 receptor agonists might be useful in schizophrenia, as this combination might promote NMDA transmission over and above the effect of D_2 receptor blockade alone. Some investigators[113–115] have argued that clozapine acts as a partial agonist at the D_1 receptor, and that this property might be important for its unique clinical profile. Another implication of this model is that the depolarization blockade of DA neurons that emerges following chronic antipsychotic treatment[116,117] might actually reduce the effectiveness of antipsychotic drugs, as this response would depress D_1 receptor activation. From there, one might speculate that the lower rate of depolarization blockade following atypical drug administration[116,118] contributes to their therapeutic advantages over typical antipsychotic drugs.

This model might also explain the efficacy of antipsychotic drugs in patients not presenting excessive stimulation of D_2 receptors. We previously observed that, in schizophrenia, excessive occupancy of D_2 receptors by DA is predictive of a rapid response of positive symptoms to D_2 receptor blockade.[36] In contrast, patients with normal levels of D_2 receptor occupancy by DA were slow responders. In these patients, the positive symptomatology might be driven by a primary NMDA deficiency rather than by D_2 receptor–induced NMDA deficiency. Antipsychotic drugs might still be helpful in these patients: by setting D_2 receptor stimulation to levels below normal and shifting the D_1 receptor/D_2 receptor equilibrium in favor of D_1 receptors, antipsychotic drugs might also promote NMDA receptor function and plasticity in these patients. However, these effects might take longer and be less reliable than in patients with NMDA deficiency primarily driven by excessive D_2 receptor activity.

DA–GLU Interactions in the Cortex

In the prefrontal cortex, D_1 receptors are localized on pyramidal cells (dendritic spines and shafts), while D_1, D_2, and D_4 receptors are localized on GABA interneurons.[119,120] Thus, DA modulates pyramidal cell excitability, both directly and indirectly via modulation of GABAergic interneurons.[121]

Stimulation of DA receptors located on GABAergic interneurons is generally viewed as promoting a GABA-mediated inhibition of pyramidal cells.[122–125] The effect of D_1 receptor stimulation on prefrontal pyramidal neurons is more complex. D_1 stimulation is neither "excitatory" nor "inhibitory," but depends on the functional status of these neurons at the time of D_1 receptor stimulation.[121] Stimulation of D_1 receptors enhances excitability of activated neurons and further stabilizes inactivated neurons. D_1 receptor augmentation of NMDA currents is most effective when the neuron is already activated (i.e., depolarized) by excitatory inputs.[126] In contrast, when excitatory inputs to a neuron are not active (in hyperpolarized neurons), D_1 receptors may decrease glutamate-mediated responses, in part by reducing sodium and N-type calcium currents.[127] These "activity-dependent" actions of DA, especially at D_1 receptors would allow the maintenance of calcium influx and spike firing in circuits activated during the processing of task-relevant information, while reducing the excitability of neurons that are not receiving sufficient excitatory input.

Thus, DA, via D_1 receptors, acts as a "reinforcer" in prefrontal cellular circuits. First, DA causes direct stimulation of D_1 receptors on pyramidal cells, leading to potentiation of the response of stimulated pyramidal neurons and silencing unstimulated neurons. Second, DA promotes activity of GABAergic interneurons, contributing to a generalized inhibitory tone or "background noise reduction" (only neurons subjected to high excitatory inputs escape from this inhibition). Via these mechanisms, DA enhances the signal-to-noise ratio in prefrontal circuits.[125,128] Thus, a deficit in D_1 receptor stimulation in the PFC in schizophrenia might contribute to the cognitive deficits presented by these patients. Conversely, atypical antipsychotic agents promote DA release in the PFC, by combining D_2 receptor blockade with 5-HT_{2A} antagonism or 5HT_{1A} agonism.[129–134] This effect, shared by atypical but not typical antipsychotic drugs, might mediate the modest improvement in WM function observed in some clinical studies.

CONCLUSIONS

In this paper, we briefly discussed recent imaging data supporting the association of schizophrenia with a dopaminergic biological marker involving deficits in cortical DA function and excesses in subcortical DA function. Animal and imaging data are consistent with the idea that both abnormalities might be secondary to a synaptic dysconnectivity involving the PFC, which is well modeled by NMDA antagonist administration. In turn, both components of this biological marker might contribute to worsen synaptic connectivity and NMDA function.

Thus, both GLU-DA and DA-GLU interactions might be very relevant to schizophrenia pathophysiology and treatment. A deficit in GLU transmission might lead to the dopaminergic biological marker associated with this illness, and these DA alterations might exacerbate GLU transmission deficits. In that sense, the view that NMDA alterations are primary and DA alterations are secondary is probably oversimplistic, as both sets of abnormalities reinforce each other. A consequence of this general model is that direct intervention to support NMDA or D_1 function might be beneficial as augmentation strategies for the treatment of schizophrenia.

ACKNOWLEDGMENTS

This work supported by the U.S. Public Health Service (K02MH01603-01, K08 MH01594-01, and K02 MH064178-01A2) and the Lieber Center for Schizophrenia Research at Columbia University.

REFERENCES

1. CARLSSON, A. & M. LINDQVIST. 1963. Effect of chlorpromazine or haloperidol on formation of 3-methoxytyramine and normetanephrine in mouse brain. Acta Pharmacol. Toxicol. **20:** 140–144.
2. SEEMAN, P. & T. LEE. 1975. Antipsychotic drugs: direct correlation between clinical potency and presynaptic action on dopamine neurons. Science **188:** 1217–1219.
3. CREESE, I., D.R. BURT & S.H. SNYDER. 1976. Dopamine receptor binding predicts clinical and pharmacological potencies of antischizophrenic drugs. Science **19:** 481–483.
4. LIEBERMAN, J.A., J.M. KANE & J. ALVIR. 1987. Provocative tests with psychostimulant drugs in schizophrenia. Psychopharmacology **91:** 415–433.
5. ANGRIST, B. & D.P. VAN KAMMEN. 1984. CNS stimulants as a tool in the study of schizophrenia. Trends Neurosci. **7:** 388–390.
6. KNABLE, M.B. & D.R. WEINBERGER. 1997. Dopamine, the prefrontal cortex and schizophrenia. J. Psychopharmacol. **11:** 123–131.
7. GOLDMAN-RAKIC, P.S., E.C. MULY, 3RD & G.V. WILLIAMS. 2000. D(1) receptors in prefrontal cells and circuits. Brain Res. Brain Res. Rev. **31:** 295–301.
8. DAVIS, K.L. et al. 1991. Dopamine in schizophrenia: a review and reconceptualization. Am. J. Psychiatry **148:** 1474–1486.
9. WEINBERGER, D.R. 1987. Implications of the normal brain development for the pathogenesis of schizophrenia. Arch. Gen. Psychiatry. **44:** 660–669.
10. OLNEY, J.W. & N.B. FARBER. 1995. Glutamate receptor dysfunction and schizophrenia. Arch. Gen. Psychiatry **52:** 998–1007.
11. JAVITT, D.C. & S.R. ZUKIN. 1991. Recent advances in the phencyclidine model of schizophrenia. Am. J. Psychiatry **148:** 1301–1308.
12. TAMMINGA, C.A. et al. 1995. Glutamate pharmacology and the treatment of schizophrenia: current status and future directions. Int. Clin. Psychopharmacol. **3:** 29–37.
13. GOFF, D.C. & J.T. COYLE. 2001. The emerging role of glutamate in the pathophysiology and treatment of schizophrenia. Am. J. Psychiatry **158:** 1367–1377.
14. JENTSCH, J.D. & R.H. ROTH. 1999. The neuropsychopharmacology of phencyclidine: from NMDA receptor hypofunction to the dopamine hypothesis of schizophrenia. Neuropsychopharmacology **20:** 201–225.
15. MILLER, D.W. & E.D. ABERCROMBIE. 1996. Effects of MK-801 on spontaneous and amphetamine-stimulated dopamine release in striatum measured with in vivo microdialysis in awake rats. Brain Res. Bull. **40:** 57–62.
16. KEGELES, L.S. et al. 2000. Modulation of amphetamine-induced striatal dopamine release by ketamine in humans: implications for schizophrenia. Biol. Psychiatry **48:** 627–640.
17. REITH, J. et al. 1994. Elevated dopa decarboxylase activity in living brain of patients with psychosis. Proc. Natl. Acad. Sci. USA **91:** 11651–11654.
18. HIETALA, J. et al. 1995. Presynaptic dopamine function in striatum of neuroleptic-naive schizophrenic patients. Lancet **346:** 1130–1131.
19. DAO-CASTELLANA, M.H. et al. 1997. Presynaptic dopaminergic function in the striatum of schizophrenic patients. Schizophrenia Res. **23:** 167–174.
20. HIETALA, J. et al. 1999. Depressive symptoms and presynaptic dopamine function in neuroleptic-naive schizophrenia. Schizophrenia Res. **35:** 41–50.
21. MEYER-LINDENBERG, A. et al. 2002. Reduced prefrontal activity predicts exaggerated striatal dopaminergic function in schizophrenia. Nat. Neurosci. **5:** 267–271.
22. LINDSTROM, L.H. et al. 1999. Increased dopamine synthesis rate in medial prefrontal cortex and striatum in schizophrenia indicated by L-(beta-11C) DOPA and PET. Biol. Psychiatry **46:** 681–688.
23. LARUELLE, M. 2000. Imaging synaptic neurotransmission with in vivo binding competition techniques: a critical review. J. Cereb. Blood Flow Metab. **20:** 423–451.
24. LARUELLE, M. et al. 1997. Microdialysis and SPECT measurements of amphetamine-induced dopamine release in nonhuman primates. Synapse **25:** 1–14.
25. BREIER, A. et al. 1997. Schizophrenia is associated with elevated amphetamine-induced synaptic dopamine concentrations: Evidence from a novel positron emission tomography method. Proc. Natl. Acad. Sci. USA **94:** 2569–2574.

26. KEGELES, L.S. et al. 1999. Stability of [123I]IBZM SPECT measurement of amphetamine-induced striatal dopamine release in humans. Synapse **31:** 302–308.
27. PICCINI, P., N. PAVESE & D.J. BROOKS. 2003. Endogenous dopamine release after pharmacological challenges in Parkinson's disease. Ann. Neurol. **53:** 647–653.
28. VILLEMAGNE, V.L. et al. 1999. GBR12909 attenuates amphetamine-induced striatal dopamine release as measured by [(11)C]raclopride continuous infusion PET scans. Synapse **33:** 268–273.
29. LARUELLE, M. et al. 1996. Single photon emission computerized tomography imaging of amphetamine-induced dopamine release in drug free schizophrenic subjects. Proc. Natl. Acad. Sci. USA **93:** 9235–9240.
30. ABI-DARGHAM, A. et al. 1998. Increased striatal dopamine transmission in schizophrenia: confirmation in a second cohort. Am. J. Psychiatry **155:** 761–767.
31. LARUELLE, M. et al. 1999. Increased dopamine transmission in schizophrenia: relationship to illness phases. Biol. Psychiatry **46:** 56–72.
32. LARUELLE, M. 1999. The role of endogenous sensitization in the pathophysiology of schizophrenia: implications from recent brain imaging studies. Brain Res. Rev. In press.
33. DEUTCH, A.Y., W.A. CLARK & R.H. ROTH. 1990. Prefrontal cortical dopamine depletion enhances the responsiveness of mesolimbic dopamine neurons to stress. Brain Res. **521:** 311–315.
34. KALIVAS, P.W. & P. DUFFY. 1995. Selective activation of dopamine transmission in the shell of the nucleus accumbens by stress. Brain Res. **675:** 325–328.
35. PARSEY, R.V. et al. 2001. Dopamine D(2) receptor availability and amphetamine-induced dopamine release in unipolar depression. Biol. Psychiatry **50:** 313–322.
36. ABI-DARGHAM, A. et al. 2000. Increased baseline occupancy of D_2 receptors by dopamine in schizophrenia. Proc. Natl. Acad. Sci. USA **97:** 8104–8109.
37. WEINBERGER, D.R., K.F. BERMAN & T.N. CHASE. 1988. Mesocortical dopaminergic function and human cognition. Ann. NY Acad. Sci. **537:** 330–338.
38. KAHN, R.S. et al. 1994. Neuropsychological correlates of central monoamine function in chronic schizophrenia: relationship between CSF metabolites and cognitive function. Schizophr. Res. **11:** 217–224.
39. DANIEL D. G. et al. 1991. The effect of amphetamine on regional cerebral blood flow during cognitive activation in schizophrenia. J. Neurosci. **11:** 1907–1917.
40. DOLAN, R.J. et al. 1995. Dopaminergic modulation of impaired cognitive activation in the anterior cingulate cortex in schizophrenia. Nature **378:** 180–182.
41. AKIL, M. et al. 1999. Lamina-specific alterations in the dopamine innervation of the prefrontal cortex in schizophrenic subjects. Am. J. Psychiatry **156:** 1580–1589.
42. HALLDIN, C. et al. 1986. Preparation of 11C-labelled SCH 23390 for the in vivo study of dopamine D_1 receptors using positron emission tomography. Appl. Radiat. Isot. **37:** 1039–1043.
43. LARUELLE, M. et al. 1991. Characterization of [125I]SCH23982 binding in human brain: comparison with [3H]SCH23390. Neurosci. Lett. **31:** 273–276.
44. SUHARA, T. et al. 1992. D_1 dopamine receptor binding in mood disorders measured by positron emission tomography. Psychopharmacology (Berl). **106:** 14–18.
45. ANDERSEN, P.H. et al. 1992. NNC-112, NNC-687 and NNC-756, new selective and highly potent dopamine D_1 receptor antagonists. Eur. J. Pharmacol. **219:** 45–52.
46. HALLDIN, C. et al. 1998. Carbon-11-NNC 112: a radioligand for PET examination of striatal and neocortical D_1-dopamine receptors. J. Nucl. Med. **39:** 2061–2068.
47. ABI-DARGHAM, A. et al. 1999. PET studies of binding competition between endogenous dopamine and the D_1 radiotracer [11C]NNC 756. Synapse **32:** 93–109.
48. ABI-DARGHAM, A. et al. 2000. Measurement of striatal and extrastriatal dopamine D_1 receptor binding potential with [11C]NNC 112 in humans: validation and reproducibility. J. Cereb. Blood Flow Metab. **20:** 225–243.
49. OKUBO, Y. et al. 1997. Decreased prefrontal dopamine D_1 receptors in schizophrenia revealed by PET. Nature **385:** 634–636.
50. KARLSSON, P. et al. 2002. PET study of D(1) dopamine receptor binding in neuroleptic-naive patients with schizophrenia. Am. J. Psychiatry **159:** 761–767.

51. ABI-DARGHAM, A. *et al.* 2002. Prefrontal dopamine D_1 receptors and working memory in schizophrenia. J. Neurosci. **22:** 3708–3719.
52. GUO, N. *et al.* 2001. The effect of chronic DA depletion on D_1 ligand binding in rodent brain. Soc. Neurosc. Abst. 27.
53. PYCOCK, C.J., R.W. KERWIN & C.J. CARTER. 1980. Effect of lesion of cortical dopamine terminals on subcortical dopamine receptors in rats. Nature **286:** 74–77.
54. KOLACHANA, B.S. *et al.* 1996. Abnormal prefrontal cortical regulation of striatal dopamine release after neonatal medial temporal-limbic lesions in rhesus monkeys. Soc. Neurosci. Abst. **22:** 1974.
55. KRYSTAL, J.H. *et al.* 1994. Subanesthetic effects of the noncompetitive NMDA antagonist, ketamine, in humans. Psychotomimetic, perceptual, cognitive, and neuroendocrine responses. Arch. Gen. Psychiatry. **51:** 199–214.
56. LAHTI, A.C. *et al.* 1995. Subanesthetic doses of ketamine stimulate psychosis in schizophrenia. Neuropsychopharmacology **13:** 9–19.
57. LAHTI, A.C. *et al.* 2001. Effects of ketamine in normal and schizophrenic volunteers. Neuropsychopharmacology **25:** 455–467.
58. JAVITT, D.C. *et al.* 1994. Amelioration of negative symptoms in schizophrenia by glycine. Am. J. Psychiatry **151:** 1234–1236.
59. JAVITT, D.C. *et al.* 2001. Adjunctive high-dose glycine in the treatment of schizophrenia. Int. J. Neuropsychopharmacol. **4:** 385–391.
60. HERESCO-LEVY, U. *et al.* 1999. Efficacy of high-dose glycine in the treatment of enduring negative symptoms of schizophrenia. Arch. Gen. Psychiatry **56:** 29–36.
61. HERESCO-LEVY, U. *et al.* 1998. Double-blind, placebo-controlled, crossover trial of D-cycloserine adjuvant therapy for treatment-resistant schizophrenia. Int. J. Neuropsychopharmcol. **1:** 131–135.
62. HERESCO-LEVY, U. *et al.* 2002. Placebo-controlled trial of D-cycloserine added to conventional neuroleptics, olanzapine, or risperidone in schizophrenia. Am. J. Psychiatry **159:** 480–482.
63. GOFF, D.C. *et al.* 1999. A placebo-controlled trial of D-cycloserine added to conventional neuroleptics in patients with schizophrenia. Arch. Gen. Psychiatry **56:** 21–27.
64. TSAI, G. *et al.* 1998. D-serine added to antipsychotics for the treatment of schizophrenia. Biol. Psychiatry **44:** 1081–1089.
65. MILLAN, M.J. 2002. N-methyl-D-aspartate receptor-coupled glycineB receptors in the pathogenesis and treatment of schizophrenia: a critical review. Curr. Drug Target CNS Neurol. Disord. **1:** 191–213.
66. HARRISON, P.J. 1999. The neuropathology of schizophrenia. A critical review of the data and their interpretation. Brain **122:** 593–624.
67. MEADOR-WOODRUFF, J.H. & D.J. HEALY. 2000. Glutamate receptor expression in schizophrenic brain. Brain Res. Brain Res. Rev. **31:** 288–294.
68. KONRADI, C. & S. HECKERS. 2003. Molecular aspects of glutamate dysregulation: implications for schizophrenia and its treatment. Pharmacol. Ther. **97:** 153–179.
69. FRANKLE, W.G., J. LERMA & M. LARUELLE. 2003. The synaptic hypothesis of schizophrenia. Neuron **39:** 205–216.
70. CARLSSON, A., N. WATERS & M.L. CARLSSON. 1999. Neurotransmitter interactions in schizophrenia—therapeutic implications. Biol. Psychiatry **46:** 1388–1395.
71. CARR, D.B. & S.R. SESACK. 2000. Projections from the rat prefrontal cortex to the ventral tegmental area: target specificity in the synaptic associations with mesoaccumbens and mesocortical neurons. J. Neurosci. **20:** 3864–3873.
72. JACKSON, M.E., A.S. FROST & B. MOGHADDAM. 2001. Stimulation of prefrontal cortex at physiologically relevant frequencies inhibits dopamine release in the nucleus accumbens. J. Neurochem. **78:** 920–923.
73. TAKAHATA, R. & B. MOGHADDAM. 2000. Target-specific glutamatergic regulation of dopamine neurons in the ventral tegmental area. J. Neurochem. **75:** 1775–1778.
74. ADAMS, B.W., C.W. BRADBERRY & B. MOGHADDAM. 2002. NMDA antagonist effects on striatal dopamine release: microdialysis studies in awake monkeys. Synapse **43:** 12–18.
75. KEGELES, L.S. *et al.* 2002. NMDA antagonist effects on striatal dopamine release: positron emission tomography studies in humans. Synapse **43:** 19–29.

76. AALTO, S. *et al.* 2002. Ketamine does not decrease striatal dopamine D_2 receptor binding in man. Psychopharmacology (Berl). **164:** 401–406.
77. BREIER, A. *et al.* 1998. Effects of NMDA antagonism on striatal dopamine release in healthy subjects: application of a novel PET approach. Synapse **29:** 142–147.
78. SMITH, G.S. *et al.* 1998. Glutamate modulation of dopamine measured in vivo with positron emission tomography (PET) and 11C-raclopride in normal human subjects. Neuropsychopharmacology **18:** 18–25.
79. VOLLENWEIDER, F.X. *et al.* 2000. Effects of (S)-ketamine on striatal dopamine: a [11C]raclopride PET study of a model psychosis in humans. J. Psychiatr. Res. **34:** 35–43.
80. BUNNEY, B.S. & G.K. AGHAJANIAN. 1978. d-Amphetamine-induced depression of central dopamine neurons: evidence for mediation by both autoreceptors and a striatonigral feedback pathway. Naunyn Schmiedebergs Arch. Pharmacol. **304:** 255–261.
81. MOGHADDAM, B. *et al.* 1997. Activation of glutamatergic neurotransmission by ketamine: A novel step in the pathway from NMDA receptor blockade to dopaminergic and cognitive disruptions associated with the prefrontal cortex. J. Neurosci. **17:** 2921–2927.
82. MOGHADDAM, B. & B.W. ADAMS. 1998. Reversal of phencyclidine effects by a group II metabotropic glutamate receptor agonist in rats. Science **281:** 1349–1352.
83. PIN, J P. *et al.* 1999. New perspectives for the development of selective metabotropic glutamate receptor ligands. Eur. J. Pharmacol. **375:** 277–294.
84. NAKANISHI, S. & M. MASU. 1994. Molecular diversity and functions of glutamate receptors. Annu. Rev. Biophys. Biomol. Struct. **23:** 319–348.
85. SCHOEPP, D.D. 2002. Metabotropic glutamate receptors. Pharmacol. Biochem. Behav. **73:** 285–286.
86. EAST, S.J., M.P. HILL & J.M. BROTCHIE. 1995. Metabotropic glutamate receptor agonists inhibit endogenous glutamate release from rat striatal synaptosomes. Eur. J. Pharmacol. **277:** 117–1121.
87. BATTAGLIA, G., J.A. MONN & D.D. SCHOEPP. 1997. In vivo inhibition of veratridine-evoked release of striatal excitatory amino acids by the group II metabotropic glutamate receptor agonist LY354740 in rats. Neurosci. Lett. **229:** 161–164.
88. DI IORIO, P. *et al.* 1996. Interaction between A1 adenosine and class II metabotropic glutamate receptors in the regulation of purine and glutamate release from rat hippocampal slices. J. Neurochem. **67:** 302–309.
89. SCHOEPP, D.D. *et al.* 1997. LY354740 is a potent and highly selective group II metabotropic glutamate receptor agonist in cells expressing human glutamate receptors. Neuropharmacology **36:** 1–11.
90. VAN BERCKEL, B.N.M. *et al.* 2001. Enhanced amphetamine-induced striatal [11C]raclopride displacement by the group II metabotropic glutamate receptor agonist L354740 in baboons. Soc. Neurosci. Abst. 454.3.
91. STARR, M.S. 1995. Glutamate/dopamine D_1/D_2 balance in the basal ganglia and its relevance to Parkinson's disease. Synapse **19:** 264–293.
92. SMITH, A.D. & J.P. BOLAM. 1990. The neural network of the basal ganglia as revealed by the study of synaptic connections of identified neurones. Trends Neurosci. **13:** 259–265.
93. KOTTER, R. 1994. Postsynaptic integration of glutamatergic and dopaminergic signals in the striatum. Prog. Neurobiol. **44:** 163–196.
94. NICOLA, S.M., J. SURMEIER & R.C. MALENKA. 2000. Dopaminergic modulation of neuronal excitability in the striatum and nucleus accumbens. Annu. Rev. Neurosci. **23:** 185–215.
95. CEPEDA, C. & M.S. LEVINE. 1998. Dopamine and N-methyl-D-aspartate receptor interactions in the neostriatum. Dev. Neurosci. **20:** 1–18.
96. LEVINE, M.S. *et al.* 1996. Neuromodulatory actions of dopamine on synaptically-evoked neostriatal responses in slices. Synapse **24:** 65–78.
97. CENTONZE, D. *et al.* 2001. Dopaminergic control of synaptic plasticity in the dorsal striatum. Eur. J. Neurosci. **13:** 1071–1077.
98. CEPEDA, C. *et al.* 2001. Facilitated glutamatergic transmission in the striatum of D_2 dopamine receptor-deficient mice. J. Neurophysiol. **85:** 659–670.

99. PERIS, J., L.P. DWOSKIN & N.R. ZAHNISER. 1988. Biphasic modulation of evoked [3H]D-aspartate release by D-2 dopamine receptors in rat striatal slices. Synapse **2**: 450–456.
100. LEVEQUE, J.C. *et al.* 2000. Intracellular modulation of NMDA receptor function by antipsychotic drugs. J. Neurosci. **20**: 4011–4220.
101. ONN, S.P., A.R. WEST & A.A. GRACE. 2000. Dopamine-mediated regulation of striatal neuronal and network interactions. Trends Neurosci. **23**: S48–56.
102. WEST, A.R. & A.A. GRACE. 2002. Opposite influences of endogenous dopamine D_1 and D_2 receptor activation on activity states and electrophysiological properties of striatal neurons: studies combining in vivo intracellular recordings and reverse microdialysis. J. Neurosci. **22**: 294–304.
103. MARTI, M. *et al.* 2002. Striatal dopamine-NMDA receptor interactions in the modulation of glutamate release in the substantia nigra pars reticulata in vivo: opposite role for D_1 and D_2 receptors. J. Neurochem. **83**: 635–644.
104. MORARI, M. *et al.* 1994. Dopamine D_1 and D_2 receptor antagonism differentially modulates stimulation of striatal neurotransmitter levels by N-methyl-D-aspartic acid. Eur. J. Pharmacol. **256**: 23–30.
105. HERNANDEZ-LOPEZ, S. *et al.* 1997. D_1 receptor activation enhances evoked discharge in neostriatal medium spiny neurons by modulating an L-type Ca^{2+} conductance. J. Neurosci. **17**: 3334–3342.
106. FLORES-HERNANDEZ, J. *et al.* 2002. Dopamine enhancement of NMDA currents in dissociated medium-sized striatal neurons: role of D_1 receptors and DARPP-32. J. Neurophysiol. **88**: 3010–3020.
107. WILSON, C.J. & Y. KAWAGUCHI. 1996. The origins of two-state spontaneous membrane potential fluctuations of neostriatal spiny neurons. J. Neurosci. **16**: 2397–2410.
108. DUNAH, A.W. & D.G. STANDAERT. 2001. Dopamine D_1 receptor-dependent trafficking of striatal NMDA glutamate receptors to the postsynaptic membrane. J. Neurosci. **21**: 5546–5558.
109. KARLSSON, P. *et al.* 1995. Lack of apparent antipsychotic effect of the D_1-dopamine receptor antagonist SCH39166 in acutely ill schizophrenic patients. Psychopharmacology (Berl). **121**: 309–316.
110. KARLE, J. *et al.* 1995. NNC 01-0687, a selective dopamine D_1 receptor antagonist, in the treatment of schizophrenia. Psychopharmacology (Berl.) **121**: 328–329.
111. DE BEAUREPAIRE, R. *et al.* 1995. An open trial of the D_1 antagonist SCH 39166 in six cases of acute psychotic states. Psychopharmacology (Berl.) **121**: 323–327.
112. DEN BOER, J.A. *et al.* 1995. Differential effects of the D_1-DA receptor antagonist SCH39166 on positive and negative symptoms of schizophrenia. Psychopharmacology (Berl.) **121**: 317–322.
113. NINAN, I. & S.K. KULKARNI. 1998. Partial agonistic action of clozapine at dopamine D_2 receptors in dopamine depleted animals. Psychopharmacology (Berl.) **135**: 311–317.
114. JACKSON, D.M., H. WIKSTROM & Y. LIAO. 1998. Is clozapine an (partial) agonist at both dopamine D_1 and D_2 receptors? Psychopharmacology (Berl.) **138**: 213–216.
115. AHLENIUS, S. 1999. Clozapine: dopamine D_1 receptor agonism in the prefrontal cortex as the code to decipher a Rosetta stone of antipsychotic drugs. Pharmacol. Toxicol. **84**: 193–196.
116. GRACE, A.A. 1992. The depolarization block hypothesis of neuroleptic action: implications for the etiology and treatment of schizophrenia. J. Neural Transm. Suppl. **36**: 91–131.
117. GRACE, A.A. *et al.* 1997. Dopamine-cell depolarization block as a model for the therapeutic actions of antipsychotic drugs. Trends Neurosci. **20**: 31–37.
118. MELTZER, H.Y. 1991. The mechanism of action of novel antipsychotic drugs. Schizophr. Bull. **17**: 263–287.
119. MRZLETAK, L. *et al.* 1996. Localization of dopamine D4 receptors in GABAergic neurons of the primate brain. Nature **381**: 245–248.

120. SMILEY, J.F. et al. 1994. D$_1$ dopamine receptor immunoreactivity in human and monkey cerebral cortex: predominant and extrasynaptic localization in dendritic spines. Proc. Natl. Acad. Sci. USA **91:** 5720–5724.
121. YANG, C.R., J.K. SEAMANS & N. GORELOVA. 1999. Developing a neuronal model for the pathophysiology of schizophrenia based on the nature of electrophysiological actions of dopamine in the prefrontal cortex. Neuropsychopharmacology **21:** 161–194.
122. DEL ARCO, A. & F. MORA. 2000. Endogenous dopamine potentiates the effects of glutamate on extracellular GABA in the prefrontal cortex of the freely moving rat. Brain Res. Bull. **53:** 339–345.
123. GROBIN, A.C. & A.Y. DEUTCH. 1998. Dopaminergic regulation of extracellular gamma-aminobutyric acid levels in the prefrontal cortex of the rat. J. Pharmacol. Exp. Ther. **285:** 350–357.
124. SEAMANS, J.K. et al. 2001. Bidirectional dopamine modulation of GABAergic inhibition in prefrontal cortical pyramidal neurons. J. Neurosci. **21:** 3628–3638.
125. GORELOVA, N., J.K. SEAMANS & C.R. YANG. 2002. Mechanisms of dopamine activation of fast-spiking interneurons that exert inhibition in rat prefrontal cortex. J. Neurophysiol. **88:** 3150–3166.
126. SEAMANS, J.K. et al. 2001. Dopamine D$_1$/D5 receptor modulation of excitatory synaptic inputs to layer V prefrontal cortex neurons. Proc. Natl. Acad. Sci. USA **98:** 301–306.
127. FIENBERG, A. et al. 1998. DARPP-32: regulator of the efficacy of dopaminergic neurotransmission. Science **281:** 838–842.
128. SEAMANS, J.K. et al. 2001. Bidirectional dopamine modulation of GABAergic inhibition in prefrontal cortical pyramidal neurons. J. Neurosci. **21:** 3628–3638.
129. WESTERINK, B.H. et al. 2001. Antipsychotic drugs classified by their effects on the release of dopamine and noradrenaline in the prefrontal cortex and striatum. Eur. J. Pharmacol. **412:** 127–138.
130. ICHIKAWA, J. et al. 2001. 5-HT(2A) and D(2) receptor blockade increases cortical DA release via 5-HT(1A) receptor activation: a possible mechanism of atypical antipsychotic-induced cortical dopamine release. J. Neurochem. **76:** 1521–1531.
131. GESSA, G.L. et al. 2000. Dissociation of haloperidol, clozapine, and olanzapine effects on electrical activity of mesocortical dopamine neurons and dopamine release in the prefrontal cortex. Neuropsychopharmacology **22:** 642–649.
132. YOUNGREN, K.D. et al. 1999. Clozapine preferentially increases dopamine release in the rhesus monkey prefrontal cortex compared with the caudate nucleus. Neuropsychopharmacology **20:** 403–412.
133. ROLLEMA, H. et al. 1997. Clozapine increases dopamine release in prefrontal cortex by 5-HT1A receptor activation. Eur. J. Pharmacol. **338:** R3–5.
134. YAMAMOTO, B.K. & M.A. COOPERMAN. 1994. Differential effects of chronic antipsychotic drug treatment on extracellular glutamate and dopamine concentrations. J. Neurosci. **14:** 4159–4166.

Glutamate-Mediated Plasticity in Corticostriatal Networks

Role in Adaptive Motor Learning

ANN E. KELLEY, MATTHEW E. ANDRZEJEWSKI, ANNE E. BALDWIN,[a] PEPE J. HERNANDEZ, AND WAYNE E. PRATT

Department of Psychiatry and Neuroscience Training Program, University of Wisconsin–Madison Medical School, 6001 Research Park Boulevard, Madison, Wisconsin 53719, USA

ABSTRACT: Little is known about how memories of new voluntary motor actions, also known as procedural memory, are formed at the molecular level. Our work examining acquisition of lever-pressing for food in rats has shown that activation of glutamate NMDA receptors, within broadly distributed but interconnected regions (e.g., nucleus accumbens core, prefrontal cortex, basolateral amygdala), is critical for such learning to occur. This receptor stimulation triggers intracellular cascades that involve protein phosphorylation and new protein synthesis. In support of this idea, we have found that posttrial inhibition of protein synthesis in the ventral striatum impairs learning, whereas posttrial NMDA receptor blockade does not. More recent data show extension of this network to the central amygdala, where infusions of NMDA antagonists also impair learning. We hypothesize that activity in this distributed network (including dopaminergic activity and perhaps muscarinic cholinergic activity) computes coincident events and thus enhances the probability that temporally related actions and events (e.g., lever pressing and delivery of reward) become associated. Such basic mechanisms of plasticity within this reinforcement learning network also appear to be profoundly affected in addiction.

KEYWORDS: nucleus accumbens, instrumental learning, dopamine; NMDA receptors

In recent years a great amount of interest has focused on the role of glutamate as well as dopamine-glutamate interactions in the control of neural plasticity, learning, and memory, and addiction. These two neurotransmitter systems are widely distributed in many regions of cortex, limbic system, and basal ganglia, where they appear to play an integrative role in motivational and associative information processing. It is currently believed that coordinated neural signaling of these systems, particularly

---

Address for correspondence: Ann E. Kelley, Department of Psychiatry and Neuroscience Training Program, University of Wisconsin-Madison Medical School, 6001 Research Park Boulevard, Madison, WI 53719. Voice: 608-262-1123; fax: 608-265-3050.

aekelley@wisc.edu

[a]Current address: Anne E. Baldwin, Pennsylvania State College of Medicine, Department Behavioral Sciences, Hershey, PA 17033-2390.

through the dopamine D1 and glutamate N-methyl-D-aspartate (NMDA) receptors, is a critical event in triggering intracellular transductional and transcriptional cascades that lead to long-term changes in gene expression, synaptic plasticity, and ultimately behavior.[1-3] Addictive drugs also induce long-term neuroadaptations at the structural, cellular, molecular, and genomic levels, primarily through their impact on dopaminergic and glutamatergic circuits. Indeed, that drugs of abuse engage D1-and NMDA-mediated neuronal cascades shared with normal reward learning and memory is one of the most important insights that has emerged in the past decade regarding the neurobiology of addiction.[4] Such drug-induced neuroadaptations may contribute to abnormal information processing and behavior, resulting in poor decision-making, loss of control, and compulsivity that characterize addiction. Thus, further information regarding the normal behavioral role of dopamine- and glutamate-mediated neural networks may help to shed light on the nature of addiction and its treatment. Our laboratory has engaged in the investigation of the role of glutamate- and dopamine-coded neural systems within corticostriatal networks in adaptive motor learning, also termed instrumental learning.

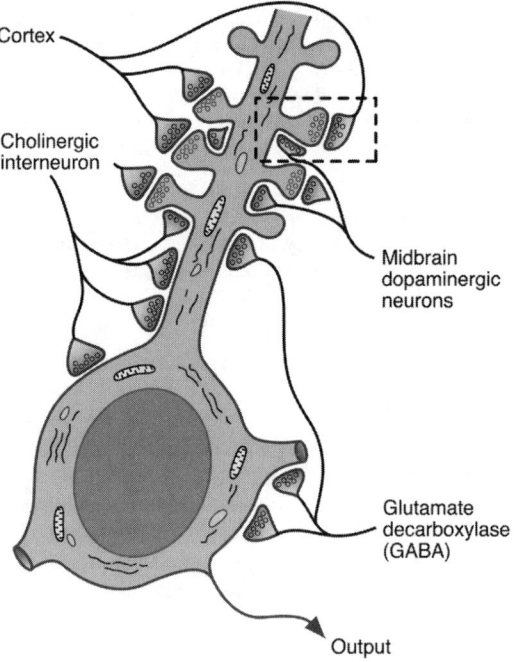

FIGURE 1. Schematic illustration of striatal medium spiny neuron showing synaptic inputs of relevance to the work discussed in text. Dendritic spines receive both glutamatergic afferents from corticothalamic regions as well as dopaminergic contacts from the midbrain (*see box*), and are thought to be the site of synaptic modification during learning. Cholinergic contacts, arising from aspiny acetylcholine-containing neurons, affect the proximal dendrite and may also play a significant role in neural processing and plasticity.

NMDA RECEPTOR–DEPENDENT PLASTICITY WITHIN A DISTRIBUTED CORTICOSTRIATAL NETWORK MEDIATES APPETITIVE INSTRUMENTAL LEARNING

We have long been interested in the role of plasticity within the medium spiny neurons of the ventral striatum, which are a major cellular component of the striatum (FIG. 1). Our earlier studies revealed quite profound learning deficits in an instrumental learning task with AMPA or NMDA antagonists infused into the core.[5–7] Specifically, infusion of AP-5, a selective competitive NMDA receptor antagonist, into the nucleus accumbens core blocked acquisition of appetitive instrumental learning. We extended these findings and wondered whether the requirement for NMDA receptor activation was specific to the accumbens core or whether other associated brain regions were also involved. Somewhat to our surprise, we found that NMDA receptor blockade in both the lateral/basolateral amygdala and medial prefrontal cortex (mPFC) also strongly disrupts acquisition of lever-pressing for food,[8] as shown in FIGURE 2. Injections into dorsal or ventral hippocampus had no effect. Several important features of the AP-5–induced impairment should be noted: first, it is only early in learning that AP-5 has any effect; infusions into active sites once learning is established have no effect. This profile suggests that NMDA receptor activation is required for plasticity only early in the learning process. Further, control experiments show that AP-5 infusions that disrupt learning have no effect on general motor behavior or on motivation for food. These data have provided novel evidence for an essential role of NMDA receptor–mediated plasticity in several key brain regions in the acquisition of new motor learning, and suggest that disruption of glutamatergic activity in any part of a distributed network is enough to prevent learning. They complement an important literature implicating NMDA-receptor mediated mechanisms in the cellular basis of learning and memory and in long-term potentiation.[9]

A KEY FEATURE OF PLASTICITY WITHIN THIS NETWORK IS COINCIDENT ACTIVATION OF DOPAMINE AND NMDA RECEPTORS

Since there has been growing evidence both in cellular and molecular models for D1-NMDA interactions in the control of learning-related plasticity, we decided to investigate the potential role of such a putative interaction in our instrumental learning model. Our first objective was to assess the effects of intra-accumbens core infusion of the D1 receptor antagonist SCH-23390 in acquisition. However, a major obstacle to investigating the role of dopamine (DA) receptors in learning and to interpreting effects on behavior is the considerable motoric impairment that often results with DA receptor blockade (unlike with AP-5). Because we indeed found evidence for a motor impairment, we examined the effects of infusion of very low doses of the D1 antagonist as well as combinations of low doses of AP-5 and SCH-23390.[6] Bilateral infusion of a relatively high dose of SCH-23390 (3 nmol or 1 µg) significantly impaired learning but also disrupted performance after the response was learned. Infusion of a much lower dose of SCH-23390 (0.3 nmol) or a much lower dose of AP-5 than used in the previous studies (0.5 nmol or 0.1 µg) had no effect on acquisition or performance. Most interestingly however, coinfusion of the low doses of the NMDA

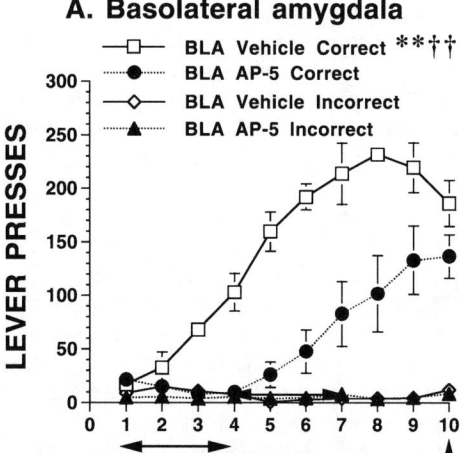
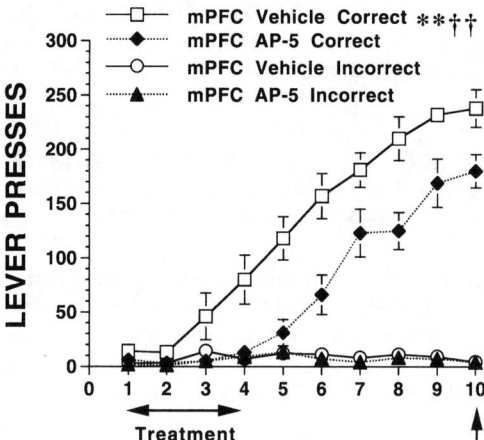

FIGURE 2. Effect of the selective, competitive NMDA receptor antagonist AP-5 on acquisition of instrumental responding for food, following bilateral infusion into the (**A**) basolateral amygdala and (**B**) medial prefrontal cortex. Animals were infused with drug or vehicle on days 1–4 just prior to training (*horizontal arrows*), and also on day 10 (*vertical arrow*), once the task was learned. Temporary blockade of NMDA receptors in these regions strongly impairs learning, although performance of the learned response is not affected. Control experiments showed no measurable effect on general motor behavior or food intake. From work by Baldwin and colleagues.[8,10]

and D1 antagonist strongly disrupted acquisition of instrumental learning. Although in control tests infusion of the higher dose of the D1 antagonist reduced spontaneous motor behaviors as might be expected, the coinfusion of low doses had no effect on motor activity or feeding.

Given that the medial prefrontal cortex (mPFC), like the striatum, receives a convergence of dopaminergic and glutamatergic inputs, we hypothesized that a similar interaction in the mPFC might underlie neural adaptation during learning. An experiment similar to the previous one was carried out with cannulae aimed at the mPFC.[10] In this study it was necessary to employ three doses of SCH-23390 as we found the mPFC to be exquisitely sensitive to D1 receptor blockade. Both the 3.0 and 0.3 nmol doses of SCH-23390 infused into the mPFC impaired acquisition of the bar-press response, and the highest dose also reduced performance of the learned response. We then infused the lowest dose of SCH-23390 with a low dose of AP-5 (0.5 nmol), which had no effect on its own. The co-infusion markedly impaired acquisition of instrumental responding. This study represents the first direct test of the effects of PFC dopamine D1 receptor antagonism and concurrent D1 and NMDA receptor antagonism on acquisition of instrumental responding. We believe these results have broad implications for the cellular basis of neuronal adaptation during motor learning and, in light of the similar profile with the nucleus accumbens, provide evidence for parallel cellular mechanisms within discrete regions of the proposed distributed network. We are currently investigating possible similar mechanisms within the amygdala.

INTRACELLULAR SIGNALING MECHANISMS IN ACCUMBENS AND mPFC ARE NECESSARY FOR INITIATION OF MOLECULAR EVENTS LEADING TO ESTABLISHED INSTRUMENTAL LEARNING

The proposed convergence of dopamine D1 and glutamatergic NMDA receptors suggests that these extracellular signals, perhaps conveying both motivational information and temporal information pertaining to sensory and motor events, trigger second messenger cascades that eventually affect transcription and translation. We have particularly focused on protein kinase A (PKA), which interacts with a number of transcription factors as well as other second messenger systems. PKA is implicated in many forms of plasticity, including long-term potentiation.[11,12] For example, intra-amygdala infusion of the selective PKA inhibitor Rp-cAMPS impairs long-term memory for contextual fear conditioning.[13] We conducted a series of experiments in which drugs interfering with protein kinase activity were infused into the nucleus accumbens in conjunction with the instrumental learning task.[14] It was demonstrated that treatment with the PKA inhibitor Rp-cAMPS impaired learning. Interestingly, infusion of an activator of PKA, Sp-cAMPS also impaired learning, suggesting that an optimal level of PKA within the accumbens is required. It was also shown that posttrial infusion of the broad-spectrum kinase inhibitor H7 dose-dependently impaired acquisition, indicating that long-term kinase activity lasting minutes or hours may be at least one important mechanism related to the plasticity involved in this type of learning.

In the recent study involving the mPFC,[10] we also conducted an experiment with infusions of Rp-cAMPS into this region. Here too we found a similar profile, in that bilateral infusion of the PKA inhibitor also impaired learning. Although not directly shown in our experiments, these results together with much data in the literature suggest that PKA may be an intracellular substrate for the D1-NMDA interaction. An example of supportive evidence is provided by the work of Gurden and colleagues,[15] who showed that long-term potentiation (LTP) at hippocampal-prefrontal synapses is dependent on NMDA and D1 receptor coactivation and on intracellular PKA.

EARLY CONSOLIDATION OF INSTRUMENTAL LEARNING REQUIRES PROTEIN SYNTHESIS IN THE NUCLEUS ACCUMBENS

It is well established that long-term memory formation is a temporally dynamic process requiring the activation of specific genes and *de novo* protein synthesis.[16,17] For example, infusion of the protein synthesis inhibitor anisomycin into the amygdala prevents consolidation of fear memories.[18] However, no studies have addressed the role of *de novo* protein synthesis within specific brain structures for consolidation of positively motivated instrumental behaviors. Given our work with D1-NMDA interactions and protein kinases, we hypothesized that posttrial blockade of protein synthesis within the nucleus accumbens would disrupt the consolidation of "instrumental memory." We recently found that posttrial infusions of anisomycin into the core but not the shell after the first 5 of 12 test sessions prevent the consolidation of long-term memory for the task.[19] Posttrial core infusions delayed by 2 or 4 hours had no effect. Once the task was learned, behavior was no longer sensitive to intra-accumbens anisomycin. Our data provide the first demonstration that a form of procedural or "habit" learning is dependent on translational events in a specific brain region. However, once the animal learns these associations and the behavior becomes firmly established, protein synthesis within accumbens is no longer required for the expression of the behavior, a profile that exactly mirrors the role of NMDA and D1 receptor activation.

POSTTRIAL BLOCKADE OF NMDA OR D1 RECEPTORS DOES NOT AFFECT ACQUISITION OF INSTRUMENTAL LEARNING

Throughout the course of our studies, we have often wondered whether posttrial infusion of AP-5 or SCH-23390 would affect acquisition of instrumental responding. According to the main hypothesis driving the work, glutamate and dopamine within the distributed network are encoding current state, that is, the temporal pattern and context of events necessary for reinforcement learning. If this were true, posttrial blockade of D1 or NMDA receptors should have no effect on learning (unlike interference with kinases or protein synthesis, whose activity has a longer timecourse). We have very recently conducted these experiments and have clear evidence that immediate posttrial infusion of AP-5 or SCH-23390 does not affect acquisition (results not yet published). These results fit nicely with the notion that dynamic and interactive activity of glutamatergic and dopaminergic circuits, only during the relevant contextual situation, is required for new learning. In contrast, within a constrained temporal window just following the context, intracellular transcription and translation contributes to long-term synaptic remodeling that is not dependent on context.

INVESTIGATION OF OTHER STRIATAL AND LIMBIC SITES SUGGESTS A BROADLY DISTRIBUTED NETWORK

Several recent sets of data have added evidence that a broadly distributed network subserves instrumental learning. First, we have been investigating the role of the central nucleus of the amygdala in this learning task. Since we previously found that AP-5 infusion into the basolateral amygdala impaired learning, it was of interest to ascertain any involvement of the closely adjoining central nucleus. An experiment was carried out in which central nucleus infusions of AP-5 were made.[20] We indeed found that the AP-5 infusions prevented acquisition of responding. Interestingly, in contrast to all previous studies, performance was also markedly impaired by the AP-5 infusions given after the animals had acquired the task. A control experiment, however, revealed that central nucleus AP-5 infusions also affected spontaneous motor behavior and patterns of food intake; the drug actually augmented motor behavior and shortened feeding bouts. We have interpreted these data in the context of the proposed role for the central nucleus in attentional or motivational functions.[21] Recent studies also show that D1 receptors in the central nucleus play a role in the network.[20]

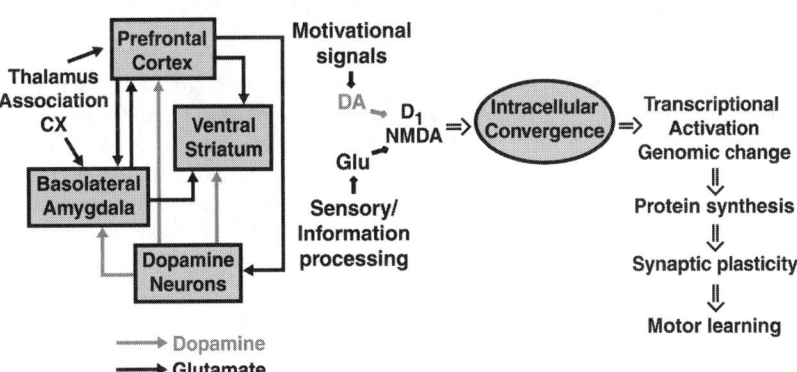

FIGURE 3. A proposed model of glutamate–dopamine interactions within a distributed network in the control of appetitive instrumental learning. Activity and plasticity in this network are hypothesized to mediate synaptic alteration at different nodes (e.g., amygdala and ventral striatum) during learning. Sensory and motivational signals converging on neurons in these nodes initiate phosphorylation events, gene transcription, and protein synthesis, resulting in the consolidation of novel adaptive motor actions. Note that this proposed model is not exclusive of other regions being involved; for example, recent evidence suggests a role for posterior striatal sites as well as central nucleus of amygdala (*see text*). It is not yet known how the thalamus participates in this network with regard to motor learning.

Finally, several recent experiments examined the effects of AP-5 infusions into additional striatal subregions, in particular in the dorsolateral sector and a posterior region. The posterior region was originally chosen as a site control for the work with the amygdala. Much to our surprise posterior, laterally placed striatal injections markedly impaired learning, and again, there was no effect on later performance of the learned response or on motor activity or feeding behavior.[20] In contrast, infusion of AP-5 into the dorsolateral sector of striatum had no effect on learning. What could possibly explain the difference in sensitivity to NMDA receptor blockade in these two sites? One hypothesis is that the critical regions within striatum involve only those regions innervated by amygdala, allocortex (prefrontal and perirhinal cortex), or mesocortex (piriform, entorhinal, and hippocampus); in effect "limbic-innervated striatum." The negative result with the dorsolateral striatum, which receives afferents only from neocortex,[22] may be very informative in this regard. In any case, our accruing results suggest that disruption of glutamatergic synapses anywhere in this network is enough to disrupt the plasticity processes that are necessary for learning. This very broad distribution suggests that glutamate-driven network synchrony or some sort of global corticostriatal population code (perhaps necessary for assessing the temporal relationship of sensory and motor events) is a critical factor underlying this form of adaptive learning. Several neural computational models emphasize the suitability of corticostriatal networks for such learning.[23,24] FIGURE 3 shows a schematic view of this corticostriatal DA- and glutamate-coded network.

ACETYLCHOLINE ALSO APPEARS TO BE A SIGNIFICANT MODULATOR OF PLASTICITY

Very recently we have begun to address the question of the role of striatal acetylcholine in instrumental learning.[25] Large aspiny cholinergic interneurons play an important role in striatal processing, presumably by modulating input to medium spiny neurons. Despite historical as well as contemporary study of the complex interaction of acetylcholine with both dopamine and glutamate within striatum, little work has been done to assess the role of striatal acetylcholine on learning within the behaving animal, or by what mechanisms it may modulate learning. We chose to begin an examination of striatal cholinergic mechanisms by first blocking nicotinic or muscarinic receptors in this region during the learning phase of lever-pressing for food reward. We have found that infusion of the muscarinic receptor blocker scopolamine into the core or shell subregions of accumbens dose-dependently impairs learning, whereas similar infusion of the nicotinic antagonist mecamylamine has no effect. Scopolamine in higher doses (10 µg bilaterally) also affected performance of the learned response, so it is unclear exactly how this dose of drug is affecting behavior. However, with a lower dose (1 µg bilaterally), we observed relatively selective effects on early acquisition of responding. Thus these results are in accordance with several recent studies suggesting that in addition to dopamine D1 and NMDA receptors, striatal muscarinic receptors also have a significant role in cellular plasticity and motor learning.[26–29] Further work will explore interrelationships among these neurotransmitter systems.

CONCLUSIONS

Instrumental learning, in which an organism learns a new motor response in order to obtain a positive outcome (procurement of food when hungry, avoidance of danger or pain), is one of the most elementary forms of behavioral adaptation.[30] Through interchange with its environment, an animal is able to learn about the consequences of its actions, and thereby modify the current environment through new behaviors to produce more favorable conditions.[31] Our work on instrumental learning, together with accruing evidence in the literature for a critical role for glutamate and glutamate-dopamine interactions in neural plasticity, suggests that activity in corticostriatal networks is an essential component of long-term molecular changes that underlie the learning of adaptive motor actions. Further work will continue to explore the nature of this activity. For example, it would be interesting to study the extent of the distribution of the active sites within this network—it is possible that the thalamus plays an important role. Additional interesting questions are whether there is communication between specific regions within the network important for acquisition of instrumental learning, and what is the nature of the relationship between local consolidation (in "nodes") and systems consolidation (reconfiguration of networks). Future experimentation will address these and other questions.

ACKNOWLEDGMENT

This work was supported by grant DA04788 from the National Institute on Drug Abuse.

REFERENCES

1. FLORESCO, S.B. *et al.* 2001. Dopamine D1 and NMDA receptors mediate potentiation of basolateral amygdala-evoked firing of nucleus accumbens neurons. J. Neurosci. **21:** 6370–6376.
2. SCOTT, L. *et al.* 2002. Selective up-regulation of dopamine D1 receptors in dendritic spines by NMDA receptor activation. Proc. Natl. Acad. Sci. USA **99:** 1661–1664.
3. KELLEY, A.E. & K.C. BERRIDGE. 2002. The neuroscience of natural rewards: relevance to addictive drugs. J. Neurosci. **22:** 3306–3311.
4. BERKE, J.D. & S.E. HYMAN. 2000. Addiction, dopamine, and the molecular mechanisms of memory. Neuron. **25:** 515–532.
5. MALDONADO-IRIZARRY, C.S. & A.E. KELLEY. 1995. Excitatory amino acid receptors within nucleus accumbens subregions differentially mediate spatial learning in the rat. Behav. Pharmacol. **6:** 527–539.
6. KELLEY, A.E., S. SMITH-ROE & M.R. HOLAHAN. 1997. Response-reinforcement learning is dependent on NMDA receptor activation in the nucleus accumbens core. Proc. Natl. Acad. Sci. USA **94:** 12174–12179.
7. SMITH-ROE, S.L., K. SADEGHIAN & A.E. KELLEY. 1999. Spatial learning and performance in the radial arm maze is impaired after N-methyl-D-aspartate (NMDA) receptor blockade in striatal subregions. Behav. Neurosci. **113:** 703–717.
8. BALDWIN, A.E. *et al.* 2000. N-methy-D-aspartate receptor-dependent plasticity within a distributed corticostriatal network mediates appetitive instrumental learning. Behav. Neurosci. **114:** 1–15.
9. ABEL, T. & K.M. LATTAL. 2001. Molecular mechanisms of memory acquisition, consolidation and retrieval. Curr. Opin. Neurobiol. **11:** 180–187.

10. BALDWIN, A.E., K. SADEGHIAN & A.E. KELLEY. 2002. Appetitive instrumental learning requires coincident activation of NMDA and dopamine D1 receptors within the medial prefrontal cortex. J. Neurosci. **22:** 1063–1071.
11. FREY, U., Y.-Y. HUANG & E.R. KANDEL. 1993. Effects of cAMP simulate a late stage of LTP in hippocampal CA1 neurons. Science **260:** 1661–1664.
12. ABEL, T. *et al.* 1997. Genetic demonstration of a role for PKA in the late phase of LTP and in hippocampus-based long-term memory. Cell **88:** 615–626.
13. SCHAFE, G.E., T.E. THIELE & I.L. BERNSTEIN. 1998. Conditioning method dramatically alters the role of amygdala in taste aversion learning. Learn. Mem. **5:** 481–492.
14. BALDWIN, A.E. *et al.* 2002. Appetitive instrumental learning is impaired by inhibition of cAMP-dependent protein kinase within the nucleus accumbens. Neurobiol. Learn. Mem. **77:** 44–62.
15. GURDEN, H., M. TAKITA & T.M. JAY. 2000. Essential role of D1 but not D2 receptors in the NMDA receptor-dependent long-term potentiation at hippocampal-prefrontal cortex synapses in vivo. J. Neurosci. **20:** RC106.
16. DAVIS, H.P. & L.R. SQUIRE. 1984. Protein synthesis and memory: a review. Psychol. Bull. **96:** 518–559.
17. DUDAI, Y. 1996. Consolidation: fragility on the road to the engram. Neuron **17:** 367–370.
18. NADER, K., G.E. SCHAFE & J.E. LE DOUX. 2000. Fear memories require protein synthesis in the amygdala for reconsolidation after retrieval. Nature **406:** 722–726.
19. HERNANDEZ, P.J., K. SADEGHIAN & A.E. KELLEY. 2002. Early consolidation of instrumental learning requires protein synthesis in the nucleus accumbens. Nat. Neurosci. **5:** 1327–1331.
20. ANDRZEJEWSKI, M.E. & A.E. KELLEY. 2002. The role of dopamine D1 and NMDA receptors on the central and basolateral amygdala on instrumental conditioning. Soc. Neurosci. Abstr. 28.
21. GALLAGHER, M. & G. SCHOENBAUM. 1999. Functions of the amygdala and related forebrain areas in attention and cognition. Ann. N.Y. Acad. Sci. **877:** 397–411.
22. MCGEORGE, A.J. & R.L.M. FAULL. 1989. The organization of the projection from the cerebral cortex to the striatum in the rat. Neuroscience **29:** 503–537.
23. HOUK, J.C., J.L. ADAMS & A.G. BARTO. 1995. A model of how the basal ganglia generate and use neural signals that predict reinforcement. *In* Models of Information Processing in the Basal Ganglia. J.C. Houk, J.L. Davis & D.G. Beiser, Eds.: 249–270. MIT Press. Cambridge, MA.
24. AMOS, A. 2000. A computational model of information processing in the frontal cortex and basal ganglia. J. Cogn. Neurosci. **12:** 505–519.
25. PRATT, W.E. 2003. Effects of muscarinic and nicotinic receptor blockade of the nucleus accumbens on operant learning, locomotion, and sucrose consumption. Soc. Neurosci. Abstr. 79.
26. SUZUKI, T. *et al.* 2001. Dopamine-dependent synaptic plasticity in the striatal cholinergic interneurons. J. Neurosci. **21:** 6492–6501.
27. BLAZQUEZ, P.M. *et al.* 2002. A network representation of response probability in the striatum. Neuron **33:** 973–982.
28. CHANG, Q. & P.E. GOLD. 2003. Switching memory systems during learning: changes in patterns of brain acetylcholine release in the hippocampus and striatum in rats. J. Neurosci. **23:** 3001–3005.
29. KITABATAKE, Y. *et al.* 2003. Impairment of reward-related learning by cholinergic cell ablation in the striatum. Proc. Natl. Acad. Sci. USA **100:** 7965–7970.
30. RESCORLA, R.A. 1991. Associative relations in instrumental learning: the eighteenth Bartlett memorial Lecture. Q. J. Exp. Psychol. **43B:** 1–23.
31. SKINNER, B.F. 1953. Science and Human Behavior. Free Press. New York.

Glutamate Transmission and Addiction to Cocaine

PETER W. KALIVAS, KRISTA McFARLAND, SCOTT BOWERS, KAREN SZUMLINSKI, ZHENG-XIONG XI, AND DAVID BAKER

Department of Physiology and Neuroscience, Medical University of South Carolina, Charleston, South Carolina 29464, USA

ABSTRACT: A variety of data point to the possibility that neuroadaptations in glutamate transmission are produced by repeated exposure to cocaine that result in the expression of behaviors characteristic of addiction, such as craving and relapse. Using the reinstatement model of relapse in rats, glutamate release in the projection from the prefrontal cortex to the nucleus accumbens has been shown to underlie cocaine- and stress-primed reinstatement. In this report, four adaptations produced by withdrawal from repeated cocaine are described that may regulate the release of glutamate underlying reinstatement of drug-seeking resulted. (1) Neurons in the prefrontal cortex have increased levels of activator of G protein signaling 3 (AGS3) that causes reduced signaling through Gi coupled receptors, and normalization of AGS3 blocked cocaine-primed reinstatement. (2) The activity of the cystine-glutamate exchanger is reduced resulting in decreased extracellular glutamate in the nucleus accumbens, and normalization of exchanger activity prevented cocaine-primed reinstatement. (3) Metobotropic glutamate receptor function is diminished after repeated cocaine administration that results in reduced regulation of glutamate release. (4) Homer1 protein is reduced in the nucleus accumbens, and Homer2 knockout mice show enhanced responsiveness to cocaine. Taken together, there appears to be both pre- and postsynaptic changes in glutamate transmission that dysregulates the glutamatergic projection from the prefrontal cortex to the nucleus accumbens. These adaptations are hypothesized to facilitate glutamate release in response to a cocaine injection or acute stress and lead to the reinstatement of drug-seeking behavior.

KEYWORDS: reinstatement; cocaine; G protein; cystine-glutamate exchange; mGluR; glutamate; prefrontal cortex; nucleus accumbens; cocaine

INTRODUCTION

Repeated exposure and withdrawal from addictive drugs, including cocaine, produces neuroadaptations that result in behaviors characteristic of addiction, such as craving and relapse.[1] Using animal models of addiction, such as behavioral sensitization, drug self-administration and the reinstatement of drug-seeking, a role for

Address for correspondence: Peter W. Kalivas, Ph.D., Medical University of South Carolina, Department of Physiology, 650 MUSC Complex, Suite 607, Charleston, SC 29425. Voice: 843-792-2005; fax: 843-792-4423.
kalivasp@musc.edu

glutamate transmission in both the development and expression of these behaviors has been identified.[2–4] Thus, antagonists of glutamate receptors and lesioning or stimulating glutamatergic cell groups in the cortex affects addiction-related behaviors.

The reinstatement of drug-seeking behavior has become a widely used animal model of craving and relapse.[5] In a series of studies it was shown that reversible inactivation of the prefrontal cortex prevents the reinstatement drug-seeking that is primed by either cocaine, stress or a cocaine-associated cue.[6] Recently it was found that there is an increase in glutamate release in the nucleus accumbens when an animal engages in drug-seeking behavior and that this release is blocked by inactivation of the prefrontal cortex.[7] The present report compiles a number of recent experiments, some published elsewhere,[8,9] that identify cellular adaptations potentially underlying the dysregulated release of glutamate in the projection from the prefrontal cortex to the nucleus accumbens that mediates drug-seeking behavior.

MODEL OF GLUTAMATERGIC ADAPTATIONS MEDIATING REINSTATEMENT

FIGURE 1 illustrates the glutamatergic projection from the medial prefrontal cortex to the nucleus accumbens. Based upon careful mapping studies it appears that the projection from the more dorsal aspects of the medial prefrontal cortex (including the prelimbic and ventral anterior cingulate cortices[10]) to the more lateral core subcompartment of the nucleus accumbens is especially critical to the primed reinstatement of drug-seeking.[6,7,11] The model illustrates four neuroadaptations of potential importance in mediating the dysregulated glutamatergic transmission in this projection that underlies primed reinstatement of drug-seeking. (1) Within the prefrontal cortex there is an upregulation of AGS3 that decreases signaling through Gi coupled receptors and may contribute to the lack of membrane bistability noted in prefrontal pyramidal cells after withdrawal from cocaine self-administration.[12] (2) In the nucleus accumbens there is a reduction in the activity of the cystine-glutamate exchanger that mediates reduced basal extracellular levels of glutamate often observed in the nucleus accumbens after withdrawal from repeated cocaine.[7,13,14] (3) In the nucleus accumbens there is decreased sensitivity of group II metabotropic glutamate receptors (mGluR2/3) that provide inhibitory feedback regulation of both synaptic glutamate release, as well as extracellular glutamate derived from cystine-glutamate exchange.[9,15–17] (4) Reduced levels of the scaffolding protein, Homer1bc, are elicited by withdrawal from repeated cocaine, which may contribute to the desensitization of group I mGluRs, as well as reported cocaine-induced alterations in the responsiveness of ionotropic glutamate receptors.[18–20]

Increased AGS3 in the Prefrontal Cortex

AGS3 is a protein that regulates G-protein signaling via binding to the GDP-bound conformation of Giα, thereby competing for the binding of Giα to Gβγ and reducing the access of Giα to surface receptor and effector proteins.[21,22] Withdrawal from repeated cocaine administration produces an increase in the level of AGS3 in the prefrontal cortex and core of the nucleus accumbens that is accompanied by reduced signaling through Giα coupled receptors, including mGluR2/3, D2, GABA$_B$

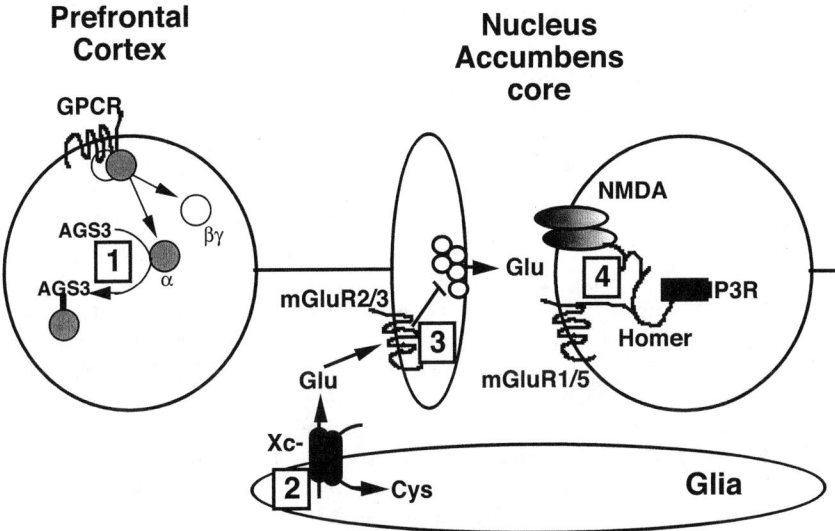

FIGURE 1. Illustration of the neuroadaptations produced by repeated cocaine in the glutamatergic projection from the prefrontal cortex to the nucleus accumbens core. The numbers refer to specific adaptations described in the text. (1) Increase AGS3 results in reduced signaling through Giα coupled receptors (GPCR) because AGS3 competes with βγ for Giα binding. (2) The the rate of cystine-glutamate exchange (Xc-) is reduced in the nucleus accumbens, which decreases the basal extracellular level of glutamate. (3) Reduced extracellular glutamate combined with desensitization of mGluR2/3 decreases inhibitory tone on synaptic glutamate release. (4) Reduced levels of Homer1bc compromise signaling in the postsynaptic density between group I mGluR (mGluR1/5), NMDA receptors and IP3 receptors.

and A1 receptors.[9,23–27] When AGS3 levels are restored to normal by infusing antisense oligonucleotides into the prefrontal cortex of rats extinguished from cocaine self-administration the reinstatement of drug-seeking by a cocaine priming injection was prevented. Moreover, when the pumps containing antisense oligonucleotide are removed and the level of AGS3 allowed to rise to the cocaine-elevated concentration, reinstatement to an acute injection of cocaine was restored.[27] These data indicate that the disruption in Giα signaling caused by increased AGS3 is a mediator of reinstatement behavior in rats trained to self-administer cocaine. Consistent with this, if an AGS3 pharmacophore containing the G protein binding domains of AGS3 fused to a Tat protein (to impart cell permeability[28]) is infused into the nucleus accumbens the capacity of an acute injection of cocaine to elevate glutamate in the nucleus accumbens is augmented.[27] Thus, mimicking cocaine-induced elevation in AGS3 in the prefrontal cortex also caused enhanced release of prefrontal-accumbens glutamate, similar to what occurs after cocaine self-administration.[7]

Decreased Cystine-Glutamate Exchange and mGluR2/3 Function

The extracellular levels of glutamate in the nucleus accumbens are maintained primarily by the activity of the cystine-glutamate exchanger, not by synaptic glutamate release.[15,29] The cystine-glutamate exchanger is a heterodimer and xCT is the unique protein constituent of the dimer.[30] The complex facilitates the exchange of one extracellular cystine for one intracellular glutamate molecule at a rate that is energy independent and determined by substrate concentrations and probably protein phosphorylation.[15,31,32] Repeated cocaine administration decreases the rate of cystine-glutamate exchange causing a reduction in the basal extracellular levels of glutamate in the nucleus accumbens.[8] When the activity of the cystine-glutamate exchanger is restored by increasing substrate via a systemic injection of the procysteine drug N-acetylcysteine, the levels of glutamate were normalized and the capacity of a cocaine-priming injection to reinstate drug-seeking behavior was abolished.

As outlined in FIGURE 1, it is hypothesized that the reduced levels of basal extracellular glutamate after discontinuing cocaine administration reduces tone on mGluR2/3 release inhibiting autoreceptors, thereby causing the facilitated synaptic release of glutamate observed during cocaine-primed reinstatement.[7] Contributing to this putative mechanism it was shown that after 3 weeks of withdrawal from repeated cocaine the amount of inactive mGluR2/3 dimer is increased in the nucleus accumbens and, consistent with receptor desensitization, the remaining mGluR monomer is more highly phosphorylated.[9,33]

Homer Proteins and Excitatory Transmission

Electrophysiological studies clearly show altered excitatory transmission in the nucleus accumbens after withdrawal from repeated cocaine.[18,19,34] The changes in excitatory transmission have not been linked to consistent changes in the expression of ionotropic glutamate receptor subunits,[35,36] posing the possibility that alterations in proteins in the postsynaptic density that modulate glutamate receptor trafficking and signaling efficiency may be affected by repeated cocaine. Consistent with this possibility the levels of Homer1bc were shown to be reduced after 3 weeks of withdrawal from repeated cocaine.[20] Homer proteins bind to a number of proteins in the postsynaptic density via an EVH1 domain, and notably are pivotal in linking group I mGluR to ionotropic glutamate receptors.[37,38] Correspondingly, the influences of mGluR receptors on glutamate transmission are reduced following repeated cocaine administration or in Homer2 knockout mice.[20,39] Moreover, the Homer2 gene deletion results in a behavioral phenotype akin to animals that have been pretreated with repeated cocaine injections. Thus, Homer2 KO mice are more sensitive to the locomotor stimulant and rewarding effects of cocaine administration.[39] While a functional link between the alteration in Homer protein by repeated cocaine and the electrophysiological effects of repeated cocaine on ionotropic glutamate receptors is not entirely clear, recently it was found that in cultured hippocampal cells from Homer2 KO mice there is a marked reduction in both NMDA and AMPA induced currents (John Woodward, Ph.D., Medical University of South Carolina, unpublished observation).

CONCLUSIONS

Repeated cocaine administration initiates a number of changes in the prefrontal cortex and nucleus accumbens. Some of these changes are present for weeks or months after the last cocaine injection. As outlined in FIGURE 1, we have recently characterized four cocaine-induced neuroadaptations that may contribute to the craving and relapse associated with cocaine addiction. There appears to be a strong association between the reinstatement of drug-seeking in an animal model of relapse and enhanced glutamate transmission in the core of the nucleus accumbens.[7,8] Each of the four adaptations outline above and illustrated in FIGURE 1 have the potential to contribute to changes in glutamate transmission in the projection from the prefrontal cortex to the nucleus accumbens that appears to mediate drug-primed reinstatement. The increase in AGS3 in the prefrontal cortex decreases $G_i\alpha$ signaling and normalization of AGS3 levels blocks cocaine primed reinstatement while increasing AGS3 in drug-naïve rats causes an increase in glutamate release by an acute injection of cocaine. The reduction in cystine-glutamate exchange in the nucleus accumbens reduces the basal extracellular levels of glutamate and may reduce inhibitory tone by extracellular glutamate on mGluR2/3 receptors. Moreover, the mGluR2/3 receptors are already desensitized following repeated cocaine. Combining these two adaptations would result in a marked decrease in inhibitory autoregulation of synaptic glutamate release, thereby increasing glutamate release. Consistent with a role for these adaptations in drug-seeking, cocaine-primed reinstatement was abolished when cystine-glutamate exchange was increased and the level of extracellular glutamate restored to normal. Finally, reduced concentration of Homer protein in the nucleus accumbens by repeated cocaine administration may contribute to reported postsynaptic changes in glutamate transmission. However, in general postsynaptic glutamate signaling appears to be reduced after repeated cocaine indicating that the reduction in Homer may serve a compensatory function to dampen the effect of the facilitated presynaptic glutamate release. The details of how molecular neuroadaptations produced by cocaine promote or inhibit drug craving and relapse remain to be fully elucidated. However, the recent findings that reversing some of these adaptations can prevent reinstatement in animals models of drug-seeking indicates that this research tact may ultimately yield novel pharmacological therapies capable of helping addicts control drug craving and relapse.

ACKNOWLEDGMENTS

This research was supported in part by U.S. Public Health Service Grants MH-40827, DA-03906, DA-06074, and DA-07288.

REFERENCES

1. NESTLER, E. 2001. Molecular basis of long-term plasticity underlying addiction. Nature Rev. **2:** 119–128.
2. EVERITT, B.J. & M.E. WOLF. 2002. Psychomotor stimulant addiction: a neural systems perspective. J. Neurosci. **22:** 3312–3320.
3. WOLF, M.E. 1998. The role of excitatory amino acids in behavioral sensitization to psychomotor stimulants. Progr. Neurobiol. **54:** 679–720.

4. VANDERSCHUREN, L.J. & P.W. KALIVAS. 2000. Alterations in dopaminergic and glutamatergic transmission in the induction and expression of behavioral sensitization: a critical review of preclinical studies. Psychopharmacology (Berl). **151:** 99–120.
5. SHALEV, U., J.W. GRIMM & Y. SHAHAM. 2002. Neurobiology of relapse to heroin and cocaine seeking: a review. Pharmacol Rev. **54:** 1–42.
6. MCFARLAND, K. & P.W. KALIVAS. 2001. The circuitry mediating cocaine-induced reinstatement of drug-seeking behavior. J. Neurosci. **21:** 8655–8663.
7. MCFARLAND, K., C.C. LAPISH & P.W. KALIVAS. 2003. Glutamate, not dopamine, in the accumbens core mediates cocaine-induced reinstatement of drug-seeking behavior. J. Neurosci. **23:** 3531–3537.
8. BAKER, D.A. et al. 2003. Neuroadaptations in cystine-glutamate exchange underlie cocaine relapse. Nat. Neurosci. **6:** 743–749.
9. XI, Z.-X. et al. 2002. Modulation of group II metabotropic glutamate receptor signaling by chronic cocaine. J. Pharmacol. Exp. Ther. **303:** 608–615.
10. PAXINOS, G. & C. WATSON. 1986. The Rat Brain in Stereotaxic Coordinates. Academic Press. New York.
11. PARK, W.K. et al. 2002. Cocaine administered into the medial prefrontal cortex reinstates cocaine-seeking behavior by increasing AMPA receptor-mediated glutamate transmission in the nucleus accumbens. J. Neurosci. **22:** 2916–2925.
12. TRANTHAM, H. et al. 2002. Repeated cocaine administration alters the electrophysiological properties of prefrontal cortical neurons. Neuroscience **113:** 749.
13. PIERCE, R. C. et al. 1996. Repeated cocaine augments excitatory amino acid transmission in the nucleus accumbens only in rats having developed behavioral sensitization. J. Neurosci. **16:** 1550–1560.
14. HOTSENPILLER, G., M. GIORGETTI & M.E. WOLF. 2001. Alterations in behaviour and glutamate transmission following presentation of stimuli previously associated with cocaine exposure. Eur. J. Neurosci. **14:** 1843–1855.
15. BAKER, D.A. et al. 2002. The origin and neuronal function of in vivo nonsynaptic glutamate. J. Neurosci. **22:** 9134–9141.
16. XI, Z.-X. et al. 2002. Inhibition of glutamate release by group II metabotropic glutamate receptors. J. Pharmacol. Exp.Ther. **300:** 162–171.
17. CARTMELL, J. & D.D. SCHOEPP. 2000. Regulation of neurotransmitter release by metabotropic glutamate receptors. J. Neurochem. **75:** 889–907.
18. THOMAS, M.J. et al. 2001. Long-term depression in the nucleus accumbens: a neural correlate of behavioral sensitization to cocaine. Nat. Neurosci. **4:** 1217–1223.
19. WHITE, F. et al. 1995. Repeated administration of cocaine or amphetamine alters neuronal responses to glutamate in the mesoaccumbens dopamine system. J. Pharmacol. Expl. Therapeutics **273**(1): 445–454.
20. SWANSON, C. et al. 2001. Repeated cocaine administration attenuates group I metabotropic glutamate receptor-mediated glutamate release and behavioral activation: A potential role for Homer 1b/c. J. Neurosci. **21:** 9043–9052.
21. NATOCHIN, M. et al. 2000. AGS3 inhibits GDP dissociation from galpha subunits of the Gi family and rhodopsin-dependent activation of transducin. J. Biol. Chem. **275:** 40981–40985.
22. BERNARD, M. et al. 2001. Selective interaction of AGS3 with G-proteins and the influence of AGS3 on the activation state of G-proteins. J. Biol. Chem. **276:** 1585–1593.
23. XI, Z.-X. et al. 2003. GABA transmission in the nucleus accumbens is altered after withdrawal from repeated cocaine. J. Neurosci. **23:** 3498–3505.
24. SHOJI, S. et al. 1997. Chronic cocaine enhances gamma-aminobutyric acid and glutamate release by altering presynaptic and not postsynaptic gamma-aminobutyric acidB receptors within the rat dorsolateral septal nucleus. J. Pharmacol. Exp. Ther. **280:** 129–137.
25. ZHANG, K. et al. 2000. GABAB receptors: altered coupling to G-proteins in rats sensitized to amphetamine. Neuroscience **101:** 5–10.
26. TODA, S., L.F. ALGUACIL & P.W. KALIVAS. 2003. Repeated cocaine administration changes the function and subcellular distribution of adenosine A1 receptor in the rat nucleus accumbens. J. Neurochem. In press.

27. BOWERS, M.S. *et al.* 2003. Activator of G-protein signaling 3: a gatekeeper of cocaine addiction. Submitted for publication.
28. NAGAHARA, H. *et al.* 1998. Transduction of full-length TAT fusion proteins into mammalian cells: TAT-p27Kip1 induces cell migration. Nat. Med. **4:** 1449–1452.
29. TIMMERMAN, W. & B.H. WESTERINK. 1997. Brain microdialysis of GABA and glutamate: what does it signify? Synapse **27:** 242–261.
30. SATO, H. *et al.* 1999. Cloning and expression of a plasma membrane cystine/glutamate exchange transporter composed of two distinct proteins. J. Biol. Chem. **274:** 11455–11458.
31. GOCHENAUER, G.E. & M.B. ROBINSON. 2001. Dibutyryl-cAMP (dbcAMP) up-regulates astrocytic chloride-dependent L-[^{3}H]glutamate transport and expression of both system xc(-) subunits. J Neurochem. **78:** 276–286.
32. WARR, O., M. TAKAHASHI & D. ATTWELL. 1999. Modulation of extracellular glutamate concentration in rat brain slices by cystine-glutamate exchange. J. Physiol. **5143:** 783–793.
33. SCHAFFHAUSER, H. *et al.* 2000. cAMP-dependent protein kinase inhibits mGluR2 coupling to G-proteins by direct receptor phosphorylation. J. Neurosci. **20:** 5663–5670.
34. ZHANG, X.-F., X.-T. HU & F. J. WHITE. 1998. Whole-cell plasticity in cocaine withdrawal: reduced sodium current in nucleus accumbens neurons. J. Neurosci. **18:** 488–498.
35. LU, W. & M. WOLF. 1999. Repeated amphetamine administration alters AMPA receptor subunit expression in rat nucleus accumbens and medial prefrontal cortex. Synapse **32:** 119–131.
36. CHURCHILL, L. *et al.* 1999. Repeated cocaine alters glutamate receptor subunit levels in the nucleus accumbens and ventral tegmental area of rats that develop behavioral sensitization. J. Neurochem. **72:** 2397–2403.
37. TU, J.C. *et al.* 1998. Homer binds a novel proline-rich motif and links group 1 metabotropic glutamate receptors with IP3 receptors. Neuron **21:** 717-726.
38. TU, J. *et al.* 1999. mGluR/Homer and PSD-95 complexes are linked by the Shank family of postsynaptic density proteins. Neuron **23:** 583–592.
39. SZUMLINSKI, K. *et al.* 2003. Altered cocaine-induced behavioral and neurochemical plasticity in Homer2 knock-out mice. Submitted for publication.

NMDA Receptor Antagonism and the Ethanol Intoxication Signal

From Alcoholism Risk to Pharmacotherapy

JOHN H. KRYSTAL,[a,b,c] ISMENE L. PETRAKIS,[a,b,c] EVGENY KRUPITSKY,[d] CHRISTIAN SCHÜTZ,[e] LOUIS TREVISAN,[a,b,c] AND D. CYRIL D'SOUZA[a,b,c]

[a]*Department of Psychiatry, Yale University School of Medicine, New Haven, Connecticut 06510, USA*

[b]*Alcohol Research Center (116-A), VA Connecticut Healthcare System, West Haven, Connecticut 06516, USA*

[c]*NIAAA Center for the Translational Neuroscience of Alcoholism, Abraham Ribicoff Research Faciltiies, Connecticut Mental Health Center, New Haven, Connecticut 06519, USA*

[d]*Department of Pharmacology, St. Petersburg Pavlov State Medical University, St. Petersburg, Russia*

[e]*Department of Psychiatry, University of Bonn, Bonn, Germany*

ABSTRACT: This paper reviews clinical evidence suggesting that antagonism of the *N*-methyl-D-aspartate subtype of glutamate receptors by ethanol may convey an important component of the ethanol intoxication signal, that is, subjective and objective responses associated with the consumption of a large amount of ethanol. It will then review recent evidence that two phenotypes associated with increased risk for heavy alcohol consumption, recovering ethanol-dependent patients, and healthy individuals with a family history of alcohol dependence, exhibit reduced sensitivity to the dysphoric consequences of administration of the NMDA receptor antagonist, ketamine. Each of these groups displays reduced sensitivity to a potentially important response that might normally trigger the cessation of ethanol consumption. These data raise the possibility that alterations in NMDA receptor function that reduce the response to the NMDA antagonist component of ethanol may increase the risk for heavy drinking. This hypothesis is consistent with growing evidence that NMDA receptor antagonists may play a role in the treatment of alcoholism by suppressing alcohol withdrawal, reducing the development or expression of alcohol tolerance, or preventing or reversing the sensitiziation to ethanol effects.

KEYWORDS: ethanol; alcoholism; alcohol dependence; phenotype; alcoholism vulnerability; pharmacotherapy; glutamate; *N*-methyl-D-aspartate receptor; ketamine

Address for correspondence: John H. Krystal, Alcohol Research Center (116-A), VA Connecticut Healthcare System, 950 Campbell Ave., West Haven, CT 06516. Voice: 203-937-4790; fax: 203-397-3468.

john.krystal@yale.edu

INTRODUCTION

Ethanol acts on ligand-gated, voltage-gated, and G-protein–regulated ion channels and metabotropic receptors in the brain.[1] The blockade of the N-methyl-D-aspartate (NMDA) glutamate receptor is among the highest potency ethanol actions in the brain.[2] In animals, NMDA receptor antagonists have ethanol-like discriminative stimulus properties that follow a characteristic pattern: NMDA antagonist effects are most similar to ethanol effects at higher subanesthetic doses, and these doses of NMDA antagonists substitute best for rather high doses of ethanol.[3] In contrast, benzodiazepines or barbiturates, drugs that facilitate $GABA_A$ receptor function, produce a different dose-related profile of generalization to the discriminative stimulus effects of ethanol: they are most similar to the effects of relatively low doses of ethanol, and their contribution to the discriminative stimulus effects of ethanol may decline at high doses.[4,5]

One potential interpretation of these preclinical data is that ethanol actions at the $GABA_A$ receptor might be most revelant to the subjective effects of low levels of ethanol intoxication. With respect to alcoholism, ethanol effects on GABA systems might pertain to the priming of ethanol consumption by the initial alcohol drinks. In contrast, the antagonism of NMDA glutamate receptors by ethanol might be expected to convey the cognitive, subjective, and physiologic responses associated with ethanol intoxication that normally constitutes a negative feedback signal for subsequent ethanol intoxication. In the face of intact mechanisms that prime or reward low levels of alcohol intoxication, one might expect that reduced responsiveness to the NMDA receptor antagonist action of ethanol might result in a deficient negative feedback signal on alcohol consumption and might thereby constitute a phenotype of risk for heavy drinking. From this perspective, it is perhaps not surprising that animals that show reduced response to NMDA receptor antagonists show increased levels of ethanol self-administration. Also, drugs that attenuate NMDA receptor function reduce alcohol self-administration (see Ref. 6 for a review).

Building on these preclinical findings, a series of recent studies conducted in healthy humans, individuals at increased familial risk for alcoholism, and recovering ethanol-dependent patients have explored the role of NMDA receptors in alcohol intoxication, dependence, the vulnerability to alcoholism, and the pharmacotherapy of alcoholism. The purpose of this review is the briefly consider these studies and their potential implications for understanding aspects of the risk for heavy drinking and the pharmacotherapy of alcoholism.

ETHANOL-LIKE EFFECTS OF NMDA RECEPTOR ANTAGONISTS IN HUMANS

In humans, the subjective effects of subanesthetic doses of NMDA receptor antagonists resemble subjective effects associated with ethanol intoxication. In studies of both phencyclidine[7] and ketamine,[8] healthy subjects spontaneously reported ethanol-like subjective effects. In the first of these studies, subjects described feeling "smashed" or "drunk" after phencyclidine administration. These anecdotal reports were followed by studies that used ketamine, dextromethorphan, or the $glycine_B$ partial agonist D-cycloserine to carefully assess the ethanol-like properties of drugs that reduce NMDA receptor function in healthy subjects and ethanol-dependent patients.[9–13] In alcohol-dependent patients, ketamine had euphoric effects that were associated with prominent sedation and cognitive impairment. The ketamine doses that were judged most similar to ethanol were associated with high levels of ethanol intoxication, approximately eight or more standard alcohol drinks.[9] Thus, the ethanol-like effects of ketamine may signal intoxication, that is, a subjective response indicative of current or impending behavioral impairment. Consistent with this interpretation, two[9,10] of the three[14] NMDA receptor antagonist studies in ethanol-dependent patients found that ketamine did not stimulate ethanol craving. The absence of alcohol craving in the face of an ethanol-like intoxication state may suggest that the NMDA receptor antagonist intoxication state conveys a negative feedback signal upon drinking in humans.

BLUNTED KETAMINE RESPONSE IN ALCOHOL-DEPENDENT PATIENTS AND HEALTHY SUBJECTS WITH A FAMILY HISTORY POSITIVE FOR ALCOHOLISM: A PHENOTYPE OF RISK FOR HEAVY DRINKING?

Ethanol dependence is associated with cross-tolerance to other NMDA receptor antagonists. It has been long known, for example, that NMDA antagonists suppress ethanol withdrawal seizures.[15] Further, the upregulation of NMDA receptors associated with ethanol dependence may contribute to ethanol tolerance and acute withdrawal symptoms, and may also contribute to withdrawal-related neurotoxicity.[16] As has been reviewed recently,[6] the upregulation of NMDA receptors associated with ethanol dependence may alter the experience of ethanol intoxication by reducing the dysphoric or behavior-impairing consequences of the NMDA receptor antagonist component of ethanol intoxication, while preserving the rewarding aspects of NMDA receptor antagonism.

The hypothesis that ethanol dependence–related upregulation of NMDA receptor function might reduce sensitivity to the NMDA receptor antagonist component of ethanol is consistent with ethanol dependence–related alterations in ketamine response.[17] Recovering ethanol-dependent patients had reduced dysphoric mood responses, perceptual changes, and cognitive impairments compared to a comparison group of healthy subjects, but the stimulant and euphoric responses to ketamine did not differ between the groups (see FIG. 1). These changes in ketamine response might signal a shift in the reward valence of ethanol, that is, it might seem to be a relatively more rewarding drug to ethanol-dependent patients. This characteristic might create expectations that might reinforce heavy drinking once it began. Also, if the capacity of ethanol to block NMDA receptors conveys an important signal to stop drinking and if this signal is blunted in patients, it would not be surprising that they might be at increased risk for heavy drinking early in recovery.

Individuals with a first degree relative with ethanol dependence (FHP), as a group, have reduced sensitivity to aspects of ethanol intoxication compared to healthy individuals without a family history of alcohol dependence (FHN).[18,19] Over time, FHP individuals appear to learn that alcohol is a relatively more rewarding drug than FHN individuals.[20,21] Perhaps as a result, reduced sensitivity to ethanol in

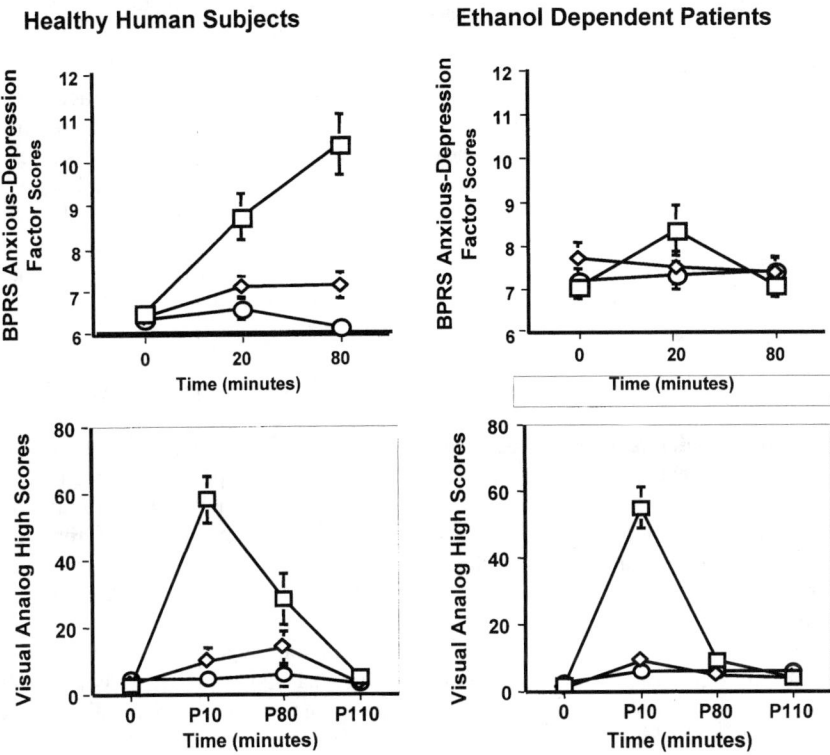

FIGURE 1. This figure presents data suggesting that there is a shift in the reward valence of ketamine effects, that is, a reduction in the dysphoric effects and preservation of the euphoric effects in recovering alcohol-dependent patients compared to healthy human subjects. *Top figures*: The effects of placebo (*circles*), ketamine 0.1 mg/kg (*diamonds*), and ketamine 0.5 mg/kg (*squares*) on the Brief Psychiatric Rating Scale anxious-depression factor scores in healthy human subjects ($n=26$; *top left figure*) and recovering ethanol-dependent patients ($n=34$; *top right figure*). *Bottom figures*: The effects of placebo (*circles*), ketamine 0.1 mg/kg (*diamonds*), and ketamine 0.5 mg/kg (*squares*) on the "high" visual analogue scale scores in healthy human subjects ($n=26$; *bottom left figure*) and recovering ethanol-dependent patients ($n=34$; *bottom right figure*). Values are presented as mean ± standard error of the mean (SEM). Figures are modified from Ref. 17.

FHP individuals appears to be one of the strongest predictors of the subsequent development of ethanol dependence in this group.[22,23]

Alterations in NMDA receptor function associated with reduced sensitivity to the actions of NMDA receptor antagonists may convey a component of the reduced sensitivity to ethanol in FHP individuals.[6] There is compelling preclinical evidence that inbred rodent strains that differ in the level of expression of NMDA receptor subunits, the functionality of NMDA receptor subunits, or aspects of NMDA receptor-related signal transduction may differ in their propensity to administer ethanol (see Ref. 6 for a review). A recent study supports the clinical relevance of these preclinical investigations. One recent report suggests that a group of FHP healthy subjects showed a pattern of reductions, relative to FHN healthy subjects, in the cognitive and dysphoric mood response to ketamine[13] that was similar to the pattern of alterations in ketamine response seen in ethanol-dependent patients, noted above. Further, there were trends for FHP ethanol-dependent patients to have more pronounced blunting of some cognitive and mood responses to ketamine relative to FHN ethanol-dependent patients (J. Krystal, unpublished observation). These findings suggest that family history of alcoholism and ethanol dependence might produce convergent, but not completely overlapping, alterations in NMDA receptor function associated with increased risk for alcoholism. If so, it would suggest a potentially important gene-environment interaction with respect to the epigenesis of alcoholism.

IMPLICATIONS FOR THE PHARMACOTHERAPY OF ALCOHOLISM

Antiglutamatergic treatments, such as NMDA receptor antagonists or anticonvulsant agents, might reduce the motivation for drinking by suppressing alcohol withdrawal symptoms. Glutamatergic activation and NMDA receptor upregulation contribute to the alcohol withdrawal syndrome.[24,25] A recent pilot study reported that lamotrigine, an anticonvulsant drug that reduces glutamate release, and memantine, a midpotency voltage-dependent antagonist of NMDA glutamate receptors, reduced acute alcohol withdrawal symptoms as effectively as benzodiazepine in alcohol-dependent patients (E. Krupitsky and J. Krystal, unpublished observation). These data are consistent with a growing literature that suggests that anticonvulsant therapies are effective for suppressing withdrawal and superior to benzodiazepines in supporting abstinence.[26,27]

However, NMDA antagonists may reduce ethanol consumption through other mechanisms. For example, recent data suggest that a drug with NMDA antagonist properties, acamprosate, was more effective when started after acute alcohol withdrawal was completed than when started during acute withdrawal,[28] even though it may have neuroprotective effects during ethanol withdrawal.[29] This finding argues against the contention that acamprosate works by suppressing acute withdrawal symptoms, and it suggests that NMDA antagonists might have other beneficial effects in alcohol-dependent patients. For example, it is possible that NMDA receptor antagonists at low doses directly convey a component of the subjective "stop drinking" subjective signal arising from the similarity of the subjective effects of NMDA antagonists to relatively high ethanol doses. In light of some preliminary evidence that NMDA antagonists might potentiate aspects of ethanol intoxication,[30] it is possible that NMDA antagonist administration might, in effect, reduce the expression of ethanol tolerance. This aspect of NMDA antagonist action might be particularly relevant to alcohol-dependent or FHP individuals who do not experience a strong intoxication signal when drinking. Further, NMDA receptor antagonists might prevent the development of tolerance to ethanol or the expression of sensitiziation to the stimulatory or rewarding effects of ethanol.[31,32] This action might help to maintain the sensitivity to the dysphoric aspects of ethanol intoxication.

There is growing interest, signaled by Project COMBINE, the NIAAA multicenter study evaluating the combination of acamprosate and naltrexone, in the possibility that the combination of opiate receptor antagonists and NMDA receptor antagonists might have particular value for the treatment of alcohol dependence. For example, a recent study provided interesting data that suggested that the combination of naltrexone and acamprosate was superior to acamprosate and numerically, but not significantly, superior to naltrexone.[33] Other data suggest that the combination of acamprsoate and naltrexone is not superior to naltrexone alone in reducing alcohol consumption.[34]

However, acamprosate is a complicated and poorly understood agent. Some recent human laboratory data might hold out hope for the combination of better understood NMDA receptor antagonists and naltrexone. In one pilot study, pretreatment with naltrexone, 25 mg, appears to promote the dysphoric effects of a subperceptual (low), but not perceptual (higher) ketamine dose in healthy subjects (S. Madonick and J. Krystal, personal communication). These data suggest that naltrexone may potentiate some negative consequences of ethanol intoxication that might be mediated by NMDA glutamate receptors. However, this effect is modest, and it appears that one could drink enough alcohol to overcome this seemingly protective effect of naltrexone. These data are complemented by another study in healthy human subjects that examined the interactive effects of memantine and naltrexone on ethanol intoxication.[35] In this study, memantine had mild euphoric effects, naltrexone had some mild dysphoric effects, and neither agent attenuated the dose-related euphoric effects of ethanol. In contrast, the combination of naltrexone and memantine had mild euphoric effects but blocked the dose-related euphoric effects of ethanol. It is not yet clear whether the combination of memantine and naltrexone is an effective strategy for the pharmacotherapy for alcoholism, but these promising results deserve follow-up in clinical trials.

SUMMARY AND FUTURE RESEARCH

The research reviewed in this paper suggests that ethanol intoxication, ethanol dependence, the vulnerability to alcohol dependence, and the pharmacotherapy of alcoholism could be linked, in part, to NMDA receptor function. The studies reviewed in this paper suggested that (1) NMDA receptor antagonists produce subjective effects that resemble ethanol intoxication in humans, (2) ethanol-dependent patients show marked reductions to the intoxicating effects of NMDA receptor antagonists compared to healthy comparison subjects, and (3) individuals with a first-degree relative with ethanol dependence (FHP) also show reduced sensitivity to the intoxicating effects of ketamine. The strong resemblance of subanesthetic doses of NMDA receptor antagonists to high levels of ethanol ingestion and the absence of ketamine-induced alcohol craving in alcohol-dependent patients suggested that the NMDA antagonist actions of ethanol contributed to the subjective sense of intoxication. A central thesis of this review is that the intoxication signal conveyed by the NMDA receptor antagonist actions of ethanol normally conveys a negative feedback on drinking. Deficits in this negative feedback signal in alcohol-dependent patients and FHP individuals suggest that genetic variation and environmental factors converge to produce phenotypes at high risk for heavy drinking. NMDA receptor antagonist drugs, alone or in combination with other agents, may suppress alcohol withdrawal symptoms. Also, they may play a role in substituting for, restoring, or maintaining this deficient negative feedback signal and thereby may be useful pharmacotherapeutic adjuncts to alcoholism treatment.

The model presented in this paper will need further empirical validation before it can be accepted as an unambiguous guide for future research. Additional unresolved research questions include (1) What genetic mutations convey the reduced responses to ketamine observed in FHP individuals? (2) To what extent does altered ketamine response account for altered ethanol response in FHP individuals? (3) How effective are NMDA receptor antagonist drugs other than acamprosate in the treatment of alcoholism? (4) Are the effects of NMDA receptor antagonists on craving in alcohol-dependent patients dose related, and, if so, are there doses of NMDA receptor antagonists that might promote rather than reduce drinking? (5) Might NMDA receptor antagonist drugs play a role in the prevention of alcohol dependence in heavy social drinkers or FHP individuals?

Overall, the pursuit of the clinical significance of the NMDA receptor antagonist actions of ethanol appears to offer many possible new insights into the neurobiology and treatment of alcoholism. It is hoped that the development of a conceptual framework for this effort may help to hasten the rational development of sorely needed pharmacotherapies for this devastating disorder.

ACKNOWLEDGMENTS

This work was supported by the NIAAA (KO2 AA 00261-01, RO1 AA11321-01A1, P50 AA99-005), the Department of Veterans Affairs (Merit Review Grant, Alcohol Research Center, Schizophrenia Biological Research Center, National Center for Post-traumatic Stress Disorder), and the U.S. Civilian Research and Development Foundation for the Independent States of the Former Soviet Union (CRDF).

REFERENCES

1. KRYSTAL, J.H., B. TABAKOFF. 2002. Ethanol abuse, dependence, and withdrawal: neurobiology and clinical implications. *In* Psychopharmacology: A Fifth Generation of Progress. K.L. Davis, D.S. Charney, J.T. Coyle & C. Nemeroff, Eds.: 1425–1443. Lippincott Williams and Wilkins. Philadelphia.
2. GRANT, K.A. & D.M. LOVINGER. 1995. Cellular and behavioral neurobiology of alcohol: receptor-mediated neuronal processes. Clin. Neurosci. **3**(3): 155–164.
3. GRANT, K.A. & G. COLOMBO. 1993. Discriminative stimulus effects of ethanol: effect of training dose on the substitution of N-methyl-D-aspartate antagonists. J. Pharmacol. Exp. Ther. **264**(3): 1241–1247.
4. GRANT, K.A. & G. COLOMBO. 1993. Pharmacological analysis of the mixed discriminative stimulus effects of ethanol. Alcohol Alcohol. Suppl. **2**: 445–449.
5. GREEN, K.L. & K.A. GRANT. 1998. Evidence for overshadowing by components of the heterogeneous discriminative stimulus effects of ethanol. Drug Alcohol Depend. **52**(2): 149–159.
6. KRYSTAL, J.H., I.L. PETRAKIS, G. MASON, *et al.* 2003. N-methyl-D-aspartate glutamate receptors and alcoholism: reward, dependence, treatment, and vulnerability. Pharmacol. Ther. **99**(1): 79–94.
7. LUBY, E.D., B.D. COHEN, G. ROSENBAUM, *et al.* 1959. Study of a new schizophrenomimetic drug—sernyl. Arch. Neurol. Psychiatry **81**: 363–369.
8. KRYSTAL, J.H., L.P. KARPER, J.P. SEIBYL, *et al.* 1994. Subanesthetic effects of the noncompetitive NMDA antagonist, ketamine, in humans. Psychotomimetic, perceptual, cognitive, and neuroendocrine responses. Arch. Gen. Psychiatry **51**(3): 199–214.

9. KRYSTAL, J.H., I.L. PETRAKIS, E. WEBB, et al. 1998. Dose-related ethanol-like effects of the NMDA antagonist, ketamine, in recently detoxified alcoholics. Arch. Gen. Psychiatry **55**(4): 354–360.
10. KRUPITSKY, E.M., A.M. BURAKOV, T.N. ROMANOVA, et al. 2001. Attenuation of ketamine effects by nimodipine in recently detoxified ethanol dependent men: psychopharmacologic implications of the interaction of NMDA and L-type calcium channel antagonists. Neuropsychopharmacology **25**: 936–947.
11. SCHUTZ, C.G. & M. SOYKA. 2000. Dextromethorphan challenge in alcohol-dependent patients and controls. Arch. Gen. Psychiatry **57**(3): 291–292.
12. KRYSTAL, J., I. PETRAKIS, S. KRASNICKI, et al. 1999. Altered responses to agonists of the strychnine-insensitive glycine NMDA coagonist (SIGLY) site in recently detoxified alcoholics. Alcohol. Clin. Exp. Res. **22**: 94A.
13. PETRAKIS, I.L., D. LIMONCELLI, R. GUERGUIEVA, et al. 2003. Altered NMDA glutamate receptor antagonist response in individuals with a family vulnerability to alcoholism. Am. J. Psychiatry. In review.
14. SOYKA, M., B. BONDY, B. EISENBURG & C.G. SCHUTZ. 2000. NMDA receptor challenge with dextromethorphan—subjective response, neuroendocrinological findings and possible clinical implications. J. Neural Transm. **107**(6): 701–714.
15. HOFFMAN, P.L. & B. TABAKOFF. 1991. The contribution of voltage-gated and NMDA receptor-gated calcium channels to ethanol withdrawal seizures. Alcohol Alcohol. Suppl. **1**: 171–175.
16. TSAI, G. & J.T. COYLE. 1998. The role of glutamatergic neurotransmission in the pathophysiology of alcoholism. Annu. Rev. Med. **49**: 173–184.
17. KRYSTAL, J.H., I.L. PETRAKIS, D. LIMONCELLI, et al. 2003. Altered NMDA glutamate receptor antagonist response in recovering ethanol dependent patients. Neuropsychopharmacology. In press.
18. SCHUCKIT, M.A. 1985. Ethanol-induced changes in body sway in men at high alcoholism risk. Arch. Gen. Psychiatry **42**(4): 375–379.
19. SCHUCKIT, M.A., J.W. TSUANG, R.M. ANTHENELLI, et al. 1996. Alcohol challenges in young men from alcoholic pedigrees and control families: a report from the COGA project. J. Stud. Alcohol **57**(4): 368–377.
20. EARLEYWINE, M. 1994. Anticipated biphasic effects of alcohol vary with risk for alcoholism: a preliminary report. Alcohol. Clin. Exp. Res. **18**(3): 711–714.
21. ERBLICH, J. & M. EARLEYWINE & B. ERBLICH. 2001. Positive and negative associations with alcohol and familial risk for alcoholism. Psychol. Addict. Behav. **15**(3): 204–209.
22. VOLAVKA, J., P. CZOBOR, D.W. GOODWIN, et al. 1996. The electroencephalogram after alcohol administration in high-risk men and the development of alcohol use disorders 10 years later. Arch. Gen. Psychiatry **53**(3): 258–263.
23. SCHUCKIT, M.A. & T.L. SMITH. 1996. An 8-year follow-up of 450 sons of alcoholic and control subjects. Arch. Gen. Psychiatry **53**(3): 202–210.
24. HOFFMAN, P.L. & B. TABAKOFF. 1996. Alcohol dependence: a commentary on mechanisms. Alcohol Alcohol. **31**(4): 333–340.
25. TSAI, G.E., P. RAGAN, R. CHANG, et al. 1998. Increased glutamatergic neurotransmission and oxidative stress after alcohol withdrawal. Am. J. Psychiatry **155**(6): 726–732.
26. MALCOLM, R., H. MYRICK, J.S. ROBERTS & R. ANTON. 2000. The effects of lorazepam and carbamazepine on single vs. multiple previous alcohol withdrawals in an outpatient randomized trial. Alcohol. Clin. Exp. Res. **24**(5): 38A.
27. MALCOLM, R., H. MYRICK, J. ROBERTS, et al. 2002. The differential effects of medication on mood, sleep disturbance, and work ability in outpatient alcohol detoxification. Am. J. Addict. **11**(2): 141–150.
28. HOPKINS, J.S., J.C. GARBUTT, C.L. POOLE, et al. 2002. Naltrexone and acamprosate: meta-analysis of two medical treatments for alcoholism. Alcohol. Clin. Exp. Res. **26**(5): 130A.
29. MAYER, S., B.R. HARRIS, D.A. GIBSON, et al. Acamprosate, MK-801, and ifenprodil inhibit neurotoxicity and calcium entry induced by ethanol withdrawal in organotypic slice cultures from neonatal rat hippocampus. Alcohol. Clin. Exp. Res. **26**(10): 1468–1478.

30. DANYSZ, W., W. DYR, E. JANKOWSKA, *et al.* 1992. The involvement of NMDA receptors in acute and chronic effects of ethanol. Alcohol. Clin. Exp. Res. **16**(3): 499–504.
31. CHESTER, J.A., N.J. GRAHAME, T.K. LI, *et al.* 2001. Effects of acamprosate on sensitization to the locomotor-stimulant effects of alcohol in mice selectively bred for high and low alcohol preference. Behav. Pharmacol. **12**(6–7): 535–543.
32. QUERTEMONT, E., C. BRABANT & P. DE WITTE. 2002. Acamprosate reduces context-dependent ethanol effects. Psychopharmacology **164**(1): 10–18.
33. KIEFER, F., H. JAHN, T. TARNASKE, *et al.* 2003. Comparing and combining naltrexone and acamprosate in relapse prevention of alcoholism: a double-blind, placebo-controlled study. Arch. Gen. Psychiatry **60**(1): 92–99.
34. STROMBERG, M.F., S.A. MACKLER, J.R. VOLPICELLI & C.P. O'BRIEN. 2001. Effect of acamprosate and naltrexone, alone or in combination, on ethanol consumption. Alcohol **23**(2): 109–116.
35. SCHÜTZ, C.G., C. MAYER, G. KOLLER, *et al.* 2002. Combining naltrexone and memantine to block the rewarding effects of alcohol—an experimental pilot study in human subjects. Alcohol. Clin. Exp. Res. **36**(5): 130A.

Short- and Long-Term Modulation of Synaptic Inputs to Brain Reward Areas by Nicotine

ZARA M. FAGEN,[a] HUIBERT D. MANSVELDER,[c] J. RUSSEL KEATH,[d] AND DANIEL S. McGEHEE[a,b]

[a]*Committee on Neurobiology, University of Chicago, Chicago, Illinois 60637, USA*

[b]*Department of Anesthesia and Critical Care, University of Chicago, Chicago, Illinois 60637, USA*

[c]*Earth and Life Science, Vrije University Amsterdam, the Netherlands*

[d]*Department of Biological Sciences, Northwestern University, Chicago, Illinois, USA*

ABSTRACT: Dopamine signaling in brain reward areas is a key element in the development of drug abuse and dependence. Recent anatomical and electrophysiological research has begun to elucidate both complexity and specificity in synaptic connections between ventral tegmental neurons and their inputs. Specifically, the activity of dopamine neurons in the ventral tegmental area relies on the combination of both excitatory and inhibitory inputs. Controlling endogenous neurotransmission to dopamine neurons is one mechanism by which drugs of abuse affect both transient and long-term changes in synaptic activity. Here, we review recent findings concerning glutamatergic, GABAergic, and cholinergic inputs to dopamine neurons, and their roles in the reinforcement associated with drug abuse. Importantly, several studies support that a single drug exposure can lead to changes in synaptic strength that are associated with learning and memory. Ultimately, these cellular changes could underlie the long-lasting effects of drugs. Furthermore, nicotinic acetylcholine receptors in the ventral tegmental area emerge as a possible common target for the behavioral and cellular actions not only of nicotine, but also of several other drugs of abuse. Finally, we explore age-related differences in nicotine sensitivity in order to understand both human epidemiological data, and laboratory animal behavioral findings that suggest adolescents are more susceptible to developing nicotine dependence.

KEYWORDS: acetylcholine; ventral tegmental area; dopamine; LTP; GABA; glutamate; sensitization

Although many factors contribute to the use and abuse of addictive drugs, the molecular interactions between drugs and neural substrates are the primary basis of behaviors such as dependence, tolerance, sensitization, and craving.[1] While addiction involves many central nervous system (CNS) effects, a common feature of many

Address for correspondence: Daniel S. McGehee, University of Chicago, Department of Anesthesia and Critical Care, 5841 S. Maryland Ave., MC 4028, Chicago, IL 60637. Voice: 773-834-0790; fax: 773-702-4791.
  dmcgehee@uchicago.edu

abused drugs is that plasma concentrations of drug achieved during self-administration increase dopamine (DA) levels in the nucleus accumbens (NAcc).[2,3] The NAcc receives DA projections from the ventral tegmental area (VTA), and 6-OHDA lesioning of these neurons, or microperfusion of DA antagonists into the NAcc reduces drug self-administration.[3] Recent observations have led to the hypothesis that VTA DA neurons signal the salience or expectation of reward.[3,4] Specifically, changes in DA release in the NAcc are detected during the learning phases of self-stimulation in rats but decline after 30 minutes of training.[5] Similarly, monkeys show increased DA neuron activity in response to unpredicted rewards and decreased activity in the absence of expected reward, suggesting that DA neurons act as reward predictors.[4] While this deviates from the simpler view that DA release is synonymous with reward, the factors that affect DA release remain a major focus for drug abuse research. In this vein, recent evidence suggests that endogenous cholinergic centers provide important control of DA release in the striatum,[6] and behavioral studies show nicotinic acetylcholine receptor (nAChR) involvement in drug-induced behavioral states, such as sensitization.[7] A growing body of evidence suggests that nAChRs and cholinergic projections may play an important role not only in nicotine addiction, but also in addiction to a wide array of abused substances through their modulation of DA neuron activity in the VTA.

Another exciting area of research implicates cellular mechanisms of learning and memory in the effects of addictive drugs.[8–12] This is reasonable given the persistence of aspects of addiction, where the threat of relapse and craving will continue for years after quitting.[1] Activation of cellular mechanisms of memory and learning within the circuitry of the reward system could contribute to the lasting behavioral and biochemical effects of drugs of abuse.

CONNECTIVITY OF BRAIN REWARD AREAS

Afferent and efferent VTA projections are complex and highly ordered (FIG. 1). The VTA includes DA and GABA projection neurons, along with local GABAergic interneurons, and sends efferent projections to cortical, striatal, thalamic, cerebellar, and limbic structures. The medial prefrontal cortex (PFC) and NAcc projections have been implicated in addiction mechanisms. There is marked similarity of VTA projections across species, and the majority are ipsilateral and reciprocal, with ~30% dopaminergic.[13]

The major VTA DA projections to cortex synapse on pyramidal neurons of the PFC. These DA terminals often oppose excitatory synapses derived from the cortex or thalamus.[14] In addition, a population of GABAergic VTA neurons projects to the PFC.[15] The VTA receives substantial glutamatergic inputs from PFC, which Carr and Sesack[16] found to selectively target the DA neurons that feed back to PFC. In the same study, PFC glutamate projections also synapse on GABAergic VTA neurons that project to the NAcc. This allows for independent modulation of mesoaccumbens and mesocortical pathways, and may implicate the PFC in different aspects of addiction.

The lateral dorsal and pedunculopontine tegmental nuclei (LDTg and PPTg, respectively) are midbrain cholinergic centers that send cholinergic and glutamatergic projections to the VTA. PPTg and LDTg glutamatergic projections to VTA synapse

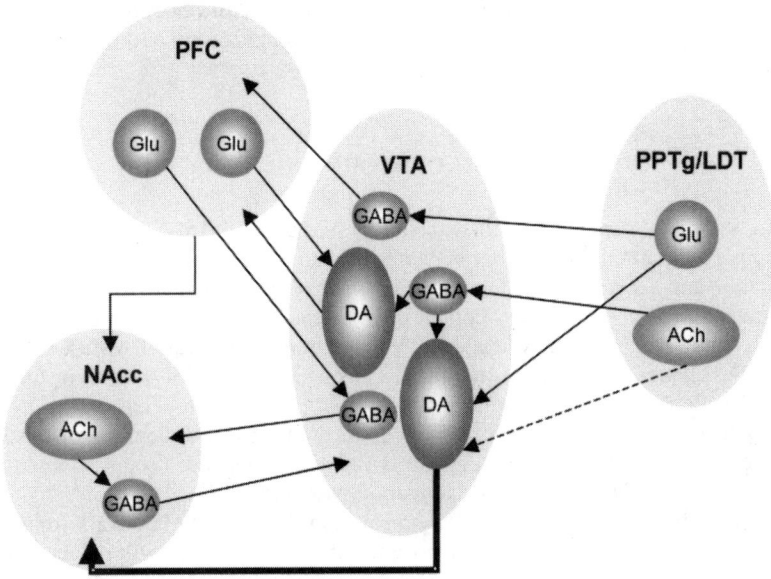

FIGURE 1. A simplified schematic of brain reward center connectivity. Dopamine (DA), GABA neurons (GABA), acetylcholine neurons (ACh), and glutamate neurons (Glu) are depicted in select brain nuclei involved in drug reward. The ventral tegmental area (VTA), nucleus accumbens (NAcc), prefrontal cortex (PFC), and laterodorsal and pedunculopontine tegmental nuclei (PPTg/LDTg) pattern of connectivity is depicted. Ultimately, the release of DA from VTA projections to NAcc and PFC is regulated by a variety of inputs. See text for supporting references.

predominately on the DA neurons that project to NAcc.[17] Cholinergic projections from PPTg and LDTg synapse preferentially on VTA neurons with low dopamine transporter expression.[18] In fact, functional studies indicate that DA neurons receive only approximately 5% of cholinergic inputs to the VTA.[19] Despite this selective targeting of cholinergic inputs, almost all DA and non-DA neurons in VTA express nAChRs.[20,21] While the PPTg and LDTg have highly segregated inputs to brain reward circuitry, the widespread expression of nAChRs suggests an important modulatory role for endogenous ACh, possibly via volume transmission.

Inhibitory GABAergic inputs to VTA DA neurons arise from local interneurons and projections from the ventral pallidum, as well as reciprocal connections from the NAcc.[22] $GABA_A$ antagonists in the VTA increase DA release in the NAcc,[23] and animals will work for intra-VTA administration of $GABA_A$ antagonists.[24] Therefore, the regulation of DA neurons by GABAergic inputs is central to our understanding of drug abuse.

In addition, numerous other neurotransmitters influence DA neuron activity, including serotonin, norepinephrine, endogenous opioids, orexins, and others.[22,25] Furthermore, electrophysiological and tracing techniques indicate excitatory inputs to VTA from the bed nucleus of the stria terminalis,[26] and *in vitro* studies suggest that DA neurons themselves may release glutamate.[27] While the roles of these inputs in addiction need further study, they provide additional control points for dopamine output.

GLUTAMATE RECEPTORS AND ADDICTION

Ionotropic glutamate receptors, including NMDA and AMPA receptors, are the principal mediators of excitatory synaptic transmission in the CNS. Long-term potentiation and depression of glutamatergic synapses are thought to underlie learning and memory. A popular theory contends that, depending upon the level of pre- and postsynaptic neuron activity, the selective insertion or removal of AMPA receptors underlies changes in synaptic strength.[28] Thus, changes in AMPA receptor function can indicate the status of cellular plasticity that may explain persistent behavioral and biochemical alterations. Furthermore, burst firing in DA neurons of the VTA efficiently increases DA release in the NAcc and is dependent on gluatamatergic transmission and NMDA receptor activation within the VTA.[29]

Using a horizontal midbrain slice preparation, Bonci and Malenka[30] found that pairing presynaptic stimulation with postsynaptic depolarization will induce NMDA receptor-dependent LTP selectively in VTA DA neurons, but not in VTA GABA neurons. With the same preparation, Mansvelder and McGehee[9] showed that a single exposure to nicotine can enhance LTP induction. Subsequent studies from the Malenka group found that a single dose of cocaine in young rats can induce LTP of glutamatergic inputs to VTA DA neurons that persists for up to 10 days.[8] Similar strengthening of glutamatergic synapses occurs following a single exposure to a variety of abused drugs, including nicotine, psychostimulants, and alcohol.[11] These studies correlate with *in vivo* evidence of increased AMPA receptor function in the VTA for up to 10 days following withdrawal from repeated psychostimulant exposure.[31]

NICOTINIC RECEPTORS AND ADDICTION

Nicotinic receptors are pentameric, ligand-gated ion channel receptors. To date, 11 neuronal subunit genes have been identified in mammals, and a subset of these, α3–7 and β2–4, are expressed in the VTA.[20] While many subunit combinations are possible, there are some restrictions. For example, α7 nAChRs form homomeric receptors predominantly, while α3, α4, and α6 form heteromeric channels with β2 or β4. α5 and β3 can contribute to functional heteromeric channels, including other α/β pairs.[32,33] *In vivo* studies of genetically modified mice demonstrate that α7 nAChRs are expressed in ~40% of VTA DA and GABA neurons.[20] All the functional heteromeric nAChRs expressed in VTA likely contain β2.[20,21] Many of the β2-containing nAChRs include α4 and possibly other subunits. β2 also combines with α6 in VTA DA neurons to form functional channels on somata as well as axon terminals in NAcc.[20] The majority of VTA GABA neurons are likely to express α4 and β2 subunits, which are selectively inhibited by either dihydro-β-erythroidine (DHBE) or mecamylamine (MEC).[10,20]

Self-administration of nicotine by rodents, nicotine-induced behaviors, and the increase in accumbal DA following nicotine exposure require the activation of VTA nAChRs. For example, intra-VTA injection of nicotine is sufficient to evoke nicotine-induced behavior, whereas injections outside the VTA are ineffective.[34] Local infusion into the VTA of nonspecific nAChR antagonists inhibits nicotine self-administration and DA release in the NAcc associated with nicotine self-administration.[35,36] Perfusion of the NAcc with these antagonists had no effect on either phenomenon. While application of nicotinic antagonists into the NAcc can alter nicotine-induced DA release, the role of these receptors in self-administration is unclear.[37] Specifically, β2-containing nAChRs are required for maintenance of nicotine self-administration in rats and mice.[38,39] However, intra-VTA infusion of the α7-selective antagonist, MLA, prevents nicotine-induced DA release in the NAcc.[40] It is still unknown which nAChR subtypes are involved in the acquisition of nicotine self-administration.

Endogenous ACh is required for the rewarding effects of nicotine. Lesions of the PPTg block nicotine's rewarding effects, enhance its aversive effects, and reduce nicotine self-administration.[41] In addition, DA release in the NAcc is modulated by cholinergic activation of the VTA from the LDTg.[42] Thus, endogenous cholinergic regulation of brain reward circuitry is important in the mechanisms of addiction.

During cigarette smoking, blood nicotine levels reach 300–500 nM within 2–3 min of initiation and are maintained close to 250 nM for ~10 minutes.[43,44] High-affinity nAChRs, including α4β2 will be activated by this concentration and then rapidly desensitized. Lower-affinity α7 nAChRs will be activated weakly at these concentrations, but this is sufficient to enhance glutamate release within the VTA due to the presynaptic localization of these receptors.[9] *In vivo* studies show that a single systemic nicotine exposure will increase accumbal DA for more than an hour.[40]

To understand the source of nicotine's long-lasting effects, we have looked at changes in excitatory and inhibitory transmission in the VTA in response to nicotine.[9,10] Glutamatergic transmission is enhanced by nicotine in several brain regions.[45] In the VTA, activation of presynaptic α7 nAChRs on glutamate terminals can replace presynaptic stimulation in LTP induction.[9] While this study did not find evidence for a contribution of postsynaptic nAChR activity in LTP induction in hippocampal slices, Ji *et al.*[46] found that a weak stimuli, which induced only STP, could induce LTP when paired with concurrent nAChR activation.

Activation of nAChRs also modulates GABAergic transmission in several areas, including the hippocampus, thalamus, cortex, and interpeduncular nucleus.[47] In the VTA, nicotine increases GABA neuron firing by activating high-affinity nAChRs.[10] This transiently increases GABAergic inputs to DA neurons and is likely to offset some of the excitatory effects of nicotine described above. Desensitization of the high-affinity nAChRs will recover in about an hour, and during this time, tonic cholinergic drive to the GABA neurons will likely be removed.[10] Non-DA neurons of the VTA are predominantly GABAergic, and these receive cholinergic input from the LDTg and PPTg.[18] In fact, blocking acetylcholinesterase, the enzyme responsible for hydrolysis of ACh in brain slices, increases GABA cell activity in the VTA.[10]

In contrast, α7 nAChRs on glutamate terminals desensitize to a lesser extent in the presence of low nicotine concentrations. Therefore, these receptors would remain active, and therefore, capable of enhancing glutamatergic transmission.[10] As a result,

DA neurons experience an increase in excitatory transmission that could lead to LTP, with a simultaneous decrease in inhibitory drive. Thus, two synaptic mechanisms contribute to the persistent effects of nicotine on VTA DA neuron excitability.[9,10] These studies correlate with *in vivo* effects of nicotine where the action potential firing rate of VTA DA neurons is first decreased and then increased in a persistent manner.[48] Ultimately, the diversity of nAChRs in the VTA, and their different biophysical characteristics, allows nicotine to optimally enhance DA neuron activity.

NICOTINIC RECEPTORS AND SENSITIZATION

In addition to their role in nicotine self-administration, nAChRs also contribute to the effects of other rewards. For example, the rewarding effects of intracranial self-stimulation are reduced by intra-VTA nAChR antagonists, such as MEC and DHBE.[49–51] The distribution of nAChRs and cholinergic projections to the VTA suggests that nicotinic receptors are poised to regulate the acquisition and/or maintenance of addiction to multiple drugs of abuse.

Sensitization, or enhanced responding to repeated drug exposure, can be observed either behaviorally as increased drug-induced locomotor activity, or biochemically as increased drug-induced dopamine release.[49,50] The induction and expression of sensitization in animal models is believed to parallel the acquisition and maintenance of drug addiction. Behavioral sensitization reflects the motivational properties of a drug, as well as the reinstatement probability of drug abuse.[49] In fact, biochemical sensitization is related to increased drug self-administration.[50] As such, the induction or expression of sensitization is used to assess the rewarding properties of a drug.

Nicotine-induced behavioral sensitization requires activation of NMDA receptors[52] along with $\alpha 4\beta 2$ nAChRs.[53] Interestingly, nAChRs in the VTA have been shown to be required for the reinforcing properties of psychostimulants, nicotine, and ethanol.[7,38,39,54] Using conditioned place preference as an indicator of drug reward, blocking nAChRs was shown to decrease cocaine-induced reward.[55] In the same study, mice lacking the $\beta 2$ subunit show decreased cocaine sensitivity in the conditioned place-preference paradigm.

A striking example of nAChR involvement in psychostimulant sensitization involved systemic applications in rats of either a broad-spectrum nicotinic antagonist, or a $\beta 2$ nAChR-selective antagonist during the induction of amphetamine and cocaine behavioral and biochemical sensitization. In both cases, inhibition of nAChRs blocked the induction of sensitization. Once established, however, the maintenance of sensitization was not dependent upon nAChR activity.[7] In contrast, systemic injection of nicotinic antagonists in mice blocked both the induction and expression of amphetamine-induced sensitization with no effect on cocaine-induced sensitization.[56] Furthermore, selective striatal exposure to nicotinic antagonists blocked only the development of amphetamine sensitization, and not its expression.[56] While systemically applied antagonists cannot reveal underlying cellular mechanisms, they highlight the involvement of nAChRs in drug sensitization. Given the importance of sensitization in understanding the motivational effects of drugs, this will likely be a busy area of investigation in the future.

Methamphetamine has been shown to interact directly with nAChRs in heterologous expression systems.[57] While this is not likely to explain the extent of amphetamine's effects *in vivo,* it does suggest a mechanism by which nAChRs affect the sensitizing effects of the drug. Not all psychostimulants interact with nAChRs directly, and an alternate possibility is the modulation of endogenous cholinergic activity. For example, amphetamine pretreatment sensitizes ACh release in the cortex and striatum.[58,59] These findings suggest enhanced release from forebrain cholinergic centers or striatal cholinergic interneurons; however, it is unclear whether similar increases in PPTg transmission occur during sensitization.

Finally, it is fascinating that behavioral, biochemical, and neuroendocrine changes associated with sensitization can occur following a single drug treatment. Vanderschuren and colleagues[60] show that biological changes associated with amphetamine sensitization develop over time following a single drug exposure. Since cellular studies show significant strengthening of synapses from a single drug treatment,[8–11] it is tempting to speculate that synaptic plasticity underlies sensitization—both behaviorally and biochemically. However, a confound to this hypothesis is the difference in the time course of observations *in vivo* relative to the induction of LTP in the VTA. For example, the enhancement of LTP by nicotine occurs in minutes *in vitro*,[9] and *in vivo* drug exposure induced LTP that is evident only one day after treatment, and that lasted up to 10 days.[8,11] In contrast, changes in biochemical, behavioral, and neuroendocrine sensitization that are marginal after three days are strongly expressed after three weeks and persist for months.[60] A potentially important consideration in this comparison is that the behavioral and biochemical studies of sensitization have been carried out almost exclusively in adult rats, while the electrophysiological studies on LTP expression have used neonatal or adolescent rats and mice. Although this is discussed below, the present findings suggest dramatic differences in the time course of sensitization relative to drug-induced synaptic plasticity. With the caveat that cellular studies in adult tissue are needed, the results from adolescent tissue suggest that synaptic plasticity in the VTA DA system may contribute to the induction of sensitization, but is not likely to mediate its long-term expression.

DEVELOPMENTAL CHANGES IN NICOTINE SENSITIVITY AND DRUG ABUSE

Evidence from human studies suggests increased vulnerability to nicotine addiction by adolescents. For example, people aged 15–24 years have been identified as those at the highest risk of dependence.[61] Nicotine use increases throughout adolescence, with one in three adolescents smoking daily by age 18.[62] Human adolescents express the initial symptoms of nicotine dependence after smoking only a few cigarettes,[63] and identify nicotine dependence as relevant to their smoking experience.[64] This epidemiological data suggests that sensitivity to drug exposure and vulnerability to drug dependence may vary with age.

Similarly, behavioral data in adolescent and adult rats exposed to nicotine show striking differences.[65] Early adolescent mice (postnatal days 24–35) show a preference for nicotine-containing drinking water and increased behavioral response to nicotine, while older adolescents (postnatal days 50–61) avoid oral nicotine administration.[66] Furthermore, nicotine-induced nAChR upregulation, cell damage, and synaptic plasticity vary significantly between fetal, adolescent, and adult animals.[67] While these *in vivo* studies cannot rule out confounding developmental processes, they suggest important differences in nicotine sensitivity of adolescents and adults.

In our laboratory, whole-cell voltage clamp recordings from adult and postnatal rat brain slices of VTA DA neurons show a greater sensitivity to nicotine in tissue from young rats (postnatal days 9–15) compared to adult animals. In these studies, we examined nicotine-induced enhancement of glutamatergic and GABAergic inputs to VTA DA neurons. The enhancement of both types of transmission was stronger in tissue from young animals. In addition, the prevalence of the nicotine-induced enhancement of GABA transmission was significantly greater in tissue from young animals (Z.M. Fagen & D.S. McGehee, unpublished observations). Since nAChR expression is developmentally regulated,[68] these data suggest that age-dependent changes in subunit expression and/or function may alter the sensitivity of DA synapses to nicotinic modulation. These alterations may underlie the behavioral differences seen in nicotine abuse in both the laboratory and the human population.

In summary, DA release in the NAcc is a key element in reinforcement and reward. As such, it remains a focus of drug abuse research. While understanding the effects of addictive drugs on VTA DA neuron activity may not be the whole story, it is helping us identify new cellular and molecular targets for possible treatments for addiction. The anatomy of the VTA is highly segregated, with specific connectivity, which provides tight regulation of DA output. Inputs from cholinergic brain stem nuclei are poised to regulate DA neuron activity through nAChRs distributed throughout reward pathways. Nicotinic receptor subunits are differentially expressed on neurons in the VTA, thereby contributing additional specificity to VTA control. In fact, nicotinic receptors are proving to be a common factor in drug reward, not only for nicotine, but for other drugs of abuse, as well. Together, these data suggest that the endogenous cholinergic drive to brain reward areas requires and will certainly receive further scrutiny by the research community.

One salient feature of recent research is that drugs of abuse induce cellular mechanisms of learning following a single exposure. Long-lasting changes in glutamate receptor function are associated not only with initial drug exposure, but also with drug-induced behaviors. In fact, cellular changes have been correlated with the expression of behavioral sensitization. Long-term changes in behavioral and biochemical responses to drugs can occur following a single exposure. Clearly, the exact links between synaptic plasticity and behavioral outcomes demand further investigation.

Finally, epidemiological data has demonstrated that adolescents are unique in their vulnerability to nicotine abuse. While behavioral studies in young animals support this idea, the underlying cellular mechanisms are largely unknown. Given that cellular electrophysiological research related to drug addiction has been conducted predominately in young tissue, it will be important to extend these studies to adult tissue to help elucidate mechanisms underlying age-dependent differences in susceptibility to drug abuse.

ACKNOWLEDGMENTS

This work was funded by the National Science Foundation DGE0202337 to Z.M.F., the Netherlands Organization for Scientific Research NWO S93-334 to

H.D.M., the National Institutes of Health DA07255 to J.R.K., DA015918 and NS 35090 to D.S.M., and by the Brain Research Foundation.

REFERENCES

1. NESTLER, E.J. & G.K. AGHAJANIAN. 1997. Molecular and cellular basis of addiction. Science **278**: 58–63.
2. RIEGEL, A.C. & E.D. FRENCH. 2002. Abused inhalants and central reward pathways: electrophysiological and behavioral studies in the rat. Ann. N.Y. Acad. Sci. **965**: 281–291.
3. DANI, J.A. 2003. Roles of dopamine signaling in nicotine addiction. Mol. Psychiatry **8**: 255–256.
4. HOLLERMAN, J.R. & W. SCHULTZ. 1998. Dopamine neurons report an error in the temporal prediction of reward during learning. Nat. Neurosci. **1**: 304–309.
5. GARRIS, P.A. et al. 1999. Dissociation of dopamine release in the nucleus accumbens from intracranial self-stimulation. Nature **398**: 67–69.
6. ZHOU, F.M., Y. LIANG & J.A. DANI. 2001. Endogenous nicotinic cholinergic activity regulates dopamine release in the striatum. Nat. Neurosci. **4**: 1224–1229.
7. SCHOFFELMEER, A.N. et al. 2002. Psychostimulant-induced behavioral sensitization depends on nicotinic receptor activation. J. Neurosci. **22**: 3269–3276.
8. UNGLESS, M.A. et al. 2001. Single cocaine exposure in vivo induces long-term potentiation in dopamine neurons. Nature **411**: 583–587.
9. MANSVELDER, H.D. & D.S. MCGEHEE. 2000. Long-term potentiation of excitatory inputs to brain reward areas by nicotine. Neuron **27**: 349–357.
10. MANSVELDER, H.D., J.R. KEATH & D.S. MCGEHEE. 2002. Synaptic mechanisms underlie nicotine-induced excitability of brain reward areas. Neuron **33**: 905–919.
11. SAAL, D. et al. 2003. Drugs of abuse and stress trigger a common synaptic adaptation in dopamine neurons. Neuron **37**: 577–582.
12. NESTLER, E.J. 2002. Common molecular and cellular substrates of addiction and memory. Neurobiol. Learn. Mem. **78**: 637–647.
13. OADES, R.D. & G.M. HALLIDAY. 1987. Ventral tegmental (A10) system: neurobiology. 1. Anatomy and connectivity. Brain Res. **434**: 117–165.
14. CARR, D.B. et al. 1999. Dopamine terminals in the rat prefrontal cortex synapse on pyramidal cells that project to the nucleus accumbens. J. Neurosci. **19**: 11049–11060.
15. CARR, D.B. & S.R. SESACK. 2000. GABA-containing neurons in the rat ventral tegmental area project to the prefrontal cortex. Synapse **38**: 114–123.
16. CARR, D.B. & S.R. SESACK. 2000. Projections from the rat prefrontal cortex to the ventral tegmental area: target specificity in the synaptic associations with mesoaccumbens and mesocortical neurons. J. Neurosci. **20**: 3864–3873.
17. OMELCHENKO, N. & S. SESACK. 2002. Projections to the rat ventral tegmental area from the pedunculopontine-laterodorsal tegmentum synapse onto dopamine neurons that project to the nucleus accumbens. Soc. Neurosci. Abstr. **28**: 4.
18. GARZON, M. et al. 1999. Cholinergic axon terminals in the ventral tegmental area target a subpopulation of neurons expressing low levels of the dopamine transporter. J. Comp. Neurol. **410**: 197–210.
19. FIORILLO, C.D. & J.T. WILLIAMS. 2000. Cholinergic inhibition of ventral midbrain dopamine neurons. J. Neurosci. **20**: 7855–7860.
20. KLINK, R. et al. 2001. Molecular and physiological diversity of nicotinic acetylcholine receptors in the midbrain dopaminergic nuclei. J. Neurosci. **21**: 1452–1463.
21. WOOLTORTON, J. et al. 2003. Differential desensitization and distribution of nicotinic acetylcholine receptor subtypes in midbrain dopamine areas. J. Neurosci. **23**: 3176–3185.
22. KALIVAS, P.W. 1993. Neurotransmitter regulation of dopamine neurons in the ventral tegmental area. Brain Res. Rev. **18**: 75–113.
23. IKEMOTO, S., R.R. KOHL & W.J. MCBRIDE. 1997. GABA(A) receptor blockade in the anterior ventral tegmental area increases extracellular levels of dopamine in the nucleus accumbens of rats. J. Neurochem. **69**: 137–143.

24. IKEMOTO, S., J.M. MURPHY & W.J. MCBRIDE. 1997. Self-infusion of GABA(A) antagonists directly into the ventral tegmental area and adjacent regions. Behav. Neurosci. **111:** 369–380.
25. KOROTKOVA, T.M. *et al.* 2003. Excitation of ventral tegmental area dopaminergic and nondopaminergic neurons by orexins/hypocretins. J. Neurosci. **23:** 7–11.
26. GEORGES, F. & G. ASTON-JONES. 2002. Activation of ventral tegmental area cells by the bed nucleus of the stria terminalis: a novel excitatory amino acid input to midbrain dopamine neurons. J. Neurosci. **22:** 5173–5187.
27. SULZER, D. *et al.* 1998. Dopamine neurons make glutamatergic synapses in vitro. J. Neurosci. **18:** 4588–4602.
28. LUSCHER, C. & M. FRERKING. 2001. Restless AMPA receptors: implications for synaptic transmission and plasticity. Trends Neurosci. **24:** 665–670.
29. COOPER, D.C. 2002. The significance of action potential bursting in the brain reward circuit. Neurochem. Int. **41:** 333–340.
30. BONCI, A. & R.C. MALENKA. 1999. Properties and plasticity of excitatory synapses on dopaminergic and GABAergic cells in the ventral tegmental area. J. Neurosci. **19:** 3723–3730.
31. GIORGETTI, M. *et al.* 2001. Amphetamine-induced plasticity of AMPA receptors in the ventral tegmental area: effects on extracellular levels of dopamine and glutamate in freely moving rats. J. Neurosci. **21:** 6362–6369.
32. MCGEHEE, D.S. 1999. Molecular diversity of neuronal nicotinic acetylcholine receptors. Ann. N.Y. Acad. Sci. **868:** 565–577.
33. CHANGEUX, J.P. *et al.* 1998. Brain nicotinic receptors: structure and regulation, role in learning and reinforcement. Brain Res. Brain Res. Rev. **26:** 198–216.
34. LAVIOLETTE, S.R. & D. VAN DER KOOY. 2003. Blockade of mesolimbic dopamine transmission dramatically increases sensitivity to the rewarding effects of nicotine in the ventral tegmental area. Mol. Psychiatry **8:** 50–59.
35. NISELL, M., G.G. NOMIKOS & T.H. SVENSSON. 1994. Systemic nicotine-induced dopamine release in the rat nucleus accumbens is regulated by nicotinic receptors in the ventral tegmental area. Synapse **16:** 36–44.
36. CORRIGALL, W.A., K.M. COEN & K.L. ADAMSON. 1994. Self-administered nicotine activates the mesolimbic dopamine system through the ventral tegmental area. Brain Res. **653:** 278–284.
37. FU, Y. *et al.* 2000. Local alpha-bungarotoxin-sensitive nicotinic receptors in the nucleus accumbens modulate nicotine-stimulated dopamine secretion in vivo. Neuroscience **101:** 369–375.
38. PICCIOTTO, M.R. *et al.* 1998. Acetylcholine receptors containing the beta2 subunit are involved in the reinforcing properties of nicotine. Nature **391:** 173–177.
39. GROTTICK, A.J. *et al.* 2000. Evidence that nicotinic alpha(7) receptors are not involved in the hyperlocomotor and rewarding effects of nicotine. J. Pharmacol. Exp. Ther. **294:** 1112–1119.
40. SCHILSTROM, B. *et al.* 1998. Nicotine and food induced dopamine release in the nucleus accumbens of the rat: putative role of alpha7 nicotinic receptors in the ventral tegmental area. Neuroscience **85:** 1005–1009.
41. LAVIOLETTE, S.R., T.O. ALEXSON & D. VAN DER KOOY. 2002. Lesions of the tegmental pedunculopontine nucleus block the rewarding effects and reveal the aversive effects of nicotine in the ventral tegmental area. J. Neurosci. **22:** 8653–8660.
42. BLAHA, C.D. *et al.* 1996. Modulation of dopamine efflux in the nucleus accumbens after cholinergic stimulation of the ventral tegmental area in intact, pedunculopontine tegmental nucleus-lesioned, and laterodorsal tegmental nucleus-lesioned rats. J. Neurosci. **16:** 714–722.
43. HENNINGFIELD, J.E. *et al.* 1993. Higher levels of nicotine in arterial than in venous blood after cigarette smoking. Drug Alcohol Depend. **33:** 23–29.
44. GOURLAY, S.G. & N.L. BENOWITZ. 1997. Arteriovenous differences in plasma concentration of nicotine and catecholamines and related cardiovascular effects after smoking, nicotine nasal spray, and intravenous nicotine. Clin. Pharmacol. Ther. **62:** 453–463.
45. MACDERMOTT, A.B., L.W. ROLE & S.A. SIEGELBAUM. 1999. Presynaptic ionotropic receptors and the control of transmitter release. Annu. Rev. Neurosci. **22:** 443–485.

46. Ji, D., R. Lape & J.A. Dani. 2001. Timing and location of nicotinic activity enhances or depresses hippocampal synaptic plasticity. Neuron **31:** 131–141.
47. Wonnacott, S. 1997. Presynaptic nicotinic ACh receptors. Trends Neurosci. **20:** 92–98.
48. Erhardt, S., L. Schwieler & G. Engberg. 2002. Excitatory and inhibitory responses of dopamine neurons in the ventral tegmental area to nicotine. Synapse **43:** 227–237.
49. De Vries, T.J. et al. 1998. Drug-induced reinstatement of heroin- and cocaine-seeking behaviour following long-term extinction is associated with expression of behavioural sensitization. Eur. J. Neurosci. **10:** 3565–3571.
50. Vezina, P. et al. 2002. Sensitization of midbrain dopamine neuron reactivity promotes the pursuit of amphetamine. J. Neurosci. **22:** 4654–4662.
51. Yeomans, J. & M. Baptista. 1997. Both nicotinic and muscarinic receptors in ventral tegmental area contribute to brain-stimulation reward. Pharmacol. Biochem. Behav. **57:** 915–921.
52. Shoaib, M. et al. 1997. Behavioural and biochemical adaptations to nicotine in rats: influence of MK801, an NMDA receptor antagonist. Psychopharmacology (Berl.) **134:** 121–130.
53. Grottick, A.J., R. Wyler & G.A. Higgins. 2000. The alpha4beta2 agonist SIB 1765F, but not the alpha7 agonist AR-R 17779, cross-sensitises to the psychostimulant effects of nicotine. Psychopharmacology (Berl.) **150:** 233–236.
54. Soderpalm, B. et al. 2000. Nicotinic mechanisms involved in the dopamine activating and reinforcing properties of ethanol. Behav. Brain Res. **113:** 85–96.
55. Zachariou, V. et al. 2001. Nicotine receptor inactivation decreases sensitivity to cocaine. Neuropsychopharmacology **24:** 576–589.
56. Karler, R., L.D. Calder & J.B. Bedingfield. 1996. A novel nicotinic-cholinergic role in behavioral sensitization to amphetamine-induced stereotypy in mice. Brain Res. **725:** 192–198.
57. Nomura, T. & T. Nishizaki. 1997. Methamphetamine modulates ACh-evoked currents in Xenopus occytes expressing the rat alpha7 receptors. Neurosci Lett. **239:** 73–76.
58. Nelson, C.L., M. Sarter & J.P. Bruno. 2000. Repeated pretreatment with amphetamine sensitizes increases in cortical acetylcholine release. Psychopharmacology (Berl.) **151:** 406–415.
59. Bickerdike, M.J. & E.D. Abercrombie. 1997. Striatal acetylcholine release correlates with behavioral sensitization in rats withdrawn from chronic amphetamine. J. Pharmacol. Exp. Ther. **282:** 818–826.
60. Vanderschuren, L.J. et al. 1999. A single exposure to amphetamine is sufficient to induce long-term behavioral, neuroendocrine, and neurochemical sensitization in rats. J. Neurosci. **19:** 9579–9586.
61. Breslau, N. et al. 2001. Nicotine dependence in the United States: prevalence, trends, and smoking persistence. Arch. Gen. Psychiatry **58:** 810–816.
62. Young, S.E. et al. 2002. Substance use, abuse and dependence in adolescence: prevalence, symptom profiles and correlates. Drug Alcohol Depend. **68:** 309–322.
63. DiFranza, J.R. et al. 2000. Initial symptoms of nicotine dependence in adolescents. Tob. Control **9:** 313–319.
64. O'Loughlin, J. et al. 2002. The hardest thing is the habit: a qualitative investigation of adolescent smokers' experience of nicotine dependence. Nicotine Tob. Res. **4:** 201–209.
65. Faraday, M.M., B.M. Elliott & N.E. Grunberg. 2001. Adult vs. adolescent rats differ in biobehavioral responses to chronic nicotine administration. Pharmacol. Biochem. Behav. **70:** 475–489.
66. Adriani, W. et al. 2002. Peculiar vulnerability to nicotine oral self-administration in mice during early adolescence. Neuropsychopharmacology **27:** 212–224.
67. Slotkin, T.A. 2002. Nicotine and the adolescent brain: insights from an animal model. Neurotoxicol Teratol. **24:** 369–384.
68. Zoli, M. et al. 1995. Developmental regulation of nicotinic ACh receptor subunit mRNAs in the rat central and peripheral nervous systems. J. Neurosci. **15:** 1912–1939.

Glutamatergic Transmission in Opiate and Alcohol Dependence

GEORGE ROBERT SIGGINS,[a] GILLES MARTIN,[a,c] MARISA ROBERTO,[a] ZHIGUO NIE,[a] SAMUEL MADAMBA,[a] AND LUIS DE LECEA[b]

[a]*Department of Neuropharmacology,* [b]*Department of Molecular Biology, The Scripps Research Institute, 10550 Torrey Pines Road, La Jolla, California 92037, USA*

ABSTRACT: Both the nucleus accumbens (NAcc) and central amygdala (CeA) are thought to play roles in tolerance to, and dependence on, abused drugs. Although our past studies in rat brain slices suggested a role for NMDA receptors (NMDARs) in NAcc neurons in the effects of acute and chronic opiate treatment, the cellular and molecular mechanisms remained unclear. Therefore, we examined the effects of morphine dependence on electrophysiological properties of NMDARs in freshly isolated NAcc neurons and on expression of mRNA coding for NR2A-C subunits using single-cell RT-PCR. Chronic morphine did not alter the affinity for NMDAR agonists glutamate, homoquinolinate, or NMDA, but decreased the affinity of the coagonist glycine. Chronic morphine altered the NMDAR inhibition by two NMDAR antagonists, 7-Cl-kynurenate and ifenprodil, but not that by d-APV or Mg^{2+}. Chronic morphine accelerated the NMDA current desensitization rate in NAcc neurons. In single-cell RT-PCR, chronic morphine predominantly reduced the number of neurons expressing multiple NR2 subunits. Ethanol also alters NMDARs. We found that low ethanol concentrations (IC_{50} = 13 mM) inhibited NMDA currents and NMDA-EPSPs in most NAcc neurons in a slice preparation. NAcc neurons from ethanol-dependent rats showed enhanced NMDA sensitivity. In CeA neurons, acute ethanol decreased (by 10–25%) non–NMDA- and NMDA-EPSPs in most neurons. In CeA neurons from ethanol-dependent rats, acute ethanol decreased the non–NMDA-EPSPs to the same extent as in naïve rats, but inhibited (by 30–40%) NMDA-EPSPs significantly more than in controls, suggesting sensitization to ethanol. Preliminary studies with microdialysis and real-time PCR analysis support this idea: local ethanol administration *in vivo* had no effect on glutamate release, but chronic ethanol nearly tripled the expression of NR2B subunits (the most ethanol sensitive) in CeA. These combined findings suggest that changes in glutamatergic transmission in NAcc and CeA may underlie the neuroadaptions that lead to opiate and ethanol dependence.

KEYWORDS: nucleus accumbens; central amygdala; chronic morphine; chronic ethanol; electrophysiology; pharmacology; RT-PCR; NMDA receptor

---

Address for correspondence: George Robert Siggins, Department of Neuropharmacology, The Scripps Research Institute, 10550 Torrey Pines Rd., La Jolla, CA. Voice: 858-784-7067; fax: 858-784-7393.

geobob@scripps.edu

[c]Present address: Department of Neurobiology, University of Massachusetts Medical School, 364 Plantation Street, Worcester, MA 01655.

BACKGROUND

The extended amygdala, composed in part of the nucleus accumbens and the central amygdala, has emerged as a key complex involved in stress and the behavioral effects of drugs of abuse.[1–3] Although behavioral studies have provided strong evidence for a key role of the extended amygdala in drug reward and dependence, relatively little is known about the cellular and molecular mechanisms responsible for acute drug effects in this complex, and even more remains to be understood about chronic effects. Much behavioral, neurochemical, and clinical evidence (see chapters in this volume by Kalivas, Krystal, and Dackis) suggests that glutamatergic transmission, and especially transmission involving NMDA receptors, is a key player in the sensitivity and tolerance to, and dependence on, various drugs of abuse. For example, Trujillo and Akil[4] and Marek et al.[5] reported that intraventricular infusion of MK801, a selective noncompetitive NMDA receptor antagonist, could attenuate morphine withdrawal, a finding subsequently confirmed by others.[6–10]

NMDA (and AMPA) receptor subunit expression also is reported to be sensitive to abused drugs and changes in brain homeostasis.[11–15] Stress or chronic treatment with several abused drugs, including ethanol, cocaine, and morphine, have been shown to increase the expression of NMDAR1 glutamate receptors in brain neurons.[12,16,17]

In situ hybridization and immunocytochemical studies have shown that the NMDAR1 (NR1) subunit is ubiquitously expressed in the brain (see Refs. 18 and 19 for a review), whereas NR2 subunits show a more specific spatial expression. Thus, in adult rats, NR2A and B subunits are preferentially located in the forebrain, whereas NR2C is mostly expressed in cerebellum, and NR2D in brain stem and spinal cord. The vast majority of accumbens medium spiny neurons express NR2A and NR2B subunits, with a clear predominance for the latter.[20–22]

In previous studies over the last two decades, we have explored in various brain slice preparations the possibility that acute and chronic morphine or ethanol treatment alter glutamate-mediated synaptic transmission in hippocampus, NAcc, and amygdala, and especially that mediated by NMDA receptor activation. For example, we found in NAcc slices that superfusion of a μ opiate receptor agonist increased the amplitude of responses to NMDA but decreased those to AMPA and kainate.[23] Chronic morphine changed several properties of NMDA receptor-mediated excitatory postsynaptic currents (NMDA-EPSCs) in complex ways.[24,25] By contrast, ethanol superfusion potently reduced NMDA-EPSCs and NMDA currents in NAcc neurons;[26,27] the NMDA receptors showed greater sensitivity to NMDA after chronic treatment with ethanol.[28]

However, uncertainty over whether the drugs were acting at pre- or postsynaptic sites in the slice preparations sometimes clouded interpretation of these results. Therefore, in some of the recent studies described here, we used freshly isolated neurons to directly examine the chronic effects of morphine on postsynaptic NAcc NMDA receptors, taking advantage of the fact that each NR2 subunit confers distinct pharmacological properties to the NMDA receptor-channel complex and acts like a regulatory subunit (for a review, see Ref. 29). We also describe electrophysiological studies of the effects of acute and chronic ethanol on the physiology and subunit expression of NMDARs in the NAcc and CeA.

OPIATES AND GLUTAMATERGIC TRANSMISSION

The following section summarizes the findings from a recent study[30] on the effects of morphine dependence on neurons freshly isolated from NAcc and subjected to patch-clamp electrophysiological recording and subsequent single-cell RT-PCR analysis for NMDAR subunit expression.

Methods Used

We prepared NAcc slices and isolated neurons from adult male Sprague Dawley rats (100–200 g), as previously reported,[31] using the methods of the Surmeier group[32,33] for cell isolation. Briefly, after cutting the slices transversely with a Vibroslicer, we incubated the slices (400 μm thick) in a gassed (95% O_2 and 5% CO_2) NaHCO3-buffered saline solution (for the composition of all solutions used, see Ref. 31). After one hour, we dissected out the nucleus accumbens and incubated it for 25 min in oxygenated HEPES-buffered solution in the inner chamber of a Cell-Stirr flask containing papain in the buffer. When the slices were obtained from chronic morphine-treated rats, 1 μM morphine was added to the solutions to prevent withdrawal. After enzymatic digestion, the tissue was transferred into a centrifuge tube and rinsed with a Na^+-isethionate solution. After 10 min, we triturated the tissue using fire-polished Pasteur pipettes and plated the supernatant onto a Petri dish placed on the stage of the inverted microscope. We allowed the cells to attach to the dish for 10 min before replacing the Na^+-isethionate solution with normal external solution[31] flowing at a rate of 1.5 mL/minute.

We used standard whole-cell recording methods[34] with patch electrodes (1.8–2.2 MΩ, 270–275 mOsM[31]), recording in voltage-clamp mode with an Axopatch 1D amplifier and an Axon Digidata 1200 interface. We administered control and drug-containing solutions by gravity at a rate of 1.5 mL/s, via a 3-barrel capillary (tip diameter: 500 μm) superfusion device. The pipette tips were placed about 200 μm from the recorded cell, and their movement was rapidly controlled by solenoid valves (drug onset time: 20 ms). We exposed NAcc medium spiny neurons for 5 s (every 30 s) to pulses of increasing concentrations of agonists (NMDA, glutamate, homoquinolinic acid) in the presence of a saturating (100 μM) concentration of glycine. Antagonists (7-Cl-kynurenic acid, ifenprodil, 5-AP) were coapplied with saturating concentrations of NMDA (200 μM) and glycine (100 μM). We fitted concentration-response curves with a Hill equation and measured NMDA current desensitization following coapplication (5 s) of 200 μM NMDA and 100 μM glycine at –60 mV holding potential; single exponential tau values were plotted by histogram with a 0.5-s bin width.

We made rats morphine dependent by subcutaneous implantation, under light halothane anesthesia, of 2 morphine pellets (75 mg of base each; provided by NIDA); control rats received placebo pellets. Electrophysiological testing was performed 4–6 d after pellet implantation.

For analysis of single-cell NMDAR subunit RNA, after recording we aspirated cell contents into the pipette by negative pressure[32,33] and ejected them into a tube for random-primed single-strand cDNA synthesis. The primers used in this study have been tested extensively in amplification reactions with whole rat brain cDNA as a template and were designed to give amplification products of 300–500 bp encompassing an intron, to allow identification of nonprocessed or genomic sequences. We amplified NMDAR cDNAs using a pair of primers common to NMDAR2A-C, and the PCR was followed by a 30-cycle amplification with nested primers specific for each of the subunits. In representative cases, the sequence of amplicons was verified.

Morphine Dependence Has Little Effect on Agonist Sensitivity of NAcc NMDA Receptors

In a previous study on NAcc slices, we found that chronic morphine treatment decreased the amplitude of NMDA receptor-mediated synaptic events, increased paired-pulse facilitation, and altered several pharmacological properties of NMDA-EPSPs,[24,25] suggesting that morphine depressed glutamate synaptic transmission either by decreasing glutamate release or altering postsynaptic NMDAR properties. Most recently,[30] we have tested the properties of postsynaptic NMDARs from freshly isolated medium spiny neurons. We first assessed the effects of morphine dependence on NMDAR glutamate sensitivity, a property shown to be controlled by NR2 subunits: the glutamate EC_{50} for NR2C is lower than for NR2A or NR2B.[35–37] In NAcc neurons of placebo rats, NMDA currents evoked with glutamate (0.01 μM to 1 mM, in the presence of 15 μM CNQX to block non-NMDARs) were maximum with 0.5 mM glutamate and steeply declined with lower concentrations. In NAcc neurons from morphine-dependent rats, NMDAR current amplitudes evoked with equivalent glutamate concentrations were nearly identical to those recorded in placebo neurons;[30] the glutamate EC_{50} for morphine-treated rats was equivalent to that from placebo rats.

We also examined the sensitivity of NAcc NMDARs for homoquinolinic acid (HQ) and NMDA, two agonists that can discriminate among NMDAR subunits: the affinity of recombinant NR1/NR2C or NR2D receptors for NMDA is higher than that of NR1/NR2A or NR2B,[35,38,39] whereas the EC_{50} for HQ is higher for NR2A and B than for NR2C. In NAcc neurons from placebo rats, HQ evoked threshold currents at around 1 μM and maximum transient current amplitudes at 0.5–1 mM, with an EC_{50} of 32 ± 4 μM. Chronic morphine had little effect on the HQ EC_{50}. With NMDA, largest current amplitudes were evoked at 1 mM. Chronic morphine had no effect on NMDA currents, as the NMDA EC_{50} was virtually identical in neurons from placebo and morphine-treated rats.[30]

Chronic Morphine Reduces Glycine Affinity for NAcc NMDARs

Glycine, the allosteric NMDA receptor coagonist, has a differential affinity for the various NMDAR subunits: the NR1/NR2A affinity for glycine is lower than that of NR2B or NR2C.[36–38,40] For glycine responses with 0.03–100 μM, in the presence of 200 μM NMDA, morphine dependence shifted the glycine EC_{50} to the right (2.24 ± 0.15 μM in placebo rats, to 5.1 ± 1.45 μM in dependent rats).[30]

Chronic morphine alters the sensitivity to NMDAR antagonists. We also assessed the inhibition of NMDA currents (evoked by 200 μM NMDA with 100 μM glycine) by the NMDAR antagonists ifenprodil, 7-Cl-kyn acid (a polyamine site antagonist), and d-APV. The affinity of NR1/NR2B recombinant receptors for ifenprodil is much stronger than NR1/NR2A and NR2C receptors.[41,42] In our study of isolated NAcc

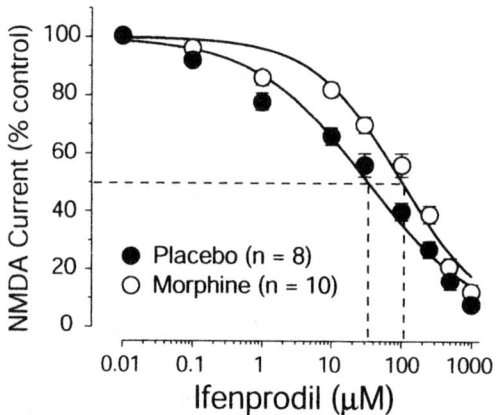

FIGURE 1. Morphine dependence reduces the sensitivity of NMDARs in isolated NAcc neurons to the NMDAR antagonist ifenprodil: concentration-response curves. *Dashed lines* indicate IC_{50}s for each condition. In this and subsequent figures, points represent mean ± SE. Drug applications as described in the text.

neurons,[30] ifenprodil blocked NMDA currents at low concentrations (1 µM) in NAcc neurons of both placebo and morphine-treated rats, but this inhibition was strongly attenuated after chronic morphine treatment for concentrations of 10 to 300 µM (FIG. 1). In placebo rats, ifenprodil blocked NMDA currents with an IC_{50} of 32 ± 2.2 µM, but in dependent rats the IC_{50} increased to 102 ± 2.4 µM. At the highest ifenprodil concentrations tested (0.5 and 1 mM), there was no detectable difference between the two groups.

In contrast to infenprodil, the affinity of NR1/NR2A recombinant NMDARs for 7-Cl-kyn is higher than that of NR1/NR2B and NR1/NR2C.[39,41] In our study,[30] the lowest effective concentration of 7-Cl-kyn in inhibiting NMDA currents was 1 µM in neurons of both placebo and morphine-dependent rats, with an almost total block reached at 1 mM 7-Cl-kyn. Morphine dependence enhanced the 7-Cl-kyn inhibition of NMDA currents at concentrations > 1 µM: the 7-Cl-kyn IC_{50} was 9.8 ± 1.58 µM in neurons of placebo rats and 27 ± 5.8 µM after chronic morphine.

We assessed the effects of chronic morphine on the action of d-APV, the selective competitive NMDAR antagonist. Although there is little difference in d-APV affinity for NR1/NR2A and NR1/NR2B recombinant NMDARs, its affinity for NR2C and NR2D is substantially lower.[38,39] In our study,[30] chronic morphine treatment did not significantly alter d-APV efficacy or potency of NMDAR antagonism for NAcc medium spiny neurons; the d-APV IC_{50} was nearly equivalent for placebo and morphine-dependent rats.

We also examined the effects of chronic morphine on the Mg^{2+}-mediated block of NMDA receptors, because we had previously shown that chronic morphine decreased this block in a NAcc slice preparation.[25] However, in the postsynaptic model of isolated NAcc neurons,[30] the Mg^{2+} IC_{50} was only slightly higher after chronic morphine compared to placebo rats, suggesting that the previous data obtained in the slice preparation may have resulted from presynaptic effects.

Chronic Morphine Alters NMDAR Desensitization

Both inactivation and desensitization of NMDARs are regulated by NR2 subunits, with a faster decay of the current when NR2A is coexpressed with the NR1 subunit. In a previous study of NAcc slices,[25] we found that chronic morphine accelerated NAcc NMDA-EPSC deactivation. We have extended this study to NAcc isolated cells, examining the effects of morphine on NMDA current desensitization (Ca^{2+} independent) by evoking NMDA currents with coapplication of 200 µM NMDA and saturating concentrations of glycine (100 µM). Current decays (fit by a single exponential) as tau values were plotted into a histogram that showed unevenly distributed tau values in neurons from both placebo and morphine-dependent rats. However, taus of NMDA currents from placebo rats were predominantly (18 of 32 neurons) distributed between 1.5 and 2.5 s, with a peak at 2 s, compared to a distribution between 0.5 and 2 s in a majority of neurons (27 out of 32 neurons; peak at 1.5 s) from morphine-dependent rats.[30] A Gaussian fit of tau distributions gave mean values of 1.99 ± 0.08 s and 1.38 ± 0.03 s for placebo and morphine-dependent rats, respectively, thus supporting the decreased deactivation times seen for NMDA-EPSCs in NAcc slices from morphinized rats.

NMDAR Subunit Expression in Single Cells

Reasoning from the literature and our electrophysiological data that morphine dependence might induce the expression of more NR2A or 2C subunits absent or scarce in NAcc neurons of placebo rats, we examined the mRNA expression of three NMDAR subunits in dissociated NAcc medium spiny neurons by single cell RT-PCR. Whole-cell aspirates were reverse transcribed and PCR amplified using primers common to the three NR2 subunits, and normalized to NR1 content. In NAcc neurons of placebo rats, the mRNAs coding for NR2A and B were detected in most NAcc medium spiny neurons, either alone or coexpressed (6 of 25 cells tested) with another NR2 subunit.[30] The only clear qualitative change was a reduction in the apparent heterogeneity of NMDAR subunit composition: the number of cells with heteromultimeric receptors decreased after chronic morphine (TABLE 1), likely resulting in NMDARs composed of either NR1/NR2A or NR1/NR2B subunits only. The expression of both NR2A (36% fell to 25% of cells) and NR2B (80 to 76%) subunits dropped in percentage,[30] even though the ratio NR2A/NR2B increased, possibly be-

TABLE 1. Morphine dependence may alter NMDA receptor composition: Single-cell RT-PCR of isolated NAcc medium spiny neurons

NR2 subunits	NR2A	NR2B	NR2C	NR2A/B	NR2A/B/C	NR2B/C
Placebo	4	14	1	4	1	1
Morphine dependent	3	8	1	0	0	0

NOTE: The chart shows the number of cells expressing each of the three NMDAR subunits: NR2A, 2B, and 2C. Note that in the morphine-dependent condition, the number of cells expressing more than one type of NR2 subunit (heteromultimeric) drops to zero for each combination (right three columns). In addition, the percentage of cells expressing NR2B subunits drops from 80% in controls to 67% in neurons of dependent rats.

cause the number of cells containing multiple NR2 types (A/B, A/B/C, or B/C) fell to zero.

In conclusion, most of our electrophysiological data predict an increase in NR2A subunit expression at the expense of NR2B subunits, yet the single-cell RT-PCR studies point to a reduction in cells with heteromultimeric NMR2s. However, because of the difficulty with the single-cell technique in sampling sufficient numbers of neurons needed to verify an increase in NR2A expression, we are now initiating quantitative RT-PCR and immunohistochemical studies of larger, multicellular samples of CeA brain tissue (see the studies of ethanol with RT-PCR below).

ETHANOL AND GLUTAMATERGIC TRANSMISSION

In concert with the data from many earlier studies showing that ethanol can inhibit glutamatergic function (for review, see Ref. 43), our past studies in NAcc

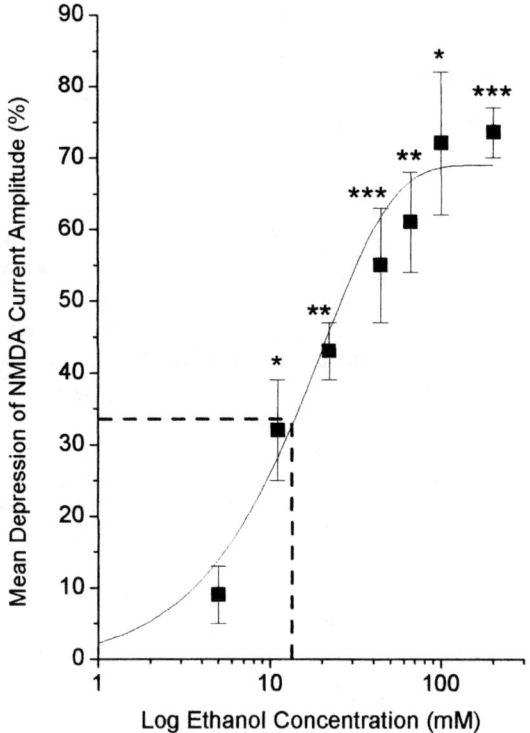

FIGURE 2. Concentration-antagonism curve: ethanol markedly depresses NMDA currents in NAcc neurons in a slice preparation. NMDA was applied either by rapid superfusion (4 mL/min) or by pressure from a small-tip pipette placed near the recorded neuron. *Dashed lines* indicate the apparent IC_{50} (13 mM) for this ethanol effect. Note the high efficacy for ethanol (maximal depression at 100 mM = ~60–83%). *significance depression at the $P < .05$ level; ** $P < .01$; ***$P < .001$. Modified and updated from Ref. 27.

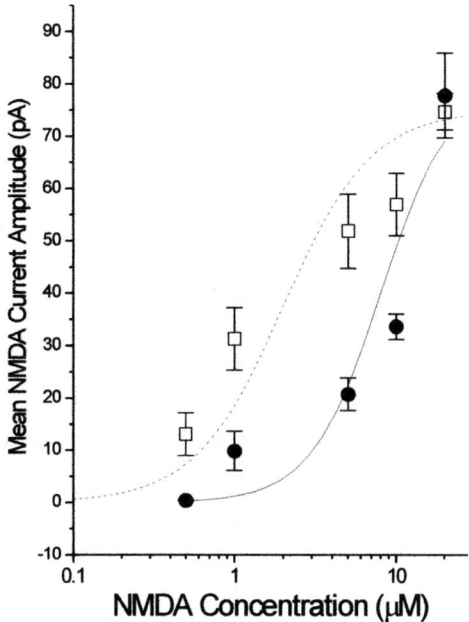

FIGURE 3. NAcc neurons (slice preparation): Chronic ethanol treatment (via vapor chamber) enhances NMDAR sensitivity to NMDA (applied by rapid superfusion); note the shift in the concentration-response relationship to the left. *Closed circles* represent slices from control animals; *open squares* represent neurons from slices of ethanol-dependent rats. The approximate EC_{50} for the controls was ~8 µM and that for the dependent animals was about 2 µM. Chronic ethanol treatment is described in the text.

neurons in a rat slice preparation found that low doses of ethanol reproducibly inhibited amplitudes of glutamatergic EPSPs and currents evoked by exogenous NMDA.[26,27] The IC_{50} for this ethanol-NMDA interaction was 13 mM; 44–66 mM ethanol reduced NMDA currents to about 20–30% of control (FIG. 2). Thus, NAcc neurons are among the most sensitive to ethanol's NMDA blocking effect. Interestingly, ethanol dependence, elicited by the standard ethanol inhalation method for chronic ethanol (CE) treatment at the Scripps Research Institute Alcohol Research Center,[44] produced an enhancement of the sensitivity of NAcc neurons to exogenous NMDA (FIG. 3),[28] suggesting an upregulation of NMDARs as a neuroadaptation induced by chronic ethanol.

In the recent study[45,46] described below, we have investigated effects of acute and chronic ethanol administration on pharmacologically isolated non-NMDA and NMDA receptor-mediated EPSP/Cs within the central amygdala (CeA) in a slice preparation.

Methods Used

We prepared and studied transverse amygdala slices (400 μm thick) from male Sprague-Dawley rats (100–300 g), as previously described.[47] Drugs were added to the ACSF from stock solutions at known concentrations. For chronic ethanol treatment we used the ethanol inhalation method, described above.[44] In the ethanol-treated group, we exposed rats to continuous ethanol vapors for 2–4 weeks before the CeA slices were obtained. Sham controls were treated similarly but without ethanol vapors. We made recordings in ethanol-free ACSF from slices of CE-treated rats 2–8 hours after cutting the slices, corresponding to the time period for the hyperexcitable (dependence) behavioral response to ethanol withdrawal. Some of the CE-treated animals were withdrawn from the ethanol exposure for 1–2 weeks prior to preparing the slices. The mean blood alcohol level of the CE-treated animals was 198.8 mg/dL.

We recorded CeA neurons intracellularly (using discontinuous voltage- or current-clamp) with sharp micropipettes containing 3M KCl, and we evoked EPSP/Cs by stimulating locally within the CeA through a bipolar stimulating electrode. We evoked pharmacologically isolated AMPA receptor-mediated EPSP/Cs using the $GABA_A$ and $GABA_B$ receptor blockers bicuculline (30 μM) and CGP55845A (1 μM), and the NMDAR antagonist d-APV (30 μM), and we isolated NMDAR-mediated EPSP/Cs in low Mg^{2+} (0.75 mM) ACSF using bicuculline, CGP55845A, and the AMPA/kainate receptor antagonist CNQX (10 μM). We also applied exogenous NMDA (2 μM in the pipette) locally near the recorded neuron by pressure (2–4 μm tip diameter; 1–10 psi; 0.5–3 s duration), while recording NMDA responses in current-clamp mode in the presence of bicuculline (30 μM), CGP55845A (1 μM), CNQX (10 μM), and 1 μM tetrodotoxin (to minimize presynaptic effects).

Acute Ethanol Reduces Synaptically Evoked Glutamatergic EPSP/Cs

Local afferent stimulation evoked compound responses in CeA neurons consisting of glutamatergic EPSP/Cs and GABAergic IPSP/Cs. At resting membrane potential (RMP), the evoked responses were composed mainly of non-NMDA glutamate (AMPA/kainate) and $GABA_A$ receptor-mediated synaptic components, because specific antagonists for these receptors completely blocked the evoked responses.

Ethanol had no significant ($P > 0.1$) effect on basic membrane properties, such as membrane potential, input resistance, or spike amplitudes (see also Ref. 47). However, acute superfusion of 44 mM ethanol for 7–10 min significantly decreased compound glutamatergic EPSP/Cs (recorded in CGP55845A and bicuculline) in all neurons tested:[45,46] the mean EPSC amplitude in ethanol was 71.9 ± 4% of control (at half-maximal stimulus), with partial recovery on washout (84 ± 9% of control). This effect was mediated primarily by a reduction of the non–NMDAR-mediated component, since d-APV (10 μM) did not significantly mitigate the inhibitory effect of ethanol on EPSPs, and, in most neurons, the addition of 10 μM CNQX totally blocked the EPSP. In direct studies of the non–NMDAR- (AMPA/kainate)-mediated component (in the presence of d-APV, CGP55845A, and bicuculline), acute ethanol (44 mM) for 7–10 min significantly decreased the mean evoked non–NMDA-EPSP amplitudes by 20 to 30% across the stimulus strengths used, with recovery on washout.[45,46] We also evoked d-APV–sensitive and voltage-dependent NMDAR-mediated EPSPs, using a low Mg^{2+} ACSF to remove the Mg^{2+} block of the NMDA channel. Acute ethanol (44 mM) also decreased these NMDA-EPSP/Cs by 15–20% ($P < 0.01$) over all the stimulus strengths used, with recovery on washout.

Ethanol Dependence Alters Effects of Acute Ethanol

Because glutamate receptors, and especially NMDARs, are thought to be involved in ethanol-related phenomena, such as tolerance and dependence, we recorded from slices taken from rats chronically exposed to ethanol vapors for 2–4 weeks.[45,46] We prepared the CeA slices from CE-treated rats, exactly as in the acute ethanol experiments, and allowed the slices to withdraw from ethanol in the recording chamber. The CE-treated animals used in our study were ethanol dependent, as shown by the alcohol withdrawal syndrome observed at the end of the CE treatment period in a separate sample of rats rated via several behavioral tests for ethanol withdrawal severity, including a common behavioral rating scale for hyperactivity (ventromedial distal limb flexion response, tail stiffness, abnormal body posture, and gait),[48,49] measured at 2, 4, 6, and 8 h of ethanol withdrawal. Behavioral signs of ethanol withdrawal were evident in all six rats tested, at 2–8 hours after termination of ethanol treatment (the equivalent time corresponding to the electrophysiological recordings).

In slices taken from CE-treated rats during acute withdrawal (2–8 h after last ethanol exposure), acute superfusion of 44 mM ethanol still significantly ($P < .001$) decreased the evoked compound EPSP/Cs and non–NMDA-EPSPs to 80 ± 6% and 81 ± 7% of control (mean from three different stimulus intensities), respectively, somewhat less than the inhibition in slices from control rats. Baseline input–output (I/O) curves (before ethanol application) for the non–NMDA-EPSPs were shifted significantly lower in slices from CE-treated rats compared to slices from control rats, whereas we observed no significant difference between the compound EPSP I/O curves in slices from control and CE-treated animals. The baseline I/O curves for NMDA EPSPs also were comparable in slices from control and CE-treated rats. In contrast, the depressive effect of acute, 44 mM ethanol on NMDA-EPSP amplitudes (over the whole I/O curve) in CE-treated slices was significantly stronger (to 60–70% of baseline) than in control (naïve or sham) slices (FIG. 4), suggesting sensitization or reverse tolerance.[45,46]

To assess whether the inhibitory effect of ethanol on NMDA-EPSPs involved presynaptic changes in neurotransmitter release, we measured paired-pulse facilitation (PPF; 50–180 ms interpulse intervals) of NMDA-EPSPs in each CeA neuron before, during, and after ethanol application. PPF is characterized by a transient increase in synaptic efficacy during the response to the second pulse of a two-pulse stimulation protocol and is thought to result primarily from residual Ca^{2+} accumulation within the terminals following the first stimulus.[50–52] Changes in PPF are thought to be *inversely* related to probability of transmitter release.[53,54] We found that baseline PPF of NMDA-EPSP/Cs was similar in neurons from control and dependent animals and that acute superfusion of 44 mM ethanol onto control slices had little effect on PPF of NMDA EPSP/Cs. However, acute ethanol significantly decreased PPF of NMDA EPSP/Cs (relative to the control; $P < .05$) in slices withdrawn within 2–8 hours, suggesting an acute ethanol-induced increase in glutamate release at this time[45,46] and indicating a neuroadaptative change after CE treatment. The in-

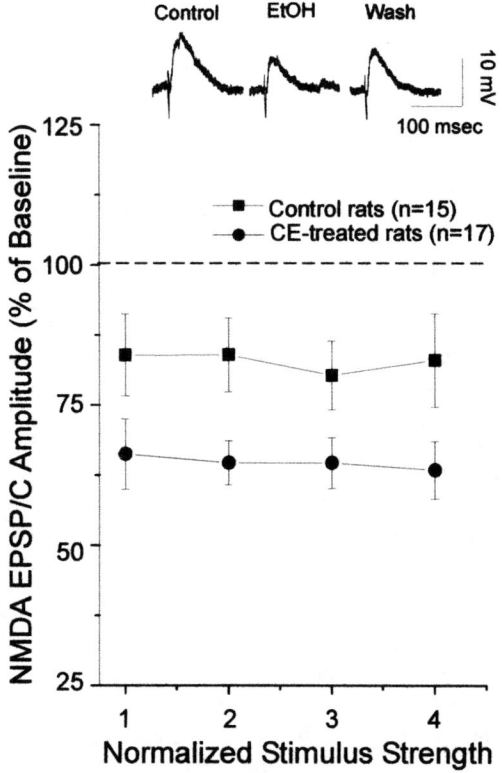

FIGURE 4. Central amygdala neurons in a slice preparation: Ethanol dependence enhances the inhibitory effect of acute ethanol (44 mM applied by superfusion) on NMDA-EPSP/C amplitudes. Note that the same concentration of acute ethanol reduces the NMDA-EPSP/C amplitudes in slices from chronically treated rats to about 65% of baseline, compared to about 85% in controls. This difference was significant at $P < .01$ (ANOVA). This effect of chronic ethanol represents a sensitization to acute ethanol, or reverse tolerance.

creased glutamate release could represent a compensatory mechanism countering the ethanol inhibition of postsynaptic NMDA receptors.

To determine the persistence of these chonic ethanol effects, we recorded from slices taken from rats withdrawn from the CE treatment for one or two weeks prior to preparing the slices. In slices from the one-week withdrawn animals, acute ethanol superfusion reduced the amplitudes of NMDA-EPSPs similarly to those of control rats, but the ethanol-induced reduction of NMDA EPSP/Cs PPF (20%) was still present in CeA neurons. However, after two weeks of withdrawal the effect of acute ethanol on PPF had totally recovered to insignificance.

Effects of Acute and Chronic Ethanol on Postsynaptic NMDA Receptors

Because acute ethanol reduced NMDA-EPSP/C amplitudes but had little effect on their PPF in CeA slices of naïve rats, we hypothesized that the ethanol effect may

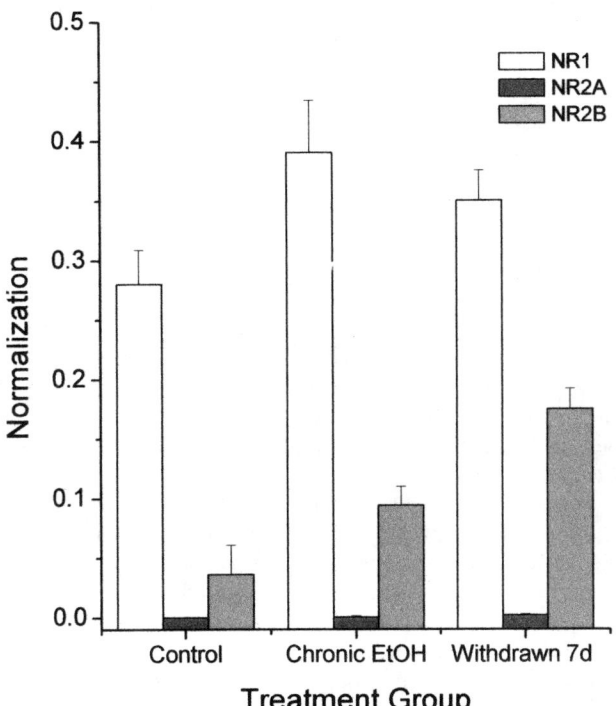

FIGURE 5. Preliminary RT-PCR data from central amygdala tissue: Comparison of NMDAR subunit expression in CeA from control rats, ethanol-dependent rats (Chronic EtOH), and those withdrawn from ethanol treatment for 7 days (Withdrawn 7d). Note that there is very little expression of NR2A subunits compared to NR2B subunits, and that NR1 subunit expression is high, as expected. However, NR2B subunit expression increases dramatically after chronic ethanol treatment, and even more after a week of withdrawal.

be exerted at postsynaptic sites, perhaps at the NMDA receptor itself. This hypothesis is supported by preliminary evidence using microdialysis in rat CeA *in vivo*: locally applied ethanol enhanced GABA release but had little effect on glutamate release (Roberto, Stouffer, Siggins, and Parsons, in preparation).

To directly test this hypothesis, we have begun preliminary studies of the effects of acute and chronic ethanol on responses of CeA neurons to locally applied exogenous NMDA, in the presence of tetrodotoxin to minimize presynaptic effects. As predicted from the NMDA-EPSP/C data, and in accord with many past studies showing ethanol inhibition of NMDA receptors in other neuron types,[27,55–58] acute ethanol decreased currents elicited by NMDA applied directly to CeA neurons from a pipette by 20–30%, equivalent to the ethanol-induced decrease of NMDA-EPSP/Cs in these neurons. In addition, as with the NMDA-EPSP/Cs, acute ethanol superfusion elicited a more pronounced inhibition (30–40%) of the NMDA currents evoked in slices taken from ethanol-dependent rats, compared to those from controls, again suggesting a sensitization or reverse tolerance but now at the postsynaptic level.

We have also begun molecular biological studies of the subunit composition of NMDARs in CeA neurons, using quantitative RT-PCR of CeA tissue dissected from naïve, CE-treated and withdrawn rats (Roberto, Bajo, Acosta, de Lecea, and Siggins, in preparation). To date, the data from these studies suggest that NR2B subunits predominate in naïve CeA and that chronic ethanol causes an even greater expression of these NR2B subunits (FIG. 5), consistent with the greater sensitivity of NMDA-EPSPs to ethanol seen in our electrophysiological studies (see Ref. 43).

CONCLUSIONS

Our studies were done primarily on neurons of the rat extended amygdala (including CeA and NAcc), because this complex is known to play a major role in several behavioral phenomena, such as emotions, stress, anxiety, and drug reinforcement and dependence.[3,59] As shown in the several studies described here, acute treatment with drugs of abuse like opiates and alcohol can have dramatic effects in this complex on glutamatergic transmission at the synaptic and molecular level. In many cases, direct drug and dependence effects on postsynaptic NMDARs were seen. Chronic treatment with, or dependence on, these abused drugs can cause changes in the NMDAR subunit composition, as well as (not studied here) transcriptional, posttranscriptional, or posttranslational regulation (e.g., by phosphorylation: see e.g., Ref. 60) of NMDAR subunits.[43] These changes could represent the neuroadaptations or altered allostasis posited to account for the changes in behavior with respect to increased drug intake.[61] For example, the complex changes in NMDAR electrophysiological properties seen after chronic morphine could cause profound changes in the excitability (e.g., reduced firing on depolarization) of the medium spiny neurons in NAcc neurons, and thus their reduced output to downstream regions. Moreover, the upregulation of NMDARs and the increase in sensitivity of NMDA-EPSPs to acute ethanol following the development of ethanol dependence could underlie the behavioral sensitization reported for several abused drugs. Changes in Ca^{2+} flux, known to occur through NMDARs, and Ca^{2+}-related phenomena, such as neurotoxicity, also should be considered as other possible neuroadaptations to drug dependence. Finally, from these diverse data, it is reasonable to assume that pharmaceutical agents acting on NMDARs may have potential for the rational treatment of opiate dependence or alcoholism.

ACKNOWLEDGMENTS

This research was supported by National Institutes of Health Grants DA03665 and AA06420. We thank Dr. J. Surmeier for help with the acutely isolated neuron preparation and Drs. W. Fröstl and A. Suter (Novartis Pharma) for the gift of CGP-55854A.

REFERENCES

1. KOOB, G.F. 1992. Drugs of abuse: anatomy, pharmacology and function of reward pathways. Trends Pharmacol. Sci. **13:** 177–184.

2. KOOB, G.F., R. MALDONADO & L. STINUS. 1992. Neural substrates of opiate withdrawal. Trends Neurosci. **15:** 186–191.
3. KOOB, G. *et al.* 1993. The mesocorticolimbic circuit in drug dependence and reward—A role for the extended amygdala? *In* Limbic Motor Circuits and Neuropsychiatry. P.W. Kalivas & C.D. Barnes, Eds.: 289–309. CRC Press, Inc. Boca Raton.
4. TRUJILLO, K.A. & H. AKIL. 1991. Inhibition of morphine tolerance and dependence by the NMDA receptor antagonist MK-801. Science **251:** 85–87.
5. MAREK, P. *et al.* 1991. Excitatory amino acid antagonists (kynurenic acid and MK-801) attenuate the development of morphine tolerance in the rat. Brain Res. **547:** 77–81.
6. TISEO, P.J. *et al.* 1994. Modulation of morphine tolerance by the competitive N-methyl-D-aspartate receptor antagonist LY274614: assessment of opioid receptor changes. J. Pharmacol. Exp. Ther. **268:** 195–201.
7. ELLIOTT, K. *et al.* 1995. N-methyl-D-aspartate (NMDA) receptors, mu and kappa opioid tolerance, and perspectives on new analgesic drug development. Neuropsychopharmacology **13:** 347–356.
8. LUTFY, K. *et al.* 1995. Blockade of morphine tolerance by ACEA-1328, a novel NMDA receptor/glycine site antagonist. Eur. J. Pharmacol. **273:** 187–189.
9. KOSTEN, T.A., J.L. DECAPRIO & M.I. ROSEN. 1995. The severity of naloxone-precipitated opiate withdrawal is attenuated by felbamate, a possible glycine antagonist. Neuropsychopharmacology **13:** 323–333.
10. HERMAN, B.H., F. VOCCI & P. BRIDGE. 1995. The effects of NMDA receptor antagonists and nitric oxide synthase inhibitors on opioid tolerance and withdrawal—medication development issues for opiate addiction. Neuropsychopharmacology **13:** 269–293.
11. FOLLESA, P. & M.K. TICKU. 1995. Chronic ethanol treatment differentially regulates NMDA receptor subunit mRNA expression in rat brain. Brain Res. Mol. Brain Res. **29:** 99–106.
12. FOLLESA, P. & M.K. TICKU. 1996. Chronic ethanol-mediated up-regulation of the N-methyl-D-aspartate receptor polypeptide subunits in mouse cortical neurons in culture. J. Biol. Chem. **271:** 13297–13299.
13. RESINK, A. *et al.* 1995. Regulation of the expression of NMDA receptor subunits in rat cerebellar granule cells: effect of chronic K^+-induced depolarization and NMDA exposure. J. Neurochem. **64:** 558–565.
14. RIVA, M.A. *et al.* 1997. Regulation of NMDA receptor subunit messenger RNA levels in the rat brain following acute and chronic exposure to antipsychotic drugs. Mol. Brain Res. **50:** 136–142.
15. SNELL, L.D. *et al.* 1996. Regional and subunit specific changes in NMDA receptor mRNA and immunoreactivity in mouse brain following chronic ethanol ingestion. Brain Res. Mol. Brain Res. **40:** 71–78.
16. FITZGERALD, L.W. *et al.* 1996. Drugs of abuse and stress increase the expression of GluR1 and NMDAR1 glutamate receptor subunits in the rat ventral tegmental area: Common adaptations among cross-sensitizing agents. J. Neurosci. **16:** 274–282.
17. TREVISAN, L. *et al.* 1994. Chronic ingestion of ethanol up-regulates NMDAR1 receptor subunit immunoreactivity in rat hippocampus. J. Neurochem. **62:** 1635–1638.
18. DUNAH, A.W. *et al.* 1999. Biochemical studies of the structure and function of the *N*-methyl-D-aspartate subtype of glutamate receptors. Mol. Neurobiol. **19:** 151–179.
19. MCBAIN, C.J. & M.L. MAYER. 1994. N-methyl-D-aspartic acid receptor structure and function. Physiol. Rev. **74:** 723–760.
20. CHEN, Q. & A. REINER. 1996. Cellular distribution of the NMDA receptor NR2A/2B subunits in the rat striatum. Brain. Res. **743:** 346–352.
21. LANDWEHRMEYER, G.B. *et al.* 1995. NMDA receptor subunit mRNA expression by projection neurons and interneurons in rat striatum. J. Neurosci. **15:** 5297–5307.
22. STANDAERT, D.G. *et al.* 1996. Expression of NMDAR2D glutamate receptor subunit mRNA in neurochemically identified interneurons in the rat neostriatum, neocortex and hippocampus. Mol. Brain Res. **42:** 89–102.
23. MARTIN, G., Z. NIE & G.R. SIGGINS. 1997. Mu-opioid receptors modulate NMDA receptor-mediated responses in nucleus accumbens neurons. J. Neurosci. **17:** 11–22.
24. MARTIN, G., R. PRZEWLOCKI & G.R. SIGGINS. 1999. Chronic morphine treatment selectively augments metabotropic glutamate receptor-induced inhibition of N-methyl-D-

aspartate receptor-mediated neurotransmission in nucleus accumbens. J. Pharmacol. Exp. Ther. **288:** 30–35.
25. MARTIN, G. *et al.* 1999. Chronic morphine treatment alters NMDA receptor-mediated synaptic transmission in the nucleus accumbens. J. Neurosci. **19:** 9081–9089.
26. NIE, Z. *et al.* 1993. Ethanol decreases glutamatergic synaptic transmission in rat nucleus accumbens in vitro: naloxone reversal. J. Pharmacol. Exp. Ther. **266:** 1705–1712.
27. NIE, Z., S.G. MADAMBA & G.R. SIGGINS. 1994. Ethanol inhibits glutamatergic neurotransmission in nucleus accumbens neurons by multiple mechanisms. J. Pharmacol. Exp. Ther. **271:** 1566–1573.
28. NIE, Z., S.G. MADAMBA & G.R. SIGGINS. 1995. Withdrawal from chronic ethanol exposure increases sensitivity to N-methyl-D-aspartate (NMDA) in rat nucleus accumbens neurons. Soc. Neurosci. Abst. **21:** 2101.
29. Yamakura, T. & K. Shimoji. 1999. Subunit- and site-specific pharmacology of the NMDA receptor channel. Prog. Neurobiol. **59:** 279–298.
30. MARTIN, G. *et al.* 2003. Chronic morphine treatment alters NMDA receptors in freshly-isolated neurons from nucleus accumbens. Submitted.
31. MARTIN, G. & G.R. SIGGINS. 2002. Electrophysiological evidence for expression of glycine receptors in freshly isolated neurons from nucleus accumbens. J. Pharmacol. Exp. Ther. **302:** 1135–1145.
32. SURMEIER, D.J., W.J. SONG & Z. YAN. 1996. Coordinated expression of dopamine receptors in neostriatal medium spiny neurons. J. Neurosci. **16:** 6579–6591.
33. YAN, Z. & D.J. SURMEIER. 1996. Muscarinic (m2/m4) receptors reduce N- and P-type Ca^{2+} currents in rat neostriatal cholinergic interneurons through a fast, membrane-delimited, G-protein pathway. J. Neurosci. **16:** 2592–2604.
34. HAMILL, O.P. *et al.* 1981. Improved patch clamp techniques for high resolution current recording from cells and cell free membrane patches. Pflugers Arch. **391:** 85–100.
35. DE CARVALHO, L.P., P. BOCHET & J. ROSSIER. 1996. The endogenous agonist quinolinic acid and the non endogenous homoquinolinic acid discriminate between NMDAR2 receptor subunits. Neurochem. Int. **28:** 445–452.
36. LAURIE, D.J. & P.H. SEEBURG. 1994. Ligand affinities at recombinant N-methyl-D-aspartate receptors depend on subunit composition. Eur. J. Pharmacol. **268:** 335–345.
37. WAFFORD, K.A. *et al.* 1993. Preferential co-assembly of recombinant NMDA receptors composed of three different subunits. Neuroreport **4:** 1347–1349.
38. BULLER, A.L. & D.T. MONAGHAN. 1997. Pharmacological heterogeneity of NMDA receptors: characterization of NR1a/NR2D heteromers expressed in *Xenopus* oocytes. Eur. J. Pharmacol. **320:** 87–94.
39. PRIESTLEY, T. *et al.* 1995. Pharmacological properties of recombinant human N-methyl-D-aspartate receptors comprising NR1a/NR2A and NR1a/NR2B subunit assemblies expressed in permanently transfected mouse fibroblast cells. Mol. Pharmacol. **48:** 841–848.
40. KUTSUWADA, T. *et al.* 1992. Molecular diversity of the NMDA receptor channel. Nature **358:** 36–41.
41. AVENET, P. *et al.* 1997. Antagonist properties of eliprodil and other NMDA receptor antagonists at rat NR1A/NR2A and NR1A/NR2B receptors expressed in *Xenopus* oocytes. Neurosci. Lett. **223:** 133–136.
42. WILLIAMS, K. 1993. Ifenprodil discriminates subtypes of the N-methyl-D-aspartate receptor: selectivity and mechanisms at recombinant heteromeric receptors. Mol. Pharmacol. **44:** 851–859.
43. KUMARI, M. & M.K. TICKU. 2000. Regulation of NMDA receptors by ethanol. Prog. Drug Res. **54:** 152–189.
44. ROGERS, J., S.G. WIENER & F.E. BLOOM. 1979. Long-term ethanol administration methods for rats: advantages of inhalation over intubation or liquid diets. Behav. Neural Biol. **27:** 466–486.
45. ROBERTO, M., S.G. MADAMBA & G.R. SIGGINS. 2002. Acute and chronic ethanol alters glutamatergic transmission in rat central amygdala. Alcoholism Clin. Exp. Res. Suppl. **26:** 63.

46. ROBERTO, M., S.G. MADAMBA & G.R. SIGGINS. 2003. Acute and chronic ethanol decreases glutamatergic transmission in rat central amygdala. Submitted.
47. ROBERTO, M. *et al.* 2003. Ethanol increases GABAergic transmission at both pre- and postsynaptic sites in rat central amygdala neurons. Proc. Natl. Acad. Sci. USA **100:** 2053–2058.
48. MACEY, D.J. *et al.* 1996. Time-dependent quantifiable withdrawal from ethanol in the rat: effect of method of dependence induction. Alcohol. **13:** 163–170.
49. SCHULTEIS, G. *et al.* 1995. Decreased brain reward produced by ethanol withdrawal. Proc. Natl. Acad. Sci. USA **92:** 5880–5884.
50. DEBANNE, D. *et al.* 1996. Paired-pulse facilitation and depression at unitary synapses in rat hippocampus: quantal fluctuation affects subsequent release. J. Physiol. **491:** 163–176.
51. JIANG, L. *et al.* 2000. Paired-pulse modulation at individual GABAergic synapses in rat hippocampus. J. Physiol. **523:** 425–439.
52. ZUCKER, R.S. 1989. Short-term synaptic plasticity. Annu. Rev. Neurosci. **12:** 13–31.
53. ANDREASEN, M. & J.J. HABLITZ. 1994. Paired-pulse facilitation in the dentate gyrus: a patch-clamp study in rat hippocampus in vitro. J. Neurophysiol. **72:** 326–336.
54. MENNERICK, S. & C.F. ZORUMSKI. 1995. Paired-pulse modulation of fast excitatory synaptic currents in microcultures of rat hippocampal neurons. J. Physiol. (Lond.) **488:** 85–101.
55. WHITE, G., D. LOVINGER & F. WEIGHT. 1990. Ethanol inhibits NMDA-activated current but does not alter GABA-activated current in an isolated adult mammalian neuron. Brain Res. **507:** 332–336.
56. WEIGHT, F.F., D.M. LOVINGER & G. WHITE. 1991. Alcohol inhibition of NMDA channel function. Alcohol Alcohol. Suppl **1:** 163–169.
57. HOFFMAN, P.L. *et al.* 1989. N-methyl-D-aspartate receptors and ethanol: inhibition of calcium flux and cyclic GMP production. J. Neurochem. **52:** 1937–1940.
58. SIMSON, P.E. *et al.* 1991. Ethanol inhibits NMDA-evoked electrophysiological activity in vivo. J. Pharmacol. Exp. Ther. **257:** 255–231.
59. HYYTIA, P. & G.F. KOOB. 1995. $GABA_A$ receptor antagonism in the extended amygdala decreases ethanol self-administration in rats. Eur. J. Pharmacol. **283:** 151–159.
60. MALDVE, R.E. *et al.* 2002. DARPP-32 and regulation of the ethanol sensitivity of NMDA receptors in the nucleus accumbens. Nat. Neurosci. **5:** 641–648.
61. KOOB, G.F. 1996. Drug addiction: the yin and yang of hedonic homeostasis. Neuron **16:** 893–896.

Exogenous and Endogenous Cannabinoids Control Synaptic Transmission in Mice Nucleus Accumbens

DAVID ROBBE,[a] GÉRARD ALONSO,[b] AND OLIVIER J. MANZONI[a]

[a]*Equipe Avenir "Plasticité synaptique: Maturation and Addiction," INSERM U378, Institut Magendie, Bordeaux, 33077 France*
[b]*CNRS UMR 5101, Montpellier 34094, cedex 5, France*

ABSTRACT: Addictive drugs are thought to alter normal brain function and cause the remodeling of synaptic functions in areas important to memory and reward. Excitatory transmission to the nucleus accumbens (NAc) is involved in the actions of most drugs of abuse, including cannabis. We have explored the functions of the endocannabinoid system at the prefrontal cortex–NAc synapses. Immunocytochemistry showed cannabinoid receptor (CB1) expression on axonal terminals making contacts with NAc neurons. In NAc slices, synthetic cannabinoids inhibit spontaneous and evoked glutamate-mediated transmission through presynaptic activation of presynaptic K^+ channels and GABA-mediated transmission most likely via a direct presynaptic action on the vesicular release machinery. How does synaptic activity lead to the production of endogenous cannabinoids (eCBs) in the NAc? More generally, do eCBs participate in long-term synaptic plasticity in the brain? We found that tetanic stimulation (mimicking naturally occurring frequencies) of prelimbic glutamatergic afferents induced a presynaptic LTD dependent on eCB and CB1 receptors (eCB-LTD). Induction of eCB-LTD required postsynaptic activation of mGlu5 receptors and a rise in postsynaptic Ca^{2+} from ryanodine-sensitive intracellular Ca^{2+} stores. This retrograde signaling cascade involved postsynaptic eCB release and activation of presynaptic CB1 receptors. In the NAc, eCB-LTD might be part of a negative feedback loop, reducing glutamatergic synaptic strength during sustained cortical activity. The fact that this new form of LTD was occluded by an exogenous cannabinoid suggested that cannabis derivatives, such as marijuana, may alter normal eCB-mediated synaptic plasticity. These data suggest a major role of the eCB system in long-term synaptic plasticity and give insights into how cannabis derivatives, such as marijuana, alter normal eCB functions in the brain reward system.

KEYWORDS: nucleus accumbens; glutamate, GABA; cannabinoid CB1 receptors; synaptic plasticity; addictive drugs

Address for correspondence: Olivier J. Manzoni, Equipe Avenir "Plasticité synaptique: Maturation and Addiction," INSERM U378, Institut Magendie, 1 rue Camille Saint Saens, Bordeaux, 33077 France. Voice: +33-557573770; fax: +33-557573776.
manzoni@bordeaux.inserm.fr

INTRODUCTION

Drug addiction can be viewed as a pathological form of memory and more specifically a permanent disorder of synaptic plasticity. Repeated consumption of addictive drugs initiates adaptive mechanisms causing remodeling of brain circuits normally used to reinforce rewarding (corticomesolimbic pathways) and/or learning behaviors (hippocampus, amygdala, cortex) that are necessary for the normal functions of the brain.[1–4] Via their initial targets, addictive drugs affect synaptic transmission at various levels and in different neurotransmitter systems. Cocaine and amphetamine directly modulate monoamine release and/or uptake and thus have a strong impact on catecholaminergic neurotransmission. Morphine, heroine, and other opiates directly bind the various opioid receptors and thus interfere with the physiological endogenous opioid systems, while cannabis derivatives substitute for endocannabinoids. Nicotine activates and desensitizes nicotinic cholinergic receptors and can influence cholinergic neuromodulation. Ethanol inhibits NMDA receptors and facilitates GABA receptor function. Addictive drugs perform actions not only on the neurons on which the initial target receptors are but also on the neural circuits in which those neurons participate.

The fact that the modifications in synaptic functions induced by addictive drugs are expressed long after complete withdrawal is strongly indicative of the key role of durable changes in specific neural systems and of gene expression in drug craving and relapse.[3,4] The mechanisms responsible for these adaptations are clearly reminiscent of mechanisms involved in nonpathological forms of activity-dependent synaptic rearrangements. Moreover, at the cellular and synaptic level, it is likely that synaptic adaptations induced by drugs of abuse have common mechanisms with other forms of activity-dependent plasticity, such as long-term potentiation (LTP) or long-term depression (LTD). This activity-dependent plasticity may result from the direct effect of addictive drugs on the excitability or transmitter release of individual cells. However, the indirect effects of addictive drugs are also important. The recent advances in understanding the molecular mechanisms underlying synaptic plasticity in the mammalian central nervous system are directly relevant to gaining an understanding of the effects of drug self-administration. In fact, the list of molecules implicated in hippocampal LTP presented by Sanes and Lichtman includes many known targets of addictive drugs.[5] Thus synaptic plasticity initiated by addictive drugs and activity-dependent processes may be linked by common mechanisms. The systematic study of the effects of chronic drug treatment on long-term potentiation (LTP) and depression (LTD)–like phenomena in structures relevant to drug addiction is crucial to the understanding of the synaptic consequences of chronic drug use. Addictive drugs may have far-reaching actions affecting synaptic plasticity by both direct and indirect mechanisms.[1–4]

Derivatives of *Cannabis sativa L.*, such as marijuana and hashish, have been used for centuries for therapeutical and recreational purposes. The psychopharmacologically active component of *Cannabis sativa L.*, (–)-trans-delta9-tetrahydrocannabinol, as well as cannabimimetics and endocannabinoids, mediate their actions in the central nervous system through specific interactions with a Gi/Go-protein–coupled receptor (CB1).[6] The CB1 receptor is widely expressed in the brain and has been shown to inhibit adenylate cyclase, activate MAP-kinases, reduce Ca^{2+} currents, and modulate several K^+ conductances.[6] Activation of CB1 receptors inhibits synaptic transmission in the hippocampus,[7,8] substantia nigra pars compacta,[9] the cerebellum,[10] and the prefrontal cortex.[11]

The mesolimbic-mesocortical dopaminergic system and particularly the nucleus accumbens (NAc) are essential to the reinforcing properties of addictive drugs.[12] Drugs of abuse, such as psychostimulants, opiates, nicotine, alcohol, and cannabinoids alter dopamine levels in the NAc.[13] Because of intrisinc and network properties, the projection cells of the NAc, the GABAergic medium-spiny neurons, depend on glutamatergic excitatory afferents to generate action potentials.[13] Glutamatergic transmission in the NAc participates in the effects of opiates and psychostimulants and is altered by chronic drug treatment.[14,15] Although cannabis derivatives are the most common illicit drugs, little is known of their actions in the mesolimbic regions and the nucleus accumbens. In particular, the potential effects of exogenous cannabinoids (cannabimimetics) in the NAc were not known. Our goal was first to identify the localization and functions of CB1 receptors in excitatory and inhibitory synapses of the NAc, second to discover the consequences of endogenous cannabinoid release on synaptic transmission and plasticity in the NAc, and finally to evaluate the consequences of *in vivo* exposure to addictive drugs on endocannbinoid-mediated synaptics in the NAc.

METHODS

Electrophysiology

Whole cell patch clamp and extracellular field recordings were made from medium spiny neurons in parasagittal slices of mouse nucleus accumbens. This method has been described previously.[15,16] In brief, mice (male C57BL6, 4–6 weeks) were anesthetized with fluorene and decapitated. The brain was sliced (300–400 m) in the parasagittal plane using a vibratome and maintained in physiological saline at 4° C. Slices containing the NAc were stored at least one hour at room temperature before being placed in the recording chamber and superfused (2 mL/min) with artificial cerebrospinal fluid (ACSF) that contained (in mM) 126 NaCl, 2.5 KCl, 1.2 $MgCl_2$, 2.4 $CaCl_2$, 18 $NaHCO_3$, 1.2 NaH_2PO_4, and 11 Glucose, and that was equilibrated with 95% O_2/5% CO_2. All experiments were done at room temperature. When recording excitatory transmission, the superfusion medium contained picrotoxin (100 M) to block GABA-A receptors. Conversely, when recording GABAergic synapses, the superfusion medium contained 1 mM kynurenate to block ionotropic glutamate receptors. All drugs were added at the final concentration to the superfusion medium.

For field potential recordings, the recording pipette was filled with a 3 M NaCl solution, and both the fEPSP slope (calculated with a least square method) and fEPSP amplitude were measured.

For patch-clamp experiments, cells were visualized using an upright microscope with infrared illumination, and recordings were made with whole cell electrodes containing the following (mM): CsGluconate 128, NaCl 20, $MgCl_2$ 1, EGTA 1, $CaCl_2$ 0.3, Mg-ATP 2, GTP 0.3, and cAMP 0.2 buffered with HEPES 10, pH 7.3. Electrode resistance was 4 MOhms, acceptable access resistance was < 15 MOhms, and the holding potential was –70 mV. An Axopatch-1D (Axon Instruments) was used to record the data, which were filtered at 1–2 kHz, digitized at 5 kHz on a Digi-

Data 1200 interface (Axon Instruments), and collected on a PC using ACQUIS-1 software (Bio-Logic, France). To evoke synaptic currents, stimuli (100–150 μs duration) were delivered at 0.033 Hz through bipolar tungsten electrodes placed at the prefrontal cortex–accumbens border.[16] Recordings were made in the rostral-medial dorsal accumbens close to the anterior commissure. Evoked excitatory/inhibitory postsynaptic current (iPSCs/EPSC) amplitudes were measured by averaging a 5-ms window around the peak and subtracting the average value obtained during a 10-ms window immediately before the stimulus. Two stimuli were applied at an interval of 50 ms, and the paired-pulse ratio was calculated by dividing the amplitude of the PSC evoked by the second stimulus by the amplitude of the first PSC evoked by the first stimulus. A change in the paired-pulse ratio is thought to result from the alteration in transmitter release caused by a presynaptic mechanism.

Miniature iPSCs/EPSCs (mIPSCs/EPSCs) were recorded in the presence of tetrodotoxin (TTX, 300 nM) using Axoscope 1.1.1; mIPSCs/EPSCs amplitude and interinterval time were measured using Axograph 3.6. For this analysis, a template of mEPSCs having the width and time course of a typical synaptic event [a double exponential: $f(t)=\exp(-t/\text{Rise})-\exp(-t/\text{Decay})$, where rise = 0.5 ms and decay = 3 ms] or of mIPSCs [a double exponential: $f(t)=\exp(-t/\text{Rise})-\exp(-t/\text{Decay})$, where rise = 0.66 ms and decay = 10 ms], was slid along the data trace one point at a time. At each position, this template is optimally scaled and offset to fit the data, and a detection criterion is calculated. The detection criterion is the template scaling factor divided by the goodness-of-fit at each position. An event is detected when this criterion exceeds a threshold and reaches a sharp maximum. The limit of detection was 2 pA.[17]

The fitting of concentration response curves was calculated according to $y = \{y_{max}-y_{min}/1+ (x/EC_{50})n\} + y_{min}$ (where y_{max} = response in the absence of agonist, y_{min} = response remaining in presence of maximal agonist concentration, x = concentration, EC_{50} = concentration of agonist producing 50% of the maximal response, and n = slope) with Kaleidagraph software (Abelbeck Software, USA). All values are given as the SEM. Statistical analyses were done with the Mann-Whitney U-test or the Kolmogorov-Smirnov tests using Statview (Abacus Concepts, Inc. USA); $P < .05$ was taken as indicating statistical significance. SR 141716A was generously provided by Sanofi-Recherche (Montpellier, France). WIN-2, WIN 55,212,3, CP 55940, and SR 141716A were prepared as (10 mM) stock solutions in DMSO. Final concentrations were < 0.1% DMSO.

RESULTS

Localization of CB1 Receptors in the Nucleus Accumbens

The general examination of CB1-immunostained (IS) structures in the brain showed that, in mice, the organization of the CB1-IS structures conformed to previous descriptions in the rat.[18] Scattered CB1-IS perikarya were observed throughout the cortex, the hippocampus, and the olfactory bulb. Numerous CB1-IS perikarya were detected in the prelimbic cortex area, which is known to strongly project to the NAc.[19] Dense plexuses of CB1-IS fibers were detected throughout the cortex, the hippocampus, the anterior olfactory nucleus, and the olfactory bulb. A number of

CB1-IS fibers were also detected throughout the striatum. Within the NAc, these CB1-IS fibers were mostly located in the core of the nucleus where they appeared as large, poorly branched fibers exhibiting a large number of intensely immunostained varicosities. Double immunostaining experiments indicated that throughout the NAc these CB1-IS fibers established frequent *en passant* or terminal synaptic-like contacts with GABA-IS perikarya or dendritic processes.[19] These results were strongly suggestive of the presence of CB1 receptors on glutamatergic terminals, making synapses with GABAergic medium spiny neurons of the NAc. In agreement with a cortical origin of these axons, intense CB1-immunostaining was located within the cytoplasm of a number of perikarya dispersed throughout perikaryas in the prefrontal cortex. This identification of presynaptic CB1 receptors on axonal fibers in the NAc prompted us to explore the effects of cannabimimetics at the glutamatergic synapses between the prelimbic cortex and the NAc (see below). Finally, we also observed few CB1-IS perikaria within the nucleus accumbens. This was in agreement with the low level of CB1-IS in the rat NAc reported by Tsou and colleagues.[18] The latter observation suggested that CB1 receptors could be expressed on the inhibitory synapses between GABAergic interneurons intrinsic to the NAc and medium spiny neurons of the NAc. Using electrophysiological techniques in the slice preparation of the nucleus accumbens, we and others directly tested these hypotheses.

Cannabimimetics Inhibit Synaptic Transmission in the Mouse NAc

Effects of Cannabimimetics on Glutamatergic Excitatory Transmission

In the NAc slice preparation we found that cannabimimetics, with a pharmacology consistent with the involvement of CB1 receptors, caused a profound inhibition of glutamatergic transmission at the synapses between the prelimbic cortex and the NAc (see FIG. 1A).[19] The electrophysiological analysis corroborated the anatomical data and pointed out a presynaptic site of action for CB1 receptors. When CB1-mediated inhibition of synaptic transmission was observed in the central nervous system, it always resulted from presynaptic actions.[20] Accordingly, electrophysiological analysis revealed an increase in the paired-pulse ratio of evoked EPSCs and a decrease of the mEPSC frequency—but not their amplitude—during CB1 inhibition. The simplest interpretation of our anatomical and electrophysiological data is that a presynaptic mechanism is responsible for the actions of the cannabinoid agonists at the excitatory synapses to the NAc. We also found that the CB1 receptor-mediated presynaptic inhibition of glutamatergic transmission is independent of the cAMP/PKA cascade and of the inhibition of voltage-sensitive calcium channels (VSCCs), in striking contrast with those showing that the modulation of presynaptic VSCCs is responsible for the CB1 receptor-mediated inhibition of transmitter release in the hippocampus[21,22] and substantia nigra pars compacta.[9]

A well-documented action of CB1 receptors in neuronal cells is their inhibitory coupling with voltage-sensitive Ca^{2+} channels.[6] Accordingly, recordings from substantia nigra pars reticulata neurons showed that cannabinoids exert a presynaptic inhibition on GABAergic transmission via the CB1 inhibition of Cd^{2+}-sensitive presynaptic Ca^{2+} channels.[9] It was also shown that Cd^{2+} prevented cannabinoid actions on spontaneous GABA release,[8] and that CB1 receptors can inhibit evoked glutamate release via the modulation of N- and P/Q-type Ca^{2+} channels.[21] Our data

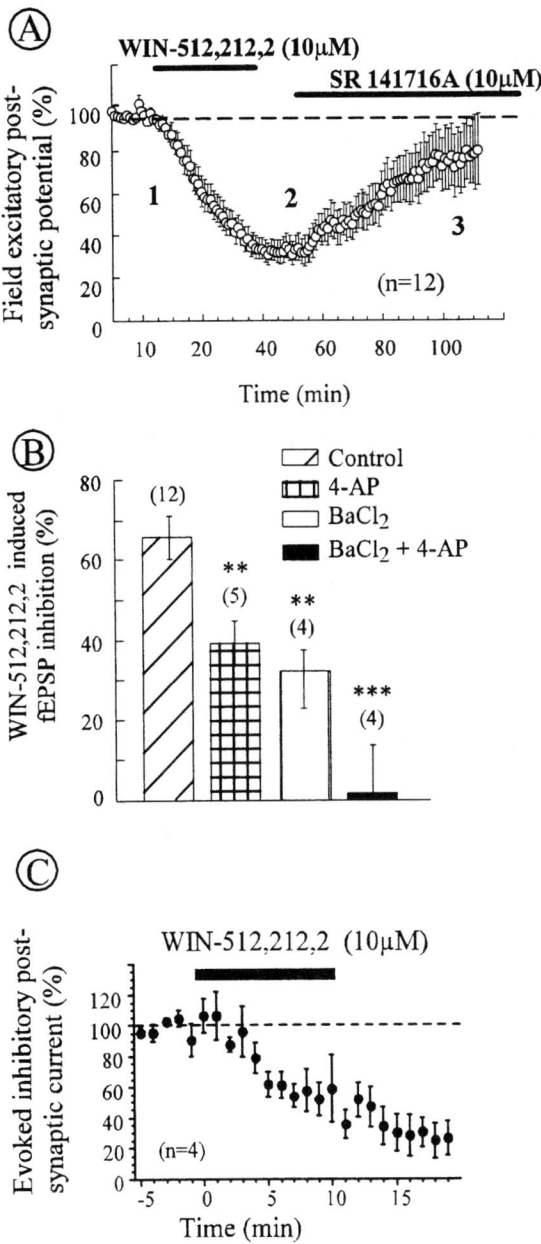

FIGURE 1. Cannabinoid CB1 receptor-mediated inhibition of evoked excitatory and inhibitory synaptic transmission in mice nucleus accumbens. (**A**) CB1 receptor-mediated inhibition of evoked excitatory synaptic transmission in mouse nucleus accumbens. 10 μM of the cannabimimetic WIN-2 reduced the fEPSP on average to $34 \pm 5\%$ ($n = 12$) of its basal value. The effects of WIN-2 were reversed by the selective CB1 antagonist SR 141716A

suggest that in the NAc, CB1 receptors inhibit spontaneous and evoked synaptic transmission independently from the modulation of presynaptic voltage-sensitive Ca^{2+} channels.

Surprisingly, we found that blockade of presynaptic K^+ channels hampered the cannabinoid effects in standard conditions, but also when $[Ca^{2+}]_o$ was lowered to reduce the size of evoked EPSPs and control for indirect actions of K^+ channel blockers on Ca^{2+}-dependent processes. Blocking voltage-dependent K^+ conductances by 4-AP (100 µM) and GIRK-like conductances with $BaCl_2$ (300 µM) both caused a clear reduction in the inhibitory actions of the cannabimimetic (FIG. 1B).

Since CB1 receptors are not present in the dopaminergic neurons of the ventral tegmental area (VTA),[18] how can cannabinoids activate mesolimbic dopamine neurons[23] and increase dopamine levels in the NAc?[24,25] The finding of CB1 receptors at the glutamatergic synapses in the NAc provides a simple explanation. The glutamatergic afferents to the NAc control the firing of the NAc GABAergic neurons, which in turn inhibit the dopaminergic neurons of the VTA. Via the reduction of excitatory transmission in the NAc, cannabinoids could disinhibit dopamine cells of the VTA, increase their firing rate, and trigger the release of dopamine in the nucleus accumbens.

Effects of Cannabimimetics on GABAergic Inhibitory Transmission

We observed that cannabinoid CB1 receptor agonists strongly inhibited evoked IPSCs recorded in the core of the nucleus accumbens.[26] Bath-applied WIN-2 (10 µM) markedly reduced evoked IPSCs ($n = 4$, FIG. 1C). The WIN-2–induced inhibition was prevented by preincubation with the selective cannabinoid CB1 receptor antagonist (SR141716A) (not shown),[27] which by itself had no effect on the IPSCs (not shown). The variation of the paired-pulse ratio of inhibitory transmission, a phenomenon sensitive to presynaptic manipulations, was measured to assess the locus of the CB1-mediated effect. At the peak of the depression induced by 10 µM WIN-2, the IPSC was reduced to $28.2 \pm 13.2\%$ of its control value, while the paired-pulse ratio was up to $272.2 \pm 88.5\%$ of its control value ($n = 4$), indicating a presynaptic site of action. The presynaptic localization of cannabinoid CB1 receptors was confirmed with the observation that WIN-2 caused a marked depression of the miniature IPSC frequency (not shown). Our results are in complete agreement with Hoffman and Lupica's detailed study.[22] It is important to note that these authors have observed

---

(10 µM). (**B**) Effects of K^+ channel blockade on the CB1-induced inhibition of glutamatergic synaptic transmission in the NAc. In standard ACSF (2.4 mM CaCl2, *hatched bar*), 4-AP (100 µM, *crossed bar*) and $BaCl_2$ (300 µM, *white bar*) reduced the WIN-2–induced fEPSP inhibition. WIN-2 reduced fEPSP by $39.1 \pm 5.5\%$ ($n = 5$, $P = .015$) and $32.3 \pm 9.4\%$ ($n = 4$, $P = .029$) in the presence of 4AP and $BaCl_2$, respectively, compared to $65.6 \pm 5.4\%$ ($n = 12$) in controls. When added together, 4-AP and $BaCl_2$ (*black bar*) completely prevented the WIN-2 (10 µM) inhibition ($1.7 \pm 11\%$, $n = 4$, $P = .002$). (**C**) CB1 receptor-mediated inhibition of evoked inhibitory synaptic transmission in mouse nucleus accumbens: the cannabimimetic WIN 55,212,2 (10 µM) reduced evoked IPSCs on average to $30.3 \pm 9.8\%$ of their basal value ($n = 4$). Summary of the time course of the depressing action of bath-applied WIN-2 in the 4 cells recorded. Each point is expressed as the percentage of inhibition of basal-evoked transmission; the error bar represents the SEM.

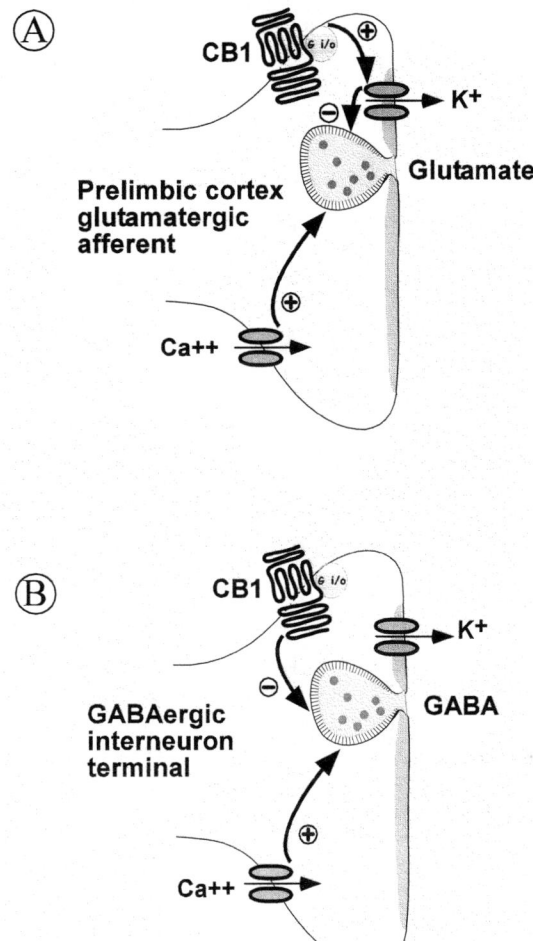

FIGURE 2. Schematic view of the presynaptic actions of CB1 receptors in the NAc: (**A**) At glutamatergic terminals, CB1 receptors reduce synaptic release via activation of presynaptic K^+ channels. (**B**) At GABAergic terminals, CB1 receptors reduce synaptic release downstream of presynaptic Ca^{2+} channels.

CB1-mediated mIPSC inhibition in the presence of the polyvalent cation ruthenium red. Because ruthenium red strongly increases spontaneous release (via direct interactions with secretory mechanisms) and blocks voltage-sensitive Ca^{2+} channels, their experiments strongly suggest that presynaptic CB1 receptors inhibit GABA release downstream of Ca^{2+} channels.

Since the vast majority of cells in the NAc are GABA containing medium-spiny neurons that may be interconnected by a dense network of recurrent collaterals[13] and innervated by inhibitory GABAergic interneurons,[28] the existence of cannabinoid CB1 receptors, reducing inhibitory synaptic GABAergic currents in addition to their presence on cortical excitatory terminals, suggests a complex role of endogenous cannabinoids in the NAc. It is possible that the total outcome of endogenous cannabinoid production may depend on the site/pattern of production of these diffusible lipids.

Endogenous Cannabinoids Mediated Long-Term Synaptic Depression in the NAc

eCBs, CB1, and eCB retrograde signaling have been involved in a short-lasting form of synaptic regulation: the depolarization-induced suppression of both inhibitory and excitatory transmission (DSI-E) observed in the cerebellum and the hippocampus (reviewed in Refs. 7, 20, and 29). In these structures, the brief depolarization of postsynaptic neurons caused eCB-release and short-living presynaptic inhibition lasting from tens of seconds to one minute.

However, the involvement of eCB in long-lasting activity-dependent synaptic plasticity remained to be documented. The finding that CB1 are localized at the excitatory/inhibitory afferents to the NAc where exogenous cannabimimetics inhibit glutamatergic/GABAergic synaptic transmission raised two questions: How does synaptic activity lead to the production of eCBs in the NAc, and What are the physiological correlates of eCB release on synaptic transmission?

To answer these questions, it was first necessary to define a synaptic stimulation protocol that releases endocannabinoids in the NAc. *In vivo* studies have shown the occurrence of low-frequency (up to 10–15 Hz) cell firing in the NAc and that long-term synaptic plasticity can be induced at the corticostriatal synapses following several minutes of stimulation at 5-Hz discharge.[30,31] We thus explored the consequences of various low-frequency stimulations in a slice preparation containing prelimbic cortex-NAc synapses.[32] By trial and error we discovered that a 10-min stimulation at 13 Hz of the prelimbic cortex afferents to the NAc reliably induced a robust long-term depression (LTD) of evoked excitatory synaptic transmission.[32]

The 13Hz-LTD was observed in the whole cell configuration (excitatory postsynaptic currents (eEPSCs) expressed were $77.9 \pm 13.1\%$ of baseline 30 min after the end of the tetanus; FIG. 3A) and when measuring field excitatory postsynaptic potentials (fEPSP measured $81.4 \pm 3.6\%$ of baseline 60 min after tetanus, $n = 39$). Repetition of the low-frequency tetanus caused saturation of 13Hz-LTD (maximum 40% depression). We tested the hypotheses that eCBs were released by the 13 Hz tetanus and induced LTD. First, the effects of the selective CB1 antagonist SR141716A (29) were tested: it completely blocked 13Hz-LTD induction (with the low concentration of 100 nM SR141716A,[27] the fEPSP was $100.9 \pm 1.9\%$ of baseline at 60 min, $n = 6$, $P < .05$ vs. control). Similar results were obtained with another CB1 antagonist AM-251 (2 µM, fEPSP was $102.3 \pm 8.9\%$ of baseline at 60 min, $n = 6$, $P < .05$ vs. control). SR141716A (1 µM) had no effects when applied once eCB-LTD was already induced (fEPSP was $79.2 \pm 5.5\%$ of baseline at 60 min, $n = 7$, $P = .7$ compared to control without SR141716A), suggesting that the late phase of 13Hz-LTD was not due to the continuous release of eCBs.

In marked contrast with their wild-type littermates, genetically altered mice lacking CB1 did not display 13Hz-induced LTD, confirming that 13Hz-LTD depends on eCB signaling and CB1. If CB1 activation and eCB-LTD shared common mechanisms to inhibit excitatory synaptic transmission, a prediction was that eCB-LTD

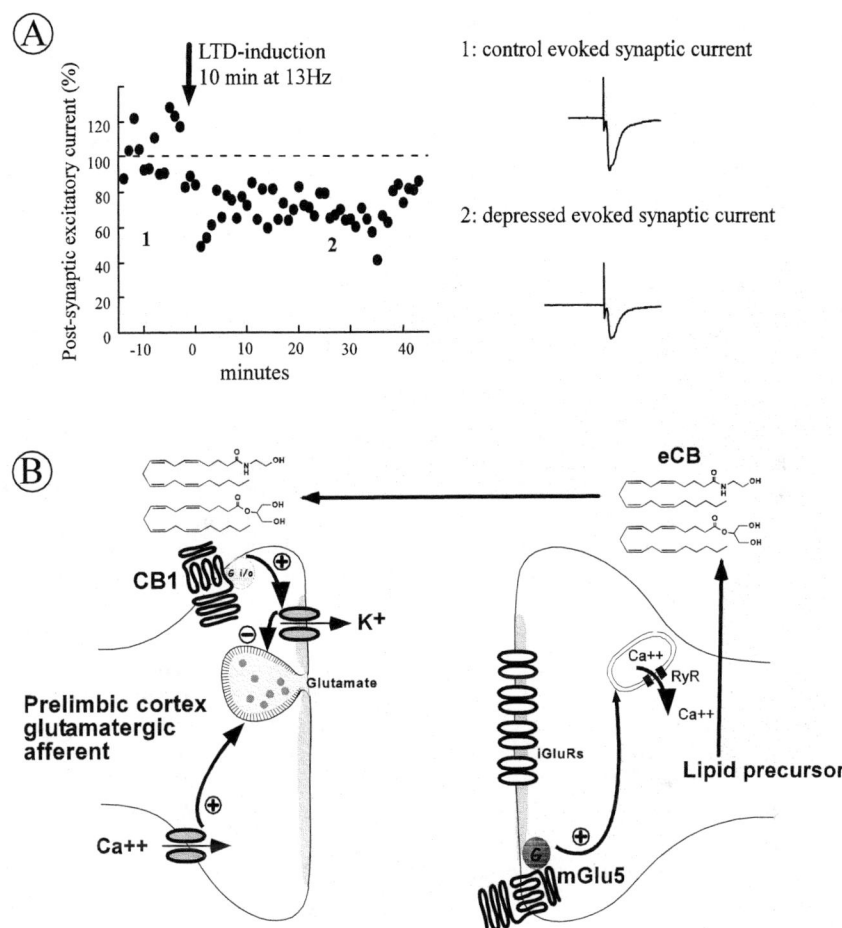

FIGURE 3. Schematic model of eCB-mediated retrograde signaling in the NAc: (**A**) Tetanic stimulation (10 min) of prelimbic afferents at a naturally occurring frequency (13 Hz) induced LTD. (**B**) Schematic model of eCB-mediated retrograde signaling in the NAc: Tetanic stimulation (10 min at 3 Hz) induced a presynaptic LTD dependent on eCB and CB1 receptors. Induction of eCB-LTD required postsynaptic activation of mGluR5 and a rise in postsynaptic Ca^{2+} mediated by ryanodine-sensitive intracellular stores. The retrograde signaling cascade led to postsynaptic eCB release, activation of presynaptic CB1 receptors, and LTD.

was occluded by exogenous cannabimimetic and endogenously released cannabinoids. Indeed, pretreatment with the CB1 agonist WIN 55,212,2 (300 µM) inhibited excitatory synaptic transmission and completely prevented eCB-LTD and bath application of 10 µM AM404, a selective inhibitor of the eCB transporter, partially occluded eCB-LTD.

Thus, eCB-LTD is occluded by exogenous cannabimimetics and naturally occurring eCBs, suggesting common pathways for these phenomena. On the basis of the previous observation of the presynaptic localization of CB1 at the prelimbic cortex-NAc synapses, we verified that eCB-LTD was also expressed at a presynaptic locus and recorded spontaneous EPSCs (sEPSC) in control conditions and 30 min after eCB-LTD: induction of eCB-LTD markedly decreased sEPSC's frequency up to 30 min after tetanus ($P < .0001$, Kolmogorov-Smirnof test). In contrast, the distribution of sEPSC amplitude was not affected ($P = .012$, Kolmogorov-Smirnof test), showing that eCB-LTD is expressed at a presynaptic locus.

Postsynaptic Ca^{2+} elevation plays a crucial role in eCB signaling and in depolarization-induced suppression of inhibition excitation (DSI/DSE) of the hippocampus and the cerebellum.[29] We decided to evaluate the Ca^{2+} requirements of eCB-LTD: when slices were preincubated with the membrane permeant Ca^{2+}-chelator BAPTA-AM (50 µM), eCB-LTD was completely prevented. Accordingly, eCB-LTD was eliminated by selectively buffering postsynaptic calcium with intrapipette BAPTA.[32]

What receptors trigger eCB release during the tetanus? Because both glutamatergic and dopaminergic afferents contact medium spiny neurons and because in the dorsal striatum eCB signaling is activated by dopamine,[33] the involvement of glutamate and dopamine receptors was evaluated. Pretreatment with the non–subtype-selective ionotropic glutamate receptor antagonist kynurenate (1 mM, 10 min prior to and during tetanus) completely blocked fEPSPs but had no effect on the induction of eCB-LTD, which was normal after kynurenate washout, suggesting the lack of involvement of postsynaptic ionotropic glutamate receptors. Similar negative results were obtained with the competitive NMDA receptor antagonist D-AP5 (50 µM), a mixture of the selective D1 antagonist SCH23390 (30 µM) and the D2 antagonist sulpiride (30 µM), and by selective blockade of mGluR2/3 receptors with a low concentration of LY341495 (200 nM) or eGlu (50 µM).

In the cerebellum, activation of postsynaptic mGlu1 receptors causes DSI.[29] Medium spiny neurons express a subtype of mGlu receptors coupled to phospholipase C—mGlu5[21,35]—which is mainly located in postsynaptic elements.[34] We reasoned that sustained stimulation of cortical glutamatergic afferents could activate postsynaptic mGlu5, which could translate the glutamate signal into eCB signaling. Indeed we found that eCB-LTD is abolished by the broad-spectrum mGluR antagonist LY341495 and by the specific mGluR5 antagonist MPEP.[35] Conversely, direct pharmacological stimulation with the specific group 1 mGluR agonist (S)-DHPG (100 µM) for 10 min was sufficient to induce LTD in wild-type mice but not in CB1$^{-/-}$ mice. These results show that stimulation of mGlu5 by synaptically released glutamate or an exogenous agonist is both necessary and sufficient for the expression of LTD at the prelimbic cortex-NAc synapses. mGluR5 receptors are classically described to be coupled to the phospholipase C and the IP3/intracellular Ca^{2+} second messenger pathways,[36] and we found that both the Ca^{2+}-ATPase inhibitor thapsigargin and ryanodine blocked eCB-LTD, indicating a role of the ryanodine-sensitive intracellular pools.

In the central nervous system, eCBs are produced by several mechanisms: depolarization-induced suppression of transmitter release is triggered by a transient postsynaptic depolarization, and interactions with group I mGlu have been proposed.[7,20,29] Our findings suggest an important physiological function for mGlu5: translating postsynaptic glutamatergic signaling to presynaptic eCB signaling.

Interestingly, at the same time, Lovinger's group made a very similar discovery at the corticostriatal synapses.[37] These authors reported that high-frequency stimulation of cortical afferents paired with postsynaptic depolarization induced a form of LTD blocked by a CB1 antagonist and absent in CB1$^{-/-}$ mice. Similar to NAc's eCB-LTD, striatal eCB-LTD required postsynaptic Ca^{2+} elevation. Although the postsynaptic receptors involved in striatal eCB-LTD were not identified, this study strongly suggests that eCB-LTD might be a common feature of central synapses.

CONCLUSIONS

The fact that in the NAc eCB-LTD could be induced with low-frequency tetanus mimicking naturally occurring frequencies suggests the existence of eCB-LTD *in vivo*. Retrograde eCB signaling could participate in the adaptative responses of many central glutamatergic and nonglutamatergic synapses. Specifically in the NAc, eCB-LTD might be part of a negative feedback loop reducing glutamatergic synaptic strength during sustained cortical activity. Moreover, on the basis of the presence of CB1 receptors on inhibitory terminals in the NAc, it is expected that in response to the appropriate naturally released stimulus pattern, eventually eCBs reach and activate these receptors. The final outcome of eCB release in the NAc will depend on the local distribution of CB1 receptors and on eCB-releasing enzymes. In accord with this idea, Marsicano and colleagues have presented compelling evidence that, in the basolateral amygdala, synaptically released eCBs induce the long-term depression of inhibitory transmission.[38] The eCB system appears to be in the unique position to participate in activity-dependent long-term plasticity of both glutamatergic and GABAergic synapses.

Thus, there is a growing body of evidence indicating a key role for the eCB system in short- or long-term synaptic plasticity in brain areas important to the behavioral effects of cannabis use. A current hypothesis is that addictive drugs participate in and alter normal brain function and cause the remodeling of synaptic functions. The finding of a form of LTD induced by eCB that is occluded by an exogenous cannabinoid gives insights into how cannabis derivatives, such as marijuana, alter normal eCB functions in the brain reward system: it is possible that cannabimimetics (such as the widely abused phytocannabinoid *Cannabis sativa L.,* (−)-trans-delta9-tetrahydrocannabinol) occlude normal eCB functions (such as LTD) and affect eCB-mediated synaptic plasticity. The importance of these phenomena to the cellular adaptations caused by marijuana abuse remains to be determined, but one can propose their participation in the complex behavioral effects of cannabis derivatives.

REFERENCES

1. HYMAN, S.E. & R.C. MALENKA. 2001. Addiction and the brain: the neurobiology of compulsion and its persistence. Nat. Rev. Neurosci. **2:** 695–703.
2. WILLIAMS, J.T., M.J. CHRISTIE & O. MANZONI. 2001. Cellular and synaptic adaptations mediating opioid dependence. Physiol. Rev. **81:** 299–343.
3. NESTLER, E.J. 2001. Molecular neurobiology of addiction. Am. J. Addict. **10:** 201–217.
4. NESTLER, E.J. 2002. Common molecular and cellular substrates of addiction and memory. Neurobiol. Learn. Mem. **78:** 637–647.

5. SANES, J.R. & J.W. LICHTMAN. 1999. Can molecules explain long-term potentiation? Nat. Neurosci. **2:** 597–604.
6. AMERI, A. 1999. The effects of cannabinoids on the brain. Prog. Neurobiol. **58:** 315–348.
7. WILSON, R.I. & R.A. NICOLL. 2002. Endocannabinoid signaling in the brain. Science **296:** 678–682.
8. HOFFMAN, A.F. & C.R. LUPICA. 2000. Mechanisms of cannabinoid inhibition of GABA(A) synaptic transmission in the hippocampus. J. Neurosci. **20:** 2470–2479.
9. CHAN, P.K., S.C. CHAN & W.H. YUNG. 1998. Presynaptic inhibition of GABAergic inputs to rat substantia nigra pars reticulata neurones by a cannabinoid agonist. Neuroreport **9:** 671–675.
10. LEVENES, C., H. DANIEL, P. SOUBRIE & F. CREPEL. 1998. Cannabinoids decrease excitatory synaptic transmission and impair long-term depression in rat cerebellar Purkinje cells. J. Physiol. **510** (Pt 3): 867–879.
11. AUCLAIR, N., S. OTANI, P. SOUBRIE & F. CREPEL. 2000. Cannabinoids modulate synaptic strength and plasticity at glutamatergic synapses of rat prefrontal cortex pyramidal neurons. J. Neurophysiol. **83:** 3287–3293.
12. HYMAN, S.E. 1996. Shaking out the cause of addiction. Science **273:** 611–612.
13. PENNARTZ, C.M., H.J. GROENEWEGEN & F.H. LOPES DA SILVA. 1994. The nucleus accumbens as a complex of functionally distinct neuronal ensembles: an integration of behavioural, electrophysiological and anatomical data. Prog. Neurobiol. **42:** 719–761.
14. PULVIRENTI, L. & M. DIANA. 2001. Drug dependence as a disorder of neural plasticity: focus on dopamine and glutamate. Rev. Neurosci. **12:** 141–158.
15. MANZONI, O., D. PUJALTE, J. WILLIAMS & J. BOCKAERT. 1998. Decreased presynaptic sensitivity to adenosine after cocaine withdrawal. J. Neurosci. **18:** 7996–8002.
16. MANZONI, O., J.M. MICHEL & J. BOCKAERT. 1997. Metabotropic glutamate receptors in the rat nucleus accumbens. Eur. J. Neurosci. **9:** 1514–1523.
17. MANZONI, O.J. & J.T. WILLIAMS. 1999. Presynaptic regulation of glutamate release in the ventral tegmental area during morphine withdrawal. J. Neurosci. **19:** 6629–6636.
18. TSOU, K., S. BROWN, M.C. SANUDO-PENA, K. MACKIE & J.M. WALKER. 1998. Immunohistochemical distribution of cannabinoid CB1 receptors in the rat central nervous system. Neuroscience **83:** 393–411.
19. ROBBE, D., G. ALONSO, F. DUCHAMP, et al. 2001. Localization and mechanisms of action of cannabinoid receptors at the glutamatergic synapses of the mouse nucleus accumbens. J. Neurosci. **21:** 109–116.
20. DOHERTY, J. & R. DINGLEDINE. 2003. Functional interactions between cannabinoid and metabotropic glutamate receptors in the central nervous system. Curr. Opin. Pharmacol. **3:** 46–53.
21. SULLIVAN, J.M. 1999. Mechanisms of cannabinoid-receptor-mediated inhibition of synaptic transmission in cultured hippocampal pyramidal neurons. J. Neurophysiol. **82:** 1286–1294.
22. HOFFMAN, A.F. & C.R. LUPICA. 2001. Direct actions of cannabinoids on synaptic transmission in the nucleus accumbens: a comparison with opioids. J. Neurophysiol. **85:** 72–83.
23. GESSA, G.L., M. MELIS, A.L. MUNTONI & M. DIANA. 1998. Cannabinoids activate mesolimbic dopamine neurons by an action on cannabinoid CB1 receptors. Eur. J. Pharmacol. **341:** 39–44.
24. TANDA, G., F.E. PONTIERI & G. DI CHIARA. 1997. Cannabinoid and heroin activation of mesolimbic dopamine transmission by a common mu1 opioid receptor mechanism. Science **276:** 2048–2050.
25. SZABO, B., T. MULLER & H. KOCH. 1999. Effects of cannabinoids on dopamine release in the corpus striatum and the nucleus accumbens in vitro. J. Neurochem. **73:** 1084–1089.
26. MANZONI, O.J. & J. BOCKAERT. 2001. Cannabinoids inhibit GABAergic synaptic transmission in mice nucleus accumbens. Eur. J. Pharmacol. **412:** R3–5.
27. BARTH, F. & M. RINALDI-CARMONA. 1999. The development of cannabinoid antagonists. Curr. Med. Chem. **6:** 745–755.

28. KOOS, T. & J.M. TEPPER. 1999. Inhibitory control of neostriatal projection neurons by GABAergic interneurons. Nat. Neurosci. **2:** 467–472.
29. KREITZER, A.C. & W.G. REGEHR. 2002. Retrograde signaling by endocannabinoids. Curr. Opin. Neurobiol. **12:** 324–330.
30. CHARPIER, S., S. MAHON. & J.M. DENIAU. 1999. In vivo induction of striatal long-term potentiation by low-frequency stimulation of the cerebral cortex. Neuroscience **91:** 1209–1222.
31. CARELLI, R.M. & S.G. IJAMES. 2000. Nucleus accumbens cell firing during maintenance, extinction, and reinstatement of cocaine self-administration behavior in rats. Brain Res. **866:** 44–54.
32. ROBBE, D., M. KOPF, A. REMAURY, et al. 2002. Endogenous cannabinoids mediate long-term synaptic depression in the nucleus accumbens. Proc. Natl. Acad. Sci. USA **99:** 8384–8388.
33. GIUFFRIDA, A. et al. 1999. Dopamine activation of endogenous cannabinoid signaling in dorsal striatum. Nat. Neurosci. **2:** 358–363.
34. SHIGEMOTO, R. et al. 1993. Immunohistochemical localization of a metabotropic glutamate receptor, mGluR5, in the rat brain. Neurosci. Lett. **163:** 53–57.
35. PIN, J.P., C. DE COLLE, A.S. BESSIS & F. ACHER. 1999. New perspectives for the development of selective metabotropic glutamate receptor ligands. Eur. J. Pharmacol. **375:** 277–294.
36. CONN, P.J. & J.P. PIN. 1997. Pharmacology and functions of metabotropic glutamate receptors. Annu. Rev. Pharmacol. Toxicol. **37:** 205–237.
37. GERDEMAN, G.L., J. RONESI & D.M. LOVINGER. 2002. Postsynaptic endocannabinoid release is critical to long-term depression in the striatum. Nat. Neurosci. **5:** 446–451.
38. MARSICANO, G. et al. 2002. The endogenous cannabinoid system controls extinction of aversive memories. Nature **418:** 530–534.

Plastic Control of Striatal Glutamatergic Transmission by Ensemble Actions of Several Neurotransmitters and Targets for Drugs of Abuse

DAVID M. LOVINGER,[a] JOHN G. PARTRIDGE,[b] AND KA-CHOI TANG[c]

[a]*Laboratory for Integrative Neuroscience,* [b]*Laboratory of Molecular Physiology, Division of Intramural Clinical and Basic Research, National Institute on Alcohol Abuse and Alcoholism, National Institutes of Health, Rockville, Maryland 20852, USA*

[c]*Ernest Gallo Clinic and Research Center, and the Department of Neurology, University of California at San Francisco, Emeryville, California 94608, USA*

> ABSTRACT: Long-lasting alterations in the efficacy of glutamatergic synapses, such as long-term potentiation (LTP) and long-term depression (LTD), are prominent models for mechanisms of information storage in the brain. It has been suggested that exposure to drugs of abuse produces synaptic plasticity at glutamatergic synapses that shares many features with LTP and LTD, and that these synaptic changes may play roles in addiction. We have examined the involvement of particular neurotransmitters in synaptic plasticity at glutamatergic synapses within the striatum, a brain region with prominent roles in initiation and sequencing of actions, as well as habit formation. Our studies indicate that multiple neurotransmitters interact to produce striatal synaptic plasticity, and that the relative strength and patterning of the afferent inputs that release the various neurotransmitters determines whether LTP or LTD is activated. Drugs of abuse interact with glutamatergic synaptic plasticity in multiple ways, including alterations in dopamine release and more direct effects on glutamate release and glutamate receptors. We hypothesize that these effects contribute to addiction by facilitating the formation of new, drug-centered habits, and by disruption of more adaptive behaviors.
>
> KEYWORDS: acetylcholine; addiction; basal ganglia; dopamine; habit formation; long-term depression; long-term potentiation; learning and memory; synaptic plasticity

INTRODUCTION

Drugs of abuse produce a range of effects on brain function and behavior after either acute or chronic ingestion. Acute effects, such as the anxiolytic effects of benzodiazepines and alcohol, temporary mood elevation by amphetamines and cocaine,

---

Address for correspondence: Dr. David M. Lovinger, National Institute on Alcohol Abuse and Alcoholism, Rockville, MD 20852. Voice: 301-443-2445; fax: 301-480-1734.
lovindav@mail.nih.gov

and pain relief by opiates, are subjectively pleasant for the users. However, prolonged use and abuse of many psychoactive drugs leads to addiction, involving drug tolerance, drug dependence, and avoidance of withdrawal syndromes. In the initial stages of use, drugs act on target proteins within key brain regions, and most often these actions result in altered synaptic transmission that produces intoxication. While many of the effects that occur during drug intoxication are reversible once the drug is metabolized or eliminated, changes in neuronal function can outlast drug exposure. In many cases actions of the abused drug on its target protein will provoke adaptive responses in brain physiology and/or neurochemistry. There is emerging evidence that drugs of abuse can activate mechanisms leading to long-lasting changes in synaptic transmission, often referred to as long-lasting synaptic plasticity.[1-3] Drugs of abuse may also act to interrupt the normal production of synaptic plasticity by other environmental stimuli. Synaptic plasticity is thought to be one essential mechanism for information storage in the brain. Thus, abused drugs may have a large impact on brain information storage by directly stimulating plastic changes in neuronal circuitry while disrupting changes in the same circuitry that normally subserves storage of information about important environmental events. These mechanisms may produce the aberrant behaviors associated with addiction, such as excessive drug seeking and neglect of other, more socially adaptive behaviors.

There is now a great deal of evidence linking both the acute and chronic effects of addictive drugs to alterations in glutamatergic transmission. As we will see from the discussion below, most, if not all, drugs of abuse either directly or indirectly alter glutamatergic transmission during acute exposure. In addition, many of the long-lasting forms of synaptic plasticity that have been characterized occur at glutamatergic synapses that connect different brain regions, and there is a growing body of studies indicating that drugs of abuse alter these forms of synaptic plasticity.

We and others are interested in the actions of drugs of abuse within the basal ganglia, a set of interconnected brain regions that regulate action production, sequencing, and habit formation. Glutamate is one of the major neurotransmitters used to communicate information between the different subregions of the basal ganglia. Plastic changes in this glutamatergic communication could play roles in habit learning and other forms of information storage. Glutamate does not work in isolation. Indeed, there are numerous other neurotransmitters that interact with and modulate the actions of glutamate in the basal ganglia. In the present paper we will focus on the striatum, a subregion of the basal ganglia known to have strong influence on the workings of the entire circuitry. We will present and discuss new evidence concerning glutamatergic transmission, synaptic plasticity at glutamatergic synapses, and interactions of various neurotransmitters and drugs of abuse with glutamatergic transmission in striatum.

STRIATAL CIRCUITRY AND PHYSIOLOGY

The striatum in the rodent brain consists largely of the caudate nucleus, while in the primate brain this structure has further differentiated into both caudate and putamen nuclei. In addition, the brains of many species contain a component of the ventral striatum known as the nucleus accumbens (NAc) that shares many cytoarchitectural, neurochemical, and physiological features with the more dorsal components of the striatum. We will mainly discuss observations made in the dorsal striatum, with occasional reference to similar or contrasting findings in the NAc. The neuronal complement of the striatum consists largely of medium spiny projection neurons that constitute more than 90% of the total neurons. These neurons are GABAergic, and, as such, they function to inhibit downstream nuclei, such as the internal pallidum and the substantia nigra. It is through these output neurons that the striatum alters basal ganglia throughput. The two different target sites form the direct and indirect pathways that have opposing effects on cortical output.[4] Exciting new studies also support previous evidence of synaptic crosstalk between medium spiny neurons via GABAergic axon collaterals.[5,6]

Synaptic input to the medium spiny neurons comes largely from extrinsic afferents. Glutamatergic inputs arising from all areas of cortex, as well as from the thalamus innervate the neurons on the spines covering the outer two-thirds of the dendrites.[7] The striatum is the largest recipient of dopaminergic synaptic innervation, which comes from the substantia nigra pars compacta. These afferents synapse on many cell types and synaptic structures within the striatum, but one of the most common sites of synapses is the necks of dendritic spines on the medium spiny neurons.[8] The close spatial arrangement of glutamatergic and dopaminergic inputs suggests strong interactions between these neurotransmitters. Glutamate most often functions as a fast-acting excitatory neurotransmitter, while dopamine (DA) has neuromodulatory effects, and thus it is most likely that DA modulates glutamatergic excitation. The striatum also receives input from other brain regions, including the serotonergic raphe nucleus, but the physiological roles of these inputs to striatum will not be considered here.

Several subtypes of striatal interneurons release different neurotransmitters with roles in intrinsic striatal communication. These include the fast-spiking, small GABAergic parvalbumin-positive neurons, the nitric oxide synthase-positive interneurons, and the large cholinergic interneurons. All of these neuronal subtypes innervate the medium spiny neurons,[9,10] and there is crosstalk among the interneurons as well.[11,12] The large cholinergic neurons are especially well characterized. These neurons receive glutamatergic input from the thalamus, as well as dopaminergic innervation from the substantia nigra. The cholinergic neuronal axons arborize extensively to form synaptic contacts throughout the striatum, with especially dense innervation in the patch (aka striosome) compartments. They innervate not only medium spiny neurons, but also the small GABAergic interneurons,[12] and probably also influence the function of dopaminergic axon terminals.[13,14] The small GABAergic interneurons provide one source of inhibitory synaptic input to medium spiny neurons, and most likely inhibit other striatal interneurons.[11] The NOS/somatostatin-positive interneurons provide rich innervation throughout the striatum and are thought to have a primarily modulatory role.

A rich variety of receptors exist for the many striatal neurotransmitters. Both ionotropic and metabotropic glutamate receptors are abundant.[15,16] The medium spiny neurons contain AMPA and NMDA ionotropic receptors, as well as group I metabotropic receptors. Group II mGluRs reside mostly on glutamatergic cortical afferents, where they appear to act as autoreceptors.[17] Both D1 and D2 dopamine receptors are highly expressed in the striatum, with D3 and D5 subtypes being expressed at lower levels. A large assortment of ACh receptors is also present, including several subunits that form nicotinic receptors, along with the m1–4 muscarinic receptors.[18]

$GABA_B$ receptors are thought to exist mainly on the cortical afferents,[19] while a number of $GABA_A$ receptor–forming subunits are expressed by the different striatal neurons. Opiate receptors are also present in abundance, with both the delta and mu subtypes being prominent.[20] In addition, the striatum is rich in CB1-type cannabinoid receptors,[21] which will be discussed more prominently in a later section of this paper. Receptors for 5-HT, somatostatin, and other neuromodulators are also present in the striatum.

INTERACTIONS AMONG STRIATAL NEUROTRANSMITTERS

The rich sources of synaptic input to striatal neurons from external afferents and local interneurons provide a situation in which multiple neurotransmitters have the potential to interact. Many such interactions have been described, with the majority being conventional neuromodulatory effects common to most brain regions. For example, corticostriatal glutamatergic synapses are inhibited by a number of neurotransmitter receptors, including presynaptic M2-class muscarinic ACh, adenosine A1, GABAB, cannabinoid, and opiate receptors.[22–25] Dopamine and other striatal neurotransmitters can also modulate the function of a variety of ion channels found on medium spiny neurons and striatal interneurons.[26]

In addition, there is good evidence for interactions among striatal neurotransmitters in determining the behavior of an organism. For example, adenosine and dopamine are thought to have opposing actions on neurophysiology, neurochemistry, and behavior.[27–29] Dopamine and ACh have long been known to have antagonistic effects on striatal output, at least in drug-naïve animals.[30,31] More recent studies indicate that this relationship may be altered by chronic amphetamine exposure, such that cholinergic output is increased by the drug.[32] This may be one of the plastic changes underlying amphetamine sensitization. Ultimately, the effects of ACh-DA interactions on circuitry and behavior will likely depend on the receptor subtypes activated by the neurotransmitters.[33] There is a growing realization that interactions among striatal neurotransmitters play key roles in determining the types of plasticity taking place at glutamatergic synapses onto striatal neurons, as discussed in the next section.

STRIATAL SYNAPTIC PLASTICITY INVOLVES COMPLEX INTERACTIONS BETWEEN MULTIPLE NEUROTRANSMITTERS

Recent years have witnessed rapid developments in our understanding of the factors that contribute to synaptic plasticity in the striatum. In particular, the role of different neurotransmitter receptors has become clearer. Early studies of plasticity using striatal slices focused primarily on long-term depression (LTD) at corticostriatal synapses, primarily due to the ease with which this form of plasticity could be evoked in the slice preparation. FIGURE 1A schematically illustrates the neurotransmitter interactions involved in striatal LTD. Calabresi and colleagues noted the need for dopaminergic transmission, a feature unique to striatal LTD, early in their analysis of this form of plasticity.[34] These investigators also made the important observation that striatal LTD requires activation of D2 dopamine receptors. A requirement

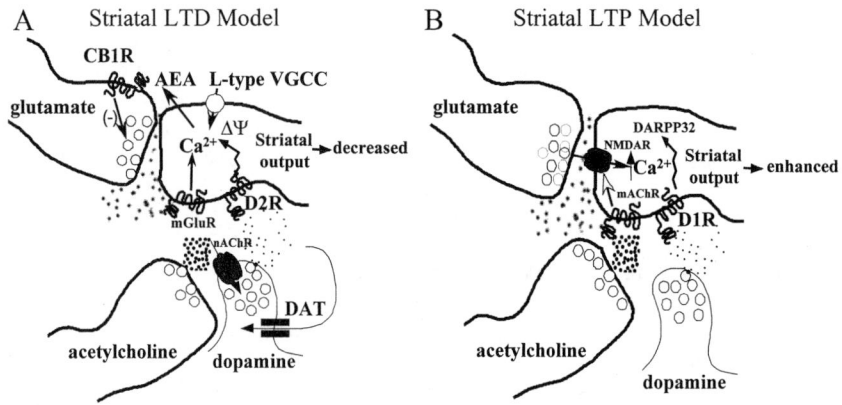

FIGURE 1. Neurotransmitter receptors involved in induction of striatal LTP and LTD. (**A**) Striatal LTD requires activation of D2 receptors, group I metabotropic glutamate receptors, and L-type calcium channels, ultimately leading to endocannabinoid release and activation of CB1 cannabinoid receptors. Nicotinic ACh receptors likely participate in LTD induction by stimulating DA release from the axon terminals of nigral neurons. AEA = arachidonyl ethanolamide (anandamide); DAT = dopamine transporter; VGCC = voltage-gated calcium channel; CB1R = cannabinoid CB1 receptor. (**B**) Striatal LTP requires coordinated activation of NMDA-type glutamate receptors, D1-like dopamine receptors, and muscarinic ACh receptors. DARPP32 = dopamine and cAMP-regulated phosphoprotein 32.

for activation of a D1-like receptor was also suggested by pharmacological studies in the Calabresi laboratory, but it is not yet clear if D1 or D5 receptors are involved in striatal LTD. One idea that has been suggested is that D1-like receptors on striatal interneurons stimulate the release of nitric oxide (NO) that plays a role in LTD induction.[35]

The actions of glutamate itself are also crucial for LTD induction. The synaptic activation of AMPA receptors provides the initial depolarization during the repetitive synaptic activation that induces LTD, and this depolarization is necessary for LTD induction.[34] Activation of the G-protein–coupled metabotropic glutamate receptors (mGluRs) has also been strongly implicated in striatal LTD induction.[36,37] It appears that the group I mGluRs that are known to reside postsynaptically on striatal medium spiny neurons[38] are involved in LTD induction but are not involved in maintenance of synaptic depression once LTD has been induced. These receptors are known to stimulate phospholipase activity and release of intracellular calcium stores in neurons, and thus it is reasonable to speculate that group I mGluRs contribute to LTD induction via stimulation of one or both of these signaling pathways. The role of NMDA receptors in striatal LTD remains unclear. While some investigators claim that LTD is blocked by NMDAR antagonists,[39] the majority of studies indicate that robust LTD can be induced during receptor blockade.[34,40] The differential findings may result from different LTD induction procedures, and the cellular locus of the NMDARs that could play a role in LTD is not known. It is possible that two forms of LTD, one NMDAR dependent and one mGluR dependent, coexist in the striatum.

As mentioned above, ACh is an abundant neurotransmitter in the striatum. Both muscarinic and nicotinic ACh receptors are highly expressed in the striatum. The nicotinic receptors have been implicated in LTD induction by virtue of the finding that antagonists of these receptors block the initiation of LTD.[41] This blockade can be overcome by application of drugs that block DA uptake, including cocaine. It appears that nicotinic receptors present on the terminals of dopaminergic neurons stimulate DA release during LTD induction. This idea is supported by the observation that nAChRs play a prominent role in DA transients produced by afferent stimulation in striatal slices.[14] The observation that DA uptake blockade restores LTD even when nAChRs are blocked indicates that nicotinic receptor activation is not strictly necessary for LTD induction. Rather, we believe that nAChRs play a permissive role.

An endocannabinoid neuromodulator has also been shown to play a crucial role in stiratal LTD. Application of blockers of the cannabinoid receptor blocks LTD induction and LTD cannot be induced in mice lacking the CB1 cannabinoid receptor.[42] Experimental evidence has been presented supporting the idea that the endocannabinoid involved in LTD is generated postsynaptically and acts presynaptically to promote LTD expression. Interestingly, similar roles for endocannabinoids and CB1 receptors are supported by studies of LTD in the nucleus accumbens and amygdala.[43,44] Thus, an endocannabinoid may be the long-sought-after retrograde messenger involved in many forms of LTD.

Until recently, most of the evidence for striatal LTP came from *in vivo* experiments, and it was generally thought that LTP induction was difficult to achieve in brain slices. However, recent experiments using slices from different aged animals and different recording techniques have revealed robust LTP in this preparation. It appears that LTP is especially prominent during field potential recordings in striatal slices from young animals. The ability to induce LTP also appears to depend on the striatal subregion examined.[39,45] FIGURE 1B schematically illustrates the neurotransmitter interactions involved in striatal LTP. Upon first glance it might appear that striatal LTP involves mechanisms already well described in the hippocampus and other brain regions. Activation of NMDARs is necessary for striatal LTP, as in the other brain regions.[39,45,46] However, further studies of striatal LTP have revealed surprising roles for the prominent striatal neurotransmitters DA and ACh. Dopamine depletion studies indicated the necessity for this neurotransmitter in LTP. Calabresi and colleagues, as well as Kerr and coworkers first demonstrated a role for D1-type DA receptors in LTP observed in a striatal slice preparation.[47,48] These experiments were carried out in slices from adult brain where LTP is not easily observed, and thus LTP was either rarely observed[48] or induced in the absence of added extracellular Mg^{2+}.[47] In brain slices from the dorsomedial subregion of P15–P21 rats, we have consistently observed LTP following delivery of 4 100 Hz stimulus trains (FIG. 2). This LTP is dependent on NMDA receptor activation.[45] The magnitude of LTP induced using this paradigm is drastically reduced in the presence of the D1-like receptor antagonist SKF83566 (FIG. 2A). This finding supports the idea that striatal LTP involves activation of D1-type DA receptors, even in the presence of extracellular Mg^{2+} and under conditions where LTP is robust and consistently observed. Some LTP remains in the presence of the D1 antagonist, and thus it remains to be determined if D1-like receptor activation is absolutely required for LTP induction, or if these receptors produce a modulatory effect that promotes LTP induction.

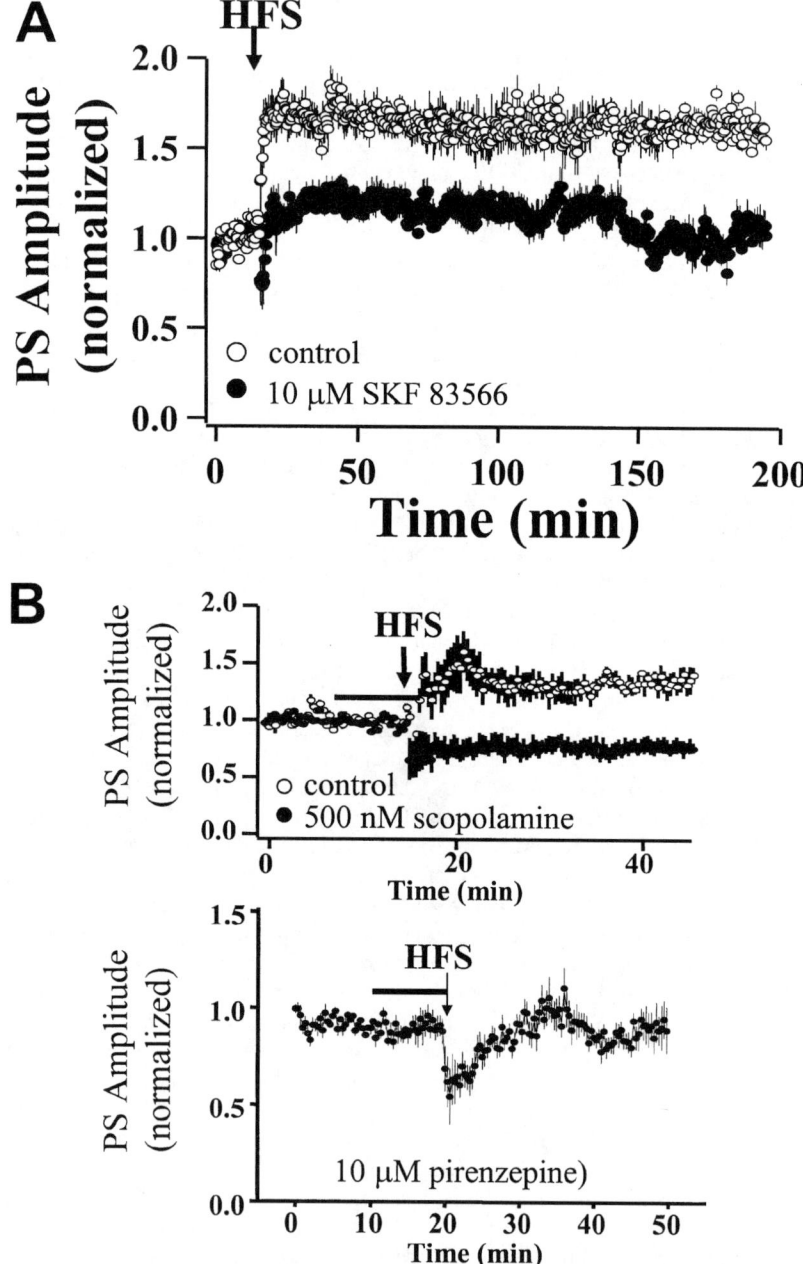

FIGURE 2. Blockade of striatal LTP by D1 dopamine receptor and mAChR antagonists. Graphs showing time course of changes in population spike responses recorded in dorsomedial striatum before and after high-frequency stimulation in the absence and presence of the indicated concentrations of a D1 dopamine receptor antagonist (**A**) and the muscarinic

Evidence has also accumulated indicating a role for muscarinic ACh receptors (mAChRs) in striatal LTP. Calabresi and coworkers have shown that scopolamine (a general mAChR antagonist) and pirenzepine (an M1-type antagonist) block LTP induction using the 0 Mg^{2+} condition.[49] We have likewise observed that scopolamine and pirenzepine block LTP induction by high-frequency stimulation in striatal slices from young animals (FIG. 2B). It is reasonable to speculate that an M1-like mAChR is involved in LTP induction, because M2-like mAChRs inhibit glutamatergic striatal synaptic transmission[23] and thus would be more likely to reduce, rather than promote, LTP induction.

At present there is no solid evidence indicating the mechanisms by which D1 DA receptors or mAChRs participate in LTP induction. D1 receptors stimulate adenylyl cyclase through G_{olf}, and this action could produce gene expression changes needed for maintained LTP expression, as has been suggested in the hippocampus.[50] D1 receptors are located primarily on postsynaptic elements of striatal medium spiny neurons,[51] and thus it is likely that their role in LTP involves postsynaptic signaling. Levine and colleagues have amassed evidence that D1 receptors potentiate NMDAR function in striatum,[52] and this mechanism is a good candidate for a D1 promotional effect on LTP induction.

The M1-like mAChRs are known to stimulate $G_{\alpha q}$-class G-proteins, and this promotes formation of second messengers that can stimulate protein kinases and rises in intracellular calcium. Thus, it is most probable that such mechanisms are activated by these receptors in medium spiny neurons,[53] and that these signals could play an important role in LTP induction in striatum. More detailed information on the induction and expression mechanisms of striatal LTP will be necessary before we can formulate better hypotheses about the roles played by D1 and mACh receptors. To date, our understanding of these mechanisms has been hampered by difficulties in eliciting striatal LTP during patch-clamp and intracellular recordings.[39,45]

One major challenge in integrating LTP and LTD into models of striatal function has been to understand the conditions under which these two forms of plasticity occur. In particular, it seems confusing that both forms of plasticity are most easily elicited by high-frequency stimulation in the slice preparation. It would seem unlikely that the same type of synaptic activation could have opposite effects on synapses impinging on the same types of neurons. However, one must bear in mind that experimenter-generated afferent stimulation in a brain slice setting, and even *in vivo*, is a poor substitute for endogenous patterns of neural activity. At best, the responses to this type of stimulation can tell us about the capabilities for plasticity at a particular synapse, but it requires deeper analysis to determine what types of plasticity will take place in response to natural patterns of afferent input to striatum.

In this respect the findings discussed above, along with evidence from recent *in vivo* studies of plasticity, begin to provide clues to the conditions under which LTP and LTD may be favored in striatum. In the case of striatal LTP, the combined requirement for activation of NMDA and D1 receptors indicates a need for strong

---

receptor antagonists scopolamine and pirenzepine (**B**). *Solid lines* indicate period of antagonist application. (HFS = high-frequency stimulation.) Experiments were carried out with standard field potential recording techniques in coronal striatal slices, as described in Partridge *et al*.[45]

glutamatergic activation and postsynaptic depolarization, coupled with strong dopaminergic input that provides the relatively high DA concentrations needed to activate D1 receptors. This idea is supported by the findings of Wickens and coworkers in the *in vivo* setting who have demonstrated that LTP is induced only during activation of both cortical and nigral afferents, while activation of cortical afferents alone results in LTD.[54] The requirement for mAChR activation suggests that low synaptic concentrations of ACh are needed for LTP induction, given the relatively high affinity of the muscarinic receptors for ACh. It is possible that the normal tonic level of activity of these neurons may be sufficient to provide the ACh needed for potentiation. The observation that tonic mACh receptor activation is often observed in striatal slices[41] even during low-frequency afferent activation supports the idea that this condition is readily attainable in the striatum.

In contrast to LTP, striatal LTD would seem to be favored under conditions where moderate-to-strong glutamatergic input from cortex, and perhaps also from thalamus, are combined with weak nigral input. The glutamatergic input to medium spiny neurons needs to be strong enough to activate group I mGluRs, which are often extrasynaptic, and thus may require some "spillover" of glutamate.[55] Glutamate depolarizes the medium spiny neuron, mostly through AMPA receptor activation, sufficiently to allow activation of L-type calcium channels that are required for LTD activation.[34,40] Glutamatergic synaptic input from thalamus can strongly activate cholinergic interneurons, and this may be one element driving these neurons during LTD induction. We believe that strong activation of the cholinergic neurons is necessary for activation of the relatively low-affinity nicotinic AChRs that help to release the DA necessary for LTD induction. With respect to DA, the consistent observation that D2R activation is necessary for LTD induction suggests that only low-to-moderate levels of dopaminergic neuronal activity are necessary for LTD induction, since D2 receptors are quite sensitive to DA.[56] It is possible that the differing patterns of dopaminergic input needed for LTP versus LTD arise not only from differences in the timing and pattern of firing of nigral neurons, but also from differences in regulation of DA release at axon terminals in the striatum. Our observations, and those made by the Dani group, support the idea that nAChR stimulation of DA release is quite strong in striatum.[14,41] It may well be the case that the pattern of DA release during LTD-inducing activity is dictated by the firing of cholinergic neurons acting locally, while the activity necessary for LTP induction is the burst firing of the nigral neurons themselves. If this is the case, then the patterns of DA release may be more important than the intrasynaptic DA concentrations in determining the type of plasticity elicited. Overall, our model for LTD includes relatively robust glutamatergic input, strong activation of cholinergic input, perhaps secondary to glutamatergic synaptic drive, and weak input from nigral dopaminergic afferents.

We believe that these proposed schemata for LTP and LTD induction fit well with the natural activity patterns of striatal medium spiny neurons and their afferent inputs. The majority of evidence indicates that medium spiny neurons spend most of their time in either the "up" or "down" states.[57] In the down state neurons are receiving little glutamatergic input and exhibit no action potential firing, while in the up state the cortical input is very strong and lasts for several seconds, during which time action potentials can fire at a regular frequency with little accommodation. The up state is dependent on continued cortical glutamatergic input.[57] Given this pattern of input and resultant postsynaptic firing, it is unlikely that medium spiny neurons would develop plasticity mechanisms that depend on low-frequency, weak cortical input. Instead, it would seem reasonable that the neurons would be poised to adjust their synaptic efficacy based on the combination of all afferent inputs during the up state. In this scenario, the pattern of nigral firing and dopaminergic input become crucial factors in determining the valence of efficacy adjustment. The nigral neurons exhibit low frequency, but steady tonic firing, interrupted by periods of higher frequency bursting, and these firing patterns have been linked to generalized behavioral activation and communication of "reward error" signals, respectively.[58] These two patterns of activity could provide the low and high concentrations of DA that help to determine what changes in efficacy occur during pairing with the up state.

The cholinergic neurons also exhibit low frequency, tonic activity even in the absence of synaptic input.[59] Glutamatergic input can activate these neurons quite strongly, presumably leading to bursts of ACh release. These patterns of activity might provide the levels of ACh needed for muscarinic and nicotinic AChR activation, respectively. Given this rich pattern of possible interactions in afferent input, it is easy to imagine how synapses onto medium spiny neurons might adjust their efficacy in response to intense cortical input in a manner that also depends on the coactivation of other synapses, which essentially provides information about the environmental and brain-state "context" in which the cortical input occurs. This context-dependent plasticity will allow medium spiny neurons to adjust their responsiveness to cortical input in an environmentally appropriate manner. We cannot rule out the possibility that adjustments in corticostriatal synaptic efficacy may occur during periods where sustained cortical input is very weak and insufficient to drive medium spiny neurons to fire, although it is not clear when such conditions occur. This could give rise to a form of LTD similar to that induced by sustained low-frequency input in the hippocampus and nucleus accumbens.[60] To date, however, there is no convincing evidence that this form of LTD occurs in the dorsal striatum.

IMPLICATIONS FOR BEHAVIOR AND ADDICTION

The striatum participates in initiation, production, and sequencing of behavior and also functions in the storage of information related to these processes. Indeed, it has been suggested that the striatum, and the basal ganglia as a whole, are involved in cognitive aspects of behavioral planning and production, although the evidence for this is sparse.[61] There has been a tendency to segregate the functions of the dorsal striatum from those of the ventral striatum (or nucleus accumbens). The more dorsal regions have been most closely linked to task-oriented motor sequence production and habit learning, whereas the ventral striatum has been implicated in working memory, reinforcement-oriented behavior, and addiction-related behaviors. However, these roles are probably different extremes of a similar role within the basal ganglia circuitry. Both structures sit in the same relative position within their respective cortico-basal ganglia-cortical circuits, and both likely have the same function within this circuitry to translate initial cortical output to inhibition of downstream structures. Although responses to neurotransmitters (e.g., dopamine) differ for medium spiny neurons in dorsal striatum versus NAc,[62] the major differences between these two striatal subregions are likely related to the specificity of their afferent and efferent connections that dictates the precise information they will be handling. The dorsal striatum will respond mainly to output from motor, sensory, and associative cortices, while the NAc will be handling information from limbic systems as well as associative cortical information. The throughput for these two striatal subregions will be similar in its physiology, to some extent, but different in its informational content.

With respect to a behavioral pattern as complex as drug addiction, it is easy to imagine that information flow through these two striatal subregions might play different but complementary roles. For example, the NAc may store information about the relative reinforcement value of a particular drug and may coordinate the cognitive and affective aspects of craving- and drug-seeking behavior. The dorsal striatum, in contrast, might drive the performance of habitual, almost stereotyped, behaviors associated with the abused substance.[2,63] It is also possible the plastic changes within the dorsal striatum contribute to decreases in behaviors that are not associated with drug seeking or use during the course of development of addiction. Synaptic plasticity in both brain regions could play key roles in these behavioral adaptations.

Studies of striatal synaptic plasticity have revealed a remarkable need for ensemble action of multiple neurotransmitters, as well as neuromodulatory agents acting in a paracrine or juxtacrine manner (e.g., the endocannabinoids) in the production of LTP or LTD. We believe that this reflects the fact that the activity of striatal output neurons is dependent on integration of a variety of lines of information coming from different afferent sources. There is good evidence that striatal output is controlled by the interactions of several neurotransmitters arising from these different afferents (i.e., glutamate, dopamine, and ACh). This integrative function, perhaps combined with metaplastic changes in the threshold for plasticity, likely determines the valence and scale of long-lasting efficacy changes at synapses onto striatal medium spiny neurons. One of the intriguing sets of results to emerge from this line of study is the idea that a single neurotransmitter can participate in multiple forms of plasticity depending on the type of receptor activated. This idea is by no means new. Indeed, in the hippocampal CA1 region it is evident that even a single neurotransmitter receptor, the NMDA receptor, can participate in both LTP and LTD. However, in the case of the striatum the complexity of these neurotransmitter roles appears to be especially rich, given the requirement for actions of multiple transmitters in both LTP and LTD. This diversity of roles of the transmitters and their receptors offers multiple potential targets for pharmacotherapies aimed at altering striatal function and ameliorating striatal disorders.

It is our hypothesis that the types of synaptic plasticity we describe in this paper are involved in striatal neuroadaptations that underlie habit learning. There is certainly good evidence that the neurotransmitter receptors implicated in these forms of plasticity play large roles in striatal-based behaviors and learning. It has long been known that blockade of NMDA receptors has a profound impact on striatal function and related motor output,[64,65] and, likewise, manipulations that alter D1 and D2R function are well known to alter behavior via alterations in the striatum. Cannabinoid drugs produce profound motor initiation disruption when injected either peripherally or directly into the striatum.[66]

With respect to habit learning and memory, elegant studies by Packard and colleagues have demonstrated that posttraining injection of NMDAR or D2R antagonists into the striatum disrupts memory for specific goal-oriented motor responses.[67] In addition, a recent study indicates that development of striatal-based memory strategies involves increased ACh release in striatum.[68] The reason why blockers of both LTP and LTD have similar effects on memory is not immediately apparent. However, if habits are shaped by combined selective tuning of the activity of different striatal efferents, for example, to enhance one set of movements while suppressing another set, then it is easy to imagine how simultaneous LTP and LTD might be necessary for proper learning and memory.

This idea is supported by information on striatal neurophysiology during habit learning paradigms. In rat, different striatal neurons undergo long-lasting increases and decreases in activity in conjunction with habit learning.[69] In monkey, similar changes can be observed during learning of a reinforcer-oriented motor task, and the direction of the change appears to depend on signals coming from the substantia nigra, implicating dopamine in this learning process.[58] Thus, there is reason to believe that concurrent LTP- and LTD-like changes in efficacy participate in striatal-based habit learning.

The crucial role of the striatum in the production and expression of behavioral repertoires could endow it with a vulnerability to foster aberrant behavioral output. A particularly compelling idea is that disruption of striatal circuitry and/or neurochemistry will lead to the overproduction of abnormal behavioral sequences or the lack of adaptive sequences. It is well known that striatal dysfunction is at the heart of the pathologies of major motor output disorders such as Huntington's and Parkinson's diseases. It has also been suggested that striatal pathology might contribute to Tourette syndrome and obsessive-compulsive disorders, two neuropathologies involving production of abnormal behavioral repertoires.[70] Thus, it is not too farfetched to speculate about roles for the striatum in production of habits that come to be viewed as maladaptive. We have recently suggested that addiction to drugs of abuse may represent one scenario in which alterations in the physiology of the striatum, both dorsal and ventral, leads to abnormal habit acquisition.[2] Indeed, it is easy to imagine how a drug of abuse, such as amphetamine or $\Delta 9$-THC, could set into motion long-lasting plasticity at corticostriatal synapses. In such a situation, the drug might substitute for some aspects of both the environmental stimuli and the reinforcing stimuli that are needed for habit acquisition. Thus, the system would learn to seek the drug of abuse as well as to reproduce the behaviors associated with drug action. Evidence supporting this idea is not overwhelming at present, especially in studies of the dorsal striatum, but we hope that this idea will provide a theoretical framework for future research in this area.

REFERENCES

1. NESTLER, E.J. 2001. Molecular basis of long-term plasticity underlying addiction. Nat. Rev. Neurosci. **2:** 119–128.
2. GERDEMAN, G.L., J.G. PARTRIDGE, C.R. LUPICA & D.M. LOVINGER. 2003. It could be habit forming: drugs of abuse and striatal synaptic plasticity. Trends Neurosci. 184–192.
3. HYMAN, S.E. & R.C. MALENKA. 2001. Addiction and the brain: the neurobiology of compulsion and its persistence. Nat. Rev. Neurosci. **2:** 695–703.
4. SCHMIDT, W.J. 1995. Balance of transmitter activities in the basal ganglia loops. J. Neural Transm. Suppl. **46:** 67–76.
5. TUNSTALL, M.J., D.E. OORSCHOT, A. KEAN & J.R. WICKENS. 2002. Inhibitory interactions between spiny projection neurons in the rat striatum. J. Neurophysiol. **88:** 1263–1269.

6. CZUBAYKO, U. & D. PLENZ. 2002. Fast synaptic transmission between striatal spiny projection neurons. Proc. Natl. Acad. Sci. USA **99:** 15764–15769.
7. KINCAID, A.E., T. ZHENG & C.J. WILSON. 1998. Connectivity and convergence of single corticostriatal axons. J. Neurosci. **18:** 4722–4731.
8. SMITH, Y., B.D. BENNETT, J.P. BOLAM, *et al.* 1994. Synaptic relationships between dopaminergic afferents and cortical or thalamic input in the sensorimotor territory of the striatum in monkey. J. Comp. Neurol. **344:** 1–19.
9. IZZO, P.N. & J.P. BOLAM. 1988. Cholinergic synaptic input to different parts of spiny striatonigral neurons in the rat. J. Comp. Neurol. **269:** 219–234.
10. BOLAM, J.P. & P.N. IZZO. 1988. The postsynaptic targets of substance P-immunoreactive terminals in the rat neostriatum with particular reference to identified spiny striatonigral neurons. Exp. Brain Res. **70:** 361–377.
11. KOOS, T. & J.M. TEPPER. 1999. Inhibitory control of neostriatal projection neurons by GABAergic interneurons. Nat. Neurosci. **2:** 467–472.
12. KOOS, T. & J.M. TEPPER. 2002. Dual cholinergic control of fast-spiking interneurons in the neostriatum. J. Neurosci. **22:** 529–535.
13. WONNACOTT, S., S. KAISER, A. MOGG, *et al.* 2000. Presynaptic nicotinic receptors modulating dopamine release in the rat striatum. Eur. J. Pharmacol. **393:** 51–58.
14. ZHOU, F.M., Y. LIANG & J.A. DANI. 2001. Endogenous nicotinic cholinergic activity regulates dopamine release in the striatum. Nat. Neurosci. **4:** 1224–1229.
15. TARAZI, F.I. & R.J. BALDESSARINI. 1999. Regional localization of dopamine and ionotropic glutamate receptor subtypes in striatolimbic brain regions. J. Neurosci. Res. **55:** 401–410.
16. ROUSE, S.T., M.J. MARINO, S.R. BRADLEY, *et al.* 2000. Distribution and roles of metabotropic glutamate receptors in the basal ganglia motor circuit: implications for treatment of Parkinson's disease and related disorders. Pharmacol. Ther. **88:** 427–435.
17. LOVINGER, D.M. & B.A. MCCOOL. 1995. Metabotropic glutamate receptor-mediated presynaptic depression at corticostriatal synapses involves mGLuR2 or 3. J. Neurophysiol. **73:** 1076–1083.
18. HERSCH, S.M., C.A. GUTEKUNST, H.D. REES, *et al.* 1994. Distribution of m1-m4 muscarinic receptor proteins in the rat striatum: light and electron microscopic immunocytochemistry using subtype-specific antibodies. J. Neurosci. **14:** 3351–3363.
19. CALABRESI, P., N.B. MERCURI, M. DE MURTAS & G. BERNARDI. 1991. Involvement of GABA systems in feedback regulation of glutamate- and GABA-mediated synaptic potentials in rat neostriatum. J. Physiol. **440:** 581–599.
20. BOWEN, W.D., S. GENTLEMAN, M. HERKENHAM & C.B. PERT. 1981. Interconverting mu and delta forms of the opiate receptor in rat striatal patches. Proc. Natl. Acad. Sci. USA **78:** 4818–4822.
21. HOHMANN, A.G. & M. HERKENHAM. 2000. Localization of cannabinoid CB(1) receptor mRNA in neuronal subpopulations of rat striatum: a double-label in situ hybridization study. Synapse **37:** 71–80.
22. JIANG, Z.G. & R.A. NORTH. 1991. Membrane properties and synaptic responses of rat striatal neurones in vitro. J. Physiol. (Lond.) **443:** 533–553.
23. MALENKA, R.C. & J.D. KOCSIS. 1988. Presynaptic actions of carbachol and adenosine on corticostriatal synaptic transmission studied in vitro. J. Neurosci. **8:** 3750–3756.
24. GERDEMAN, G. & D.M. LOVINGER. 2001. CB1 cannabinoid receptor inhibits synaptic release of glutamate in rat dorsolateral striatum. J. Neurophysiol. **85:** 468–471.
25. HUANG, C.C., S.W. LO & K.S. HSU. 2001. Presynaptic mechanisms underlying cannabinoid inhibition of excitatory synaptic transmission in rat striatal neurons. J. Physiol. **532:** 731–748.
26. NICOLA, S.M., J. SURMEIER & R.C. MALENKA. 2000. Dopaminergic modulation of neuronal excitability in the striatum and nucleus accumbens. Annu. Rev. Neurosci. **23:** 185–215.
27. AOYAMA, S., H. KASE & E. BORRELLI. 2000. Rescue of locomotor impairment in dopamine D2 receptor-deficient mice by an adenosine A2A receptor antagonist. J. Neurosci. **20:** 5848–5852.

28. ZAHNISER, N.R., J.K. SIMOSKY, R.D. MAYFIELD, et al. 2000. Functional uncoupling of adenosine A(2A) receptors and reduced response to caffeine in mice lacking dopamine D2 receptors. J. Neurosci. **20:** 5949–5957.
29. CHEN, J.F., R. MORATALLA, F. IMPAGNATIELLO, et al. 2001. The role of the D(2) dopamine receptor (D(2)R) in A(2A) adenosine receptor (A(2A)R)-mediated behavioral and cellular responses as revealed by A(2A) and D(2) receptor knockout mice. Proc. Natl. Acad. Sci. USA **98:** 1970–1975.
30. GONZALEZ, L.P. & E.H. ELLINWOOD, JR. 1984. Cholinergic modulation of stimulant-induced behavior. Pharmacol. Biochem. Behav. **20:** 397–403.
31. COSTALL, B. & R.J. NAYLOR. 1972. Modification of amphetamine effects by intracerebrally administered anticholinergic agents. Life Sci. I **11:** 239–253.
32. BICKERDIKE, M.J. & E.D. ABERCROMBIE. 1997. Striatal acetylcholine release correlates with behavioral sensitization in rats withdrawn from chronic amphetamine. J. Pharmacol. Exp. Ther. **282:** 818–826.
33. ZHOU, F.M., C.J. WILSON & J.A. DANI. 2002. Cholinergic interneuron characteristics and nicotinic properties in the striatum. J. Neurobiol. **53:** 590–605.
34. CALABRESI, P., R. MAJ, A. PISANI, et al. 1992. Long-term synaptic depression in the striatum: physiological and pharmacological characterization. J. Neurosci. **12:** 4224–4233.
35. CALABRESI, P., P. GUBELLINI, D. CENTONZE, et al. 1999. A critical role of the nitric oxide/cGMP pathway in corticostriatal long-term depression. J. Neurosci. **19:** 2489–2499.
36. BATTAGLIA, G., V. BRUNO, A. PISANI, et al. 2001. Selective blockade of type-1 metabotropic glutamate receptors induces neuroprotection by enhancing gabaergic transmission. Mol. Cell. Neurosci. **17:** 1071–1083.
37. SUNG, K.W., S. CHOI & D.M. LOVINGER. 2001. Activation of group I mGluRs is necessary for induction of long-term depression at striatal synapses. J. Neurophysiol. **86:** 2405–2412.
38. TESTA, C.M. I.K. FRIBERG, S.W. WEISS & D.G. STANDAERT. 1998. Immunohistochemical localization of metabotropic glutamate receptors mGluR1a and mGluR2/3 in the rat basal ganglia. J. Comp. Neurol. **390:** 5–19.
39. SPENCER, J.P. & K.P. MURPHY. 2000. Bi-directional changes in synaptic plasticity induced at corticostriatal synapses in vitro. Exp. Brain Res. **135:** 497–503.
40. CHOI, S. & D.M. LOVINGER. 1997. Decreased probability of neurotransmitter release underlies striatal long-term depression and postnatal development of corticostriatal synapses. Proc. Natl. Acad. Sci. USA **94:** 2665–2670.
41. PARTRIDGE, J.G., S. APPARSUNDARAM, G.A. GERHARDT, et al. 2002. Nicotinic acetylcholine receptors interact with dopamine in induction of striatal long-term depression. J. Neurosci. **22:** 2541–2549.
42. GERDEMAN, G.L., J. RONESI & D.M. LOVINGER. 2002. Postsynaptic endocannabinoid release is critical to long-term depression in the striatum. Nat. Neurosci. **5:** 446–451.
43. ROBBE, D., M. KOPF, A. REMAURY, et al. 2002. Endogenous cannabinoids mediate long-term synaptic depression in the nucleus accumbens. Proc. Natl. Acad. Sci. USA **99:** 8384–8388.
44. MARSICANO, G., C.T. WOTJAK, S.C. AZAD, et al. 2002. The endogenous cannabinoid system controls extinction of aversive memories. Nature **418:** 530–534.
45. PARTRIDGE, J.G., K.C. TANG & D.M. LOVINGER. 2000. Regional and postnatal heterogeneity of activity-dependent long-term changes in synaptic efficacy in the dorsal striatum. J. Neurophysiol. **84:** 1422–1429.
46. CALABRESI, P., A. PISANI, N.B. MERCURI & G. BERNARDI. 1992. Long-term potentiation in the striatum is unmasked by removing the voltage-dependent magnesium block of NMDA receptor channels. Eur. J. Neurosci. **4:** 929–935.
47. CENTONZE, D., B. PICCONI, P. GUBELLINI, et al. 2001. Dopaminergic control of synaptic plasticity in the dorsal striatum. Eur. J. Neurosci. **13:** 1071–1077.
48. KERR, J.N. & J.R. WICKENS. 2001. Dopamine D-1/D-5 receptor activation is required for long-term potentiation in the rat neostriatum in vitro. J. Neurophysiol. **85:** 117–124.

49. CALABRESI, P., D. CENTONZE, P. GUBELLINI & G. BERNARDI. 1999. Activation of M1-like muscarinic receptors is required for the induction of corticostriatal LTP. Neuropharmacology **38:** 323–326.
50. HUANG, Y.Y. & E.R. KANDEL. 1995. D1/D5 receptor agonists induce a protein synthesis-dependent late potentiation in the CA1 region of the hippocampus [see comments]. Proc. Natl. Acad. Sci. USA **92:** 2446–2450.
51. HERSCH, S.M., B.J. CILIAX, C.A. GUTEKUNST, et al. 1995. Electron microscopic analysis of D1 and D2 dopamine receptor proteins in the dorsal striatum and their synaptic relationships with motor corticostriatal afferents. J. Neurosci. **15:** 5222–5237.
52. FLORES-HERNANDEZ, J., C. CEPEDA, E. HERNANDEZ-ECHEAGARAY, et al. 2002. Dopamine enhancement of NMDA currents in dissociated medium-sized striatal neurons: role of D1 receptors and DARPP-32. J. Neurophysiol. **88:** 3010–3020.
53. HERSCH, S.M. & A.I. LEVEY. 1995. Diverse pre- and post-synaptic expression of m1-m4 muscarinic receptor proteins in neurons and afferents in the rat neostriatum. Life Sci. **56:** 931–938.
54. REYNOLDS, J.N. & J.R. WICKENS. 2002. Dopamine-dependent plasticity of corticostriatal synapses. Neural Netw. **15:** 507–521.
55. LUJAN, R., J.D. ROBERTS, R. SHIGEMOTO, et al. 1997. Differential plasma membrane distribution of metabotropic glutamate receptors mGluR1 alpha, mGluR2 and mGluR5, relative to neurotransmitter release sites. J. Chem. Neuroanat. **13:** 219–241.
56. RICHFIELD, E.K., J.B. PENNEY & A.B. YOUNG. 1989. Anatomical and affinity state comparisons between dopamine D1 and D2 receptors in the rat central nervous system. Neuroscience **30:** 767–777.
57. WILSON, C.J. & Y. KAWAGUCHI. 1996. The origins of two-state spontaneous membrane potential fluctuations of neostriatal spiny neurons. J. Neurosci. **16:** 2397–2410.
58. SCHULTZ, W. 2002. Getting formal with dopamine and reward. Neuron **36:** 241–263.
59. BENNETT, B.D. & C.J. WILSON. 1999. Spontaneous activity of neostriatal cholinergic interneurons in vitro. J. Neurosci. **19:** 5586–5596.
60. KOMBIAN, S.B. & R.C. MALENKA. 1994. Simultaneous LTP of non-NMDA- and LTD of NMDA-receptor-mediated responses in the nucleus accumbens. Nature **368:** 242–246.
61. GRAYBIEL, A.M. 1998. The basal ganglia and chunking of action repertoires. Neurobiol. Learn. Mem. **70:** 119–136.
62. NICOLA, S.M. & R.C. MALENKA. 1998. Modulation of synaptic transmission by dopamine and norepinephrine in ventral but not dorsal striatum. J. Neurophysiol. **79:** 1768–1776.
63. EVERITT, B.J. & M.E. WOLF. 2002. Psychomotor stimulant addiction: a neural systems perspective. J. Neurosci. **22:** 3312–3320.
64. BROBERGER, C., D. BLACKER, L. GIMENEZ-LLORT, et al. 1998. Modulation of motor behaviour by NMDA- and cholecystokinin-antagonism. Amino Acids **14:** 25–31.
65. Schmidt, W.J. & B.D. Kretschmer. 1997. Behavioural pharmacology of glutamate receptors in the basal ganglia. Neurosci. Biobehav. Rev. **21:** 381–392.
66. GOUGH, A.L. & J.E. OLLEY. 1978. Catalepsy induced by intrastriatal injections of delta9-THC and 11-OH- delta9-THC in the rat. Neuropharmacology **17:** 137–144.
67. PACKARD, M.G. & B.J. KNOWLTON. 2002. Learning and memory functions of the basal ganglia. Annu. Rev. Neurosci. **25:** 563–593.
68. CHANG, Q. & P.E. GOLD. 2003. Switching memory systems during learning: changes in patterns of brain acetylcholine release in the hippocampus and striatum in rats. J. Neurosci. **23:** 3001–3005.
69. JOG, M.S., Y. KUBOTA, C.I. CONNOLLY, et al. 1999. Building neural representations of habits. Science **286:** 1745–1749.
70. GRAYBIEL, A.M. & S.L. RAUCH. 2000. Toward a neurobiology of obsessive-compulsive disorder. Neuron **28:** 343–347.

Mechanisms by which Dopamine Receptors May Influence Synaptic Plasticity

MARINA E. WOLF, SIMONA MANGIAVACCHI, AND XIU SUN

Department of Neuroscience, Finch University of Health Sciences/The Chicago Medical School, North Chicago, Illinois 60064-3095, USA

ABSTRACT: While dopamine (DA) receptors mediate acute effects of amphetamine and cocaine, chronic drug administration produces many glutamate-dependent adaptations, including LTP in reward-related neuronal circuits. An important question presents itself: How do DA receptors influence glutamate-dependent synaptic plasticity? Alterations in AMPA receptor phosphorylation and trafficking are critical for LTP. We hypothesize that D1 DA receptors modulate these processes, that chronic drug-induced adaptations in D1 receptor signaling, therefore, trigger compensatory changes in AMPA receptor function, and that this ultimately contributes to inappropriate plasticity in addiction-related neuronal circuits. Postnatal rat nucleus accumbens (NAc) cultures were used to study D1 receptor regulation of the AMPA receptor subunit GluR1. We found that D1 receptor stimulation enhances phosphorylation of GluR1 at the protein kinase A (PKA) site. Furthermore, D1 receptor stimulation increases GluR1 surface expression by increasing the rate of GluR1 externalization. The latter effect is prevented by the PKA inhibitors KT5720 and RpcAMPS, whereas the PKA activator SpcAMPS increases the rate of GluR1 externalization. These findings indicate that PKA phosphorylation is important in determining AMPA receptor surface expression and suggest a mechanism by which DA-releasing drugs of abuse may directly tap into fundamental mechanisms that enable synaptic plasticity. A limitation of our current model is that there are no intrinsic glutamate neurons in the NAc and thus no glutamate synapses in NAc cultures. To address this problem, we have restored excitatory synaptic inputs to NAc neurons by co-culturing them with prefrontal cortex (PFC) neurons. We are also studying GluR1 trafficking in PFC cultures. In both systems, synaptic AMPA receptors can be defined based on colocalization of GluR1 and the synaptic marker synaptobrevin. Preliminary results suggest that D1 receptor stimulation or PKA activation leads to increased surface GluR1 expression in PFC neurons but not to insertion into synaptic sites.

KEYWORDS: AMPA receptor; D1 dopamine receptor; cell culture; LTP; nucleus accumbens; phosphorylation; prefrontal cortex; receptor trafficking

---

Marina E. Wolf, Department of Neuroscience, FUHS/The Chicago Medical School, 3333 Green Bay Rd., North Chicago, IL 60064-3095. Voice: 847-578-8659; fax: 847-578-8515.
marina.wolf@finchcms.edu

INTRODUCTION

The nucleus accumbens (NAc) is an important interface between the limbic system, which controls motivational aspects of behavior, and the motor system, which controls the execution of behavior.[1] NAc neurons receive convergent inputs from midbrain dopamine (DA) neurons and glutamate-containing neurons originating from cortical and limbic brain regions.[2] Postsynaptic interactions between DA and glutamate in the NAc are critical for motivation, reward, and other behavioral functions that are disrupted after chronic exposure to drugs of abuse.[3] Yet, little is known about the mechanisms that enable DA receptors to modulate glutamate transmission or glutamate-dependent forms of plasticity, such as long-term potentiation (LTP) and long-term depression (LTD). These forms of plasticity are increasingly recognized as contributing to long-term changes in the brain during exposure to drugs of abuse.[4]

Recent studies suggest that two processes are critical for altering the strength of excitatory synapses during LTP and LTD—regulation of AMPA receptor phosphorylation and regulation of AMPA receptor trafficking in and out of synapses.[5,6] We hypothesized that DA receptors influence these basic mechanisms of synaptic plasticity in NAc neurons. As a first step toward testing this hypothesis, we have conducted studies using primary cultures prepared from the NAc of postnatal day 1 rats. We have also studied DA/AMPA receptor interactions in rat prefrontal cortex (PFC) cultures and in co-cultures of mouse NAc and PFC neurons.

NAc CELL CULTURES

Neurons in the NAc can be divided into two major groups. Medium spiny GABA-containing neurons comprise ~90% of cells in the intact NAc and are its projection neurons. The remaining 10% is composed of interneurons that can be divided into at least four populations based on transmitter phenotype and other properties.[7] For our analysis, we classified neurons as either medium spiny neurons or interneurons based on previously defined morphological criteria.[8,9] Consistent with characteristics of the intact NAc, ~80% of cells in our cultures exhibit the morphology of medium spiny neurons (soma diameter ~10 mm, with 2–4 relatively closely projecting processes), while ~20% exhibit morphology of interneurons (soma diameter ≥ 15 mm, with extended processes over 10 × the length of the soma). Using immunocytochemical methods, we found that nearly all neurons express glutamic acid decarboxylase, a marker for GABAergic neurons, as would be predicted based on the fact that medium spiny neurons and many interneurons in the intact NAc use GABA as a transmitter.[7] We also found that nearly all neurons express the AMPA receptor subunit GluR1, ~80% express D1 DA receptors, and ~80% express D2 DA receptors. These findings imply considerable overlap of DA and AMPA receptor expression at the single cell level, making our cultures a useful model system for studying DA/AMPA receptor interactions. A more detailed characterization of the NAc cultures has been published previously.[10]

D1 RECEPTOR STIMULATION INCREASES GluR1 PHOSPHORYLATION

Our first goal was to test the hypothesis that D1 DA receptor stimulation increases the phosphorylation of the AMPA receptor subunit GluR1 at the protein kinase A (PKA) phosphorylation site (ser-845). For these studies, NAc cells from postnatal day 1 rats were plated at a high density (120,000 cells/cm^2) and cultured for two weeks prior to experiments.[10] To determine the effect of D1 receptor stimulation on GluR1 phosphorylation, we added the D1 receptor selective agonist SKF 81297 to NAc cultures, extracted membrane protein, and performed Western analysis using an antibody recognizing GluR1 phosphorylated at ser-845.[11] Results were normalized to total GluR1 (detected using a phosphorylation-independent antibody) and to actin.

SKF 81297 produced a robust increase in GluR1 phosphorylation that was maximal within 5 min of incubation. This effect was concentration dependent, with significant increases produced by concentrations of SKF 81297 as low as 10 nM and maximal increases (~2- to 3-fold) produced by 10 μM SKF 81297. The effect of SKF 81297 was blocked by the D1 receptor antagonist SCH 23390 and the PKA inhibitor H89, and was reproduced by forskolin. The D2 receptor agonist quinpirole attenuated the response to D1 receptor stimulation. Neither D1 nor D2 receptor agonists altered GluR1 phosphorylation at ser-831, the site phosphorylated by protein kinase C and calcium/calmodulin-dependent protein kinase II.[10]

GluR1 phosphorylation by PKA has been shown to increase AMPA receptor currents in several brain regions (see Ref. 10 for more discussion). For example, in dorsal striatum, D1 receptor stimulation enhances AMPA receptor currents via a PKA-dependent mechanism.[12–14] Thus, the present results suggest that AMPA receptor transmission in the NAc may be augmented by concurrent D1 receptor stimulation. Indeed, recent *in vivo* findings support the idea that D1 receptors enhance excitatory transmission in the intact NAc and dorsal striatum.[15,16] It should be noted, however, that both excitatory and inhibitory effects of DA receptor stimulation on striatal neurons have been found, depending on experimental parameters, including the membrane potential of the striatal neuron, the concentration of DA, and the subtypes of DA and glutamate receptors activated (see Ref. 10 for more discussion).

D1 RECEPTOR STIMULATION INCREASES GluR1 SURFACE EXPRESSSION

Our next goal was to determine if D1 DA receptors also modulate cell surface expression of GluR1, detected using antibody to the extracellular portion of GluR1 and fluorescence microscopy. For these studies, NAc neurons from postnatal day 1 rats were plated at low density (12,000 cells/cm^2) to facilitate image analysis. To promote healthy cell growth under low-density conditions, cells were plated in media that had been conditioned with astrocyte cultures for 8–24 hours.[17] Experiments were performed after ~3 weeks in culture, as preliminary studies showed that this time point was associated with the presence of extensive processes and robust GluR1 expression. Surface GluR1 was labeled by incubating live cultures for 30 min with antibody recognizing the extracellular N-terminus portion of GluR1 (N-GluR1; Oncogene, Cambridge, MA; 1:10). Then, cultures were fixed with 4% paraformaldehyde, rinsed, blocked, and incubated for 1–2 h at room temperature with donkey anti-rabbit secondary antibody conjugated to Cy3 (1:250, Jackson ImmunoResearch Laboratories, West Grove, PA). Images were acquired and analyzed using a Nikon inverted microscope, an ORCA ER digital camera (Hamamatsu Photonics, Japan), and MetaMorph Imaging software (Universal Imaging, West Chester, PA, USA). To eliminate bias, neurons were chosen for analysis under phase contrast imaging, and fluorescence images were then collected without further adjustment. Methods for quantifying surface GluR1 labeling were adapted from those described previously for hippocampal cultures.[18–20] The number of fluorescent puncta was counted for a fixed length of a neuronal process located at least one soma diameter away from the soma. Thresholds were set at least two times higher than unlabeled neurites (imaged from control cultures incubated with secondary antibody only). Neurons were classified as either medium spiny neurons or interneurons based on morphological criteria described above, but we did not attempt to distinguish between subtypes of interneurons.

Previous studies in cultured hippocampal neurons have shown that glutamate agonists rapidly downregulate GluR1 surface expression.[18] To verify that GluR1 trafficking is under normal regulatory control in our cultures, cells were incubated with 10 µM or 50 µM glutamate for 15 minutes. Glutamate produced a concentration-dependent decrease in GluR1 surface expression in both medium spiny neurons and interneurons, although the maximal decrease was greater in interneurons (~40% decrease) than in medium spiny neurons (~15% decrease).[17] Subsequent studies, using a method that selectively detects newly internalized GluR1,[21,22] have shown that AMPA and NMDA also decrease GluR1 surface expression in medium spiny neurons and interneurons (Mangiavacchi and Wolf, unpublished observations).

Having established that regulated GluR1 trafficking can be demonstrated in NAc cultures, we examined the effect of D1 receptor stimulation. GluR1 labeling on processes of both medium spiny neurons and interneurons was increased by brief (5–15 min) incubation with the D1 agonist SKF 81297 (1 µM). This effect was blocked by the D1 receptor antagonist SCH 23390 (10 µM) and reproduced by the adenylyl cyclase activator forskolin (10 µM).[17] The latter result suggests that PKA may be involved in the response to D1 receptor stimulation (see next section). These results are the first to demonstrate modulation of AMPA receptor surface expression by a nonglutamatergic G protein–coupled receptor. They suggest a novel mechanism by which DA receptors may modulate excitatory synaptic transmission and glutamate-dependent synaptic plasticity.

RELATIONSHIP BETWEEN GluR1 PHOSPHORYLATION AND ITS SURFACE EXPRESSION

As described above, D1 receptor stimulation increases both GluR1 phosphorylation and its surface expression. The apparent involvement of PKA in both processes suggests that they may be related. Indeed, results in other systems support a relationship between PKA phosphorylation of GluR1 and AMPA receptor trafficking,[23,24] and the nature of this relationship has become a topic of intense research. Our NAc culture system is well suited to exploring this important issue, as we have identified a receptor (D1 DA receptor) that regulates both processes. Furthermore, our approach enables us to examine signal transduction pathways activated through physiological routes in real neurons.

To examine the role of PKA in regulating GluR1 surface expression, we have used an immunocytochemical method that selectively detects newly externalized GluR1.[25] First, GluR1 already present on the cell surface is preblocked by incubating with antibody to the extracellular N-terminus of GluR1 and unlabeled secondary antibody. Then, cells are incubated at room temperature to allow insertion of new GluR1 subunits into the plasma membrane. After fixation, cultures are incubated with primary antibody again, followed by fluorescent secondary antibody. Because cells are not permeabilized, the second round of immunostaining detects only GluR1 subunits that have been newly inserted onto the cell surface. This method offers a greater signal-to-noise ratio than the approach used in our prior studies.[17] The prior method measured the total density of surface GluR1 and therefore detected newly externalized GluR1 against the background of preexisting surface GluR1.

By selectively visualizing newly externalized GluR1, the method described above enables us to determine the rate of GluR1 externalization. The first step was to define the rate of constitutive GluR1 externalization using cultures that were brought to room temperature for 15 min in normal media (media control group). Then, we showed that cultures incubated with the D1 agonist SKF 81297 during the 15 min of room temperature incubation exhibited a greatly increased rate of GluR1 externalization. This effect of SKF 81297 was concentration dependent and observed in both medium spiny neurons and interneurons. More detailed analysis, using MetaMorph software, revealed that SKF 81297 produced a robust increase in both the density of GluR1 puncta and the area of GluR1 puncta. The simplest interpretation is that D1 receptor stimulation caused the formation of new surface GluR1-containing puncta (increased puncta density) and also resulted in the addition of GluR1 to existing puncta (increased puncta area). Both effects were blocked by the D1 receptor antagonist SCH 23390.

To test whether PKA mediates the effect of D1 receptor stimulation on GluR1 externalization, two different cell-permeable PKA inhibitors, KT5720 (2 μM and 10 μM) or RpcAMPS (10 μM), were added 5 min before SKF 81297 (1 μM). At the 10-μM concentration, both PKA inhibitors produced a near-complete block of the D1 agonist-induced increase in GluR1 puncta area and puncta density. Conversely, the PKA activator SpcAMPS increased GluR1 externalization in medium spiny neurons and interneurons in a concentration-dependent manner. Finally, SKF 81297 had no additional effect on GluR1 externalization in the presence of a maximally effective dose of SpcAMPS, further implicating PKA in the action of the D1 agonist. These results indicate that phosphorylation of GluR1 at the PKA site is required for its externalization in response to D1 receptor stimulation and that constitutive externalization of GluR1 is also regulated by its phosphorylation.[26,27]

DEVELOPMENT OF A CO-CULTURE SYSTEM TO ENABLE STUDIES OF SYNAPTIC AMPA RECEPTORS

The NAc contains no intrinsic glutamatergic neurons, although it receives glutamate inputs from cortical and limbic brain regions. Thus, a major limitation of our current "NAc only" culture system is that we are studying AMPA receptors in the absence of glutamate synapses. A new goal is therefore to develop a co-culture system in which cortical neurons are cultured with NAc neurons in order to restore excitatory synaptic input to the NAc neurons. However, we cannot simply mix freshly dissociated cortical and NAc cells because morphological criteria are not sufficient to distinguish cells from these regions. Pyramidal neurons might be identifiable, but cortical interneurons would not be distinguishable from NAc neurons. It is important to definitively identify cell populations to preserve the ability to selectively analyze surface GluR1 on processes of NAc neurons.

After trying several strategies, we had success with the approach of preparing co-cultures using cortical neurons from enhanced green fluorescent protein (EGFP) transgenic mice (Jackson Laboratory) and NAc neurons from littermates that do not express EGFP.[28] In preliminary studies, we found that cortical and NAc neurons could be readily distinguished based on GFP expression, even after three weeks in culture. Next, we established a method to detect synaptic GluR1 based on double immunostaining for cell surface GluR1 and the synaptic marker synaptobrevin/VAMP2. Using this method, we were able to quantify the number of synaptic and extrasynaptic AMPA receptors on both NAc and cortical neurons in the co-cultures. NAc/cortex co-cultures will provide an improved model system for our studies of DA receptor regulation of AMPA receptor phosphorylation and trafficking. In addition, this co-culture strategy can be widely used in other studies that require the ability to distinguish between two sources of cells.

AMPA RECEPTOR TRAFFICKING IN RAT PREFRONTAL CORTEX CULTURES

Although this chapter has focused on the NAc, the prefrontal cortex (PFC) is also an important area in which to study DA and glutamate receptor interactions, as interactions between these transmitter systems are important in many disorders believed to involve PFC dysfunction, including schizophrenia and drug addiction. Although studying synaptic AMPA receptors on NAc neurons requires co-culturing with a source of glutamate neurons (see previous section), the intact PFC contains many intrinsic glutamate neurons. Accordingly, primary cultures prepared from the PFC contain glutamate synapses.

Using primary cultures prepared from the PFC of postnatal day 1 rats, we have found that approximately 50% of synaptic sites (identified with antibody to the synaptic marker synaptobrevin/VAMP2) also contain GluR1, confirming the existence of AMPA receptor-containing synapses. As observed in NAc cultures, GluR1 surface expression is increased by incubation with either the D1 agonist SKF 81297 or the PKA activator SpcAMPS. However, preliminary results suggest that neither treatment increases the extent to which GluR1 is colocalized with the synaptic marker. This suggests that PKA activation is sufficient to produce GluR1 externalization onto the cell membrane, but not sufficient to produce GluR1 insertion into synaptic sites. This is consistent with prior studies showing that GluR1 synaptic insertion is a two-step process, consisting of initial insertion into extrasynaptic sites followed by lateral movement into the synapse.[29–32] Similar to our results, a recent study of activity-dependent incorporation of GluR1 into synaptic sites in organotypic slices from rat hippocampus found that PKA phosphorylation was not sufficient to drive GluR1 into synapses—both PKA phosphorylation and activation of calcium/calmodulin-dependent kinase II (CaMKII) were required.[24]

One interpretation of our findings is that D1 receptor stimulation increases the nonsynaptic cell surface pool of GluR1 in PFC neurons and thus increases the GluR1 pool available for synaptic insertion during LTP. This interpretation is consistent with results indicating that D1 receptors promote LTP in the PFC through a mechanism that involves PKA.[33–36]

SIGNIFICANCE OF RESEARCH FOR DEVELOPMENT OF TREATMENTS

DA and glutamate are two of the most important transmitter systems in schizophrenia and addiction. Neurons in most brain regions important in these disorders receive convergent DA and glutamate inputs. However, very little is known about the mechanisms that enable interactions between these transmitters at the single cell level. Our studies provide novel information about the signaling mechanisms that enable DA receptors to modulate glutamate transmission and glutamate-dependent plasticity. This kind of information is essential in order to develop rational hypotheses to account for DA/glutamate interactions observed in whole animal models or clinical populations.

Our results suggest that D1 DA receptors (and perhaps other monoamine receptors coupled to PKA) can control the fundamental mechanisms for regulating excitatory synaptic transmission and glutamate-dependent forms of plasticity, such as LTP and LTD, both in the NAc and the PFC. Under normal conditions, this may enable ongoing regulation of glutamate transmission in response to changes in the activity of DA inputs. In animals receiving repeated treatment with cocaine or amphetamine, abnormally high levels of DA receptor stimulation lead to compensatory changes in monoamine transmission.[37] Abnormal monoamine transmission will disrupt regulation of glutamate transmission, conceivably leading to inappropriate glutamate-dependent plasticity in neuronal circuits governing motivation and reward. This, in turn, may underlie behavioral changes in addiction. Similarly, synaptic plasticity may be altered as a result of perturbations of monoamine transmission associated with the onset of psychiatric disorders and as a result of drugs administered to treat these disorders (because most such drugs work by altering monoamine transmission).

This perspective, which focuses attention on the contribution of glutamate-dependent plasticity to addiction and to psychiatric disorders such as schizophrenia, may encourage the development of novel therapeutic strategies. For example, rather than targeting an "endpoint" associated with a disorder (e.g., an increase or decrease in transmission at a particular receptor), it may be useful to target the plasticity mechanisms that enable the transition from normal to abnormal function. More generally, our results show that insight into basic mechanisms of synaptic plasticity can help us understand the pathology that occurs in neurobiological disorders.

ACKNOWLEDGMENTS

This work was supported by USPHS Grants DA13006 and DA09621, Independent Scientist Award DA00453, and a NARSAD Independent Investigator Award (MEW).

REFERENCES

1. KELLEY, A.E. 1999. Neural integrative activities of nucleus accumbens subregions in relation to learning and motivation. Psychobiol. **27:** 198–213.
2. GROENEWEGEN, H.J., C.I. WRIGHT, A.V.J. BEIJER & P. VOORN. 1999. Convergence and segregation of ventral striatal inputs and outputs. Ann. N.Y. Acad. Sci. **877:** 49–64.
3. EVERITT, B.J. & M.E. WOLF. 2002. Psychomotor stimulant addiction: a neural systems perspective. J. Neurosci. **22:** 3312–3320.
4. WOLF, M.E. 2002. Addiction: making the connection between behavioral changes and neuronal plasticity in specific circuits. Molecular Interventions **2:** 146–157.
5. LÜSCHER, C. & M. FRERKING. 2001. Restless AMPA receptors: implications for synaptic transmission and plasticity. Trends Neurosci. **24:** 665–670.
6. MALINOW, R. & R.C. MALENKA. 2002. AMPA receptor trafficking and synaptic plasticity. Annu. Rev. Neurosci. **25:** 103–126.
7. MEREDITH, G.E. & S. TOTTERDELL. 1999. Microcircuits in nucleus accumbens' shell and core involved in cognition and reward. Psychobiology **27:** 165–186.
8. SHI, W.-X. & S. RAYPORT. 1994. GABA synapses formed in vitro by local axon collaterals of nucleus accumbens neurons. J. Neurosci. **14:** 4548–4560.
9. SHETREAT, M.E., L. LIN, A.C. WONG & S. RAYPORT. 1996. Visualization of D1 dopamine receptors on living nucleus accumbens neurons and their colocalization with D2 receptors. J. Neurochem. **66:** 1475–1482.
10. CHAO, S.Z., W. LU, H.-K. LEE, R.L. HUGANIR & M.E. WOLF. 2002. D1 dopamine receptor stimulation increases GluR1 phosphorylation and AMPA receptor currents in postnatal nucleus accumbens cultures. J. Neurochem. **81:** 984–992.
11. MAMMEN, A.L., K. KAMEYAMA, K.W. ROCHE & R.L. HUGANIR. 1997. Phosphorylation of the alpha-amino-3-hydroxy-5-methyl-isoxazole-4-propionic acid receptor GluR1 subunit by calcium/calmodulin-dependent protein kinase II. J. Biol. Chem. **272:** 32582–32593.
12. PRICE, C.J., P. KIM & L.A. RAYMOND. 1999. D1 dopamine receptor-induced cyclic AMP-dependent protein kinase phosphorylation and potentiation of striatal glutamate receptors. J. Neurochem. **73:** 2441–2446.
13. YAN, Z., L. HSIEH-WILSON, J. FENG, et al. 1999. Protein phosphatase 1 modulation of neostriatal AMPA channels: regulation by DARPP-32 and spinophilin. Nat. Neurosci. **2:** 13–17.
14. SNYDER, G.L., P.B. ALLEN, A.A. FIENBERG, et al. 2000. Regulation of phosphorylation of the GluR1 AMPA receptor in the neostriatum by dopamine and psychostimulants in vivo. J. Neurosci. **20:** 4480–4488.
15. GONON, F. & L. SUNDSTROM. 1996. Excitatory effects of dopamine released by impulse flow in the rat nucleus accumbens in vivo. Neuroscience **75:** 13–18.
16. WEST, A.R. & A.A. GRACE. 2002. Opposite influences of endogenous dopamine D_1 and D_2 receptor activation on activity states and electrophysiological properties of striatal neurons: studies combining in vivo intracellular recordings and reverse microdialysis. J. Neurosci. **22:** 294–204.
17. CHAO, S.Z., M.A. ARIANO, D.A. PETERSON & M.E. WOLF. 2002. Dopamine D1 receptor stimulation increases GluR1 surface expression in nucleus accumbens neurons. J. Neurochem. **83:** 704–412.
18. LISSIN, D.V., R.C. CARROLL, R.A. NICOLL, et al. 1999. Rapid, activation-induced redistribution of ionotropic glutamate receptors in cultured hippocampal neurons. J. Neurosci. **19:** 1263–1272.

19. CARROLL, R.C., D.V. LISSIN, M. VON ZASTROW, et al. 1999. Rapid redistribution of glutamate receptors contributes to long-term depression in hippocampal cultures. Nat. Neurosci. **2:** 454–460.
20. LIAO D., X. ZHANG, R. O'BRIEN, et al. 1999. Regulation of morphological postsynaptic silent synapses in developing hippocampal neurons. Nat. Neurosci. **2:** 37–43.
21. CARROLL, R.C., E.C. BEATTIE, H. XIA, et al. 1999. Dynamin-dependent endocytosis of ionotropic glutamate receptors. Proc. Natl. Acad. Sci. USA **96:** 14112–14117.
22. BEATTIE, E.C., R.C. CARROLL, X. YU, et al. 2000. Regulation of AMPA receptor endocytosis by a signaling mechanism shared with LTD. Nat. Neurosci. **3:** 1291–1300.
23. EHLERS, M.D. 2000. Reinsertion or degradation of AMPA receptors determined by activity-dependent endocytic sorting. Neuron **28:** 511–525.
24. ESTEBAN, J.A., S.H. SHI, C. WILSON, et al. 2003. PKA phosphorylation of AMPA receptor subunits controls synaptic trafficking underlying plasticity. Nat. Neurosci. **6:** 136–143.
25. LU, W.-Y., H.-Y. MAN, W. JU, et al. 2001. Activation of synaptic NMDA receptors induces membrane insertion of new AMPA receptors and LTP in cultured hippocampal neurons. Neuron **29:** 243–254.
26. MANGIAVACCHI, S. & M.E. WOLF. 2002. D1 receptor stimulation increases GluR1 externalization in nucleus accumbens neurons through a protein kinase A (PKA)-dependent mechanism. Soc. Neurosci. Abstr. **28:** #502.16.
27. MANGIAVACCHI, S. & M.E. WOLF. D1 dopamine receptor stimulation increases GluR1 surface expression in cultured nucleus accumbens neurons through a pathway dependent on protein kinase A. J. Neurochem. Submitted.
28. SUN, X., S. MANGIAVACCHI & M.E. WOLF. 2002. Use of fluorescent tracers to distinguish between cortical and nucleus accumbens neurons in a co-culture system. Soc. Neurosci. Abstr. **28:** #502.17.
29. PASSAFARO, M., V. PIECH & M. SHENG. 2001. Subunit-specific temporal and spatial patterns of AMPA receptor exocytosis in hippocampal neurons. Nat. Neurosci. **4:** 917–926.
30. CHEN, L., D.M. CHETKOVICH, R.S. PETRALIA, et al. 2000. Stargazin regulates synaptic targeting of AMPA receptor by two distinct mechanisms. Nature **408:** 936–943.
31. CHETKOVICH, D.M., L. CHEN, T.J. STOCKER, et al. 2002. Phosphorylation of the postsynaptic density-95 (PSD-95)/Discs large/zona occludens-1 binding site of stargazing regulates binding to PSD-95 and synaptic targeting of AMPA receptors. J. Neurosci. **22:** 5791–5796.
32. SCHNELL, E., M. SIZEMORE, S. KARIMZADEGAN, et al. 2002. Direct interactions between PSD-95 and stargazing control synaptic AMPA receptor number. Proc. Natl. Acad. Sci. USA **99:** 13902–13907.
33. JAY, T.M., H. GURDEN & T.YAMAGUCHI. 1998. Rapid increase in PKA activity during long-term potentiation in the hippocampal afferent fibre system to the prefrontal cortex in vivo. Eur. J. Neurosci. **10:** 3302–3306.
34. GURDEN, H., J.P. TASSIN & T.M. JAY. 1999. Integrity of the mesocortical dopaminergic system is necessary for complete expression of in vivo hippocampal-prefrontal cortex long-term potentiation. Neuroscience **94:** 1019–1027.
35. GURDEN, H., M. TAKITA & T.M. JAY. 2000. Essential role of D1 but not D2 receptors in the NMDA receptor-dependent long-term potentiation at hippocampal-prefrontal cortex synapses in vivo. J. Neurosci. **20:** RC106.
36. BLOND, O., F. CREPEL & S. OTANI. 2002. Long-term potentiation in rat prefrontal slices facilitated by phased application of dopamine. Eur. J. Pharmacol. **438:** 115–116.
37. WHITE, F.J. & P.W. KALIVAS. 1998. Neuroadaptations involved in amphetamine and cocaine addiction. Drug Alcohol Depend. **51:** 141–153.

Glutamate and Depression

Clinical and Preclinical Studies

IAN A. PAUL[a] AND PHIL SKOLNICK[b]

[a]*Laboratory of Neurobehavioral Pharmacology and Immunology, Division of Neurobiology and Behavior Research, Departments of Psychiatry and Pharmacology, University of Mississippi Medical Center, Jackson, Mississippi 39216, USA*

[b]*DOV Pharmaceuticals, 433 Hackensack Avenue, Hackensack, New Jersey 07601, USA*

ABSTRACT: The past decade has seen a steady accumulation of evidence supporting a role for the excitatory amino acid (EAA) neurotransmitter, glutamate, and its receptors in depression and antidepressant activity. To date, evidence has emerged indicating that N-methyl-D-aspartate (NMDA) receptor antagonists, group I metabotropic glutamate receptor (mGluR$_1$ and mGluR$_5$) antagonists, as well as positive modulators of α-amino-3-hydroxy-5-methyl-4-isoxazolepropionic acid (AMPA) receptors have antidepressant-like activity in a variety of preclinical models. Moreover, antidepressant-like activity can be produced not only by drugs modulating the glutamatergic synapse, but also by agents that affect subcellular signaling systems linked to EAA receptors (e.g., nitric oxide synthase). In view of the extensive colocalization of EAA and monoamine markers in nuclei such as the locus coeruleus and dorsal raphe, it is likely that an intimate relationship exists between regulation of monoaminergic and EAA neurotransmission and antidepressant effects. Further, there is also evidence implicating disturbances in glutamate metabolism, NMDA, and mGluR$_{1,5}$ receptors in depression and suicidality. Finally, recent data indicate that a single intravenous dose of an NMDA receptor antagonist is sufficient to produce sustained relief from depressive symptoms. Taken together with the proposed role of neurotrophic factors in the neuroplastic responses to stressors and antidepressant treatments, these findings represent exciting and novel avenues to both understand depressive symptomatology and develop more effective antidepressants.

KEYWORDS: excitatory amino acid transmitter; NMDA receptor; AMPA receptor; metabotropic glutamate receptor; nitric oxide synthase; neuroplastic responses; antidepressant treatment

INTRODUCTION

Overview of Major Depressive Disorders

Major depression is a multifaceted disorder that features some combination of depressed mood, anhedonia, feelings of worthlessness, guilt and/or hopelessness.

Address for correspondence: Ian A. Paul, Ph.D., Department of Psychiatry and Human Behavior, Box 127, University of Mississippi Medical Center, 2500 North State Street, Jackson, MS 39216. Voice: 601-984-5883; fax: 601-984-5884.
ipaul@psychiatry.umsmed.edu

Major depression usually produces a constellation of disturbances of appetite, sleep, sexual activity, and concentration, as well as agitation or lethargy/fatigue. In addition, major depression is strongly associated with suicidality and increased risk of death from all causes.[1–5] Conservative estimates indicate that 7.1% of the population between 18 and 54 years of age will suffer a serious mood disorder in a given year.[6] Most adults will suffer one or more depressive episodes over the course of their lifetime, and as many as 15% will suffer recurrent bouts of depression of variable severity throughout their lives. The immense societal burden of this disease has made it a major focus of therapeutic research for more than 50 years.

Overview of Antidepressants

The discovery of effective antidepressants among monoamine oxidase inhibitors and tricyclic inhibitors of norepinephrine (NE) and serotonin (5HT) reuptake provided a stimulus and drive for research on the actions of antidepressants and the pathophysiology of major depression for nearly two decades.[7] The demonstration that selective inhibitors of 5HT are antidepressant has driven this research in recent years owing to their greatly diminished toxicity and side effect profile compared with first-generation antidepressants.[8–10] However, the response rate with any given antidepressant treatment has remained essentially unchanged since the early 1960s. Thus, regardless of whether patients are prescribed monoamine oxidase inhibitors, tricyclic compounds, serotonin reuptake inhibitors, or newer 5HT and NE specific reuptake inhibitors (SNRIs), response rates hover between 65–70%. Moreover, in most double-blind, placebo-controlled studies, 3–6 weeks of drug administration is required to produce a therapeutic response. These data suggest an agency of factors other than, or in addition to, the acute effects of these drugs is required to effect therapeutic response.[10,11] Given the difficulties in maintaining patients on treatment after unsuccessful drug trials and the risk of self-injurious behavior and suicide until remission of depressive symptoms, the need for rapidly acting antidepressants with high response rates remains as pressing now as it was 40 years ago.

The failure to identify rapidly acting antidepressants that are effective in a large majority of depressed patients stems from the largely serendipitous nature of antidepressant drug development. Rational drug development depends on understanding either the etiology of the targeted disease or the mechanism of therapeutic/palliative action of successful treatments. Drugs so designed are specifically targeted at points along an etiological or therapeutic cascade. In the case of major depressive disorders we clearly do not understand the etiology of the disease(s) and our understanding of the mechanism of therapeutic action is comparatively conjectural.

Most clinically effective antidepressants interfere with monoamine reuptake or in some other fashion increase synaptic cleft concentrations of the monoamines, norepinephrine and serotonin. A detailed account of these interactions is beyond the scope of this review, and excellent reviews of the subject have recently been published.[12,13] To summarize, most clinically effective antidepressants initially result in increases in synaptic concentrations of one or more monoamines (norepinephrine, serotonin, or dopamine). However, the onset of therapeutic action for these drugs is quite slow, appearing only after 3–5 weeks of continuous administration.[11,14] In contrast, stable plasma and brain concentrations of these drugs are usually achieved within a week of treatment. This separation of stable drug concentrations from the onset of therapeutic response has led to the conclusion that pharmacodynamic or adaptive neurobiological responses to continuous treatment underlie the mechanism of action of antidepressants.

Long-term antidepressant administration is accompanied by adaptive regulation of monoaminergic receptors (β-adrenoceptors, $5\text{-HT}_{4,6,7}$, α-adrenoceptors), second-messenger systems, transporters (NET and SERT), and synthetic enzyme systems (tyrosine and tryptophan hydroxylase). Taken together, these data indicate a significant role for adaptation of monoaminergic neurotransmitter systems, especially norepinephrine and serotonin, in the actions of antidepressants. Nonetheless, the existence of antidepressant medications that do not obviously interact with monoaminergic neurons, non-pharmacological treatments (such as electroconvulsive shock and REM sleep deprivation), and the relatively modest success rate for any given antidepressant (estimated at 65–70% in most studies) have led many investigators to conclude that actions at monoaminergic synapses alone cannot completely account for the actions of antidepressants.[15]

Overview of Excitatory Amino Acid Neurotransmission

Although glutamate was first proposed as a neurotransmitter as early as 1959, only in the past two decades have pharmacological tools in the form of selective agonists and antagonists become available to enable investigation of this system.[16] It is now known that neurons employing excitatory amino acids for neurotransmission comprise some 30% the neurons in forebrain and that these neurons are non-uniformly distributed in the CNS. High densities of glutamate-reactive neurons are found in cortex, both as interneurons and as the primary neurotransmitter for pyramidal neurons of layers IV and V.[17] Likewise, glutamate-containing neurons are found in subcortical structures, such as hippocampus, caudate nucleus, thalamic nuclei, and in the cerebellum.[17]

Glutamate receptors have been pharmacologically classified as ionotropic and metabotropic. The former include N-methyl-D-aspartate (NMDA), AMPA and kainate families of receptors, while the latter include the g-protein coupled family of glutamate receptors. The ionotropic receptors gate cations while the metabotropic receptors use both adenylyl cyclase and phosphoinositide activity to transmit signals across the post-synaptic membrane.

MAJOR DEPRESSION AND EXCITATORY AMINO ACIDS

Glutamate and Aspartate

As early as 1982, reports appeared suggesting that glutamate metabolism differed significantly between depressed patients and controls.[18,19] More recent studies have indicated these differences can be resolved with chronic antidepressant treatments.[20,21] Moreover, Berk and colleagues have reported that platelet sensitivity to glutamate is increased in major depressives.[22] In brain, Auer and colleagues have reported reduced glutamate levels in anterior cingulate cortex using pMRI.[23] However, frontal cortical samples obtained from neurosurgical procedures evinced no significant overall effect of depression on brain glutamate and aspartate levels, although they did report an increase in aspartate levels in a subgroup of depressives.[24] Murck and colleagues report that total REM sleep suppression, which is an effective antidepressive treatment, produces robust increases in pontine glutamine levels.[25] Most recently, Michael and colleagues have reported an increase in left amygdalar glutamate/glutamine (Glx) after successful resolution of depressive symptoms with electroconvulsive shock therapy.[26] In a parallel study, Pfleiderer and colleagues report that Glx levels in left anterior cingulate cortex are reduced in major depressives compared to controls and that this reduction is reversed with effective ECT.[27]

Calcium Homeostasis

There is an extensive clinical literature suggesting a role for calcium homeostasis in the pathophysiology of human depression.[28] For example, hypoparathyriodism has long been associated with affective disturbance.[29–31] The etiology of these disturbances has been hypothesized to be related to errors in calcium metabolism. In addition, changes in serum or CSF calcium levels have been reported in patients with depression not directly associated with endocrine dysfunction.[32–40] Conversely, reduced plasma, serum, or CSF calcium levels accompany improvement of depressive symptomatology after either electroconvulsive shock or antidepressant drug treatment.[33–35,39,41–43] Finally, calcium channel antagonists have been reported to be antidepressant in an anecdotal report.[44,45]

ANTIDEPRESSANTS AND EXCITATORY AMINO ACIDS

Early Hints

In vitro studies have shown that some antidepressant drugs bind to NMDA receptors[46] and inhibit the binding of NMDA receptor ligands.[47] Similarly, several laboratories have reported that antidepressants can modulate the release and/or reuptake of glutamate.[48–52] Likewise, at least one group reported that conditions resulting in antidepressant-sensitive behavioral changes also disrupted long-term potentiation, a neurobiological process dependent upon release of glutamate and activation of several glutamate receptor subtypes.[53–55]

NMDA Receptors

Based on these findings, Trullas and Skolnick proposed that antagonists of the NMDA receptor would display antidepressant-like properties in acute, preclinical screening procedures such as the forced swim and tail suspension tests.[56,57] Since this seminal report, studies in several laboratories have demonstrated that functional antagonists of the NMDA receptor, including ligands at the glutamate, glycine, polyamine, bivalent cation (Zn^{2+}) and ionophore recognition sites are as efficacious as tricyclic antidepressants in preclinical antidepressant screening procedures in mice[56,58–61] and rats.[62–68] Likewise, chronic treatment with NMDA receptor antagonists results in behavioral effects analogous to those produced by chronic antidepressant treatments in the chronic mild stress,[69] learned helplessness,[70,71] footshock-induced aggression[67,72,73] and olfactory bulbectomy models.[74] Moreover, a single dose of the NMDA receptor ionophore ligand, ketamine, has long-lasting effects on the behavior of rats in the forced swim test.[75]

Congruent with the behavioral effects of chronic treatment with NMDA receptor antagonists are the observations that chronic, but not acute, treatment with these drugs results in a reduction in the density of forebrain β-adrenoceptors[76,77] and 5-HT$_2$ receptors.[78] Similarly, as observed with other antidepressants, administration of NMDA receptor antagonists to animals processed in the forced swim test results in a rapid down-regulation of forebrain β-adrenoceptors.[66] Taken together, this body of evidence provides strong support for the hypothesis that NMDA receptor antagonists have antidepressant properties.

Antidepressant drugs produce time- and dose-dependent changes in the radioligand binding properties of the NMDA receptor.[70,74,79–90] These changes display nearly complete within-class generality and are not obtained with closely related non-antidepressant treatments (between-class specificity).[84] These effects may be related to the ability of antidepressant administration to alter the expression of NMDA receptor subunits[91] perhaps through expression of brain-derived neurotrophic factor (BDNF).[15,92] Thus, in both behavioral and biochemical screening procedures, antagonists at the NMDA receptor complex behave in ways comparable to clinically active antidepressants.

Likewise, preclinical procedures, such as the forced swim test and chronic mild stress paradigm, which produce antidepressant-sensitive changes in behavior, also produce alterations in the radioligand binding properties of the NMDA receptor opposite to those produced by antidepressants.[93,94] In addition, antidepressant treatment in the chronic mild stress paradigm that normalizes sucrose consumption deficits also normalizes the radioligand binding properties of the NMDA receptor.[93,94]

NMDA receptor abnormalities are observed in human suicide victims[95] and major depressives.[96] In homogenates of normal human frontal cortical (Brodmann Area 10), glycine displaces the binding of the glutamate receptor antagonist [^{3}H]CGP-39653 in a biphasic manner, similar to its effects in rodents. In contrast, the high affinity component of glycine displacement of [^{3}H]CGP-39653 is significantly reduced in frontal cortical homogenates from the majority of suicide victims. Moreover, specific binding of 10 nM [^{3}H]CGP-39653 was significantly lower (40 ± 14%) in cortices from suicide victims relative to that from age- and postmortem interval-matched control subjects. No differences between control and suicide tissue were observed in the IC$_{50}$ for either component of glycine displacement of [^{3}H]CGP-39653 binding. In addition, no differences between groups were observed in either specific or glycine-displaceable [^{3}H]5,7-DCKA binding or in specific, glycine- or glutamate-stimulated [^{3}H]dizocilpine binding.

Similarly, Law and Deakin have recently reported a reduction in the expression of NMDAR1 mRNA in the hippocampus of depressed subjects.[96] Inasmuch as this reduction was also observed in samples from both bipolar and schizophrenic subjects, the specificity of this observation is open to question. However, while laterality was observed in samples from schizophrenic subjects, no such laterality was observed with samples from major depressives or bipolars. Thus, in addition to fundamental differences in NMDA receptor subunit expression, an absence of laterality in the reductions in NMDA receptor expression may be a significant feature of affective disorders.

Although not known at the time to be a functional NMDA receptor antagonist, as early as 1959, Crane had noted the antidepressant properties of D-cycloserine, a glycine recognition site partial agonist, in human major depressives.[97,98] Similarly, while the NMDA antagonist properties of amantadine were not initially appreciated, this drug has been reported to relieve depressive symptoms in humans[99–102,103] and to augment conventional antidepressant therapies.[104] In addition, although no longer in clinical use due to unacceptable non-CNS toxicity, metapramine showed promise as an antidepressant with no obvious or dramatic effects on monoamine transport or metabolism.[105–108] Recently, this drug was identified as a low affinity NMDA receptor antagonist.[109] The related low potency NMDA receptor antagonist memantine has been proposed as a potential antidepressant[110] and early clinical studies are consistent with its efficacy as an antidepressant[111,112] Recently, Berman and co-workers have demonstrated that intravenous administration of low doses of the NMDA receptor antagonist ketamine results in robust and rapid relief of depressive symptoms in a double-blind crossover study.[113] These effects appear some 4 h after the disappearance of the transient dissociative effects of ketamine and continue to improve over three days post-infusion. Berman and co-workers noted an average reduction in Hamilton Depression Rating Scale scores of ~15 points and reported that these effects persisted for up to two weeks post-infusion in the absence of any other antidepressant treatment. The antidepressant effect of ketamine was recently confirmed in a larger study of surgical patients.[114] While these studies must be replicated, the effects of ketamine in depressed humans are precisely those predicted by studies in animal screening paradigms and analogs of depression.

These data lend significant support to the hypothesis that dysfunction of the NMDA receptor complex is involved in the psychopathology of major depression. Moreover, the data from both rodents and humans indicate that antagonists of the NMDA receptor complex possess antidepressant-like properties. The robustness, consistency, and reproducibility of these studies provide a clear impetus for additional studies to explore the hypothesis that excitatory amino acid receptor-linked neuronal signaling will hold the potential for both antidepressant development and specific relevance to the pathophysiology of major depression.

Metabotropic Glutamate Receptors

The initial studies of Pilc and Legutko demonstrated that prolonged antidepressant treatment reduces the stimulation of adenylyl cyclase by activation of group I metabotropic glutamate receptors (mGluR$_1$ and mGluR$_5$).[86,87] Similarly, chronic antidepressant treatment reduces phosphoinositol production by mGluR1 receptors in the CA1 region of the hippocampus.[115,116] This is accompanied by a reduction in hippocampal neuronal responsiveness to mGluR$_1$ receptor activation.[117] Together with the data described above for the NMDA receptor, these data indicate that antidepressant treatments induce adaptive reductions in the sensitivity of both ionotropic and metabotropic glutamate receptors.

Likewise, Tatarczynska and colleagues have demonstrated that an mGluR5 antagonist displays antidepressant-like activity in mice.[118] However, this effect was not observed in rats. Whether this represents a difference in species selectivity or is indicative of limitations on the antidepressant effects of these compounds remains to be determined. Nonetheless, these data are congruent with previous studies of the antidepressant-like activity of NMDA receptor antagonists described above.

Most recently, Smialowska and co-workers have demonstrated that both pharmacological (imipramine) and nonpharmacological (electroconvulsive shock) antidepressant treatments, administered chronically, increased $mGluR_{1a}$ and $mGluR_{5a}$ immunoreactivity in rat hippocampus. The most pronounced effects were observed in the CA1 and CA3 regions.[119,120] Similarly, Matrisciano and colleagues have reported that chronic imipramine amplifies both the expression and the function of $mGluR_{2/3}$ receptors in hippocampal slices.[121] Thus, like NMDA receptors, metabotropic glutamate receptors appear to respond adaptively to chronic antidepressant treatments.

AMPA Receptors

Several studies have demonstrated that AMPA receptor activation can increase expression of brain-derived neurotrophic factor (BDNF) both *in vitro* and *in vivo*.[122–124] Since converging lines of evidence point to a critical role for BDNF in the actions of antidepressants,[125] Skolnick and co-workers hypothesized that an AMPA receptor potentiator would possess antidepressant-like properties. Recently, this group demonstrated that LY392098, an AMPA receptor potentiator, was as active as imipramine in both the rat and the mouse forced swim test at doses that were without effect on locomotor activity.[126] Likewise, LY392098 markedly increased BDNF mRNA in primary neuronal culture.[127] Thus, activation of AMPA receptors may be of therapeutic benefit in the treatment of depression.

Most recently, Martinez-Turrillas and co-workers have reported that chronic antidepressant treatment increases the membrane expression of AMPA receptors in rat hippocampus.[128] There was no change in expression in the total hippocampal extract, indicating that chronic antidepressant treatments specifically increase trafficking of AMPA receptors from intracellular pools to synaptic sites. In addition, at least for the $GluR_{2/3}$ subunit, expression remained elevated 72 h after the last antidepressant treatment. Moreover, chronic treatment with the serotonin reuptake inhibitor, fluoxetine, increased phosphorylation of the Ser^{831}-$GluR_1$ and Ser^{845}-$GluR_1$ sites of the AMPA receptor in extracts of cortex, hippocampus, and striatum.[129] Thus, as for the NMDA and metabotropic glutamate receptors, AMPA receptors appear responsive to chronic antidepressant treatment, consistent with a target for neural adaptation believed to underlie the therapeutic response to these treatments.

ANTIDEPRESSANTS AND EXCITATORY AMINO ACID–LINKED SUBCELLULAR SIGNALING

Overview

Activation of glutamate receptors often results in increases in Ca^{2+} levels at the postsynaptic neuron. The Ca^{2+} binds to and stimulates a Ca^{2+}-calmodulin complex, which in turn stimulates nitric oxide (NO) synthesis[130] to convert L-arginine to L-citrulline and liberate NO. The released NO can then stimulate an NO-sensitive guanylyl cyclase to convert GTP to cGMP (FIG. 1), which in turn can stimulate phosphodiesterase II and protein kinase G. Alternately, NO can serve as a transsynaptic signal either at the presynaptic neuron or nearby cells. NO synthase, the enzyme responsible for the production of NO, can be exogenously stimulated with compounds such as sodium nitroprusside, molsidomine, S-nitroso-*N*-acetylpenicillamine, and peripherally administered L-arginine. Conversely, NO synthase activity can be inhibited with N^G-nitro-L-arginine (L-NNA), N^G-nitro-L-arginine methyl ester (L-NAME), N^G-monomethyl-L-arginine (L-NMMA), and 7-nitroindizole (7NI).

In addition to its effects on guanylyl cyclase, NO is a membrane-permeable molecule involved in signaling processes and cellular communication in a variety of systems. Previous studies have suggested that NO may be able to stimulate the release of neurotransmitters (dopamine, norepinephrine)[131] and that this effect is dependent on NMDA receptor stimulation. In fact, both an increase[131,132] and decrease[133,134] in basal monoamine efflux after pretreatment with NO precursors or donors has been reported. Recent reports also indicate that NO plays the role of an inhibitory endogenous substance in discriminative effects of the psychostimulants in rats, because in-

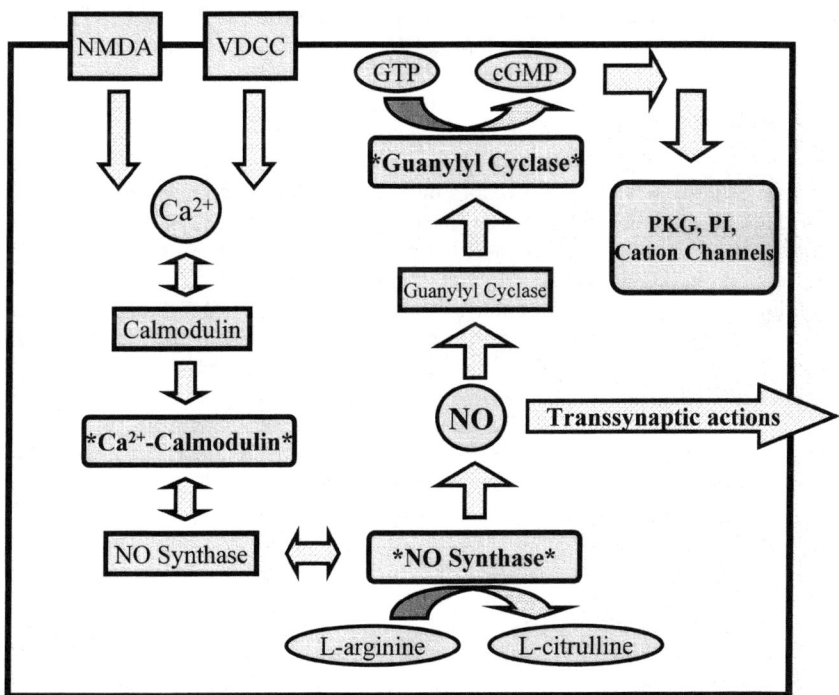

FIGURE 1. Ca^{2+}-calmodulin-nitric oxide synthase-guanylyl cyclase (CC-NOS-GC) cascade. Abbreviations: VDCC, voltage-dependent calcium channel; NMDA, *N*-methyl-D-aspartate receptor; CAM, calmodulin; NO, nitric oxide; NO Synthase, nitric oxide synthase; GTP, guanosine triphosphate; cGMP, cyclic guanosine monophosphate; PKG, Protein kinase G; PI, Phosphoinositol; Asterisk brackets (e.g., *Ca^{2+}-Calmodulin*) denote activated form of enzymes.

hibition of NOS enhances the effects of amphetamine and cocaine while an increased NO level attenuates them.[135] Moreover microdialysis data have shown that NO has an inhibitory influence on DA release in the stratum.[136] Likewise Silva and colleagues[137] demonstrated increased DA release *in vivo* following intrastriatal administration of 7-nitroindazole that was antagonized by coperfusion with the NOS substrate, L-arginine. NMDA-mediated neurotransmitter release is linked to NO production and the NOS inhibitors L-NNA and 7-nitroindazole (7NI) blocked the NMDA-mediated release of neurotransmitters in dose-dependent fashion.[138]

Nitric Oxide Synthase

A potential role for NO in affective disorder has also recently been proposed.[139,140] NO is a regulator of both short- and long-term neuronal adaptive changes and consequently may play a role in neuronal adaptation to antidepressant drugs. Targets of NO include guanylyl cyclase, G proteins, amino acid, amine and neuropeptide release and transport. NO-mediated cGMP synthesis also mediates induction of immediate early gene expression, which is implicated in long-term synaptic changes and more recently in the mechanism of action of antidepressant drugs.[12,139] Selective inhibitors of NO synthase isoforms are under development, which raises the possibility of developing NO synthase inhibitors that may be targeted for the treatment of CNS disorders where NO has been implicated.

Antidepressants can inhibit the activity of neuronal NO synthase (nNOS), albeit for the most part with low affinity.[141,142] The first compelling biochemical evidence of linkage between antidepressant medication and NOS activity was described by Finkel and colleagues.[143] In this report, the serotonin reuptake inhibitor, paroxetine, was shown to inhibit constitutive nitric oxide synthase (cNOS) activity in animal models and humans (FIG. 1) at concentrations comparable to those achieved in clinical therapy. It is of interest to note that, in the earlier study of Lee and colleagues, only the serotonin reuptake inhibitor, fluoxetine, inhibited nNOS at a clinically relevant concentration.[141] Taken together, these studies raise the possibility that NOS inhibition may be a clinically relevant feature of that class of antidepressants. NOS inhibition could theoretically cause hypertension, coronary artery spasm, and platelet aggregation, and thus would not be appropriate for patients with ischemic heart disease and/or coronary insufficiency.[144] However, paroxetine has an extraordinary safety profile and all examined patients with ischemic heart disease tolerated this NOS inhibitor without any serious adverse cardiac events.

The first report of NO synthase antagonist activity in an antidepressant-sensitive paradigm was from Jefferys and Funder[145] using the rat forced swim test. Two reports have recently demonstrated that inhibitors of NO synthase are active in both acute and chronic preclinical antidepressant screening procedures in mice.[146,147] Yildiz and colleagues have presented evidence in the rat forced swim test to indicate that these effects may be specific to the neuronal isoforms of NO synthase.[148] These effects may also generalize to other elements of the NMDA receptor-linked subcellular signaling pathway. A preliminary study by Eroğlu and Cağlayan indicated that methylene blue, an inhibitor both NO synthase and soluble guanylyl cyclase, is active in the rat forced swim test.[149] Similar effects have also been reported for the inhibitor of soluble guanylyl cyclase, ^{1}H-[1,2,4]oxadiazole[4,3-a]quinoxalin-1-one (ODQ).[150] Thus, at least nitric oxide synthase and possibly other elements of the

Ca^{2+}-calmodulin-NOS-guanylyl cyclase subcellular signaling pathway appear to play a role in antidepressant-sensitive behavior and neurochemical response.

Recently, Suzuki and colleagues have demonstrated that chronic antidepressant drug treatment increases the expression of inducible NOS in several areas of forebrain but is without effect on neuronal isoforms of NOS.[151] Conversely, subjecting rodents to uncontrollable forced swimming results in rapid and selective enhancement of NADPH-diaphorase (a marker for NO synthase) staining in the paraventricular nucleus of the hypothalamus.[152] Similarly, exposure of rats to chronic mild stress markedly increases NADPH-diaphorase staining in hippocampus—an effect blocked or reversed by chronic treatment with fluoxetine.[153]

Antidepressants, Stress, and Neurogenesis

It has long been recognized that at least half of all patients suffering from major depression display elevated 24-h plasma cortisol levels and an inability to suppress cortisol after exogenous administration of corticosteroids such as dexamethasone.[154] Abnormally high circulating levels of corticosteroids exert neurotoxic effects in several areas of CNS that are implicated in the symptomatology of major depressive disorders (such as hippocampus and nucleus accumbens[155]). Recent postmortem evidence supports neuronal and glial loss and/or shrinkage and in several areas of frontal cortex, most notably orbitofrontal and cingulate cortex.[156-158] Together, these data suggest that morphometric changes in neurons and glia attend major depressive disorders and chronic stress conditions.[14]

During the past decade, it has been shown that phosphorylation of the cyclic AMP response-element binding protein (CREB) and expression of brain-derived nerve growth factor (BDNF), and its receptor trkB can be altered in response to both stress and antidepressant treatments.[12,159,160] These studies have demonstrated that stress reduces CREB phosphorylation and BDNF and trkB expression, while chronic antidepressant treatments increase phosphorylation of CREB and expression of BDNF and trkB. These effects appear selective and perhaps capable of reliably detected possible antidepressants. Notably, these effects are regionally restricted and appear primarily in cortex and hippocampus.[14,159]

It is thus of considerable interest to note that accumulating evidence indicates that major depressive disorders are accompanied by alterations in brain morphology and stereology indicative of reduced cortical volume[157,158] in prefrontal and anterior cingulate cortex. For an excellent review of this literature, see the recent review by Manji and colleagues.[161] Finally, recent reports have provided evidence that chronic antidepressant administration can induce neurogenesis in rat hippocampus.[162]

SYNTHESIS

Glutamate Signaling Cascades as Sites of Antidepressant Action

The data described above provide compelling evidence that subcellular signaling processes linked to modulation of glutamate receptors are involved in antidepressant actions. Modulation of the activity at multiple loci within these signaling processes results in antidepressant-like actions, at least in preclinical models. Moreover, several of these sites appear to respond adaptively to long-term exposure to clinically effective antidepressant treatments.

We now hypothesize (1) that diminishing or interfering with signaling at multiple points along the excitatory amino acid signaling pathway will produce antidepressant-like effects; (2) that the acute actions of at least some antidepressants in preclinical screening procedures are a consequence of the ability to disrupt excitatory amino acid signaling; and (3) that chronic but, not acute treatment with antidepressants results in adaptation of excitatory amino acid signaling at multiple loci. We also hypothesize (4) that the critical neuroanatomical sites for these effects are in neuronal circuits involving the prefrontal cortex, hippocampus and brainstem monoaminergic nuclei. We further suggest (5) that by directly regulating the influx of calcium and the production of NO free radicals and by indirectly regulating CREB phosphorylation and BDNF production, antidepressant-induced regulation of excitatory amino acid signaling stimulates neurogenesis and neurite outgrowth, which ultimately result in amelioration of depressive symptomatology.

Integration with Monoamine Hypotheses

Glutamatergic neurons provide the major excitatory neurotransmitter input to the locus coeruleus. Glutamatergic innervation of the locus coeruleus derives largely from the nucleus paragigantocellularis.[163] Glutamate activates the LC through activation of both NMDA and non-NMDA receptors.[164] Handling and immobilization stress increase glutamate measured in the rat locus coeruleus by microdialysis.[165,166] Interestingly, noise stress–induced enhancement of glutamate release in the locus coeruleus is abolished by superfusion of the locus coeruleus with a corticotropin releasing factor (CRF) antagonist,[167] indicating an important interaction between CRF and glutamate systems at the level of the locus coeruleus.[168] It is tempting to speculate that a deficit in noradrenergic transmission in major depression is secondary to a chronic elevation in glutamatergic input into the LC and a resulting depletion of central norepinephrine. Thus, agents that act to diminish excitatory input to the locus coeruleus might well normalize noradrenergic firing.

The raphe nuclei also receive glutamatergic input. At least part of the glutamatergic input to the dorsal raphe nuclei is transmitted through the habenula.[169] As is the case for the LC, glutamate is excitatory in the raphe nuclei. There is also evidence for an indirect inhibitory role of glutamate via intervening GABAergic neurons.[170] The activity of dopamine (DA) neurons in the mesolimbic and mesocortical circuitry can also be modulated by excitatory amino acids.[171] DA neurons in the ventral tegmental area receive direct glutamatergic innervation from the prefrontal cortex.[171] Glutamate excites DA cell activity via ionotropic and metabotropic receptors.[172]

Integration with the Role of Neurotrophic Factors

As shown in FIGURE 2, the subcellular signaling of all three glutamate receptors converges on nuclear cyclic AMP response element (CRE) and the associated CRE binding protein (CREB), which in turn regulates synthesis of many proteins, including BDNF. For example, the NMDA receptor, in particular the heteromer using the NR2A subunit, is coupled via the subcellular protein PSD95 to nitric oxide synthase.[173–175] Gating of calcium regulates mitochondrial associated protein kinase

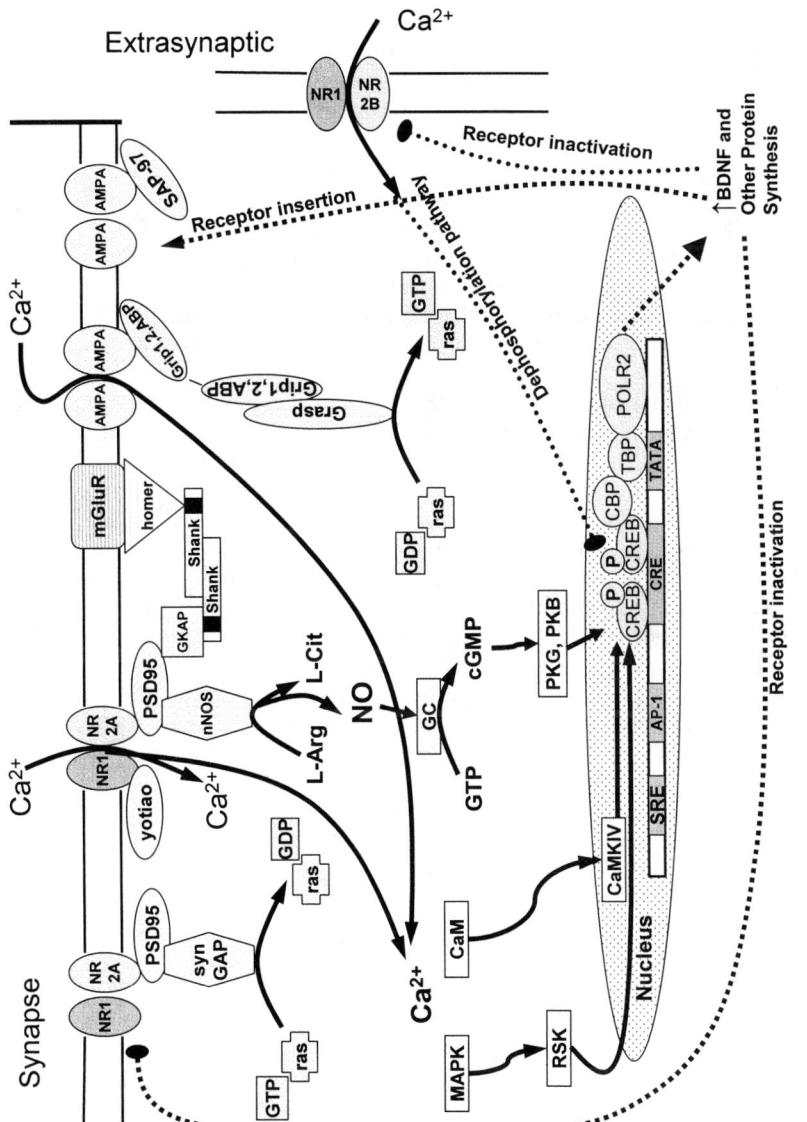

FIGURE 2. See following page for legend.

(MAPK) and ribosomal S6 kinase (RSK) as well as calmodulin (CaM) and calmodulin kinase IV (CaMKIV), which in turn regulate phosphorylation of nuclear CREB.[176,177] Similarly, activation of the NMDA receptor increases production of nitric oxide by nitric oxide synthase, which also regulates phosphorylation of nuclear CREB.[178,179] In addition, AMPA receptors gate calcium, which can activate MAPK and CaM and hence regulate phosphorylation of CREB.[180,181] Finally, group I mGluR are linked to and regulate NMDA receptor function via several subcellular anchor proteins, which provides yet another means by which glutamatergic neurotransmission can ultimately regulate phosphorylation of nuclear CREB and ultimately, synthesis of BDNF and other proteins.[182,183]

An apparent contradiction in this system is the fact that stimulation of both AMPA receptors and synaptic NMDA receptors strongly activates CREB phosphorylation and gene expression. Given the data presented above, then, we might logically infer that facilitating AMPA receptor gating of calcium should activate CREB and hence facilitate the production of BDNF and other proteins. This makes good sense given the data that indicate that exogenously administered BDNF has antidepressant-like properties in rodents, similar to those observed with AMPA potentiators.

Likewise, stimulation of synaptic NMDA receptors activates CREB and, presumably, BDNF production in a manner similar to AMPA receptor activation. Thus, antagonists of the NMDA receptor should inhibit CREB activation and BDNF production, which would be expected to produce a prodepressant-like effect. However, as noted above, NMDA receptor antagonists are potent antidepressant-like compounds both in rodents and in humans. This presents a paradox in our interpretation of the interaction between NMDA receptors and neurotrophic factors.

Recent evidence indicates that the site of NMDA receptor activation or inhibition is a critical factor in determining its synaptic effects. Hardingham and colleagues have demonstrated that activation of intrasynaptic NMDA receptors strongly activates nuclear CREB.[184] However, activation of extrasynaptic NMDA receptor triggers a dephosphorylation cascade, which results in inactivation of CREB and the protein synthesis, including BDNF, stimulated by phosphorylated CREB.[185,186] Thus, we can hypothesize that by inhibiting the CREB dephosphorylation cascade, NMDA receptor antagonists binding to extrasynaptic NMDA receptors may be responsible for the antidepressant-like effects of these compounds.

FIGURE 2. Subcellular linkage between glutamatergic receptors and nuclear transcription factors. NR1, NR2A, NR2B = NMDA receptor NR1, NR2A and NR2B subunits. mGluR = Group I metabotropic glutamate receptor. AMPA = AMPA receptor. PSD95 = post-synaptic density 95 protein. Membrane-associated guanylate kinase = MAGUK) associated with NR2A subunit. Yotiao = PDZ domain-carrying anchoring protein associated with NR1 subunit. Homer = anchoring protein associated with mGluR. Grip1,2, ABP = glutamate receptor xx protein 1,2, AMPA receptor binding protein. SAP-97 = synapse-associated protein-97 (MAGUK) associated with AMPA receptors. synGAP = synaptic glutamate receptor associated protein. nNOS = neuronal nitric oxide synthase. GKAP = guanylate kinase-associated protein. Shank = PDZ protein associated with GKAP and homer. GTP = guanosine triphosphate. GDP = guanosine diphosphate. Grasp = PDZ domain carrying protein associated with AMPA receptor. NO = nitric oxide. MAPK = mitogen associate protein kinase. CaM = calmodulin. RSK = ribosomal S6 kinase. CaMKIV = calmodulin kinase IV. CREB = cAMP response element binding protein.

PROPOSED ANATOMICAL CIRCUITRY OF RELEVANCE TO THE INTERACTION OF ANTIDEPRESSANTS AND EXCITATORY AMINO ACID NEUROTRANSMISSION

Noradrenergic projections from the locus coeruleus and serotonergic projections from the raphe nuclei modulate the activity of GABAergic interneurons and the apical dendrites of glutamatergic pyramidal neurons of prefrontal cortex and modulate neuronal activity in hippocampus. An extensive glutamatergic projection from frontal cortex also modulates hippocampal neuronal activity and feeds back both directly and indirectly via the nucleus paragigantocellularis (n. PGi) and habenula to the locus coeruleus[187–189] and raphe nuclei,[190] respectively. The connection to the locus coeruleus from the n. PGi appears also to be glutamatergic, while the connection to the raphe nuclei from the habenula appears to be GABAergic. In addition, there exist reciprocal modulatory projections from locus coeruleus and raphe nuclei (or periraphe areas). Glutamatergic neurons in prefrontal cortex and hippocampus thus appear to (1) be critically situated both to receive modulatory input from noradrenergic and serotonergic projections and (2) feed back upon these projection areas potentially relevant for antidepressant activity. As can be seen in FIGURE 3, most known an-

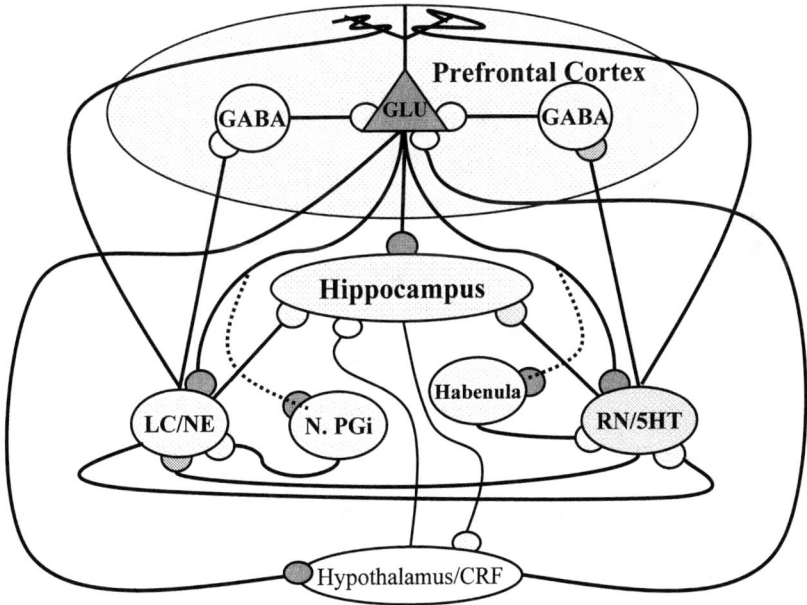

FIGURE 3. Proposed neural circuitry involved in the activity of antidepressant-like drugs. N. Pgi = Nuc. Paragigantocellularis. LC = Locus coeruleus. RN = Raphe nuclei. Projections described from LC are noradrenergic; those from RN are serotonergic; those from habenula are GABAergic. Not shown are BDNF/TrkB modulatory presynaptic connections on glutamatergic terminals in prefrontal cortex and hippocampus. Noradrenergic and serotonergic fibers synapse/modulate GABAergic neurons and apical dendrites of glutamatergic neurons in prefrontal cortex.

tidepressants would be expected to exert effects on this circuit by increasing synaptic concentrations of NE or 5HT in the frontal cortex. This would initiate a cascade of modulatory effects via GABAergic interneurons, glutamatergic pyramidal cells, and hippocampal neurons. This schema also provides an plausible mechanism for the antidepressant-like effects of NMDA receptor antagonists by direct action at frontal cortical pyramidal or hippocampal glutamatergic neurons.

CONCLUSION

Modulation of excitatory amino acid signaling represents a hitherto underexploited means of understanding affective disorders and developing novel antidepressant treatments. Clearly, additional study is needed. Controlled clinical trials are needed to determine if the antidepressant-like properties of NMDA receptor antagonists, metabotropic glutamate receptor antagonists, AMPA receptor potentiators, and modulators of the Ca^{2+}-calmodulin-NOS-guanylyl cyclase subcellular signaling pathway observed in animals obtain as well in humans suffering from major depression. Moreover, it is clear that potential toxic liabilities may limit the application of these compounds in a clinical setting. While the clinical impact of modulators of this cascade remains unknown, the emerging body of evidence described here makes clear the potential impacts of glutamatergic pathways in antidepressant-like actions and the etiology of major depression.

ACKNOWLEDGMENTS

This publication was supported in part by an Independent Investigator award from the National Alliance for Research in Schizophrenia and Depression to IAP and by research funds from the Center of Psychiatric Neuroscience at the University of Mississippi Medical Center, which is supported by National Institutes of Health Grant Number RR-P20 RR17701 from the Institutional Development Award (IDeA) Program of the National Center for Research Resources.

REFERENCES

1. 1994. Diagnostic and statistical manual of mental disorders, Fourth edition (DSM-IV). American Psychiatric Association. Washington, D.C.
2. MUSSELMAN, D.L., D.L. EVANS & C.B. NEMEROFF. 1998. The relationship of depression to cardiovascular disease: epidemiology, biology, and treatment. Arch. Gen. Psychiatry **55:** 580–592.
3. SCHULZ, R. et al. 2000. Association between depression and mortality in older adults: the Cardiovascular Health Study. Arch. Intern. Med. **160:** 1761–1768.
4. MICHELSON, D., et al.1996. Bone mineral density in women with depression. N. Engl. J. Med. **335:** 1176–1181.
5. CIECHANOWSKI, P.S., W.J. KATON & J.E. RUSSO. 2000. Depression and diabetes: impact of depressive symptoms on adherence, function, and costs. Arch. Intern. Med. **160:** 3278–3285.
6. REGIER, D.A. et al. 1993. The de facto US mental and addictive disorders service system. Epidemiologic catchment area prospective 1-year prevalence rates of disorders and services. Arch. Gen. Psychiatry. **50:** 85–94.

7. STONE, E.A. 1983. Problems with the current catecholamine hypothesis of antidepressant agents: speculations leading to a new hypothesis. Behav. Brain Sci. **6:** 555–577.
8. MAJ, J., E. PRZEGALINSKI & E. MOGILNICKA. 1984. Hypotheses concerning the mechanism of action of antidepressant drugs. Rev. Physiol. Biochem. Pharmacol. **100:** 1–74.
9. CALDECOTT-HAZARD, S. et al. 1991. Clinical and biochemical aspects of depressive disorders—II. Transmitter/receptor theories. Synapse **9:** 251–301.
10. CHARNEY, D.S. 1998. Monoamine dysfunction and the pathophysiology and treatment of depression. J. Clin. Psychiatry. **14:** 11–14.
11. OSWALD, I., V. BREZINOVA & D.L.F. DUNLEAVY. 1972. On the slowness of action of tricyclic antidepressant drugs. Br. J. Psychiat. **120:** 673–677.
12. DUMAN, R.S., G.R. HENINGER & E.J. NESTLER. 1997. A molecular and cellular theory of depression. Arch. Gen. Psychiatry. **54:** 597–606.
13. ROSSBY, S.P. & F. SULSER. 1997. Antidepressants: Beyond the synapse. *In* Antidepressants: New Pharmacological Strategies. P. Skolnick, Ed.: 195–212. Humana Press. Totowa, NJ.
14. MANJI, H.K., W.C. DREVETS & D.S. CHARNEY. 2001. The cellular neurobiology of depression. Nat. Med. **7:** 541–547.
15. SKOLNICK, P. 1999. Antidepressants for the new millennium. Eur. J. Pharmacol. **375:** 31–40.
16. WATKINS, J.C. 2000. L-glutamate as a central neurotransmitter: looking back. Biochem. Soc. Trans. **28:** 297–309.
17. RAJKOWSKA, G. 2000. Histopathology of the prefrontal cortex in major depression: what does it tell us about dysfunctional monoaminergic circuits? Prog. Brain Res. **126:** 397–412.
18. KIM, J.S. et al. 1982. Increased serum glutamate in depressed patients. Arch. Psychiatr. Nervenkr. **232:** 299–304.
19. ALTAMURA, C.A. et al. 1993. Plasma and platelet excitatory amino acids in psychiatric disorders. Neuroscience **55:** 511–519.
20. MAURI, M.C. et al. 1998. Plasma and platelet amino acid concentrations in patients affected by major depression and under fluvoxamine treatment. Neuropsychobiology **37:** 124–129.
21. MAES, M. et al. 1998. Serum levels of excitatory amino acids, serine, glycine, histidine, threonine, taurine, alanine and arginine in treatment-resistant depression: modulation by treatment with antidepressants and prediction of clinical responsivity. Acta Psychiatr. Scand. **97:** 302–308.
22. BERK, M., H. PLEIN & D. FERREIRA. 2001. Platelet glutamate receptor supersensitivity in major depressive disorder. Clin. Neuropharmacol. **24:** 129–132.
23. AUER, D.P. et al. 2000. Reduced glutamate in the anterior cingulate cortex in depression: an in vivo proton magnetic resonance spectroscopy study. Biol. Psychiatry. **47:** 305–313.
24. FRANCIS, P.T. et al. 1989. Brain amino acid concentrations and Ca^{2+}-dependent release in intractable depression assessed antemortem. Brain Res. **494:** 315–324.
25. MURCK, H. et al. 2002. Increase in amino acids in the pons after sleep deprivation: a pilot study using proton magnetic resonance spectroscopy. Neuropsychobiology **45:** 120–123.
26. MICHAEL, N. et al. 2003. Neurotrophic effects of electroconvulsive therapy: a proton magnetic resonance study of the left amygdalar region in patients with treatment-resistant depression. Neuropsychopharmacology **28:** 720–725.
27. PFLEIDERER, B. et al. 2003. Effective electroconvulsive therapy reverses glutamate/glutamine deficit in the left anterior cingulum of unipolar depressed patients. Psychiatry Res. **122:** 185–192.
28. ORTOLANO, G.A. et al. 1983. A calcium hypothesis of antidepressant action. Med. Hypoth. **10:** 207–221.
29. CLARK, J.A., L.J. DAVIDSON & H.C. FERGUSON. 1962. Psychosis in hypoparathyroidism. J. Mental Sci. **108:** 811–815.
30. FOURMAN, P. et al. 1967. Effect of calcium on mental symptoms in partial parathyroid insufficiency. Lancet **2:** 914–915.

31. WHYBROW, P.C. & T. HURWITZ. 1976. Psychological disturbances associated with endocrine disease and hormone therapy. *In* Hormones, Behavior, and Psychopathology. E.J. Sachar, Ed.: 125–143. Raven Press. New York.
32. GOUR, K.N. & H.M. CHAUDRY. 1957. Study of calcium metabolism in electric convulsive therapy (ECT) in certain. J. Mental Sci. **103:** 275–285.
33. FLACH, F.F., E. LIANG & P.E. STOKES. 1960. The effects of electric convulsive treatments on nitrogen, calcium, and phosphorus. J. Mental Sci. **106:** 638–647.
34. FLACH, F.F. 1964. Calcium metabolism in states of depression. Br. J. Psychiatr. **110:** 588–593.
35. FARAGALLA, F.F. & F.F. FLACH. 1970. Studies of mineral metabolism in mental depression. I. The effects of imipramine and electric convulsive therapy on calcium balance and kinetics. J. Nerv. Ment. Dis. **151:** 120–129.
36. JIMERSON, D.C. *et al.* 1979. CSF Calcium: Clinical correlates in affective illness and schizophrenia. Biol. Psychiatry **14:** 37–51.
37. CARMAN, J.S. *et al.* 1977. Calcium and electroconvulsive therapy in severe depressive illness. Biol. Psychiatry **12:** 5–17.
38. CARMAN, J.S. & R.J. WYATT. 1979. Calcium: Bivalent cation in the bivalent psychoses. Biol. Psychiatry **14:** 295–336.
39. LINDER, J. *et al.* 1989. Calcium and magnesium concentrations in affective disorder: difference between plasma and serum in relation to symptoms. Acta Psychiatr. Scand. **80:** 527–537.
40. DUBOVSKY, S.L. *et al.* 1989. Increased platelet intracellular calcium concentration in patients with bipolar affective disorders. Arch. Gen. Psychiatry **46:** 632–638.
41. EIDUSON, S., N.Q. BRILL & E. CRUMPTON. 1960. The effect of electro-convulsive therapy on spinal fluid constituents. J. Mental Sci. **106:** 692–698.
42. CARMAN, J.S. & R.J. WYATT. 1977. Alterations in cerebrospinal fluid and serum total calcium with changes in psychiatric state. *In* Neuroregulators and Psychiatric Disorders. E. Usdin *et al.*, Eds. Oxford University Press. New York.
43. MELLERUP, E.T. *et al.* 1979. Calcium and electroconvulsive therapy of depressed patients. Biol. Psychiatry **14:** 711–714.
44. DUBOVSKY, S.L., R.D. FRANKS & D. SCHRIER. 1985. Phenelzine-induced hypomania: Effect of verapamil. Biol. Psychiatry **20:** 1009–1014.
45. DUBOVSKY, S.L. 1993. Calcium antagonists in manic-depressive illness. Neuropsychobiology **27:** 184–192.
46. SILLS, M.A. & P.S. LOO. 1988. Tricyclic antidepressants and dextromethorphan bind with higher affinity to the phencyclidine receptor in the absence of magnesium and L-glutamate. Mol. Pharmacol. **36:** 160–165.
47. REYNOLDS, I.J. & R.J. MILLER. 1988. Tricyclic antidepressants block N-methyl-D-aspartate receptors: similarities to the action of zinc. Br. J. Pharmacol. **95:** 95–102.
48. KIM, J.S. *et al.* 1982. Effects of amitriptyline on serum glutamate and free tryptophan in rats. Arch. Psychiatr. Nervenkr. **232:** 391–394.
49. PRIKHOZHAN, A.V., G.I. KOVALEV & K.S. RAEVSKII. 1990. [Effects of antidepressive agents on glutamatergic autoregulatory presynaptic mechanism in the rat cerebral cortex]. Biull. Eksp. Biol. Med. **110:** 624–626.
50. MCCASLIN, P.P. *et al.* 1992. Amitriptyline prevents N-methyl-D-aspartate (NMDA)-induced toxicity, does not prevent NMDA-induced elevations of extracellular glutamate, but augments kainate-induced elevations of glutamate. J. Neurochem. **59:** 401–405.
51. BOURON, A. & J.Y. CHATTON. 1999. Acute application of the tricyclic antidepressant desipramine presynaptically stimulates the exocytosis of glutamate in the hippocampus. Neuroscience **90:** 729–736.
52. GOLEMBIOWSKA, K. & A. ZYLEWSKA. 1999. Effect of antidepressant drugs on veratridine-evoked glutamate and aspartate release in rat prefrontal cortex. Pol. J. Pharmacol. **51:** 63–70.
53. BORTOLOTTO, Z.A., S.M. FITZJOHN & G.L. COLLINGRIDGE. 1999. Roles of metabotropic glutamate receptors in LTP and LTD in the hippocampus. Curr. Opin. Neurobiol. **9:** 299–304.
54. COLLINGRIDGE, G.L. & T.V. BLISS. 1995. Memories of NMDA receptors and LTP. Trends Neurosci. **18:** 54–56.

55. SHORS, T.J. et al. 1989. Inescapable versus escapable shock modulates long-term potentiation in the rat hippocampus. Science **244:** 224–226.
56. TRULLAS, R. & P. SKOLNICK. 1990. Functional antagonists at the NMDA receptor complex exhibit antidepressant actions. Eur. J. Pharmacol. **185:** 1–10.
57. PETRIE, R.X., I.C. REID & C.A. STEWART. 2000. The N-methyl-D-aspartate receptor, synaptic plasticity, and depressive disorder. A critical review. Pharmacol. Ther. **87:** 11–25.
58. TRULLAS, R., B. JACKSON & P. SKOLNICK. 1989. 1-aminocyclopropanecarboxylic acid, a ligand of the strychnine-insensitive glycine receptor. Pharmacol. Biochem. Behav. **34:** 313–316.
59. TRULLAS, R. et al. 1991. 1-aminocyclopropanecarboxylates exhibit antidepressant and anxiolytic actions in animal models. Eur. J. Pharmacol. **203:** 379–385.
60. LAYER, R.T. et al. 1995. Antidepressant-like actions of the polyamine site NMDA antagonist, eliprodil (SL-82.0715). Pharmacol. Biochem. Behav. **52:** 621–627.
61. TRULLAS, R. 1997. Functional NMDA antagonists: A new class of antidepressant agents. In Antidepressants: New Pharmacological Strategies. P. Skolnick, Ed.: 103–124. Humana Press. Totowa, NJ.
62. MAJ, J., Z. ROGOZ & G. SKUZA. 1992. The effects of combined treatment with MK-801 and antidepressant drugs in the forced swimming test in rats. Pol. J. Pharmacol. Pharm. **44:** 217–226.
63. MAJ, J. et al. 1992. The effect of CGP 37849 and CGP 39551, competitive NMDA receptor antagonists, in the forced swimming test. Pol. J. Pharmacol. Pharm. **44:** 337–346.
64. CAI, Z. & P.P. MCCASLIN. 1992. Amitriptyline, desipramine, cyproheptadine and carbamazepine, in concentrations used therapeutically, reduce kainate- and N-methyl-D- aspartate-induced intracellular Ca^{2+} levels in neuronal culture. Eur. J. Pharmacol. **219:** 53–57.
65. MAJ, J. et al. 1992. Effects of MK-801 and antidepressant drugs in the forced swimming test in rats. Eur. Neuropsychopharmacol. **2:** 37–41.
66. WEDZONY, K., V. KLIMEK & G. NOWAK. 1995. Rapid down-regulation of beta-adrenergic receptors evoked by combined forced swimming test and CGP 37849—a competitive antagonist of NMDA receptors. Pol. J. Pharmacol. **47:** 537–540.
67. PRZEGALINSKI, E. et al. 1997. Antidepressant-like effects of a partial agonist at strychnine- insensitive glycine receptors and a competitive NMDA receptor antagonist. Neuropharmacology **36:** 31–37.
68. KROCZKA, B. et al. 2001. Antidepressant-like properties of zinc in rodent forced swim test. Brain Res. Bull. **55:** 297–300.
69. PAPP, M. & E. MORYL. 1994. Antidepressant activity of non-competitive and competitive NMDA receptor antagonists in a chronic mild stress model of depression. Eur. J. Pharmacol. **263:** 1–7.
70. MJELLEM, N., A. LUND & K. HOLE. 1993. Reduction of NMDA-induced behaviour after acute and chronic administration of desipramine in mice. Neuropharmacology **32:** 591–595.
71. MELONI, D. et al. 1993. Dizocilpine antagonizes the effect of chronic imipramine on learned helplessness in rats. Pharmacol. Biochem. Behav. **46:** 423–426.
72. MAJ, J. et al. 1995. Some central effects of GYKI 52466, a non-competitive AMPA receptor antagonist. Pol. J. Pharmacol. **47:** 501–507.
73. OSSOWSKA, G., B. KLENK-MAJEWSKA & G. SZYMCZYK. 1997. The effect of NMDA antagonists on footshock-induced fighting behavior in chronically stressed rats. J. Physiol. Pharmacol. **48:** 127–135.
74. KELLY, J.P., A.S. WRYNN & B.E. LEONARD. 1997. The olfactory bulbectomized rat as a model of depression: an update. Pharmacol. Ther. **74:** 299–316.
75. YILMAZ, A. et al. 2002. Prolonged effect of an anesthetic dose of ketamine on behavioral despair. Pharmacol. Biochem. Behav. **71:** 341–344.
76. PAUL, I.A. et al. 1992. Down-regulation of cortical beta-adrenoceptors by chronic treatment with functional NMDA antagonists. Psychopharmacology **106:** 285–287.
77. MAJ, J. et al. 1993. Central effects of repeated treatment with CGP 37849, a competitive NMDA receptor antagonist with potential antidepressant activity. Pol. J. Pharmacol. **45:** 455–466.

78. PAPP, M., V. KLIMEK & P. WILLNER. 1994. Effects of imipramine on serotonergic and beta-adrenergic receptor binding in a realistic animal model of depression. Psychopharmacology (Berl). **114:** 309–314.
79. MAJ, J., Z. ROGOZ & G. SKUZA. 1991. Antidepressant drugs increase the locomotor hyperactivity induced by MK- 801 in rats. J. Neural Transm. Gen. Sect. **85:** 169–179.
80. MAJ, J. *et al.* 1992. The effect of antidepressant drugs on the locomotor hyperactivity induced by MK-801, a non-competitive NMDA receptor antagonist. Neuropharmacology **31:** 685–691.
81. PAUL, I.A. *et al.* 1993. Adaptation of the NMDA receptor in rat cortex following chronic electroconvulsive shock or imipramine. Eur. J. Pharmacol. **247:** 305–311.
82. MASSICOTTE, G., J. BERNARD & M. OHAYON. 1993. Chronic effects of trimipramine, an antidepressant, on hippocampal synaptic plasticity. Behav. Neural Biol. **59:** 100–106.
83. NOWAK, G. *et al.* 1993. Adaptive changes in the N-methyl-D-aspartate receptor complex after chronic treatment with imipramine and 1-aminocyclopropanecarboxylic acid. J. Pharmacol. Exp. Ther. **265:** 1380–1386.
84. PAUL, I.A. *et al.* 1994. Adaptation of the N-methyl-D-aspartate receptor complex following chronic antidepressant treatments. J. Pharmacol. Exp. Ther. **269:** 95–102.
85. BERNARD, J., M. OHAYON & G. MASSICOTTE. 1994. Modulation of the AMPA receptor by phospholipase A2: effect of the antidepressant trimipramine. Psychiatry Res. **51:** 107–114.
86. PILC, A. & B. LEGUTKO. 1995. Antidepressant treatment influences cyclic AMP accumulation induced by excitatory amino acids in rat brain. Pol. J. Pharmacol. **47:** 359–361.
87. PILC, A. & B. LEGUTKO. 1995. The influence of prolonged antidepressant treatment on the changes in cyclic AMP accumulation induced by excitatory amino acids in rat cerebral cortical slices. Neuroreport **7:** 85–88.
88. SKOLNICK, P. *et al.* 1996. Adaptation of N-methyl-D-aspartate (NMDA) receptors following antidepressant treatment: implications for the pharmacotherapy of depression. Pharmacopsychiatry **29:** 23–26.
89. WATKINS, C.J., Q. PEI & N.R. NEWBERRY. 1998. Differential effects of electroconvulsive shock on the glutamate receptor mRNAs for NR2A, NR2B and mGluR5b. Brain Res. Mol. Brain Res. **61:** 108–113.
90. NOWAK, G. *et al.* 1998. Adaptation of cortical NMDA receptors by chronic treatment with specific serotonin reuptake inhibitors. Eur. J. Pharmacol. **342:** 367–370.
91. BOYER, P.A., P. SKOLNICK & L.H. FOSSOM. 1998. Chronic administration of imipramine and citalopram alters the expression of NMDA receptor subunit mRNAs in mouse brain. A quantitative in situ hybridization study. J. Mol. Neurosci. **10:** 219–233.
92. BRANDOLI, C. *et al.* 1998. Brain-derived neurotrophic factor and basic fibroblast growth factor downregulate NMDA receptor function in cerebellar granule cells. J. Neurosci. **18:** 7953–7961.
93. NOWAK, G. *et al.* 1995. Swim stress increases the potency of glycine at the N-methyl-D- aspartate receptor complex. J. Neurochem. **64:** 925–927.
94. PAUL, I.A. 1997. NMDA receptors and affective disorders. *In* Antidepressants: New Pharmacological Strategies. P. Skolnick, Ed.: 145–158. Humana Press. Totowa, NJ.
95. NOWAK, G., G.A. ORDWAY & I.A. PAUL. 1995. Alterations in the N-methyl-D-aspartate (NMDA) receptor complex in the frontal cortex of suicide victims. Brain Res. **675:** 157–164.
96. LAW, A.J. & J.F. DEAKIN. 2001. Asymmetrical reductions of hippocampal NMDAR1 glutamate receptor mRNA in the psychoses. Neuroreport **12:** 2971–2974.
97. CRANE, G.E. 1961. The psychotropic effects of cycloserine: A new use for an antibiotic. Compr. Psychiatry. **2:** 51–59.
98. CRANE, G.E. 1959. Cycloserine as an antidepressant agent. Am. J. Psychiat. **115:** 1025–1026.
99. VALE, S., M.A. ESPEJEL & J.C. DOMINGUEZ. 1971. Amantadine in depression. Lancet **2:** 437.
100. BODE, L. *et al.* 1997. Amantadine and human Borna disease virus in vitro and in vivo in an infected patient with bipolar depression. Lancet **349:** 178–179.

101. RIZZO, M. & P.L. MORSELLI. 1972. Amantadine-induced aggressiveness. Br. Med. J. **3:** 50.
102. PARKES, J.D. *et al.* 1970. Amantadine dosage in treatment of Parkinson's disease. Lancet **1:** 1130–1133.
103. HUBER, T.J., D.E. DIETRICH & H.M. EMRICH. 1999. Possible use of amantadine in depression. Pharmacopsychiatry **32:** 47–55.
104. STRYJER, R. *et al.* 2003. Amantadine as augmentation therapy in the management of treatment-resistant depression. Int. Clin. Psychopharmacol. **18:** 93–96.
105. BOUGEROL, T. *et al.* 1990. [Plasma levels of metapramine and its 3 major metabolites in patients with depression. Results and preliminary interpretations]. Encephale **16:** 35–40.
106. AZORIN, J.M. *et al.* 1988. Plasma 3,4-dihydroxyphenylethyleneglycol and 3-methoxy-4-hydroxyphenylethyleneglycol as indicators of central noradrenergic activity. A comparative study on control subjects and depressed patients. Neuropsychobiology **20:** 67–73.
107. GAYRAL, L.F. *et al.* 1978. [Therapeutic trial of the antidepressive agent, 19560 RP metapramine]. Therapie **33:** 541–542.
108. DICK, P. 1978. [Comparative study of the pharmacological and clinical effects of clorimipramine and metaprimine (19560 R.P.)]. Encephale **4:** 41–51.
109. BOIREAU, A. *et al.* 1996. The antidepressant metapramine is a low-affinity antagonist at N-methyl- D-aspartic acid receptors. Neuropharmacology **35:** 1703-1707.
110. MORYL, E., W. DANYSZ & G. QUACK. 1993. Potential antidepressive properties of amantadine, memantine and bifemelane. Pharmacol. Toxicol. **72:** 394–397.
111. GORTELMEYER, R. & H. ERBLER. 1992. Memantine in the treatment of mild to moderate dementia syndrome. A double-blind placebo-controlled study. Arzneimittelforschung **42:** 904–913.
112. AMBROZI, L. & W. DANIELCZYK. 1988. Treatment of impaired cerebral function in psychogeriatric patients with memantine--results of a phase II double-blind study. Pharmacopsychiatry **21:** 144–146.
113. BERMAN, R.M. *et al.* 2000. Antidepressant effects of ketamine in depressed patients. Biol. Psychiatry **47:** 351-354.
114. KUDOH, A. *et al.* 2002. Small-dose ketamine improves the postoperative state of depressed patients. Anesth. Analg. **95:** 114-118.
115. PILC, A. *et al.* 1998. Antidepressant treatment influences group I of glutamate metabotropic receptors in slices from hippocampal CA1 region. Eur. J. Pharmacol. **349:** 83–78.
116. PALUCHA, A. *et al.* 1997. Influence of imipramine treatment on the group I of metabotropic glutamate receptors in CA1 region of hippocampus. Pol. J. Pharmacol. **49:** 495–497.
117. ZAHORODNA, A. & M. BIJAK. 1999. An antidepressant-induced decrease in the responsiveness of hippocampal neurons to group I metabotropic glutamate receptor activation. Eur. J. Pharmacol. **386:** 173–179.
118. TATARCZYNSKA, E. *et al.* 2001. Potential anxiolytic- and antidepressant-like effects of MPEP, a potent, selective and systemically active mGlu5 receptor antagonist. Br. J. Pharmacol. **132:** 1423–1430.
119. BAJKOWSKA, M. *et al.* 1999. Effect of chronic antidepressant or electroconvulsive shock treatment on mGLuR1a immunoreactivity expression in the rat hippocampus. Pol. J. Pharmacol. **51:** 539–541.
120. SMIALOWSKA, M. *et al.* 2002. Effect of chronic imipramine or electroconvulsive shock on the expression of mGluR1a and mGluR5a immunoreactivity in rat brain hippocampus. Neuropharmacology **42:** 1016–1023.
121. MATRISCIANO, F. *et al.* 2002. Imipramine treatment up-regulates the expression and function of mGlu2/3 metabotropic glutamate receptors in the rat hippocampus. Neuropharmacology **42:** 1008–10015.
122. ZAFRA, F. *et al.* 1990. Activity dependent regulation of BDNF and NGF mRNAs in the rat hippocampus is mediated by non-NMDA glutamate receptors. EMBO J. **9:** 3545–3550.
123. HAYASHI, T. *et al.* 1999. The AMPA receptor interacts with and signals through the protein tyrosine kinase Lyn. Nature **397:** 72–76.

124. LAUTERBORN, J.C. *et al.* 2000. Positive modulation of AMPA receptors increases neurotrophin expression by hippocampal and cortical neurons. J. Neurosci. **20:** 8–21.
125. ALTAR, C.A. 1999. Neurotrophins and depression. Trends Pharmacol. Sci. **20:** 59–61.
126. LI, X. *et al.* 2001. Antidepressant-like actions of an AMPA receptor potentiator (LY392098). Neuropharmacology **40:** 1028–1033.
127. LEGUTKO, B., X. LI & P. SKOLNICK. 2001. Regulation of BDNF expression in primary neuron culture by LY392098, a novel AMPA receptor potentiator. Neuropharmacology **40:** 1019–10127.
128. MARTINEZ-TURRILLAS, R., D. FRECHILLA & J. DEL RIO. 2002. Chronic antidepressant treatment increases the membrane expression of AMPA receptors in rat hippocampus. Neuropharmacology **43:** 1230–1237.
129. SVENNINGSSON, P. *et al.* 2002. Involvement of striatal and extrastriatal DARPP-32 in biochemical and behavioral effects of fluoxetine (Prozac). Proc. Natl. Acad. Sci. USA **99:** 3182–3187.
130. SOUTHAM, E. & J. GARTHWAITE. 1993. The nitric oxide-cyclic GMP signalling pathway in rat brain. Neuropharmacology **32:** 1267–1277.
131. ZHU, X.Z. & L.G. LUO. 1992. Effect of nitroprusside (nitric oxide) on endogenous dopamine release from rat striatal slices. J. Neurochem. **59:** 932–935.
132. STRASSER, A. *et al.* 1994. L-arginine induces dopamine release from the striatum in vivo. Neuroreport **5:** 2298–300.
133. BOWYER, J.F. *et al.* 1995. Nitric oxide regulation of methamphetamine-induced dopamine release in caudate/putamen. Brain Res. **699:** 62–70.
134. LIN, A.M., L.S. KAO & C.Y. CHAI. 1995. Involvement of nitric oxide in dopaminergic transmission in rat striatum: an in vivo electrochemical study. J. Neurochem. **65:** 2043–2049.
135. FILIP, M. & E. PRZEGALINSKI. 1998. The role of the nitric oxide (NO) pathway in the discriminative stimuli of amphetamine and cocaine. Pharmacol. Biochem. Behav. **59:** 703–708.
136. GUEVARA-GUZMAN, R., P.C. EMSON & K.M. KENDRICK. 1994. Modulation of in vivo striatal transmitter release by nitric oxide and cyclic GMP. J. Neurochem. **62:** 807–810.
137. SILVA, M.T. *et al.* 1995. Increased striatal dopamine efflux in vivo following inhibition of cerebral nitric oxide synthase by the novel monosodium salt of 7-nitro indazole. Br. J. Pharmacol. **114:** 257–258.
138. MONTAGUE, P.R. *et al.* 1994. Role of NO production in NMDA receptor-mediated neurotransmitter release in cerebral cortex. Science **263:** 973–977.
139. HARVEY, B.E. 1996. Affective disorders and nitric oxide: A role in pathways to relapse and refractoriness? Human Psychopharmacol. **11:** 309–319.
140. VAN AMSTERDAM, J.G. & A. OPPERHUIZEN. 1999. Nitric oxide and biopterin in depression and stress. Psychiatry Res. **85:** 33–38.
141. LEE, S.-Y., O.-K. KIM & E.E. EL-FAKAHANY. 1995. Inhibition of neuronal nitric oxide synthase by antidepressants. Pharmacol. Comm. **7:** 1–9.
142. WEGENER, G. *et al.* 2003. Local, but not systemic, administration of serotonergic antidepressants decreases hippocampal nitric oxide synthase activity. Brain Res. **959:** 128–134.
143. FINKEL, M.S. *et al.* 1996. Paroxetine is a novel nitric oxide synthase inhibitor. Psychopharmacol. Bull. **32:** 653–658.
144. MONCADA, S. & A. HIGGS. 1993. The L-arginine-nitric oxide pathway. N. Engl. J. Med. **329:** 2002–2012.
145. JEFFERYS, D. & J. FUNDER. 1996. Nitric oxide modulates retention of immobility in the forced swimming test in rats. Eur. J. Pharmacol. **295:** 131–135.
146. HARKIN, A.J. *et al.* 1999. Nitric oxide synthase inhibitors have antidepressant-like properties in mice. 1. Acute treatments are active in the forced swim test. Eur. J. Pharmacol. **372:** 207–213.
147. KAROLEWICZ, B. *et al.* 1999. Nitric oxide synthase inhibitors have antidepressant-like properties in mice. 2. Chronic treatment results in downregulation of cortical beta-adrenoceptors. Eur. J. Pharmacol. **372:** 215–220.
148. YILDIZ, F. *et al.* 2000. Antidepressant-like effect of 7-nitroindazole in the forced swimming test in rats. Psychopharmacology (Berl). **149:** 41–44.

149. EROGLU, L. & B. CAGLAYAN. 1997. Anxiolytic and antidepressant properties of methylene blue in animal models. Pharmacol. Res. **36:** 381–385.
150. HEIBERG, I.L., G. WEGENER & R. ROSENBERG. 2002. Reduction of cGMP and nitric oxide has antidepressant-like effects in the forced swimming test in rats. Behav. Brain Res. **134:** 479–484.
151. SUZUKI, E. *et al.* 2002. Antipsychotic, antidepressant, anxiolytic, and anticonvulsant drugs induce type II nitric oxide synthase mRNA in rat brain. Neurosci. Lett. **333:** 217–219.
152. SANCHEZ, F. *et al.* 1999. Swim stress enhances the NADPH-diaphorase histochemical staining in the paraventricular nucleus of the hypothalamus. Brain Res. **828:** 159-62.
153. LUO, L. & R.X. TAN. 2001. Fluoxetine inhibits dendrite atrophy of hippocampal neurons by decreasing nitric oxide synthase expression in rat depression model. Acta Pharmacol. Sin. **22:** 865–870.
154. PLOTSKY, P.M., M.J. OWENS & C.B. NEMEROFF. 1998. Psychoneuroendocrinology of depression. Hypothalamic-pituitary-adrenal axis. Psychiatr. Clin. North Am. **21:** 293–307.
155. SAPOLSKY, R.M. 2000. The possibility of neurotoxicity in the hippocampus in major depression: a primer on neuron death. Biol. Psychiatry **48:** 755–765.
156. RAJKOWSKA, G. 2000. Postmortem studies in mood disorders indicate altered numbers of neurons and glial cells. Biol. Psychiatry **48:** 766–777.
157. RAJKOWSKA, G. *et al.* 1999. Morphometric evidence for neuronal and glial prefrontal cell pathology in major depression. Biol. Psychiatry **45:** 1085–1098.
158. ONGUR, D., W.C. DREVETS & J.L. PRICE. 1998. Glial reduction in the subgenual prefrontal cortex in mood disorders. Proc. Natl. Acad. Sci. U S A **95:** 13290–13295.
159. NIBUYA, M., S. MORINOBU & R.S. DUMAN. 1995. Regulation of BDNF and trkB mRNA in rat brain by chronic electroconvulsive seizure and antidepressant drug treatments. J. Neurosci. **15:** 7539–7547.
160. NIBUYA, M., E.J. NESTLER & R.S. DUMAN. 1996. Chronic antidepressant administration increases the expression of cAMP response element binding protein (CREB) in rat hippocampus. J. Neurosci. **16:** 2365–2372.
161. MANJI, H.K. *et al.* 2003. Enhancing neuronal plasticity and cellular resilience to develop novel, improved therapeutics for difficult-to-treat depression. Biol. Psychiatry **53:** 707–742.
162. MALBERG, J.E. *et al.* 2000. Chronic antidepressant treatment increases neurogenesis in adult rat hippocampus. J. Neurosci. **20:** 9104–9110.
163. ASTON-JONES, G. *et al.* 1986. The brain nucleus locus coeruleus: restricted afferent control of a broad efferent network. Science **234:** 734–737.
164. OLPE, H.R. *et al.* 1989. Excitatory amino acid receptors in rat locus coeruleus. An extracellular in vitro study. Naunyn Schmiedebergs Arch. Pharmacol. **339:** 312–314.
165. SINGEWALD, N., G.Y. ZHOU & C. SCHNEIDER. 1995. Release of excitatory and inhibitory amino acids from the locus coeruleus of conscious rats by cardiovascular stimuli and various forms of acute stress. Brain Res. **704:** 42–50.
166. TIMMERMAN, W. *et al.* 1999. Effects of handling on extracellular levels of glutamate and other amino acids in various areas of the brain measured by microdialysis. Brain Res. **833:** 150–160.
167. SINGEWALD, N. *et al.* 1996. Corticotropin-releasing factor modulates basal and stress-induced excitatory amino acid release in the locus coeruleus of conscious rats. Neurosci. Lett. **204:** 45–48.
168. LEVINE, J. *et al.* 2000. Increased cerebrospinal fluid glutamine levels in depressed patients. Biol. Psychiatry. **47:** 586–593.
169. KALEN, P. *et al.* 1986. Further evidence for excitatory amino acid transmission in the lateral habenular projection to the rostral raphe nuclei: lesion-induced decrease of high affinity glutamate uptake. Neurosci. Lett. **68:** 35–40.
170. TAO, R. & S.B. AUERBACH. 2000. Regulation of serotonin release by GABA and excitatory amino acids. J. Psychopharmacol. **14:** 100–113.
171. KALIVAS, P.W., L. CHURCHILL & M.A. KLITENICK. 1993. The circuitry mediating the translation of motivational stimuli into adaptive motor responses. *In* Limbic Motor Circuits and Psychiatry. P.W. Kalivas & C.D. Barnes, Eds.: 237–287. CRC Press. Boca Raton, FL.

172. SWERDLOW, N.R. & G.F. KOOB. 1987. Dopamine, schizophrenia, mania and depression: Toward a unified hypothesis of cortico-striato-pallido-thalamic function. Behav. Brain Sci. **10:** 197–245.
173. WATANABE, Y. et al. 2003. Post-synaptic density-95 promotes calcium/calmodulin-dependent protein kinase II-mediated Ser847 phosphorylation of neuronal nitric oxide synthase. Biochem. J. **372:** 465–471.
174. CHRISTOPHERSON, K.S. et al. 1999. PSD-95 assembles a ternary complex with the N-methyl-D-aspartic acid receptor and a bivalent neuronal NO synthase PDZ domain. J. Biol. Chem. **274:** 27467–27473.
175. SATTLER, R. et al. 1999. Specific coupling of NMDA receptor activation to nitric oxide neurotoxicity by PSD-95 protein. Science **284:** 1845–1848.
176. XING, J., D.D. GINTY & M.E. GREENBERG. 1996. Coupling of the RAS-MAPK pathway to gene activation by RSK2, a growth factor-regulated CREB kinase. Science **273:** 959–963.
177. PENDE, M. et al. 1997. Neurotransmitter- and growth factor-induced cAMP response element binding protein phosphorylation in glial cell progenitors: role of calcium ions, protein kinase C, and mitogen-activated protein kinase/ribosomal S6 kinase pathway. J. Neurosci. **17:** 1291–1301.
178. DAS, S. et al. 1997. NMDA and D1 receptors regulate the phosphorylation of CREB and the induction of c-fos in striatal neurons in primary culture. Synapse **25:** 227–233.
179. DING, J.M. et al. 1997. Resetting the biological clock: mediation of nocturnal CREB phosphorylation via light, glutamate, and nitric oxide. J. Neurosci. **17:** 667–675.
180. PERKINTON, M.S. et al. 2002. Phosphatidylinositol 3-kinase is a central mediator of NMDA receptor signalling to MAP kinase (Erk1/2), Akt/PKB and CREB in striatal neurones. J. Neurochem. **80:** 239–254.
181. PERKINTON, M.S., T.S. SIHRA & R.J. WILLIAMS. 1999. Ca(2+)-permeable AMPA receptors induce phosphorylation of cAMP response element-binding protein through a phosphatidylinositol 3-kinase-dependent stimulation of the mitogen-activated protein kinase signaling cascade in neurons. J. Neurosci. **19:** 5861–5874.
182. EHLERS, M.D. 1999. Synapse structure: glutamate receptors connected by the shanks. Curr Biol. **9:** R848-50.
183. TU, J.C. et al. 1999. Coupling of mGluR/Homer and PSD-95 complexes by the Shank family of postsynaptic density proteins. Neuron **23:** 583–592.
184. HARDINGHAM, G.E., F.J. ARNOLD & H. BADING. 2001. Nuclear calcium signaling controls CREB-mediated gene expression triggered by synaptic activity. Nat. Neurosci. **4:** 261–267.
185. HARDINGHAM, G.E., Y. FUKUNAGA & H. BADING. 2002. Extrasynaptic NMDARs oppose synaptic NMDARs by triggering CREB shut-off and cell death pathways. Nat. Neurosci. **5:** 405–414.
186. HARDINGHAM, G.E. & H. BADING. 2002. Coupling of extrasynaptic NMDA receptors to a CREB shut-off pathway is developmentally regulated. Biochim. Biophys. Acta **1600:** 148–153.
187. JODO, E., C. CHIANG & G. ASTON-JONES. 1998. Potent excitatory influence of prefrontal cortex activity on noradrenergic locus coeruleus neurons. Neuroscience **83:** 63–79.
188. JODO, E. & G. ASTON-JONES. 1997. Activation of locus coeruleus by prefrontal cortex is mediated by excitatory amino acid inputs. Brain Res. **768:** 327–332.
189. SINGEWALD, N. & A. PHILIPPU. 1998. Release of neurotransmitters in the locus coeruleus. Prog Neurobiol. **56:** 237-67.
190. PEYRON, C. et al. 1998. Forebrain afferents to the rat dorsal raphe nucleus demonstrated by retrograde and anterograde tracing methods. Neuroscience **82:** 443–468.

Regulation of Cellular Plasticity Cascades in the Pathophysiology and Treatment of Mood Disorders

Role of the Glutamatergic System

CARLOS A. ZARATE JR., JING DU, JORGE QUIROZ, NEIL A. GRAY, KIRK D. DENICOFF, JASKARAN SINGH, DENNIS S. CHARNEY, AND HUSSEINI K. MANJI

Laboratory of Molecular Pathophysiology, Mood and Anxiety Disorders Program, National Institute of Mental Health, Bethesda, Maryland 20892, USA

ABSTRACT: There is increasing evidence from a variety of sources that mood disorders are associated with regional reductions in brain volume, as well as reductions in the number, size, and density of glia and neurons in discrete brain areas. Although the precise pathophysiology underlying these morphometric changes remains to be fully elucidated, the data suggest that severe mood disorders are associated with impairments of structural plasticity and cellular resilience. In this context, it is noteworthy that a growing body of data suggests that the glutamatergic system—which is known to play a major role in neuronal plasticity and cellular resilience—may be involved in the pathophysiology and treatment of mood disorders. Preclinical studies have shown that the glutamatergic system represents targets (often indirect) for the actions of antidepressants and mood stabilizers. There are a number of glutamatergic "plasticity enhancing" strategies that may be of considerable utility in the treatment of mood disorders. Among the most immediate ones are NMDA antagonists, inhibitors of glutamate-release agents, and AMPA potentiators; this research progress holds much promise for the development of novel therapeutics for the treatment of severe, refractory mood disorders.

KEYWORDS: antidepressant; bipolar disorder; depression; glutamate; memantine; mood stabilizer; NMDA/AMPA/kainate; riluzole

INTRODUCTION

Despite the devastating impact that mood disorders have on the lives of millions worldwide, there is still a dearth of knowledge concerning their underlying etiology and pathophysiology. The brain systems that have heretofore received the greatest attention in neurobiologic studies of mood disorders have been the monoaminergic neurotransmitter systems that are extensively distributed throughout the network of

---

Address for correspondence: Carlos A. Zarate Jr., M.D., National Institute of Mental Health, Laboratory of Molecular Pathophysiology, Mood and Anxiety Disorders Program, National Institute of Mental Health, 9000 Rockville Pike, Building 10, Unit 3 West, Room 3s250, Bethesda, MD 20892. Voice: 301-402-9359; fax: 301-402-9360.
zaratec@intra.nimh.nih.gov

limbic, striatal, and prefrontal cortical neuronal circuits and are thought to support the behavioral and visceral manifestations of mood disorders.[1–3] Thus, clinical studies over the past 40 years have attempted to uncover the specific defects of these neurotransmitter systems in mood disorders by using a variety of biochemical and neuroendocrine strategies.

While such investigations have been heuristic over the years, they have been of limited value in elucidating the unique biology of mood disorders, which must include an understanding of the underlying basis for the predilection to episodic and often-profound mood disturbance, which can become progressive over time. Thus, mood disorders likely arise from the complex interaction of multiple susceptibility (and protective) genes and environmental factors, and the phenotypic expression of these diseases includes not only episodic and often profound mood disturbance, but also a constellation of cognitive, motoric, autonomic, endocrine, and sleep/wake abnormalities. Furthermore, while most antidepressants exert their initial effects by increasing the intrasynaptic levels of serotonin and/or norepinephrine, their clinical antidepressant effects are only observed after chronic (days to weeks) administration, suggesting that a cascade of downstream effects are ultimately responsible for their therapeutic effects. These observations have led to the appreciation that while dysfunction within the monoaminergic neurotransmitter systems is likely to play important roles in mediating some facets of the pathophysiology of mood disorders, they likely represent the downstream effects of other, more primary abnormalities.[4,5]

In addition to the growing appreciation that investigations into the pathophysiology of complex mood disorders have been excessively focused on monoaminergic systems, there has been a growing appreciation that progress in developing truly novel and improved antidepressant medications has consequently also been limited. The SSRIs, for example, have a better side effect profile for many patients, and are easier for physicians to prescribe. However, these newer medications have essentially the same mechanism of action as the tricyclic antidepressants and, as a result, the efficacy of the newer agents and the range of depressed patients they treat are no better than the older medications. Moreover, today's treatments remain suboptimal for many patients afflicted with depressive syndromes, and they continue to suffer protracted illnesses.

A recognition of the clear need for better treatments and the lack of significant advances in our ability to develop novel, improved therapeutics for these devastating illnesses has led to the investigation of the putative roles of intracellular signaling cascades and nonaminergic systems in the pathophysiology and treatment of mood disorders. Consequently, recent evidence demonstrating that impairments of neuroplasticity and cellular resilience may underlie the pathophysiology of mood disorders, and that antidepressants and mood stabilizers exert major effects on signaling pathways that regulate neuroplasticity and cell survival, have generated considerable excitement among the clinical neuroscience community, and are reshaping views about the neurobiological underpinnings of these disorders.[1,2,6–8]

EVIDENCE FOR IMPAIRMENTS OF STRUCTURAL PLASTICITY AND CELLULAR RESILIENCE IN MOOD DISORDERS

Positron emission tomography (PET) imaging studies have revealed multiple abnormalities of regional cerebral blood flow (CBF) and glucose metabolism in limbic and prefrontal cortical (PFC) structures in mood disorders. These abnormalities implicate limbic-thalamic-cortical and limbic-cortical-striatal-pallidal-thalamic circuits, involving the amygdala, orbital and medial PFC, and anatomically related parts of the striatum and thalamus in the pathophysiology of mood disorders. Interestingly, recent morphometric magnetic resonance imaging (MRI) and postmortem investigations have also demonstrated abnormalities of brain structure that persist independently of mood state and may contribute to the corresponding abnormalities of metabolic activity.[1,9]

Structural imaging studies have demonstrated reduced gray matter volumes in areas of the orbital and medial PFC, ventral striatum, and hippocampus, and enlargement of third ventricle in patients with mood disorders compared to healthy control samples.[3,10,11] Also consistent is the presence of white matter hyperintensities (WMH) in the brains of elderly depressed patients, and patients with bipolar disorder; these lesions may be associated with poor treatment response.[12] There is a growing awareness of the genetic influences on WMH, and their possible impact on neuropsychological functioning,[13] but WMH may have multiple causes including cerebrovascular accidents, demyelination, loss of axons, dilated perivascular space, minute brain cysts, and necrosis.[12] In this context, it is noteworthy that recent studies have used diffusion tensor imaging (DTI) of brain tissue to study possible white matter tract disruption in mood disorders.[14] This procedure measures the apparent diffusion coefficient (ADC), or isotropic diffusion, and anisotropy, or diffusion as influenced by tissue structure. These authors showed that WMH showed higher ADC and lower anisotropy than normal regions, with gray matter showing similar trends. Together, these results support the contention that WMH damage the structure of brain tissue, and likely disrupt the neuronal connectivity necessary for normal affective functioning. Although the cause of WMH in mood disorders is unknown, their presence—particularly in the brains of young bipolar patients—suggests importance in the pathophysiology of the disorder.[12,15]

Complementary postmortem neuropathological studies have shown abnormal reductions in cortex volume, glial cell counts, and/or neuron size in the subgenual PFC, orbital cortex, dorsal anterolateral PFC, amygdala and in basal ganglia and dorsal raphe nuclei.[9,16,17] It is not known whether these deficits constitute developmental abnormalities that may confer vulnerability to abnormal mood episodes, compensatory changes to other pathogenic processes, or the sequelae of recurrent affective episodes per se. Understanding these issues will partly depend upon experiments that delineate the onset of such abnormalities within the illness course and determine whether they antedate depressive episodes in individuals at high familial risk for mood disorders. While there is not total reproducibility among either the neuroimaging or postmortem studies, the differences likely represent variations of experimental design (including medication effects, *vide infra*), and—as would be expected in heterogeneous disorders such as mood disorders—in patient populations. Thus, research is required in order to understand if more rigorously defined subtypes of depression, or mood disorders, are associated with any particular abnormality.[12] Nevertheless, the marked reduction in glial cells in these regions has been particularly intriguing in view of the growing appreciation that glia play critical roles in regulating synaptic glutamate concentrations and CNS energy homeostasis, and in releasing trophic factors that participate in the development and maintenance of synaptic networks formed by neuronal and glial processes.[18–23] Abnormalities of glial function could thus prove integral to the impairments of structural plasticity and overall pathophysiology of mood disorders.

STRESS AND GLUCOCORTICOIDS MODULATE NEURAL PLASTICITY: IMPLICATIONS FOR MOOD DISORDERS

In developing hypotheses regarding the pathogenesis of these histopathological changes in mood disorders, the alterations in cellular morphology resulting from various stressors have been the focus of considerable recent research.[7] Thus, although mood disorders undoubtedly have a strong genetic basis, considerable evidence has shown that severe stressors are associated with a substantial increase in risk for the onset of mood disorders in susceptible individuals. Most studies of atrophy and survival of neurons in response to stress, as well as hormones of the hypothalamic-pituitary-adrenal (HPA) axis, have focused on the hippocampus. This is due, in part, to the well-defined and easily studied neuronal populations of this limbic brain region, including the dentate gyrus granule cell layer, and the CA1 and CA3 pyramidal cell layers. These cell layers and their connections (mossy fiber pathway and Schaffer collateral) have also been used as cellular models of learning and memory (i.e., long-term potentiation). Another major reason that the hippocampus has been the focus of stress research is that the highest levels of glucocorticoid receptors are expressed in this brain region.[24] However, it is clear that stress and glucocorticoids also influence the survival and atrophy of neurons in other brain regions (e.g. prefrontal cortex, *vide infra*) that have not yet been studied in the same detail as the hippocampus.

One of the most consistent effects of stress on cellular morphology is atrophy of hippocampal neurons.[25,26] This atrophy is observed in the CA3 pyramidal neurons, but not in other hippocampal cell groups (i.e., CA1 pyramidal and dentate gyrus granule neurons). The stress-induced atrophy of CA3 neurons (i.e., decreased number and length of the apical dendritic branches), occurs after two to three weeks of exposure to restraint stress or more long-term social stress, and has been observed in rodents and tree shrews.[25,26] Atrophy of CA3 pyramidal neurons also occurs upon exposure to high levels of glucocorticoids, suggesting that activation of the HPA axis likely plays a major role in mediating the stress-induced atrophy.[25–27] The hippocampus has a very high concentration of glutamate and expresses both Type I and Type II corticosteroid receptors, though Type II receptors may be relatively scarce in the hippocampus of primates,[28,29] and more abundant in cortical regions. Mineralocorticoid or Type I (MR) receptor activation in the hippocampus (CA1) is associated with reduced calcium currents, whereas activation of glucocorticoid or Type II receptors (GR) causes increased calcium currents and enhanced responses to excitatory amino acids. Very high levels of Type II receptor activation markedly increases calcium currents and lead to increased NMDA receptor throughput that could predispose to neurotoxicity. Indeed, as we discuss in greater detail below, a growing body of data has implicated glutamatergic neurotransmission in stress-induced hippocampal atrophy and death.[26]

THE ROLE OF THE GLUTAMATERGIC SYSTEM IN MEDIATING THE STRESS-INDUCED MORPHOMETRIC CHANGES

Microdialysis studies have shown that stress increases extracellular levels of glutamate in hippocampus, and *N*-methyl-D-aspartate (NMDA) glutamate receptor antagonists attenuate stress-induced atrophy of CA3 pyramidal neurons.[26,30] Although a variety of methodological issues remain to be fully resolved, the preponderance of the evidence to date suggests that the atrophy, and possibly death, of CA3 pyramidal neurons arises, at least in part, from increased glutamate neurotransmission.[26,30] It should be noted, however, that although NMDA antagonists block stress-induced hippocampal atrophy, there have not been any studies demonstrating that they are able to block the cell death induced by severe stress. This suggests that the mechanisms underlying atrophy and death may lie on a continuum, with severe (or prolonged) stresses "recruiting" additional pathogenic pathways in addition to enhanced NMDA-mediated neurotransmission. As discussed, stress increases extracellular levels of glutamate and sustained activation of NMDA, as well as non-NMDA ionotropic receptors could result in high intracellular levels of calcium. Overactivation of the glutamate ionotropic receptors is known to contribute to the neurotoxic effects of a variety of insults, including repeated seizures and ischemia. Neurotoxicity follows as a response to overactivation of calcium-dependent enzymes and the generation of oxygen free radicals. Stress or glucocorticoid exposure also compromises the metabolic capacity of neurons, thereby increasing the vulnerability to other types of neuronal insults.

DIRECT CLINICAL EVIDENCE OF GLUTAMATERGIC DYSFUNCTION IN MOOD DISORDERS

It has to be acknowledged that, at this point, there is not very much direct clinical evidence for abnormalities of the glutamatergic system in the pathophysiology of mood disorders; this undoubtedly reflects, in large part, the difficulty in assessing CNS glutamatergic function in humans. The clinical findings of abnormalities in glutamate function in mood disorders reported in the majority of the studies, consist of increases in glutamate and/or glutamine levels in plasma[31–35] in CSF[36] or platelet intracellular calcium in response to glutamate stimulation[37] (TABLE 1). In addition, reductions in high-affinity glycine displaceable [^{3}H]CGP-39653 binding to NMDA receptors[38] and reductions in glutamate/glutamine (Glx) levels in a patient with suicidal ideation,[39] reductions in Glx levels in the anterior cingulate cortex[40] and left cingulum[41] have been found in adult depressed patients. Conversely, an elevated Glx level in the frontal lobe and basal ganglia in children with bipolar disorder[42] has been reported. In acute mania, increased Glx levels in the dorsolateral PFC have been reported[43] (TABLE 1).

Effects of Chronic Antidepressants on the Glutamatergic System

The effects of antidepressants on the glutamatergic system are reviewed extensively elsewhere in this volume (Paul and Skolnick, this volume).

TABLE 1. Preclinical and clinical data supporting the role of the glutamatergic system in the pathophysiology and treatment of mood disorders

Relevant preclinical findings	
Reynolds and Miller[123]	Tricyclic antidepressants slowed the dissociation rate of the NMDA antagonist [^{3}H]MK-801 in a similar manner to Zn^{2+}
Trullas and Skolnick[125]	Competitive NMDA antagonist AP-7, noncompetitive antagonist dizocilpine, and a partial agonist at strychnine-insensitive glycine receptors ACPC, mimicked the effects of clinically effective antidepressants in inescapable-stress model in rats
Paul et al.[47]	Chronic administration of ACPC and MK-801 in mice reduces the density of cortical β-adrenoreceptors with a magnitude comparable to that produced by imipramine
Moryl et al.[96]	Chronic administration of amantadine decreased the immobility time in the forced swimming test in rats
Nowak et al.[48]	Chronic citalopram in mouse, lowered 6.2-fold high affinity glycine displaceable [^{3}H]CGP-39653 binding to glutamate receptors and reduced 1.5-fold the potency of glycine to inhibit [^{3}H]DCKA binding in cortex. Also increases aspartate concentration 110% in cortex and 33% in hippocampus
Papp and Moryl[121]	Chronic administration of the glycine partial agonist ACPC was associated with significant antidepressant properties in the mild stress model and occurred more rapidly than with imipramine
Dixon and Hokin[58]	Chronic lithium upregulated and stabilized glutamate uptake by presynaptic nerve endings in mouse cerebral cortex
Boyer et al.[49]	Chronic (16 days) of citalopram in mouse lowered NMDA ε1-subunit mRNA level in frontal cortex, CA2 of hippocampus, and amygdala, whereas imipramine does it only in amygdala. Imipramine lowered NMDA ε2-subunit mRNA level in cortex, CA1-4 of hippocampus, and amygdala, whereas citalopram does it only in amygdala. Both drugs reduced transcript levels of ζ-subunit in cortex, thalamus, striatum, and cerebellum
Bouron and Chatton[44]	Desipramine enhanced spontaneous vesicular release of glutamate in hippocampal neurons dissociated from neonatal rats
Michael-Titus et al.[45]	Imipramine and phenelzine decreased potassium stimulated glutamate outflow in rat prefrontal cortex and not in striatum
Chen et al.[82]	Ketamine pretreatment attenuated ECS-induced mossy fiber sprouting in dentate gyrus and BDNF expression in medial prefrontal cortex and the dentate gyrus in rats
Li et al.[69]	AMPA receptor potentiator LY392098 (a biarylpropylsulfonamide) produced antidepressant-like effects in animal models of depression in rats and mice
Rogoz et al.[97]	Fluoxetine, which was inactive in the forced swimming test in rats when given alone, showed a positive effect when combined with the NMDA receptor antagonists amantadine (10 and 20 mg/kg) and memantine (2.5 and 5 mg/kg)

TABLE 1. Preclinical and clinical data supporting the role of the glutamatergic system in the pathophysiology and treatment of mood disorders (*Continued*)

Relevant clinical findings
Drug Trials

Crane et al.[118]	D-Cycloserine in tuberculosis	Antidepressant patients with tuberculosis and depression
Parkes et al.[122]	Antiparkinsonian response to amantadine (double-blind crossover study)	Significant improvement in mood and relief of parkinsonian symptoms in 37 patients
Vale et al.[126]	Antidepressant response to amantadine (randomized placebo-controlled study)	Significant antidepressant efficacy in 40 depressed patients
Calabrese et al[117]	Antidepressant response to lamotrigine (double-blind placebo-controlled study)	Significant antidepressant efficacy in 195 depressed BPD I patients
Berman et al.[116]	Antidepressant response to ketamine (double-blind placebo-controlled study)	Improved depressive symptoms in depressed patients (8 MDD, 1 BPD) lasting longer (3 days) than euphoric effects (hours)
Stryjer et al.[124]	Antidepressant response to amantadine (open-label, add-on)	Significant antidepressant efficacy in 8 patients with treatment-resistant depression
Zarate et al.[127]	Riluzole (open-label study)	Riluzole was effective in treatment-resistant unipolar depression

Glutamate Receptor Studies

Holemans et al.[119]	Binding of [^{3}H]dizocilpine sites	No changes of binding in 22 depressed, medication-free suicide victims in cortex, hippocampus, thalamus, and basal ganglia. Negative correlation between age and NMDA receptor binding in frontal cortex suicide victims
Nowak et al.[38]	High affinity glycine displaceable [^{3}H]CGP-39653 binding to Glu receptors	Reduced in suicide victims (50% of them depressed) vs. controls, in frontal cortex
Meador-Woodruff et al.[120]	NMDA mRNA subunit levels in striatum	Postmortem analysis. Only NR2D (a subunit of NMDA receptor) mRNA is higher in BPD vs. MDD. Only GluR1 (a subunit of AMPA receptor) mRNA is lower in BPD vs. controls. [^{3}H]AMPA binding was higher in BPD vs. MDD

TABLE 1. Preclinical and clinical data supporting the role of the glutamatergic system in the pathophysiology and treatment of mood disorders (*Continued*)

Glutamate levels: MRS		
Cousins and Harper[39]	Glu/Gln	A 22–39% drop in Glu/Gln metabolites in a patient with metastatic breast cancer was associated with depression and suicidal ideation
Castillo et al.[42]	Glx/Gln ratio levels in frontal and temporal cortex and basal ganglia	Elevated in frontal lobe and basal ganglia in BPD medication-free children vs. controls
Auer et al.[40]	Glx levels	Decreased in anterior cingulate cortex of depressed patients vs. controls (7 patients were medication free and 12 under antidepressant treatment; 1 BPD and 18 MDD)
Pfleiderer et al.[41]	Glx levels in cingulum	Reduced Glx levels in left cingulum in patients with MDD vs. controls. Normalization of Glx levels in ECT responders
Michael et al.[43]	Glx levels in dorsolateral prefrontal cortex	Increased Glx levels in dorsolateral prefrontal cortex inpatients with acute mania
Glutamate, glutamine, and glycine: CSF, plasma, and platelet levels		
Mathis et al.[32]	Gln plasma level	Gln level higher in 59 depressive patients (MDD, BPD) vs. controls
Altamura et al.[33]	Glu plasma and platelet level	Increased plasmatic and decreased platelet level in medication-free depressed patients (4 MDD, 11 BPD) vs. controls
Altamura et al.[34]	Glu and Gly plasma levels	Lower in 25 MDD medication-free patients vs. controls
Mauri et al.[31]	Glu plasma levels	Higher Glu plasma level in 29 MDD patients vs. controls, not altered by fluvoxamine
Maes et al.[35]	Glu, Gln plasma levels	No differences between patients and control. Antidepressant treatment for 5 weeks reduced Glu and increased Gln plasmatic levels
Levine et al.[36]	CSF Gln levels	Elevated in medication-free depressed patients vs. control (2 BPD, 16 MDD) and correlated with CSF magnesium level
Berk et al.[37]	Platelet intracellular calcium response to Glu stimulation	Greater in 15 MDD medication-free patients vs. controls

ABBREVIATIONS: AMPA, α-amino-3-hydroxy-5-methyl-4-isoxazole propionic acid; ACPC, 1-aminocyclopropanecarboxylic acid; BDNF, brain derived neurotrophic factor; BPD, bipolar disorder patients; CSF, cerebrospinal fluid; DCKA, dichlorokynurenic acid; ECS, electroconvulsive shock; ECT, electroconvulsive therapy; Glx, peak on magnetic resonance spectroscopy containing glutamate, but also other compounds (i.e., not exclusively glutamate); Glu, glutamate; Gly, glycine; Gln, glutamine; MDD, major depressive disorder patients; MRS, magnetic resonance spectroscopy; NMDA, *N*-methyl-D-aspartate. (Modified and reproduced with permission from Shirayama and colleagues.[114])

In addition, antidepressants affect release of glutamate,[44] decrease potassium stimulated glutamate outflow in rat prefrontal cortex,[45] NMDA receptor function[38,46] and receptor binding profiles.[47,48] Moreover, chronic administration of antidepressants regionally alters expression of mRNA that encodes multiple NMDA receptor subunits[49] and radioligand binding to these receptors in localized areas of the brain.[50]

Effects of Lithium and Valproate on the Glutamatergic System

Patients with Bipolar Disorder (manic-depressive illness) are generally treated with a class of medications referred to as "mood stabilizers." Mood stabilizers are agents that have antimanic effects, exert prophylactic effects in preventing recurrent manic or depressive episodes, and may also possess some antidepressant properties. The prototypic agent of this class is lithium, a seemingly simple monovalent cation. More recently, a variety of anticonvulsant agents, most notably valproic acid (a substituted pentanoic acid) have also been utilized as mood stabilizing agents.[51,52]

Lithium

Nonaka and colleagues[53] found that chronic treatment with therapeutically relevant concentrations of lithium chloride in cultured rat cerebellar, cortical, and hippocampal neurons protected against glutamate-induced excitotoxicity involving apoptosis mediated by NMDA receptors. Furthermore they found that chronic treatment of mice with therapeutically relevant concentrations of lithium (0.7 mM) upregulated synaptosomal uptake of glutamate. The investigators found that the protection in cerebellar neurons is specific for glutamate induced excitotoxicity and can be attributed to inhibition of NMDA receptor-mediated Ca^{2+} influx and is not the result of down-regulation of NMDA receptor subunit proteins (NR1, NR2A, NR2B, or NR2C)[53,54] or the ability to block inositol monophosphatase activity. However, while lithium has been reported not to alter the total protein levels of NR1, NR2A and NR2B subunits of NMDA receptors, a recent study suggests that lithium protection against glutamate excitotoxicity in rat cerebral neurons may be a result of reducing the level of NR2B phosphorylation at Tyr1472.[54] NR2B phosphorylation correlates with NMDA receptor-mediated synaptic activity and excitotoxicity. It is also possible that lithium modulates glutamatergic neurotransmission by affecting the non-NMDA receptor-sensitive channels. A recent study by Karkanias and Papke,[55] reported that lithium in frog oocytes (*Xenopus laevis*) produces an increase of inward and outward currents of AMPA receptors and a decrease in the currents of KA and NMDA receptors.

Valproate

Valproate regulates excitatory amino acid neurotransmission[56] and increases concentrations of glutamate in neuronal cultures and in animal brains, possibly by regulation of glutaminase and glutamine synthetase.[57] Valproate also stimulated glutamate release in mouse cerebral cortex,[58] and potentiated glutamate-evoked intracellular calcium activity.[59] Ueda and Willmore[60] recently reported the effect of sodium valproate on glutamate transporter expression in the hippocampus. With a dose of 100 mg/kg/day of sodium valproate given for 14 days, they found an increase in EAAT1 levels and a decrease in EAAT2 levels. Hassel and colleagues[61] reported that chronic treatment of rats with sodium valproate (200 mg or 400 mg/kg, for 90 days) leads to a dose-dependent increase in hippocampal glutamate uptake capacity as measured by uptake of [^{3}H]glutamate into proteoliposomes via increasing the levels of the glutamate transporters EAAT1 and EAAT2 in the hippocampus.

At the receptor level, valproate modulates physiological responses by glutamate receptors including NMDA,[62] and non-NMDA receptors AMPA,[63] KA receptor,[64] and quisqualate metabotropic glutamate receptor. In rodent models, valproate has been shown to reduce seizure activity induced by AMPA glutamate receptor agonists.[65,66] In postmortem human brain tissue, Kunig and colleagues,[63] showed that therapeutic levels of valproate decreased binding of AMPA to the AMPA glutamate receptors- thus effectively blocking them.

LITHIUM AND VALPROATE ALSO REGULATE AMPA RECEPTOR SUBUNIT TRAFFICKING

Recent studies suggest that AMPA receptor trafficking may be associated with the pharmacological effects of drugs, which affect mood. AMPA receptors appear to be common targets of structurally dissimilar mood stabilizers when administered in therapeutically relevant paradigms (dose, duration, critical brain regions); the functional consequences and therapeutic relevance are under extensive investigation in our laboratory. Overall, the two antimanic agents, lithium and valproate reduce GluR1 level in synaptosomal preparation from hippocampus of chronically treated animals.[67] In contrast, very recent studies have found that chronic antidepressant treatment to rats increases membrane expression of AMPA receptors in the hippocampus, which may lead to an increase in excitatory synaptic strength. Furthermore, Malenka's group have recently shown that the psychostimulants cocaine and amphetamine, enhance the AMPA/NMDA EPSC ratio at excitatory synapses on midbrain dopamine cells, also suggesting an enhancement in AMPA synaptic strength in the midbrain.[68] Also consistent with the potential mood elevating properties of enhancing AMPA "throughput" are the observations that Ampakines (which are small benzamide compounds that allosterically produce positive modulation of AMPA receptors) have been demonstrated to have antidepressant effects in animal models of depression (including the application of inescapable stressors, forced-swim test, and tail-suspension-induced immobility tests), in learned-helplessness models of depression, and in animals exposed to chronic mild stress procedure.[69] In addition, Ampakines have recently been reported to enhance the neurotrophic activity of brain derived neurotrophic factor (BDNF);[70] as we discuss below, enhancement of neurotrophic signaling cascades may be a functional mechanism to counter the deleterious effects of excessive NMDA signaling.

LITHIUM AND VALPROATE ROBUSTLY ACTIVATE NEUROTROPHIC SIGNALING CASCADES

Several endogenous growth factors—including nerve growth factor (NGF) and BDNF—exert many of their neurotrophic effects via the MAP kinase-signaling cascade. In view of the important role of MAP kinases in mediating long term neuroplastic events, it is noteworthy that lithium and valproate, at therapeutically relevant concentrations, have recently been demonstrated to robustly activate the ERK MAP kinase cascade in human neuroblastoma SH-SY5Y cells and in critical limbic and limbic-related areas of rodent brain.[71] Interestingly, neurotrophic factors are now known to promote cell survival by activating MAP kinases to suppress intrinsic, cellular apoptotic machinery, not by inducing cell survival pathways (*vide supra*).[72–79] Thus, a downstream target of the MAP kinase cascade, ribosomal S-6 kinase (Rsk) phosphorylates the cAMP response element binding protein (CREB) and this leads to induction of bcl-2 gene expression. Consistent with an activation of neurotrophic signaling cascades, chronic treatment of rats with "therapeutic" doses of lithium produces a doubling of bcl-2 levels in frontal cortex, effects that were primarily due to a marked increase in the number of bcl-2 immunoreactive cells in layers II and III of frontal cortex.[80–82]

As discussed already, lithium administration has also been demonstrated to enhance cellular resiliency by attenuating aberrant NMDA-mediated glutamatergic functioning. Consistent with the cytoprotective effects that would be expected by attenuation of NMDA-mediated excitotoxicity—and, more importantly, bcl-2 upregulation—lithium, at therapeutically relevant concentrations, has been shown to exert neuroprotective effects in a variety of preclinical paradigms. Thus, lithium has been demonstrated to protect against the deleterious effects of glutamate, NMDA receptor activation, aging, serum/nerve growth factor deprivation, ouabain, thapsigargin (which mobilizes intracellular MPP^+), Ca^{2+} and β-amyloid *in vitro*.[81] In addition to the demonstration of protective effects *in vitro*, a number of studies have also investigated lithium's neuroprotective effects *in vivo*. Most notably, chronic lithium has also been shown to exert dramatic protective effects against middle cerebral artery occlusion, reducing not only the infarct size (56%), but also the neurological deficits (abnormal posture and hemiplegia).[53] More recently, the same research group has demonstrated that chronic *in vivo* lithium treatment robustly protects neurons in the striatum from quinolinic acid-induced toxicity, in a putative model of Huntington's disease.[83]

HUMAN EVIDENCE FOR THE NEUROTROPHIC EFFECTS OF LITHIUM

While the body of preclinical data demonstrating neurotrophic and neuroprotective effects of mood stabilizers is striking, considerable caution must clearly be exercised in extrapolating to the clinical situation with humans. To investigate the potential neurotrophic effects of lithium in humans more definitively, a longitudinal clinical study was recently undertaken using proton magnetic resonance spectroscopy (MRS) to quantitate *N*-acetyl-aspartate (NAA, a putative marker of neuronal viability) levels.[84] Four weeks of lithium treatment produced a significant increase in NAA levels, effects that were localized almost exclusively to gray matter.[85] These findings provide intriguing indirect support for the contention that chronic lithium increases neuronal viability/function in the human brain. Furthermore, a ~0.97 correlation between lithium-induced NAA increases and regional voxel gray matter content was observed. A follow-up volumetric MRI study has demonstrated that four weeks of lithium treatment also significantly increased total gray matter content in the human brain,[86] suggesting an increase in the volume of the neuropil (the moss-like layer, comprising axonal and dendritic fibers, that occupies much of the cortex gray matter volume). A finer grained subregional analysis of this brain imaging data is ongoing, and clearly shows that lithium produces a regionally selective increase in gray matter, with prominent effects being observed in hippocampus and caudate.[87] Furthermore, no changes in overall gray matter volume are observed in healthy volunteers treated chronically with lithium, suggesting that lithium is truly producing a reversal of illness-related atrophy, rather than nonspecific gray matter increases.

Glutamate-Modulating Agents as Putative Treatments for Mood Disorders

In view of the preclinical and clinical data supporting a role for the glutamatergic system in the pathophysiology and treatment of mood disorders, there is considerable recent interest in investigating the putative efficacy of glutamate modulating agents; here we discuss two such agents currently undergoing clinical trials.

Memantine

Memantine has in double-blind placebo controlled studies been shown to be effective in reducing clinical deterioration in Alzheimer's, vascular and mixed dementia of all severities.[88–90] Chemically, memantine is a 1-amino-3,5-dimethyl-adamantane of the adamantine class, a compound that can easily cross the blood-brain barrier. Memantine is a potent noncompetitive voltage-dependent NMDA antagonist with a receptor effect comparable to MK-801[91] and has been demonstrated to inhibit [^{3}H]MK-801 binding to human hippocampal NMDA receptors.[92] Memantine acts on the NMDA receptor at a fairly low but clinically relevant dose (brain concentrations of 1–2 mM at a median dose of 20 mg/day with a dose range of 10–60 mg/day) while interactions with σ-site, AMPA and KA receptors have been reported to take place only at considerably higher concentrations.[93] This is important in that the therapeutic effects of memantine may occur at much lower doses than the psychotomimetic effects unlike ketamine and MK-801. In addition, memantine blocks the NMDA receptor-associated ion channel only when it is open for pathological periods of time. Increasing concentrations of glutamate or other NMDA agonists can cause NMDA channels to remain open on average for a greater fraction of time. Under such conditions, an open-channel blocking drug such as memantine has a better chance to enter the channel and block it. Because of this mechanism of action, the untoward effects of pathological concentrations of glutamate are prevented to a greater extent than the effects of physiological concentrations, which are relatively spared.[94,95]

Memantine has been used since 1978 in Germany for the treatment of mild and moderate cerebral performance disorders with the following cardinal symptoms: concentration and memory problems, loss of interest and drive, premature fatigue and depressive mood (dementia syndromes) as well as in diseases that affect attention and alertness (vigilance). Such symptoms overlap with those seen in mood disorders and as such, it is possible that memantine may also prove to have beneficial effects in patients with mood disorders. In preclinical studies, Moryl and colleagues[96] described a dose-dependent decrease in the immobility time in the forced swimming test in rats following administration of memantine. A synergistic effect was seen when imipramine and fluoxetine was given jointly with memantine in the forced swim test in rats.[97] Compared to the other NMDA antagonists, memantine has been reported to have the greatest effective potency for binding at the MK-801 receptor site in mouse brain as demonstrated by fluorine-18 PET[98] and in human brain tissue. Memantine binds to the MK-801 binding site of the NMDA receptor in postmortem human frontal cortex at therapeutic concentrations,[99] and reduces membrane currents.[91]

In summary, memantine is one of the few NMDA antagonists available for use in humans and is ideal for testing in mood disorders as it has been in clinical use for many years with minimal side effects[100] and has a very favorable pharmacological profile.

Riluzole

Riluzole, a neuroprotective agent with anticonvulsant properties, is a member of the benzothiazole class. Chemically, riluzole is 2-amino-6-(trifluoromethoxy) benzothiazole, that can easily cross the blood brain barrier 101. Riluzole is the only drug currently approved (by the FDA in US, CPMP in Europe and MHW in Japan) for the treatment of amyotrophic lateral sclerosis (ALS). Evidence from a variety of studies with experimental animals and with humans indicate that riluzole is devoid of the psychotomimetic or other behavioral side-effects commonly associated with excitatory amino acid antagonists.[102,103] The mechanism of action by which riluzole exerts its therapeutic effects on the glutamatergic system has been reported to be through the inhibition of voltage-dependent sodium channels in mammalian CNS neurons[104,105] and the reduction of glutamate release.[106] However, tetrodotoxin, a sodium channel blocker, failed to block the inhibitory effect of riluzole on glutamate, indicating that the effects of the drug are not exclusively mediated by its action on sodium channels.[107] In contrast to memantine, riluzole does not act directly on the NMDA receptor.[108] The anticonvulsant activity of riluzole may also be due to its effect on neurotransmission mediated by the AMPA/KA receptors,[108–112] however the mechanism by which riluzole interacts with these receptors remains unclear. Recent studies have shown that riluzole may exert its neurotrophic effects by stimulating the synthesis of BDNF in cultured mouse astrocytes.[113] This finding is of interest in that previous studies demonstrated that antidepressant treatments increase the expression of BDNF in rat hippocampus and that infusion of BDNF into the dentate gyrus of hippocampus produces antidepressant effects in behavioral models of depression.[114]

In summary, riluzole is different from memantine in that it inhibits the release of glutamate instead of producing NMDA antagonism and is one of the few antiglutamatergic agents available for use in humans. It is available for clinical use in severely debilitated medically ill patients (ALS) and it has a very favorable side effect profile; these characteristics make it an ideal candidate to test in patients with mood disorders.[115]

SUMMARY

Regionally selective impairments of structural plasticity and cellular resiliency, which have been postulated to contribute to the development of classical neurodegenerative disorders, may also exist in mood disorders. It remains unclear whether these impairments correlate with the magnitude or duration of the biochemical perturbations extant in mood disorders, reflect an enhanced vulnerability to the deleterious effects of these perturbations (e.g., due to genetic factors and/or early life events), or indeed represent the fundamental etiological process in mood disorders. It is thus noteworthy that there is growing evidence from preclinical and clinical research that the glutamatergic system is involved in the pathophysiology and treatment of mood disorders. The development of new modulators of glutamatergic signaling as plasticity enhancers for the treatment of mood disorders may lead to improved therapeutics for these devastating disorders.

REFERENCES

1. MANJI, H.K., W.C. DREVETS & D.S. CHARNEY. 2001. The cellular neurobiology of depression. Nat. Med. **7:** 541–547.
2. NESTLER, E.J., M. BARROT, R.J. DILEONE, et al. 2002. Neurobiology of depression. Neuron **34:** 13–25.
3. DREVETS, W.C. 2001. Neuroimaging and neuropathological studies of depression: implications for the cognitive-emotional features of mood disorders. Curr. Opin. Neurobiol. **11:** 240–249.
4. MANJI, H.K. & R.H. LENOX. 2000. Signaling: cellular insights into the pathophysiology of bipolar disorder. Biol. Psychiatry **48:** 518–530.
5. PAYNE, J.L., J.A. QUIROZ, C.A. ZARATE & H.K. MANJI. 2002. Timing is everything: Does the robust upregulation of noradrenergically regulated plasticity genes underlie the rapid antidepressant effects of sleep deprivation? Biol. Psychiatry. In press.
6. MANJI, H.K., G.J. MOORE, G. RAJKOWSKA & G. CHEN. 2000. Neuroplasticity and cellular resilience in mood disorders. Millennium Article. Mol. Psychiatry. **5:** 578–593.
7. D'SA, C. & R. DUMAN. 2002. Antidepressants and neuroplasticity. Bipolar Disorder **4:** 183.
8. YOUNG, L.T. 2002. Neuroprotective effects of antidepressant and mood stabilizing drugs. J. Psychiatry Neurosci. **27:** 8–9.
9. MANJI, H. & R. DUMAN. 2001. Impairments of neuroplasticity and cellular resilience in severe mood disorder: implications for the development of novel therapeutics. Psychopharmacol. Bull. **35:** 5–49.
10. STRAKOWSKI, S.M., C.M. ADLER & M.P. DELBELLO. 2002. Volumetric MRI studies of mood disorders: do they distinguish unipolar and bipolar disorder? Bipolar Disord. **4:** 80–88.
11. BEYER, J.L. & K.R. KRISHNAN. 2002. Volumetric brain imaging findings in mood disorders. Bipolar Disord. **4:** 89–104.
12. LENOX, R.H., T.D. GOULD & H.K. MANJI. 2002. Endophenotypes in bipolar disorder. Am. J. Med. Genet. **114:** 391–406.
13. CARMELLI, D., T. REED & C. DECARLI. 2002. A bivariate genetic analysis of cerebral white matter hyperintensities and cognitive performance in elderly male twins. Neurobiol. Aging **23:** 413–420.
14. TAYLOR, W.D., M.E. PAYNE, K.R. KRISHNAN, et al. 2001. Evidence of white matter tract disruption in MRI hyperintensities. Biol. Psychiatry **50:** 179–183.
15. STOLL, A.L., P.F. RENSHAW, D.A. YURGELUN-TODD & B.M. COHEN. 2000. Neuroimaging in bipolar disorder: what have we learned? Biol. Psychiatry **48:** 505–517.
16. COTTER, D., D. MACKAY, S. LANDAU, et al. 2001. Reduced glial cell density and neuronal size in the anterior cingulate cortex in major depressive disorder. Arch. Gen. Psychiatry **58:** 545–553.
17. RAJKOWSKA, G. 2002. Cell pathology in bipolar disorder. Bipolar Disord. **4:** 105–116.
18. COYLE, J.T. & R. SCHWARCZ. 2000. Mind glue: implications of glial cell biology for psychiatry. Arch. Gen. Psychiatry **57:** 90–93.

19. HAYDON, P.G. 2001. GLIA: listening and talking to the synapse. Nat. Rev. Neurosci. **2:** 185–193.
20. ONGUR, D., W.C. DREVETS & J.L. PRICE. 1998. Glial reduction in the subgenual prefrontal cortex in mood disorders. Proc. Natl. Acad. Sci. USA **95:** 13290–13295.
21. RAJKOWSKA, G., J.J. MIGUEL-HIDALGO, J. WEI, et al. 1999. Morphometric evidence for neuronal and glial prefrontal cell pathology in major depression. Biol. Psychiatry **45:** 1085–1098.
22. RAJKOWSKA, G. 2000. Postmortem studies in mood disorders indicate altered numbers of neurons and glial cells. Biol. Psychiatry. **48:** 766–777.
23. ULLIAN, E.M., S.K. SAPPERSTEIN, K.S. CHRISTOPHERSON & B.A. BARRES. 2001. Control of synapse number by glia. Science **291:** 657–661.
24. LOPEZ, J.F., D.T. CHALMERS, K.Y. LITTLE, et al. 1998. Regulation of serotonin1A, glucocorticoid, and mineralocorticoid receptor in rat and human hippocampus: implications for the neurobiology of depression. Biol. Psychiatry. **43:** 547–573.
25. SAPOLSKY, R.M. 2000. Glucocorticoids and hippocampal atrophy in neuropsychiatric disorders. Arch. Gen. Psychiatry **57:** 925–935.
26. MCEWEN, B.S. 1999. Stress and hippocampal plasticity. Annu. Rev. Neurosci. **22:** 105–122.
27. SAPOLSKY, R.M. 1996. Stress, glucocorticoids, and damage to the nervous system: The current state of confusion. Stress **1:** 1–19.
28. SANCHEZ, M.M., L.J. YOUNG, P.M. PLOTSKY & T.R. INSEL. 2000. Distribution of corticosteroid receptors in the rhesus brain: relative absence of glucocorticoid receptors in the hippocampal formation. J. Neurosci. **20:** 4657–4668.
29. PATEL, P.D., J.F. LOPEZ, D.M. LYONS et al. 2000. Glucocorticoid and mineralocorticoid receptor mRNA expression in squirrel monkey brain. J. Psychiatr. Res. **34:** 383–392.
30. SAPOLSKY, R.M. 2000. The possibility of neurotoxicity in the hippocampus in major depression: a primer on neuron death. Biol. Psychiatry **48:** 755–765.
31. MAURI, M.C., A. FERRARA, L. BOSCATI, et al. 1998. Plasma and platelet amino acid concentrations in patients affected by major depression and under fluvoxamine treatment. Neuropsychobiology **37:** 124–129.
32. MATHIS, P., L. SCHMITT, M. BENATIA, et al. 1988. [Plasma amino acid disturbances and depression]. Encephale **14:** 77–82.
33. ALTAMURA, C.A., M.C. MAURI, A. FERRARA, et al. 1993. Plasma and platelet excitatory amino acids in psychiatric disorders. Am. J. Psychiatry **150:** 1731–1733.
34. ALTAMURA, C., M. MAES, J. DAI & H.Y. MELTZER. 1995. Plasma concentrations of excitatory amino acids, serine, glycine, taurine and histidine in major depression. Eur. Neuropsychopharmacol. **5** (Suppl.): 71–75.
35. MAES, M., R. VERKERK, E. VANDOOLAEGHE, et al. 1998. Serum levels of excitatory amino acids, serine, glycine, histidine, threonine, taurine, alanine and arginine in treatment-resistant depression: modulation by treatment with antidepressants and prediction of clinical responsivity. Acta Psychiatr. Scand. **97:** 302–308.
36. LEVINE, J., K. PANCHALINGAM, A. RAPOPORT, et al. 2000. Increased cerebrospinal fluid glutamine levels in depressed patients. Biol. Psychiatry **47:** 586–593.
37. BERK, M., H. PLEIN & D. FERREIRA. 2001. Platelet glutamate receptor supersensitivity in major depressive disorder. Clin. Neuropharmacol. **24:** 129–132.
38. NOWAK, G., G.A. ORDWAY & I.A. PAUL. 1995. Alterations in the N-methyl-D-aspartate (NMDA) receptor complex in the frontal cortex of suicide victims. Brain Res. **675:** 157–164.
39. COUSINS, J.P. & G. HARPER. 1996. Neurobiochemical changes from Taxol/Neupogen chemotherapy for metastatic breast carcinoma corresponds with suicidal depression. Cancer Lett. **110:** 163–167.
40. AUER, D.P., B. PUTZ, E. KRAFT, et al. 2000. Reduced glutamate in the anterior cingulate cortex in depression: an in vivo proton magnetic resonance spectroscopy study. Biol Psychiatry. **47:** 305–313.
41. PFLEIDERER, B., N. MICHAEL, A. ERFURTH, et al. 2003. Effective electroconvulsive therapy reverses glutamate/glutamine deficit in the left anterior cingulum of unipolar depressed patients. Psychiatry Res. **122:** 185–192.

42. CASTILLO, M., L. KWOCK, H. COURVOISIE & S.R. HOOPER. 2000. Proton MR spectroscopy in children with bipolar affective disorder: preliminary observations. Am. J. Neuroradiol. **21:** 832–838.
43. MICHAEL, M. A. ERFURTH, P. OHRMANN, et al. 2003. Acute mania is accompanied by elevated glutamate/glutamine levels within the left dorsolateral prefrontal cortex. Psychopharmacology (Berl). 2003.
44. BOURON, A. & J.Y. CHATTON. 1999. Acute application of the tricyclic antidepressant desipramine presynaptically stimulates the exocytosis of glutamate in the hippocampus. Neuroscience **90:** 729–736.
45. MICHAEL-TITUS, A.T., S. BAINS, J. JEETLE & R. WHELPTON. 2000. Imipramine and phenelzine decrease glutamate overflow in the prefrontal cortex—a possible mechanism of neuroprotection in major depression? Neuroscience **100:** 681–684.
46. NOWAK, G., I.A. PAUL, P. POPIK, et al. 1993. Ca^{2+} antagonists effect an antidepressant-like adaptation of the NMDA receptor complex. Eur. J. Pharmacol. **247:** 101–102.
47. PAUL, I.A., R. TRULLAS, P. SKOLNICK & G. NOWAK. 1992. Down-regulation of cortical beta-adrenoceptors by chronic treatment with functional NMDA antagonists. Psychopharmacology (Berl). **106:** 285–287.
48. NOWAK, G., Y. LI & I.A. PAUL. 1995. Adaptation of cortical but not hippocampal NMDA receptors after chronic citalopram treatment. Eur. J. Pharmacol. **295:** 75–85.
49. BOYER, P.A., P. SKOLNICK & L.H. FOSSOM. 1998. Chronic administration of imipramine and citalopram alters the expression of NMDA receptor subunit mRNAs in mouse brain. A quantitative *in situ* hybridization study. J. Mol. Neurosci. **10:** 219–233.
50. SKOLNICK, P. 1999. Antidepressants for the new millennium. Eur. J. Pharmacol. **375:** 31–40.
51. POPE, H.G., JR., S.L. MCELROY, P.E. KECK, JR. & J.I. HUDSON. 1991. Valproate in the treatment of acute mania. A placebo-controlled study. Arch. Gen. Psychiatry **48:** 62–68.
52. BOWDEN, C.L., A.M. BRUGGER, A.C. SWANN, et al. 1994. Efficacy of divalproex vs lithium and placebo in the treatment of mania. The Depakote Mania Study Group. J. Am. Med. Assoc. **271:** 918–924.
53. NONAKA, S. & D.M. CHUANG. 1998. Neuroprotective effects of chronic lithium on focal cerebral ischemia in rats. Neuroreport **9:** 2081–2084.
54. HASHIMOTO, R., C. HOUGH, T. NAKAZAWA, et al. 2002. Lithium protection against glutamate excitotoxicity in rat cerebral cortical neurons: involvement of NMDA receptor inhibition possibly by decreasing NR2B tyrosine phosphorylation. J. Neurochem. **80:** 589–597.
55. KARKANIAS, N.B. & R.L. PAPKE. 1999. Lithium modulates desensitization of the glutamate receptor subtype gluR3 in *Xenopus* oocytes. Neurosci. Lett. **277:** 153–156.
56. LOSCHER, W. 1993. Effects of the antiepileptic drug valproate on metabolism and function of inhibitory and excitatory amino acids in the brain. Neurochem. Res. **18:** 485–502.
57. COLLINS, R.M., JR., H.R. ZIELKE & R.C. WOODY. 1994. Valproate increases glutaminase and decreases glutamine synthetase activities in primary cultures of rat brain astrocytes. J. Neurochem. **62:** 1137–1143.
58. DIXON, J,F, & L.E. HOKIN. 1997. The antibipolar drug valproate mimics lithium in stimulating glutamate release and inositol 1,4,5-trisphosphate accumulation in brain cortex slices but not accumulation of inositol monophosphates and bisphosphates. Proc. Natl. Acad. Sci. USA **94:** 4757–4760.
59. NILSSON, M., E. HANSSON & L. RONNBACK. 1992. Agonist-evoked Ca^{2+} transients in primary astroglial cultures—modulatory effects of valproic acid. Glia **5:** 201–209.
60. UEDA, Y. & L.J. WILLMORE. 2000. Molecular regulation of glutamate and GABA transporter proteins by valproic acid in rat hippocampus during epileptogenesis. Exp. Brain Res. **133:** 334–339.
61. HASSEL, B., E.G. IVERSEN, L. GJERSTAD & E. TAUBOLL. 2001. Up-regulation of hippocampal glutamate transport during chronic treatment with sodium valproate. J. Neurochem. **77:** 1285–1292.
62. LOSCHER, W. 1999. Valproate: a reappraisal of its pharmacodynamic properties and mechanisms of action. Prog. Neurobiol. **58:** 31–59.

63. KUNIG, G., B. NIEDERMEYER, J. DECKERT, et al. 1998. Inhibition of [^{3}H]alpha-amino-3-hydroxy-5-methyl-4-isoxazole-propionic acid [AMPA] binding by the anticonvulsant valproate in clinically relevant concentrations: an autoradiographic investigation in human hippocampus. Epilepsy Res. **31:** 153–157.
64. BOLANOS, A.R., M. SARKISIAN, Y. YANG, et al. 1998. Comparison of valproate and phenobarbital treatment after status epilepticus in rats. Neurology **51:** 41–48.
65. TURSKI, L. 1990. [The N-methyl-D-aspartate receptor complex. Various sites of regulation and clinical consequences]. Arzneimittelforschung **40:** 511–514.
66. STEPPUHN, K.G. & L. TURSKI. 1993. Modulation of the seizure threshold for excitatory amino acids in mice by antiepileptic drugs and chemoconvulsants. J. Pharmacol. Exp. Ther. **265:** 1063–1070.
67. JING, D., N.A. GRAY, C.S. FALKE, et al. 2002. The mood stabilizer lithium regulates the synaptic and total protein expression of AMPA glutamate receptors in vitro and in vivo. Soc. Neurosci. Vol. Program No. **308.** Washington, D.C.
68. SAAL, D., Y. DONG, A. BONCI & R.C. MALENKA. 2003. Drugs of abuse and stress trigger a common synaptic adaptation in dopamine neurons. Neuron **37:** 577–582.
69. LI, X., J.P. TIZZANO, K. GRIFFEY, et al. 2001. Antidepressant-like actions of an AMPA receptor potentiator (LY392098). Neuropharmacology **40:** 1028–1033.
70. LAUTERBORN, J.C., G. LYNCH, P. VANDERKLISH, et al. 2000. Positive modulation of AMPA receptors increases neurotrophin expression by hippocampal and cortical neurons. J. Neurosci. **20:** 8–21.
71. YUAN, P.X., L.D. HUANG, Y.M. JIANG, et al. 2001. The mood stabilizer valproic acid activates mitogen-activated protein kinases and promotes neurite growth. J Biol Chem. **276:** 31674–31683.
72. ADAMS, J.M. & S. CORY, 1998. The Bcl-2 protein family: arbiters of cell survival. Science **281:** 1322–1326.
73. BONNI, A., A. BRUNET, A.E. WEST, et al. 1999. Cell survival promoted by the Ras-MAPK signaling pathway by transcription-dependent and -independent mechanisms. Science **286:** 1358–1362.
74. CHEN, D.F., G.E. SCHNEIDER, J.C. MARTINOU & S. TONEGAWA. 1997. Bcl-2 promotes regeneration of severed axons in mammalian CNS. Nature **385:** 434–439.
75. FINKBEINER, S. 2000. CREB couples neurotrophin signals to survival messages. Neuron **25:** 11–14.
76. PETTMANN, B. & C.E. HENDERSON. 1998. Neuronal cell death. Neuron **20:** 633–647.
77. RICCIO, A., S. AHN, C.M. DAVENPORT, et al. 1999. Mediation by a CREB family transcription factor of NGF-dependent survival of sympathetic neurons. Science **286:** 2358–2361.
78. THOENEN, H. 1995. Neurotrophins and neuronal plasticity. Science **270:** 593–598.
79. CHEN, G., H. EINAT, P. YUAN & H. MANJI. 2002. Evidence for the involvement of the MAP/ERK signaling pathway in mood modulation. Biol. Psychiatry **51:** 126S.
80. CHEN, G., W.Z. ZENG, P.X. YUAN, et al. 1999. The mood-stabilizing agents lithium and valproate robustly increase the levels of the neuroprotective protein bcl-2 in the CNS. J. Neurochem. **72:** 879–882.
81. MANJI, H., J. DU & T. GOULD. 2002. The role of neurotrophic signaling pathways in mood disorders. *In* Signal Transduction and Human Disease. J. Gutkind, Ed. In press.
82. CHEN, G., L.D. HUANG, W.Z. ZENG & H.K. MANJI. 2001. Mood stabilizers regulate cytoprotective and mRNA-binding proteins in the brain: long-term effects on cell survival and transcript stability. Int. J. Neuropsychopharmacol. **4:** 47–64.
83. WEI, H., Z.H. QIN, V.V. SENATOROV, et al. 2001. Lithium suppresses excitotoxicity-induced striatal lesions in a rat model of Huntington's disease. Neuroscience **106:** 603–612.
84. TSAI, G. & J.T. COYLE. 1995. N-acetylaspartate in neuropsychiatric disorders. Prog. Neurobiol. **46:** 531–540.
85. MOORE, G.J., J.M. BEBCHUK, K. HASANAT, et al. 2000. Lithium increases N-acetylaspartate in the human brain: in vivo evidence in support of bcl-2's neurotrophic effects? Biol .Psychiatry **48:** 1–8.

86. MOORE, G.J., J.M. BEBCHUK, I.B. WILDS, *et al.* 2000. Lithium-induced increase in human brain grey matter. Lancet **356:** 1241–1242.
87. MOORE, G., R. RAJARETHINAM, B. CORTESE, *et al.* 2001. Regionally specific increases in human brain gray matter with chronic lithium treatment. Soc. Neurosci. (Abstr.) **27:** (Program No.111.118).
88. AREOSA, S.A. & F. SHERRIFF. 2003. Memantine for dementia. Cochrane Database Syst Rev. VCD003154.
89. ORGOGOZO, J.M., A.S. RIGAUD, A. STOFFLER, *et al.* 2002. Efficacy and safety of memantine in patients with mild to moderate vascular dementia: a randomized, placebo-controlled trial (MMM 300). Stroke **33:** 1834–1839.
90. REISBERG, B., R. DOODY, A. STOFFER, *et al.* 2003. Memantine in moderate-to-severe Alzheimer's disease. N. Engl. J. Med. **348:** 1333–1341.
91. BORMANN, J. 1989. Memantine is a potent blocker of N-methyl-D-aspartate (NMDA) receptor channels. Eur. J. Pharmacol. **166:** 591–592.
92. BERGER, W., J. DECKERT, J. HARTMANN, *et al.* 1994. Memantine inhibits [^{3}H]MK-801 binding to human hippocampal NMDA receptors. Neuroreport **5:** 1237–1240.
93. KORNHUBER, J., K. SCHOPPMEYER & P. RIEDERER. 1993. Affinity of 1-aminoadamantanes for the sigma binding site in post-mortem human frontal cortex. Neurosci. Lett. **163:** 129–131.
94. CHEN, H.S., Y.F. WANG, P.V. RAYUDU, *et al.* 1998. Neuroprotective concentrations of the N-methyl-D-aspartate open-channel blocker memantine are effective without cytoplasmic vacuolation following post-ischemic administration and do not block maze learning or long-term potentiation. Neuroscience **86:** 1121–1132.
95. CHEN, H.S. & S.A. LIPTON. 1997. Mechanism of memantine block of NMDA-activated channels in rat retinal ganglion cells: uncompetitive antagonism. J. Physiol. **499:** 27–46.
96. MORYL, E., W. DANYSZ & G. QUACK. 1993. Potential antidepressive properties of amantadine, memantine and bifemelane. Pharmacol. Toxicol. **72:** 394–397.
97. ROGOZ, Z., G. SKUZA, J. MAJ & W. DANYSZ. 2002. Synergistic effect of uncompetitive NMDA receptor antagonists and antidepressant drugs in the forced swimming test in rats. Neuropharmacology **42:** 1024–1030.
98. AMETAMEY, S.M., S. SAMNICK, K.L. LEENDERS, *et al.* 1999. Fluorine-18 radiolabelling, biodistribution studies and preliminary PET evaluation of a new memantine derivative for imaging the NMDA receptor. J. Recept. Signal Transduct. Res. **19:** 129–141.
99. KORNHUBER, J., J. BORMANN, W. RETZ, *et al.* 1989. Memantine displaces [^{3}H]MK-801 at therapeutic concentrations in postmortem human frontal cortex. Eur. J. Pharmacol. **166:** 589–590.
100. KORNHUBER, J., M. WELLER, K. SCHOPPMEYER & P. RIEDERER. 1994. Amantadine and memantine are NMDA receptor antagonists with neuroprotective properties. J. Neural Transm. Suppl. **43:** 91–104.
101. BENAVIDES, J., J.C. CAMELIN, N. MITRANI, *et al.* 1985. 2-Amino-6-trifluoromethoxy benzothiazole, a possible antagonist of excitatory amino acid neurotransmission—II. Biochemical properties. Neuropharmacology **24:** 1085–1092.
102. KRETSCHMER, B.D., U. KRATZER & W.J. SCHMIDT. 1998. Riluzole, a glutamate release inhibitor, and motor behavior. Naunyn Schmiedebergs Arch. Pharmacol. **358:** 181–190.
103. DOBLE, A. 1996. The pharmacology and mechanism of action of riluzole. Neurology **47:** S233–241.
104. URBANI, A. & O. BELLUZZI. 2000. Riluzole inhibits the persistent sodium current in mammalian CNS neurons. Eur. J. Neurosci. **12:** 3567–3574.
105. BENOIT, E. & D. ESCANDE D. 1991. Riluzole specifically blocks inactivated Na channels in myelinated nerve fibre. Pflug. Arch. **419:** 603–609.
106. CHERAMY, A., L. BARBEITO, G. GODEHEU & J. GLOWINSKI. 1992. Riluzole inhibits the release of glutamate in the caudate nucleus of the cat in vivo. Neurosci. Lett. **147:** 209–212.
107. MARTIN, D., M.A. THOMPSON & J.V. NADLER. 1993. The neuroprotective agent riluzole inhibits release of glutamate and aspartate from slices of hippocampal area CA1. Eur. J. Pharmacol. **250:** 473–476.

108. DEBONO, M.W., J. LE GUERN, T. CANTON, et al. 1993. Inhibition by riluzole of electrophysiological responses mediated by rat kainate and NMDA receptors expressed in *Xenopus* oocytes. Eur. J. Pharmacol. **235:** 283–289.
109. SINISCALCHI, A., A. BONCI, N.B. MERCURI & G. BERNARDI. 1997. Effects of riluzole on rat cortical neurones: an in vitro electrophysiological study. Br. J. Pharmacol. **120:** 225–230.
110. DE SARRO, G., A. SINISCALCHI, G. FERRERI, et al. 2000. NMDA and AMPA/kainate receptors are involved in the anticonvulsant activity of riluzole in DBA/2 mice. Eur. J. Pharmacol. **408:** 25–34.
111. HUBERT, J.P., M.C. BURGEVIN, F. TERRO, et al. 1998. Effects of depolarizing stimuli on calcium homeostasis in cultured rat motoneurones. Br. J. Pharmacol. **125:** 1421–1428.
112. Keita, H., C. Lepouse, D. Henzel, et al. 1997. Riluzole blocks dopamine release evoked by N-methyl-D-aspartate, kainate, and veratridine in the rat striatum. Anesthesiology **87:** 1164–1171.
113. MIZUTA, I., M. OHTA, K. OHTA, et al. 2001. Riluzole stimulates nerve growth factor, brain-derived neurotrophic factor and glial cell line-derived neurotrophic factor synthesis in cultured mouse astrocytes. Neurosci. Lett. **310:** 117–120.
114. SHIRAYAMA, Y., A.C. CHEN, S. NAKAGAWA, et al. 2002. Brain-derived neurotrophic factor produces antidepressant effects in behavioral models of depression. J. Neurosci. **22:** 3251–3261.
115. ZARATE, C.A., JR., J. QUIROZ, J. PAYNE & H.K. MANJI. 2002. Modulators of the glutamatergic system: implications for the development of improved therapeutics in mood disorders. Psychopharmacol. Bull. **36:** 35–83.
116. BERMAN, R.M., A. CAPPIELLO, A. ANAND, et al. 2000. Antidepressant effects of ketamine in depressed patients. Biol. Psychiatry **47:** 351–354.
117. CALABRESE, J.R., C.L. BOWDEN, G.S. SACHS, et al. 1999. A double-blind placebo-controlled study of lamotrigine monotherapy in outpatients with bipolar I depression. Lamictal 602 Study Group. J. Clin. Psychiatry **60:** 79–88.
118. CRANE, G. 1959. Cycloserine as an antidepressant agent. Am. J. Psychiatry **115:** 1025–1026.
119. HOLEMANS, S., F. DE PAERMENTIER, R.W. HORTON, et al. 1993. NMDA glutamatergic receptors, labelled with [^{3}H]MK-801, in brain samples from drug-free depressed suicides. Brain Res. **616:** 138–143.
120. MEADOR-WOODRUFF, J.H., A.J. HOGG JR. & R.E. SMITH. 2001. Striatal ionotropic glutamate receptor expression in schizophrenia, bipolar disorder, and major depressive disorder. Brain Res. Bull. **55:** 631–640.
121. PAPP, M. & E. MORYL. 1996. Antidepressant-like effects of 1-aminocyclopropanecarboxylic acid and D-cycloserine in an animal model of depression. Eur. J. Pharmacol. **316:** 145–151.
122. PARKES, J.D., D.M. CALVER, K.J. ZILKHA & R.P. KNILL-JONES. 1970. Controlled trial of amantadine hydrochloride in Parkinson's disease. Lancet **1:** 259–262.
123. REYNOLDS, I.J. & R.J. MILLER. 1988. Tricyclic antidepressants block N-methyl-D-aspartate receptors: similarities to the action of zinc. Br. J. Pharmacol. **95:** 95–102.
124. STRYJER, R., R.D. STROUS, G. SHAKED, et al. 2003. Amantadine as augmentation therapy in the management of treatment-resistant depression. Int. Clin. Psychopharmacol. **18:** 93–96.
125. TRULLAS, R. & P. SKOLNICK. 1990. Functional antagonists at the NMDA receptor complex exhibit antidepressant actions. Eur. J. Pharmacol. **185:** 1–10.
126. VALE, S., M.A. ESPEJEL & J.C. DOMINGUEZ. 1971. Amantadine in depression. Lancet **2:** 437.
127. ZARATE, C.A.J., J.L. PAYNE, J. QUIROZ, et al. An open-label trial of riluzole in treatment-resistant major depression. Am. J. Psychiatry. In press.

Clinical Studies Implementing Glutamate Neurotransmission in Mood Disorders

GERARD SANACORA, DOUGLAS L. ROTHMAN,[b] GRAEME MASON, AND JOHN H. KRYSTAL

Department of Psychiatry and [b]Department of Radiology, Yale University, School of Medicine, New Haven, Connecticut 06519, USA

ABSTRACT: Emerging evidence suggests that the amino acid neurotransmitter systems are associated with the pathophysiology and treatment of mood disorders. Recent advances in the areas of molecular neurobiology, pharmacology, and magnetic resonance spectroscopy (MRS) now provide better tools to probe the function of the amino acid neurotransmitter systems and are affording us the opportunity to better investigate the relationship of these systems to mood disorders. Here we review the available literature in the field and suggest a possible pathophysiological model that may account for the many of the findings.

KEYWORDS: mood disorders; glutamate neurotransmission

INTRODUCTION

Over the past four decades mood disorders research has been dominated by the monoamine hypothesis of depression; however, there is now increasing recognition of the role played by the amino acid neurotransmitter systems in the pathophysiology and treatment of these disorders. Not surprisingly, glutamate, being the principal excitatory neurotransmitter in the mammalian brain, contributes to a wide array of both normal and abnormal brain functions. Mounting preclinical evidence suggests that both the glutamatergic and GABAergic neurotransmitter systems may also contribute to the pathophysiology of mood disorders and the mechanisms of antidepressant and mood stabilizer action[1,2] (see also Paul and Skolnick and Zarate and colleagues articles in this volume). However, relatively little attention has been given to studies probing the role of the amino acid neurotransmitter systems in relation to mood disorders in humans. We will review the evidence supporting a role for the amino acid neurotransmitters in the pathophysiology and treatment of mood disorders, focusing on human studies, and attempt to synthesize a model that may help us understand the glutamatergic mechanisms related to the neurobiology of mood disorders.

Address for correspondence: Gerard Sanacora, M.D., Ph.D., CNRU @CMHC, 34 Park Street, New Haven CT 06519. Voice: 203-974-7535; fax: 203-974-7662.
gerard.sanacora@yale.edu

HUMAN STUDIES OF GLUTAMATE IN POSTMORTEM AND PERIPHERAL MEASURES

Glutamate is a ubiquitous compound with multiple biological roles. While it is best known for its involvement in intermediary metabolism and neurotransmission within the central nervous system, it is present in high concentrations in many tissues of the body including blood and cerebral spinal fluid (CSF). Studies attempting to identify abnormalities in plasma glutamate levels from individuals with mood disorders have not offered conclusive results. Kim[3] reported elevated glutamate levels in a group of depressed patients, but could not separate the effect from the use of antidepressant medications. Altamura and colleagues[4] also initially reported elevated plasma glutamate levels in a group of mood disorder patients but was unable to replicate the findings when comparing unmedicated patients with major depression to healthy controls.[5] More recently, Mauri[6] found elevated plasma and platelet levels of glutamate in depressed patients compared to controls. However, in contrast, Maes and colleagues[7] found no difference in serum glutamate levels from depressed patients compared to age- and sex-matched controls, but they did find significantly reduced serum glutamate levels following a five-week period of treatment with antidepressants.

There have been no direct studies of glutamate in the CSF, but two studies have explored the relationship between CSF glutamine levels and mood disorders. Significantly higher CSF glutamine concentrations were found in 18 hospitalized patients with acute unmedicated severe depression compared to 22 control subjects using *in vitro* magnetic resonance spectroscopy (MRS).[8] In a second study Hiraoka and colleagues[9] reported that mean CSF glutamine levels in functional psychosis (schizophrenia, manic-depressive illness, and other psychoses) did not differ from those of neurotic patients, patients with cerebrovascular disorders, or patients with metabolic neuropathy.

No significant difference in frontal cortex glutamate concentrations was found in a single study examining neurosurgical samples from chronically depressed subjects.[10] In another postmortem study no difference in NMDA receptor binding was found among depressed suicide victims and an age-matched comparison group.[11] However, the proportion of high affinity, glycine-displaceable [^{3}H]CGP-39653 binding to glutamate receptors was found to be reduced in age- and postmortem interval-matched suicide victims.[12] This apparent change in NMDA receptor binding is likely related to changes in subunit composition and is consistent with observations in animals models of chronic stress.[13] Interestingly, changes in NMDA receptor binding have also been observed following chronic administration of 17 different antidepressants and electroconvulsive shock in a mouse model.[14]

HUMAN MRS STUDIES OF GLUTAMATE

^{1}H-MRS can be used to make *in vivo* measures of excitatory amino acids in the brain. The limited ability to assign unequivocal resonance peaks to these metabolites at lower field strengths has led to the use of a combined measure termed Glx that contains several compounds including glutamate, glutamine, homocarnosine, and GABA. Using this methodology, Cousins and colleagues demonstrated temporary decreases in Glx levels coinciding with a patient's transient experience of suicidal depression following taxol and neupogen chemotherapy.[15] Auer and colleagues[16] reported decreased Glx measures in the anterior cingulate cortex of severely depressed subjects. In addition, a recently published study by Michael also found reduced measures of Glx in the left amygdala that increased following ECT.[17] A preliminary study by the same group also showed a decrease in baseline anterior cingulate Glx levels that later increased following treatment with ECT.[18] Interestingly, no decrease was seen in a group of bipolar depressed subjects participating in this study. Consistent with the idea that Glx measures may differ between unipolar and bipolar depressed patients, elevated levels of Glx were found in both the frontal lobe and basal ganglia of depressed bipolar children compared to a control group.[19] In a recent study using a macromolecule nulling sequence to better identify the components of the Glx peak, we found significantly elevated glutamate levels in the occipital cortex (the only region examined) of a subgroup of unipolar depressed compared to control subjects (manuscript in preparation).

In summary, while not always showing consistent differences, the assemblage of studies in humans do support the idea that glutamatergic function (glutamate concentrations and receptor function) is altered in depressed individuals. The fact that the system appears sensitive to pharmacological and environmental manipulations suggests that it is the regulation of the system that is altered during depressive episodes. These findings are consistent with preclinical studies demonstrating significant effects of stress on the function of the glutamate system.[20–25]

ROLE OF GLUTAMATERGIC DRUGS IN THE TREATMENT OF DEPRESSION

Accompanying the reports of the altered glutamatergic function in depressed individuals are studies suggesting that several classes of glutamatergic drugs possess antidepressant activity. NMDA receptor antagonists exhibit antidepressant-like effects in several animal models that are used to screen for antidepressant activity.[2,26] In humans, Crane unknowingly was the first to report an antidepressant action for an agent with NMDA receptor activity when he suggested that high-dose D-cycloserine (DCS), an antibiotic developed to treat tuberculosis, possessed antidepressant efficacy.[27,28] He reported beneficial effects with respect to depressed mood, insomnia, and reduced appetite in depressed tuberculosis patients treated with DCS in doses of 500 mg/day. The rapid onset of antidepressant activity was a striking feature of DCS effects in these patients. At doses similar to those used in these studies, DCS has mild acute dose-related euphoric and amnestic effects in humans that resemble those seen with low doses of noncompetitive NMDA antagonists.[29] In preclinical studies DCS acts like a partial agonist, producing roughly half of the maximal facilitation of NMDA receptor function associated with glycine at maximally stimulating doses.[30,31] Therefore, under conditions where the occupancy of glycine-B sites by agonists exceeds 50%, glycine-B antagonist-like effects emerge.[32]

A series of separate studies suggest that amantadine, a low-affinity noncompetitive antagonist of the NMDA receptor developed initially as an antiviral agent, also possesses antidepressant activity in depressed patients with Parkinson's disease.[33,34] Other case reports and preliminary studies suggest it may also have clinical efficacy

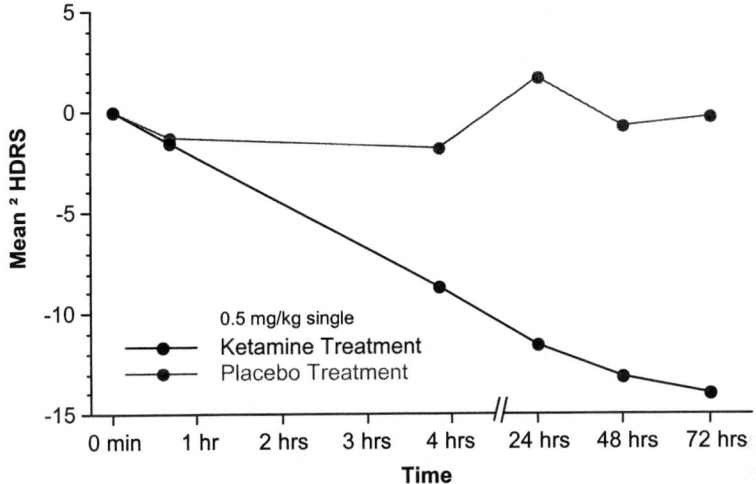

FIGURE 1. Adapted from Berman and colleagues.[45]

in treating depressive symptoms in patients without Parkinson's disease providing further evidence of the antidepressant activity of NMDA receptor antagonists.[35–39] Additional studies using memantine, a related noncompetitive voltage-dependent NMDA receptor antagonist, show that it is also associated with behavioral improvement across a number of dimensions in patients with neuropsychiatric disorders, including mood and motor activity.[40,41]

Berman and colleagues,[45] demonstrated that a single intravenous dose of ketamine 0.5 mg/kg (an uncompetitive NMDA receptor antagonist) had rapid and prolonged antidepressant effects in depressed patients (FIG. 1). In these subjects, ketamine infusion produced mild psychosis and euphoria that dissipated within 2 h, while the antidepressant effects of ketamine infusion emerged over the first 3 h and persisted for more than 72 hours. A similar prolonged antidepressant effect has also been shown following the use of ketamine anesthesia in depressed patients undergoing orthopedic surgery.[46] A recent animal study using ketamine in the behavior despair animal model of depression yielded results consistent with these findings of prolonged antidepressant action by demonstrating an enduring effect on immobility measures following a single dose of ketamine.[47]

A collection of clinical studies suggests that other (non-NMDA antagonist) classes of glutamate-inhibiting drugs may also produce antidepressant effects. Lamotrigine, an agent believed to inhibit the excessive release of glutamate via inhibition of use-dependent sodium channels, P-type and N-type calcium channels, and effects on potassium channels, is effective in the treatment bipolar depression.[48–54] Limited preliminary studies suggest that it may also have potential benefit in the treatment of unipolar depression, however this is much less clear.[55–57] In addition, a recent case report[58] and an open-label study[59] also suggest riluzole, an agent with antiglutamatergic properties,[60] is associated with antidepressant activity.

TABLE 1.

Medication	Putative Action on Glutamatergic System	Effect on Depression	Studies for or against; References
D-Cycloserine	NMDA partial agonist	Antidepressant at concentrations >250 mg/day	27, 28
Amantadine	Low-affinity noncompetitive antagonist of the NMDA receptor	Antidepressant activity when given to patients with Parkinson's Early reports of clinical efficacy in treating MDD	33–38, 39
Memantine	Low-affinity noncompetitive voltage-dependent NMDA receptor antagonist	Antidepressant activity when given to subjects with dementia Early reports of efficacy in MDD	40, 41
Ketamine	Noncompetitive NMDA receptor antagonist that binds within the open receptor channel	Extended antidepressant action following single infusions	42, 43
Riluzole	Reduces presynaptic conduction in glutamatergic nerve fibers	Case report and small open-label trial suggest antidepressant activity	58, 59
Lamotrigine	Reduces glutamate release via inhibition of use-dependent sodium channels, P-type and N-type calcium channels, and potassium channels	Effective in the treatment of depressive episodes associated with bipolar disorder and possibly unipolar depression	49–53, 55–57

Appearing somewhat incongruous with the hypothesis that general inhibition glutamatergic transmission confers antidepressant action, emerging evidence suggests that positive allosteric modulators of AMPA receptors are effective in animal models of behavioral despair that possess considerable predictive validity for antidepressant activity.[26,61,62] Similar agents have been shown to enhance some aspects of memory in humans,[63,64] however no studies have yet been done to evaluate this class of drugs for antidepressant action in humans.

In summary, there is increasing data suggesting drugs that affect glutamatergic transmission possess antidepressant properties (TABLE 1). However the mechanism by which these agents produce their effect remains uncertain. The potential for the NMDA receptor antagonists and glutamate release inhibitors to protect against glutamate-related excitotoxicity and provide neuroprotection by decreasing calcium transmission and increasing the expression of neurotrophic factors has been proposed as a mechanism of action.[26,65] While this mechanism could contribute to the long-term benefits of these medications, it is unlikely to explain the studies reporting acute mood-elevating effects of ketamine administration in healthy controls and recently detoxified alcoholics, or the rapid antidepressant response to a single infusion of ketamine.[29,66–71] These acute and subacute effects of NMDA receptor antagonism appear better accounted for by events that rapidly alter the system, in addition to generating longer lasting physiological adaptations. Evidence indicates that some of the more acute neuropsychiatric effects of NMDA antagonists are mediated via increased glutamate release.[72] NMDA receptor antagonism by ketamine and phencyclidine has been previously shown to increase glutamatergic transmission through non-NMDA receptor mechanisms (i.e., AMPA and kinate receptors).[72,73] Moreover, the co-administration of metabotropic glutamate type II receptor agonist (+)-2-aminobicyclo-(3.1.0)-hexam-2,6-dicarboxylate monohydrate (LY354740) has been shown to attenuate many of the motor and cognitive effects associated with NMDA antagonism in rats[73] and to attenuate the disruptive effects of ketamine upon working memory in humans.[74] This acute increase in non-NMDA receptor fast glutamate transmission associated with NMDA receptor antagonist may contribute to the acute antidepressant effects and may help to explain the seemingly paradoxical antidepressant effects of the AMPA receptor potentiators in animal models of depression. The fact that AMPA receptor potentiators induce BDNF expression[75,76] suggests that they may also have effects on long-term plasticity. In actuality, both non-NMDA mediated responses and effects mediated through NMDA antagonists, such as stimulation of cell proliferation and neurogenesis,[77,78] may be working in conjunction to produce the rapid and sustained antidepressant effect.

EVIDENCE OF GABAergic ABNORMALITIES IN DEPRESSION

Accumulating evidence also suggests GABA, the other major amino acid neurotransmitter in the mammalian brain, may also contribute to the pathophysiology associated with mood disorders (see Lloyd,[79] Petty,[80] Shiah,[81] and Sanacora[82] for complete reviews). Rodent models suggest adaptive changes in GABAergic function contribute to the stress response. Decreased GABA concentrations and synthesis, and decreased $GABA_A$ receptor binding have been reported in response to both acute and chronic stress.[83–91] Furthermore, stress can have long lasting effects on GABAergic function that appear related to altered expression of adult behaviors.[92] GABA enhancing agents have been shown to have antidepressant-like actions in several rodent models used as tests of antidepressant activity.[79,87,93,94] In humans two GABA-mimetic compounds, progabide and fengabine (SL 79229), also appear to possess antidepressant properties in preliminary studies (see Bartholini and colleagues for compilation of studies).[95,96]

Similar to what is seen with the glutamatergic system, there is also evidence that standard antidepressant agents have significant effects on GABA function. Elevated GABA levels, enhanced GABA release in rat brains, and enhanced $GABA_B$ receptor binding following chronic administration of all three classes of classic antidepressant agents (TCA, SSRI, MAOI) have been reported.[79,97–101] However, other studies give inconsistent findings, adding some confusion to the picture.[102–107]

In contrast to the inconclusive studies of plasma and CSF glutamate, multiple studies demonstrate consistently decreased GABA measures in the CSF and plasma of depressed individuals. Eight studies have compared CSF GABA concentrations from depressed subjects to those of a control group. Four found significantly lower GABA concentrations in depressed subjects compared to various control groups, and the other four found reduced GABA at trend levels. A meta-analysis of this data shows the finding of GABA reductions in depression to be highly significant.[108] Multiple reports of decreased plasma GABA levels in individuals with mood disorders are also in the literature.[109–113] The plasma findings show a marked shift in the frequency distribution of plasma GABA levels of depressed subjects with 40% of depressed subjects having levels below 100 pmol/ml, while only 6% of healthy subjects had plasma GABA levels below 100 pmol/ml. Others have attempted to measure GABA concentrations and glutamic acid decarboxylase (GAD) activity in postmortem studies. Unfortunately, rapid changes in glutamic acid decarboxylase (GAD) activity following death greatly limit the ability to interpret the results from these studies[114–121] and lead to conflicting results. This apparent inconsistency in the postmortem studies serves to highlight the need to conduct *in vivo* measures of the GABAergic system.

MRS FINDINGS OF GABAergic INVOLVEMENT IN DEPRESSION

We have been able to use ^{1}H-MRS to explore the relationship between *in vivo* cortical GABA concentrations and depression. In our initial study, occipital cortex GABA levels were measured in 14 medication-free, severely depressed patients and eighteen healthy comparison subjects (FIG. 2). The depressed patients had markedly lower occipital cortex GABA concentrations (mean±SD, 0.71±0.27 mmoles/kg)

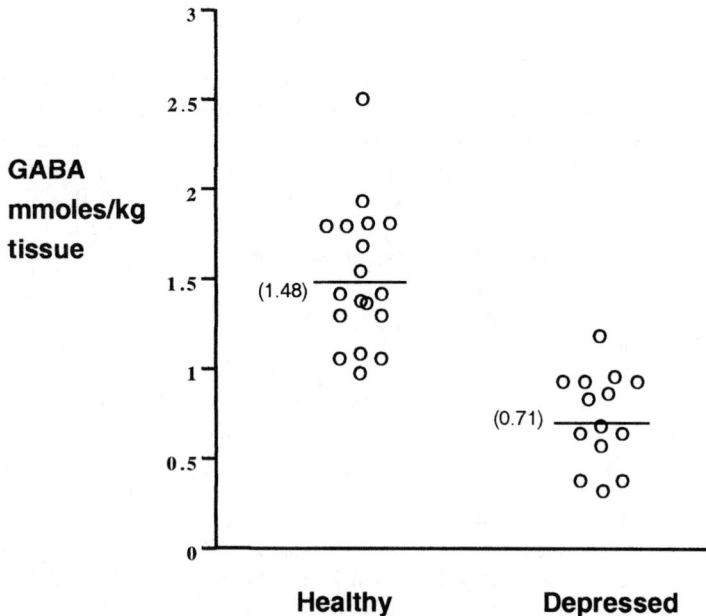

FIGURE 2. Occipital cortex GABA concentration in the brains of depressed and healthy control subjects as measured by ^{1}H-MRS.

than healthy comparison subjects ($N=18$, 1.48 ± 0.39 mmoles/kg).[122] The difference remained highly significant in an ANCOVA covarying for age and gender ($F[1,28] = 83.0$; $P = .0001$).

We have recently completed a replication study with a new sample of depressed subjects and healthy controls showing findings consistent with those of the initial study (data in preparation for publication). This larger study included primarily mild to moderately depressed outpatient subjects and healthy control subjects. We again found the depressed subjects to have significantly lower occipital cortex GABA concentrations than the healthy controls employing an ANCOVA controlling for age and gender. The mean difference in cortical GABA concentrations between the depressed and healthy subjects was not as great in this sample, and there was an increase in the variability observed with depressed GABA concentrations compared to the initial study. The striking feature, however, was the extremely rare healthy subject with GABA concentrations below 1 mmole/kg tissue, in contrast to ~50% of the depressed subjects who have GABA concentrations below 1 mmole/kg tissue. Preliminary data suggest that the differences may be due to the greater diagnostic subtype heterogeneity in this sample of depressed subjects.[123]

We have also conducted two studies examining the effects of electroconvulsive therapy (ECT) and selective serotonin reuptake inhibitor (SSRI) medications on cortical GABA concentrations to determine if the treatment of major depression is associated with changes in occipital cortex GABA concentrations. Post-ECT GABA concentrations were significantly elevated compared to pre-ECT concentrations.[124] Mean GABA concentrations rose from 0.85 (SD, 0.34) to 1.51 (SD, 0.48) mmoles/ kg brain tissue following treatment. Seven of the eight subjects demonstrated increased cortical GABA concentrations at the time of the post-ECT study and four showed increases of greater than 85% over the pre-ECT levels. Similar increases were seen following treatment with SSRI medications. Post-SSRI treatment occipital cortex GABA concentrations (mean=1.70, SD=.37) were significantly increased compared to pretreatment concentrations (mean=1.27, SD=.30) ($P<.03$, $N=11$ paired t test), with nine of the eleven subjects demonstrating elevated posttreatment concentrations.[125]

In sum, The results of our ^{1}H-MRS studies suggest that both the GABAergic and glutamatergic neurotransmitter systems are altered in the brains of a subgroup of depressed individuals, and at least the GABAergic system appears to normalize following treatment of the depressive episodes with either ECT or SSRI antidepressants. The reduced GABA concentrations in the brains of depressed subjects are consistent with both the preclinical studies and earlier clinical findings. However, we are left to speculate on what mechanisms could account for the apparent widespread effects on both amino acid neurotransmitter systems. Preliminary work by Mason and colleagues using ^{13}C-MRS suggests that the rate of GABA synthesis is decreased in depressed subjects.[126] The reduced GABA synthesis does not appear to be the result of genetic differences in GABA's major synthetic enzyme GAD-67, since we recently failed to demonstrate any significant findings in a preliminary mutation screen of the GAD-67 gene and a haplotype association to unipolar depression.[127] Although, it remains possible that allelic variation in GAD-65 gene may be related to the baseline differences; the fact that GABA concentrations appear to normalize following antidepressant treatments suggests it is more likely that the regulation of the amino acid neurotransmitter system is aberrant during depressive episodes.

A TRIPARTITE SYNAPSE MODEL FOR THE DISRUPTION OF GLUTAMATE AND GABA NEURONAL FUNCTION IN DEPRESSION

In review of the available literature it appears that the regulation of both the GABAergic and glutamatergic neurotransmitter systems are simultaneously altered during depressive episodes, and that both systems contribute to the mechanism of antidepressant action. Presence of both GABAergic and glutamatergic abnormalities in depression is consonant with the close physiological coupling of the systems. Recent work has highlighted the importance of glial cells in the functioning and regulation of the amino acid neurotransmitter systems. In addition to supplying the primary source of energy to neurons,[128] glial cells also provide the major pathway for neuronal glutamate and GABA synthesis.[129,130] The potential role of glial cells in the regulation of amino acid neurotransmitter function is of added interest after considering the recent post mortem studies demonstrating decreased glial cell number and density in depressed individuals. This finding has been replicated in several studies over varied brain regions.[131–139] Additional reports of abnormally elevated levels of S-100B, a calcium binding peptide produced by astroglial cells of the central nervous system in melancholic depressed subjects,[140–142] further supports a potential relationship between glial pathology, altered amino acid neurotransmitter function, and major depression.

Reduced glial function could account for several of the differences observed between depressed and healthy controls and possibly some of the antidepressant effects of agents targeting the glutamatergic system (FIG. 3). For example, impaired astrocyte function would be expected to decrease synaptic glutamate uptake, resulting in increased spillover and elevated levels of extrasynaptic glutamate. NMDA antagonists, such as ketamine, DCS, and amantadine, may produce their antidepressant effects in part by decreasing the amount of NMDA activation that results from the increased extrasynaptic concentrations of glutamate. However it is likely that the elevated extrasynaptic glutamate resulting from impaired uptake would ultimately reduce glutamate release by the presynaptic glutamatergic neuron due to feedback. If this occurs, the NMDA receptor antagonists may in fact be acting somewhat paradoxically in the depressed individuals by increasing transmission through AMPA and kinate receptors, as previously reported by Moghaddam.[72,73] This model of elevated extrasynaptic glutamate concentrations but decreased synaptic glutamate release may help reconcile the seemingly contradictory preclinical reports suggesting that increased glutamate transmission via AMPA potentiator agents is associated anti-depressant activity.[61,62]

Decreased glutamate release and impaired glial function would also be expected to result in decreased flux through the glutamate/glutamine cycle. The decreased glutamine synthesis could lead to reduced glutamine supplies to GABAergic neurons,[130] which has been shown to be the main precursor for GABA synthesis. Thus, the reduced glutamine availability could account for the consistent finding of reduced GABA concentrations in the brains and CSF of depressed subjects, as well as our preliminary finding of reduced GABA synthesis. However, predicting what effect this would have on static glutamate and glutamine total tissue levels is more difficult since release may also be significantly reduced in this situation, resulting in increased intracellular stores.

This hypothesis can be directly tested employing ^{13}C-MRS to specifically investigate whether glu/gln cycling is impaired in depressed subjects. The prediction that impaired cycling is secondary to a disruption of astrocyte function can also be tested using acetate-labeled ^{13}C-MRS studies.[129] Using ^{13}C-labeled acetate allows for the direct interrogation of glial cell metabolism and should demonstrate reduced rates in depressed subjects if glial function is significantly impaired.

The proposed mechanism by which NMDA receptor antagonists act as antidepressant agents via lifting inhibition of glutamate release could also be directly tested. Use of these agents in combination with measures of glu/gln cycling, specific AMPA/kinate, and/or metabotropic glutamate receptor modulators could be used to further explore the total effect of the NMDA antagonists on overall glutamate function. The excitatory amino acid transporters would also serve as another site of po-

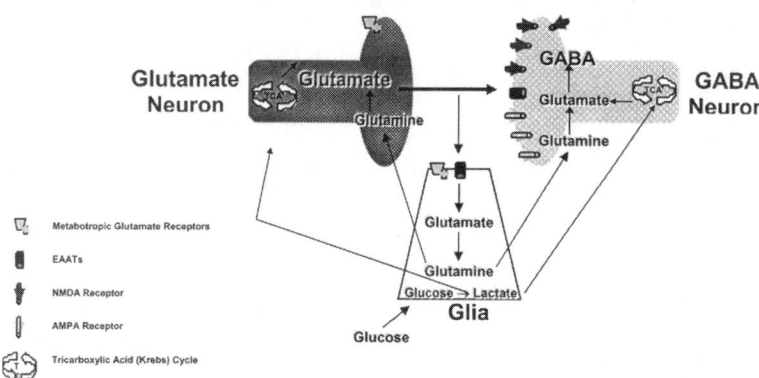

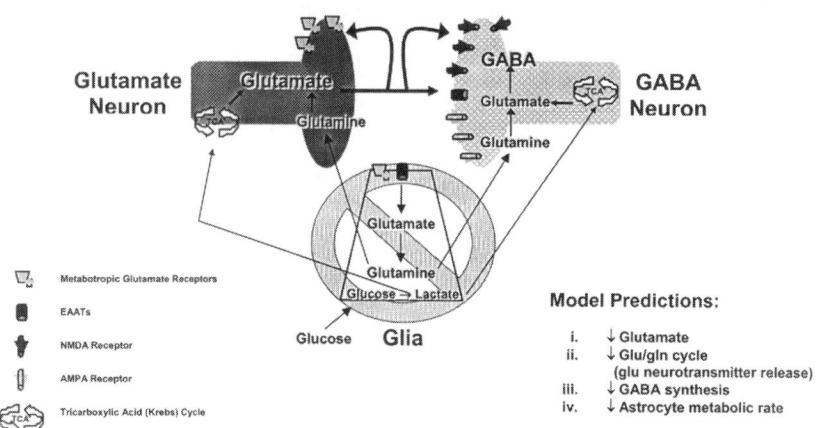

FIGURE 3. (*Top*) The role of the tripartite synapse in regulation of glutamate and GABA neurotransmission and (*bottom*) the potential effects of glial disruption on glutamate neurotransmission and GABA metabolism.

tential interest. Recent reports of decreased excitatory amino acid transporter expression in a small postmortem sample of depressed subjects bring additional interest to this line of pursuit.[143]

In conclusion, mounting evidence suggests that the amino acid neurotransmitter systems contribute to both the pathophysiology and treatment of major depressive disorder. More recent advances in our understanding of the physiology associated with these systems are now providing potential new targets for investigation and new drug development.

REFERENCES

1. KRYSTAL, J.H. *et al.* 2002. Glutamate and GABA systems as targets for novel antidepressant and mood-stabilizing treatments. Mol. Psychiatry **7:** S71–80.
2. SKOLNICK, P. 1999. Antidepressants for the new millennium. Eur. J. Pharmacol. **375:** 31–40.
3. KIM, J.S. *et al.* 1982. Increased serum glutamate in depressed patients. Arch. Psychiatr. Nervenkr. **232:** 299–304.
4. ALTAMURA, C.A. *et al.* 1993. Plasma and platelet excitatory amino acids in psychiatric disorders. Am. J. Psychiatry **150:** 1731–1733.
5. ALTAMURA, C. *et al.* 1995. Plasma concentrations of excitatory amino acids, serine, glycine, taurine and histidine in major depression. Eur. Neuropsychopharmacol. **5:** 71–75.
6. MAURI, M.C. *et al.* 1998. Plasma and platelet amino acid concentrations in patients affected by major depression and under fluvoxamine treatment. Neuropsychobiology **37:** 124–129.
7. MAES, M. *et al.* 1998. Serum levels of excitatory amino acids, serine, glycine, histidine, threonine, taurine, alanine and arginine in treatment-resistant depression: Modulation by treatment with antidepressants and prediction of clinical responsivity. Acta Psychiatr. Scand. **97:** 302–308.
8. LEVINE, J. *et al.* 2000. Increased cerebrospinal fluid glutamine levels in depressed patients. Biol. Psychiatry **47:** 586–593.
9. HIRAOKA, A. *et al.* 1989. Capillary-isotachophoretic determination of glutamine in cerebrospinal fluid of various neurological disorders. Clin. Biochem. **22:** 293–296.
10. FRANCIS, P.T. *et al.* 1989. Brain amino acid concentrations and Ca^{2+}-dependent release in intractable depression assessed antemortem. Brain Res. **494:** 315–324.
11. HOLEMANS, S. *et al.* 1993. NMDA glutamatergic receptors, labeled with [^{3}H]MK-801, in brain samples from drug-free depressed suicides. Brain Res. **616:** 138–143.
12. NOWAK, G., G.A. ORDWAY & I.A. PAUL. 1995. Alterations in the N-methyl-D-aspartate (NMDA) receptor complex in the frontal cortex of suicide victims. Brain Res. **675:** 157–164.
13. NOWAK, G. *et al.* 1998. Strychnine-insensitive glycine/NMDA sites are altered in two stress models of depression. Pol. J. Pharmacol. **50:** 365–369.
14. SKOLNICK, P. *et al.* 1996. Adaptation of N-methyl-D-aspartate (NMDA) receptors following antidepressant treatment: Implications for the pharmacotherapy of depression. Pharmacopsychiatry **29:** 23–26.
15. COUSINS, J.P. & G. HARPER. 1996. Neurobiochemical changes from Taxol/Neupogen chemotherapy for metastatic breast carcinoma corresponds with suicidal depression. Cancer Lett. **110:** 163–167.
16. AUER, D.P. *et al.* 2000. Reduced glutamate in the anterior cingulate cortex in depression: An in vivo proton magnetic resonance spectroscopy study. Biol. Psychiatry **47:** 305–313.
17. MICHAEL, N. *et al.* 2003. Neurotrophic effects of electroconvulsive therapy: A proton magnetic resonance study of the left amygdalar region in patients with treatment-resistant depression. Neuropsychopharmacology **28:** 720–725.

18. MICHAEL, N. et al. 2001. Clinical response to electroconvulsive therapy (ECT) restores reduced glutamate/glutamine levels in the left anterior cingulum of severely depressed patients. In World Congress of Biol. Psychiatry **2**: 191S World Congress of Biological Psychiatry. Berlin, Germany.
19. CASTILLO, M. et al. 2000. Proton MR spectroscopy in children with bipolar affective disorder: Preliminary observations. Am. J. Neuroradiol. **21**: 832–838.
20. MOGHADDAM, B. et al. 1994. Glucocorticoids mediate the stress-induced extracellular accumulation of glutamate. Brain Res. **655**: 251–254.
21. MOGHADDAM, B. 1993. Stress preferentially increases extraneuronal levels of excitatory amino acids in the prefrontal cortex: Comparison to hippocampus and basal ganglia. J. Neurochem. **60**: 1650–1657.
22. BAGLEY, J. & B. MOGHADDAM. 1997. Temporal dynamics of glutamate efflux in the prefrontal cortex and in the hippocampus following repeated stress: Effects of pretreatment with saline or diazepam. Neuroscience **77**: 65–73.
23. LOWY, M.T., L. WITTENBERG & B.K. YAMAMOTO. 1995. Effect of acute stress on hippocampal glutamate levels and spectrin proteolysis in young and aged rats. J. Neurochem. **65**: 268–274.
24. NOWAK, G. et al. 1995. Swim stress increases the potency of glycine at the N-methyl-D-aspartate receptor complex. J. Neurochem. **64**: 925–927.
25. VIRGIN, C.E., JR. et al. 1991. Glucocorticoids inhibit glucose transport and glutamate uptake in hippocampal astrocytes: Implications for glucocorticoid neurotoxicity. J. Neurochem. **57**: 1422–1428.
26. SKOLNICK, P. et al. 2001. Current perspectives on the development of non-biogenic amine-based antidepressants. Pharmacol. Res. **43**: 411–423.
27. CRANE, G. 1959. Cycloserine as an antidepressant agent. Am. J. Psychiatry **115**: 1025–1026.
28. CRANE, G. 1961. The psychotropic effect of cycloserine: A new use of an antibiotic. Compr. Psychiatry **2**: 51–59.
29. KRYSTAL, J. et al. 1994. Subanesthetic effects of the NMDA antagonist, ketamine, in humans: Psychotomimetic, perceptual, cognitive, and neuroendocrine effects. Arch. Gen. Psychiatry **51**: 199–214.
30. MONAHAN, J.B. et al. 1989. D-cycloserine, a positive modulator of the N-methyl-D-aspartate receptor, enhances performance of learning tasks in rats. Pharmacology Biochem. Behav. **34**: 649–653.
31. EMMETT, M.R. et al. 1991. Actions of D-cycloserine at the N-methyl-D-aspartate-associated glycine receptor site in vivo. Neuropharmacology **30**: 1167–1171.
32. HENDERSON, G., J.W. JOHNSON & P. ASCHER. 1990. Competitive antagonists and partial agonists at the glycine modulatory site of the mouse N-methyl-D-aspartate receptor. J. Physiol. **430**: 189–212.
33. PARKES, J.D. et al. 1970. Controlled trial of amantadine hydrochloride in Parkinson's disease. Lancet **1**: 259–262.
34. MINDHAM, R.H., C.D. MARSDEN & J.D. PARKES. 1976. Psychiatric symptoms during L-dopa therapy for Parkinson's disease and their relationship to physical disability. Physiol. Med. **6**: 23–33.
35. VALE, S., M.A. ESPEJEL & J.C. DOMINGUEZ. 1971. Amantadine in depression. Lancet **2**: 437.
36. DIETRICH, D.E. et al. 2000. Word recognition memory before and after successful treatment of depression. Pharmacopsychiatry **33**: 221–228.
37. FERSZT, R. et al. 1999. Amantadine revisited: An open trial of amantadinesulfate treatment in chronically depressed patients with Borna disease virus infection. Pharmacopsychiatry **32**: 142–147.
38. HUBER, T.J., D.E. DIETRICH & H.M. EMRICH. 1999. Possible use of amantadine in depression. Pharmacopsychiatry **32**: 47–55.
39. STRYJER, R. et al. 2003. Amantadine as augmentation therapy in the management of treatment-resistant depression. Int. Clin. Psychopharmacol. **18**: 93–96.
40. GORTELMEYER, R. & H. ERBLER. 1992. Memantine in the treatment of mild to moderate dementia syndrome: A double-blind placebo-controlled study. Arzneimittelforschung **42**: 904–913.

41. AMBROZI, L. & W. DANIELCZYK. 1988. Treatment of impaired cerebral function in psychogeriatric patients with memantine—results of a phase II double-blind study. Pharmacopsychiatry **21:** 144–146.
42. YAMAKURA, T. *et al.* 1993. Different sensitivities of NMDA receptor channel subtypes to non-competitive antagonists. Neuroreport **4:** 687–690.
43. ANIS, N.A. *et al.* 1983. The dissociative anaesthetics, ketamine and phencyclidine, selectively reduce excitation of central mammalian neurones by N-methyl-aspartate. Br. J. Pharmacol. **79:** 565–575.
44. KRYSTAL, J.H. *et al.* 1999. NMDA agonists and antagonists as probes of glutamatergic dysfunction and pharmacotherapies in neuropsychiatric disorders. Harv. Rev. Psychiatry **7:** 125–143.
45. BERMAN, R.M. *et al.* 2000. Antidepressant effects of ketamine in depressed patients. Biol. Psychiatry **47:** 351–354.
46. KUDOH, A. *et al.* 2002. Small-dose ketamine improves the postoperative state of depressed patients. Anesth. Analg. **95:** 114–118.
47. YILMAZ, A. *et al.* 2002. Prolonged effect of an anesthetic dose of ketamine on behavioral despair. Pharmacol. Biochem. Behav. **71:** 341–344.
48. FOGELSON, D.L. & H. STERNBACH. 1997. Lamotrigine treatment of refractory bipolar disorder.[comment]. J. Clin. Psychiatr. **58:** 271–273.
49. BOWDEN, C.L. & N.U. KARREN. 2002. Lamotrigine in the treatment of bipolar disorder. Expert Opin. Pharmacother. **3:** 1513–1519.
50. HURLEY, S.C. 2002. Lamotrigine update and its use in mood disorders. Ann. Pharmacother. **36:** 860–873.
51. BOWDEN, C.L. 2001. Novel treatments for bipolar disorder.[erratum appears in Expert Opin. Investig. Drugs 2001 10(7):following 1205]. Expert Opin. Investig. Drugs **10:** 661–671.
52. FRYE, M.A. *et al.* 2000. A placebo-controlled study of lamotrigine and gabapentin monotherapy in refractory mood disorders. J. Clin. Psychopharmacol. **20:** 607–614.
53. CALABRESE, J.R. *et al.* 2000. A double-blind, placebo-controlled, prophylaxis study of lamotrigine in rapid-cycling bipolar disorder: Lamictal 614 Study Group. J. Clin. Psychiatr. **61:** 841–850.
54. BOWDEN, C.L. *et al.* 2003. A placebo-controlled 18-month trial of lamotrigine and lithium maintenance treatment in recently manic or hypomanic patients with bipolar I disorder. Arch. Gen. Psychiatry **60:** 392–400.
55. BARBEE, J.G. & N.J. JAMHOUR. 2002. Lamotrigine as an augmentation agent in treatment-resistant depression. J. Clin. Psychiatr. **63:** 737–741.
56. NORMANN, C. *et al.* 2002. Lamotrigine as adjunct to paroxetine in acute depression: A placebo-controlled, double-blind study. J. Clin. Psychiatr. **63:** 337–344.
57. OBROCEA, G.V. *et al.* 2002. Clinical predictors of response to lamotrigine and gabapentin monotherapy in refractory affective disorders. Biol. Psychiatry **51:** 253–260.
58. CORIC, V. *et al.* 2003. Beneficial effects of the antiglutamatergic agent riluzole in a patient diagnosed with obsessive-compulsive disorder and major depressive disorder. Psychopharmacology **167:** 219–220.
59. ZARATE, C.A. JR., *et al.* 2002. Riluzole: An investigation of the antidepressant efficacy of an antiglutamatergic agent with neruotrophic properties. American College of Neuropsychopharmacology, 41st Annual Meeting, San Juan, Puerto Rico, 2002. p. 132.
60. MACIVER, M.B. *et al.* 1996. Riluzole anesthesia: Use-dependent block of presynaptic glutamate fibers. Anesthesiology **85:** 626–634.
61. LI, X. *et al.* 2001. Antidepressant-like actions of an AMPA receptor potentiator (LY392098). Neuropharmacology **40:** 1028–1033.
62. QUIRK, J.C. & E.S. NISENBAUM. 2002. LY404187: A novel positive allosteric modulator of AMPA receptors. CNS Drug Rev. **8:** 255–282.
63. INGVAR, M. *et al.* 1997. Enhancement by an ampakine of memory encoding in humans. Exp. Neurol. **146:** 553–559.
64. LYNCH, G. *et al.* 1997. Evidence that a positive modulator of AMPA-type glutamate receptors improves delayed recall in aged humans. Exp. Neurol. **145:** 89–92.

65. MANJI, H.K. et al. 2003. Enhancing neuronal plasticity and cellular resilience to develop novel, improved therapeutics for difficult-to-treat depression. Biol. Psychiatry **53:** 707–742.
66. DOMINO, E., P. CHODOFF & G. CORSSEN. 1965. Pharmacologic effects of CI-581, a new dissociative anesthetic, in man. Clin. Pharm. Ther. **6:** 279–291.
67. OYE, I., O. PAULSEN & A. MAURSET. 1992. Effects of ketamine on sensory perception: Evidence for a role of N-methyl-D-aspartate receptors. J. Pharmacol. Exp. Ther. **260:** 1209–1213.
68. GHONEIM, M.M. et al. 1985. Ketamine: Behavioral effects of subanesthetic doses. J. Clin. Psychopharmacol. **5:** 70–77.
69. MALHOTRA, A.K. et al. 1996. NMDA receptor function and human cognition: The effects of ketamine in healthy volunteers. Neuropsychopharmacology **14:** 301–307.
70. NEWCOMER, J.W. et al. 1999. Ketamine-induced NMDA receptor hypofunction as a model of memory impairment and psychosis. Neuropsychopharmacology **20:** 106–118.
71. KRYSTAL, J.H. et al. 1998. Dose-related ethanol-like effects of the NMDA antagonist, ketamine, in recently detoxified alcoholics.[comment]. Arch. Gen. Psychiatry **55:** 354–360.
72. MOGHADDAM, B. et al. 1997. Activation of glutamatergic neurotransmission by ketamine: A novel step in the pathway from NMDA receptor blockade to dopaminergic and cognitive disruptions associated with the prefrontal cortex. J. Neurosci. **17:** 2921–2927.
73. MOGHADDAM, B. & B.W. ADAMS. 1998. Reversal of phencyclidine effects by a group II metabotropic glutamate receptor agonist in rats. [see comments]. Science **281:** 1349–1352.
74. KRYSTAL, J.H. et al. 2003. Attenuation of the disruptive effects of the NMDA glutamate receptor antagonist, ketamine, on working memory by pretreatment with the group II metabotropic glutamate receptor (mGluR) agonist, LY354740, in healthy human subjects. In review.
75. MACKOWIAK, M. et al. 2002. An AMPA receptor potentiator modulates hippocampal expression of BDNF: An in vivo study. Neuropharmacology **43:** 1–10.
76. LEGUTKO, B., X. LI & P. SKOLNICK. 2001. Regulation of BDNF expression in primary neuron culture by LY392098, a novel AMPA receptor potentiator. Neuropharmacology **40:** 1019–1027.
77. NACHER, J. et al. 2003. NMDA receptor antagonist treatment increases the production of new neurons in the aged rat hippocampus. Neurobiol. Aging **24:** 273–284.
78. NACHER, J. et al. 2001. NMDA receptor antagonist treatment induces a long-lasting increase in the number of proliferating cells, PSA-NCAM-immunoreactive granule neurons and radial glia in the adult rat dentate gyrus. Eur. J. Neurosci. **13:** 512–520.
79. LLOYD, K.G., P.L. MORSELLI & G. BARTOLINI. 1987. GABA and affective disorders. Med. Biol. **65:** 159–165.
80. PETTY, F. 1995. GABA and mood disorders: A brief review and hypothesis. J Affect. Disord. **34:** 275–281.
81. SHIAH, I. & L.N. YATHAM. GABA function in mood disorders: An update and critical review. Life Sci. **63:** 1289–1303.
82. SANACORA, G., G.F. MASON & J.H. KRYSTAL. 2000. Impairment of GABAergic function in depression: New insights from neuroimaging. Crit. Rev. Neurobiol. **14:** 23–45.
83. BIGGO, G. et al. 1984. Stress and beta-carbolines decrease the density of low affinity GABA binding sites: An effect reversed by diazepam. Brain Res. **305:** 13–18.
84. CONCAS, A. et al. 1988. Foot-shock and anxiogenic beta-carbolines increase t[^{35}S] butylbicyclophosphorothionate binding in the rat cerebral cortex, an effect opposite to anxiolytics and gamma-aminobutyric acid mimetics. Pharmacol. Ther. **48:** 1868–1876.
85. SERRA, M. et al. 1991. Foot-shock stress enhances the increase of [^{35}S]TBPS binding in the rat cerebral cortex and the convulsions induced by isoniazid. Neurochemical. Res. **16:** 17–22.
86. SANNA, E. et al. 1992. Carbon dioxide inhalation reduces the function of GABA$_A$ receptors in the rat brain. Eur. J. Pharmacol. **216:** 457–458.

87. BORSINI, F. et al. 1988. On the role of endogenous GABA in the forced swimming test in rats. Pharmacol. Biochem. Behav. **29:** 275–279.
88. ACOSTA, G.B., M.E.O. LOSADA & M.C. RUBIO. 1993. Area-dependent changes in GABAergic function after acute and chronic cold stress. Neurosci. Lett. **154:** 175–178.
89. INSEL, T.R. 1989. Decreased in vivo binding to brain benzodiazepine receptors during social isolation. Psychopharmacology 97: 142–144.
90. ACOSTA, G.B. & M.C. RUBIO. 1994. GABA$_A$ receptors mediate the changes produced by stress on GABA function and locomotor activity. Neurosci. Lett. **176:** 29–31.
91. WEIZMAN, R. et al. 1989. Repeated swim stress alters brain benzodiazepine receptors measured in vivo. J. Pharmacol. Exp. Ther. **249:** 701–707.
92. CALDJI, C. et al. 2000. The effects of early rearing environment on the development of GABA$_A$ and central benzodiazepine receptor levels and novelty-induced fearfulness in the rat. Neuropsychopharmacology **22:** 219–229.
93. SHERMAN, A.D. & F. PETTY. 1980. Neurochemical basis of the action of antidepressants on learned helplessness. Behav. Neural. Biol. **30:** 119–134.
94. DELINI-STULA, A. & A. VASSOUT. 1978. Influence of baclofen and GABA-mimetic agents on spontaneous and olfactory-bulb-ablation-induced muricidal behavior in the rat. Arzneimittelforschung **28:** 1508–1509.
95. BARTHOLINI, G. 1985. GABA receptor agonists: Pharmacological spectrum and therapeutic actions. Med. Res. Revs. **5:** 55–75.
96. BARTHOLINI, G., K.G. LLOYD & P.L. MORSELLI, EDS. 1986. GABA and Mood Disorders: Experimental and Clinical Research. Raven Press. New York, NY.
97. POPOV, N. & H. MATTHIES. 1969. Some effects of monoamine oxidase inhibitors on the metabolism of gamma-aminobutyric acid in rat brain. J. Neurochem. **16:** 899–907.
98. SCHATZ, R.A. & H. LAL. 1971. Elevation of brain GABA by pargyline: A possible mechanism for protection against oxygen toxicity. J. Neurochem. **18:** 2553–2555.
99. KORF, J. & K. VENEMA. 1983. Desmethylimipramine enhances the release of endogenous GABA and other neurotransmitter amino acids from the rat thalamus. J. Neurochem. **40:** 946–950.
100. BOWLER, J.M. et al. 1983. Regional GABA concentration and ^{3}H-diazepam binding in rat brain following repeated electroconvulsive shock. J. Neurol. Transm. **56:** 3–12.
101. PASLAWSKI, T. et al. 1996. The antidepressant drug phenelzine produces antianxiety effects in the plus-maze and increases in rat brain GABA. Psychopharmacology (Berl) **127:** 19–24.
102. PILC, A. & K.G. LLOYD. 1984. Chronic antidepressants and GABA$_B$ receptors: A GABA hypothesis of antidepressant drug action. Life Sci. **35:** 2149–2154.
103. OLSEN, R.W. et al. 1978. Effects of drugs on gamma-aminobutyric acid receptors, uptake, release and synthesis in vitro. Brain Res. **139:** 277–279.
104. CROSS, J.A. & R.W. HORTON. 1988. Effects of chronic oral administration of the antidepressants desmethylimipramine and zimeldine on rat cortical GABA$_B$ binding sites: A comparison with 5-HT2 binding site changes. Br. J. Pharmacol. **93:** 331–336.
105. MCMANUS, D.J. & A.J. GREENSHAW. 1991. Differential effects of antidepressants on GABA$_B$ and beta-adrenergic receptors in rat brain. Biochem. Pharmacol. **42:** 1525–1528.
106. SZEKELY, A.M., M.L. BARBACCIA & E. COSTA. 1987. Effect of protracted antidepressant treatment on signal transduction and [^{3}H](–)-baclofen binding at GABA$_B$ receptors. J. Pharmacol. Exp. Ther. **243:** 155–159.
107. MONTELEONE, P. et al. 1990. GABA, depression and the mechanism of antidepressant action: A neuroendocrine approach. Affect. Disord. **20:** 1–5.
108. PETTY, F., G.L. KRAMER & W. HENDRICKSE. 1993. GABA and depression. In Biology of Depressive Disorders, Part A: A systems perspective. J.J. Mann & D.J. Kupler, Eds.: 79–108. Plenum Press. New York, NY.
109. PETTY, F. & M.A. SCHLESSER. 1981. Plasma GABA in affective illness: A preliminary investigation. J. Affect. Disord. **3:** 339–343.

110. PETTY, F. & A.D. SHERMAN. 1984. Plasma GABA in psychiatric illness. J. Affect. Disord. **6:** 131–138.
111. PETTY, F. *et al.* 1990. Plasma GABA in mood disorders. Psychopharmacol. Bull. **26:** 157–161.
112. PETTY, F. *et al.* 1992. Low plasma gamma-aminobutyric acid levels in male patients with depression. Biol. Psychiatry **32:** 354–363.
113. BERRETTINI, W.H. *et al.* 1982. Plasma and CSF GABA in affective illness. Brit. J. Psychiat. **141:** 483–487.
114. KORPI, E.R., J.E. KLEINMAN & R.J. WYATT. 1988. GABA concentrations in forebrain areas of suicide victims. Biol. Psychiatry **23:** 109–114.
115. HONIG, A. *et al.* 1989. Amino acid levels in depression: A preliminary investigation. J. Psychiatr. Res. **22:** 159–164.
116. PERRY, E. *et al.* 1977. Neurotransmitter abnormalities in senile dementia. J. Neurol. Sci. **34:** 247–265.
117. KAIYA, H. *et al.* 1982. Plasma glutamate decarboxylase activity in neuropsychiatry. Psychiatry Res. **6:** 335–343.
118. HECKERS, S. *et al.* 2002. Differential hippocampal expression of glutamic acid decarboxylase 65 and 67 messenger RNA in bipolar disorder and schizophrenia. Arch. Gen. Psychiatry **59:** 521–529.
119. BENES, F.M. *et al.* 2000. Glutamate decarboxylase(65)-immunoreactive terminals in cingulate and prefrontal cortices of schizophrenic and bipolar brain. J. Chem. Neuroanat. **20:** 259–269.
120. CHEETHAM, S. *et al.* 1988. Brain GABAA/benzodiazepine binding sites and glutamic acid decarboxylase activity in depressed suicide victims. Brain Res. **460:** 114–123.
121. TOTH, Z. *et al.* 1999. Gene expression for glutamic acid decarboxylase is increased in prefrontal cortex of depressed patients. Soc. Neurosci. **29:** 2097.
122. SANACORA, G. *et al.* 1999. Reduced cortical gamma-aminobutyric acid levels in depressed patients determined by proton magnetic resonance spectroscopy. Arch. Gen. Psychiatry **56:** 1043–1047.
123. SANACORA, G. *et al.* 2003. Occipital cortex GABA concentrations differentiate depressive subtypes. Biol. Psychiatry **53:** 171–172.
124. SANACORA, G. *et al.* 2003. Increased occipital cortex GABA concentrations following electroconvulsive therapy in depressed patients. Am. J. Psychiatry **160:** 577–579.
125. SANACORA, G. *et al.* 2002. Increased occipital cortex GABA concentrations in depressed patients after therapy with selective serotonin reuptake inhibitors. Am. J. Psychiatry **159:** 663–665.
126. MASON, G.F. *et al.* 2001. Preliminary evidence of reduced cortical GABA synthesis rate in major depression. Soc. Neurosci. **142:** 6.
127. LAPPALAINEN, J. *et al.* 2003. Mutation screen of the glutamate decarboxylase-67 gene and haplotype association to unipolar depression. Neuropsychiatric Gen. In press.
128. MAGISTRETTI, P.J. *et al.* 1999. Energy on demand. Science **283:** 496–497.
129. LEBON, V., *et al.* 2002. Astroglial contribution to brain energy metabolism in humans revealed by ^{13}C nuclear magnetic resonance spectroscopy: Elucidation of the dominant pathway for neurotransmitter glutamate repletion and measurement of astrocytic oxidative metabolism. J. Neurosci. **22:** 1523–1531.
130. PATEL, A.B. *et al.* 2001. Glutamine is the major precursor for GABA synthesis in rat neocortex in vivo following acute GABA-transaminase inhibition. Brain Res. **919:** 207–220.
131. RAJKOWSKA, G. *et al.* 1999. Morphometric evidence of neuronal and glial prefrontal cell pathology in major depression. Biol. Psychiatry **45:** 1085–1098.
132. RAJKOWSKA, G. 2000. Postmortem studies in mood disorders indicate altered numbers of neurons and glial cells. Biol. Psychiatry **48:** 766–777.
133. BOWLEY, M.P. *et al.* 2002. Low glial numbers in the amygdala in major depressive disorder. Biol. Psychiatry **52:** 404–412.
134. RAJKOWSKA, G., A. HALARIS & L.D. SELEMON. 2001. Reductions in neuronal and glial density characterize the dorsolateral prefrontal cortex in bipolar disorder.[comment]. Biol. Psychiatry **49:** 741–752.

135. MIGUEL-HIDALGO, J.J. *et al.* 2002. Glia pathology in the prefrontal cortex in alcohol dependence with and without depressive symptoms. Biol. Psychiatry **52:** 1121–1133.
136. WEBSTER, M.J. *et al.* 2001. Immunohistochemical localization of phosphorylated glial fibrillary acidic protein in the prefrontal cortex and hippocampus from patients with schizophrenia, bipolar disorder, and depression. Brain, Behav., Immun. **15:** 388–400.
137. COTTER, D.R., C.M. PARIANTE & I.P. EVERALL. 2001. Glial cell abnormalities in major psychiatric disorders: The evidence and implications. Brain Res. Bull. **55:** 585–595.
138. MIGUEL-HIDALGO, J.J. *et al.* 2000. Glial fibrillary acidic protein immunoreactivity in the prefrontal cortex distinguishes younger from older adults in major depressive disorder. Biol. Psychiatry **48:** 861–873.
139. ONGUR, D., W.C. DREVETS & J.L. PRICE. 1998. Glial reduction in the subgenual prefrontal cortex in mood disorders. Proc. Natl. Acad. Sci. USA **95:** 13290–13295.
140. SCHROETER, M.L. *et al.* 2002. S100B is increased in mood disorders and may be reduced by antidepressive treatment. NeuroReport **13:** 1675–1678.
141. ROTHERMUNDT, M. *et al.* 2001. S-100B is increased in melancholic but not in nonmelancholic major depression. J. Affective Disorders **66:** 89–93.
142. GRABE, H.J. *et al.* 2001. Neurotrophic factor S100 beta in major depression. Neuropsychobiology **44:** 88–90.
143. MCCULLUMSMITH, R.E. & J.H. MEADOR-WOODRUFF. 2002. Striatal excitatory amino acid transporter transcript expression in schizophrenia, bipolar disorder, and major depressive disorder. Neuropsychopharmacology **26:** 368-375.

A Role for Noradrenergic Transmission in the Actions of Phencyclidine and the Antipsychotic and Antistress Effects of mGlu2/3 Receptor Agonists

CHAD J. SWANSON AND DARRYLE D. SCHOEPP

Neuroscience Research, Eli Lilly and Company, Indianapolis, Indiana 46285 USA

ABSTRACT: Evidence suggests that glutamatergic neuronal transmission is involved in psychiatric and neurological disorders and that drugs that target glutamate systems may serve as novel therapeutics in humans. For example, agonists for group II mGlu receptors (mGlu2 and mGlu3) have been shown to be anxiolytic in certain animal models and have shown promise in early human trials. mGlu2/3 receptor agonists also block the neurochemical and behavioral actions of psychotogens, such as phencyclidine and amphetamine in rodents, suggesting that they may be useful to treat psychosis in humans. Recently, we have used *in vivo* microdialysis and behavioral methods to further explore the potential antipsychotic and antistress actions of mGlu2/3 receptor agonists in rats. In subjects undergoing brain microdialysis of the nucleus accumbens shell, we have shown that LY379268 (3 mg/kg s.c.) (a systemically active mGlu2/3 receptor agonist) blocks PCP-induced locomotor activations for ~3 hours. In these animals, PCP-induced dopamine release was reduced, but only in a transient fashion (15–75 min). PCP-induced norepinephrine release was also reduced, but unlike dopamine, in a manner that was temporally correlated with the reduction of PCP-induced behaviors. In separate experiments in rats not undergoing microdialysis, the α_2-adrenergic receptor agonist, clonidine, was shown to block PCP behaviors, and the norepinephrine reuptake inhibitor reboxetine was shown to exacerbate PCP-induced ambulations. In the latter study, LY379268 pretreatment effectively reversed the PCP behaviors in both control and reboxetine-treated animals. These data support a role for noradrenergic neurotransmission in the actions of drugs such as phencyclidine and suggest that stress pathways associated with these drugs can be normalized by mGlu2/3 receptor activation.

KEYWORDS: LY379268; metabotropic glutamate receptors; phencyclidine; dopamine; norepinephrine; microdialysis; psychosis

Excessive or pathologically enhanced glutamate neurotransmission has been implicated in a number of neurological and psychiatric disorders, including acute and chronic neurodegenerative diseases[7,22] pain/migraine,[13] epilepsy,[9,24] depression,[35]

Address for correspondence: Dr. Darryle D. Schoepp, Neuroscience Research Division, Lilly Research Laboratories, Eli Lilly and Company, Indianapolis, IN 46285. Voice: 317-276-6316; fax: 317-276-7600.
  schoepp@lilly.com

anxiety,[40] and psychosis.[21,30] Thus, novel drugs that modulate glutamate neurotransmission may have broad application as novel therapies for CNS disorders.[20,33] Much of the focus in the field this past decade has been on NMDA or AMPA receptor antagonists as neuroprotective agents. However, clinical experience for finding novel, effective, and safe drugs that specifically target the glutamate system has not produced many useful therapies to date.[28]

More recent approaches that specifically target the modulation of molecular subtypes of ionotropic receptors, metabotropic receptors, and transporters are now leading to new systemically active pharmacological agents with promising pharmacology in both animal models and human patients (FIG. 1). This new generation of compounds now includes potentiators of AMPA receptors (cognition),[2,14,26] AMPA and/or kainate receptor antagonists (epilepsy, pain, migraine),[3,4,22] inhibitors of glycine transport to enhance NMDA receptor function (cognition/psychosis),[17,28] antagonists for mGlu5 receptors (anxiety/pain),[5] agonists for metabotropic glutamate receptors (mGlu2/3) (anxiety/psychosis),[23,27,30,31,34] and potentiators of mGlu2 receptors (anxiety/migraine).[18,19] Thus, the discovery of metabotropic glutamate receptors as modulators of glutamatergic transmission and the exploration of novel mGlu receptor ligand pharmacology in animals and humans offers new approaches to safely and effectively treat these disorders.

Agonists for group II mGlu receptors (mGlu2 and mGlu3) include compounds such as LY354740 and LY379268.[32] LY354740 has been shown to be anxiolytic in certain animal models[31] and this compound has shown promise in early human studies for anxiety disorders.[23,34] As a class of agents, mGlu2/3 receptor agonists also block certain biochemical manifestations and behavioral actions of psychotogens, such as phencyclidine and amphetamine in rodents, suggesting that they may also be useful to treat psychosis in humans.[8,27,30] For example, LY354740 and/or LY379268 have been shown to prevent PCP induced increases in glutamate release in the prefrontal cortex or nucleus accumbens in rats, while having no inhibitory effects on PCP-induced dopamine release in these brain regions.[6,27]

However, studies indicate that the nucleus accumbens includes at least two distinct subfields (core and shell) distinguished by their anatomical connectivity, neurochemistry, and pharmacology.[11,16] The core of the nucleus accumbens is thought to be an extension of the dorsal striatum and is thus linked to extrapyramidal motor functions. In contrast, the shell of the nucleus accumbens is thought be integrated with limbic functions such as stress, arousal, and reinforcement.

With this in mind, the effects of mGlu2/3 receptor activation on monoamine release in subregions of the nucleus accumbens have been investigated by Mitchell.[25] They have observed that reverse dialysis of LY379268 into the shell, but not core, of the nucleus accumbens reduces basal dopamine levels and prevents PCP-induced dopamine release in rats. Thus, these data suggest a role for nucleus accumbens (shell) dopamine in the pharmacological effects of mGlu2/3 receptor agonists in the PCP model.

To further explore the role of monoamines in the actions of PCP and the putative antipsychotic actions of mGlu2/3 receptor agonists, we used α-methyl-DL-p-tyrosine methyl ester (α-MPT; 400 mg/kg, i.p.) administration to selectively deplete rats of catecholamines (dopamine and norepinephrine) and DL-p-chlorophenyl-alanine methyl ester (PCPA; 300 mg/kg, i.p.) administration to deplete brain serotonin (5-HT) 24 h prior to testing.[37] α-MPT treatment was shown to completely prevent

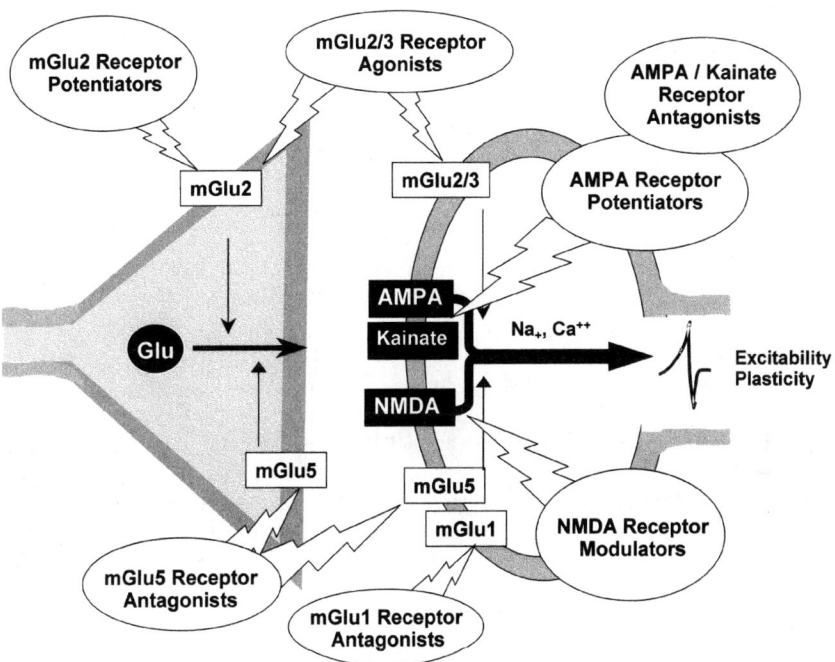

FIGURE 1. Targets for novel drugs that modulate glutamate neuronal transmission. New approaches to modulate excitatory neuronal transmission at the glutamate synapse have produced several new classes of drugs that may be useful to treat psychiatric and/or neurological disorders. Small molecule compounds for these targets have been discovered and have shown promising pharmacology in animal models and/or have shown early clinical efficacy. For example, agonists for mGlu2/3 receptors decrease the presynaptic release of glutamate and subsequently suppress postsynaptic excitations. These agents have shown activity in animal models of anxiety, stress, psychosis, pain, and neurotoxicity. Other promising new pharmacological agents include allosteric potentiators of mGlu2 receptors (migraine/anxiety), AMPA receptor potentiators (cognition/depression), AMPA/kainate selective antagonists (pain, epilepsy), mGlu1 receptor antagonists (pain, neuroprotection), mGlu5 receptor antagonists (anxiety, pain, neuroprotection), and modulators of NMDA receptor function such as NR2B-selective antagonists (neuroprotection), glycine site antagonists (neuroprotection, pain), and inhibitors of glycine transport (GLT1) inhibitors (cognition, psychosis).

the locomotor stimulant actions of amphetamine in rats, suggesting that no functionally relevant release of catecholamines occurs after this treatment paradigm. In contrast to amphetamine, locomotor stimulations were produced by PCP, but were appreciably attenuated (by approximately one-half) following α-MPT pretreatment. These findings indicate that PCP-induced motor behaviors are not only mediated by catecholamines, but other neurotransmitters as well. Accordingly, certain PCP-mediated behaviors in catecholamine-depleted (α-MPT treated) animals were significantly reduced by systemic pretreatment with the AMPA/kainite selective antagonist, LY293558, but not haloperidol. Previous work from other laboratories has indicated that neurotransmitters such as glutamate, GABA, and 5-HT are also involved in the pharmacological actions of PCP[15,21]; these findings are consistent with that literature. Our data also support a role for 5-HT in the behavioral actions of PCP, as the 5-HT2 receptor antagonist, ketanserin, blocked PCP-induced behaviors in control and α-MPT-treated animals only. Interestingly, PCPA pretreatment did not affect the ability of the mGlu2/3 receptor agonist LY379268 or the atypical antipsychotic clozapine to block PCP behaviors. As $5-HT_2$ receptor antagonists such as MDL-100,907 have not shown robust efficacy in treating schizophrenia, critics might suggest that the PCP model does not offer predictive validity for identifying novel therapeutic agents. However, our experiments in 5-HT-depleted rats clearly distinguish the mechanism of action of mGlu2/3 receptor agonists and/or atypical antipsychotics from $5-HT_2$ receptor antagonists using variations of the PCP model of psychosis. Like clozapine, but not ketanserin, LY379268 produced a significant blockade of PCP-induced hyper-locomotion in control, catecholamine-, and 5-HT-depleted animals. Collectively, these data suggest a role for catecholamines and other neurotransmitters in mediating the anti-PCP actions of LY379268. Furthermore, the profile of activity in these different paradigms shows that LY379268 displays more similar pharmacology to the atypical antipsychotic clozapine than either haloperidol or the 5HT2 receptor antagonist ketanserin. These data further support the hypothesis that mGlu2/3 receptor agonists may be useful antipsychotic agents in humans.

To further examine the role of catecholamines in mediating the behavioral actions of mGlu2/3 receptor agonists in the PCP model, we have examined PCP behaviors in rats undergoing simultaneous *in vivo* brain microdialysis. In free-moving animals probed in the nucleus accumbens shell, systemically administered LY379268 (3 mg/kg s.c.) blocks PCP-induced locomotor activations for up to 180 min post-PCP injection.[38] In agreement with Mitchell,[25] PCP-induced increases in extracellular dopamine were reduced in these animals, but only in a relatively transient manner (15–75 min after PCP administration). Interestingly, PCP-induced norepinephrine release was also reduced, but unlike dopamine, in a manner more temporally correlated with the reduction of PCP-induced behaviors (for up to 180 min post-PCP administration). In separate behavioral experiments, the α_2-adrenergic receptor agonist clonidine was shown to block PCP-induced behaviors, indicating a role for norepinephrine in mediating some of the behavioral effects of PCP.

Norepinephrine is posited to play an important role in modulating behavioral processes associated with stress and arousal.[12] Noradrenergic innervation of the nucleus accumbens appears to originate primarily from the nucleus tractus solitarius where it is confined to the shell subregion, while little or no innervation is observed in the core.[1,10] This anatomical distinction supports a role for noradrenergic function in the nucleus accumbens shell in modulating emotionally relevant behaviors, such as the response to stress and possibly its contribution to drug seeking behavior. Our data indicate a role for noradrenergic as well as dopaminergic neuronal transmission in the actions of PCP. Furthermore, our results suggest that noradrenergic-associated "stress" pathways are activated by phencyclidine administration and this response can be normalized by mGlu2/3 receptor activation. These data are consistent with other recent studies from our laboratory where we have shown that immobilization stress-induced prefrontal cortical norepinephrine release can be prevented by systemic administration of LY354740.[39]

On the basis of the above observations, we set forth to further explore the potential role of noradrenergic transmission in mediating certain behavioral aspects of PCP administration, as well as the role that mGlu2/3 receptor agonist pharmacology plays in modulating these effects. To do this, we examined the effects of the selective norepinephrine reuptake inhibitor, reboxetine, on PCP-induced behavioral activation and investigated the ability of LY379268 to block PCP-induced behaviors in the presence of reboxetine.

In these experiments, behavioral parameters were monitored in transparent, shoebox cages that measured 45×25×20 cm, with a 1 cm depth of wood chips on the cage floor and a metal grill on top of the cage. Rectangular photocell monitors (Hamilton Kinder, Poway, CA) with a bank of 12 photocell beams (8×4 formation) surrounded each test cage. A lower rack of photocell beams was positioned 5 cm above the cage floor to enable detection of both body and head movements. Ambulations (locomotor activity), fine motor movements (an estimate of stereotyped behavior), and time at rest (total number of seconds in a 60 minute session in which no beams were broken; taken at 1-s intervals) were recorded by computer and stored for each test session. Rats were pretreated with vehicle (sterile water) or reboxetine (3 mg/kg, i.p.) in home cages thirty minutes prior to testing. Following the 30-min pretreatment period, subjects were placed in the test cage for an additional 30-min habituation period to allow for acclimation to the testing environment, at the beginning of which they received LY379268 (3 mg/kg, s.c.) or vehicle (sterile saline) injections. Following this habituation period, animals were administered subcutaneous injections of phencyclidine (PCP, 5 mg/kg) or vehicle (sterile saline) and testing began immediately following PCP administration and continued for 120 minutes. Thus, this study consisted of the following four pretreatment groups: (1) vehicle/vehicle (2) reboxetine/vehicle (3) vehicle/LY379268, and (4) reboxetine/LY379268. Animals were then administered either vehicle or PCP within each respective pretreatment group, for a total of eight treatment groups in all. Statistical analysis was carried out using the GraphPad PRISM statistical/graphing package (GraphPad, San Diego, CA). Data were analyzed using a one-way analysis of variance (ANOVA), and upon discovery of statistical significance ($P<.05$), post-hoc comparisons were performed using Tukey's multiple comparisons test.

As shown in FIGURE 2, PCP-induced ambulations (locomotor activity) were selectively enhanced by reboxetine pretreatment, while fine movements and rest time were not significantly affected. These data are in support of previous work indicating a role for nucleus accumbens norepinephrine in producing locomotor activation and promoting open-field exploration in rats.[29,36] LY379268 pretreatment effectively reversed PCP-mediated behaviors to a similar degree in both control and reboxetine-pretreated animals.

Collectively, our previous studies and the present data further support a role for noradrenergic neurotransmission in the actions of drugs such as phencyclidine. Furthermore, it would appear that activation of noradrenergic pathways associated with PCP can be normalized by mGlu2/3 receptor activation. Overall, the actions of mGlu2/3 receptor agonists in animal models of psychiatric disorders appears to involve not only negative modulation of glutamatergic transmission, but may also include direct or indirect modulation of catecholamine (norepinephrine, dopamine) release and function.

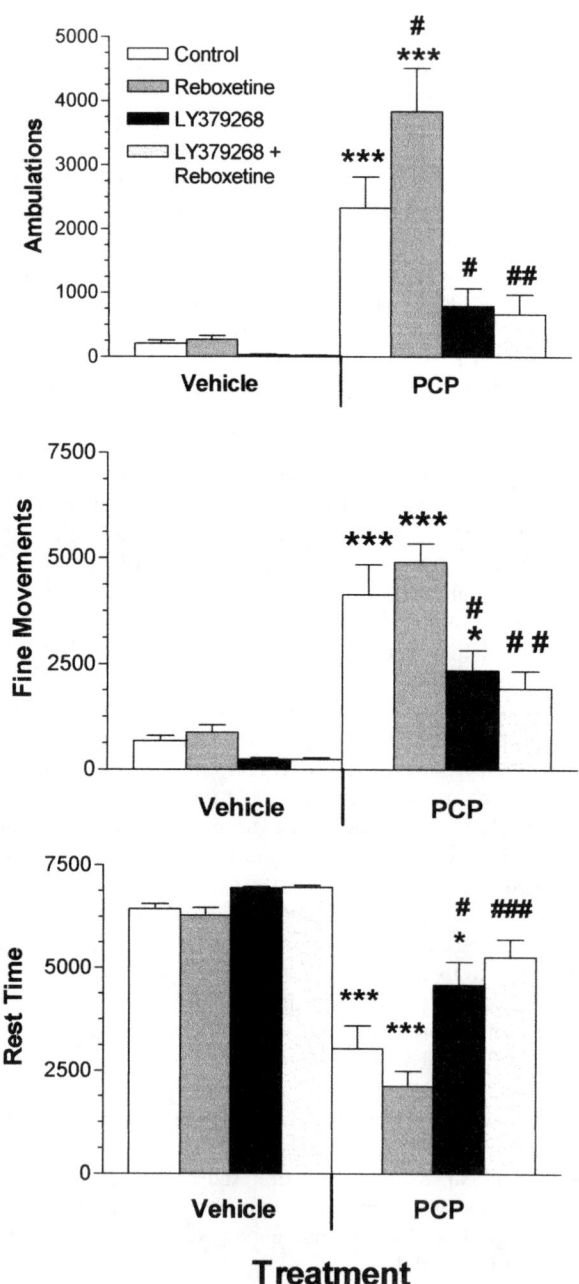

FIGURE 2. *See following page for legend.*

REFERENCES

1. BRIDGE, C.W., T.L. STRATFORD, S.L. FOOTE & A.E. KELLEY. 1997. Distribution of dopamine B-hydroxyls-like immunoreactive fibers within the shell subregion of the nucleus accumbens. Synapse **27:** 230–241.
2. BLEAKMAN, D. & D. LODGE. 1998. Neuropharmacology of AMPA and kainate receptors. Neuropharmacology **37:** 1187–1204.
3. BLEAKMAN, D. 1999. Kainate receptor pharmacology and physiology. Cell. Mol. Life Sci. **56:** 558–566.
4. BLEAKMAN, D., M.R. GATES, A.M. OGDEN & M. MACKOWIAK. 2002. Kainate receptor agonists, antagonists and allosteric modulators. Curr. Pharmaceutical Design **8:** 873–885.
5. BORDI, F. & A. UGOLINI. 1999. Group I metabotropic glutamate receptors: implications for brain diseases. Progr. Neurobiol. **59:** 55–79.
6. BRISTOW, L.J., D. LORRAIN, U. CAMPBELL, et al. 2002. The mGlu2/3 agonist LY379268 attenuates the hyperactivity and increased prefrontal cortex and Hippocampal glutamate release induced by NMDA receptor antagonists in the rat. Neuropharmacology **43:** 279.
7. BRUNO, V., G. BATTAGLAI, A. COPANI, et al. 2001. Metabotropic glutamate receptor subtypes as targets for neuroprotective drugs. J. Cerebral Blood Flow Metabolism **21:** 1013–1033.
8. CARTMELL, J., J.A. MONN & D.D. SCHOEPP. 1999. The mGlu2/3 agonists, LY354740 and LY379268, selectively attenuate phencyclidine versus d-amphetamine motor behaviors in rats. J. Pharmacol. Exp. Ther. **291:** 161–170.
9. CHAPMAN, A. 2000. Glutamate and epilepsy. J. Nutr. **130:** 1043S–1045S.
10. DELFS, J.M., Y. ZHU, J.P. DRUHAN & G.S. ASTON-JONES. 1998. Origin of noradrenergic afferents to the shell subregion of the nucleus accumbens: anterograde and retrograde tract-tracing studies in the rat. Brain Res. **28:** 127–140
11. DI CHIARA, G. 2002. Nucleus accumbens shell and core dopamine: differential role in behavior and addiction. Behav. Brain Res. **137:** 75–114.
12. FOOTE, S.L., G. ASTON-JONES & F.E. BLOOM. 1980. Impulse activity of locus coeruleus neurons in awake rats and monkeys is a function of sensory stimulation and arousal. Proc. Natl. Acad. Sci. USA **77:** 3033–3037.
13. FUNDYTUS, M. 2001. Glutamate receptors and nociception implications for the drug treatment of pain. CNS Drugs **15:** 29–58.
14. GATES, M., A. OGDEN & D. BLEAKMAN. 2001. Pharmacological effects of AMPA receptor potentiators LY392098 and LY404187 on rat neuronal AMPA receptors in vitro. Neuropharmacology **40:** 984–991.
15. HALBERSTADT, A.L. 1995. The phencyclidine-glutamate model of schizophrenia. Clin. Neuropsychopharmacol. **18:** 237–249.
16. HERRERO, M.T., C. BARCIA & J.M. NAVARRO. 2002. Functional anatomy of thalamus and basal ganglia. Childs Nervous System **18:** 386–404.

FIGURE 2. mGlu2/3 receptor modulation of reboxetine effects on PCP-induced behavioral activation in the rat. Rats were pretreated 60 minutes prior to PCP administration with either vehicle (1 mL/kg, s.c.) or reboxetine (3 mg/kg, i.p.) administration. At 30 min prior to PCP administration, rats were administered either vehicle (1 mL/kg, s.c.) or LY379268 (3 mg/kg, s.c.). PCP (5 mg/kg, s.c) or vehicle was then administered to these four groups of pretreated animals. Behavioral monitoring began immediately following PCP or vehicle administration and continued over a 120-min period. Data are expressed as mean (± SEM) of total photocell counts over the 120-min test period. $^*P < .05$ compared to corresponding vehicle/control group. $^{**}P < .01$ compared to corresponding vehicle/control group. $^{***}P < .005$ compared to corresponding vehicle/control group. $^{\#}P < .05$ compared to corresponding vehicle/PCP group. $^{\#\#}P < .01$ compared to corresponding vehicle/PCP group. $^{\#\#\#}P < .005$ compared to corresponding vehicle/PCP group.

17. ISHIMARU, M.J. & M. TORU. 1997. The glutamate hypothesis of schizophrenia. CNS Drugs **7:** 47–67.
18. JOHNSON, K., D. DIECKMAN, T. BRITTON, *et al.* 2002. Selective non-amino acid allosteric mGlu2 receptor potentiators inhibit dural plasma extravasation: a potential role in the treatment of migraine. Neuropharmacology **43:** 291.
19. JOHNSON, M., M. BAEZ, T. BRITTON, *et al.* 2002. Subtype-selective positive allosteric modulators of the metabotropic glutamate 2 receptor: In vivo and in vivo characterization of novel mGlu2 potentiators. Neuropharmacology **43:** 291.
20. KNOPFEL, T., R. KUHN & H. ALLGEIER. 1995. Metabotropic glutamate receptors novel targets for drug development. J. Med. Chem. **38:** 1417–1426.
21. KRYSTAL, J.H., A. BELGER, C. D'SOUZA, *et al.* 1999. Therapeutic implications of the hyperglutamatergic effects of NMDA antagonists. Neuropsychopharmacology **21** (S6): S143–S157.
22. LEES, G.J. 2000. Pharmacology of AMPA / kainate receptor ligands and their therapeutic potential in neurological and psychiatric disorders. Drugs **59:** 33–78.
23. LEVINE, L., B. GAYDOS, D. SHEENAN, *et al.* 2002. The mGlu2/3 receptor agonist, LY354740, reduces panic anxiety induced by CO_2 challenge in patients diagnosed with panic disorder. Neuropharmacology **43:** 294.
24. LIN, Z. & P.K. KADABA. 1997. Molecular targets for the rational design of antieplieptic drugs and related neuroprotective agents. Medicinal Res. Rev. **17:** 537–572.
25. MITCHELL, S. 2003. LY379268 reduction in phencycline induced dopamine release. Neuropharmacology, in press.
26. MIU, P., K.R. JARVIE, V. RADHAKRISHNAN, *et al.* 2001. Novel AMPA receptor potentiators LY392098 and LY404187: effects on recombinant human AMPA receptors in vitro. Neuropharmacology **40:** 976–983.
27. MOGHADDAM, B. & B.W. ADAMS. 1998. Reversal of phencyclidine effects by a group II metabotropic glutamate receptor agonist in rats. Science **281:** 1349–1352.
28. PARSONS, C.G., W. DANYSZ & G. QUACK. 1998. Glutamate in CNS disorders as a target for drug development: an update. Drug News Perspectives **11:** 523–569.
29. PIJNENBURG, A.J., W.M. HONIG & J.M. VAN ROSSUM. 1975. Effects of antagonists upon locomotor stimulation induced by injection of dopamine and noradrenaline into the nucleus accumbens of nialamide-pretreated rats. Psychopharmacology **41:** 175–180.
30. SCHOEPP, D.D. & G.J. MAREK. 2002. Preclinical pharmacology of mGlu2/3 receptor agonists: novel agents for schizophrenia? Current Drug Targets–CNS and Neurological Disorders **1:** 215–225.
31. SCHOEPP, D.D., J.A. MONN, G.J. MAREK, *et al.* 1999. LY354740: A systemically active mGlu2/3 receptor agonist. CNS Drug Rev. **5:** 1–12
32. SCHOEPP, D.D., D.E. JANE & J.A. MONN. 1999. Pharmacological agents acting at subtypes of metabotropic glutamate receptors. Neuropharmacology **38:** 1431–1476.
33. SCHOEPP, D.D. 2001. Unveiling the functions of presynaptic metabotropic glutamate receptors in the central nervous system. Perspective: J. Pharmacol. Exp. Ther. **299:** 12–20.
34. SCHOEPP, D.D., L.R. LEVINE, B. GAYDOS & W.Z. POTTER. 2003. Metabotropic glutamate receptor agonists as a novel approach to treat anxiety/stress. Stress, in press.
35. SKOLNICK, P. 1999. Antidepressants of the new millennium. Eur. J. Pharmacol. **375:** 31–40.
36. SVENSSON L. & S. AHLENIUS. 1982. Functional importance of nucleus accumbens noradrenaline in the rat. Acta Pharmacol. Toxicol. (Copenh). **50:** 22–24.
37. SWANSON C.J. & D.D. SCHOEPP. 2002. The group II metabotropic glutamate receptor agonist (−)-2-oxa-4-aminobicyclo[3.1.0.]hexane-4,6-dicarboxylate (LY379268) and clozapine reverse phencyclidine-induced behaviors in monoamine-depleted rats. J. Pharmacol. Exp. Ther. **303:** 919–927.
38. SWANSON C.J., K.W. PERRY & D.D. SCHOEPP. 2003. Blockade of phencyclidine-induced behavioral activation by the group II metabotropic glutamate receptor agonist, LY379268, correlates with effect on extracellular noradrenaline but not dopamine levels in the nucleus accumbens shell of the rat. Pharmacologist. In press.

39. SWANSON, C.J., K. PERRY & D.D. SCHOEPP. 2003. The group II metabotropic glutamate receptor agonist, LY354740, blocks immobilization-induced increases in catecholamine release in the rat medial prefrontal cortex. Pharmacologist. In press.
40. WILEY, J. 1997. Behavioral pharmacology of N-methyl-D-aspartate antagonists: implications for the study and pharmacotherapy of anxiety and schizophrenia. Exp. Clin. Psychopharmacology **5:** 365–374.

Converging Evidence of NMDA Receptor Hypofunction in the Pathophysiology of Schizophrenia

JOSEPH T. COYLE,[a,c] GUOCHUAN TSAI,[a,c] AND DONALD GOFF[b,c]

[a]McLean Hospital, Belmont, Massachusetts 02478, USA

[b]Department of Psychiatry, Massachusetts General Hospital, Boston, Massachusetts, USA

[c]Department of Psychiatry, Harvard Medical School, Boston, Massachusetts, USA

> ABSTRACT: Numerous clinical studies demonstrate that subanesthetic doses of dissociative anesthetics, which are noncompetitive antagonists at the NMDA receptor, replicate in normal subjects the cognitive impairments, negative symptoms, and brain functional abnormalities of schizophrenia. Postmortem and genetic studies have identified several abnormalities associated with schizophrenia that would interfere with the activation of the glycine modulatory site on the NMDA receptor. Placebo-controlled clinical trials with agents that directly or indirectly activate the glycine modulatory site consistently reduce negative symptoms and frequently improve cognition in patients with chronic schizophrenia who are receiving concurrent typical antipsychotics. Thus, there is convincing evidence that hypofunction of a subset of NMDA receptors may contribute to the symptomatic features of schizophrenia.
>
> KEYWORDS: NMDA receptor; schizophrenia; glycine modulatory site; glycine; D-serine; D-cycloserine; negative symptoms; GlyT1

INTRODUCTION

Schizophrenia is a disorder that affects multiple domains resulting in positive symptoms (delusions and hallucinations), negative symptoms (apathy, social isolation), and cognitive symptoms (e.g., impairments in memory and problem solving). While positive symptoms wax and wane, negative symptoms and cognitive impairments are more enduring, cause greater disability, and correlate with cortical atrophy/ventricular enlargement.[1] Schizophrenia is highly heritable (80%) with the risk in first-degree relatives being 12-fold greater than that in the general population (1%).[2] The lack of complete concordance in identical twins (approximately 60%) indicates that environmental and epigenetic factors play a role in the phenotype, with most evidence pointing toward perinatal insults.[3] Linkage studies have identified a number of sites on the human genome associated with risk for schizophrenia. The

---

Address for correspondence: Joseph T. Coyle, M.D., McLean Hospital, 115 Mill Street, Belmont, MA 02478. Voice: 617-855-2101; fax: 617-855-2705.
joseph_coyle@hms.harvard.edu

disorder likely involves complex genetics with multiple risk alleles of small effect interacting to create the phenotype. Given the variation in phenotypic features of the disorder, it is likely that certain genes may involve the cognitive aspects of the disorder while others may contribute to the psychotic features.

NMDA RECEPTOR HYPOFUNCTION

For the last 40 years the dominant hypothesis about the pathophysiology of schizophrenia has been the dopamine hypothesis, which is based on the findings that dopamine-releasing stimulants cause psychosis and antipsychotics act by blocking D2 dopamine receptors.[4] The limitation of this hypothesis is that negative symptoms, cognitive impairments, and cortical atrophy are poorly responsive to typical antipsychotics.[5] An alternative hypothesis for the pathophysiology of schizophrenia arose from clinical observations that dissociative anesthetics, such as phencyclidine (PCP) and ketamine, produce a syndrome that is clinically indistinguishable from schizophrenia.[6,7] These phenomena were first linked to dysfunction of glutamatergic neurotransmission in 1991 when Javitt and Zukin[8] noted that the serum concentration of PCP associated with psychiatric symptoms corresponded to the level that blocks NMDA receptors (NMDAR). Dissociative anesthetics bind to sites within the channel of NMDAR where they act as noncompetitive, use-dependent antagonists.

The hypothesis that schizophrenia might result from NMDAR hypofunction gained support from subsequent studies in normal individuals under laboratory conditions.[1,9] The infusion of subanesthetic doses of ketamine produced a syndrome characterized by withdrawal, blunted affect, psychomotor retardation, delusions, and illusions (but not hallucinations). Furthermore, ketamine impaired performance on a number of cognitive tasks that require frontal cortex and/or hippocampal involvement including the Wisconsin Card Sorting Test, verbal declarative memory, delayed word call, and verbal fluency in a manner similar to what is observed in schizophrenia.[10,11] Dissociative anesthetics also reproduce some of the physiologic abnormalities, such as abnormal event related potentials (ERPs),[12] disrupted eye tracking,[13] and hypofrontality,[14] in functional imaging studies. Thus, dissociative anesthetics replicated the negative symptoms and cognitive impairments of schizophrenia, unlike the amphetamine model that recreates only the positive symptoms.

Preclinical studies indicate that dissociative anesthetics increase the release of dopamine in frontal cortex and ventral striatum, producing behaviors consistent with psychotomimetic effects.[15] Furthermore, in PET scans in normal humans, subanesthetic doses of ketamine produce an increased striatal release of dopamine with an amphetamine challenge similar to that observed in schizophrenic subjects.[16] These findings suggest that psychosis (i.e., positive symptoms) may be secondary to a primary defect in corticolimbic NMDAR hypofunction.

What is extremely important to appreciate about these challenge studies is that the impairments in cognition and cortical function caused by subanesthetic doses of ketamine were observed under conditions in which the individuals exhibited normal performance on the Mini-Mental Status Exam (a test for delirium/dementia), as is the case for subjects with schizophrenia.[9] This indicates that only a discrete subpopulation of NMDAR are affected by the low dose of ketamine.

GLUTAMATE RECEPTORS

Tremendous advances have been made in recent years in characterizing the dynamics of the glutamatergic synapse.[1] Two classes of ionotropic glutamate receptors are the AMPA/kainate receptors (AMPAR) and the NMDAR. The AMPAR (GluR 1–4) are the primary mediators of excitatory postsynaptic currents (EPSCs). While the NMDAR (NR1; NR2A–D) may contribute to the EPSC, they play a more fundamental role in coincidence detection because of their peculiar biophysical characteristic of being voltage dependent.[17] Thus, at the resting membrane potential, the channel of the NMDAR is blocked by Mg^{2+}, which is removed upon depolarization. The channels of the NMDAR are sufficiently large to readily transduce Ca^{2+}, which activates intracellular kinases that ultimately regulate gene expression. The recruitment of NMDAR during high presynaptic glutamatergic activity results in the permanent increase in synaptic efficacy known as long-term potentiation (LTP).[17] Influx of Ca^{2+} through the NMDAR during LTP causes the recruitment of AMPAR to the synapse from intracellular stores. Persistent hyperactivity through a glutamatergic pathway can cause sprouting of post-synaptic spines via NMDAR activation, further strengthening synaptic connections. NMDAR activation has trophic effects, especially during development, with inactivity of NMDAR resulting in neuronal apoptosis.

Another unique characteristic of the NMDAR is that in addition to the binding site for the agonist, glutamate, there is a second glycine modulatory site (GMS) to which glycine and/or D-serine bind. The GMS must be occupied for glutamate to open the channel. The availability of D-serine depends upon the activities of serine racemase (SR) and the degrading enzyme D-amino acid oxidase (DAAO),[18] whereas the availability of glycine is determined by the activity of the glycine transporter, GlyT 1.[19] Notably, both SR and GlyT1 as well as the glutamate transporters that protect against excitotoxicity (EAAT 1 and 2), are expressed exclusively in astrocytes, indicating a vital role of astroglia in modulating glutamatergic neurotransmission. Our research and that of others on the hippocampal Schaffer collateral synapse indicate that the GMS is not saturated by glycine/D-serine.[20–22] Thus, regulation of the availability of glycine/D-serine at the GMS plays a critical role in optimal NMDAR function.

NEGATIVE MODULATION OF NMDAR

In addition to the negative modulation of the GMS on the NMDAR by GlyT 1 and DAAO, there are endogenous antagonists at the GMS that behave as noncompetitive antagonists for glutamate at the NMDAR analogous to the dissociative anesthetics. Kynurenic acid, a metabolite of tryptophan, is a functionally relevant GMS antagonist. Postmortem studies of the brains from a large cohort of subjects with schizophrenia have revealed a significant elevation of kynurenic acid in prefrontal cortex but not motor cortex in the schizophrenic subjects.[23] N-Acetyl aspartyl glutamate (NAAG), an agonist at the mGluR 3 receptor, which inhibits glutamate release, and an antagonist at the NMDAR GMS, is degraded by glutamate carboxypeptidase II (GCP II), an enzyme restricted to astroglia in brain.[24] Postmortem studies in schizophrenic subjects from three different brain banks have revealed decreased expression of GCP II in prefrontal cortex and temporal cortex as determined by enzyme activity,[25] DNA micro array,[26] and quantitative PCR (S. Bahn, personal communication). Thus elevated endogenous antagonists could account for NMDAR hypofunction in some subjects with schizophrenia.

Recent postmortem and genetic findings lend further support to the hypothesis of dysregulation NMDAR modulation. In related findings, mGluR 3, at which NAAG acts[27] has been linked to schizophrenia.[28] A mutation at 13q34 linked to schizophrenia affects the gene G72, a primate-specific modulator of DAAO; this mutation increases the activity of DAAO. Four SNPs for the DAAO gene have been linked to schizophrenia in Canadian pedigrees.[29] Reduced levels of blood D-serine have been reported in schizophrenia.[30] The neuregulin 1 gene, which among its many actions regulates NMDAR function,[31] has been linked to schizophrenia.[32]

NMDAR–GABA CONNECTION

We have proposed a heuristic model for the pathophysiology of schizophrenia that attempted to reconcile the neuropathologic, synaptic chemical, and neurocognitive features of the disorder (FIG. 1).[33] We hypothesize an hypofunction of the

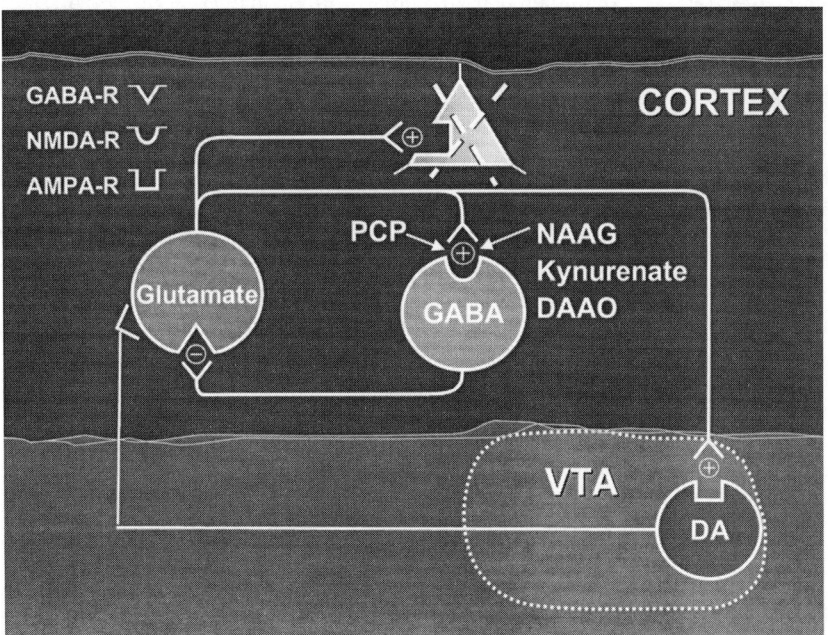

FIGURE 1. Schematic representation of critical synaptic relationships hypothesized to be involved in pathophysiology of schizophrenia. NMDAR on cortico-limbic GABAergic interneurons are differentially sensitive to exogenous (ketamine) or endogenous antagonists (NAAG, kynurenate) or low D-serine (DAAO). This results in disinhibition of glutamatergic pyramidal cells, disrupting cortical processing and enhancing cortico-limbic dopamine release. Persistent activation of AMPA receptors could contribute to neuropil atrophy.

NMDAR on corticolimbic GABAergic interneurons that disrupts columnar functional integrity causing cognitive impairments and negative symptoms,[34] a disinhibition of the glutamatergic efferents, which drives increased subcortical dopamine release resulting in positive symptoms,[15] and cortical atrophy.[35] Neurophysiologic studies suggest that the NMDAR on the GABAergic interneurons may be disproportionately sensitive to NMDAR antagonists.[34,36] This inference is supported by postmortem findings of decreased GAD67,[37,38] decreased GABA Transporter (GAT);[39] and increased GABA-A receptor[40,41] in the frontal cortex/hippocampus in schizophrenia. These very same alterations in GABAergic markers are replicated in rat cortex by chronic MK-801 treatment.[42] The recent findings of decreased serum D-serine, genetic linkages to G72 and DAAO, and reduced GCP II expression in schizophrenia emphasize the validity of considering the GMS as a final common pathway for one family of risk genes contributing to the schizophrenia phenotype.

GMS CLINICAL TRIALS

If NMDAR hypofunction is involved in the pathophysiology of schizophrenia, then interventions directed at enhancing NMDAR function should reduce symptoms. To avoid the excitotoxic consequences of direct activation of the NMDAR, clinical trials have focused on agents that directly or indirectly enhance GMS activation. Preclinical studies indicate that GMS agonists can attenuate PCP-induced behaviors but do not promote excitotoxicity. Over the last seven years, more than six placebo-controlled studies of GMS agonists effects on the symptoms of schizophrenia have been carried out and were reviewed by Coyle and colleagues.[43] These studies have focused primarily on patients with chronic schizophrenia and prominent negative symptoms, who were on a stable regimen of antipsychotics, primarily typical antipsychotics. The reason for this design was that the basic hypothesis was that GMS agonists would affect primarily negative symptoms and cognition, which are relatively insensitive to typical antipsychotics.

Javitt and his colleagues[44,45] carried out double-blind placebo-controlled studies of high-dose glycine added to typical antipsychotics ranging from approximately 30–60 g/day (0.4–0.8 g/kg/day) (TABLE 1). These studies revealed significant improvements in negative symptoms and cognition with negligible effects on positive symptoms. Maximum effects occurred between four and six weeks of treatment. Notably, in crossover arms, those randomized to placebo from glycine in the second half did not deteriorate, suggesting that the glycine treatment caused some persistent change.

The authors have carried out several metabolic studies on their subjects. The chronic glycine treatment caused a 3.5-fold increase in serum glycine and a much more prolonged elevation than observed with single dose administration in naive subjects. Furthermore, serum serine levels were also significantly elevated which may be therapeutically relevant as it is a precursor to D-serine, a potent agonist at the GMS. Notably, the improvements in negative symptoms ($P = .002$), cognition ($P = .01$), and depression ($P = .002$) on the Positive and Negative Symptom Scale (PANSS) inversely correlated with pretreatment serum glycine levels. In other words, those schizophrenics with the lowest serum glycine levels at baseline exhib-

TABLE 1. Effects of glycine modulatory site agonists in schizophrenic patients receiving antipsychotics

Agent	Reference	Mechanism	N	Negative	Cognitive	Positive
Glycine	44	Agonist	14^a	S	S	—
Glycine	45	Agonist	22^b	S	S	—
D-Cycloserine	49	Partial agonist	9^b	S	S	—
D-Cycloserine	50	Partial agonist	47^a	S	—	—
D-Cycloserine	51	Partial agonist	24^b	S	NT	—
D-Serine	46	Agonist	28^a	S	S	S
Sarcosine	52	GlyT1 inhibitor	38^a	S	S	S

N = number of patients; S=significant; ($P<.05$); NT=not tested; ^{a}Parallel placebo control; ^{b}Placebo controlled crossover.

ited the greatest improvement with glycine, a finding independently reported in studies with the partial agonist, D-cycloserine.

D-serine is a full agonist at the GMS on the NMDA receptor and is threefold more potent than glycine and more readily crosses the blood-brain barrier. Tsai and colleagues[46] carried out a double-blind placebo-controlled trial of D-serine (30 mg/kg/day) on a cohort of 28 patients with chronic schizophrenia, who were poorly responsive to neuroleptics (none receiving clozapine) and had prominent negative symptoms (Scale for the Assessment of Negative Symptoms (SANS) > 40). Six weeks of D-serine treatment was associated with significant reductions in negative symptoms, cognitive symptoms and positive symptoms. Cognitive improvement imputed from the PANSS was corroborated by a significantly better performance on the Wisconsin Card Sort Test. Metabolic studies on the subjects revealed that serum D-serine levels increased nearly 50-fold, whereas serum glycine, glutamate, and aspartate levels were unchanged. Symptom improvement observed at six weeks of D-serine treatment correlated significantly with serum D-serine levels.

D-Cycloserine was shown in *Xenopus* oocyte studies to act as a partial agonist at the GMS on the NMDAR with 60% of the efficacy of glycine; it inhibited the effects of saturating concentrations of glycine by 40%.[47] Furthermore, at high concentrations, D-cycloserine inhibits serine racemase, resulting in decreased synthesis of D-serine,[48] which should produce a U-shaped dose-response curve. Consistent with these neurophysiologic characteristics, Goff and colleagues[49] found a U-shaped dose response in a dose-finding study in which D-cycloserine was added to typical antipsychotics in a cohort of patients with schizophrenia characterized by prominent negative symptoms. A 21% reduction in negative symptoms occurred at 50 mg/day, the same dose associated with significant improvement on the Sternberg item recognition paradigm, a task that involves frontal lobe functions.

Goff and colleagues[50] carried out a parallel double-blind placebo-controlled study of D-cycloserine at 50 mg/day in 46 schizophrenic patients, who were receiving stable doses of typical antipsychotics for four months and who had SANS scores of 30 or greater. The eight-week trial revealed a 23% reduction in SANS total score, which was substantially represented by an improvement in blunted affect. No effects on positive symptoms, mood, or extrapyramidal symptoms were observed. Heresco-Levy and colleagues[51] replicated these findings on negative symptoms in 24 chronic schizophrenic patients receiving typical antipsychotics, resperidol or olanzapine in a placebo-controlled crossover design study.

N-methyl glycine (sarcosine) is a potent inhibitor of GlyT-1. Tsai and colleagues[52] have carried out a double-blind placebo-controlled study of sarcosine (30 mg/kg/day) involving 38 patients with chronic schizophrenia, who were receiving concurrent resperidone or typical antipsychotics. They observed significant reductions in negative symptoms, cognitive symptoms, and positive symptoms that became apparent by the fourth week of this six week trial.

Thus, these placebo-controlled trials of four different agents that have in common only one known property—direct or indirect activation of the GMS on the NMDAR—consistently reduced negative symptoms and frequently improved cognitive function in patients with chronic schizophrenia, who were receiving concurrent antipsychotics. Notably, both D-serine and sarcosine also affected positive symptoms, which may reflect their greater intrinsic potency than glycine or the partial agonist D-cycloserine.

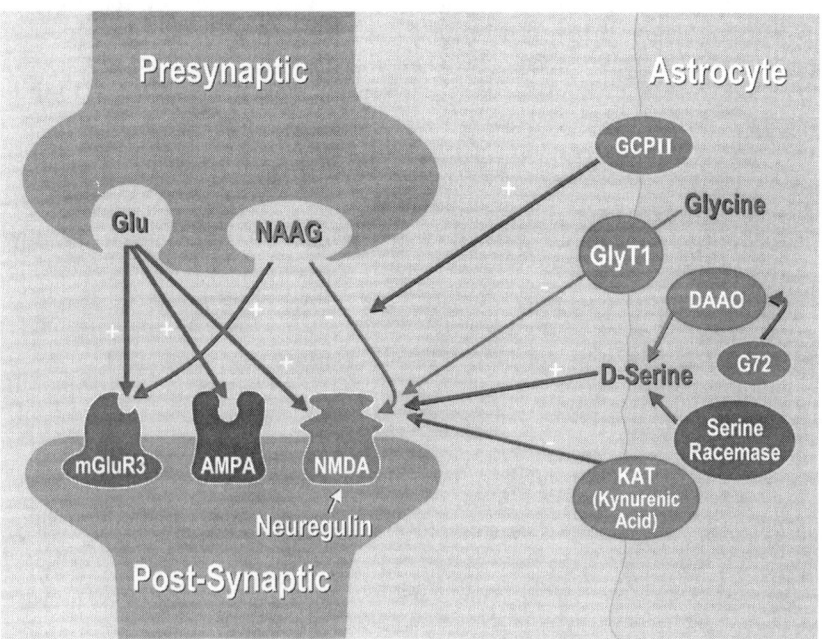

FIGURE 2. NMDA receptor modulation relevant to schizophrenia. NAAG, a GMS antagonist and mGluR3 agonist, is degraded by GCP II so that reduced GCP II increases GMS inhibition. Kynurenic acid is a GMS antagonist. GlyT1 reduces glycine availability at the GMS. Serine racemase synthesizes D-serine, a GMS agonist, whereas DAAO, which is activated by mutant G72, decreases D-serine availability at the GMS.

CONCLUSION

Clinical studies of the neurocognitive and functional effects of subanesthetic doses of the dissociative anesthetics in normal subjects have demonstrated compelling parallels to the negative symptoms, cognitive deficits, and brain functional abnormalities observed in individuals with schizophrenia. Recent postmortem and genetic studies have identified abnormalities in schizophrenia that would result in hypofunction of the NMDAR. These findings include reduced GCP II expression, elevated kynurenic acid levels, reduced D-serine levels, increased DAAO activity, and mutations in neuregulin 1. Thus, the NMDAR, especially those located in corticolimbic GABAergic interneurons, may represent a converging target of several risk genes associated with the cognitive dysfunction and negative symptoms found with schizophrenia (FIG. 2). That hypofunction of the NMDA may contribute to the symptoms of schizophrenia has received convincing support from placebo-controlled clinical trials of four agents that directly or indirectly enhance GMS function—D-cycloserine, glycine, D-serine, and sarcosine. Thus, the GMS on the NMDAR, which is not saturated by endogenous agonists, appears to be a potential site for therapeutic intervention in schizophrenia.

Brain imaging studies have consistently revealed reduced cortical volume in schizophrenia, which may be progressive during the early stages of the disorder. This volume loss may result from atrophy of the neuropil and shrinkage of pyramidal cells in critical cortico-limbic regions.[52] Given the central role of the NMDAR in the functional and structural plasticity of brain neurons, the effect of chronic treatment with GMS agonists may be more profound than simply acute enhancement of cognitive function.[17] Rather, coupled with cognitive rehabilitation, GMS therapy could potentially reverse cortical atrophy and promote the development of corrective synaptic connections.

ACKNOWLEDGMENTS

We thank Frances MacNeil for assistance in preparing this manuscript. Some of the research described in this article was supported by an National Institute of Mental Health Conte Center on the Neurobiology Schizophrenia and MH-572901 and grants from the Stanley Foundation and National Alliance for Research on Schizophrenia and Affective Disorders (NARSAD) to Joseph T. Coyle.

REFERENCES

1. GOFF, D. & J.T. COYLE. 2001. The emerging role of glutamate in the pathophysiology and treatment of schizophrenia. Am. J. Psychiatry **158**(9): 1367–1377.
2. HARRISON, P.J. & M.J. OWEN. 2003. Genes for schizophrenia? Recent findings and their pathophysiological implications. Lancet **361**(9355): 417–419.
3. TSUANG, M. 2000. Schizophrenia: genes and environment. Biol. Psychiatry **47**(3): 210–220.
4. SNYDER, S.H. 1981. Dopamine receptors, neuroleptics, and schizophrenia. Am. J. Psychiatry **138**(4): 460–464.
5. MELTZER, H.Y. 1997. Treatment-resistant schizophrenia—the role of clozapine. Curr Med. Res. Opin. **14**(1): 1–20.

6. ITIL, T. *et al.* 1967. Effect of phencyclidine in chronic schizophrenics. Can. Psychiatr. Assoc. J. **12**(2): 209–212.
7. LUBY, E.D. *et al.* 1959. Study of a new schizophrenomimetic drug—sernyl. Arch. Neurol. Psychiatry **81**: 363–369.
8. JAVITT, D.C. & S.R. ZUKIN. 1991. Recent advances in the phencyclidine model of schizophrenia. Am. J. Psychiatry **148**(10): 1301–1308.
9. KRYSTAL, J.H. *et al.* 1994. Subanaesthetic effects of the noncompetitive NMDA antagonist, ketamine, in humans. Psychotomimetic, perceptual, cognitive, and neuroendocrine responses. Arch. Gen. Psychiatry **51**(3): 199–214.
10. MALHOTRA, A.K. *et al.* 1997. Ketamine-induced exacerbation of psychotic symptoms and cognitive impairment in neuroleptic-free schizophrenics. Neuropsychopharmacology **17**(3): 141–150.
11. NEWCOMER, J. *et al.* 1999. Ketamine-induced NMDA receptor hypofunction as a model of memory impairment and psychosis. Neuropsychopharmacology **20**(2): 106–118.
12. UMBRICHT, D. *et al.* 2002. Mismatch negativity predicts psychotic experiences induced by NMDA receptor antagonist in healthy volunteers. Biol. Psychiatry **51**(5): 400–406.
13. AVILA, M.T. *et al.* 2002. Effects of ketamine on leading saccades during smooth-pursuit eye movements may implicate cerebellar dysfunction in schizophrenia. Am. J. Psychiatry **159**(9): 1490–1496.
14. VOLLENWEIDER, F.X. *et al.* 1997. Metabolic hyperfrontality and psychopathology in the ketamine model of psychosis using positron emission tomography (PET) and [^{18}F] fluorodeoxyglucose (FDG). Eur. Neuropsychopharmacol. **7**(1): 9–24.
15. MOGHADDAM, B. *et al.* 1997. Activation of glutamatergic neurotransmission by ketamine: a novel step in the pathway from NMDA receptor blockade to dopaminergic and cognitive disruptions associated with the prefrontal cortex. J. Neurosci. **17**(8): 2921–2927.
16. KEGELES, L.S. *et al.* 2000. Modulation of amphetamine-induced striatal dopamine release by ketamine in humans: implications for schizophrenia. Proc. Natl. Acad. Sci. USA **48**(7): 627–640.
17. TSIEN, J.Z. 2000. Linking Hebb's coincidence-detection to memory formation. Curr. Opin. Neurobiol. **10**(2): 266–273.
18. MOTHET, J.-P. *et al.* 2000. D-Serine is an endogenous ligand for the glycine site of the N-methyl-D-aspartate receptor. Proc. Natl. Acad. Sci. USA **97**(9): 4926–4931.
19. KIM, K.M. *et al.* 1994. Cloning of the human glycine transporter type 1: molecular and pharmacological characterization of novel isoform variants and chromosomal localization of the gene in the human and mouse genomes. Mol. Pharmacol. **45**(4): 608–617.
20. BERGERON, R. *et al.* 1998. Modulation of N-methyl-D-aspartate receptor function by glycine transport. Proc. Natl. Acad. Sci. USA **95**(26): 15730–15734.
21. BERGER, A.J. *et al.* 1998. Glycine uptake governs glycine site occupancy at NMDA receptors of excitatory synapses. J. Neurophysiol. **80**(6): 3336–3340.
22. CHEN, L. *et al.* 2003. Glycine transporter-1 blockade potentiates NMDA-mediated responses in rat prefrontal cortical neurons in vitro and in vivo. J. Neurophysiol. **89**(2): 691–703.
23. SCHWARCZ, R. *et al.* 2001. Increased cortical kynurenate content in schizophrenia. Biol. Psychiatry **50**(7): 521–530.
24. BERGER, U.V. *et al.* 1999. Glutamate carboxypeptidase II is expressed by astrocytes in the rat nervous system. J. Comp. Neurol. **415**: 52–64.
25. TSAI, G. *et al.* 1995. Abnormal excitatory neurotransmitter metabolism in schizophrenic brains. Arch. Gen. Psychiatry **52**(10): 829–836.
26. HAKAK, Y. *et al.* 2001. Genome-wide expression analysis reveals dysregulation of myelination-related genes in chronic schizophrenia. Proc. Natl. Acad. Sci. USA **98**(8): 4746–4751.
27. LEA, P.M. IV *et al.* 2001 beta-NAAG rescues LTP from blockade by NAAG in rat dentate gyrus via the type 3 metabotropic glutamate receptor. J. Neurophysiol. **85**(3): 1097–1106.
28. MARTI, S.B. *et al.* 2002. Metabotropic glutamate receptor 3 (GRM3) gene variation is not associated with schizophrenia or bipolar affective disorder in the German population. Am. J. Med. Genet. **114**(1): 46–50.

29. CHUMAKOV, I. *et al.* 2002 Genetic and physiological data implicating the new human gene G72 and the gene for D-amino acid oxidase in schizophrenia. Proc. Natl. Acad. Sci. USA **99**(21): 13675–13680.
30. HASHIMOTO, K. *et al.* 2003. Decreased serum levels of D-serine in patients with schizophrenia: evidence in support of the N-methyl-D-aspartate receptor hypofunction hypothesis of schizophrenia. Arch. Gen. Psychiatry **60**(6): 572–576.
31. BUONANNO, A. *et al.* 2001 Receptor signaling pathways in the nervous system. Curr. Opin. Neurobiol. **11**(3): 287–296.
32. STEFANSSON, H. *et al.* 2002. Neuregulin 1 and susceptibility to schizophrenia. Am. J. Hum. Genet. **71**(4): 877–892.
33. COYLE, J.T. 1996. The glutamatergic dysfunction hypothesis for schizophrenia. Harvard Rev. Psychiatry **3**(5): 241–253.
34. GRUNZE, H.C. *et al.* 1996. NMDA-dependent modulation of CA1 local circuit inhibition. J. Neurosci. **16**(6): 2034–2043.
35. FARBER, N.B. *et al.* 1995. Age-specific neurotoxicity in the rat associated with NMDA receptor blockade: potential relevance to schizophrenia? Biol. Psychiatry **15**: 788–796.
36. LI, Q. *et al.* 2002. NMDA receptor antagonists disinhibit rat posterior cingulate and retrosplenial cortices: a potential mechanism of neurotoxicity. J. Neurosci. **22**: 3070–3080.
37. GUIDOTTI, A. *et al.* 2000. Decrease in reelin and glutamic acid decarboxylase67 (GAD 67) expression in schizophrenia and bipolar disorder: A postmortem study. Arch. Gen Psychiatry 57(11): 1061-1069.
38. VOLK, D. *et al.* 2000. Decreased glutamic acid decarboxylase67 messenger RNA expression in a subset of prefrontal cortical gamma-aminobutyric acid neurons in subjects with schizophrenia. Arch. Gen Psychiatry **57**: 237–245.
39. VOLK, D. *et al.* 2001. GABA transporter-1 mRNA in prefrontal cortex in schizophrenia: decreased expression in a subset of neurons. Am. J. Psychiatry **158**(2): 256–265.
40. VOLK, D.W. *et al.* 2002. Reciprocal alterations in pre- and post-synaptic inhibitory markers at chandler cell inputs to pyramidal neurons in schizophrenia. Cereb. Cortex **12**(10): 1063–1070.
41. BENES, F.M. *et al.* 1996. Up-regulation of GABAA receptor binding on neurons of the prefrontal cortex in schizophrenic subjects. Neuroscience **75**(4): 1021–1031.
42. PAULSON, L. *et al.* 2003. Comparative genome- and proteome analysis of cerebral cortex from MK-801-treated rats. J. Neurosci. Res. **71**(4): 526–533.
43. COYLE, J.T. et al. 2002. Ionotropic glutamate receptors as therapeutic targets in schizophrenia. Curr. Drug Targets CNS Neurol. Disorders **1**: 183–189.
44. JAVITT, D.C. *et al.* 1999. AE Bennett Research Award. Reversal of phencyclidine-induced effects by glycine and glycine transport inhibitors. Biol. Psychiatry **45**(6): 668–679.
45. HERESCO-LEVY, U. *et al.* 1999. Efficacy of high-dose glycine in the treatment of enduring negative symptoms of schizophrenia. Arch. Gen. Psychiatry **56**(1): 29–36.
46. TSAI, G. *et al.* 1998. D-Serine added to antipsychotics for the treatment of schizophrenia. Biol. Psychiatry **44**(11): 1081–1089.
47. SHEININ, A. *et al.* 2001. Subunit specificity and mechanism of action of NMDA partial agonist D-cycloserine. Neuropharmacology **41**(2): 151–158.
48. COOK, S.P. *et al.* 2002. Direct calcium binding results in activation of brain serine racemase. J. Biol. Chem. **277**: 27782–27792.
49. GOFF, D.C, *et al.* 1995. Dose-finding trial of D-cycloserine added to neuroleptics for negative symptoms in schizophrenia. Am. J. Psychiatry **152**(8): 1213–1215.
50. GOFF, D.C. *et al.* 1999. A placebo-controlled trial of D-cycloserine added to conventional neuroleptics in patients with schizophrenia. Arch. Gen. Psychiatry **56**(1): 21–27.
51. HORESCO-LEVY, U. *et al.* 2002. Placebo-controlled trial of D-cycloserine added to conventional neuroleptics, olanzapine, or risperidone in schizophrenia. Am. J. Psychiatry **159**: 480–482.
52. TSAI, G. *et al.* 2003. Glycine transporter I inhibitor, N-methylglycine (sarcosine), added to antipsychotics for treatment of schizophrenia. Biol. Psychiatry. In press.
53. KASAI, K. *et al.* 2003. Progressive decrease of left Heschl gyrus and planum temporale gray matter volume in first-episode schizophrenia: a longitudinal magnetic resonance imaging study. Arch. Gen. Psychiatry **60**(8): 766–775.

Glutamatergic Agents for Cocaine Dependence

CHARLES DACKIS AND CHARLES O'BRIEN

Treatment Research Center, University of Pennsylvania,
3900 Chestnut Street, Philadelphia, Pennsylvania 19104, USA

ABSTRACT: Effective medications for cocaine dependence are needed to improve outcome in this chronic, relapsing disorder. Medications affecting glutamate function are reasonable candidates for investigation, given the involvement of glutamate circuits in reward-related brain regions and evdence of cocaine-induced glutamatergic dysregulation. In addition, it is increasingly apparent that glutamatergic mechanisms underlie several clinical aspects of cocaine dependence, including euphoria, withdrawal, craving, and hedonic dysfunction. Even denial, traditionally viewed as purely psychological, may result, in part, from dysfunctional glutamate-rich cortical regions. We review the involvement of glutamate in reward-related circuits, the acute and chronic effects of cocaine on these pathways, and glutamatergic mechanisms that contribute to the neurobiology of cocaine dependence. We also present preliminary data from our research of modafinil, a glutamate-enhancing agent with promise in the treatment of cocaine-addicted individuals.

KEYWORDS: cocaine; glutamate; modafinil; pharmacotherapy; craving

CLINICAL ASPECTS OF COCAINE DEPENDENCE

About 16% of individuals who try cocaine become addicted,[1] often experiencing a rapid loss of control over drug intake and cocaine-seeking behaviors. Cocaine dependence is largely driven by a cycle of euphoria and craving that exerts positive and negative reinforcement of continued use. The brief "rush" of cocaine euphoria surpasses the normal range of human hedonic experience, and its extraordinary lure should not be underestimated in the clinical area. After only a few minutes, cocaine euphoria gives way to cocaine-induced craving[2,3] that typically leads to a binge pattern of use and exhausts any available supply. Protracted binges (hours or days) are often followed by periods of cocaine withdrawal, characterized by hypersomnia, anergia, depressed mood, poor concentration, overeating, and psychomotor retardation.[4,5] Severe cocaine withdrawal occurs in a subset of cocaine users and predicts poor clinical outcome,[6,7] suggesting that its pharmacological reversal might be clin-

Address for correspondence: Charles Dackis, Treatment Research Center, University of Pennsylvania, 3900 Chestnut Street, Philadelphia, PA 19104. Voice: 215-662-8752; fax: 215-243-4665.
Dackis@mail.med.upenn.edu

ically advantageous.[8] The presence of these neurovegetative symptoms may also identify patients with hedonic dysregulation resulting from cocaine-induced neuroadaptations.[9]

Aside from cocaine-induced craving, patients also describe withdrawal[10] and cue-induced craving.[3] Withdrawal craving gradually emerges after the previous binge and exerts a dysphoric pull to cocaine. Cue-induced craving is dramatically evoked by environmental cues (people, places, and things) that have been paired with cocaine use through conditioned learning. Hypermetabolic responses during cue-induced craving have been reported in the orbitofrontal cortex, anterior cingulate, and amygdala by several positron emission tomography (PET)[11–15] and functional magnetic resonance imaging (fMRI)[16–19] studies. Interestingly, similar anatomical activation in these glutamate-rich regions occurs when normal subjects view sexually explicit videos,[17] linking sexual arousal and cocaine craving to common neuronal substrates. Cue-induced craving is persistent, lasting months or years, and frequently leads to relapse in the clinical setting. Consequently, rehabilitative strategies for cocaine dependence advocate the vigorous avoidance of cues that trigger craving.[20] A medication capable of dampening cue reactivity would therefore provide significant value in the clinical management of these patients.

Individuals addicted to cocaine are remarkably willing to risk death, incarceration, medical and psychiatric complications, job loss, and family turmoil in the pursuit of cocaine. Denial, a hallmark of addiction, often shields patients from perceiving these risks and interferes with their motivation for change. While denial undoubtedly involves psychological factors, cocaine-induced alterations in the medial prefrontal cortex (PFC) might also contribute to this critical phenomenon. The PFC mediates conscious executive functions, including decision making, weighing risk against benefit, assigning emotional valence to experiences, and suppressing limbic impulses. On PET, cocaine-addicted patients have persistently reduced metabolism (hypofrontality) in the orbitofrontal cortex and anterior cingulate[12,21] that correlates with low D2 binding[22] and the extent of prior cocaine use.[23] On MRI testing, cocaine-dependent subjects also have reductions in orbitofrontal gray matter density[24] and diminished prefrontal volume.[25] Furthermore, several neuropsychological studies have identified prefrontal deficits in cocaine-dependent subjects.[26–28] While it is unclear whether PFC dysfunction precedes or results from cocaine use, it might contribute to clinical manifestations of cocaine dependence that have traditionally been viewed as purely psychological, including denial, poor judgment, impulsivity, and pathological risk tolerance.[20]

Since cocaine dependence is associated with poor clinical outcome and recidivism,[29,30] pharmacological agents for this disorder have been intensely investigated to improve treatment response. The involvement of dopamine (DA) neurotransmission in cocaine reward[31] prompted numerous trials of DA agonists and antagonists that yielded disappointing results. In fact, more than two decades of intense research has failed to identify any agent with proven robust efficacy for cocaine dependence. This elusive search has recently gained impetus from studies that further elaborate many neuronal mechanisms underlying cocaine dependence. Particular attention has recently been directed toward agents affecting glutamate and GABA systems that are functionally linked to reward-related DA neurons. While the rationales for testing glutamatergic and GABAergic agents are speculative, preliminary studies are encouraging.

PRELIMINARY STUDIES WITH MODAFINIL

Since severe cocaine withdrawal reliably predicts poor clinical outcome, we decided to test an agent with properties that might reverse withdrawal symptoms. Detoxification is standard treatment for alcohol and heroin withdrawal,[32] and represents a reasonable approach to cocaine withdrawal. Modafinil, recently approved for the treatment of narcolepsy, has clinical effects in nonaddicted patients that are opposite to the cocaine-withdrawal syndrome. In particular, modafinil is alerting, increases energy, enhances mood, reduces appetite, and improves concentration and attention.[8,33] Modafinil also enhances glutamate function through an unknown mechanism that leads to an increase in glutamate synthesis and striatal glutamate brain levels,[34,35] and reduced striatal GABA release.[36,37]

Before testing modafinil in outpatients, we conducted a double-blind drug interaction study of modafinil and intravenous cocaine that found no additive effects on cocaine-induced vital sign or electrocardiogram changes.[8] This study also found that modafinil attenuated cocaine euphoria under controlled conditions. We subsequently studied 17 (4 female, 13 male; mean age 36.7, SD 5.2) cocaine-dependent subjects who were randomized to receive 200 mg/day ($n=10$) or 400 mg/day ($n=7$) of open-label modafinil for an 8-week outpatient trial that included twice-weekly cognitive behavioral therapy (CBT). Sixteen subjects smoked cocaine (crack), and one used intravenously. Written informed consent was obtained after procedures were fully explained, and an appropriate institutional review board approved the project. All subjects had severe baseline cocaine withdrawal symptoms, quantified by high scores (≥ 22) on the Cocaine Selective Severity Assessment (CSSA) (mean 45.1, SD

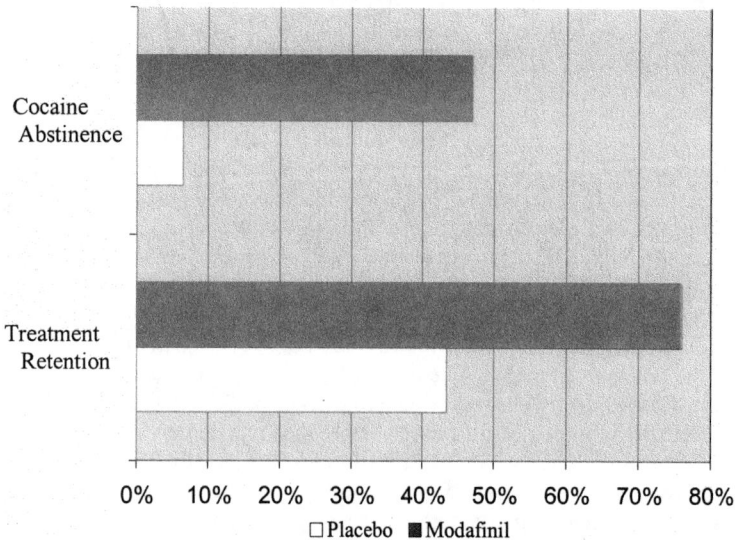

FIGURE 1. Cocaine abstinence (47% vs. 6.5%) and clinic attendance (76% vs. 43%) were superior in modafinil-treated subjects versus placebo-treated subjects from another study.

12.5). As noted previously, high baseline CSSA scores predict poor outcome[7] and may differentiate patients with cocaine-induced neuroadaptations who are more likely to benefit from pharmacotherapy.[38,39] Abstinence was assessed twice weekly by urine testing (16 samples) for the cocaine metabolite, benzoylecgonine (BE), and treatment retention was based on attendance logs. The average cocaine abstinence rate for the modafinil-treated subjects (number of urine samples that were BE-negative divided by 16) was 47%; the average treatment retention rate (percent of 16 scheduled CBT treatments attended) was 76%. These results compare very favorably to average rates of abstinence (6.5%) and retention (43%) in a benchmark group ($n = 22$) of placebo-treated, high CSSA (mean 37.0, SD 14.4) cocaine subjects from another clinical trial that was conducted concurrently in our center (see FIG. 1).[39] Furthermore, abstinence was probably understated in both groups because failing to provide a urine sample was counted the same as a BE-positive urine. We found that both modafinil dosages were well tolerated, and there was no evidence of modafinil overuse, which is consistent with its reportedly low abuse potential.[40] Subjects anecdotally reported diminished cue-induced craving, and four subjects who used cocaine while taking modafinil spontaneously reported absent or diminished euphoria. These preliminary, open-label findings encouraged us to launch a controlled clinical trial of modafinil for cocaine dependence.

GLUTAMATE AND REWARD-RELATED CIRCUITRY

Glutamate-containing pyramidal cells are heavily represented in the PFC, a cortical region that mediates such executive functions as the anticipation, pursuit, and enjoyment of rewarding activities, the recognition of survival-related environmental stimuli, and the execution of goal-directed activities. Neuronal activity within the orbitofrontal cortex, a PFC region, is associated with unpleasant emotional states and embarrassment.[27] Antisocial males have impaired orbitofrontal function on neuropsychological testing,[41] and orbitofrontal lesions can produce profound deficits in social behavior thought to result from inadequate self-punishment.[42] The orbitofrontal cortex is also activated during pleasurable activities, and in response to reward-paired cues processed from visual, auditory, olfactory, somatic, and gustatory thalamic input.[43] Specialized orbitofrontal taste-related neurons fire in response to sweet stimuli only when an animal is hungry, suggesting that this region also responds to natural drive states.[43]

Several other glutamate-containing regions are involved in reward function. The anterior cingulate, by virtue of its extensive brain stem connections, participates in many conscious functions, including wakefulness, attention, learning, and the processing of pleasure and pain.[44] The insula is an important target of visceral projections to the cortex and processes the conscious recognition of emotion and feelings. Other reward-related glutamatergic regions include the thalamus, ventral subiculum region of the hippocampus, the pedunculopontine tegmentum, and the basolateral amygdala, which is implicated in the recognition of emotionally relevant stimuli, goal-directed behavior (including cocaine seeking), and conditioned fear.[45]

The PFC sends massive glutamatergic bundles to the nucleus accumbens (NAc), a reward-related limbic structure that is the primary target of mesolimbic DA neurons. Glutamate depolarizes target neurons by activating N-methyl-D-aspartate

(NMDA), alpha-amino-3-hydroxy-5-methyl-4-isoxazole propionate (AMPA), and kainate (KA) ionotropic receptors, and stimulates three groups of metabotropic receptors (mGluRs) that, linked to G-proteins, activate phosphoinositide (Group I) or inhibit cAMP (Groups II and III) intraneuronal cascades.[46] Corticostriatal terminals express κ-opioid receptors that inhibit NAc glutamate release.[47] GABA-containing medium spiny cells comprise about 90% of NAc neurons, and some of the remaining 10% include large, slow-firing cholinergic interneurons that inhibit glutamate release and express D_5 (D1 family) receptors.[48] The medium spiny cell is named for its dense dendritic spines that receive projections from about four-hundred thousand individual pyramidal cells, largely from the PFC, but also from the thalamus, amygdala, and hippocampus.[49] This morphology enables the NAc neurons to funnel, process, and convey extensive cortical information to distant reward-related regions. Consequently, the firing characteristics and long axon projections of these output neurons are germane to the neurobiology of hedonic function.

In the striatum, medium spiny cells separate loosely into two opposing populations that contain dynorphin (and substance P) or enkephalin[50] and predominantly express D1 or D2 receptors, respectively.[51,52] Some spiny cells contain both D1 and

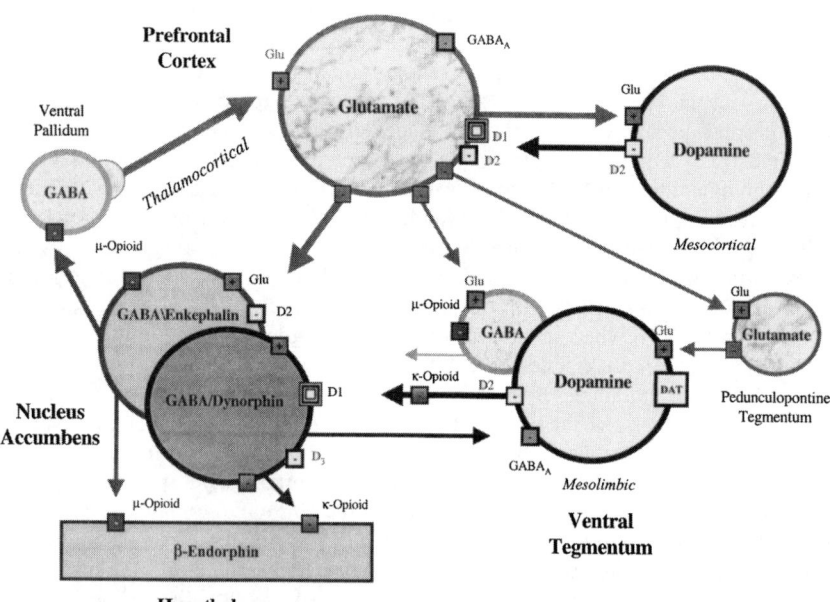

FIGURE 2. Reward-related pathways affected by cocaine are depicted. Glutamate and DA circuits are activated during cocaine administration and dysregulated after chronic cocaine exposure.

D2 receptors,[53] and GABA/dynorphin cells notably express D_3 (D2 family) receptors.[54] Although both subpopulations are similarly innervated by excitatory cortical and inhibitory pallidal axons,[55] their outflow is different (see FIG. 2). GABA/dynorphin cells project directly to the ventral tegmental area (VTA) and inhibit DA neurons via $GABA_A$[56] and κ-opioid[57] receptors, whereas GABA/enkephalin cells project to the ventral pallidum (and activate μ-opioid receptors) as part of the indirect pathway that continues to the mediodorsal thalamus and anterior cingulate.[50] This multisynaptic circuit, which includes glutamatergic bundles connecting the thalamus, anterior cingulate, and NAc, is implicated in reward-related learning[58] and cocaine craving.[21] GABA/dynorphin and GABA/enkephalin cells also project directly to reward regions in the hypothalamus[59] and preoptic area,[60] and their terminals express κ-opioid and μ-opioid autoreceptors, respectively.[61]

DA modulates PFC pyramidal cells at the dendritic and terminal levels. Mesocortical DA axons synapse on the dendrites of PFC neurons that release glutamate into the NAc.[62] DA inhibits these corticostriatal cells through D2 receptors,[63] but D1 neurotransmission can be excitatory or inhibitory, depending on the bistable resting membrane potential.[64] Mesocortical DA projections also synapse on local circuit GABA neurons in the PFC and release GABA through a D2-dependent mechanism.[65] At the terminal level, DA modulates glutamate excitation of NAc neurons at convergent synapses formed by mesolimbic and corticostriatal projections.[66]

Spike formation in the NAc is intrinsically involved in hedonic function[67] and results from the simultaneous stimulation of glutamate and D1 receptors. As with cortical cells, medium spiny cells move between two stable membrane states. Usually found in a quiescent, hyperpolarized "down" state, medium spiny cells are activated to a stable "up" state by glutamate stimulation. When in the up state, the membrane potential is close to the spike generation threshold, and medium spiny cells are excited by D1 neurotransmission.[68] When in the down state, medium spiny cells are inhibited by D1 neurotransmission as a result of intrinsic membrane conductances reviewed elsewhere.[54] Phasic NAc signaling is closely associated with hedonic function, occurring during and in anticipation of natural reward (sex, feeding, and drinking) in animals.[67] D1 neurotransmission also produces long-term metabotropic effects that are implicated in the development of addiction, including the activation of cAMP[69] and the production of dynorphin.[50] Also, the concomitant activation of NMDA and D1 receptors in the NAc is involved in synaptic plasticity related to appetitive instrumental learning.[70]

PFC projections to the VTA provide basal and stimulus-induced activation of mesocortical DA neurons[71] and make no synaptic contact with mesolimbic DA neurons.[72] The release of DA in the NAc by direct PFC stimulation has been proposed to result from the activation of mesolimbic neurons by the glutamatergic pedunculopontine tegmentum.[72] VTA burst-firing occurs during novel rewarding stimuli[73] but later habituates and shifts to the presentation of reward-paired cues.[74] Interestingly, when anticipated reward does not materialize, DA neuronal firing plunges below basal levels.[74] This degree of environmental interactivity seems remarkable for a midbrain structure, and it is possible that glutamatergic projections from the PFC drive VTA burst-firing during anticipated reward. In particular, the orbitofrontal cortex receives sensory information from the thalamus, projects to the VTA, and fires in anticipation of reward.[43] Orbitofrontal firing during anticipated reward might explain its metabolic activation during cue-induced craving.[12,15]

EFFECTS OF COCAINE ON REWARD-RELATED CIRCUITRY

Several lines of evidence link increased DA neurotransmission with cocaine-induced reward in animals[75,76] and humans.[77] However, glutamate neurotransmission has also been implicated by recent reports that mGluR5[78] and NMDA[79–82] antagonists reduce cocaine reward. Furthermore, mGluR5-deficient mice do not self-administer cocaine at all, despite showing increased levels of DA in the NAc.[83] The role of glutamate in cocaine reward is complex. Many, but not all, studies report that cocaine acutely increases extracellular glutamate levels in the NAc,[84–87] VTA,[88] and PFC,[85] particularly in cocaine-preexposed rats.[86,89] After repeated cocaine exposure, there is an enhancement of glutamate release by cocaine that may be D1 dependent.[85,88] Cocaine administration also increases β-endorphin levels in the NAc[90] and throughout the subcortex,[91] which has been linked to cocaine reward by several studies.[92–95] Cocaine stimulates corticotropin-releasing hormone (thereby releasing β-endorphin) and also directly stimulates neurons in the arcuate nucleus[96] that express Group 1 mGluRs[97] and release β-endorphin in response to NMDA stimulation.[98] Hypothalamic β-endorphin is subsequently dispersed to distant brain regions through cerebrospinal fluid.[99] These findings suggest that glutamatergic neurotransmission contributes to the release of β-endorphin by cocaine, which has been linked to cocaine reward.[92–95]

As opposed to its acute effect on NAc glutamate levels,[84–87] chronic exposure to cocaine reduces extracellular glutamate levels in the NAc.[100–102] Repeated cocaine administration also reduces mGluR2/3 autoreceptor function in the NAc,[103] which might represent a compensatory response to reduced extracellular glutamate levels. In addition, chronic cocaine exposure has been reported to alter the electrophysiological properties,[104] dendritic branching,[105] and actual number of pyramidal cells in animal studies.[106] These findings of cocaine-induced glutamate dysregulation are consistent with previously cited studies reporting reduced metabolism and neuronal density in the glutamate-rich frontal lobes of humans exposed to cocaine.

Cocaine also has different acute and chronic effects on the DA system. Opposite to its acute activation of DA, repeated cocaine administration reduces extracellular DA levels in the NAc[107–111] and desensitizes D2 autoreceptors.[108,112–118] Desensitization of somatodendritic D2 autoreceptors may lead to increased DA excitability and contribute to sensitization. It might also be speculated that the upregulation of D1 receptors, as reported by a number of studies,[113,119–123] represents a compensatory response to DA dysregulation. Reduced DA function in cocaine-addicted humans is supported by several lines of research, including neuroendocrine,[10,124–126] PET,[127,128] electroretinogram,[129] and autopsy studies.[130,131] DA dysregulation may be clinically significant because hyperprolactinemia (a marker for DA dysregulation) is associated with poor clinical outcome,[125,126,132] Interestingly, DA depletion has also been reported after chronic exposure to alcohol, opioids, and amphetamine,[107] and has been theorized to underlie craving and hedonic dysregulation associated with cocaine dependence.[10]

Cocaine-induced DA dysregulation contributes to the phenomenon of sensitization. Repeated, intermittent administration of cocaine produces persistent behavioral sensitization,[133] classically in the form of a progressive accentuation of cocaine-induced locomotion. Related to this phenomenon (although sometimes disassociated) is the enhancement of cocaine-induced NAc DA efflux in cocaine-pretreated animals.[134,135] It was recently reported that sensitized animals with enhanced cocaine-induced DA efflux also have low baseline DA release.[108] Furthermore, chemical ablation of DA projections to the PFC produces behavioral sensitization and enhances cocaine-induced DA efflux in the NAc.[136] These findings suggest that DA dysregulation may be one of many alterations associated with sensitization.

Glutamate circuits are also intrinsically involved in sensitization, as evidenced by the fact that sensitization can be prevented by interventions that reduce glutamatergic input to the VTA, including PFC ablation,[137] and glutamate receptor blockade.[138,139] Also, sensitized rats show enhanced glutamate release after cocaine when compared to cocaine-naive rats.[84] Glutamatergic receptors in the NAc and VTA appear to be altered with repeated cocaine exposure. In cocaine-pretreated animals, glutamatergic synaptic strength in the NAc is persistently depressed,[138,140,141] and Group I mGluR-mediated glutamate release is reduced.[142] Decreased excitatory synaptic strength (long-term depression) in the NAc would inhibit spike generation and potentially dysregulate hedonic function by curtailing reward-related NAc output. In the VTA, cocaine pretreatment increases glutamate receptor function,[138] and recent work suggests that long-term potentiation (LTP)-like mechanisms may be involved.[143]

Cocaine-induced hedonic dysregulation is supported by animal studies measuring intracranial self-stimulation (ICSS) of electrical current into reward structures. These studies demonstrate that more current is required to support ICSS after repeated cocaine administration and in proportion to the amount consumed.[144,145] Although hedonic dysregulation in cocaine-addicted patients has not been adequately studied, these patients often report difficulty experiencing pleasure without cocaine. In addition to DA and glutamate dysregulation, cocaine may affect hedonic function by upregulating dynorphin circuits. Autopsy studies of cocaine-dependent humans report elevated brain dynorphin-related mRNA[146] and increased κ-opioid density,[146–148] and animals treated with cocaine develop elevations in brain dynorphin,[149] striatal dynorphin-related mRNA,[50,150–157] and κ-opioid receptor binding.[92] Dynorphin inhibits DA and glutamate release in the NAc by binding κ-opioid receptors expressed by DA[50] and glutamate[47] terminals, and dampens D1 transduction via κ-opioid receptors expressed by GABA/dynorphin cells.[158,159] Repeated κ-opioid stimulation downregulates D2 receptors[160,161] and might contribute to low D2 binding on PET that has been reported in cocaine-dependent subjects.[162–164] Cocaine-induced upregulation of dynorphin may contribute to hedonic dysregulation that could be targeted by selective κ-opioid antagonists.

Dynorphin upregulation might also interfere with hedonic function by affecting other reward-related regions. GABA/dynorphin and GABA/enkephalin are thought to represent functionally opposed opioid systems that mediate aversion and reward.[165] Enkephalin release by striatopallidal spiny cells is strongly implicated in cocaine reward,[166,167] while dynorphin is aversive. Dynorphin and enkephalin-containing axons project directly from the NAc to pleasure regions of the hypothalamus and preoptic area[59] that are involved in consummatory reward related to sex[168] and feeding.[169] Relative predominance of dynorphin over enkephalin in the hypothalamus or other reward-related regions, as a result of cocaine pretreatment, might affect hedonic function.

Another glutamatergic contribution to cocaine dependence involves the previously reviewed metabolic activation of glutamate-rich limbic regions during cue-induced craving. Animal models substantiate these human studies and furnish direct evidence that cocaine-paired stimuli are capable of increasing NAc glutamate, even from cocaine-depleted levels.[102] Cocaine conditioned stimuli likewise increase DA levels in the NAc,[74] and the combined effect of DA and glutamate neurotransmission might produce the NAc firing that is associated with anticipated natural and cocaine-induced reward.[67,170] Agents that block the release of glutamate or DA might therefore attenuate cue reactivity and reduce its ability to produce relapse.

GLUTAMATERGIC STRATEGIES FOR FUTURE CLINICAL TRIALS

We reviewed several neuronal mechanisms involved in the clinical aspects of cocaine dependence, including euphoria, craving, withdrawal, hedonic dysregulation, and disturbed executive function. Pharmacological agents that reverse cocaine-induced neuroadaptations might be effective in treating associated clinical manifestations. Cocaine-induced dysregulation of glutamate, DA, and dynorphin circuits might reduce NAc firing, which is an important mechanism in hedonic function.[67] Glutamate dysregulation by cocaine might also reduce β-endorphin release that, combined with dynorphin upregulation, could produce clinically significant imbalances in reward-related endogenous opioid function. These neuroadaptations might contribute to hedonic dysregulation by cocaine, which has been demonstrated in animal studies and anecdotally encountered in the clinical management of cocaine-dependent patients.

Glutamate-enhancing agents provide a direct means of addressing clinical manifestations that might result from glutamate dysregulation. Modafinil increases glutamate synthesis and glutamate brain levels in the striatum,[34,35] and improves neuropsychological task performance.[33] Acamprosate is a partial NMDA agonist that has been tested in Europe for alcoholism[171] but is not approved in the United States. Glycine is an obligatory coagonist at NMDA receptors and could be tested as a means of enhancing glutamatergic neurotransmission. These agents might theoretically improve PFC dysfunction and hedonic dysregulation in cocaine-dependent patients by enhancing glutamate function.

Opioid and DA agents might also activate glutamatergic circuits. Selective κ-opioid antagonists release striatal glutamate,[172] and their eventual development would produce additional candidates for investigation. Naltrexone nonselectively blocks κ-opioid receptors and is a reasonable choice for testing in cocaine-dependent patients, although its action on μ-opioid and δ-opioid receptors might dampen any κ-opioid effect. D2 agonists have not been found effective in cocaine dependence, and selective D1 agonists are not currently approved for humans.[9] However, DA agonists only provide constant receptor occupancy, which would not replace environmentally interactive burst-firing that is associated with normal VTA functional integrity.[9] Selective D_3 agonists, if developed, might provide clinical improvement by specifically downregulating GABA/dynorphin neurons that express inhibitory D_3 receptors.[54] Interestingly, increased D_3 density has been reported in human cocaine users at autopsy.[173]

Many investigators have also proposed glutamate blockade for cocaine dependence. NMDA antagonists reduce cocaine self-administration[79,80,82] and may blunt cocaine-induced β-endorphin release[98] that is associated with cocaine reward.[95]

NMDA antagonists might also interfere with the development of cue-reactivity, as they block the acquisition of place preference and behavioral sensitization,[81] and NMDA neurotransmission is implicated in behaviorally relevant learning.[70] Cue-induced craving is associated with increased metabolic activity in glutamate-rich brain regions, and cocaine-paired cues release NAc glutamate in animal studies. Glutamate antagonists might therefore be effective in cocaine dependence by dampening cue-induced craving, which is a common relapse precipitant.[20] Dextromethorphan and amantadine are currently available NMDA antagonists, and the latter has been reported to be effective in cocaine dependence.[7] Memantine is a noncompetitive NMDA antagonist that prevents the acquisition and expression of place preference produced by cocaine[174] and attenuates sensitized motor responses to cocaine-paired cues.[175] However, memantine (which is not approved in the United States) has also been reported to increase the euphoric effect of smoked cocaine in a human study.[176] Glutamate antagonists might also benefit cocaine users by providing neuroprotection. Cocaine-induced disruptions of pyramidal cells (e.g., hypofrontality or neuronal cell loss) might result from glutamate-induced excitotoxicity, providing rationale for neuroprotective agents in cocaine dependence. Neuroprotective agents include NMDA antagonists, Group II and III mGluR agonists, and Group I mGluR antagonists.[177] Interestingly, although modafinil has glutamate-enhancing properties, it has been reported to be neuroprotective[178] and prevents glutamate excitotoxicity in cultured cortical neurons.[179]

GABA-activating agents provide an alternative means of reducing glutamatergic neurotransmission. Allosteric $GABA_A$ agents, such as benzodiazepines or barbiturates, reduce glutamate activity but have addictive potential.[20] $GABA_B$ agents, such as baclofen, are not addicting and are currently under investigation for cocaine dependence based on their ability to dramatically reduce cocaine self-administration.[180] In addition to attenuating DA release, baclofen acts at $GABA_B$ receptors on glutamate terminals to depress glutamate transmission. Other approved GABA agonists that might be tested as relapse prevention agents in cocaine dependence include tiagabine, a GABA uptake inhibitor, topiramate, and gabapentin. Preliminary data from our center found topiramate to be more effective than placebo as a relapse prevention agent. The theoretical effectiveness of glutamate blockade in early recovery might be limited by its potential ability to exacerbate hedonic dysregulation and hypofrontality.

CONCLUSIONS

We have reviewed glutamatergic mechanisms related to cocaine dependence. Glutamate is intrinsically involved in cocaine reward, and repeated cocaine administration alters glutamate-dependent plasticity in ways that dysregulate reward circuitry. Cocaine also dysregulates DA circuits that act synergistically with glutamate in many reward-related functions. Extensive reciprocal interconnections link DA neurons in the VTA and glutamate neurons in the PFC, and their axons converge on common NAc neurons that synthesize GABA and endogenous opioids. Consequently, glutamate neurons are affected by cocaine-induced neuroadaptations involving DA, GABA, and endogenous opioid systems. Glutamatergic agents that reverse clinically significant cocaine-induced neuroadaptations in these functionally linked neurotransmitter systems, or dampen the acute effects of cocaine and cocaine cues, might have roles in the management of cocaine dependence. It is anticipated that effective pharmacological strategies for cocaine dependence will become increasingly evident as we refine our understanding of reward function and cocaine neurobiology.

REFERENCES

1. ANTHONY, J.C. & L.A. WARNER. 1994. Comparative epidemiology of dependence on tobacco, alcohol, controlled substances, and inhalants: basic findings from the National Comorbidity Survey. Exp. Clin. Psychopharmacol. **2:** 244–268.
2. JAFFE, J.H. et al. 1989. Cocaine-induced cocaine craving. Psychopharmacology **97:** 59–64.
3. O'BRIEN, C.P. et al. 1992. Classical conditioning in drug-dependent humans. Ann. N.Y. Acad. Sci. **654:** 400–415.
4. WEDDINGTON, W.W. et al. 1990. Changes in mood, craving, and sleep during short-term abstinence reported by male cocaine addicts. A controlled, residential study. Arch. Gen. Psychiatry **47:** 861–868.
5. GAWIN, F.H. & H.D. KLEBER. 1986. Abstinence symptomatology and psychiatric diagnosis in cocaine abusers. Clinical observations. Arch. Gen. Psychiatry **43:** 107–113.
6. KAMPMAN, K.M. et al. 2001. Cocaine withdrawal symptoms and initial urine toxicology results predict treatment attrition in outpatient cocaine dependence treatment. Psychol. Addict. Behav. **15:** 52–59.
7. KAMPMAN, K.M. et al. 2000. Amantadine in the treatment of cocaine-dependent patients with severe withdrawal symptoms. Am. J. Psychiatry **157:** 2052–2054.
8. DACKIS, C.A. et al. 2003. Modafinil and Cocaine: A double-blind, placebo-controlled drug interaction study. Drug Alcohol Depend. **70:** 29–37.
9. DACKIS, C.A. & C.P. O'BRIEN. 2002. Cocaine dependence: the challenge for pharmacotherapy. Curr. Opin. Psychiatry **15:** 261–267.
10. DACKIS, C.A. & M.S. GOLD. 1985. New concepts in cocaine addiction: the dopamine depletion hypothesis. Neurosci. Biobehav. Rev. **9:** 469–477.
11. KILTS, C.D. et al. 2001. Neural activity related to drug craving in cocaine addiction. Arch. Gen. Psychiatry **58:** 334–341.
12. CHILDRESS, A.R. et al. 1999. Limbic activation during cue-induced cocaine craving. Am J Psychiatry. **156:** 11–18.
13. WANG, G.J. et al. 1999. Regional brain metabolic activation during craving elicited by recall of previous drug experiences. Life Sci. **64:** 775–784.
14. VOLKOW, N.D. et al. 1999. Association of methylphenidate-induced craving with changes in right striato-orbitofrontal metabolism in cocaine abusers: implications in addiction. Am. J. Psychiatry **156:** 19–26.
15. GRANT, S. et al. 1996. Activation of memory circuits during cue-elicited cocaine craving. Proc. Natl. Acad. Sci. USA **93:** 12040–12045.
16. BREITER, H.C. et al. 1997. Acute effects of cocaine on human brain activity and emotion. Neuron **19:** 591–611.
17. GARAVAN, H. et al. 2000. Cue-induced cocaine craving: neuroanatomical specificity for drug users and drug stimuli. Am. J. Psychiatry **157:** 1789–1798.
18. MAAS, L.C. et al. 1998. Functional magnetic resonance imaging of human brain activation during cue-induced cocaine craving. Am. J. Psychiatry **155:** 124–126.
19. WEXLER, B.E. et al. 2001. Functional magnetic resonance imaging of cocaine craving. Am. J. Psychiatry **158:** 86–95.
20. DACKIS, C.A. & C.P. O'BRIEN. 2001. Cocaine dependence: a disease of the brain's reward centers. J. Subst. Abuse Treat. **21:** 111–117.
21. VOLKOW, N.D. & J.S. FOWLER. 2000. Addiction, a disease of compulsion and drive: involvement of the orbitofrontal cortex. Cereb. Cortex **10:** 318–325.

22. VOLKOW, N.D. *et al.* 2001. Low level of brain dopamine D2 receptors in methamphetamine abusers: association with metabolism in the orbitofrontal cortex. Am. J. Psychiatry **158:** 2015–2021.
23. VOLKOW, N.D. *et al.* 1992. Long-term frontal brain metabolic changes in cocaine abusers. Synapse **11:** 184–190.
24. FRANKLIN, T.R. *et al.* 2002. Decreased gray matter concentration in the insular, orbitofrontal, cingulate, and temporal cortices of cocaine patients. Biol. Psychiatry **51:** 134–142.
25. FEIN, G., V. DI SCLAFANI & D.J. MEYERHOFF. 2002. Prefrontal cortical volume reduction associated with frontal cortex function deficit in 6-week abstinent crack-cocaine dependent men. Drug Alcohol Depend. **68:** 87–93.
26. MAJEWSKA, M.D. 1996. Cocaine addiction as a neurological disorder: implications for treatment. NIDA Res. Monogr. **163:** 1–26.
27. BECHARA, A., H. DAMASIO & A.R. DAMASIO. 2000. Emotion, decision making and the orbitofrontal cortex. Cereb. Cortex **10:** 295–307.
28. MONTEROSSO, J. *et al.* 2001. Three decision-making tasks in cocaine-dependent patients: do they measure the same construct? Addiction **96:** 1825–1837.
29. KANG, S.Y. *et al.* 1991. Outcomes for cocaine abusers after once-a-week psychosocial therapy. Am J Psychiatry **148:** 630-635.
30. ALTERMAN, A.I. *et al.* 1994. Effectiveness and costs of inpatient versus day hospital cocaine rehabilitation. J. Nerv. Ment. Dis. **182:** 157–163.
31. WISE, R.A. 1984. Neural mechanisms of the reinforcing action of cocaine. NIDA Res. Monogr. **50:** 15–33.
32. DACKIS, C.A. & M.S. GOLD. 1992. Psychiatric hospitals for treatment of dual diagnosis. *In* Substance Abuse, A Comprehensive Textbook. J.H. Lowinson, Ed.: 467–485. Williams & Wilkins. Baltimore.
33. TURNER, D.C. *et al.* 2003. Cognitive enhancing effects of modafinil in healthy volunteers. Psychopharmacology (Berl.) **165:** 260–269.
34. TOURET, M. *et al.* 1994. Effects of modafinil-induced wakefulness on glutamine synthetase regulation in the rat brain. Brain Res. Mol. Brain Res. **26:** 123–128.
35. PIERARD, C. *et al.* 1995. Effects of a vigilance-enhancing drug, modafinil, on rat brain metabolism: a 2D COSY 1H-NMR study. Brain Res. **693:** 251–256.
36. FERRARO, L. *et al.* 1996. The vigilance promoting drug modafinil increases dopamine release in the rat nucleus accumbens via the involvement of a local GABAergic mechanism. Eur. J. Pharmacol. **306:** 33–39.
37. FERRARO, L. *et al.* 1998. The effects of modafinil on striatal, pallidal and nigral GABA and glutamate release in the conscious rat: evidence for a preferential inhibition of striato-pallidal GABA transmission. Neurosci. Lett. **253:** 135–138.
38. DACKIS, C.A. & C.P. O'BRIEN. 2002. Cocaine dependence: the challenge for pharmacotherapy. Curr. Opin. Psychiatry **15:** 261–268.
39. KAMPMAN, K.M. *et al.* 2001. Effectiveness of propranolol for cocaine dependence treatment may depend on cocaine withdrawal symptom severity. Drug Alcohol Depend. **63:** 69–78.
40. JASINSKI, D.R. & R. KOVACEVIC-RISTANOVIC. 2000. Evaluation of the abuse liability of modafinil and other drugs for excessive daytime sleepiness associated with narcolepsy. Clin. Neuropharmacol. **23:** 149–156.
41. BLAIR, R.J., E. COLLEDGE & D.G. MITCHELL. 2001. Somatic markers and response reversal: is there orbitofrontal cortex dysfunction in boys with psychopathic tendencies? J. Abnorm. Child Psychol. **29:** 499–511.
42. VAN HONK, J. *et al.* 2002. Defective somatic markers in sub-clinical psychopathy. Neuroreport **13:** 1025–1027.
43. ROLLS, E.T. 2000. The orbitofrontal cortex and reward. Cereb. Cortex **10:** 284–294.
44. VAN VEEN, V. *et al.* 2001. Anterior cingulate cortex, conflict monitoring, and levels of processing. Neuroimage **14:** 1302–1388.
45. WHITELAW, R.B. *et al.* 1996. Excitotoxic lesions of the basolateral amygdala impair the acquisition of cocaine-seeking behaviour under a second-order schedule of reinforcement. Psychopharmacology (Berl.) **127:** 213–224.
46. SCHOEPP, D.D., D.E. JANE & J.A. MONN. 1999. Pharmacological agents acting at subtypes of metabotropic glutamate receptors. Neuropharmacology **38:** 1431–1476.

47. MESHUL, C.K. & J.F. MCGINTY. 2000. Kappa opioid receptor immunoreactivity in the nucleus accumbens and caudate-putamen is primarily associated with synaptic vesicles in axons. Neuroscience **96:** 91–99.
48. MCGINTY, J.F. 1999. Regulation of neurotransmitter interactions in the ventral striatum. Ann. N. Y. Acad. Sci. **877:** 129–139.
49. KINCAID, A.E., T. ZHENG & C.J. WILSON. 1998. Connectivity and convergence of single corticostriatal axons. J. Neurosci. **18:** 4722–4731.
50. STEINER, H. & C.R. GERFEN. 1998. Role of dynorphin and enkephalin in the regulation of striatal output pathways and behavior. Exp. Brain Res. **123:** 60–76.
51. LE MOINE, C. & B. BLOCH. 1995. D1 and D2 dopamine receptor gene expression in the rat striatum: sensitive cRNA probes demonstrate prominent segregation of D1 and D2 mRNAs in distinct neuronal populations of the dorsal and ventral striatum. J. Comp. Neurol. **355:** 418–426.
52. WASZCZAK, B.L. *et al.* 1998. Expression of a dopamine D2 receptor-activated K^+ channel on identified striatopallidal and striatonigral neurons. Proc. Natl. Acad. Sci. USA **95:** 11440–11444.
53. SURMEIER, D.J., W.J. SONG & Z. YAN. 1996. Coordinated expression of dopamine receptors in neostriatal medium spiny neurons. J. Neurosci. **16:** 6579–6591.
54. NICOLA, S.M., J. SURMEIER & R.C. MALENKA. 2000. Dopaminergic modulation of neuronal excitability in the striatum and nucleus accumbens. Annu. Rev. Neurosci. **23:** 185–215.
55. YUNG, K.K. *et al.* 1996. Synaptic connections between spiny neurons of the direct and indirect pathways in the neostriatum of the rat: evidence from dopamine receptor and neuropeptide immunostaining. Eur. J. Neurosci. **8:** 861–869.
56. IKEMOTO, S., R.R. KOHL & W.J. MCBRIDE. 1997. GABA(A) receptor blockade in the anterior ventral tegmental area increases extracellular levels of dopamine in the nucleus accumbens of rats. J. Neurochem. **69:** 137–143.
57. MITROVIC, I. & T.C. NAPIER. 2002. Mu and kappa opioid agonists modulate ventral tegmental area input to the ventral pallidum. Eur. J. Neurosci. **15:** 257–268.
58. KALIVAS, P.W., L. CHURCHILL & A. ROMANIDES. 1999. Involvement of the pallidal-thalamocortical circuit in adaptive behavior. Ann. N. Y. Acad. Sci. **877:** 64–70.
59. FALLON, J.H. & F.M. LESLIE. 1986. Distribution of dynorphin and enkephalin peptides in the rat brain. J. Comp. Neurol. **249:** 293–336.
60. ZAHM, D.S. & L. HEIMER. 1993. Specificity in the efferent projections of the nucleus accumbens in the rat: comparison of the rostral pole projection patterns with those of the core and shell. J. Comp. Neurol. **327:** 220–232.
61. GARZON, M. & V.M. PICKEL. 2002. Ultrastructural localization of enkephalin and mu-opioid receptors in the rat ventral tegmental area. Neuroscience **114:** 461–474.
62. CARR, D.B. *et al.* 1999. Dopamine terminals in the rat prefrontal cortex synapse on pyramidal cells that project to the nucleus accumbens. J. Neurosci. **19:** 11049–11060.
63. GODBOUT, R. *et al.* 1991. Inhibitory influence of the mesocortical dopaminergic neurons on their target cells: electrophysiological and pharmacological characterization. J. Pharmacol. Exp. Ther. **258:** 728–738.
64. LEWIS, B.L. & P. O'DONNELL. 2000. Ventral tegmental area afferents to the prefrontal cortex maintain membrane potential "up" states in pyramidal neurons via D(1) dopamine receptors. Cereb. Cortex **10:** 1168–1175.
65. GROBIN, A.C. & A.Y. DEUTCH. 1998. Dopaminergic regulation of extracellular gamma-aminobutyric acid levels in the prefrontal cortex of the rat. J. Pharmacol. Exp. Ther. **285:** 350–357.
66. SESACK, S.R. & V.M. PICKEL. 1992. Prefrontal cortical efferents in the rat synapse on unlabeled neuronal targets of catecholamine terminals in the nucleus accumbens septi and on dopamine neurons in the ventral tegmental area. J. Comp. Neurol. **320:** 145–160.
67. CARELLI, R.M. 2002. Nucleus accumbens cell firing during goal-directed behaviors for cocaine vs. "natural" reinforcement. Physiol Behav. **76:** 379–387.
68. HERNANDEZ-LOPEZ, S. *et al.* 1997. D1 receptor activation enhances evoked discharge in neostriatal medium spiny neurons by modulating an L-type $Ca2^+$ conductance. J. Neurosci. **17:** 3334–3342.

69. SELF, D.W. et al. 1998. Involvement of cAMP-dependent protein kinase in the nucleus accumbens in cocaine self-administration and relapse of cocaine-seeking behavior. J. Neurosci. **18:** 1848–1859.
70. SMITH-ROE, S.L. & A.E. KELLEY. 2000. Coincident activation of NMDA and dopamine D1 receptors within the nucleus accumbens core is required for appetitive instrumental learning. J. Neurosci. **20:** 7737–7742.
71. TONG, Z.Y., P.G. OVERTON & D. CLARK. 1996. Stimulation of the prefrontal cortex in the rat induces patterns of activity in midbrain dopaminergic neurons which resemble natural burst events. Synapse. **22:** 195–208.
72. CARR, D.B. & S.R. SESACK. 2000. Projections from the rat prefrontal cortex to the ventral tegmental area: target specificity in the synaptic associations with mesoaccumbens and mesocortical neurons. J. Neurosci. **20:** 3864–3873.
73. WISE, R.A. 2000. Addiction becomes a brain disease. Neuron **26:** 27–33.
74. SCHULTZ, W. 2001. Reward signaling by dopamine neurons. Neuroscientist **7:** 293–302.
75. KOOB, G.F., P.P. SANNA & F.E. BLOOM. 1998. Neuroscience of addiction. Neuron **21:** 467–476.
76. WISE, R.A. 1996. Neurobiology of addiction. Curr. Opin. Neurobiol. **6:** 243–251.
77. VOLKOW, N.D., J.S. FOWLER & G.J. WANG. 1999. Imaging studies on the role of dopamine in cocaine reinforcement and addiction in humans. J. Psychopharmacol. **13:** 337–345.
78. MCGEEHAN, A.J. & M.F. OLIVE. 2003. The mGluR5 antagonist MPEP reduces the conditioned rewarding effects of cocaine but not other drugs of abuse. Synapse. **47:** 240–242.
79. PULVIRENTI, L., C. BALDUCCI & G.F. KOOB. 1997. Dextromethorphan reduces intravenous cocaine self-administration in the rat. Eur. J. Pharmacol. **321:** 279–283.
80. HYYTIA, P., P. BACKSTROM & S. LILJEQUIST. 1999. Site-specific NMDA receptor antagonists produce differential effects on cocaine self-administration in rats. Eur. J. Pharmacol. **378:** 9–16.
81. PAPP, M., P. GRUCA & P. WILLNER. 2002. Selective blockade of drug-induced place preference conditioning by ACPC, a functional NDMA-receptor antagonist. Neuropsychopharmacology **27:** 727–743.
82. PULVIRENTI, L., R. MALDONADO-LOPEZ & G.F. KOOB. 1992. NMDA receptors in the nucleus accumbens modulate intravenous cocaine but not heroin self-administration in the rat. Brain Res. **594:** 327–330.
83. CHIAMULERA, C. et al. 2001. Reinforcing and locomotor stimulant effects of cocaine are absent in mGluR5 null mutant mice. Nat. Neurosci. **4:** 873–874.
84. REID, M.S. & S.P. BERGER. 1996. Evidence for sensitization of cocaine-induced nucleus accumbens glutamate release. Neuroreport **7:** 1325–1329.
85. REID, M.S., K. HSU JR. & S.P. BERGER. 1997. Cocaine and amphetamine preferentially stimulate glutamate release in the limbic system: studies on the involvement of dopamine. Synapse **27:** 95–105.
86. PIERCE, R.C. et al. 1996. Repeated cocaine augments excitatory amino acid transmission in the nucleus accumbens only in rats having developed behavioral sensitization. J. Neurosci. **16:** 1550–1560.
87. SMITH, J.A. et al. 1995. Cocaine increases extraneuronal levels of aspartate and glutamate in the nucleus accumbens. Brain Res. **683:** 264–269.
88. KALIVAS, P.W. & P. DUFFY. 1995. D1 receptors modulate glutamate transmission in the ventral tegmental area. J. Neurosci. **15:** 5379–5388.
89. REID, M.E. et al. 1999. Group I mGlu receptors potentiate synaptosomal [3H]glutamate release independently of exogenously applied arachidonic acid. Neuropharmacology **38:** 477–485.
90. OLIVE, M.F. et al. 2001. Stimulation of endorphin neurotransmission in the nucleus accumbens by ethanol, cocaine, and amphetamine. J. Neurosci. **21:** RC184.
91. GERRITS, M.A., V.M. WIEGANT & J.M. VAN REE. 1999. Endogenous opioids implicated in the dynamics of experimental drug addiction: an in vivo autoradiographic analysis. Neuroscience **89:** 1219–1227.
92. UNTERWALD, E.M. 2001. Regulation of opioid receptors by cocaine. Ann. N.Y. Acad. Sci. **937:** 74–92.

93. BAIN, G.T. & C. KORNETSKY. 1987. Naloxone attenuation of the effect of cocaine on rewarding brain stimulation. Life Sci. **40:** 1119–1125.
94. CORRIGALL, W.A. & K.M. COEN. 1991. Opiate antagonists reduce cocaine but not nicotine self-administration. Psychopharmacology (Berl.) **104:** 167–170.
95. KIYATKIN, E.A. & P.L. BROWN. 2003. Naloxone depresses cocaine self-administration and delays its initiation on the following day. Neuroreport **14:** 251–255.
96. HAYASE, T. *et al.* 1998. Brain beta-endorphin immunoreactivity as an index of cocaine and combined cocaine-ethanol toxicities. Pharmacol. Biochem. Behav. **60:** 263–270.
97. KISS, J. *et al.* 1997. Metabotropic glutamate receptor in GHRH and beta-endorphin neurones of the hypothalamic arcuate nucleus. Neuroreport **8:** 3703–3707.
98. BACH, F.W. & T.L. YAKSH. 1995. Release of beta-endorphin immunoreactivity from brain by activation of a hypothalamic N-methyl-D-aspartate receptor. Neuroscience **65:** 775–783.
99. MACMILLAN, S.J., M.A. MARK & A.W. DUGGAN. 1998. The release of beta-endorphin and the neuropeptide-receptor mismatch in the brain. Brain Res. **794:** 127–136.
100. KEYS, A.S. *et al.* 1998. Reduced glutamate immunolabeling in the nucleus accumbens following extended withdrawal from self-administered cocaine. Synapse **30:** 393–401.
101. BELL, K., P. DUFFY & P.W. KALIVAS. 2000. Context-specific enhancement of glutamate transmission by cocaine. Neuropsychopharmacology **23:** 335–344.
102. HOTSENPILLER, G., M. GIORGETTI & M.E. WOLF. 2001. Alterations in behaviour and glutamate transmission following presentation of stimuli previously associated with cocaine exposure. Eur. J. Neurosci. **14:** 1843–1855.
103. XI, Z.X. *et al.* 2002. Modulation of group II metabotropic glutamate receptor signaling by chronic cocaine. J. Pharmacol. Exp. Ther. **303:** 608–615.
104. TRANTHAM, H. *et al.* 2002. Repeated cocaine administration alters the electrophysiological properties of prefrontal cortical neurons. Neuroscience **113:** 749–753.
105. ROBINSON, T.E. *et al.* 2001. Cocaine self-administration alters the morphology of dendrites and dendritic spines in the nucleus accumbens and neocortex. Synapse **39:** 257–266.
106. LIDOW, M.S. & Z.M. SONG. 2001. Primates exposed to cocaine in utero display reduced density and number of cerebral cortical neurons. J. Comp. Neurol. **435:** 263–275.
107. ROSSETTI, Z.L. *et al.* 1992. Dramatic depletion of mesolimbic extracellular dopamine after withdrawal from morphine, alcohol or cocaine: a common neurochemical substrate for drug dependence. Ann. N.Y. Acad. Sci. **654:** 513–516.
108. CHEFER, V.I. & T.S. SHIPPENBERG. 2002. Changes in basal and cocaine-evoked extracellular dopamine uptake and release in the rat nucleus accumbens during early abstinence from cocaine: quantitative determination under transient conditions. Neuroscience **112:** 907–919.
109. PARSONS, L.H., A.D. SMITH & J.B. JUSTICE JR. 1991. Basal extracellular dopamine is decreased in the rat nucleus accumbens during abstinence from chronic cocaine. Synapse **9:** 60–65.
110. ROBERTSON, M.W., C.A. LESLIE & J.P. BENNETT JR. 1991. Apparent synaptic dopamine deficiency induced by withdrawal from chronic cocaine treatment. Brain Res. **538:** 337–339.
111. IMPERATO, A. *et al.* 1992. Chronic cocaine alters limbic extracellular dopamine. Neurochemical basis for addiction. Eur. J. Pharmacol. **212:** 299–300.
112. HENRY, D.J., M.A. GREENE & F.J. WHITE. 1989. Electrophysiological effects of cocaine in the mesoaccumbens dopamine system: repeated administration. J. Pharmacol. Exp. Ther. **251:** 833–839.
113. HENRY, D.J., X.T. HU & F.J. WHITE. 1998. Adaptations in the mesoaccumbens dopamine system resulting from repeated administration of dopamine D1 and D2 receptor-selective agonists: relevance to cocaine sensitization. Psychopharmacology (Berl.) **140:** 233–242.
114. ACKERMAN, J.M. & F.J. WHITE. 1990. A10 somatodendritic dopamine autoreceptor sensitivity following withdrawal from repeated cocaine treatment. Neurosci. Lett. **117:** 181–187.

115. JONES, S.R. et al. 1996. Effects of intermittent and continuous cocaine administration on dopamine release and uptake regulation in the striatum: in vitro voltammetric assessment. Psychopharmacology (Berl.) **126:** 331–338.
116. DAVIDSON, C., E.H. ELLINWOOD & T.H. LEE. 2000. Altered sensitivity of dopamine autoreceptors in rat accumbens 1 and 7 days after intermittent or continuous cocaine withdrawal. Brain Res. Bull. **51:** 89–93.
117. YI, S.J. & K.M. JOHNSON. 1990. Chronic cocaine treatment impairs the regulation of synaptosomal 3H-DA release by D2 autoreceptors. Pharmacol. Biochem. Behav. **36:** 457–461.
118. PIERCE, R.C., P. DUFFY & P.W. KALIVAS. 1995. Sensitization to cocaine and dopamine autoreceptor subsensitivity in the nucleus accumbens. Synapse **20:** 33–36.
119. UNTERWALD, E.M., J. FILLMORE & M.J. KREEK. 1996. Chronic repeated cocaine administration increases dopamine D1 receptor-mediated signal transduction. Eur. J. Pharmacol. **318:** 31–35.
120. HENRY, D.J. & F.J. WHITE. 1991. Repeated cocaine administration causes persistent enhancement of D1 dopamine receptor sensitivity within the rat nucleus accumbens. J. Pharmacol. Exp. Ther. **258:** 882–890.
121. BEURRIER, C. & R.C. MALENKA. 2002. Enhanced inhibition of synaptic transmission by dopamine in the nucleus accumbens during behavioral sensitization to cocaine. J. Neurosci. **22:** 5817–5822.
122. NADER, M.A. et al. 2002. Effects of cocaine self-administration on striatal dopamine systems in rhesus monkeys: initial and chronic exposure. Neuropsychopharmacology **27:** 35–46.
123. UNTERWALD, E.M. et al. 1994. Time course of the development of behavioral sensitization and dopamine receptor up-regulation during binge cocaine administration. J. Pharmacol. Exp. Ther. **270:** 1387–1396.
124. MENDELSON, J.H. et al. 1988. Anterior pituitary, adrenal, and gonadal hormones during cocaine withdrawal. Am. J. Psychiatry **145:** 1094–1098.
125. KRANZLER, H.R. & D.J. WALLINGTON. 1992. Serum prolactin level, craving, and early discharge from treatment in cocaine-dependent patients. Am. J. Drug Alcohol Abuse **18:** 187–195.
126. PATKAR, A.A. et al. 2002. Serum prolactin and response to treatment among cocaine-dependent individuals. Addict. Biol. **7:** 45–53.
127. VOLKOW, N.D. et al. 1997. Decreased striatal dopaminergic responsiveness in detoxified cocaine-dependent subjects. Nature **386:** 830–833.
128. WU, J.C. et al. 1997. Decreasing striatal 6-FDOPA uptake with increasing duration of cocaine withdrawal [see comments]. Neuropsychopharmacology **17:** 402–409.
129. SMELSON, D.A. et al. 1998. Electroretinogram in withdrawn cocaine-dependent subjects. Relationship to cue-elicited craving. Br. J. Psychiatry **172:** 537–539.
130. LITTLE, K.Y. et al. 1999. Striatal dopaminergic abnormalities in human cocaine users. Am. J. Psychiatry **156:** 238–245.
131. WILSON, J.M. et al. 1996. Striatal dopamine, dopamine transporter, and vesicular monoamine transporter in chronic cocaine users. Ann. Neurol. **40:** 428–439.
132. TEOH, S.K. et al. 1990. Hyperprolactinemia and risk for relapse of cocaine abuse. Biol. Psychiatry **28:** 824–828.
133. ROBINSON, T.E. & K.C. BERRIDGE. 1993. The neural basis of drug craving: an incentive-sensitization theory of addiction. Brain Res. Brain Res. Rev. **18:** 247–291.
134. STEKETEE, J.D., B.A. SORG & P.W. KALIVAS. 1992. The role of the nucleus accumbens in sensitization to drugs of abuse. Prog. Neuropsychopharmacol. Biol. Psychiatry **16:** 237–246.
135. KALIVAS, P.W. & P. DUFFY. 1990. Effect of acute and daily cocaine treatment on extracellular dopamine in the nucleus accumbens. Synapse **5:** 48–58.
136. BEYER, C.E. & J.D. STEKETEE. 1999. Dopamine depletion in the medial prefrontal cortex induces sensitized-like behavioral and neurochemical responses to cocaine. Brain Res. **833:** 133–141.
137. LI, Y. et al. 1999. Both glutamate receptor antagonists and prefrontal cortex lesions prevent induction of cocaine sensitization and associated neuroadaptations. Synapse **34:** 169–180.

138. WOLF, M.E. 1998. The role of excitatory amino acids in behavioral sensitization to psychomotor stimulants. Prog. Neurobiol. **54:** 679–720.
139. KARLER, R. et al. 1989. Blockade of "reverse tolerance" to cocaine and amphetamine by MK-801. Life Sci. **45:** 599–606.
140. WHITE, F.J. et al. 1995. Repeated administration of cocaine or amphetamine alters neuronal responses to glutamate in the mesoaccumbens dopamine system. J. Pharmacol. Exp. Ther. **273:** 445–454.
141. THOMAS, M.J. et al. 2001. Long-term depression in the nucleus accumbens: a neural correlate of behavioral sensitization to cocaine. Nat. Neurosci. **4:** 1217–1223.
142. SWANSON, C.J. et al. 2001. Repeated cocaine administration attenuates group I metabotropic glutamate receptor-mediated glutamate release and behavioral activation: a potential role for Homer. J. Neurosci. **21:** 9043–9052.
143. THOMAS, M.J. & R.C. MALENKA. 2003. Synaptic plasticity in the mesolimbic dopamine system. Philos. Trans. R. Soc. Lond. B Biol. Sci. **358:** 815–819.
144. MARKOU, A. & G.F. KOOB. 1991. Postcocaine anhedonia. an animal model of cocaine withdrawal. Neuropsychopharmacology **4:** 17–26.
145. KOKKINIDIS, L. & B.D. MCCARTER. 1990. Postcocaine depression and sensitization of brain-stimulation reward: analysis of reinforcement and performance effects. Pharmacol. Biochem. Behav. **36:** 463–471.
146. HURD, Y.L. & M. HERKENHAM. 1993. Molecular alterations in the neostriatum of human cocaine addicts. Synapse **13:** 357–369.
147. MASH, D.C. & J.K. STALEY. 1999. D3 dopamine and kappa opioid receptor alterations in human brain of cocaine-overdose victims. Ann. N.Y. Acad. Sci. **877:** 507–522.
148. STALEY, J.K. et al. 1997. Kappa2 opioid receptors in limbic areas of the human brain are upregulated by cocaine in fatal overdose victims. J. Neurosci. **17:** 8225–8233.
149. SIVAM, S.P. 1989. Cocaine selectively increases striatonigral dynorphin levels by a dopaminergic mechanism. J. Pharmacol. Exp. Ther. **250:** 818–824.
150. HURD, Y.L. & M. HERKENHAM. 1992. Influence of a single injection of cocaine, amphetamine or GBR 12909 on mRNA expression of striatal neuropeptides. Brain Res. Mol. Brain Res. **16:** 97–104.
151. CARTA, A.R., C.R. GERFEN & H. STEINER. 2000. Cocaine effects on gene regulation in the striatum and behavior: increased sensitivity in D3 dopamine receptor-deficient mice. Neuroreport **11:** 2395–2399.
152. STEINER, H. & C.R. GERFEN. 1999. Enkephalin regulates acute D2 dopamine receptor antagonist-induced immediate-early gene expression in striatal neurons. Neuroscience **88:** 795–810.
153. GERFEN, C.R., K.A. KEEFE & H. STEINER. 1998. Dopamine-mediated gene regulation in the striatum. Adv. Pharmacol. **42:** 670–673.
154. STEINER, H. & C.R. GERFEN. 1996. Dynorphin regulates D1 dopamine receptor-mediated responses in the striatum: relative contributions of pre- and postsynaptic mechanisms in dorsal and ventral striatum demonstrated by altered immediate-early gene induction. J. Comp. Neurol. **376:** 530–541.
155. DRAGO, J. et al. 1996. D1 dopamine receptor-deficient mouse: cocaine-induced regulation of immediate-early gene and substance P expression in the striatum. Neuroscience **74:** 813–823.
156. STEINER, H. & C.R. GERFEN. 1995. Dynorphin opioid inhibition of cocaine-induced, D1 dopamine receptor-mediated immediate-early gene expression in the striatum. J. Comp. Neurol. **353:** 200–212.
157. STEINER, H. & C.R. GERFEN. 1993. Cocaine-induced c-fos messenger RNA is inversely related to dynorphin expression in striatum. J. Neurosci. **13:** 5066–5081.
158. MINAMI, M. et al. 1993. In situ hybridization study of kappa-opioid receptor mRNA in the rat brain. Neurosci. Lett. **162:** 161–164.
159. MANSOUR, A. et al. 1994. Kappa 1 receptor mRNA distribution in the rat CNS: comparison to kappa receptor binding and prodynorphin mRNA. Mol. Cell. Neurosci. **5:** 124–144.
160. ACRI, J.B., A.C. THOMPSON & T. SHIPPENBERG. 2001. Modulation of pre- and postsynaptic dopamine D2 receptor function by the selective kappa-opioid receptor agonist U69593. Synapse **39:** 343–350.

161. IZENWASSER, S. et al. 1998. Repeated treatment with the selective kappa opioid agonist U-69593 produces a marked depletion of dopamine D2 receptors. Synapse **30:** 275–283.
162. VOLKOW, N.D. et al. 2002. Brain DA D2 receptors predict reinforcing effects of stimulants in humans: replication study. Synapse **46:** 79–82.
163. VOLKOW, N.D. et al. 1990. Effects of chronic cocaine abuse on postsynaptic dopamine receptors. Am. J. Psychiatry **147:** 719–724.
164. VOLKOW, N.D. et al. 1993. Decreased dopamine D2 receptor availability is associated with reduced frontal metabolism in cocaine abusers. Synapse **14:** 169–177.
165. AKIL, H. et al. 1998. Endogenous opioids: overview and current issues. Drug Alcohol Depend. **51:** 127–140.
166. ROBLEDO, P. & G.F. KOOB. 1993. Two discrete nucleus accumbens projection areas differentially mediate cocaine self-administration in the rat. Behav. Brain Res. **55:** 159–166.
167. OLIVE, M.F. et al. 1997. Presynaptic versus postsynaptic localization of mu and delta opioid receptors in dorsal and ventral striatopallidal pathways. J. Neurosci. **17:** 7471–7479.
168. KATO, A. & Y. SAKUMA. 2000. Neuronal activity in female rat preoptic area associated with sexually motivated behavior. Brain Res. **862:** 90–102.
169. BERNARDIS, L.L. & L.L. BELLINGER. 1996. The lateral hypothalamic area revisited: ingestive behavior. Neurosci. Biobehav. Rev. **20:** 189–287.
170. SCHULTZ, W., L. TREMBLAY & J.R. HOLLERMAN. 2000. Reward processing in primate orbitofrontal cortex and basal ganglia. Cereb. Cortex **10:** 272–284.
171. AL QATARI, M., O. BOUCHENAFA & J. LITTLETON. 1998. Mechanism of action of acamprosate. Part II. Ethanol dependence modifies effects of acamprosate on NMDA receptor binding in membranes from rat cerebral cortex. Alcohol Clin. Exp. Res. **22:** 810–814.
172. RAWLS, S.M. & J.F. MCGINTY. 1998. Kappa receptor activation attenuates L-transpyrrolidine-2,4-dicarboxylic acid-evoked glutamate levels in the striatum. J. Neurochem. **70:** 626–634.
173. STALEY, J.K. & D.C. MASH. 1996. Adaptive increase in D3 dopamine receptors in the brain reward circuits of human cocaine fatalities. J. Neurosci. **16:** 6100–6106.
174. KOTLINSKA, J. & G. BIALA. 2000. Memantine and ACPC affect conditioned place preference induced by cocaine in rats. Pol. J. Pharmacol. **52:** 179–185.
175. BESPALOV, A.Y. et al. 2000. Effects of NMDA receptor antagonists on cocaine-conditioned motor activity in rats. Eur. J. Pharmacol. **390:** 303–311.
176. COLLINS, E.D. et al. 1998. The effects of memantine on the subjective, reinforcing and cardiovascular effects of cocaine in humans. Behav. Pharmacol. **9:** 587–598.
177. FLOR, P.J. et al. 2002. Neuroprotective activity of metabotropic glutamate receptor ligands. Adv. Exp. Med. Biol. **513:** 197–223.
178. AGUIRRE, J.A. et al. 1999. A stereological study on the neuroprotective actions of acute modafinil treatment on 1-methyl-4-phenyl-1,2,3,6-tetrahydropyridine-induced nigral lesions of the male black mouse. Neurosci. Lett. **275:** 215–218.
179. ANTONELLI, T. et al. 1998. Modafinil prevents glutamate cytotoxicity in cultured cortical neurons. Neuroreport **9:** 4209–4213.
180. COUSINS, M.S., D.C. ROBERTS & H. DE WIT. 2002. GABA(B) receptor agonists for the treatment of drug addiction: a review of recent findings. Drug Alcohol Depend. **65:** 209–220.

Dopamine–Acetylcholine Interactions in the Modulation of Glutamate Release

MARCO ATZORI,[a] PATRICK KANOLD,[b] JUAN CARLOS PINEDA,[c] AND JORGE FLORES-HERNANDEZ[d]

[a]*Blanchette Rockefeller Neuroscience Institute, Rockville, Maryland 20850, USA*

[b]*Harvard Medical School, Department of Neurobiology, Boston Massachusetts 02115, USA*

[c]*Centro de Investigaciones Regionales Hideyo Noguchi, Merida, Yucatan 97135 Mexico*

[d]*Benemerita Universidad Autonoma de Puebla, Instituto de Fisiologia, Puebla 72000 Mexico*

KEYWORDS: dopamine; acetylcholine; glutamate; cortex; schizophrenia; brain slice; patch-clamp

Schizophrenia is a neurological disease whose precise anatomic substrate and cellular mechanisms are largely unknown. Internally generated voices and auditory hallucinations are among the recurrent symptoms characteristic of schizophrenia, suggestive of an impaired function of the temporal lobes.[1,2] Several neurotransmitters appear to be involved in the pathophysiology of schizophrenia, including dopamine,[3] acetylcholine,[4] and glutamate.[5] The "dopamine hypothesis," supported by pharmacological and clinical studies, proposes that an excess of dopamine or dopamine sensitivity might be associated with the disease.[3] Acetylcholine, secreted by a complex of nuclei in the basal forebrain, is an important regulator of cerebral functions such as learning and memory, sleep and wake cycles, and attention.[6] Among its cellular effects, acetylcholine binds to muscarinic receptors depressing the release of glutamate, the main excitatory neurotransmitter in the brain. Muscarinic reduction of glutamate release has been observed in many brain areas including the amygdala,[7] the hypothalamus,[8] basal ganglia,[9] and visual cortex.[10] These data suggest that the reduction of glutamate release is a general mechanism for limiting other potent excitatory effects of acetylcholine such as blockage of K^+ channels[11] and potentiation of NMDAR-mediated currents.[12] Both dopamine and acetylcholine act on complex cascades involving multiple intracellular pathways potentially interacting with each other. We considered the possibility that dopamine affects the capability of acetylcholine to depress glutamate release in the temporal cortex. Using patch-clamp recording in a thin slice preparation, we measured pharmacologically isolated AMPAR-mediated currents from visually identified pyramidal cells of layer II/III. We evoked monosynaptic glutamatergic currents (EPSCs) stimulating the neighboring axons with two current pulses at 50 ms, delivered every 6 seconds.

Address for correspondence: Marco Atzori, Blanchette Rockefeller Neuroscience Institute, Rockville, MD 20850. Voice: 301-294-7184; fax: 301-294-7007.

marco@brni-jhu.org

Dopamine and the muscarinic agonist oxotremorine were used at 20 μM and 10 μM, respectively.

We first separately determined the direct effect of dopamine and oxotremorine on the glutamatergic currents. Consistent with previous results, oxotremorine depressed the amplitude of EPSCs (A) and increased pair pulse facilitation (PPF) both in the prefrontal cortex ($A_{oxo}/A_{ctrl} = 0.37 \pm 0.16$, $PPF_{oxo}/PPF_{ctrl} = 1.55 \pm 0.21$), taken as control, and in the temporal cortex ($A_{oxo}/A_{ctrl} = 0.48 \pm 0.04$, $PPF_{oxo}/PPF_{ctrl} = 1.53 \pm 0.31$). On the contrary, dopamine had different effects in the two brain areas, since it depressed glutamatergic currents in the prefrontal cortex ($A_{DA}/A_{ctrl} = 0.55 \pm 0.09$) but left the signal unchanged in the temporal cortex ($A_{DA}/A_{ctrl} = 1.12 \pm 0.14$). In order to test possible interactions between the two neurotransmitters in the temporal cortex, we tested the effect of acetylcholine in the presence of dopamine. In the presence of dopamine, acetylcholine failed to reduce the amplitude of the glutamatergic synaptic current ($A_{DA+oxo}/A_{ctrl} = 0.87 \pm 0.05$), as well as to change pair pulse facilitation ($PPF_{oxo}/PPF_{ctrl} = 1.0 \pm 0.11$). Inverting the order of application (oxotremorine preceding dopamine) resulted in the expected depression of the glutamatergic signal ($A_{oxo}/A_{DA} = 0.69 \pm 0.12$) but was not followed by a recovery of the glutamatergic signal ($A_{oxo}+DA/A_{ctrl} = 0.67 \pm 0.12$). Our results confirm that acetylcholine depresses glutamatergic signals, probably acting through presynaptic receptors, and suggest that dopamine has distinct effects on the release of glutamate in different brain areas. More importantly, they indicate that the presence of dopamine in the temporal cortex prevents acetylcholine from depressing the glutamatergic signals, presumably via a presynaptic interaction with muscarinic receptors. The failure of dopamine to revert the muscarinic-induced depression of glutamate release once it began suggests that dopamine acts upstream or at an intermediate level in the muscarinic second-messenger cascade.

We speculate that an increase in the release of dopamine due to an altered interaction between the frontal cortex and dopaminergic nuclei can compromise the physiologic reduction of glutamatergic currents following activation of the cholinergic nuclei. An enduring impairment of the cholinergic reduction of glutamate release might result in multiple long-term consequences, such as glutamate-induced toxicity, under- or overexpression of glutamate and monoamine transporters, and redistribution of glutamate receptors. These alterations could compromise the function of the neuronal circuitry leading eventually to a temporal cortex component in schizophrenia.

Further investigation is required to establish the biochemical nature of the muscarinic–dopaminic interaction as well as the clinical implications of our finding.

ACKNOWLEDGMENTS

This investigation has been funded by the Blanchette Rockefeller Neuroscience Institute, by NARSAD (to MA), and by CONACYT 34424-N (to JCP).

REFERENCES

1. McKay, C.M., D.M Headlam & D.L. Copolov. 2000. Central auditory processing in patients with auditory symptoms. Am. J. Psychiatry **157:** 759–766.

2. KESHAVAN, M.S. 1999. Development, disease and degeneration in schizophrenia: a unitary pathophysiological model. J. Psychiatr. Res. **33:** 513–521.
3. MOORE, H. *et al.* 1999. The regulation of the forebrain dopamine transmission: relevance to the pathophysiology and psychopathology of schizophrenia. Biol. Psychiatry **46:** 40–55.
4. TANDON, R. 1999. Cholinergic aspects of schizophrenia. Br. J. Psychiatry Suppl.: 7–11.
5. TAMMINGA, C. 1999. Glutamatergic aspects of schizophrenia. Br. J. Psychiatry Suppl.: 12–15.
6. KILGARD, M.P. & M.M. MERZENICH. 1998. Cortical map reorganization enabled by nucleus basalis activity. Science **279:** 1714–1718.
7. YAJEYA, J. *et al.* 2000. Muscarinic agonist carbachol depresses excitatory synaptic transmission in the rat basolateral amygdala in vitro. Synapse **38:** 151–160.
8. BELLINGHAM, M.C. & A.J. BERGER. 1996. Presynaptic depression of excitatory synaptic inputs to rat hypoglossal motoneurons by muscarinic M2 receptors. J. Neurophysiol. **76:** 3758–3770.
9. GRILLNER, P. *et al.* 1999. Presynaptic muscarinic (M3) receptors reduce excitatory transmission in dopamine neurons of the rat mesencephalon. Neuroscience **91:** 557–565.
10. KIMURA, F. & R.W. BAUGHMAN. 1997. Distinct muscarinic receptor subtypes suppress excitatory and inhibitory synaptic responses in cortical neurons. J. Neurophysiol. **77:** 709–716.
11. KRNJEVIC, K. 1993. Central cholinergic mechanisms and function. Prog. Brain Res. **98:** 285–292.
12. ARAMAKIS, V.B. & R. METHERATE. 1998. Nicotine selectively enhances NMDAR-mediated synaptic transmission during postnatal development in sensory neocortex. J. Neurosci. **18:** 8485–8495.
13. GAO, W.J. *et al.* 2001. Presynaptic regulation of recurrent excitation by D1 receptors in prefrontal circuits. Proc. Natl. Acad. Sci. USA **98:** 295–300.

N-Acetyl Cysteine–Induced Blockade of Cocaine-Induced Reinstatement

DAVID A. BAKER, KRISTA McFARLAND, RUSSELL W. LAKE, HUI SHEN, SHIGENOBU TODA, AND PETER W. KALIVAS

Department of Physiology and Neuroscience, Medical University of South Carolina, Charleston, South Carolina 29425, USA

KEYWORDS: reinstatement; cystine-glutamate antiporter; xCT; cocaine; glutamate; microdialysis

INTRODUCTION

Repeated cocaine treatment produces persistent neuroplasticity in extracellular glutamate levels in the nucleus accumbens, including a reduction in basal levels and an augmented glutamate response following a cocaine challenge.[1,2] The rise in extracellular glutamate may be of particular importance since glutamate release in the nucleus accumbens is critical for drug-seeking behavior produced by cocaine.[3,4] Nonvesicular glutamate release from cystine-glutamate antiporters has been shown to be the primary source of extracellular glutamate levels in the nucleus accumbens and modulates the release of vesicular glutamate and dopamine via stimulation of glutamate group 2/3 metabotropic glutamate receptors.[5–7] The goal of the present study was to determine whether targeting cystine-glutamate antiporters using the cysteine prodrug *N*-acetyl cysteine (NAC) would block cocaine-induced reinstatement.

First, the capacity of NAC (0 or 60 mg/kg, i.p.) to increase the activity of cystine-glutamate antiporters was examined using *in vivo* microdialysis in rats withdrawn for 21 days from 7 daily cocaine injections. NAC administration (6–600 mg/kg, i.p.) produced a dose-dependent increase in extracellular glutamate in the nucleus accumbens. Fours hours postinjection, rats receiving 60 (mean ± SEM; baseline: 20±7; postNAC: 58±17 pmoles/sample) and 600 mg/kg NAC (baseline: 28±5; postNAC: 57±17 pmoles/sample) exhibited significant increases in extracellular accumbal glutamate levels, whereas rats treated with 6 mg/kg NAC did not (baseline: 23±6; postNAC: 31±8 pmoles/sample). Next, the capacity of NAC to elevate extracellular glutamate levels in the presence or absence of the cystine-glutamate antiporter inhibitor (S)-4-carboxyphenylglycine (CPG; 50 μM) was examined to verify the involvement of cystine-glutamate antiporters in NAC-induced glutamate release. As expected, NAC (60 mg/kg, s.c.) significantly increased accumbal extracellular

Address for correspondence: David A. Baker: Department of Biomedical Sciences, Marquette University, Schroeder Health Complex Suite 426, P.O. Box 1881, Milwaukee, WI 53201-1881. Voice: 414-288-6634; fax: 414-288-6564.
david.baker@marquette.edu

**Ann. N.Y. Acad. Sci. 1003: 349–351 (2003). © 2003 New York Academy of Sciences.
doi: 10.1196/annals.1300.023**

TABLE 1. N-acetyl cysteine–induced blockade of cocaine reinstatement

	Vehicle	NAC 6 mg/kg	NAC 60 mg/kg	NAC 600 mg/kg
Lever presses extinction	12 ± 1.4	13 ± 1.9	8 ± 1.2	15 ± 1.6
Lever presses reinstatement	132 ± 18^a	176 ± 38^a	18 ± 6	15 ± 4

^{a}Indicates a significant increase in responding relative to extinction training.

glutamate (baseline: 31±11; post-NAC: 76±22 pmoles/sample) in the absence, but not the presence of CPG (baseline: 37±14; postNAC: 49±9 pmoles/sample).

The capacity of NAC to block cocaine-induced reinstatement was then examined. Rats were trained to self-administer cocaine (0.25 mg/kg). Following self-administration, rats underwent 2-h daily extinction sessions until responding varied by 10% or less across three consecutive sessions. After extinction, subjects received a NAC injection (0–600 mg/kg, s.c.), and 4 h later were tested for reinstatement of extinguished lever pressing following a cocaine injection (10 mg/kg, i.p.). NAC dose-dependently blocked cocaine-induced reinstatement (TABLE 1). Next, the capacity of NAC to block reinstatement produced by a low (5 mg/kg) or high (10 mg/kg) dose of cocaine was examined. NAC was equally effective in blocking reinstatement produced by the low (mean lever presses±SEM; vehicle controls: 60±9.5; NAC: 15±1.6) or the high dose of cocaine (vehicle controls: 132±17.8; NAC 15±3.8).

In conclusion, systemic administration of cysteine prodrugs increases the activity of cystine-glutamate antiporters in the nucleus accumbens and blocks cocaine-induced reinstatement produced by a low or high dose of cocaine. One mechanism whereby cystine-glutamate antiporters may alter reinstatement involves increased stimulation of group 2/3 metabotropic glutamate receptors,[6] which have previously been shown to regulate the release of vesicular dopamine and glutamate.[5–8] Collectively, these data imply that cystine-glutamate antiporters represent a novel therapeutic target for cocaine addiction.

ACKNOWLEDGMENT

This research was supported by United States Public Health Service Grants MH40817, DA03906, DA12513, DA07288, and DA06074.

REFERENCES

1. PIERCE, R.C. et al. 1996. Repeated cocaine augments excitatory amino acid transmission in the nucleus accumbens only in rats having developed behavioral sensitization. J. Neurosci. **16:** 1550–1560.
2. REID, M.S. & S.P. BERGER. 1996. Evidence for sensitization of cocaine-induced nucleus accumbens glutamate release. Neuroreport **7:** 1325–1329.
3. CORNISH, J.L. & P.W. KALIVAS. 2000. Glutamate transmission in the nucleus accumbens mediates relapse in cocaine addiction. J. Neurosci (Online) **20:** RC89.

4. DI CIANO, P. & B.J. EVERITT. 2001. Dissociable effects of antagonism of NMDA and AMPA/KA receptors in the nucleus accumbens core and shell on cocaine-seeking behavior. Neuropsychopharmacology **25:** 341–360.
5. HU, G. *et al.* 1999. The regulation of dopamine transmission by metabotropic glutamate receptors. J. Pharmacol. Exp. Ther. **289:** 412–416.
6. BAKER, D.A. *et al.* 2002. The origin and neuronal function of *in vivo* non-synaptic glutamate. J. Neurosci. **22:** 9134–9141.
7. XI, Z.X. *et al.* 2002. Group II metabotropic glutamate receptors modulate extracellular glutamate in the nucleus accumbens. J. Pharmacol. Exp. Ther. **300:** 162–171.
8. BASKYS, A. & R.C. MALENKA. 1991. Agonists at metabotropic glutamate receptors presynaptically inhibit EPSCs in neonatal rat hippocampus. J. Physiol. **444:** 687–701.

AMPA- and NMDA-Associated Postsynaptic Protein Expression in the Human Dorsolateral Prefrontal Cortex

MONICA BENEYTO AND JAMES H. MEADOR-WOODRUFF

Mental Health Research Institute and Department of Psychiatry, University of Michigan, Ann Arbor, Michigan 48109, USA

INTRODUCTION

Pharmacological and anatomical data suggest that abnormal glutamate neurotransmission may be associated with the pathophysiology of schizophrenia and mood disorders. Glutamate is the primary excitatory neurotransmitter in the central nervous system. It regulates numerous cellular signaling cascades and controls the excitability of central synapses both pre- and postsynaptically. Region-specific localization, cell surface expression, and activity-dependent regulation of neurotransmitter receptors in neurons are essential for their function. In addition, there is a complex network of postsynaptic density proteins (PSD) that act via specific protein-protein interactions to serve as scaffold/adaptor proteins targeting, anchoring, and spatially organizing synaptic proteins in the cell membrane. Proteins that are specific for individual subunits that form the ionotropic glutamate receptors have been identified. Recently, we have shown that there are abnormalities in NMDA receptor expression as well as the expression of some of these PSD proteins in the thalamus in schizophrenia.[1] The distribution and levels of expression of these molecules in the human cerebral cortex is largely unknown, however, and characterization of the cortical expression of these molecules in human brain was the goal of this study.

METHODS

We used *in situ* hybridization to determine the laminar distribution and transcript expression levels of NMDA (PSD95, NF-L)- and AMPA (NSF, PICK-1)-associated PSD proteins in the dorsolateral prefrontal cortex. We used specific [^{35}S]-labeled riboprobes designed for human PSD95, NF-L, NSF, and PICK-1 as we have previously described in detail.[2–4] Images were acquired from digitized X-ray films with a CCD imaging system.

Address for correspondence: Monica Beneyto, Ph.D., Mental Health Research Institute and Department of Psychiatry, University of Michigan, 205 Zina Pitcher Place, Ann Arbor, MI 48109. Voice: 734-936-2056; fax: 734-647-4130.
mbeneyto@umich.edu

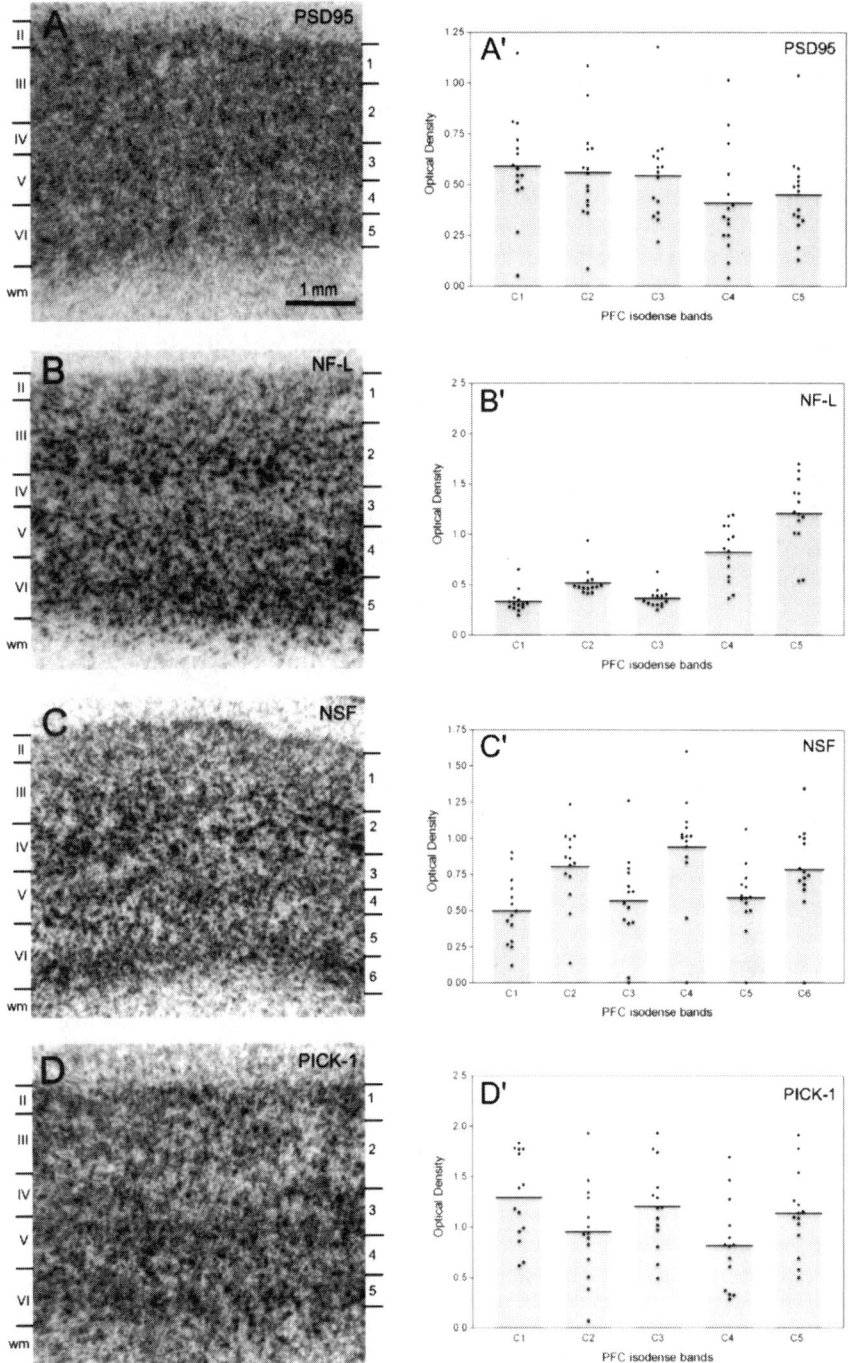

FIGURE 1. *See following page for legend.*

RESULTS AND COMMENT

We found specific laminar patterns of expression for each of the PSD proteins studied. Transcripts encoding PSD95, which binds to both NMDA (NR2) and kainate (GluR6/KA2) subunits, were expressed at similar levels in all laminae, with slight enrichment in layers II, IV, and VI. NF-L, a protein exclusively related to NMDA receptors (binding to exon 21 of NR1) had a different pattern of laminar expression. NF-L mRNA was highly expressed in two thin bands corresponding to layers III and the superficial aspect of layer V, as well as in a wider band associated with layer VI. A granular appearing hybridization signal was distinct from the more homogeneous pattern seen for PSD95.

For the AMPA-associated PSD proteins, transcripts encoding NSF (which binds GluR1-4) were moderately expressed in layers II, IV, and the deep aspect of layer V, and high in layers III and V-VI. PICK-1 mRNA was expressed at moderate to high levels throughout the entire cortical depth, with maximum levels of expression in layers II, IV, and VI.

Previous publications have reported changes in the expression of some of these molecules in schizophrenia, either by *in situ* hybridization (PSD95),[5] or by microarray (NSF).[6] On the other hand, NF-L and PICK-1 have thus far not been studied in schizophrenia. Previous work has been done in tissue homogenates, but has not yet focused on laminar or cellular changes. Our present study provides the normal patterns of expression in the prefrontal cortex of four of these molecules, and given the heterogeneity of expression, suggests that subsequent studies of these molecules in schizophrenia should target laminar and cellular resolution.

ACKNOWLEDGMENT

This work has been supported by Grant MH53327 from the NIMH (JMW).

REFERENCES

1. CLINTON, S.M., V. HAROUTUNIAN, K.L. DAVIS & J.H. MEADOR-WOODRUFF. 2002. Am. J. Psychiatry. 2003. In press.
2. HEALY, D.J., V. HAROUTUNIAN, P. POWCHIK, *et al.* 1998. AMPA receptor binding and subunit mRNA expression in prefrontal cortex and striatum of elderly schizophrenics. Neuropsychopharmacology **19:** 278–286.
3. IBRAHIM, H.J., A.J. HOGG, D.J. HEALY, *et al.* 2000. Ionotropic glutamate receptor binding and subunit mRNA expression in thalamic nuclei in schizophrenia. Am. J. Psychiatry **157:** 1811–1823.

FIGURE 1. (A–D) Photomicrographs of film autoradiograms from adjacent sections of the dorsolateral prefrontal cortex. Laminar distribution of transcripts encoding PSD-95 (**A**), NF-L (**B**), NSF (**C**), and PICK-1 (**D**). On the *left side of each panel* are the boundaries of the cortical layers identified by Nissl staining and on the *right* isodense bands that were measured. (**A′–D′**) Optical density values of isodense bands. Data are expressed as means as well as individual values from cortical layer I (*left*) to VI (*right*).

4. IBRAHIM, H.J., D.J. HEALY, A.J. HOGG & J.H. MEADOR-WOODRUFF. 2000. Nucleus-specific expression of ionotropic glutamate receptor subunit mRNAs and binding sites in primate thalamus. Mol Brain Res. **79:** 1–17.
5. OHNUMA, T., H. KATO, H. ARAI, *et al.* 2000. Gene expression of PSD95 in prefrontal cortex and hippocampus in schizophrenia. Neuroreport 28(11): 3133–3177.
6. MIRNICS, K., F.A. MIDDLETON, A. MARQUEZ, *et al.* 2000. Molecular characterization of schizophrenia viewed by microarray analysis of gene expression in prefrontal cortex. Neuron **28:** 53–67.

AGS3: A G-Protein Regulator of Addiction-Associated Behaviors

M. S. BOWERS,[a] R. W. LAKE,[a] K. McFARLAND,[a] Y. K. PETERSON,[b]
S. M. LANIER,[b] C. C. LAPISH,[a] AND P. W. KALIVAS[a]

[a]*Department of Physiology and Neuroscience, Medical University of South Carolina, Charleston, South Carolina 29425, USA*

[b]*Department of Pharmacology and Experimental Therapeutics, Louisiana State University Health Sciences Center, New Orleans, Louisiana 70112, USA*

KEYWORDS: G-protein regulator; Giα; addiction; AGS3

Signaling through the inhibitory G-protein, Giα, is dysregulated during protracted withdrawal from repeated cocaine. This imbalance may explain aberrant protein expression patterns, such as Δ-Fos B and P-CREB, which are associated with addiction.[1,2] G-protein signaling kinetics and specificity are influenced by cell type or addition of purified proteins,[3,4] which implicates accessory protein involvement. The activator of G-protein signaling 3 (AGS3) is the first G-protein dissociation inhibitor described for heterotrimeric G-proteins.[5] AGS3 specifically inhibits signaling through Giα by competing with the βγ dimer to render AGS3-bound Giα incapable of recognition by receptor or effector.[6,7] Interestingly, protracted withdrawal from repeated administration of cocaine upregulated AGS3 roughly 60% in the rat prefrontal cortex (PFC) and nucleus accumbens core (NAcore). Both the PFC and NAcore are required for expression of behaviors associated with addiction.[8,9] The PFC is dysfunctional in addicts[10] and Giα signaling is dysregulated during cocaine withdrawal.[11] The rise in AGS3 observed during withdrawal was prevented by continuous infusion of antisense oligonucleotides into the PFC for 2 weeks. When AGS3 expression is knocked down behaviors that model components of cocaine-induced plasticity or relapse, for example, locomotor sensitization or cocaine-induced reinstatement of cocaine seeking, were selectively blocked. Conversely, transduction of the consensus Giα recognition motif of AGS3 into the PFC induced behavioral sensitization to acute cocaine. Blockade of addiction-associated behaviors coincided with normalized Giα signaling. These data reveal that upregulated prefrontal cortical AGS3 allows for the expression of addiction-associated behaviors during withdrawal by blunting signaling through Giα.

Address for correspondence: M. Scott Bowers, Ernest Gallo Clinic and Research Center, 5858 Horton St., Suite 200, Emeryville, CA 94608. Voice: 510-985-3136.
sbowers@egcrc.net

ACKNOWLEDGMENTS

This work was supported by NIDA Grants DA03906, DA12513 (PWK) and a NASA predoctoral fellowship (MSB).

REFERENCES

1. NESTLER, E.J., M. BARROT & D.W. SELF. 2001. DeltaFosB: a sustained molecular switch for addiction. Proc. Natl. Acad. Sci. USA **98:** 11042–11046.
2. HOPE, B.T. 1998. Cocaine and the AP-1 transcription factor complex. Ann. NY Acad. Sci. **844:** 1–6.
3. MARJAMAKI, A. et al. 1997. Factors determining the specificity of signal transduction by guanine nucleotide-binding protein-coupled receptors. Integration of stimulatory and inhibitory input to the effector adenylyl cyclase. J. Biol. Chem. **272:** 16466–16473.
4. SATO, M. et al. 1995. Factors determining specificity of signal transduction by G-protein-coupled receptors. Regulation of signal transfer from receptor to G-protein. J. Biol. Chem. **270:** 15269–15276.
5. TAKESONO, A. et al. 1999. Receptor-independent activators of heterotrimeric G-protein signaling pathways. J. Biol. Chem. **274:** 33202–33205.
6. NATOCHIN, M. et al. 2000. AGS3 inhibits GDP dissociation from Gα subunits of the Gi family and rhodopsin-dependent activation of transducin. J. Biol. Chem. **275:** 40981–40985.
7. PETERSON, Y.K. et al. 2000. Stabilization of the GDP-bound conformation of Gialpha by a peptide derived from the G-protein regulatory motif of AGS3. J. Biol. Chem. **275:** 33193–33196.
8. PIERCE, R.C. et al. 1998. Ibotenic acid lesions of the dorsal prefrontal cortex disrupt the expression of behavioral sensitization to cocaine. Neuroscience **82:** 1103–1114.
9. MCFARLAND, K. & P.W. KALIVAS. 2001. The circuitry mediating cocaine-induced reinstatement of drug-seeking behavior. J. Neurosci. **21:** 8655–8663.
10. GOLDSTEIN, R.Z. & N.D. VOLKOW. 2002. Drug addiction and its underlying neurobiological basis: neuroimaging evidence for the involvement of the frontal cortex. Am. J Psychiatry **159:** 1642–1652.
11. XI, Z.X. et al. 2002. Modulation of group II metabotropic glutamate receptor signaling by chronic cocaine. J. Pharmacol. Exp. Ther. **303:** 608–615.

Changes in Electrophysiological Properties of Nucleus Accumbens Neurons Depend on the Extent of Behavioral Sensitization to Chronic Methamphetamine

ANNE MARIE BRADY, STANLEY D. GLICK, AND PATRICIO O'DONNELL

Center for Neuropharmacology and Neuroscience, Albany Medical College, Albany, New York 12208, USA

KEYWORDS: nucleus accumbens; prefrontal cortex; ventral tegmental area; dopamine; glutamate; methamphetamine; behavioral sensitization; intracellular recording

INTRODUCTION

Behavioral sensitization, which occurs following chronic administration of psychostimulant drugs in rats, is characterized by an enhanced locomotor response to a low challenge dose of the drug and has been suggested to model some aspects of addictive behavior, such as drug craving.[1] Behavioral sensitization is thought to be mediated by neurophysiological changes in the nucleus accumbens (NAcc) and its inputs from the prefrontal cortex (PFC) and ventral tegmental area (VTA). In particular, the development of sensitization appears to depend on dopaminergic and glutamatergic transmission in the VTA,[2,3] putatively via activation of descending PFC inputs.[4] In contrast, the long-term expression of behavioral sensitization is hypothesized to be mediated in the NAcc.[5,6] Here, we used *in vivo* intracellular recordings to characterize the electrophysiological properties of NAcc neurons after chronic administration of methamphetamine (METH). In addition, we investigated the effects of individual differences in the extent of behavioral sensitization on NAcc physiology.

METHODS

Behavioral Testing

Adult male rats (Sprague-Dawley, Charles River, 200–250 g, N=30) were tested for baseline locomotor activity in response to a challenge dose of METH (0.5 mg/kg, i.p.).

Address for correspondence: Anne Marie Brady, Center for Neuropharmacology and Neuroscience, Albany Medical College, 47 New Scotland Ave., MC-136, Albany, NY 12208. Voice: 518-262-5903; fax: 518-262-0687.
bradyam@mail.amc.edu

Ann. N.Y. Acad. Sci. 1003: 358–363 (2003). © 2003 New York Academy of Sciences.
doi: 10.1196/annals.1300.026

Activity counts (beam breaks) were recorded for 2 h postinjection in opaque, cylindrical (60 cm diameter) activity chambers. Three days later, daily injections of METH (5.0 mg/kg, i.p.) in the home cages were begun, and were administered for five consecutive days. After a 13-day withdrawal period, rats were again tested for locomotor activity in response to METH challenge (0.5 mg/kg, i.p.). Rats that exhibited at least a 50% increase above baseline in total activity counts were classified as "sensitized;" rats with less than a 20% increase from baseline were classified as "not sensitized." A separate group of rats received saline at all time points (baseline, chronic, and test).

Electrophysiology

In vivo intracellular recordings were performed 3 to 7 days after the test challenge. Rats were initially anesthetized with chloral hydrate (400 mg/kg, i.p.) and anesthesia was maintained throughout recording (20–30 mg/kg/h, i.p.) via the use of a syringe pump. Rats were placed in a stereotaxic apparatus (David Kopf, Tujunga, CA) and bipolar stimulating electrodes were placed in the prelimbic/infralimbic region of the medial PFC (mm from bregma: AP +3.0, L +1.6, DV −4.3 at a 30° angle) and the VTA (AP −6.0, L −0.5, DV −7.4 from dura). Intracellular recording electrodes were filled with 2% neurobiotin in 2 M potassium acetate and lowered into the NAcc (AP 1.2 to 1.8, L +1.2 to 1.5, DV −6.0 to −7.5 from cortical surface) with the aid of a hydraulic microdrive.

Following recordings, cells were injected with neurobiotin tracer and animals were perfused with saline followed by 4% paraformaldehyde. Filled cells were visualized and located in the shell, core, or rostral pole regions of the NAcc based on calbindin immunohistochemistry.

RESULTS

Successful intracellular recordings were made from 15 cells in 14 sensitized animals (METH-S; mean±SD change from baseline: +157±101%), 7 cells in 7 nonsensitized animals (METH-N; −2±15%), and 9 cells in 9 saline control animals. Cells in all groups exhibited a bistable membrane potential, with spontaneous fluctuations between depolarized *up* states and hyperpolarized *down* states (FIG. 1). No group differences were observed in the resting membrane potentials of either state (*down* state: METH-S, −84.6±8.5 mV; METH-N, −88.2 ± 3.4 mV; saline, −83.0±6.9 mV; *up* state: METH-S, −67.4±9.1 mV; METH-N, −73.8±6.8 mV; saline, −70.0±8.6 mV), nor for the frequency of *up* states (METH-S, 0.85±0.21 Hz; METH-N, 0.85± 0.21 Hz; saline, 0.82±0.11 Hz). However, NAcc cells from sensitized animals displayed longer *up* states than cells from nonsensitized animals (METH-S, 607±167 ms; METH-N, 426±134 ms; $P=.021$) and, accordingly, spent more time in the *up* state (METH-S, 50.2±12.6%; METH-N, 36.9±15.3%; $P =.043$). Preliminary examination of data classified by NAcc cell location suggested that these differences between sensitized and nonsensitized animals were more pronounced in the shell and rostral pole regions, as compared to the core. However, the low number of cells in each group precluded further analysis by subregion. Despite the increase in *up* state duration, cells from sensitized animals were less excit-

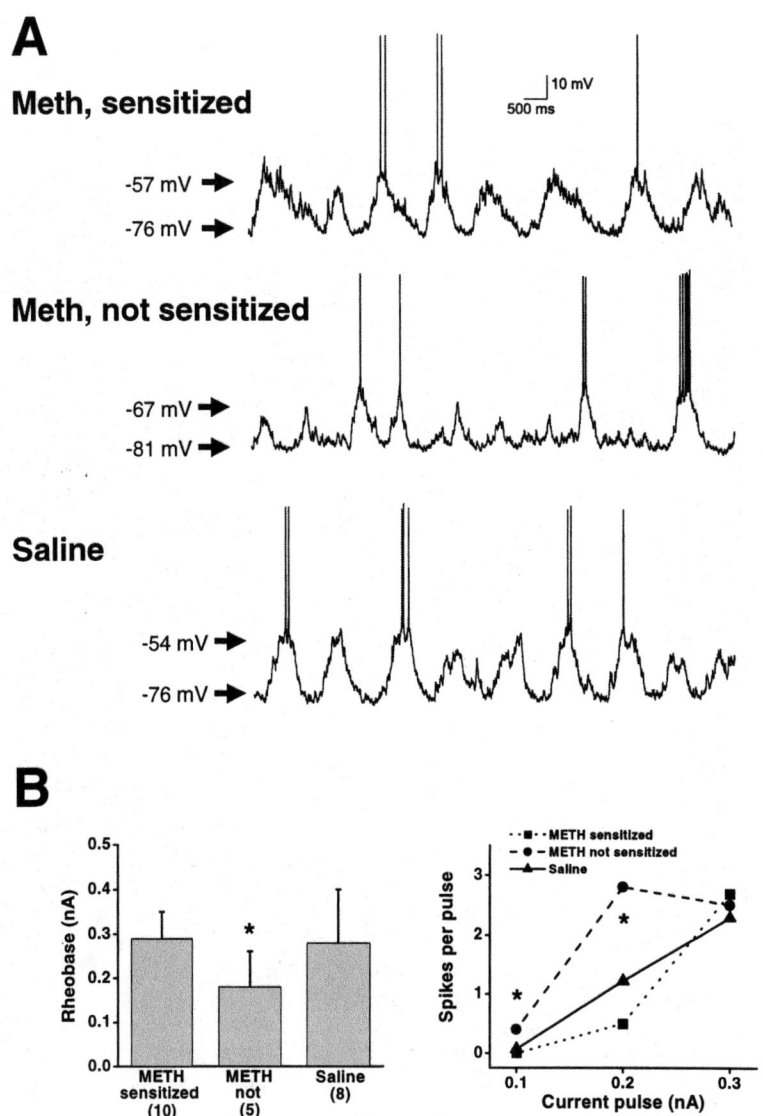

FIGURE 1. (**A**) NAcc cells from sensitized, nonsensitized, and saline control animals all exhibited a bistable membrane potential, with *up* and *down* states. *Up* states were prolonged in cells from sensitized animals. (**B**) Excitability was reduced in NAcc cells from sensitized animals compared to nonsensitized animals, as evidenced by increased rheobase value (*left*, means±SD) and decreased number of spikes to current injection (*right*, group means). Numbers of cells per group are given in parentheses. *$P<.02$, sensitized vs. nonsensitized.

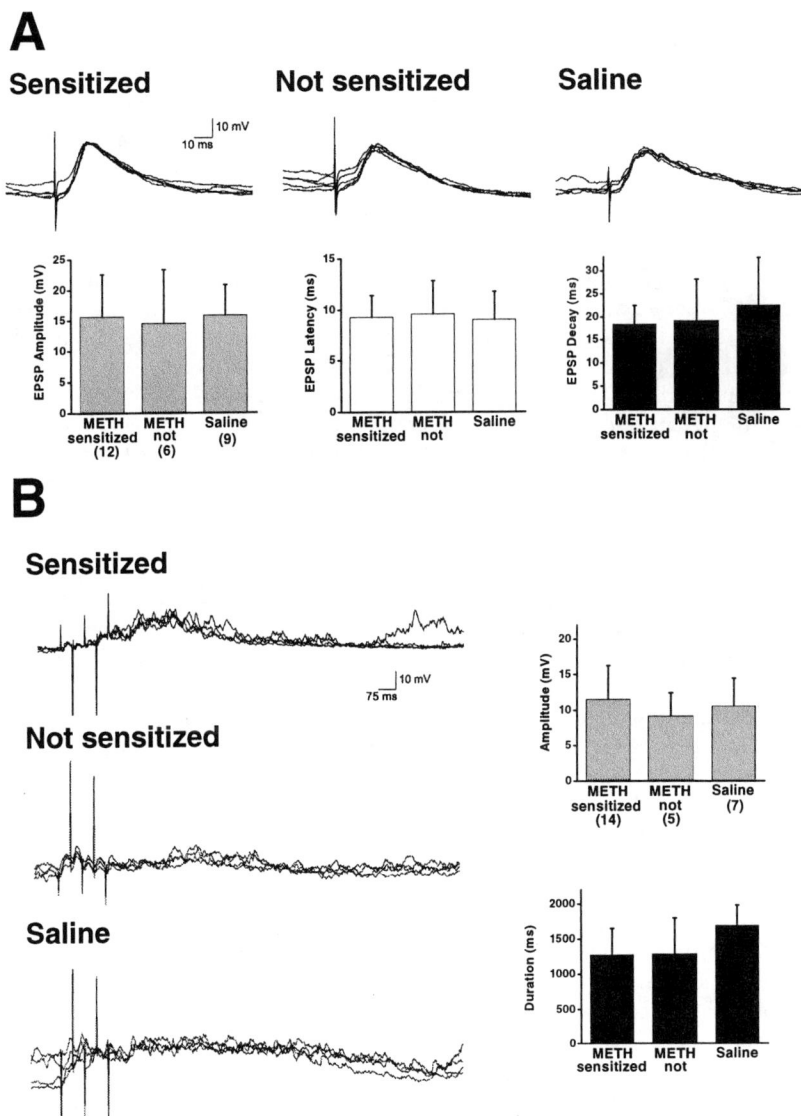

FIGURE 2. The responses of NAcc cells to stimulation of the PFC (**A**) or VTA (**B**) were not altered by chronic methamphetamine administration. Traces represent overlays of 4–5 sweeps per cell. Means±SD are represented graphically. Numbers of cells per group are given in parentheses.

able than cells from nonsensitized animals, as evidenced by a higher rheobase value to positive current injection and a decreased number of spikes to low intensity current pulses (FIG. 1). The spontaneous firing rates were unchanged among all groups.

Unexpectedly, no group differences were observed in response to any afferent stimulation (FIG. 2). Single-pulse stimulation of the PFC (0.5–1.0 mA, 0.5 msec) evoked an EPSP in all neurons tested, with no differences between groups in amplitude ($P=.926$), onset latency ($P=.921$), or decay time ($P=.469$). Train stimulation of the VTA (5 pulses, 20 Hz, 1.0 mA) elicited a sustained depolarization that also did not differ in amplitude ($P=.573$) or duration ($P=.080$) among all groups.

DISCUSSION

The first main finding of this preliminary investigation is that NAcc neurons continue to exhibit *up* and *down* states following chronic methamphetamine administration, but show further differences depending on whether the drug produced behavioral sensitization. Cells from sensitized animals were biased toward the *up* state when compared with non-sensitized animals. As *up* states in NAcc cells are driven by glutamatergic inputs from the hippocampus,[7] it may be speculated that hippocampal input to these cells is enhanced in behaviorally sensitized animals. However, cells from sensitized animals were also less excitable. These seemingly paradoxical effects resemble the effects of dopamine *in vitro*, where NAcc cells are depolarized and yet exhibit a reduced excitability.[8] Together, these findings suggest that tonic mesolimbic dopaminergic transmission may be enhanced following chronic METH, but only in animals that exhibit behavioral sensitization.

The second major finding is that NAcc cells from sensitized animals did not respond differently to PFC or VTA stimulation. This was unexpected, given that repeated amphetamine administration has previously been shown to render NAcc cells less responsive to glutamate activation,[9] and more sensitive to dopamine receptor stimulation.[10] However, these previous studies differed from the current study in that they did not specifically activate the afferent pathways to the NAcc.

The current study represents a preliminary investigation of the electrophysiological responses of NAcc neurons following chronic METH treatment, and was limited to recordings done under baseline (i.e., METH-free) conditions. As the behavioral and neurochemical changes associated with sensitization appear to emerge only under subsequent drug challenge,[11] further experiments assessing the effects of METH challenge during recording are likely to be important in fully characterizing the effects of behavioral sensitization on NAcc cell physiology. Further analysis of cells located in shell, core, and rostral pole subregions may also reveal additional differences between groups. However, the current results support the following speculations. The prolonged *up* states and reduced excitability in cells from sensitized animals suggest that basal dopamine levels may be selectively enhanced in animals that exhibit behavioral sensitization following chronic METH. However, the lack of changes in response to artificial electrical stimulation of the VTA suggests that this effect is unlikely to be mediated by alterations in dopamine terminals impinging on NAcc neurons. Similarly, there do not appear to be significant changes in glutamatergic PFC terminals in the NAcc. Instead, changes in the excitability of NAcc afferents such as the PFC and VTA are speculated to occur at the cell body level rather than locally in the NAcc. Such changes in mesocorticolimbic circuitry may underlie the observed effects on spontaneous NAcc activity, and mediate the expression of behavioral sensitization.

ACKNOWLEDGMENTS

This work was supported by DA14821 (AMB), DA07307 (SDG), and DA14020 and a NARSAD Independent Investigator Award (PO'D).

REFERENCES

1. ROBINSON, T.E. & K.C. BERRIDGE. 1993. The neural basis of drug craving: an incentive-sensitization theory of addiction. Brain Res. Rev. **18:** 247–291.
2. KALIVAS, P.W. & J.E. ALESDATTER. 1993. Involvement of N-methyl-D-aspartate receptor stimulation in the ventral tegmental area and amygdala in behavioral sensitization to cocaine. J. Pharmacol. Exp. Ther. **267:** 486–495.
3. VEZINA, P. 1996. D1 dopamine receptor activation is necessary for the induction of sensitization by amphetamine in the ventral tegmental area. J. Neurosci. **16:** 2411–2420.
4. CADOR, M., Y. BJIJOU, S. CAILHOL & L. STINUS. 1999. D-amphetamine-induced behavioral sensitization: implication of a glutamatergic medial prefrontal cortex-ventral tegmental area innervation. Neuroscience **94:** 705–721.
5. WOLF, M.E., F.J. WHITE, R. NASSAR, et al. 1993. Differential development of autoreceptor subsensitivity and enhanced dopamine release during amphetamine sensitization. J. Pharmacol. Exp. Ther. **264:** 249–255.
6. CADOR, M., Y. BJIJOU & L. STINUS. 1995. Evidence of a complete independence of the neurobiological substrates for the induction and expression of behavioral sensitization to amphetamine. Neuroscience **65:** 385–395.
7. O'DONNELL, P. & A.A. GRACE. 1995. Synaptic interactions among excitatory afferents to nucleus accumbens neurons: hippocampal gating of prefrontal cortical input. J. Neurosci. **15:** 3622–3639.
8. O'DONNELL, P. & A.A. GRACE. 1996. Dopaminergic reduction of excitability in nucleus accumbens neurons recorded in vitro. Neuropsychopharmacology **15:** 87–97.
9. WHITE, F.J., X.T. HU, X.F. ZHANG & M.E. WOLF. 1995. Repeated administration of cocaine or amphetamine alters neuronal responses to glutamate in the mesoaccumbens dopamine system. J. Pharmacol. Exp. Ther. **273:** 445–454.
10. WOLF, M.E., F.J. WHITE & X.T. HU. 1994. MK-801 prevents alterations in the mesoaccumbens dopamine system associated with behavioral sensitization to amphetamine. J. Neurosci. **14:** 1735–1745.
11. PAULSON, P.E. & T.E. ROBINSON. 1995. Amphetamine-induced time-dependent sensitization of dopamine neurotransmission in the dorsal and ventral striatum: a microdialysis study in behaving rats. Synapse **19:** 56–65.

Evaluation of NMDA Receptors *in Vivo* in Schizophrenic Patients with [^{123}I]CNS 1261 and SPET

Preliminary Findings

RODRIGO A. BRESSAN,[a] KJELL ERLANDSSON,[b] RACHEL S. MULLIGAN,[b] ROGER N. GUNN,[d] VINCENT J. CUNNINGHAM,[c] JONATHAN OWENS,[e] PETER J. ELL,[b] AND LYN S. PILOWSKY[a,b]

[a]*Institute of Psychiatry, King's College, London, WC2R 2LS, United Kingdom*

[b]*Institute of Nuclear Medicine, University College London, United Kingdom*

[c]*IRSL, Hammersmith Hospital, London, United Kingdom*

[d]*McConnell Brain Imaging Centre, McGill University, Montreal, Canada*

[e]*West of Scotland Radionuclide Dispensary, Glasgow, United Kingdom*

KEYWORDS: NMDA receptor dysfunction; schizophrenia; SPET technique

INTRODUCTION

Emerging data points toward a central role for the glutamatergic *N*-methyl-D-aspartate (NMDA) receptor dysfunction in the pathophysiology of schizophrenia. SPET techniques could provide relevant information about the basic neurochemistry of schizophrenia and the mechanism of action of antipsychotic drugs.

OBJECTIVE

We have tested the hypothesis that NMDA receptor dysfunction is present in schizophrenia and is reversed by antipsychotic treatment *in vivo* with [^{123}I]CNS 1261, a SPET ligand highly selective for the MK801 intrachannel site of the NMDA receptor.

METHODS

Subjects

Healthy controls (HC, $N=13$) and schizophrenic patients: drug free (DF, $N=5$, >2 months), typical antipsychotic treated (TA, $N=7$, 173 mg/d CPZ equivalent), and clozapine treated (CZ, $N=10$, 372 mg/day with plasma concentration 0.4 mg/mL).

Address for correspondence: Rodrigo A. Bressan, King's College London, Strand, London WC2R 2LS, United Kingdom. Voice: +44-0-20-7848-0807; fax: +44-0-20-7848-0053.
r.bressan@iop.kcl.ac.uk

SPET Data Acquisition

[^{123}I]-CNS1261-1(1-naphthyl)-N'-(3-[^{123}I]iodo-phenyl)-N'methylguanidine) is a SPET ligand for the PCP/MK801 intrachannel site of the NMDA receptor, with high affinity (K_i = 4.2 nM; [^{3}H]MK801 K_i=1.6 nM) and high selectivity.[1]

The tracer can only access the binding site when the receptor is activated and the channel is opened. NMDA receptor binding was quantified *in vivo* with dynamic [^{123}I]CNS 1261 SPET scans.[2] Bolus infusion (B/I) [^{123}I]-CNS1261 SPET scans were performed in a Prism 3000XP (Philips Medical systems) camera.[3]

Intravenous bolus of 71 MBq [^{123}I]-CNS1261 (SD 5) and infusion of [^{123}I]-CNS1261 via a saline drip into an antecubital vein was performed using a volumetric infusion pump (IVAC 572, Allaris Medical) for ~4 h at a rate of 80 mL/h (total dose ~185 MBq). Image acquisition: 0–30 min; 90–135 min; 185–225 min p.i. Data from 5 HC and 3 patients (2 TA and 1 CZ) scanned with the bolus protocol[2] were included in the analysis (inclusion or exclusion of this subjects did change the findings).

Imaging Reconstruction

Imaging reconstruction was performed according to the method previously reported.[4] Tomographic images were reconstructed by filtered backprojection. Scatter correction was performed using the triple energy window method.[4] Attenuation correction was performed with the transmission based on second order Chang method. Realignment of images from different time frames was done based on fiducial markers.

Blood Analysis

Blood and plasma clearance of radioactivity was determined from blood samples. Venous samples were taken at 5, 10, 20, 30, and up to 60 min p.i., as well as 6 samples in the following scans. Parent radioligand and metabolites in plasma were measured by HPLC.

Imaging Analysis

Images from the equilibrium phase of the scans (90–240 min) were integrated and transformed into Talairach space using SPM-99. A set of regions of interest (ROIs) were drawn in Talairach space and used for all subjects: thalamus, striatum, and cerebellum (thalamus and striatum were chosen because these brain regions represent the areas where [^{123}I]-CNS1261 has the best signal-to-background ratios; cerebellum was used as gray matter region representing low binding). Equilibrium analysis was performed, and V_T was calculated based on the total volume of distribution (V_T) [mL plasma/mL tissue]. It was calculated as follows:

$$V_r = \frac{C_t}{C_p} f_s$$

where C_t is the activity concentration in tissue, C_p is the activity concentration of unchanged tracer in venous plasma and f_s is the slope correction factor.[3] To access the influence of plasma-free fraction of the tracer (f_1) and the tissue-free fraction of the tracer (f_2) in the results, two outcome measures respectively independent of f_2 and f_1,[5] V_T difference (V_{TD}), where,

$$V_{TD} = V_T ROI - V_T Cer$$

and V_T ratio,

$$V_{TR} = \frac{(V_T ROI - V_T Cer)}{(V_T Cer)}.$$

Statistical Analysis

Repeated measures analysis of variance (ANOVA) with planned comparisons using "group" differences as a between subject factor and "brain region" as within subject factor was used. Simple planned comparisons were performed using the HC as reference group: (1) HC vs. DF, (2) HC vs. TA, and (3) HC vs. CZ. Since significant interaction between "group" and "brain region" on V_T was observed (F(3, 30) = 3.0, $P = .04$), separate one-way ANOVAs with the same planned comparisons were performed for thalamus and striatum and adjusted for two regional comparisons by using the Bonferroni correction (adjusted significance level = 0.025). Age was significantly different between groups and was included in the ANOVAs as a covariate. The same analyses were performed with V_{TD} and V_{TR} to check the impact of f_1 and f_2 in the V_T results.

RESULTS

No significant differences were observed respectively in thalamic and striatal V_T (SD) between DF or TA patients versus HC. CZ patients had significant lower V_T values in thalamus and striatum than HC. The difference in V_{TD} and V_{TR} between HC and CZ patients was statistically significant in thalamus and striatum (results not shown) suggesting the findings are not due to f_1 or f_2.

DISCUSSION

Clozapine treatment was associated with a >35% decline in thalamic [^{123}I]-CNS1261 V_T. Functional modulation of NMDA receptor binding is governed by a number of complex allosteric influences and is unlikely to be a direct effect of clozapine at the MK801 intrachannel site. The data suggest a specific effect of clozapine on glutamatergic systems *in vivo* and may help account for its unique antipsychotic efficacy.

ACKNOWLEDGMENTS

R.A. Bressan was supported by a charitable educational grant from by CAPES Foundation, Janssen Ltd. Sanofi Synthelabo during the data collection. K. Erlandsson is supported on a UK MRC Senior clinical research fellowship. L.S. Pilowsky is a UK Medical Research Council (MRC) Senior Clinical Research Fellow.

REFERENCES

1. OWENS, J. et al. 2000. Synthesis and binding characteristics of N-(1-naphthyl)-N'-(3-[(^{125})I]- iodophenyl)-N'-methylguanidine ([(^{125})I]-CNS 1261): a potential SPECT agent for imaging NMDA receptor activation. Nucl. Med. Biol. **27:** 557–564.
2. ERLANDSSON, K., R.A. BRESSAN, R.S. MULLIGAN, et al. 2003. Kinetic modelling of [^{123}I]CNS 1261—A potential SPET tracer for the NMDA receptor. Nucl. Med. Biol. **30:** 441–454.
3. BRESSAN, R.A., K. ERLANDSSON, R.S. MULLIGAN, et al. 2003. A bolus/infusion paradigm for the novel NMDA receptor SPET tracer [^{123}I]CNS 1261. Nucl. Med. Biol. In press.
4. BRESSAN, R.A., K. ERLANDSSON, H.M. JONES, et al. 2003. Is regionally selective D2/D3 dopamine occupancy sufficient for atypical antipsychotic effect? An in vivo quantitative [^{123}I]epidepride SPET study of amisulpride-treated patients. Am. J. Psychiatry **160:** 1413–1420.
5. LARUELLE, M. 2000. Imaging synaptic neurotransmission with in vivo binding competition techniques: a critical review. J. Cereb. Blood Flow Metab. **20:** 423–451.

Effects of Naloxone-Precipitated Morphine Withdrawal on Glutamate-Mediated Signaling in Striatal Neurons *in Vitro*

ELENA H. CHARTOFF, MARIA PAPADOPOULOU, CHRISTINE KONRADI, AND WILLIAM A. CARLEZON JR.

Department of Psychiatry, Harvard Medical School at McLean Hospital, Belmont, Massachusetts 02478, USA

KEYWORDS: SKF 82958; primary striatal culture; GluR1; AMPA; CREB; naloxone; morphine withdrawal

Repeated administration of morphine leads to behavioral and molecular adaptations mediated by mu opiate receptors.[1] The ventral striatum (nucleus accumbens [NAc]) plays a key role in mediating both the rewarding effects of morphine,[2] as well as aversive states associated with withdrawal.[3] The NAc integrates dopaminergic and glutamatergic inputs via activation of dopamine D1 and D2 receptors and several classes of glutamate receptors, including ionotropic AMPA and NMDA receptors.

It has been established that dopamine mediates, at least in part, critical aspects of morphine actions within mesolimbic systems.[2,3] There is increasing evidence that alterations in glutamatergic transmission can also regulate the effects of morphine.[4,5] In striatum, D1 receptor activation can phosphorylate GluR1 at Ser845 (P-GluR1), resulting in increased AMPA currents.[6,7] We have developed primary cultures of prenatal rat striatal cells as a model system to investigate molecular adaptations that occur in response to chronic morphine and morphine withdrawal. Chronic morphine leads to a compensatory upregulation of cAMP pathways,[8] and we have shown that D1 receptor-mediated activation of cAMP response element binding protein (CREB) in striatal cultures during naloxone-precipitated morphine withdrawal is enhanced.[9] Here we tested the hypothesis that, in striatal cultures undergoing morphine withdrawal, D1 receptor activation would superinduce P-GluR1 and that glutamate receptor activation would potentiate CREB phosphorylation (P-CREB).

Primary cultures of striatal neurons were treated chronically with morphine, and withdrawal was precipitated by the addition of naloxone prior to treatment with the D1 receptor agonist, SKF 82958. We observed a robust and long-lasting increase in SKF 82958-induced P-GluR1 levels in striatal cells during precipitated morphine withdraw-

---

Address for correspondence: William A. Carlezon Jr., Department of Psychiatry, Harvard Medical School at McLean Hospital, Belmont, MA 02478. Voice: 617-855-2021; fax: 617-855-2023.
carlezon@mclean.harvard.edu

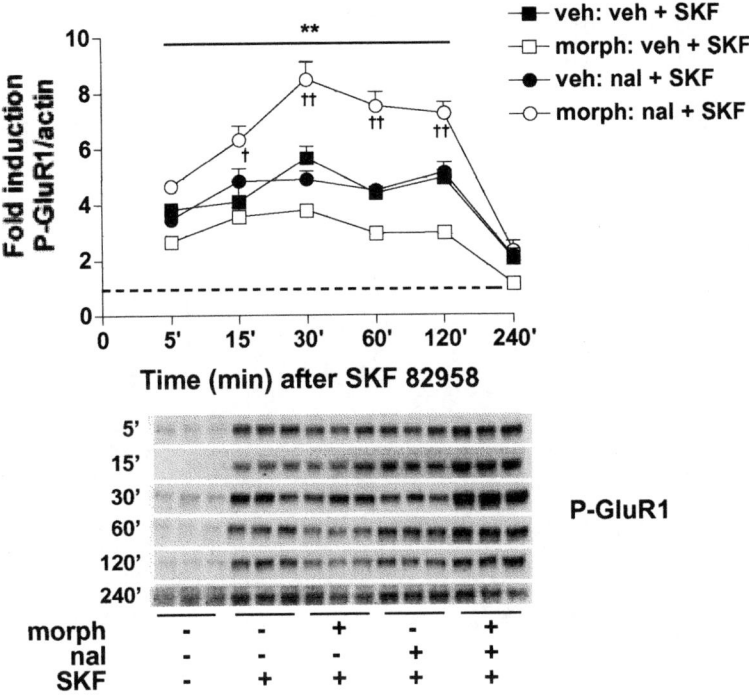

FIGURE 1. SKF 82958-induced P-GluR1 levels are superinduced in primary striatal cultures during precipitated morphine withdrawal. Cultures were treated with vehicle or morphine (10 μM) for 6 days, followed by a 15-min treatment with vehicle or naloxone prior to administration of vehicle or SKF 82958 (50 μM) for 5–240 min. Immunoblotting with antibodies specific for P-GluR1 (Ser845) and actin was conducted. The ratio of P-GluR1/actin was determined for each sample and normalized to the control group ratio to yield an index of "fold induction." Data are plotted as the mean fold induction ± SEM. **$P<.01$ compared to vehicle (indicated by dashed line). ††$P<.01$ compared to veh+nal+SKF 82958, Newman-Keul's *post hoc* tests.

al, compared to cells treated only with SKF 82958 (FIG. 1). This superinduction of P-GluR1 was PKA dependent and was not due to changes in GluR1 levels, which were unaffected by chronic morphine (data not shown). The prolonged time course of GluR1 phosphorylation during morphine withdrawal suggests a concomitant increase in glutamate transmission via AMPA receptors and perhaps an increase in Ca^{2+} influx.

Since both Ca^{2+}- and cAMP-dependent kinases can phosphorylate CREB (P-CREB),[10] we examined whether glutamate-mediated CREB activation was altered in response to chronic morphine. We treated primary cultures of striatal neurons with glutamate, AMPA, or NMDA in the presence of chronic morphine and measured P-CREB. Although there was a slight trend towards enhanced P-CREB in response to glutamate and NMDA during withdrawal, the effects were not significant (FIG. 2). These data suggest that D1 receptor-mediated phosphorylation of GluR1 may be re-

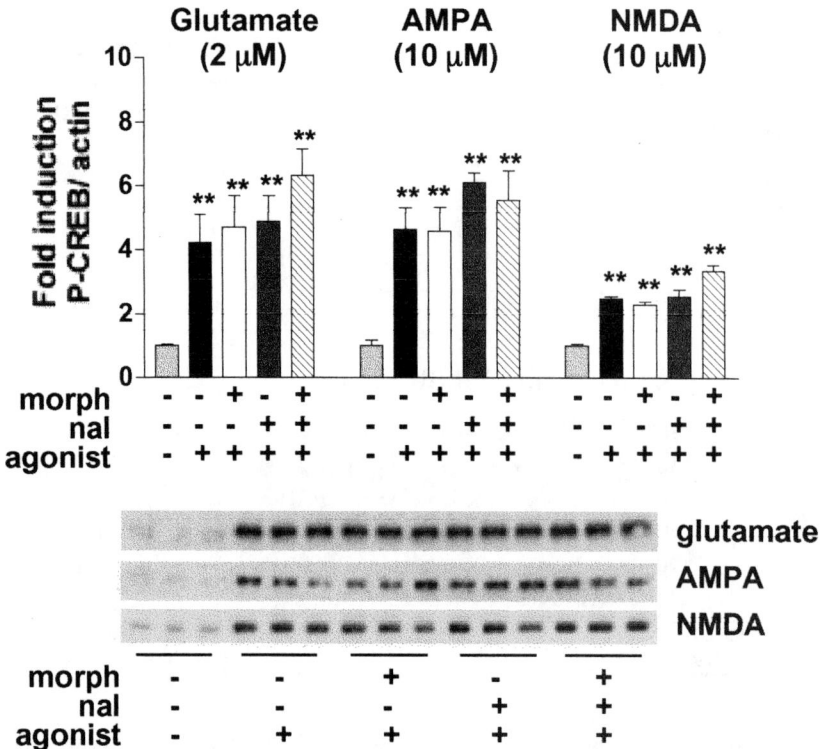

FIGURE 2. Induction of P-CREB by glutamate receptor agonists in primary striatal cultures is not affected by precipitated morphine withdrawal. Cultures were treated with vehicle or morphine (10 μM) for 6 days, followed by a 15-min treatment with vehicle or naloxone prior to administration of vehicle or glutamate (2 μM), AMPA (10 μM), or NMDA (10 μM) for 15 min. Data are plotted as the mean fold induction ± SEM. **$P<.01$ compared to vehicle, Newman-Keul's *post hoc* tests.

quired to observe a change in glutamatergic signaling during morphine withdrawal. In summary, our data suggest that glutamatergic signaling in the NAc may be potentiated during opiate withdrawal and this could contribute to aversive states.

REFERENCES

1. MATTHES, H.W. *et al.* 1996. Loss of morphine-induced analgesia, reward effect and withdrawal symptoms in mice lacking the mu-opioid-receptor gene. Nature **383:** 819–823.
2. WISE, R.A. 1989. Opiate reward: sites and substrates. Neurosci. Biobehav. Rev. **13:** 129–133.
3. KOOB, G.F., T.L. WALL & F.E. BLOOM. 1989. Nucleus accumbens as a substrate for the aversive stimulus effects of opiate withdrawal. Psychopharmacologia **98:** 530–534.
4. CARLEZON, W.A., JR. & E.J. NESTLER. 2002. Elevated levels of GluR1 in the midbrain: a trigger for sensitization to drugs of abuse? Trends Neurosci. **25:** 610–615.

5. VEKOVISCHEVA, O.Y. *et al.* 2001. Morphine-induced dependence and sensitization are altered in mice deficient in AMPA-type glutamate receptor-A subunits. J. Neurosci. **21:** 4451–4459.
6. PRICE, C.J., P. KIM & L.A. RAYMOND. 1999. D1 dopamine receptor-induced cyclic AMP-dependent protein kinase phosphorylation and potentiation of striatal glutamate receptors. J. Neurochem. **73:** 2441–2446.
7. CHAO, S.Z. *et al.* 2002. D(1) dopamine receptor stimulation increases GluR1 phosphorylation in postnatal nucleus accumbens cultures. J. Neurochem. **81:** 984–992.
8. NESTLER, E.J. & G.K. AGHAJANIAN. 1997. Molecular and cellular basis of addiction. Science **278:** 58-63.
9. CHARTOFF, E.H., M. PAPADOPOULOU, C. KONRADI & W.A. CARLEZON JR. 2003. Dopamine-dependent increases in phosphorylation of cAMP response element binding protein (CREB) during precipitated morphine withdrawal in primary cultures of rat striatal neurons. J. Neurochem. **87:** 107–118.
10. LONZE, B.E. & D.D. GINTY. 2002. Function and regulation of CREB family transcription factors in the nervous system. Neuron **35:** 605–623.

Opposite Effects of GluR1 and PKA-Resistant GluR1 Overexpression in the Ventral Tegmental Area on Cocaine Reinforcement

KWANG-HO CHOI, ZIA RAHMAN, SCOTT EDWARDS, STEPHANIE HALL, RACHAEL L. NEVE,[a] AND DAVID W. SELF

Department of Psychiatry, The University of Texas Southwestern Medical Center at Dallas, Dallas, Texas 75390, USA

[a]*Department of Psychiatry, Harvard Medical School, Belmont, Massachusetts 02178, USA*

KEYWORDS: GuR1; ventral tegmental area; cocaine; neural plasticity; AMPA receptor

Drugs of abuse may cause a maladaptive form of neural plasticity and experience-dependent behavioral changes.[1,2] Previous studies reported that repeated administration of psychostimulant cocaine either increases or has no effect on AMPA glutamate receptor subunit GluR1 levels in the ventral tegmental area (VTA) of rats.[3–5] Our studies found that chronic intravenous cocaine self-administration increases GluR1 levels in the VTA. Protein kinase A (PKA) pathways in the VTA have been implicated in behavioral sensitization to psychostimulants and synaptic plasticity.[6,7] Several lines of evidence suggest that PKA-mediated phosphorylation of GluR1 may be necessary for activity-dependent AMPA receptor delivery and long-term potentiation in synapse.[8–10] In the present study, we investigated the behavioral consequences of upregulation of GluR1 wildtype and PKA-resistant GluR1 (S845A) in the VTA by using herpes simplex viral (HSV)-mediated gene transfer in animals self-administering cocaine.

Male Sprague-Dawley rats (300–350 g) were implanted with bilateral guide cannulae in the VTA and chronic indwelling intravenous catheters. Animals were trained to lever-press for intravenous cocaine (0.5 mg/kg/injection; FR1-FR5, 4 h daily session). Following stabilization in within-session dose-response tests (0, 0.03, 0.1, 0.3, and 1 mg/kg/injection; FR5), animals received bilateral infusions of HSV vectors (LacZ, GluR1 wildtype and GluR1 S845A mutant; 2×10^4 IU/µL) in the VTA, and the effects of GluR1 and PKA-resistant GluR1 overexpression on cocaine self-administration were tested.

Address for correspondence: Kwang-Ho Choi, Department of Psychiatry, 5323 Harry Hines Blvd., University of Texas Southwestern Medical Center, Dallas, TX 75390. Voice: 214-648-1444; fax: 214-648-4182.

Kwang-ho.Choi@UTSouthwestern.edu

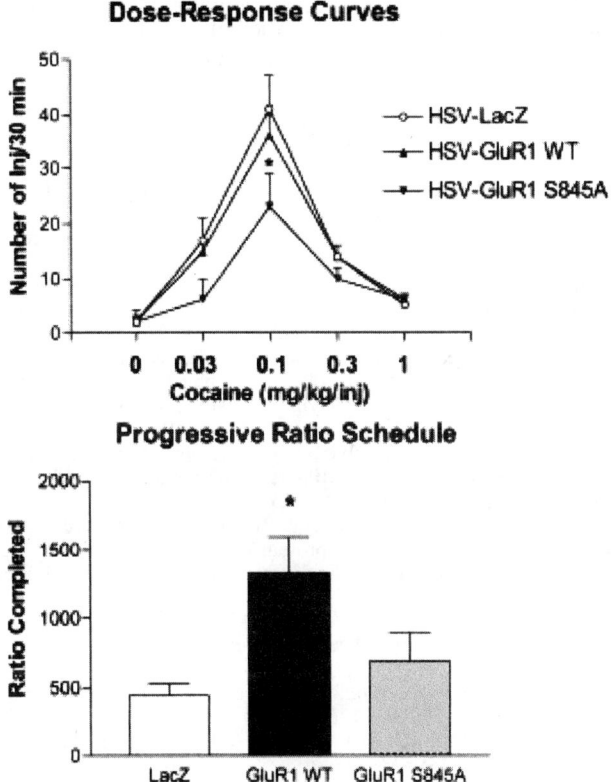

FIGURE 1. Effects of viral-mediated overexpression of GluR1 wildtype and PKA-resistant GluR1 (S845A) in the VTA on cocaine self-administration in rats. *Upper panel* shows that overexpression of PKA-resistant GluR1 in the VTA shifts dose-response curve downward, suggesting reduced reinforcing efficacy of cocaine ($N=5-7$ per group, *Significant from LacZ, $P<.05$). *Lower panel* shows that overexpression of GluR1 wildtype in the VTA increases the highest ratio of responses/injection completed (breaking point) in progressive ratio schedule ($N=5-7$ per group, *Significant from LacZ, $P<.05$).

Overexpression of PKA-resistant GluR1 in the VTA reduced reinforcing efficacy of cocaine when compared to baseline cocaine self-administration or to LacZ control group, suggesting PKA-resistant GluR1 prevented delivery of AMPA receptors to synaptic membranes (dominant negative), thereby reducing VTA excitability (FIG. 1). Following the dose-response test, animals were stabilized on a progressive ratio schedule of reinforcement at a higher cocaine dose (1 mg/kg/injection), where each successive self-injection required a greater amount of effort (lever-presses). Overexpression of GluR1 wildtype in the VTA increased the motivation for cocaine on the progressive ratio schedule, as indicated by an increase in the highest ratio of responses/injection completed. The results suggest that levels of the AMPA receptor subunit GluR1 in the VTA play a significant role in regulating the reinforcing effects of cocaine, and that PKA-mediated phosphorylation of GluR1 is essential for this regulation. Moreover, drug-induced upregulation of GluR1 levels in the VTA would contribute to increased motivation for cocaine as addiction develops.

REFERENCES

1. WOLF, M.E. 2002. Addiction and glutamate-dependent plasticity. B.H. Herman *et al.*, Eds.: 127–141. Humana Press Inc., Totowa, NJ.
2. SUTTON, M.A., E.F. SCHMIDT, K-H. CHOI, *et al.* 2003. Extinction-induced upregulation in AMPA receptors reduces cocaine-seeking behaviour. Nature **421:** 70–75.
3. FITZGERALD, L.W., J. ORTIZ, A.G. HAMEDANI & E.J. NESTLER. 1996. Regulation of glutamate receptor subunit expression by drugs of abuse and stress: Common adaptations among cross-sensitizing agents. J. Neurosci. **16:** 274–282.
4. CHURCHILL, L., C.J. SWANSON, M. URBINA & P.W. KALIVAS. 1999. Repeated cocaine alters glutamate receptor subunit levels in the nucleus accumbens and ventral tegmental area of rats that develop behavioral sensitization. J. Neurochem. **72:** 2397–2403.
5. LU, W., L.M. MONTEGGIA & M.E. WOLF. 2001. Repeated administration of amphetamine and cocaine does not alter AMPA receptor subunit expression in the rat midbrain. Neuropsychopharmacology **26:** 1–13.
6. TOLLIVER, B.K., L.B. HO, L.M. FOX & S.P. BERGER. 1999. Necessary role for ventral tegmental area adenylate cyclase and protein kinase A in induction of behavioral sensitization to intraventral tegmental area amphetamine. J. Pharmacol. Exp. Ther. **289:** 38–47.
7. GUTLERNER, J.L., E.C. PENICK, E.M. SNYDER & J.A. KAUER. 2002. Novel protein kinase A-dependent long-term depression of excitatory synapses. Neuron **36:** 921–931.
8. MALINOW, R. & R.C. MALENKA. 2002. AMPA receptor trafficking and synaptic plasticity. Ann. Rev. Neurosci. **25:** 103–126
9. SONG, I. & R.L. HUGANIR. 2002. Regulation of AMPA receptors during synaptic plasticity. Trends Neurosci. **25:** 578–588.
10. ESTEBAN, J.A., S-H. SHI, C. WILSON, *et al.* 2003. PKA phosphorylation of AMPA receptor subunits controls synaptic trafficking underlying plasticity. Nature Neurosci. **6:** 136–143.

Expression of ARHGEF11 mRNA in Schizophrenic Thalamus

GENOVEVA DAVIDKOVA, ROBERT E. McCULLUMSMITH, AND JAMES H. MEADOR-WOODRUFF

Mental Health Research Institute and Department of Psychiatry, University of Michigan, Ann Arbor, Michigan 48109, USA

KEYWORDS: Rho guanine nucleotide exchange factor 11; schizophrenia; thalamus

Human Rho guanine nucleotide exchange factor 11 (ARHGEF11) is a novel PDZ domain-containing Rho GTPase, which belongs to a family of small GTP-binding proteins that has a role in remodeling of the cytoskeleton and signal transduction processes.[1,2] The rat homologue of ARHGEF11, GTRAP48, has been shown to interact with the C terminus of the neuronal excitatory amino acid transporter EAAT4 *in vitro*, and to enhance glutamate transport *in vivo*, possibly by stabilizing and/or anchoring EAAT4 to the cell membrane, and thereby making it less likely to be internalized and subsequently degraded.[3] In view of the participation of ARHGEF11 in glutamate transport, we studied of the expression of ARHGEF11 mRNA in postmortem human thalamus from nonpsychiatric controls and schizophrenic individuals obtained from the Mount Sinai Medical Center Brain Bank.

We used a riboprobe *in situ* hybridization procedure established in our laboratory.[4] The published sequence for human ARHGEF11 (NCBI Accession NM_014784) was used to design PCR primers (located at nucleotide positions 1935-1955 for the 5′ primer and 2288-2309 for the 3′ primer) to subclone this portion of the ARHGEF11 into the Zero Blunt TOPO vector (Invitrogen). Antisense and sense ARHGEF11 riboprobes were radiolabeled with [^{35}S]UTP. Initially, these antisense and sense ARHGEF11 riboprobes were hybridized to sections from several different regions of the macaque brain (thalamus, cerebellum, prefrontal cortex, striatum) to determine the specificity of the probes. Subsequently, the expression of ARHGEF11 mRNA was evaluated in the human postmortem samples. For each individual, duplicate sections were hybridized to the antisense riboprobe and the slides were apposed to Kodak Biomax film for 1 month. In each thalamic section, six discrete nuclei were identified: anterior (A), ventral anterior (VA), dorsomedial (DM), central medial (CM), pooled ventral tier (V), and reticular (R). The levels of ARHGEF11mRNA

Address for correspondence: Genoveva Davidkova, Mental Health Research Institute and Department of Psychiatry, University of Michigan, 205 Zina Pitcher Place, Ann Arbor, MI 48109. Voice: 734-936-2061; fax: 734-647-4130.
gpuzunov@umich.edu

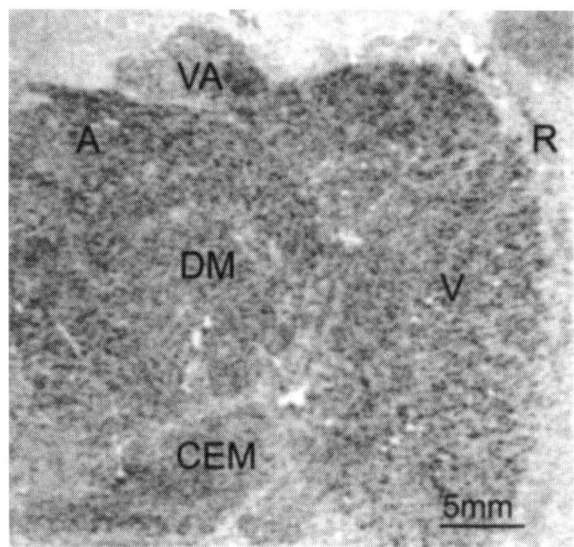

FIGURE 1. ARHGEF11 mRNA expression in human thalamus from a schizophrenic individual. Shown is a photomicrograph of a representative section with the six analyzed thalamic nuclei abbreviated as follows: A = anterior; VA = ventral anterior; V = ventral; DM = dorsomedial; CEM = central medial; R = reticular. Scale bar = 5 mm.

were measured in the individual nuclei and expressed as optical density (OD) units. Statistical analyses were performed using analyses of variance (ANOVA) with nucleus and diagnosis as independent variables and optical density as the dependent variable. For all tests $\alpha=0.05$.

ARHGEF11 mRNA was expressed in a heterogeneous pattern in the human thalamus (FIG. 1). The overall expression levels of ARHGEF11 mRNA were found to be significantly higher in the thalamic nuclei from the schizophrenic subjects as compared to the nonpsychiatric control individuals ($F_{diagnosis}=11.7$, $P=.001$)(FIG. 2). Highest levels of ARHGEF11 mRNA were found in the dorsomedial, followed by the anterior nucleus ($F_{nuclei}=4.1$, $P=.0021$). The lowest levels of ARHGEF11 mRNA were found in the reticular nucleus in both nonpsychiatric controls and schizophrenic individuals. The expression levels of ARHGEF11 varied to a greater extent among the six examined nuclei in schizophrenic subjects, whereas in the nuclei of the control subjects the expression levels were more homogeneous.

To our knowledge, this is the first *in situ* hybridization study of the expression of ARHGEF11 in postmortem human thalamus. Since the highest levels of ARHGEF11 mRNA were found in the dorsomedial and in the anterior thalamic nuclei of schizophrenic individuals, and these nuclei have connections to the limbic system, it is reasonable to speculate that ARHGEF11 is involved in the pathologic changes and associated with decreased glutamatergic neurotransmission in schizophrenia. These same subjects were previously found in our laboratory to express altered levels of certain NMDA receptor subunit and postsynaptic density (PSD) protein mRNAs.[5] To

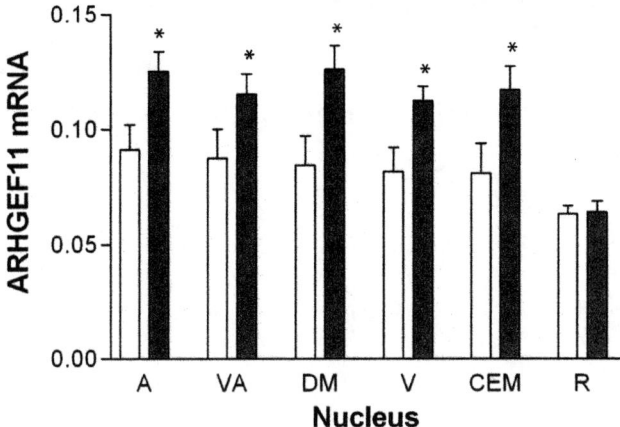

FIGURE 2. ARHGEF11 levels in thalamic nuclei from nonpsychiatric control and schizophrenic individuals. Each bar represents optical density (OD) values expressed as mean ± SEM from six control and six schizophrenic individuals. The abbreviations of the nuclei have been indicated in the legend to FIGURE 1. Statistical analysis showed that ARHGEF11 levels are significantly higher in the schizophrenic individuals as compared to the nonpsychiatric controls ($F_{diagnosis} = 11.7$, $P = .001$). *Open bars*, control; *filled bars*, schizophrenic. $P = .05$.

evaluate further the significance of ARHGEF11 for the complex molecular glutamatergic changes in schizophrenia, we plan to compare the cellular localization of ARHGEF11 mRNA with the localization of EAAT4 mRNA, as well as examine the expression of similar molecules in these subjects.

REFERENCES

1. FUKUHARA, S. *et al.* 1999. A novel PDZ domain containing guanine nucleotide exchange factor links heterotrimeric G proteins to Rho. J. Biol. Chem. **274:** 5868–5879.
2. RUMENAPP, U. *et al.* 1999. Rho-specific binding and guanine nucleotide exchange catalysis by KIAA0380, a Dbl family member. FEBS Lett. **459:** 313–318.
3. JACKSON, M. *et al.* 2001. Modulation of the neuronal glutamate transporter EAAT4 by two interacting proteins. Nature **410:** 89–93.
4. MEADOR-WOODRUFF, J.H *et al.* 1996. Differential regulation of hippocampal AMPA and kainite receptor subunit expression by haloperidol and clozapine. Mol. Psychiatry **1:** 41–53.
5. MEADOR-WOODRUFF, J.H. & J.E. KLEINMAN. 2002. Neurochemistry of schizophrenia: glutamatergic abnormalities. In Neuropsychopharmacology: The Fifth Generation of Progress. K.L. Davis, *et al.*, Eds.: 717–728. American College of Neuropsychopharmacology.

Structurally Dissimilar Antimanic Agents Modulate Synaptic Plasticity by Regulating AMPA Glutamate Receptor Subunit GluR1 Synaptic Expression

JING DU, NEIL A. GRAY, CYNTHIA FALKE, PEIXIONG YUAN, STEVEN SZABO, AND HUSSEINI K. MANJI

Laboratory of Molecular Pathophysiology, National Institutes of Mental Health, National Institutes of Health, Bethesda, Maryland 20892-4405, USA

ABSTRACT: A growing body of data from clinical and preclinical studies suggests that the glutamatergic system may represent a novel therapeutic target for severe recurrent mood disorders. Since synapse-specific glutamate receptor expression/localization is known to play critical roles in synaptic plasticity, we investigated the effects of mood stabilizers on AMPA receptor expression. Rats were treated chronically with lithium or valproate, hippocampal synaptosomes were isolated, and GluR1 levels were determined. Additionally, hippocampal neurons were prepared from E18 rat embryos and treated with lithium or valproate. Surface expression of GluR1 was determined using a biotinylation assay, and double-immunostaining with anti-GluR1 and anti-synaptotagmin antibodies was used to determine synaptic GluR1 levels. The AMPA receptor subunit GluR1 expression in hippocampal synaptosomes was significantly reduced by both chronic lithium and valproate. Overall, these studies show that AMPA receptor subunit GluR1 is a common target for two structurally highly dissimilar, but highly efficacious, mood stabilizers, lithium and valproate. These studies suggest that regulation of glutamatergically mediated synaptic plasticity may play a role in the treatment of mood disorders, and raise the possibility that agents more directly affecting synaptic GluR1 may represent novel therapies for this devastating illness.

KEYWORDS: lithium; valproate; GluR1; mood disorders

INTRODUCTION

Bipolar disorder (manic-depressive illness) is a severe, chronic, and often life-threatening illness. The disease has a lifetime incidence of about 1% and is characterized by episodes of mania and depression.[1] The most prescribed medications used to treat bipolar disorder are lithium (a monovalent cation) and valproic acid (VPA, an eight-carbon branched fatty acid).[1] Preliminary data suggest that modulation of

Address for correspondence: Husseini K. Manji, M.D., Laboratory of Molecular Pathophysiology, National Institutes of Mental Health, National Institutes of Health, Bethesda, MD 20892-4405. Voice: 301-451-8441; fax: 301-480-0123.
Manjih@intra.nimh.nih.gov

the glutamatergic system by noncompetitive NMDA receptor antagonist ketamine or lamotrigine may have efficacy in the treatment of mood disorders.[1] Furthermore, emerging data suggest that AMPA receptor subunit trafficking plays a critical role in regulating various forms of plasticity. Furthermore, AMPA receptor trafficking is regulated by the very same signaling pathways that lithium and VPA affect, including PKC, PKA, and MAPK.[1,2] We have therefore sought to determine if AMPA receptor trafficking may play a role in the therapeutic effects of mood stabilizers.

MATERIALS AND METHODS

Synaptosomal Preparation and Western Blot Analysis

Male Wistar rats were treated with chow containing lithium (2.4 g/kg) or valproate (20 g/kg) (Agway Lab Chow 3000) for 4 weeks. Synaptosomal samples were prepared from hippocampal tissue via differential and discontinuous Ficoll gradient centrifugation method. Equal amounts of proteins (2 µg) were analyzed by Western blot analysis with anti-GluR1 and anti-synaptophysin antibodies.

Culture Preparations

Hippocampal cultures was prepared from embryonic day 18 (E18) rats, cultured first in 10% fetal bovine serum/DMEM, and one day later switched to serum-free medium neurobasal plus B27 (Life Technologies). Cultures were grown in serum-free medium for 8–10 days before used for experiments.

Surface Biotinylation and Western Blot Analysis of GluR1

Surface GluR1 receptors were detected by biotinylation assay, followed by Western blot analysis. At the end of VPA or lithium treatment, cell surface protein were biotinylated, precipitated with 100 µL of ImmunoPure Immobilized Streptavidin and analyzed by Western blot analysis with a polyclonal anti-GluR1 antibody.

RESULTS

Structurally Dissimilar Antimanic Agents Lithium and Valproate Downregulate Synaptic GluR1 Levels in Vivo

We investigated whether two structurally dissimilar antimanic agents bring about changes in glutamate receptor expression at neuronal synapses. Chronic lithium and valproate treatment decreases protein levels of the AMPA-glutamate receptor subunits GluR1 in rat hippocampal synaptosomes by 35 and 25%, respectively (Jing Du et al.[3]).

Surface GluR1 Are Attenuated after Lithium and Valproate Treatment in Cultured Hippocampal Neurons

We found that surface localization of GluR1 subunits is decreased in cultured hippocampal neurons following long-term but not acute lithium (1.0 mM) or valproate (1.0 mM) treatment both by about 40% (Jing Du et al.[3]). These results are consistent with a decreased immunostaining of GluR1 on the neuronal surface after lithium (1.0 mM) treatment. Furthermore, long-term (4 days), but not short term (1 day, data not shown) lithium or VPA attenuates GluR1 expression at synaptotagmin-positive synapses (data not shown).

Total GluR1 Protein Levels Do not Change Significantly after Lithium and Valproate Treatments in Vitro and in Vivo

After chronic treatment with lithium and valproate, GluR1 levels in hippocampal tissue remain unchanged *in vivo*. Furthermore, in cultured hippocampal neurons, GluR1 level do not change significantly after lithium (1.0 mM) and valproate (1.0 mM) treatment for 4 days (data not shown).

CONCLUSIONS

Two clinically effective, but structurally highly dissimilar antimanic agents, lithium and VPA, reduce the hippocampal synaptosomal levels of GluR1 after chronic administration. Lithium and VPA reduce surface levels of GluR1 in cultured hippocampal neurons upon chronic administration. The effects do not reverse immediately after drug discontinuation and are not due to the change in total GluR1 protein level. Taken together, these results suggest that mood stabilizers attenuate AMPA receptor-mediated synaptic strength. This regulation of synaptic strength may contribute to the communication of critical circuits involved in affective functioning and buffering and therefore, affect mood-associated behavior.

REFERENCES

1. MANJI, H.K., W.C. DREVETS, *et al.* 2001. The cellular neurobiology of depression. Nat. Med. **7:** 541.
2. SHENG, M. & M.J. KIM. 2001. Postsynaptic signaling and plasticity mechanisms. Science **298:** 776.
3. DU, J., N.A. GRAY, C. FALKE, *et al.* 2003. Common targeting of mood stabilizers on AMPA glutamate receptor subunit GluR1 expression at synapses. Submitted.

Cocaine-Induced Expression Differences in Glutamate Receptor Subunits and Transporters in Amygdalae of Taste Aversion–Prone and Taste Aversion–Resistant Rats

R. L. ELKINS,[a,d] T. E. ORR,[a,d] J. L. RAUSCH,[a,d] Y. J. FEI,[b] G. F. CARL,[d] S. H. HOBBS,[f] J. J. BUCCAFUSCO,[a,c,d] AND G. L. EDWARDS[e]

[a]*Department Psychiatry and Health Behavior,* [b]*Department of Biochemistry and Molecular Biology,* [c]*Department of Pharmacology and Toxicology, Medical College of Georgia, Augusta, Georgia 30912, USA*

[d]*VA Medical Center, Augusta, Georgia 30904, USA*

[e]*Department of Physiology and Pharmacology, University of Georgia, Athens, Georgia 30602, USA*

[f]*Department of Psychology, Augusta State University, Augusta, Georgia 30904, USA*

KEYWORDS: cocaine; taste aversion; gene chip; gene expression: glutamate; amygdalae

INTRODUCTION

Taste aversion (TA)–prone (TAP) and TA-resistant (TAR) rats were selectively bred from a parental stock of Sprague Dawley–derived rats. During TA conditioning, ingestion of a novel 0.1% saccharin solution–conditioned stimulus was paired with an injection of the emetic class agent, cyclophosphamide. TA conditioning was determined by subsequent saccharin preference scores using a two-bottle choice paradigm (saccharin solution vs. plain water). Breeding pairs for TAP and TAR rats were chosen from rats exhibiting high and low conditioned aversions, respectively. The selective breeding has resulted in TAP and TAR lines that differ markedly TA conditioning (see FIG. 1).[1]

The selectively bred line differences in TA conditionability were maintained when drugs of abuse, including cocaine, were substituted for the cyclophosphamide.[1,2] Drugs of abuse have been shown to have both rewarding and aversive properties, and the aversive effects are most commonly demonstrated using a TA paradigm.[3] The lines do not differ with respect to food-reinforced learning, despite their marked difference in TA conditionability.[4] Therefore, these lines offer a suitable model for studies of the aversive properties of drugs of abuse.

Address for correspondence: Ralph L. Elkins, Ph.D., Mental Health and Behavioral Science 26, VA Medical Center, 1 Freedom Way, Augusta, GA 30904-6385. Voice: 706-733-0188 ext. 6255; fax: 706-731-7190.

ralph_elkins@msn.com

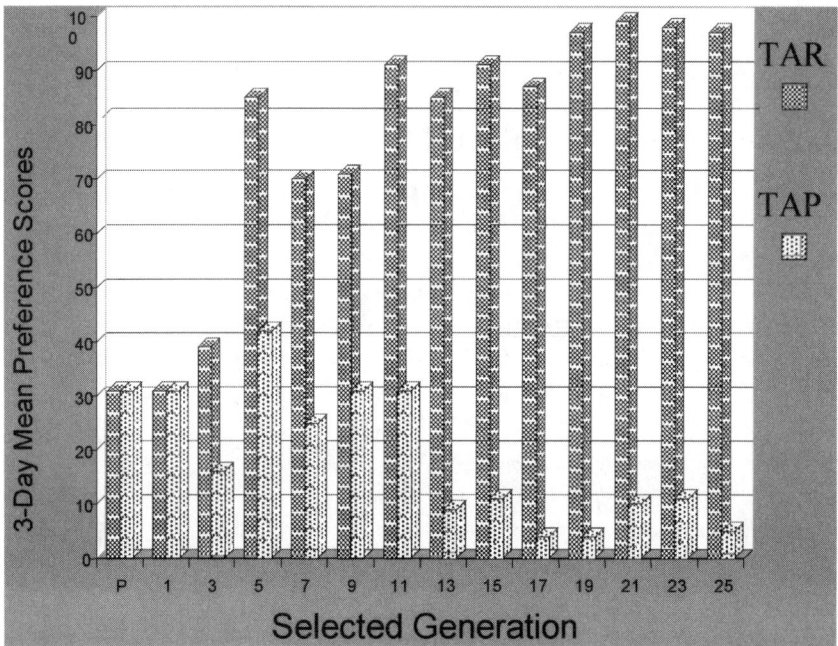

FIGURE 1. CTA line separation.

METHODOLOGY AND GENE EXPRESSION MEASUREMENTS

Amygdalae tissues from TAP and TAR rats brains were harvested 6 hours after a subcutaneous cocaine injection (25 mg/kg) or a saline control injection. Tissue from two male rat brains was pooled as one sample for each gene chip analysis. The analysis used an Affymetrix Corporation Commercial RN-U34 oligonucleotide probe array to quantify gene expression measurements, as previously described.[5] The Affymetrix software categorically designates fold changes in gene expression as Increase, Marginal Increase, Decrease, Marginal Decrease, or No Call. For present purposes, only fold changes that (1) were equal to or greater than 2.0 and (2) generated an Affymetrix designation other than No Call are reported.

RESULTS AND DISCUSSION

The findings include cocaine–induced diminished gene expressions in TAR rats as compared to TAP rats of a glutamate/aspartate transporter (GLAST) (−50.6 fold), a glutamate excitatory amino acid carrier 1 transporter (EAAC1) (−10.0 fold), an alternatively spliced NMDA receptor (−22.8 fold), and an AMPA receptor binding protein identified as the glutamate receptor interacting protein2 (GRIP2) (−22.1 fold). GLAST is distributed throughout the cerebrum and is reported to be specific for L-glutamate and L-aspartate.[6] Proposed EAAC1 functions include the sequesteration of glutamate at glutamatergic synapses and the protection of neurons from glutamate excitotoxicity.[7] The alternatively spliced NMDA receptor is one of a family of such previously described receptors.[8] GRIP2 is widely expressed in brain and is enriched in synaptic plasma membrane and postsynaptic density fractions. GRIP2 is expressed relatively late in embryonic development. It parallels the expression of AMPA receptors and is potentially involved in the targeting of AMPA receptors to synapses.[9]

TABLE 1. Cocaine-induced mRNA gene expression fold change differences between TAR and TAP rats

Gene	Description	TAR saline vs. TAP saline	TAR cocaine vs. TAP cocaine	TAR cocaine vs. TAR saline	TAP cocaine vs. TAP saline
S75687	Glutamate/aspartate transporter (GLAST)	No difference	−50.6 Decrease	−36.3 Decrease	No difference
X96790	Metabotropic glutamate receptor subunit 7b	No difference	−24.1 Decrease	−21.7 Decrease	3.6 Increase
S39221	NMDA receptor (alternatively spliced)	No difference	−22.8 Decrease	−25.3 Decrease	−2.1 Decrease
AF090113	Glutamate receptor interacting protein 2 (GRIP2) AMPA receptor binding protein	No difference	−22.1 Decrease	No Call	No Call
M92075	Metabotropic glutamate receptor 2	−2.8 Decrease	−18.3 Decrease	No Call	No difference
D16817	Metabotropic glutamate receptor mGluR-7	No difference	−17.6 Decrease	No Call	No Call
AF038571	Glutamate excitaory amino acid carrier 1 (EAAC1) transporter	No difference	−10.0 Decrease	−9.2 Decrease	No difference
M36420	Glutamate receptor GluR-C	No difference	−9.7 Decrease	No Call	3.7 Increase
M36421	Glutamate receptor GluR-D	No difference	−7.1 Decrease	No Call	No Call
M36418	Glutamate receptor GluR-A	No difference	−4.8 Decrease	−2.1 Decrease	No difference
U08256	Glutamate receptor delta-2 subunit	No difference	−4.5 Decrease	−4.0 Decrease	No difference

Seven other genes for glutamate receptors or receptor subunits also displayed cocaine-induced expression decreases in TAR rats as compared to TAP rats. The fold decreases ranged from –24.1 to –4.5 (see TABLE 1). The present findings identify glutamate-related candidate genes involving both transporters and receptors that may be found to play causal roles in individual differences in TA conditionability and in responsiveness to the aversive attributes of cocaine. This prospect is consistent with numerous recent reports linking glutamate transmission with experience-dependent synaptic plasticity.[10]

The gradual bidirectional response to the breeding of the TAP and TAR lines suggests polygenetic control of TA conditionability. In accord with this conclusion, the presently reported glutamate results were coincident with cocaine-induced TAR versus TAP gene expression differences in the GABA, acetylcholine, and dopamine neurotransmitter systems (unpublished data). Moreover, TAR rats in comparison to TAP rats have decreased gene expression of the 5-HT_3 receptor in amygdala,[5] lower level of whole brain 5-HT,[11] and less efficient 5-HT transporter function.[12]

The present gene expression findings are based on the use of a single cocaine dose and a single time interval. The results need to be replicated or confirmed by other methods. In particular, the use of a range of cocaine doses in combination with different time delays between cocaine injections and tissue harvesting should produce useful information concerning possible early gene expression changes that may be driving subsequent expression differences. The research also should be extended to other brain regions that have been implicated in TA conditioning and cocaine responsivity.

ACKNOWLEDGMENTS

This work was supported by the Department of Veterans Affairs Medical Research Service, and by the Department of Psychiatry and Health Behavior and the School of Medicine of the Medical College of Georgia.

REFERENCES

1. ELKINS, R.L., P.A. WALTERS & T.E. ORR. 1992. Continued development and unconditioned stimulus characterization of selectively-bred lines of taste aversion prone and resistant rats. Alcohol. Clin. Exp. Res. **16:** 928–934.
2. ELKINS, R.L., P.A. WALTERS, T.E. ORR, *et al.* 1991. Taste aversion inducing effects of cocaine in selectively-bred taste aversion prone and resistant rats. Neurosci. Abst. **17:** 662.
3. ORR, T.E., P.A. WALTERS & R.L. ELKINS. 2001. Fundamentals, methodologies, and uses of taste aversion learning. *In* Behavioral Methods in Neurological Research. J. Buccafusco, Ed.: 51–69. CRC Press. Boca Raton.
4. HOBBS, S.H., P.A. WALTERS, E.F. SHEALY & R.L. ELKINS. 1993. Radial maze learning by strains of taste aversion prone and resistant rats. Psychonomic Sci. **31:** 171-174.
5. ELKINS, R.L., T.E. ORR, G.L. EDWARDS, *et al.* 2003. Cocaine-induced expression differences of 5-HT_3 receptors and Na^+/K^+-APTase pump subunits in amygdalae of taste-aversion-prone and taste-aversion-resistant rats. Ann. N.Y Acad. Sci. **985:** 519–521.
6. STORCK, T., S. SCHULTE, K. HOFMANN & W. STOFFEL. 1992. Structure, expression, and functional analysis of a Na^+-dependent glutamate/aspartate transporter from rat brain. Proc. Natl. Acad. Sci. USA **89:** 10955–10959.

7. KANAI, Y., P/G. BHIDE, M. DIFIGLIA & M.A. HEDIGER. 1995. Neuronal high-affinity glutamate transport in the rat central nervous system. Neuroreport **6:** 2357–2362.
8. SUGIHARA, H., K. MORIYOSHI, T. ISHII, *et al.* 1992. Structures and properties of seven isoforms of the NMDA receptor generated by alternative splicing. Biochem. Biophys. Res. Commun. **185:** 826–832.
9. DONG, H., P. ZHANG, I. SONG, *et al.* 1999. Characterization of the glutamate receptor-interacting proteins GRIP1 and GRIP2. J. Neurosci. **15:** 6930–6941.
10. CARROLL, R.C. & R.S. ZUKIN. 2002. NMDA-receptor trafficking and targeting: implications for synaptic transmission and plasticity. Trends Neurosci. **25:** 571–577.
11. ORR, T.E., P.A. WALTERS, G.F. CARL & R.L. ELKINS. 1993. Brain levels of amines and amino acids in taste aversion-prone and -resistant rats. Physiol. & Behav. **53:** 495–500.
12. ELKINS, R.L., T.E. ORR, J.Q. LI, *et al.* 2000. Serotonin reuptake is less efficient in taste aversion-resistant than in taste aversion-prone rats. Pharmacol. Biochem. Behav. **66:** 609–614.

Cocaine-Induced Expression Differences in PSD-95/SAP-90–Associated Protein 4 and in Ca^{2+}/Calmodulin-Dependent Protein Kinase Subunits in Amygdalae of Taste Aversion–Prone and Taste Aversion–Resistant Rats

R. L. ELKINS,[a,d] T. E. ORR,[a,d] J. L. RAUSCH,[a,d] Y. J. FEI,[b] G. F. CARL,[d] S. H. HOBBS,[f] J. J. BUCCAFUSCO,[a,c,d] AND G. L. EDWARDS[e]

[a]*Department Psychiatry and Health Behavior,* [b]*Department of Biochemistry and Molecular Biology,* [c]*Department of Pharmacology and Toxicology, Medical College of Georgia, Augusta, Georgia 30912, USA*

[d]*VA Medical Center, Augusta, Georgia 30904, USA*

[e]*Department of Physiology and Pharmacology, University of Georgia, Athens, Georgia 30602, USA*

[f]*Department of Psychology, Augusta State University, Augusta, Georgia 30904, USA*

KEYWORDS: cocaine; taste aversion; gene chip; gene expression; glutamate; amygdalae

INTRODUCTION

Taste aversion (TA)–prone (TAP) and TA-resistant (TAR) rats were selectively bred from a parental stock of Sprague Dawley–derived rats.[1] During the standard TA conditioning, ingestion of a novel 0.1% saccharin solution–conditioned stimulus (CS) was paired with an injection of cyclophosphamide, an emetic class agent–unconditioned stimulus (US). After a three-day recovery period, conditioned aversions were assessed by providing simultaneous ad lib home cage access to separate bottles containing the CS saccharin solution and plain water. Breeders for subsequent generations of TAP and TAR rats were selected from those rats demonstrating, respectively, low and high postconditioning preferences for the saccharin solution. Sibling matings were prohibited to minimize inbreeding. The breeding program has produced TAP and TAR lines having markedly different TA learning propensities as depicted graphically elsewhere in this volume.[2] However, the learning capabilities of the TAP and TAR lines have diverged only with respect to TA conditionability, and the lines remain equivalent learn-

Address for correspondence: Ralph L. Elkins, Ph.D., Mental Health and Behavioral Science 26, VA Medical Center, 1 Freedom Way, Augusta, GA 30904-6385. Voice: 706-733-0188 ext. 6255; fax: 706-731-7190.
ralph_elkins@msn.com

Ann. N.Y. Acad. Sci. 1003: 386–390 (2003). © 2003 New York Academy of Sciences.
doi: 10.1196/annals.1300.068

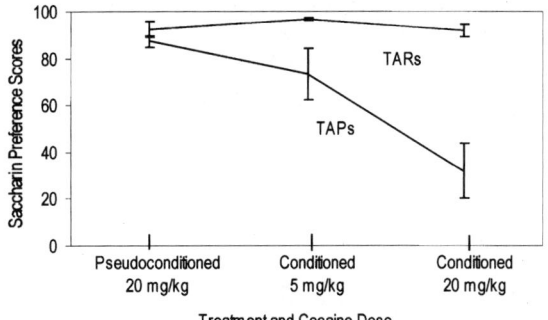

FIGURE 1. Taste-aversion conditioning to saccharin using subcutaneously injected cocaine. Pseudoconditioned refers to a control procedure that is needed to demonstrate between-line differences in TA conditioning. The pseudoconditioned control rats were injected with 20 mg/kg of cocaine following access to only plain water instead of the saccharin CS solution consumed by conditioned subjects. The pseudoconditioned rats first encountered the saccharin solution as a novel flavor during the subsequent two-bottle preference testing of all subjects. One day of 24-h preference testing was conducted 10 days after cocaine administration.

ers within other conventional paradigms.[3] These findings are in accord with the major objective of the selective breeding that was to produce TAP and TAR rat lines that were outbred to the maximum degree possible, except those genes that subserve individual differences in TA conditionability.

Drugs of abuse have both rewarding and aversive properties, and the aversive effects are most commonly demonstrated using a TA paradigm.[3] The TAP vs. TAR differences in TA conditionability were maintained when drugs of abuse, including cocaine, were substituted for the cyclophosphamide US.[1,4] Intraperitoneal cocaine injections produced dose-dependent TAs in TAP but not TAR rats of the 22nd selected generations.[4] Subcutaneous US cocaine injections produced comparable findings within the 29th generations of TAP and TAR subjects, as summarized graphically in FIGURE 1.

The dose-dependent differences in TA conditioning among TAP and TAR conditioned subjects in the absence of a between-lines difference among pseudoconditioned control subjects confirms that the cocaine-induced TA conditioning differences result from an associative learning process as opposed to some sensitization to the effects of cocaine. The TAP and TAR lines are, therefore, a viable model of studies of the genetic underpinnings of individual differences in reactivity to the aversive attributes of cocaine.

METHODOLOGY AND GENE EXPRESSION MEASUREMENTS

TAP and TAR rats brains were harvested 6 hours following a subcutaneous cocaine injection (25 mg/kg) or a saline control injection. Amygdalae tissues from the brains of two male rats were pooled as one sample for each gene chip analysis. The analysis used an Affymetrix Corporation Commercial RN-U34 oligonucleotide

TABLE 1. Cocaine-induced mRNA gene expression differences between TAP and TAR rats in a PSD-95/SAP-90–associated protein-4 and in Ca^{2+}/calmodulin-dependent protein kinase subunits. Number entries report fold change differences in gene expression

Gene	Description	TAR saline vs. TAP saline	TAR cocaine vs. TAP cocaine	TAR cocaine vs. TAR saline	TAP cocaine vs. TAP saline
U67140	PSD-95/SAP-90–associated protein-4	No difference	−15.2 Decrease	−13.3 Decrease	No call
M74488	Ca^{2+}/calmodulin-dependent protein kinase	No difference	−12.7 Decrease	−11.2 Decrease	No measurable difference
M16112	Ca^{2+}/calmodulin-dependent protein kinase II beta subunit	No difference	−6.8 Decrease	−5.2 Decrease	−1.8 Decrease
S71570	Ca^{2+}/calmodulin-dependent protein kinase II isoform gamma-b	No difference	−5.7 Decrease	−5.3 Decrease	No measurable difference
AB004267	Ca^{2+}/calmodulin-dependent protein kinase I beta 2	No difference	−4.6 Decrease	−4.8 Decrease	−2.9 Decrease
L24907	Ca^{2+}/calmodulin-dependent protein kinase I	No difference	−2.7 Decrease	−1.7 Decrease	1.9 Increase

probe array to quantify gene expression measurements as previously described.[5] The RN-U34 array represents more that 1200 rat genes relevant to neuroscience research. The Affymetrix software yields a fold change as reflective of 1 of 5 categorical designations, which include: Increase, Marginal Increase, Decrease, Marginal Decrease, and No Call. For present purposes we report only fold changes in gene expression that (1) were equal to or greater than 2.0 following a comparison of TAR cocaine vs. TAP cocaine injections and (2) generated an Affymetrix designation other than No Call.

RESULTS AND DISCUSSION

Cocaine injections produced decreased expressions of the PSD-95/SAP-90–associated protein-4 mRNA and several Ca^{2+}/calmodulin-dependent related genes in TAR vs. TAP rats. Saline-injected control animals did not display any between-lines differences in the presented genes (TABLE 1).

PSD-95 appears to be important in coupling the NMDA receptor to pathways that control bidirectional synaptic plasticity and learning.[6] Ca^{2+} influx through NMDA receptor channels has been functionally coupled to neurotoxic signaling pathways and, specifically, to nitric oxide neurotoxicity. Moreover, the functioning of PSD-95 as a scaffolding protein downstream to the NMDA receptor appears to be essential to the NMDA-mediated toxic effects of nitric oxide.[7] Nitric oxide synthesis is stimulated by Ca^{2+}/calmodulin. Furthermore, the 15 mg/kg cocaine injection also produced marked TAR vs. TAP expression decreases in NMDA transporters, in NMDA receptors, and in NMDA receptor subunits, as reported elsewhere in this volume.[2] The likelihood of toxic nitric oxide effects in TAR vs. TAP rats may be lessened by decreased gene expressions of cocaine-responsive NMDA species, of the PSD-95/SAP-90 Associated Protein-4, and of multiple species of Ca^{2+}/calmodulin-dependent protein kinases.

Although causal linkages remain to be demonstrated, the pattern of present findings is consistent with the following hypothesis: Cocaine-induced neurotoxic effects are involved in differences in reactivity to the aversive attributes of cocaine and in differences in TA conditioning of TAR vs. TAP rats. It is further hypothesized that, following cocaine injection, TAR but not TAP rats are protected from NMDA receptor-mediated neurotoxicity via the protein expression consequences of decreased expression (1) of cocaine-responsive NMDA species, (2) of the PSD-95/SAP-90–associated protein-4, and (3) of Ca^{2+}/calmodulin species.

Present findings are based on the use of a single cocaine dose and a single time interval. The findings therefore are in need of replication or confirmation by other methods. Studies of multiple cocaine doses and time intervals as well as of other brain structures that may be involved in cocaine responsivity and TA conditioning also are indicated.

ACKNOWLEDGMENTS

This work was supported by the Department of Veterans Affairs Medical Research Service, and by the Department of Psychiatry and Health Behavior and the School of Medicine of the Medical College of Georgia.

REFERENCES

1. ELKINS, R.L., P.A. WALTERS & T.E. ORR. 1992. Continued development and unconditioned stimulus characterization of selectively-bred lines of taste aversion prone and resistant rats. Alcohol. Clin. Exp. Res. **16:** 928–934.
2. ELKINS, R.L., T.E. ORR, J.L. RAUSCH, et al. Cocaine-induced expression differences in glutamate receptor subunits and transporters in amygdalae of taste aversion–prone and taste aversion–resistant rats. Ann. N.Y. Acad. Sci. This volume.
3. ORR, T.E., P.A. WALTERS & R.L. ELKINS. 2001. Fundamentals, methodologies, and uses of taste aversion learning. In Behavioral Methods in Neurological Research. J. Buccafusco, Ed.: 51–69. CRC Press. Boca Raton.
4. ELKINS, R.L., P.A. WALTERS, T.E. ORR, et al. 1991. Taste aversion inducing effects of cocaine in selectively-bred taste aversion prone and resistant rats. Neurosci. Abst. **17:** 662.
5. ELKINS, R.L., T.E. ORR, G.L. EDWARDS, et al. 2003. Cocaine-induced expression differences of 5-HT$_3$ receptors and Na$^+$/K$^+$-APTase pump subunits in amygdalae of taste-aversion-prone and taste-aversion-resistant rats. Ann. N.Y. Acad. Sci. **985:** 519–521.

6. MIGAUD, M., P. CHARLESWORTH, M. DEMPSTER, *et al.* 1998. Enhanced long-term potentiation and impaired learning in mice with mutant postsynaptic density-95 protein. Nature **369:** 433–439.
7. SATTLER, R., Z. XIONG, W.-Y. LU, *et al.* 1999. Specific coupling of NMDA receptor activation to nitric oxide neurotoxicity by PSD-95 protein. Science **284:** 1845–1848.

Rapid AMPAR/NMDAR Response to Amphetamine

A Detectable Increase in AMPAR/NMDAR Ratios in the Ventral Tegmental Area Is Detectable after Amphetamine Injection

L. J. FALEIRO,[a] S. JONES,[b] AND J. A. KAUER[a]

[a]*Department of Molecular Physiology, Pharmacology and Biotechnology, Brown University, Providence, Rhode Island 02912, USA*

[b]*Anatomy Department, University of Cambridge, Cambridge, United Kingdom*

KEYWORDS: ventral tegmental area; VTA; plasticity; AMPAR/NMDAR ratio; amphetamine; addiction

Repeated exposure to drugs of abuse produces progressive hyperlocomotion or behavioral sensitization thought to parallel changes in the neural circuitry that underlie the intensification of drug craving.[1,2] Drugs of abuse activate the mesocorticolimbic system, a circuit also activated by naturally rewarding stimuli. This circuit includes: (1) glutamatergic projections from the prefrontal cortex (PFC) that project to the ventral tegmental area (VTA) and (2) dopaminergic projections from the VTA that project to the PFC and the nucleus accumbens (NAcc). Repeated local amphetamine micro-injections into the VTA of rats result in behavioral sensitization to a systemic amphetamine challenge.[3–6] This finding suggests that the VTA is the locus of sensitization.

Amphetamine activates the reward circuit by increasing extracellular levels of dopamine and glutamate in the VTA,[7,8] which activate their cognate receptors. One mechanistic model of behavioral sensitization is that molecular and cellular sequelae of dopamine and glutamate receptor activation contribute to the altered behavior of glutamate synapses on dopamine (DA) neurons in sensitized animals.[9] Such molecular and cellular changes are hypothesized to co-opt the basal reward circuitry and contribute to addiction by promoting addictive behaviors and actions.[10,11]

We have focused our attention on glutamatergic synapses on VTA dopamine neurons and asked whether exposure to amphetamine produces plasticity at these synapses. We found previously that amphetamine blocks long-term depression (LTD) at these synapses in acute slices prepared from young rats on postnatal days (P) 16–23.[12] In addition, plasticity at glutamatergic synapses on VTA dopamine neurons as

Address for correspondence: J.A. Kauer, Department of Molecular Physiology, Pharmacology and Biotechnology, Brown University, Providence, RI 02912. Voice: 401-863-9803; fax: 401-863-1595.

Julie_Kauer@Brown.edu

Ann. N.Y. Acad. Sci. 1003: 391–394 (2003). © 2003 New York Academy of Sciences.
doi: 10.1196/annals.1300.032

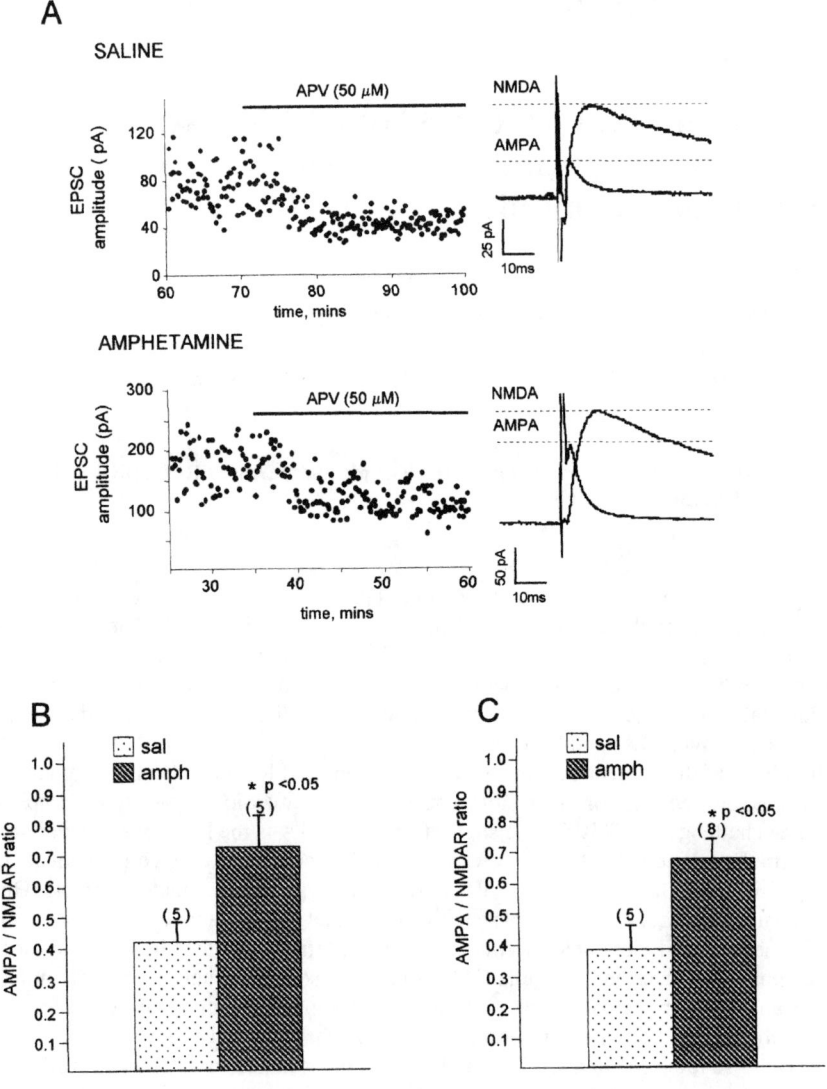

FIGURE 1. AMPAR/NMDAR ratios are augmented at glutamatergic synapses on dopamine neurons in the VTA following exposure to amphetamine. (**A**) Examples of EPSCs from dopamine neurons in the VTA of saline- or amphetamine-treated animals 24 h following injections. EPSCs were evoked with bipolar stimulating electrodes from neurons voltage-clamped at +40 mV. D-APV (50 μM) was added to the bath to isolate the NMDAR component (**A**) from the AMPAR component. AMPAR/NMDAR ratios were calculated by computing average EPSCs at +40 mV (30 EPSCs, 5 min) before and after application of APV and then subtracting the AMPAR response (+APV) from the total response (–APV) to get the NMDAR-only response. AMPAR/NMDAR ratios are significantly higher 24 h following amphetamine exposure (**B**). This change occurs as soon as 2 h following amphetamine exposure (**C**).

measured by changes in AMPAR/NMDAR ratios has been reported 24 h after a single dose of cocaine, which is consistent with long-term potentiation (LTP) induction at these synapses.[13] More recently, similar responses were observed with other drugs of abuse and in response to stressors,[14] demonstrating that drugs of abuse alter the behavior of glutamatergic synapses on VTA dopamine neurons.

Here, we examined behavioral sensitization to amphetamine in young rats (P15–P19). We quantified the AMPAR/NMDAR (A/N) ratio in sensitized versus control animals to determine whether this physiological parameter can be correlated with the behavioral response. Once we established a correlation, we determined how rapidly following amphetamine exposure a change in plasticity could be detected at the glutamatergic synapse.

Young rats (P16–P18) were administered either amphetamine (2.5 mg/kg) or saline on two consecutive days. Horizontal ambulation was measured for 40–60 min after each injection. Consistent with a sensitized response, ambulation was greater after the second amphetamine injection (1,093 ± 199 counts, $N = 7$) compared to the saline/amphetamine control (122 ± 61, $N = 5$; $P < .005$). Ambulation was progressively enhanced in rats (P11–P15) given a daily injection of amphetamine (2.5 mg/kg) for 5 days (600 ± 57 counts day 5 vs. 165±37 counts day 1, $N = 10$; $P < .00001$).

To examine the effect of a single dose of amphetamine on the A/N ratio of glutamatergic synapses on DA cells in the VTA, we administered amphetamine (2.5 mg/kg) or saline to rats (P16–18) and prepared acute horizontal midbrain slices (250 mm) 24 h later. We made whole-cell recordings from DA neurons in the presence of picrotoxin (100 μM). EPSCs composed of both AMPAR and NMDAR components were observed when the neurons were voltage-clamped to +40mV. D-APV (50 μM) was then added to the bath to isolate the NMDAR component (FIG. 1, A). The A/N ratios were significantly increased in animals administered a single dose of amphetamine (0.67 ± 0.08, $N = 5$) compared to saline controls (0.41 ± 0.06, $N = 5$, $P < .05$ (FIG. 1, B). These data demonstrate that behavioral sensitization to amphetamine is correlated with an increase in A/N ratio.

To determine how soon following amphetamine exposure a change in the A/N ratio could be detected, we administered amphetamine (2.5 mg/kg) or saline to rats (P15–P19). We observed that not all animals exhibited increased locomotor activity to amphetamine; therefore, two hours following the injections, we prepared acute slices from only those animals with a behavioral response. Surprisingly, we found that as soon as two hours following amphetamine exposure, A/N ratios were significantly altered (in animals administered a single dose of amphetamine (0.69 ± 0.085, $N = 8$) compared to saline controls (0.38 ± 0.051, $N = 5$, $P < .05$) (FIG. 1, C).

Taken together, our data demonstrate that young rats exhibit behavioral sensitization to amphetamine and that this behavioral effect is correlated with the physiological response of augmented AMPAR/NMDAR ratios at the glutamatergic synapses on VTA dopamine neurons. It is of particular significance that this physiological response is observed within two hours following amphetamine exposure.

ACKNOWLEDGMENT

This work was supported by National Institutes of Health Grants DA11289 (J.A.K.) and DA15588 (L.J.F.).

REFERENCES

1. ROBINSON, T.E. & K.C. BERRIDGE. 1993. Brain Res. Rev. **18:** 247–291.
2. SELF, D.W. & E.J. NESTLER. 1995. Ann. Rev. Neurosci. **18:** 463–495.
3. KALIVAS, P.W. & B. WEBER. 1988. J. Pharmacol. Exp. Ther. **245:** 1095–1101.
4. DOUGHERTY, G.G., JR. & E.H. ELLINWOOD, JR. 1981. Life Sci. **28:** 2295–2298.
5. VEZINA, P. 1993. Brain Res. **605:** 332–337.
6. CADOR, M. *et al.* 1995. Neuroscience **65:** 385–395.
7. KALIVAS, P. & P. DUFFY. 1995. J. Neurosci. **15:** 5379–5388.
8. XUE, C.-J. *et al.* 1996. J. Neurochem. **67:** 352–363.
9. WOLF, M.E. 1998. Prog. Neurobiol. **54:** 1–42
10. SPANAGEL, R. & F. WEISS. 1999. Trends Neurosci. **22:** 521–527.
11. HYMAN, S.E. & R.C. MALENKA. 2001. Nat. Rev. Neurosci. **2:** 695–703.
12. JONES, S.J. & J.A. KAUER. 1999. J. Neurosci. **19:** 9780–9787.
13. UNGLESS, M.A. *et al.* 2001. Nature **411:** 583–587.
14. SAAL, D. *et al.* 2003. Neuron **37:** 577–582.

Nucleus Accumbens Homer Proteins Regulate Behavioral Sensitization to Cocaine

M. BEHNAM GHASEMZADEH, LINDSAY K. PERMENTER, RUSSELL W. LAKE, AND PETER W. KALIVAS

Department of Physiology and Neuroscience, Medical University of South Carolina, Charleston, South Carolina 29425, USA

KEYWORDS: Homer proteins; cocaine; behavioral sensitization; AMPA; nucleus accumbens; antisense

Locomotor sensitization is an enduring form of behavioral plasticity produced by the repeated administration of cocaine or amphetamine. Glutamatergic neurotransmission is intimately involved in this neuroplasticity.[1,2] Recent studies indicate that synaptic scaffolding proteins regulate glutamate receptor trafficking and intracellular signaling, and have implicated these proteins in establishing neuroplastic changes at glutamatergic synapses.[3] The Homer family of proteins contribute to the excitatory synaptic scaffold.[4–6] All members of the Homer protein family selectively bind group 1 metabotropic glutamate receptors. The short form, Homer1a, lacks the coiled-coil domain and is significantly expressed only in response to an increase in synaptic activity, including an acute injection of cocaine.[4] Withdrawal from repeated exposure to cocaine reduces Homer1b/c expression selectively in the nucleus accumbens.[7] In the present study, it was hypothesized that the reduced Homer1b/c proteins in the nucleus accumbens produced by repeated cocaine administration mediates the expression of locomotor sensitization. To test this hypothesis, an antisense (AS) oligonucleotide strategy was employed to mimic the selective reduction of Homer1 gene products in the nucleus accumbens and evaluate the behavioral effect of repeated exposure to cocaine.

Rats were infused for seven days with an antisense oligonucleotide sequence that spanned the initiation codon (AS1) of Homer1 gene or a random sequence into the nucleus accumbens. This treatment reduced the level of Homer1b/c by about 35%. In contrast, AS1 did not affect the structurally related Homer3 protein, nor did it alter levels of other cytoskeletal proteins, including actin or GFAP. The levels of other proteins related to Homer, including IP3R1, Shank1a, mGluR1, mGluR5, GKAP, PSD95, NMDAR1, NMDAR2A, NMDAR2B, were also not affected. However, AS1 reduced GluR1 protein level in nucleus accumbens by about 40%. A second inde-

---

Address for correspondence: M. Behnam Ghasemzadeh, Department of Biomedical Sciences, Marquette University, 561 N. 15th Street, SC 426, Milwaukee, WI 53233. Voice: 414-288-6636; fax: 414-288-6564.

behnam.ghasemzadeh@marquette.edu

pendent Homer1 antisense sequence (AS2) was used to test the possibility of a nonspecific antisense effect. AS2 lowered the level of Homer1b/c protein levels by about 40% without affecting the level of the GluR1 protein, suggesting that the effect of AS1 on GluR1 protein was independent of the reduction in Homer1b/c.

Rats were injected with cocaine (15 mg/kg) after one week of oligonucleotide infusion. The two AS treatment groups exhibited significantly greater motor stimulant effect than the random oligonucleotide group. In order to determine if the reduction in Homer1b/c affected the development of behavioral sensitization, these animals were administered daily cocaine injections for one week in the presence of AS1, AS2, or random oligonucleotide infusion into the accumbens. They were challenged with an injection of cocaine (15 mg/kg, i.p.) at 3 weeks after the last cocaine injection and removal of the oligonucleotide minipumps. The random treatment group showed a sensitized motor response compared to day 1 of cocaine administration. The AS1 treatment group demonstrated significantly less motor activity than the random group on day 28, while the AS2 treatment group showed motor activity equivalent to the random group.

As indicated above, AS1 infusion into nucleus accumbens reduced both Homer1b/c and GluR1 content. In contrast, reducing Homer1b/c with AS2 did not alter GluR1. While both antisense sequences reduced Homer1b/c and caused a sensitized motor response to an acute injection of cocaine, only AS1 prevented the development of behavioral sensitization. To determine if changes in AMPA receptor function may mediate the capacity of AS1 to inhibit the development of behavioral sensitization to repeated cocaine treatment, rats were pretreated with the AMPA/kainate antagonist NBQX into the nucleus accumbens prior to each daily injection of cocaine. At three weeks after the last daily NBQX/cocaine administration, the behavioral response to cocaine alone (15 mg/kg) in the NBQX/cocaine group was equivalent to the response in the daily NBQX/saline and saline/saline groups and significantly lower than the daily cocaine treatment group that was pretreated with daily saline (saline/cocaine). The blockade of the development of sensitization was selective for AMPA/kainate receptors since blockade of neither NMDA and group 1 mGluRs, nor D1/D2 receptors prevented the development of behavioral sensitization to cocaine.

The present study indicates that a reduction of Homer1 gene products in nucleus accumbens produces a sensitized behavioral response to cocaine. Therefore, the previously reported decrease in Homer1b/c elicited by withdrawal from repeated cocaine administration may underlie the expression of behavioral sensitization to cocaine.[7] However, the induction of cocaine-mediated behavioral plasticity is independent of the Homer1b/c protein levels in nucleus accumbens and requires excitatory neurotransmission through AMPA/kainate receptors.

ACKNOWLEDGMENT

This study was funded by NIH Grants DA03906 and MH40817 (P.W.K.) and DA11482 and DA14328 (M.B.G.).

REFERENCES

1. WOLF, M.E. 1998. The role of excitatory amino acids in behavioral sensitization to psychomotor stimulants. Progr. Neurobiol. **54:** 679–720.
2. NESTLER, E. 2001. Molecular basis of long-term plasticity underlying addiction. Nature Rev. **2:** 119–128.
3. SCANNEVIN, R.H. & R.L. HUGANIR. 2000. Postsynaptic organization and regulation of excitatory synapses. Nature Rev. Neurosci. **1:** 133–141.
4. BRAKEMAN, P.R. et al. 1997. Homer: a protein that selectively binds metabotropic glutamate receptors. Nature **386:** 221-223.
5. KATO, A. et al. 1998. Novel members of the Vesl/Homer family of PDZ proteins that bind metabotropic glutamate receptors. J. Biol. Chem. **273:** 23969–23975.
6. TU, J.C. et al. 1999. mGluR/Homer and PSD-95 complexes are linked by the Shank family of postsynaptic density proteins. Neuron **23:** 583–592.
7. SWANSON, C. et al. 2001. Repeated cocaine administration attenuates group I metabotropic glutamate receptor-mediated glutamate release and behavioral activation: A potential role for Homer 1b/c. J. Neurosci. **21:** 9043–9052.

Altered Prefrontal Cortex–Nucleus Accumbens Information Processing in a Developmental Animal Model of Schizophrenia

YUKIORI GOTO AND PATRICIO O'DONNELL

Center for Neuropharmacology and Neuroscience, Albany Medical College, Albany, New York 12208, USA

KEYWORDS: prefrontal cortex; nucleus accumbens; developmental animal model; schizophrenia

INTRODUCTION

Neonatal ventral hippocampal (VH) damage results in postpubertal emergence of abnormal behaviors in rodents and monkeys.[1,2] This procedure has been proposed as a developmental model of hippocampal deficits in schizophrenia. We have shown that prefrontal cortical (PFC) and nucleus accumbens (NAcc) neural responses to electrical stimulation of the ventral tegmental area (VTA) are altered in adult, but not prepubertal, rats with a neonatal VH lesion.[3] In these animals, VTA activation with trains of pulses mimicking dopamine (DA) cell burst firing resulted in increased action potential firing in both PFC[4] and NAcc[3] neurons. This contrasts with the reduced firing observed in sham rats,[4] which is also typical of naïve rats.[5] Given the important PFC projection to the NAcc, the enhanced firing in the NAcc may be secondary to enhanced glutamate transmission driven by the abnormal PFC firing. To test this hypothesis, *in vivo* intracellular recordings were performed in rats that had received neonatal VH and adult PFC lesions.

METHODS

A neonatal VH lesion was performed in male rats at postnatal day (PD) 6-8 with bilateral injections of 0.3 μL ibotenic acid (10 μg/μL) in artificial cerebrospinal fluid (ACSF).[1] Adult medial PFC lesion or sham treatment was performed in these animals at PD56 with either 0.5 mL of ibotenic acid or ACSF.[6] *In vivo* intracellular recordings from NAcc neurons were conducted after at least 2 weeks of recovery from PFC operations. Concentric bipolar electrodes were placed in the VTA for electrical activation of these neurons by delivering five pulses at 20 Hz at 1.0 mA.[3]

Address for correspondence: Patricio O'Donnell, Center for Neuropharmacology and Neuroscience, Albany Medical College, Albany, NY 12208. Voice: 518-262-5904; fax: 518-262-0687. odonnep@mail.amc.edu

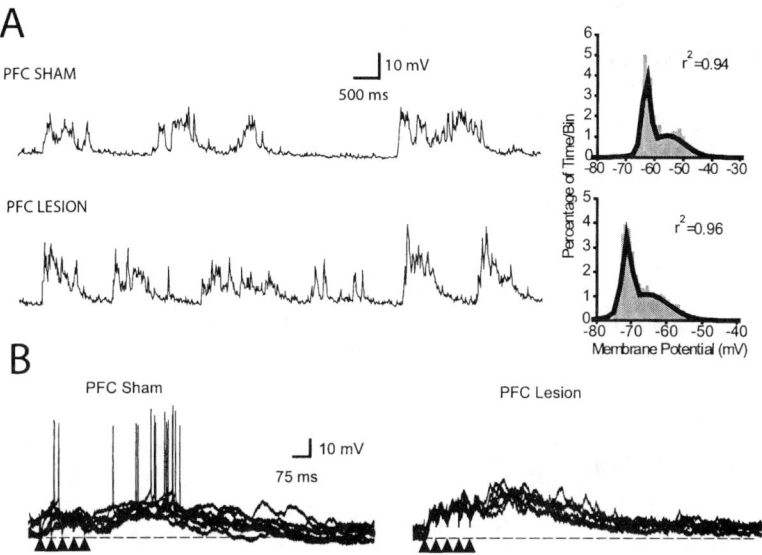

FIGURE 1. Representative traces of recordings from LESION and SHAM animals. (**A**) Spontaneous activity (*left*) and bimodal gaussian membrane potential distribution histograms (*right*). (**B**) Responses to VTA stimulation.

RESULTS

Both animals that had received neonatal VH and PFC lesions (LESION group) and a neonatal VH lesion and PFC sham treatment (SHAM group) exhibited similar membrane potential fluctuations (FIG. 1, A), suggesting that PFC lesion has little effect on spontaneous membrane activity in NAcc neurons in these animals. VTA stimulation evoked membrane depolarizations with an increase of spike firing in most SHAM groups ($N = 7/11$; FIG. 1, B), as previously described for animals with a neonatal VH lesion.[3] On the other hand, LESION groups exhibited NAcc responses with a reduction in spike firing, a response similar to that observed in naïve animals ($N = 2/27$; FIG. 1, B).[7] The proportion of NAcc neurons exhibiting such increased spike firing in SHAM animals was significantly higher than in LESION animals (Fisher exact test; $P = .0007$).

DISCUSSION

A PFC lesion prevented the delayed effects of a neonatal VH lesion on mesolimbic responses. Since PFC neurons also exhibit increased action potential firing in response to VTA stimulation in these animals,[4] the VTA-induced increase of NAcc firing in rats with a neonatal VH lesion may be mediated by increased glutamate release from the PFC. In our previous study, we have also shown that subchronic antipsychotic treatment prevented increased firing in response to VTA stimulation.[3] This treatment also reduced PFC synaptic responses in the NAcc, further supporting the evidence of enhanced excitatory neurotransmission in the PFC-NAcc pathway. Thus, an altered PFC response to the mesocortical projection is transferred to the NAcc in the form of excessive firing in animals with a neonatal VH lesion.

Abnormal PFC cell firing upon mesocortical activation could be induced by a number of factors. A loss of cortical interneurons would result in this altered response. Indeed, both neonatal VH lesioned animals and schizophrenia patients exhibit a loss of selective interneurons in the PFC.[8] Another possibility is endogenous sensitization to dopamine (DA).[9] A decreased tonic DA cell activity has been proposed in schizophrenia.[10] With decreased tonic DA release in the PFC, DA receptors in the PFC would be upregulated, resulting in increased D1 receptor binding.[11] Upon DA cell burst firing, an enhanced phasic DA release would be expected. Increased D1 receptors would render PFC neurons more excitable, and this is likely to cause excessive glutamate release in the NAcc and VTA. Interaction between the PFC and VTA would further enhance DA release in the NAcc, resulting in appearance of positive symptoms.

ACKNOWLEDGMENTS

This work was supported by U.S. Public Health Service grants and a National Alliance for Research on Schizophrenia and Depression (NARSAD) Independent Investigator Award.

REFERENCES

1. LIPSKA, B.K., G.E. JASKEW & D.R. WEINBERGER. 1993. Postpubertal emergence of hyperresponsiveness to stress and to amphetamine after neonatal excitotoxic hippocampal damage: a potential animal model of schizophrenia. Neuropsychopharmacology **9:** 67–75.
2. BACHEVALIER, J., M.C. ALVARADO & L. MALKOVA. 1999. Memory and socioemotional behavior in monkeys after hippocampal damage incurred in infancy or in adulthood. Biol. Psychiatry **46:** 329–339.
3. GOTO, Y. & P. O'DONNELL. 2002. Delayed mesolimbic system alteration in a developmental animal model of schizophrenia. J. Neurosci. **22:** 9070–9077.
4. O'DONNELL, P. et al. 2002. Neonatal hippocampal damage alters electrophysiological properties of prefrontal cortical neurons in adult rats. Cereb. Cortex **12:** 975–982.
5. LEWIS, B.L. & P. O'DONNELL. 2000. Ventral tegmental area afferents to the prefrontal cortex maintain membrane potential "up" states in pyramidal neurons via D(1) dopamine receptors. Cereb. Cortex **10:** 1168–1175.
6. LIPSKA, B.K., H.A. AL-AMIN & D.R. WEINBERGER. 1998. Excitotoxic lesions of the rat medial prefrontal cortex. Effects on abnormal behaviors associated with neonatal hippocampal damage. Neuropsychopharmacology **19:** 451–464.
7. GOTO, Y. & P. O'DONNELL. 2001. Network synchrony in the nucleus accumbens in vivo. J. Neurosci. **21:** 4498-4504.
8. LIPSKA, B.K. & D.R. WEINBERGER. 2000. To model a psychiatric disorder in animals: schizophrenia as a reality test. Neuropsychopharmacology **23:** 223–239.
9. LARUELLE, M. 2000. The role of endogenous sensitization in the pathophysiology of schizophrenia: implications from recent brain imaging studies. Brain Res. Rev. **31:** 371–384.

10. GRACE, A.A. 1991. Phasic versus tonic dopamine release and the modulation of dopamine system responsivity: a hypothesis for the etiology of schizophrenia. Neuroscience **41:** 1–24.
11. ABI-DARGHAM, A. *et al.* 2002. Prefrontal dopamine D1 receptors and working memory in schizophrenia. J. Neurosci. **22:** 3708–3719.

Lithium Regulates Total and Synaptic Expression of the AMPA Glutamate Receptor GluR2 *in Vitro* and *in Vivo*

NEIL A. GRAY, JING DU, CYNTHIA S. FALKE, PEIXIONG YUAN, AND HUSSEINI K. MANJI

National Institute of Mental Health, National Institutes of Health, Department of Health and Human Services, Bethesda, Maryland 20892, USA

KEYWORDS: lithium; GluR2; bipolar disorder; expression; localization

INTRODUCTION

Bipolar disorder (manic-depressive illness) is a severe, chronic, and life-threatening illness, characterized by episodes of mania and depression. The disease has a prevalence of approximately 1%, and it is believed that as many as 15% of bipolar patients will eventually die by suicide. While there is ample evidence that the disorder is highly genetic, the underlying pathophysiology remains poorly characterized.

The two most common medications used in the treatment of bipolar disorder are lithium, a monovalent cation, and valproate, an anticonvulsant. A growing body of data suggests that these agents may regulate intracellular signaling, leading to therapeutic alterations in neurotransmission in critically involved brain regions. Moreover, recent studies have suggested that the glutamatergic system may be a useful target in the treatment of affective disorders.[1]

Glutamate acts on ionotropic (AMPA, NMDA, and kainate) and metabotropic receptors. AMPA and NMDA receptors function in distinct but coordinated manners: AMPA receptors depolarize the postsynaptic terminal primarily by allowing an influx of sodium ions, while NMDA receptors trigger calcium transients. Increasing evidence has shown that AMPA receptor trafficking is dynamically regulated during synaptic plasticity and is targeted by a variety of major intracellular signaling pathways (such as, e.g., PKA, CaMKII, MAPK). We therefore sought to determine the effects of lithium, as well as valproate and imipramine, on the expression and localization of the AMPA receptor GluR2.

Address for correspondence: Husseini K. Manji, M.D., Laboratory of Molecular Pathophysiology, National Institute of Mental Health, MSC 4405, Building 49, Room B1EE16, Bethesda, MD 20892. Voice: 301-496-9802; 301-480-0123.
manjih@intra.nimh.nih.gov

METHODS

Ex Vivo *Studies*

Wistar-Kyoto rats were fed control, lithium-, or valproate-containing chow for 4 weeks, or given twice-daily injections of imipramine (10 mg/kg, i.p.) or saline for four weeks, prior to sacrifice. Separate control groups were used for each drug. For lithium and valproate, the presence of therapeutic drug levels was confirmed by serum test (MedTox), and animals above or below the appropriate range were excluded (0.5–1.2 mEq, 50–125 mg/L). Both crude homogenates and synaptosomal fractions were prepared from the hippocampus. Western blots were performed using antibodies from Chemicon (GluR2) and Santa Cruz Biotech (synaptophysin). Luminescence was detected using Kodak film, and signal strength was quantified using the Kodak Image System. Values were compared using the Student's t test.

Primary Hippocampal Culture

Hippocampi from embryonic day–18 rats were prepared as described and cultured for 8–10 days prior to being used for experiments. Medium was changed every three days. For total expression studies, cells were harvested in lysis buffer containing Triton X-100 and protease and phosphatase inhibitors, and then used for Western blotting as above.

Surface Expression

Surface GluR2 receptors were detected using a biotinylation assay. Following treatment, cells were incubated with sulfo-NHS-LC-biotin to label all surface proteins. Following harvest, biotinylated proteins were precipitated using ImmunoPure Immobilized Streptavidin (Pierce). Western blots were performed as above.

Immunostaining

Hippocampal cells were cultured on glass cover slips as described above. Following treatment, primary hippocampal neurons were fixed and incubated in primary antibody (GluR2 and synapsin, Chemicon). Cy3 and FITC secondary antibodies were used. Slides were imaged using the Zeiss 510 Meta confocal microscope.

RESULTS

Four-week treatment with lithium led to a consistent downregulation of both synaptosomal and total GluR2 protein in the rat hippocampus. Similarly, rats treated with valproic acid for 4 weeks also experienced a significant reduction in synaptosomal GluR2 level. In striking contrast, however, rats treated for 14 days with imipramine displayed an increase in synaptosomal GluR2. Results are summarized in TABLE 1.

In cultures of primary hippocampal neurons, both lithium and valproate induced a 30–40% downregulation of surface GluR2. Both agents required chronic administration to produce their effects; thus, lithium's effect was not apparent after six hours of treatment, but was achieved by 24 hours, and maintained for at least 4 days.

TABLE 1. The effects of lithium, valproate, and imipramine on GluR2 expression

		Lithium– Total protein	Lithium– Synaptosomes	Valproate– Synaptosomes	Imipramine– Synaptosomes
GluR2 immunoreactivity (percentage of control value)	Control	100% (2.6%) $N = 5$	100% (4.4%) $N = 5$	100% (7.2%) $N = 8$	100% (11.2) $N = 8$
	Treatment	90.78% (3.0%)* $N=6$	80.7% (1.6%)* $N=6$	71.5% (7.0%)* $N = 7$	131.6% (7.2)* $N = 7$

Each drug treatment was conducted as a separate experiment, with a separate control group.
*$P<.05$.

Valproate's effect failed to achieve significance by 24 hours, but was comparable to lithium's effect at 4 days. Immunolabeling by confocal microscopy indicated a reduction in GluR2 level at synapses.

DISCUSSION

The observation that chronic treatment with both lithium and valproate reduce synaptic GluR2 expression suggests that this phenomenon may be related to their common therapeutic effect. Because both drugs are effective antimanic agents, it is furthermore suggestive that a drug that can trigger mania in bipolar patients, imipramine, has the opposite effect. Although speculative, it is therefore possible that manipulation of AMPA-mediated glutamatergic neurotransmission within the hippocampus can modulate manic behavior.

The GluR2 subunit is of particular interest, in that its presence determines the calcium permeability of the AMPA receptor in which it is incorporated: AMPA receptors excluding GluR2 are able to increase intracellular calcium concentrations when stimulated. It is therefore possible that the observed actions of lithium, valproate, and imipramine on this receptor might confer an alteration in glutamatergic calcium signaling.

REFERENCE

1. KRYSTAL J.H., G. SANACORA, H. BLUMBERG, et al. 2002. Glutamate and GABA systems as targets for novel antidepressant and mood-stabilizing treatments. Mol. Psychiatry 7(S1): S71–80.

Prefrontal Group II Metabotropic Glutamate Receptor Activation Decreases Performance on a Working Memory Task

MARY L. GREGORY, NICHOLAS E. STECH, RUSSELL W. OWENS, AND PETER W. KALIVAS

Department of Physiology and Neuroscience, Medical University of South Carolina, Charleston, South Carolina, 29425, USA

KEYWORDS: glutamate; group II metabotropic glutamate receptor; working memory; prefrontal cortex

Working memory is a transient form of memory used for holding information "on-line" and using this information to perform intricate cognitive functions—specifically to form appropriate responses to environmental stimuli.[1–3] The prefrontal cortex (PFC) is integral to the functioning of many aspects of personality, awareness, and cognition, including working memory.[4,5] Damage to the PFC results in significant deficits in tasks that are designed to measure working memory.[6–8]

The PFC receives innervation from multiple sources, including glutamatergic efferents from the mediodorsal thalamus, hippocampus, and amygdala. An interference of glutamatergic neurotransmission in the PFC results in decreased performance on working memory tasks.[9,10,15] In the PFC, presynaptically located group II metabotropic glutamate receptors (mGluR 2/3) have an inhibitory action on slow asynchronous glutamate (Glu) release[11,12] when activated, regulating extracellular glutamate concentrations by a negative feedback mechanism.[9,13]

While other studies have demonstrated that systemic administration of mGluR 2/3 agonists selectively impairs performance on working memory tasks,[14] the drug effect responsible for this decrease in working memory has not been localized to the PFC. We hypothesized that localized injection into the PFC of an mGluR 2/3 agonist, aminopyrrolidine dicarboxylate (APDC) would decrease performance on working memory tasks. Additionally we hypothesized that localized injection of the mGluR 2/3 antagonist LY341495 into the PFC would increase glutamate release by blocking the negative feedback mechanism and, at levels producing moderate increases in extracellular glutamate, would facilitate working memory performance.

Address for correspondence: Mary Lee Gregory, Department of Physiology and Neuroscience, Basic Science Building, Suite 403, Medical University of South Carolina, 173 Ashley Avenue, Charleston, SC 29425. Voice: 843-792-1838; fax: 843-792-4423.
gregorml@musc.edu

METHODS

Rats were trained to perform a forced choice delayed-alteration task[12,14] on a T-maze. Animals were initially trained using five intramaze delays: 3, 10, 20, 30, and 50 seconds. A 20-s intramaze delay was used for the remainder of the study. Microinjection cannulae were surgically implanted into the PFC. Rats were retrained and entered a series of experiment days during which each received bilateral microinjections (0.5 μL/min) into the PFC of APDC (1 or 0.1 nmol/0.5 μL), LY341495 (0.1 or 0.01 nmol/0.5 μL), or 0.9% saline ten minutes prior to the working memory task. Experiment days were separated by practice days to insure recovery from prior injections. On test days, animals were videotaped to measure latency to leave the start box and traverse the maze. A separate group of control animals was trained on a spatial discrimination task and received a series of bilateral microinjections of 1.0 nmol/ 0.5 μL APDC, 0.1 nmol/0.5 μL LY341495, and 0.9% saline. Apart from the task, control animals were handled and housed in identical conditions to the test group.

RESULTS AND DISCUSSION

The rats' performance levels on the variable delay forced-choice delayed-alteration T-maze task were inversely proportional to the length of the intramaze delay (FIG. 1). At the 20-s intramaze delay, the animals performed at submaximal but

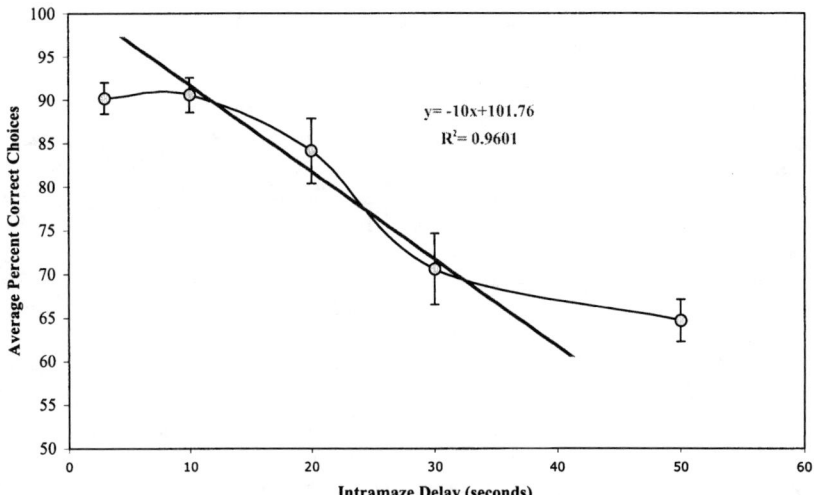

FIGURE 1. Working memory task performance at variable intramaze delays. Delay response tests reflect the ability of a subject to correctly perform a task based on information previously acquired but no longer available at the time the decision must be made. The forced choice delayed-alteration task requires that the rat remember which arm to enter on the free choice while returned to the starting box for the intramaze delay period. Animals were trained to perform a forced choice delayed-alteration task at a range of intramaze delays. As the intramaze delay period increased, performance on the maze decreased (*$P < .05$, 3-s delay; #$P < .05$, previous delay, $N = 17$).

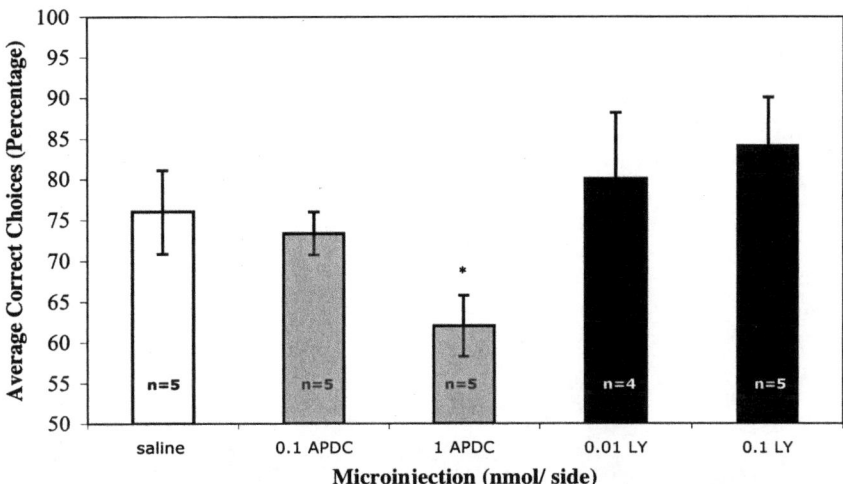

FIGURE 2. Effects of group II metabotropic receptor activation and blockade on working memory. Bilateral microinjection (0.5 µL/side) of the mGluR 2/3 agonist APDC (1 nmol) into the PFC significantly (*$P<.05$, repeated measures ANOVA) impaired working memory performance compared with 0.9% saline. The mGluR 2/3 antagonist LY341495 showed a nonsignificant trend toward increased performance. The latency of the animals to exit the start box and traverse the long arm of the T-maze was not altered at any of the injected doses.

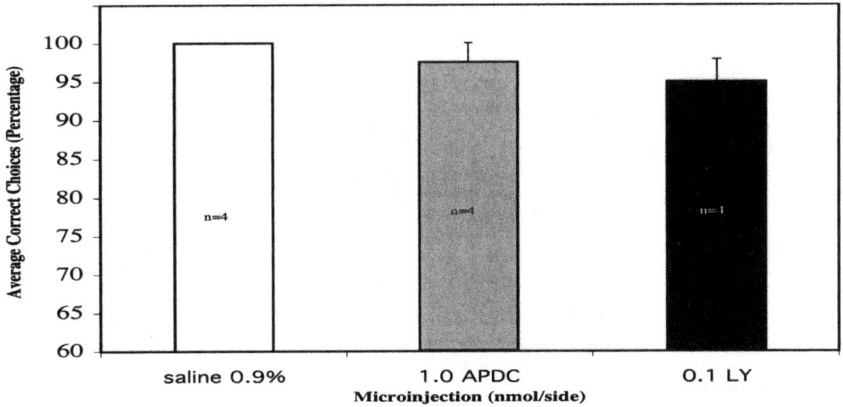

FIGURE 3. Effects of microinjection of APDC or LY341495 into the PFC on spatial discrimination. Bilateral PFC microinjections of 1.0 nmol/ 0.5 µL APDC or 0.1 nmol/ 0.5 µL LY341495 did not affect performance on a spatial discrimination task compared with vehicle.

above chance levels, allowing both increases and decreases in performance to be observed. This delay was used throughout the remainder of the study. Microinjection into the PFC of aminopyrrolidine dicarboxylate (APDC), an mGluR-2/3 agonist, prior to working memory task performance resulted in significant impairment in working memory compared with saline. Conversely, microinjection of mGluR-2/3 antagonist LY341495 showed a trend towards increased performance on the task (FIG. 2). Drug microinjection resulted in no differences in latency (data not shown.) Microinjection of 0.1 nmol/0.5 µL LY341495 and 1 nmol/0.5 µL APDC did not result in impairment on a spatial discrimination task (FIG. 3). These results confirm the previously reported involvement of mGluR-2/3 in working memory and demonstrate that this arises from a direct action in the PFC. Thus, stimulating mGluR2/3 decreased performance, while blocking mGluR-2/3 tended to increase performance.

ACKNOWLEDGMENT

This work was supported by National Institute of Mental Health grant MH-40817 (P.W.K).

REFERENCES

1. BADDELEY, A. 1992. Working memory. Science **255:** 556–559.
2. KALIVAS, P.W., L. CHURCHILL & A. ROMANIDES. 1999. Involvement of the pallidal-thalamocortical circuit in adaptive behavior. Ann. N.Y. Acad. Sci. **877:** 64–70.
3. KALIVAS, P.W. & M. NAKAMURA. 1999. Neural systems for behavioral activation and reward. Curr. Opin. Neurobiol. **9:** 223–227.
4. BADDELEY, A. 1981. The concept of working memory: a view of its current state and probable future development. Cognition **10:** 17–23.
5. GOLDMAN-RAKIC, P.S. 1987. Circuitry of primate prefrontal cortex and regulation of behavior by representational memory. *In* Handbook of Physiology. F. Plum & E. Mountcastle, Eds.: 373–417. American Physiological Society. Bethesda, MD.
6. BROERSEN, L.M., R.P.W. HEINSBROEK, J.P.C. DE BRUIN, *et al.* 1995. The role of the medial prefrontal cortex of rats in short-term memory functioning: further support for involvement of cholinergic, rather than dopaminergic mechanisms. Brain Res. **674:** 221–229.
7. GRANON, S., C. VIDAL, C. THINUS-BLANC, *et al.* 1994. Working memory, response selection, and effortful processing in rats with medial prefrontal lesions. Behav. Neurosci. **108:** 883–891.
8. KOLB, B. 1984. Functions of the frontal cortex of the rat: A comparative review. Brain Res. Rev. **8:** 65–98.
9. SCHOEPP, D., D. JANE & J. MONN. 1999. Pharmacological agents acting as subtypes of metabotropic glutamate receptors. Neuropharmacology **38:** 1431–1476.
10. VERMA, A. & B. MOGHADDAM. 1996. NMDA receptor antagonists impair prefrontal cortex function as assessed via spatial delayed alternation performance in rats: Modulation by dopamine. J. Neurosci. **16:** 373–379.
11. AGHAJANIAN, G.K. & G.J. MAREK. 1999. Serotonin, via 5-HT$_{2A}$ receptors increases EPSCs in layer V pyramidal cells of prefrontal cortex by an asynchronous mode of glutamate release. Brain Res. **825:** 161–171.
12. MAREK, G. & G. AGHAJANIAN. 1998. The electrophysiology of prefrontal serotonin systems: therapeutic implications for mood and psychosis. Biol. Psychiatry **44:** 1118–1127.
13. KILBRIDE, J., L. HUANG, M. ROWAN & R. ANWYL. 1998. Presynaptic inhibitory action of the group II metabotropic glutamate receptor agonists, LY354740 and DCG-IV. Eur. J. Pharmacol. **356:** 149–157.

14. AULTMAN, J.M. & B. MOGHADDAM. 2001. Distinct contributions of glutamate and dopamine receptors to temporal aspects of rodent working memory using a clinically relevant task. Psychopharm. **153:** 353–364.
15. ROMANIDES, A.J., P. DUFFY & P.W. KALIVAS. 1999. Glutamatergic and dopaminergic afferents to the prefrontal cortex regulate spatial working memory in rats. Neuroscience **92:** 97–106.

The Effects of Selective Orbitofrontal Cortex Lesions on the Acquisition and Performance of Cue-Controlled Cocaine Seeking in Rats

DANIEL M. HUTCHESON AND BARRY J. EVERITT

*Department of Experimental Psychology,
University of Cambridge, Cambridge, United Kingdom*

KEYWORDS: orbitofrontal cortex lesions; drug-craving behavior; cocaine

In functional imaging studies it has been demonstrated that the orbitofrontal cortex (OFC), anterior cingulate cortex, and amygdala are activated during drug-craving elicited by drug-associated stimuli.[1] Previously we have shown using selective excitotoxic lesions that the basolateral amygdala[2] and nucleus accumbens core[3] are critical structures mediating the process of conditioned reinforcement. Moreover, rats with lesions of these structures cannot acquire behavior that depends upon the contingent presentation of drug cues as assessed in a second-order schedule of cocaine-reinforcement.[2,3] However, the involvement of the OFC in drug-seeking and taking remains unclear. We therefore sought to investigate the involvement of this frontal cortical area in an animal model of cocaine-taking and -seeking behavior, especially in the context of the rich interconnections between the OFC and both the basolateral amygdala and nucleus accumbens core.

Male Lister-Hooded rats with bilateral excitotoxic lesions of the orbitofrontal cortex were trained to self-administer cocaine (0.75 mg/kg) under a fixed ratio schedule (FR1) for 10 days. Each drug infusion was paired with the illumination of a light conditioned stimulus (CS) for 20 seconds. The rats were then required to acquire a second-order $FRx(FRy:S)$ schedule of cocaine reinforcement in which the number of fixed ratios of responses for each drug infusion (x) was set at 10 throughout the training whilst the number of responses for each presentation of the CS (y) before drug, was gradually increased between sessions from 1 to 10 in the following steps: FR10(FR1:S), FR10(FR2:S), FR10(FR4:S), FR10(FR7:S), and finally FR10(FR10:S). The CS in this case was presented for a 2-s duration. This second-order schedule of cocaine reinforcement therefore provides a means of measuring cocaine-seeking under the control of a cocaine-associated conditioned reinforcer.[4,5]

Address for correspondence: Daniel M. Hutcheson, Department of Experimental Psychology, University of Cambridge, Cambridge, UK. Voice: +44-1223-339549; fax: +44-1223-333864.
dmh24@cam.ac.uk

The findings show that under FR1 those rats with OFC lesions attained the maximum number of infusions (50) more rapidly and with erratic response patterns compared to the slower, regulated pattern of self-administration seen in the sham control group. In contrast, rats with OFC lesions were markedly impaired in their acquisition of cue-controlled cocaine seeking. Indeed, only a minority of rats in the OFC-lesioned group made sufficient responses to receive intravenously self-administered infusions of cocaine at higher schedule requirements.

To investigate whether this impairment was due to a deficit in learning about the drug-predictive nature of the CS, an identical experiment was performed, but with the OFC lesions being made after initial FR1 training (again 10 days with CS-drug pairings).

The results showed that the acquisition of the second-order schedule was also significantly impaired in rats with OFC lesions, suggesting that this cortical area has an especially important role in cue-controlled drug seeking. In the same animals, responding under a FR1 schedule, a within-session dose-response experiment was performed, with the dose varied every 40 min (doses from 0.01 to 1.5 mg/kg). The results indicated no difference in the cocaine dose-response function between the lesioned and sham control groups, indicating that the deficits in the acquisition of the second-order schedule were not a consequence of any change in the rewarding effects of cocaine.

Taken together, these findings indicate that the OFC is a critical structure underlying drug cue–conditioned influences over cocaine-seeking behavior. Further studies will define both the psychological processes subserved by the OFC and also investigate the nature of the interaction between the basolateral amygdala and striatum, including the nucleus accumbens core.

REFERENCES

1. BREBNER, K., A.R. CHILDRESS & D.C. ROBERTS. 2002. A potential role for GABA(B) agonists in the treatment of psychostimulant addiction. Alcohol Alcohol. **37:** 478–484.
2. WHITELAW, R.B. *et al.* 1996. Excitotoxic lesions of the basolateral amygdala impair the acquisition of cocaine-seeking behaviour under a second-order schedule of reinforcement. Psychopharmacology (Berl) **127:** 213–224.
3. HUTCHESON, D.M. *et al.* 2001. The effects of nucleus accumbens core and shell lesions on intravenous heroin self-administration and the acquisition of drug-seeking behaviour under a second-order schedule of heroin reinforcement. Psychopharmacology (Berl) **153:** 464–472.
4. ARROYO, M. *et al.* 1998. Acquisition, maintenance and reinstatement of intravenous cocaine self-administration under a second-order schedule of reinforcement in rats: effects of conditioned cues and continuous access to cocaine. Psychopharmacology (Berl) **140:** 331–344.
5. EVERITT, B.J. & T.W. ROBBINS. 2000. Second-order schedules of drug reinforcement in rats and monkeys: measurement of reinforcing efficacy and drug-seeking behaviour. Psychopharmacology (Berl) **153:** 17–30.

In Vivo Characterization of Changes in Glycine Levels Induced by GlyT1 Inhibitors

KIRK W. JOHNSON, AMY CLEMENS-SMITH, GEORGE NOMIKOS, RICHARD DAVIS, LEE PHEBUS, HARLAN SHANNON, PATRICK LOVE, KEN PERRY, JASON KATNER, FRANK BYMASTER, HONG YU, AND BETH J. HOFFMAN

Neuroscience Research Division, Eli Lilly and Company, Indianapolis, Indianapolis 46285, USA

KEYWORDS: glycine level changes; GlyT1 inhibitors; schizophrenia

Hypofunction of the glutamatergic system has been suggested to contribute to the complex pathophysiology of schizophrenia. In particular, the noncompetitive NMDA antagonist phencyclidine (PCP) induces psychotomimetic and cognitive effects in humans, suggesting decreased activation of NMDA receptors in individuals with schizophrenia.[1] NMDA receptors contain multiple modulation sites including a strychnine-insensitive glycine-B (GlyB) site on the NR1 subunit that binds extracellular glycine as an obligatory coagonist. Glycine and GlyB site agonists are efficacious as adjunctive therapies for the treatment of schizophrenic symptoms.[2–5] Two related glycine transporter have been cloned (GlyT1 and GlyT2). GlyT1 mRNA is present in frontal cortex, hippocampus, brain stem, and spinal cord with a distribution that overlaps that of NMDA receptors.[6] Extracellular levels of glycine near NMDA receptors are regulated by glial GlyT1.[6] Thus, inhibition of glycine reuptake at GlyT1 should increase extracellular glycine levels, potentiate NMDA receptor function, and potentially represent an efficacious therapy for the symptoms of schizophrenia.

High-performance liquid chromatography with electrochemical detection (HPLC-EC) was used to quantify extracellular levels of glycine, glutamate, glutamine, serine, threonine, taurine, and alanine in microdialysate samples collected from the lateral hippocampus of rats treated with either ALX 5407 or Org 24461-HCl (10 mg/kg, s.c.), selective GlyT1 inhibitors. Glycine levels were also measured in microdialysate samples collected from the prefrontal cortex (PFC) and striatum of rats dosed with ALX 5407 (10 mg/kg, s.c.). Amino acid and drug levels were quantified in cerebrospinal fluid (CSF) samples from rats treated with ALX5407 or Org 24461-HCl (1–10 mg/kg, s.c.) using HPLC-EC and HPLC-mass spectrometry.

Address for correspondence: Kirk W. Johnson, Neuroscience Research Division, Eli Lilly and Company, Indianapolis, IN 46285. Voice: 317-276-8680; fax: 317-276-5546.
JOHNSON_KIRK_W@Lilly.com

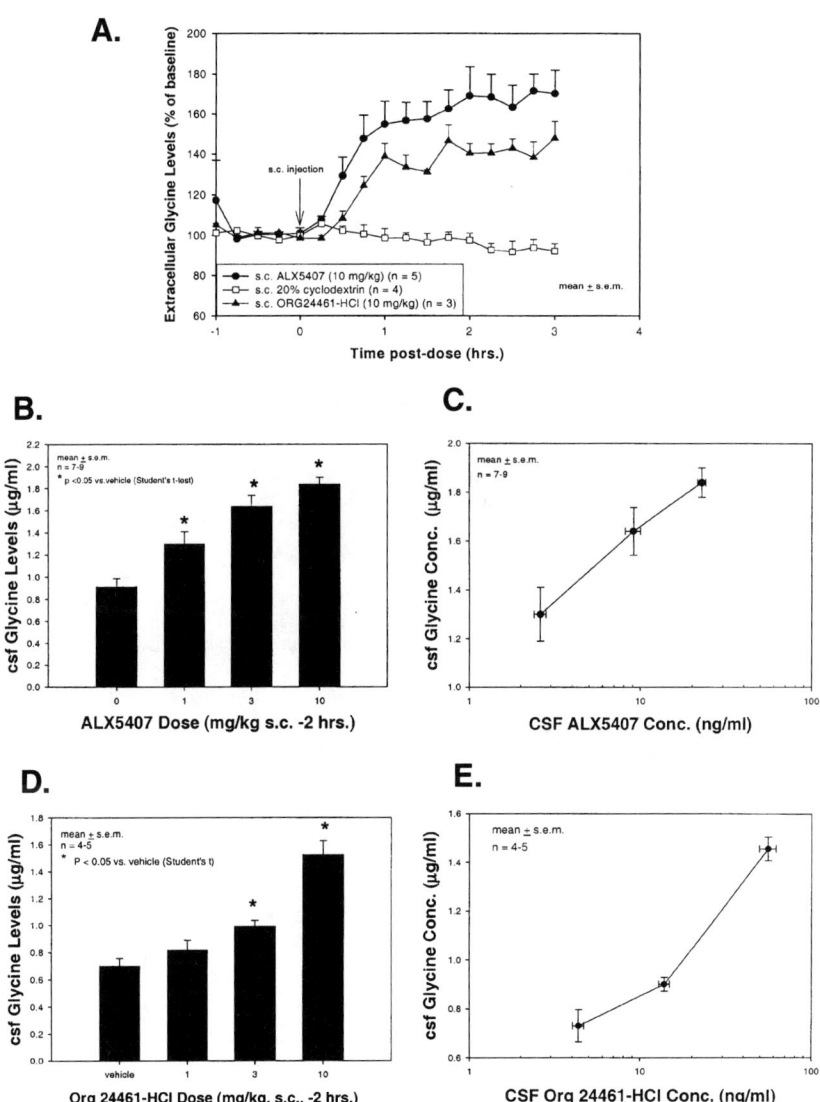

FIGURE 1. (**A**) Effect of ALX547 and Org 24461-HCl (10 mg/kg, s.c.) on glycine levels in microdialysate samples from rat hippocampus. (**B**) Effect of ALX5407 (1, 3, and 10 mg/kg, s.c.) on glycine levels in rat cerebral spinal fluid (csf). (**C**) Correlation of csf ALX5407 concentrations to csf glycine concentrations following administration of ALX5407 (1, 3, and 10 mg/kg, s.c.). (**D**) Effect of Org 24461-HCl (1, 3, and 10 mg/kg, s.c.) on glycine levels in rat csf. (**E**) Correlation of csf Org 24461-HCl concentrations to csf glycine concentrations following administration of Org 24461-HCl (1, 3, and 10 mg/kg, s.c.).

ALX 5407 and Org 24461-HCl significantly increased extracellular levels of glycine, but not other amino acids, in the lateral hippocampus (170% and 145% of baseline, respectively) (FIG. 1A). ALX 5407 also caused a significant increase in extracellular glycine levels in the PFC and striatum (180% of baseline in both regions) (data not shown). Subcutaneous injection of either ALX 5407 or Org 24461-HCl resulted in statistically significant, dose-dependent increases in CSF levels of glycine and compound 2 h post-dose (FIGS. 1B–E). Thus, ALX 5407 and Org 24461-HCl are efficacious and specific GlyT1 inhibitors. Additionally, the increases in CSF glycine levels following administration of GlyT1 inhibitors parallels changes in extracellular glycine levels and, as such, may represent a clinical surrogate for assessing efficacy of GlyT1 inhibitors.

REFERENCES

1. JAVITT, D.C. & S.R. ZUKIN. 1991. Recent advances in the phencyclidine model of schizophrenia. Am. J. Psychiatry **148:** 1301–1308.
2. JAVITT, D.C. et al. 1994. Amelioration of negative symptoms in schizophrenia by glycine. Am. J. Psychiatry **151:** 1234–1236.
3. HERESCO-LEVY, U. et al. 1999. Efficacy of high-dose glycine in the treatment of enduring negative symptoms of schizophrenia. Arch. Gen. Psychiatry **56:** 29–36.
4. TSAI, G.E. et al. 1999. D-serine added to clozapine for the treatment of schizophrenia. Am. J. Psychiatry **156:** 1822–1825.
5. GOFF, D.C. et al. 1999. A placebo-controlled trial of D-cycloserine added to conventional neuroleptics in patients with schizophrenia. Arch. Gen. Psychiatry **56:** 21–27.
6. BOROWSKY, B. et al. 1993. Two glycine transporter variants with distinct localization in the CNS and peripheral tissues are encoded by a common gene. Neuron **10:** 851–863.

Metabotropic Glutamate 5 Receptor Antagonist MPEP Decreased Nicotine and Cocaine Self-Administration but Not Nicotine and Cocaine-Induced Facilitation of Brain Reward Function in Rats

P. J. KENNY, N. E. PATERSON, B. BOUTREL, S. SEMENOVA, A. A. HARRISON, F. GASPARINI,[a] G. F. KOOB, P. D. SKOUBIS, AND A. MARKOU

The Scripps Research Institute, Department of Neuropharmacology, 10550 North Torrey Pines Road, La Jolla, California 92037, USA

[a]*Nervous System Research, Novartis-Pharma AG, 4002 Basel, Switzerland*

KEYWORDS: cocaine; nicotine; glutamate; self-administration; intracranial self-stimulation

Most drugs of abuse have been shown to increase excitatory glutamatergic transmission throughout brain reward circuitries.[1–5] Although the precise function of glutamatergic transmission in reward remains controversial and unclear,[6] it is likely that drug-induced increases in excitatory glutamatergic transmission contribute to the positive reinforcing properties of addictive drugs. Indeed, blockade of glutamatergic transmission has been shown to decrease the rewarding actions of cocaine and other drugs of abuse. More recently, metabotropic glutamate 5 (mGlu5) receptors have been shown to play a crucial role in regulating the positive reinforcing properties of addictive drugs. For example, mice in which the gene for the mGlu5 receptor was deleted did not acquire cocaine self-administration behavior, even though the acquisition of food self-administration was unaffected in these mice.[7]

The present studies investigated the behavioral mechanisms by which mGlu5 receptors regulated the reinforcing properties of cocaine and nicotine. Systemic administration of the mGlu5 receptor antagonist 6-methyl-2-(phenylethynyl)-pyridine (MPEP; 1–9 mg/kg) dose-dependently decreased cocaine (0.25 mg/infusion) self-administration in rats, and nicotine (rats: 0.01 or 0.03 mg/kg/infusion; mice: 0.048 mg/infusion) self-administration in rats and mice, while having no effects on responding for food in rats (FIG. 1).

Address for correspondence: P.J. Kenny, The Scripps Research Institute, Department of Neuropharmacology, 10550 N. Torrey Pines Rd., La Jolla, CA 92037. Voice: 858-784-7305; fax 858-784-7405.

pjkenny@scripps.edu

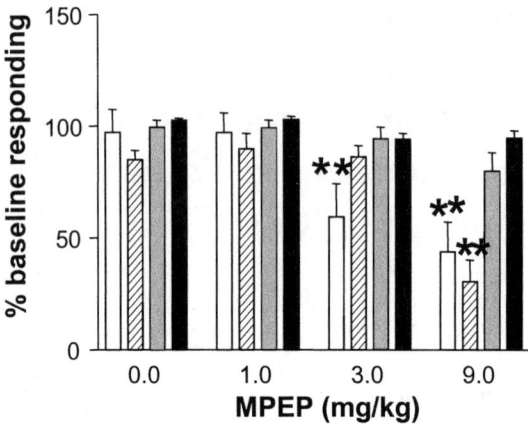

FIGURE 1. MPEP dose-dependently decreased nicotine self-administration and had no effect on responding for food in rats. Data were analyzed using a two-way ANOVA with reinforcer as the between-subjects factor, and MPEP dose as a within-subjects factor. The ANOVA revealed a main effect of both MPEP dose ($F_{3,102} = 23.33$; $P < .001$) and reinforcer ($F_{3,34} = 14.07$; $P < .001$), as well as a significant interaction ($F_{9,102} = 4.22$; $P < .001$). Post hoc analyses demonstrated that 3 mg/kg MPEP selectively decreased self-administration of the lower unit dose of nicotine, while 9.0 mg/kg MPEP significantly decreased self-administration of both unit doses of nicotine. There were no effects of MPEP on responding for food. *Open bars*, 0.01 mg/kg/inf ($n = 9$); *hatched bars*, 0.03 mg/kg/inf ($n = 9$); *gray bars*, food: TO20 s ($n = 10$); *black bars*, food: TO210 s ($n = 10$). (Reproduced with permission from Paterson *et al.*[12])

These data suggest that mGlu5 receptors regulate the reinforcing properties of cocaine and nicotine, and that the attenuation of drug self-administration was not secondary to performance-disrupting actions of MPEP. The positive affective state associated with acute cocaine and nicotine administration plays a crucial role in establishing and maintaining self-administration behavior.[8,9] Cocaine and nicotine lower intracranial self-stimulation (ICSS) reward thresholds, and this action is considered a measure of the positive affective state associated with acute cocaine and nicotine consumption.[10,11] MPEP, at doses that decreased cocaine and nicotine self-administration (1–9 mg/kg), did not attenuate the magnitude of the lowering in ICSS reward thresholds observed after acute cocaine (10 mg/kg) or nicotine (0.25 mg) administration, nor the decrease in response latency observed after cocaine administration that most likely reflects cocaine's psychomotor stimulant effects. Further, MPEP (3 mg/kg) did not alter the duration of cocaine's (10 mg/kg) lowering action on ICSS thresholds (FIG. 2).

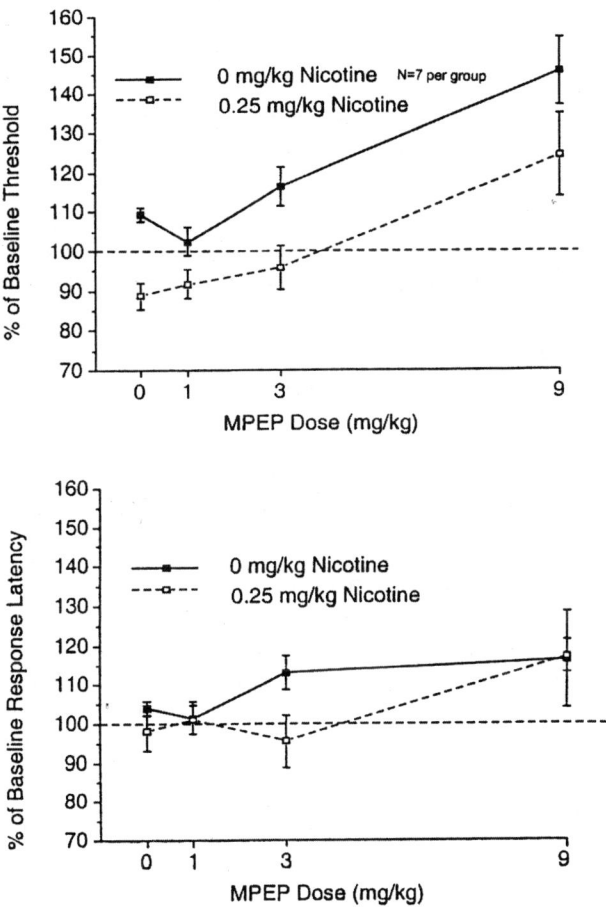

FIGURE 2. MPEP did not reverse nicotine-induced potentiation of brain stimulation reward. A two-way repeated measures ANOVA revealed a main effect of MPEP ($F_{3,18}$ = 11.864; P <.01). Post hoc analysis demonstrated that 9 mg/kg MPEP significantly elevated reward thresholds compared to all other doses (P <.05). Nicotine significantly lowered brain stimulation thresholds ($F_{1,16}$ = 27.648; P <.01), however MPEP had no effect on the reward potentiating effects of nicotine, as seen by a lack of an interaction effect (P >.05). Post hoc analysis of a main effect of MPEP on response latencies revealed that 9 mg/kg significantly increased response latencies compared to all other doses ($F_{3,18}$ = 9.104; P <.01). (Reproduced with permission from Harrison et al.[11])

Taken together, these observations suggest that mGlu5 receptors may not regulate the hedonic impact of cocaine and nicotine or the psychomotor stimulant properties of cocaine, at least as reflected in the intracranial self-stimulation reward procedure. As stated above, deletion of the mGlu5 receptor gene in mice blocked the acquisition of cocaine self-administration behavior. However, mGlu5 receptor mutant mice responded to food reinforcement similar to wild-type control mice.[7] Similarly, we have shown here that the mGlu5 receptor antagonist MPEP decreased cocaine and nicotine self-administration, but did not effect responding for food reinforcement. Overall, these observations suggest that mGlu5 receptors may not be involved in regulating "natural" (i.e., food) reinforcement, but instead may selectively regulate the motivation to consume drugs of abuse, such as cocaine and nicotine.

ACKNOWLEDGMENTS

This work was supported by the Peter F. McManus Charitable Trust (P.J.K.), grants from the National Institute on Drug Abuse (DA04398 to G.F.K. and DA11946 to A.M.), and a Novartis Research grant (A.M.).

REFERENCES

1. UNGLESS, M.A. *et al.* 2001. Single cocaine exposure in vivo induces long-term potentiation in dopamine neurons. Nature **411:** 583–587.
2. KALIVAS, P.W. & P. DUFFY. 1998. Repeated cocaine administration alters extracellular glutamate in the ventral tegmental area. J. Neurochem. **70:** 1497–1502.
3. WOLF, M.E. 2003. Effects of psychomotor stimulants on glutamate receptor expression. Methods Mol. Med. **79:** 13–31.
4. MANSVELDER, H.D., J.R. KEATH & D.S. MCGEHEE. 2002. Synaptic mechanisms underlie nicotine-induced excitability of brain reward areas. Neuron **33:** 905–919.
5. MANSVELDER, H.D. & D.S. MCGEHEE. 2000. Long-term potentiation of excitatory inputs to brain reward areas by nicotine. Neuron **27:** 349–357.
6. HARRIS, G.C. & G. ASTON-JONES. 2003. Critical role for ventral tegmental glutamate in preference for a cocaine-conditioned environment. Neuropsychopharmacology **28:** 73–76.
7. CHIAMULERA, C. *et al.* 2001. Reinforcing and locomotor stimulant effects of cocaine are absent in mGluR5 null mutant mice. Nat. Neurosci. **4:** 873–874.
8. STEWART, J., H. DE WIT & R. EIKELBOOM. 1984. Role of unconditioned and conditioned drug effects in the self-administration of opiates and stimulants. Psychol. Rev. **91:** 251–268.
9. KENNY, P.J. *et al.* 2003. Low dose cocaine self-administration transiently increases but high dose cocaine persistently decreases brain reward function in rats. Eur. J. Neurosci. **17:** 191–195.
10. MARKOU, A. & G.F. KOOB. 1992. Construct validity of a self-stimulation threshold paradigm: effects of reward and performance manipulations. Physiol. Behav. **51:** 111–119.
11. HARRISON, A.A., F. GASPARINI & A. MARKOU. 2002. Nicotine potentiation of brain stimulation reward reversed by DHβE and SCH 23390, but not by eticlopride, LY 314582 or MPEP in rats. Psychopharmacology (Berl) **160:** 56–66.
12. PATERSON, N.E. *et al.* 2003. The mGluR5 antagonist MPEP decreased nicotine self-administration in rats and mice. Psychopharmacology (Berl.) **167**(3): 257–264.

Elucidation of Homer 1a Function in the Nucleus Accumbens Using Adenovirus Gene Transfer Technology

M. S. BOWERS,[a] R. W. LAKE,[a] S. RUBINCHIK,[b] J-Y. DONG,[b] AND P. W. KALIVAS[a]

[a]*Department of Physiology and Neuroscience,*
[b]*Department of Microbiology and Immunology, Medical University of South Carolina, Charleston, South Carolina 29425, USA*

KEYWORDS: Homer 1a; nucleus accumbens; cocaine; glutamate

Glutamatergic transmission in the nucleus accumbens is necessary for the expression of behavioral sensitization to repeated administration of cocaine.[1] Additionally, Group I metabotropic glutamate transmission is strongly implicated in the conditioning, reinforcing, and locomotor stimulating effects of cocaine.[2–4] The Homer family of proteins selectively bind Group I metabotropic glutamate receptors (mGluRs). The long form of Homer 1, Homer 1b/c, is constitutively expressed as a homodimer linking Group I mGluRs to calcium-selective endoplasmic inositol trisphosphate (IP3) receptors through the C-terminal coiled-coil domain of Homer 1b/c. The immediate early gene (IEG) form, Homer 1a, which lacks the coiled-coil domain, has been shown to disrupt this complex in response to synaptic activity, thereby reducing glutamate-induced endoplasmic Ca^{2+} release.[5] To elucidate the potential role of Homer 1a in glutamate-induced locomotion, we designed a replication-deficient adenovirus carrying *Homer 1a* and *green fluorescent protein* (*GFP*) transgenes under the control of the strong cytomegalovirus (CMV) promoter (FIG. 1) in titers up to 7×10^7 pfu. It was found that rat nucleus accumbens (NA) can be stably transfected for at least one month and inflammation, as well as immune response to the vector, was similar to saline microinjection after day 3. Further, the ubiquitous CMV promoter yielded 54% neuronal transfection. The locomotor response to the Group I metabotropic glutamate receptor selective agonist (S)-3,5-dihydroxyphenylglycine (DHPG) was significantly reduced in animals overexpressing Homer 1a as compared to controls expressing only GFP (FIG. 2). Therefore, viral-mediated overexpression of Homer 1a was able to selectively modulate behaviors induced by the Group I agonist DHPG. However, in a cocaine sensitization paradigm (15 mg/kg, days 1 and 7; 30 mg/kg, days 2–6; i.p.), Homer 1a transduction showed no change in cocaine-me-

Address for correspondence: M. Scott Bowers, Ernest Gallo Clinic and Research Center, 5858 Horton Street, Suite 200, Emeryville, CA 94608. Voice: 510-985-3136.
sbowers@egcrc.net

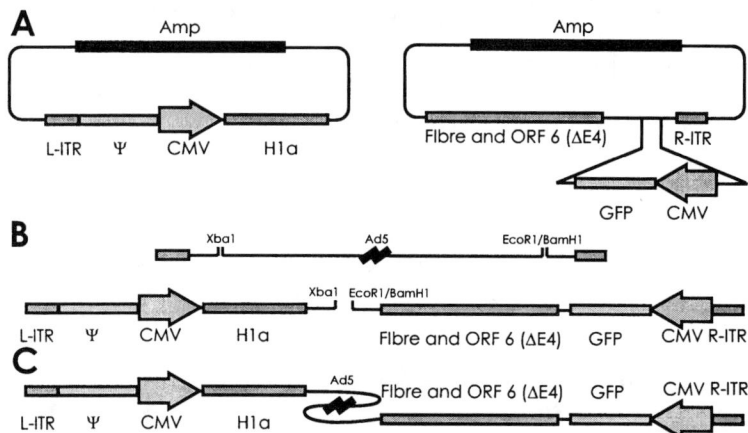

FIGURE 1. Virus construction. (**A**) Plasmids containing green fluorescent protein (GFP) and Homer 1a (H1a) transgenes under the strong cytomegalovirus (CMV) promoter are separately amplified in bacteria. (**B**) Transgenes contained in the left and right arms are ligated with the replication-deficient adenovirus backbone. (**C**) Final construct expressing both genes. Amp, ampicillin terminal repeat; L-TR, left inverted terminal repeat; R-ITR, right inverted terminal repeat; ORF, open reading frame; ψ, packing sequence.

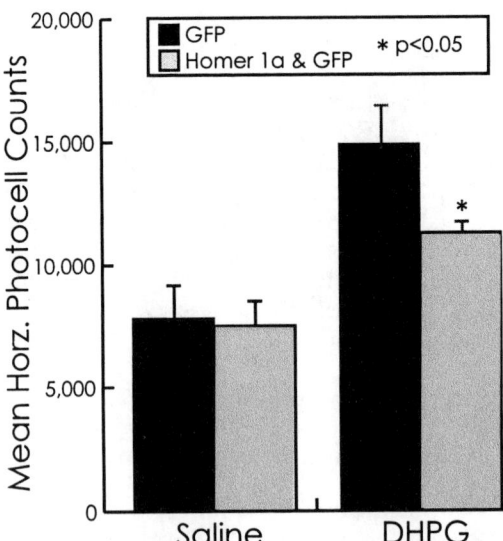

FIGURE 2. Group I mGluR stimulation: Homer 1a overexpression in the NA significantly blunts locomotion induced by specific Group I stimulation with DHPG.

diated locomotion over GFP controls. This lack of behavioral modification after repeated exposure to cocaine fits with the hypothesized role of IEG-like products, such as Homer 1a, and supports data obtained from Homer knockdown work with rats and with Homer knockout mice in our laboratory. These works illustrate the utility of adenovirus in the study of neuronal plasticity.

ACKNOWLEDGMENTS

This work was supported by NIDA Grants DA12514, MH40817, and DA03096 (P.W.K).

REFERENCES

1. CORNISH, J.L. & P.W. KALIVAS. 2001. Cocaine sensitization and craving: Differing roles for dopamine and glutamate in the nucleus accumbens. J. Addic. Dis. **20**: 43–54.
2. MCGEEHAN, A.J. & M.F. OLIVE. 2003. The mGluR5 antagonist MPEP reduces the conditioned rewarding effect of cocaine but not other drugs of abuse. Synapse **47**: 240–242.
3. CHIAMULERA, C., M.P. EPPING-JORDAN, A. ZOCCHI, et al. Reinforcing and locomotor stimulating effects of cocaine are absent In mGluR5 null mutant mice. Nat. Neurosci. **4**: 873–874.
4. SWANSON, C.J., D.A. BAKER, D. CARSON, et al. 2001. Repeated cocaine administration attenuates group I metabotropic glutamate receptor-mediated glutamate release and behavioral activation: a potential role for homer. J. Neurosci. **21**: 9043–9052.
5. TU, J.C., B. XIAO, J.P. YUAN, et al. 1998. Homer binds a novel proline-rich motif and links group I metabotropic glutamate receptors with IP3 receptors. Neuron **21**: 717–725.

Glutamate/Monoamine Interactions in the Limbic Thalamus

ANTONIETA LAVIN

Department of Physiology, Medical University of South Carolina, Charleston, South Carolina 29425, USA

KEYWORDS: dopamine; glutamate; thalamic activity

The dorsal thalamus, the prefrontal cortex (PFC), and the forebrain dopamine (DA) system each play a major role in cognitive processes and emotional states. However, despite strong clinical and behavioral evidence for the importance of the thalamocortical circuit and the mesocortical DA system for cognitive function, relatively little is known about the dopaminergic contributions to similar processes in the thalamus. A dopaminergic innervation to the dorsal thalamus has been shown[1,2] and although this innervation is not as extensive as the innervation to the striatum, the presence of DAergic receptors in the dorsal thalamus has been reported.[3,4] Moreover, the midline thalamic nuclei send projections to the prelimbic and infralimbic cortices, as well as to the nucleus accumbens, the extended amygdala, and the hippocampus, thus giving the midline thalamic nuclei broad control over cognitive and emotional processes. Despite strong clinical and behavioral evidence for the importance of the thalamocortical circuit and the mesocortical DA system for normal cognitive functioning and the implications of some of the deficits associated with several psychiatric disorders, relatively little is known about the effects of DA on thalamic neurons or the possible interactions between DA and glutamate in the limbic thalamus. By using whole-cell clamp recordings *in vitro,* we intended to address the question of whether dopamine plays a role in modulating thalamic activity.

In vitro whole cell clamp recordings were performed in slices containing the dorsal thalamus. Young male rats (18–25 days postnatal) were deeply anesthetized with an intraperitoneal injection of chloral hydrate (400 mg/kg) and perfused transcardially with oxygenated ice-cold sucrose-enriched aCSF saturated with 95%:5% O_2:CO_2. Brains were quickly removed and 4-mm sagittal blocks were cut. The blocks were sectioned in ice-cold physiological saline using a Vibratome into 350-µm slices and contained the limbic thalamus. Sections were incubated for 1.5 h at room temperature in continuously oxygenated physiological saline before placing them in a submersion-type chamber superfused with oxygenated physiological saline delivered at a flow rate of 1–2 mL/min with a peristaltic pump.

Address for correspondence: Antonieta Lavin, Department of Physiology, Medical University of South Carolina, Charleston, SC 29425. Voice: 843-792-6799; fax: 843-792-4423. lavina@musc.edu

Ann. N.Y. Acad. Sci. 1003: 422–425 (2003). © 2003 New York Academy of Sciences.
doi: 10.1196/annals.1300.042

Whole-cell recordings were performed at room temperature. Electrodes were filled with a solution containing 130 mM K$^+$ gluconate, 10 mM KCl, 10 mM HEPES, 2 mM MgCl$_2$, 3 mM Na$_2$ATP, 0.3 mM GTP, 2 mM EGTA pH 7.2–7.3, sometimes adding 1.5% biocytin. Using infrared differential interference contrast (IR-DIC) microscopy (Leica, DMS LFS microscope), thalamic neurons were visualized. DA and other drugs were either bath applied or locally applied close to the soma or the apical dendrites by a puffer pipette. Custom-made software was used for data collection, storage, and analysis.

With regard to thalamocortical co-cultures in P2–P6 rats, 350 µm sections containing the thalamus or the prelimbic and infralimbic regions of the PFC were obtained using a Vibratome with tissue submerged in oxygenated sucrose-enriched aCSF. Slices were placed next to each other on a Millipore millicell insert in a 6-well culture dish. The plating of the media contained: 50% basal medium Eagle, 25% Earle's balanced salt solution, 25% for the first 3 days of 5% horse serum 6.5 mg/mL glucose, 25 mM HEPES-NaOH (pH 7.2), and streptomycin. Solutions were exchanged every day and recordings were performed after 14 days.

Bath administration of DA (10 µM) in acute slices elicited an increase in intrinsic excitability in 4 of 7 (57%) thalamic neurons recorded in aCSF (control = 1.8±1.1 spikes, DA = 2.9±0.8 spikes, $P<.07$). When recordings were performed in aCSF + CNQX (AMPA blocker, 10 µM), clamping the membrane potential (MP) at −50 mV, DA elicited a significant increase in intrinsic excitability in 6 of 16 (37.5%) of the neurons recorded (control = 1.5±1.5 spikes, DA = 3.6±1.5, $P<.0012$, FIG. 1, A and B). The increase in excitability observed in cells held to −50 mV seems to include a decrease in anomalous rectification.

In additional recordings, NMDA receptors were stimulated by puffing NMDA (5 µM) into the thalamic cell recorded while the slice was perfused with CNQX (10 µM). In those cases, NMDA, as expected, increased cell excitability, however, the addition of DA (10 µM) to the perfusate produced and additional enhancement of the cell excitability (FIG. 1, C and D). In a small group of thalamic neurons, the blocking of NMDA receptors by APV (50 µM) seems to affect the ability of DA to increase cell excitability (FIG. 1, E and F).

When whole-cell recordings were performed in thalamic-PFC co-cultures, bath perfusion of DA (10 µM) increases the duration and amplitude of the PFC-evoked response (FIG. 1, G).

These results indicate that DA can modulate intrinsic and synaptic activity in the dorsal thalamus and that the effects of DA in thalamus may be mediated through NMDA receptors. The present findings suggest that drugs that affect DAergic activity, such as antipsychotics and psychostimulants, could also be affecting thalamic activity, and, given the crucial role that the thalamus plays in setting thalamocortical rhythms, the study of the role of DA and monoamines in modulating thalamocortical interactions is extremely important.

ACKNOWLEDGMENT

This work was supported by National Institutes of Health Grant DA 14698.

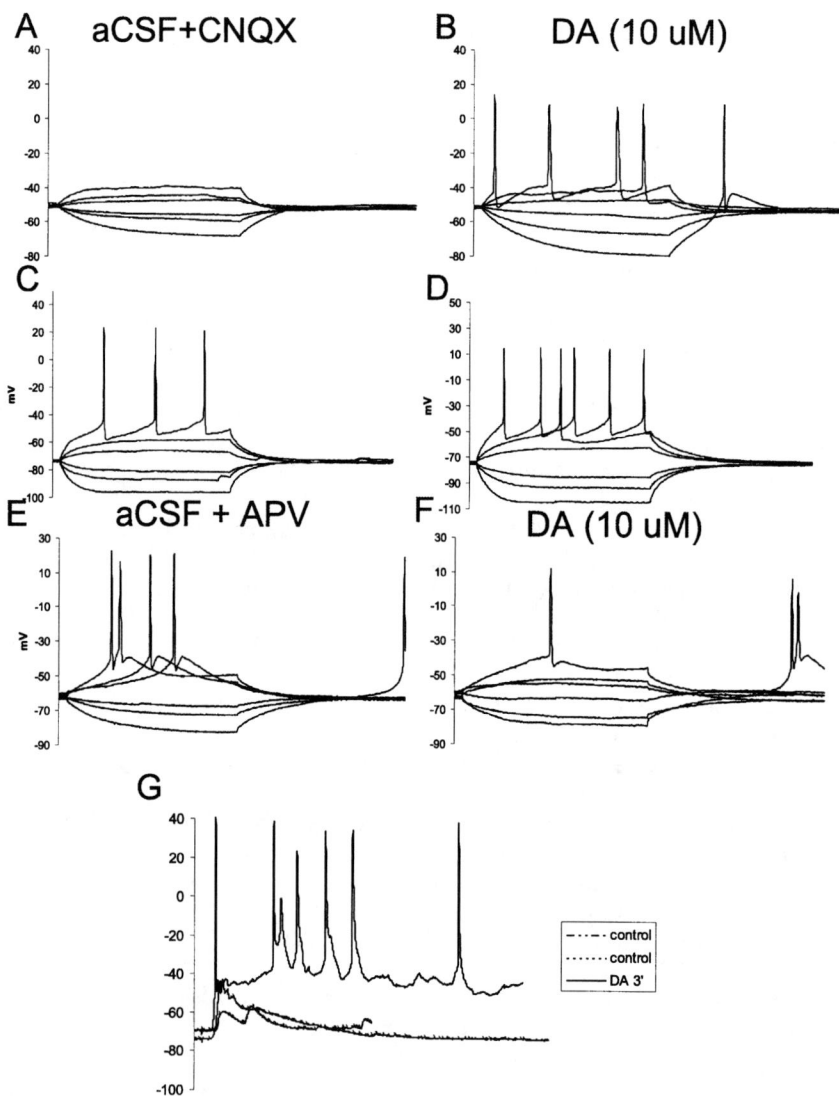

FIGURE 1. Dopamine modulates intrinsic and synaptic activity in limbic thalamus. (**A**) Whole-cell recording of a thalamic neuron in control conditions. The perfusate contained CNQX (10 µM) and the neuron was clamped at –50 mV. (**B**) Following perfusion with DA (10 µM) there is an increase in intrinsic excitability. (**C**) When NMDA was puffed into a thalamic neuron, action potentials were evoked. (**D**) DA administration enhanced the intrinsic excitability produced by a puff of NMDA. (**E**) Baseline recording of a thalamic neuron perfusated with the NMDA channel blocker APV (50 µM). (**F**) Administration of DA (10 µM) did not increase thalamic excitability when NMDA channels are blocked. (**G**) DA (10 µM) increases synaptic excitability in thalamic-PFC co-cultures.

REFERENCES

1. BECKSTEAD, R.M. 1979. An autoradiographic examination of cortico-cortical and subcortical projections of the mediodorsal projection (prefrontal) cortex in the rat. J. Comp. Neurol. **184:** 43–62.
2. GROENEWEGEN, H. 1988. Organization of the afferent connections of the mediodorsal thalamic nucleus in the rat, related to the mediodorsal-prefrontal topography. Neuroscience **24:** 379–431.
3. MANSOUR, A., J.H. MEADOR-WOODRUFF, J.R. BUNZOW, *et al.* 1992. A comparison of D1 receptor binding and mRNA in rat brain using receptors autoradiography and in situ hybridization techniques. Neuroscience **46:** 959–971.
4. SEDVALL, G. & L. FRADE. 1996. Dopamine receptors in schizophrenia. Lancet **347:** 264.

Changes in NMDA Receptor Subunit mRNAs and Cyclophilin mRNA during Development of the Human Hippocampus

AMANDA J. LAW, CYNTHIA SHANNON WEICKERT,[a] MAREE J. WEBSTER,[b] MARY M. HERMAN,[a] JOEL E. KLEINMAN,[a] AND PAUL J. HARRISON

Department of Psychiatry, University of Oxford, OX3 7JX, United Kingdom

[a]*Clinical Brain Disorders Branch, National Institute of Mental Health, National Institutes of Health, Bethesda, Maryland 20892-1385, USA*

[b]*Stanley Laboratory of Brain Research, Uniformed Services University for the Health Sciences, Bethesda, Maryland 20814-4799, USA*

KEYWORDS: glutamate receptor; cyclophilin; neonate; hippocampus; *in situ* hybridization; ontogeny

INTRODUCTION

The N-methyl-D-aspartate receptor (NMDAR) is critically involved in early neuronal development, synaptogenesis, and neurotoxicity.[1] Functional NMDARs are heteromeric assemblies of the NR1 and NR2(A–D) subunits, and the subunit composition in part determines the pharmacological properties. In the developing rat brain, changes in subunit expression, assembly, and function are seen in the early postnatal period.[2,3] In comparison, little is known about NMDAR subunit expression during human brain development.

NMDAR SUBUNIT GENE EXPRESSION IN THE DEVELOPING HUMAN HIPPOCAMPUS

Brain tissue was obtained from 34 individuals at five stages of life (TABLE 1). NR1, NR2A and NR2B subunit mRNA abundance was determined in the hippocampus and adjacent temporal cortex, along with that of the housekeeping gene cyclophilin. Quantitative radioactive *in situ* hybridization was performed using oligonucleotide (NR1, NR2A, and NR2B) and cRNA probes (cyclophilin).

Address for correspondence: Dr. Amanda J. Law, Department of Psychiatry, University of Oxford, Neurosciences Building, Warneford Hospital, Oxford, United Kingdom OX3 7JX. Voice: +44-1865-223784; fax: +44-1865-251076.
amanda.law@psych.ox.ac.uk

TABLE 1. Deomographic information about subjects studied

	Age	Number	Sex	PMI	pH
Neonate	5–20 wk	6	4 F, 2 M	43.1	6.5
Infant	22–51 wk	4	2 F, 2 M	41.8	6.6
Adolescent	14–18 yr	9	9 M	32.2	6.4
Young adult	21–24 yr	6	6 M	30.1	6.4
Adult	34–55 yr	9	9 M	28.8	6.2

TABLE 2. NMDAR subunit mRNA epxression in the human hippocampus

	Neonate	Infant	Adolescent	Young Adult	Adult
Dentate gyrus					
NR1	45 (13)	71 (4)	65 (5)	65 (55)	63 (4)
NR2A	85 (7)	116 (16)	108 (9)	87 (6)	94 (11)
NR2B[a]	205 (22)	135 (26)[1]	140 (14)[2]	155 (12)	117 (16)[3]
CA4					
NR1[b]	23 (2)	50 (4)[1]	48 (5)[3]	44 (5)[2]	44 (5)[2]
NR2A	55 (4)	61 (14)	75 (7)	54 (5)	58 (10)
NR2B	61 (10)	46 (9)	41 (6)	44 (9)	42 (8)
CA3					
NR1[c]	22 (4)	57 (4)[1]	60 (7)[3]	54 (9)[1]	54 (6)[3]
NR2A	62 (6)	63 (14)	97 (8)	81 (8)	88 (16)
NR2B	77 (15)	86 (19)	66 (11)	72 (13)	57 (10)
CA2					
NR1[d]	29 (6)	53 (4)[1]	47 (4)[1]	59 (9)[3]	55 (4)[3]
NR2A	55 (2)	45 (4)	71 (8)	65 (7)	74 (12)
NR2B	77 (13)	87 (31)	67 (9)	61 (9)	58 (10)
CA1					
NR1	33 (9)	59 (6)	44 (4)	46 (5)	47 (3)
NR2A	64 (4)	66 (12)	61 (6)	52 (7)	46 (5)
NR2B	80 (9)	82 (21)	49 (9)	55 (9)	49 (7)
Subiculum					
NR1	32 (5)	50 (3)	44 (7)	44 (2)	46 (3)
NR2A	63 (6)	55 (6)	58 (6)	47 (7)	55 (5)
NR2B[e]	93 (15)	49 (7)[3]	37 (5)[3]	39 (7)[3]	52 (8)[3]
PHG					
NR1	38 (10)	56 (3)	55 (7)	40 (5)	54 (4)
NR2A	53 (8)	47 (7)	60 (8)	52 (10)	56 (7)
NR2B[f]	85 (20)	71 (10)	32 (7)[3]	33 (7)[3]	47 (8)[1]
PRC					
NR1	38 (7)	52 (5)	56 (4)	40 (4)	42 (5)
NR2A	61 (11)	97 (9)	93 (10)	88 (6)	81 (9)
NR2B	66 (11)	76 (13)	71 (11)	57 (3)	56 (7)

Values are mean ^{35}S nCi/g (SEM). ANOVA, effect of age group: [a]$P=.02$; [b]$P=.007$; [c]$P=.01$; [d]$P=.02$; [e]$P=.001$; [f]$P=.005$. Post-hoc (LSD test) comparisons between neonates and other age groups, [1]$P<.05$, [2]$P<.01$, [3]$P<.005$. Abbreviations: PHG, parahippocampal gyrus; PRC, perirhinal cortex.

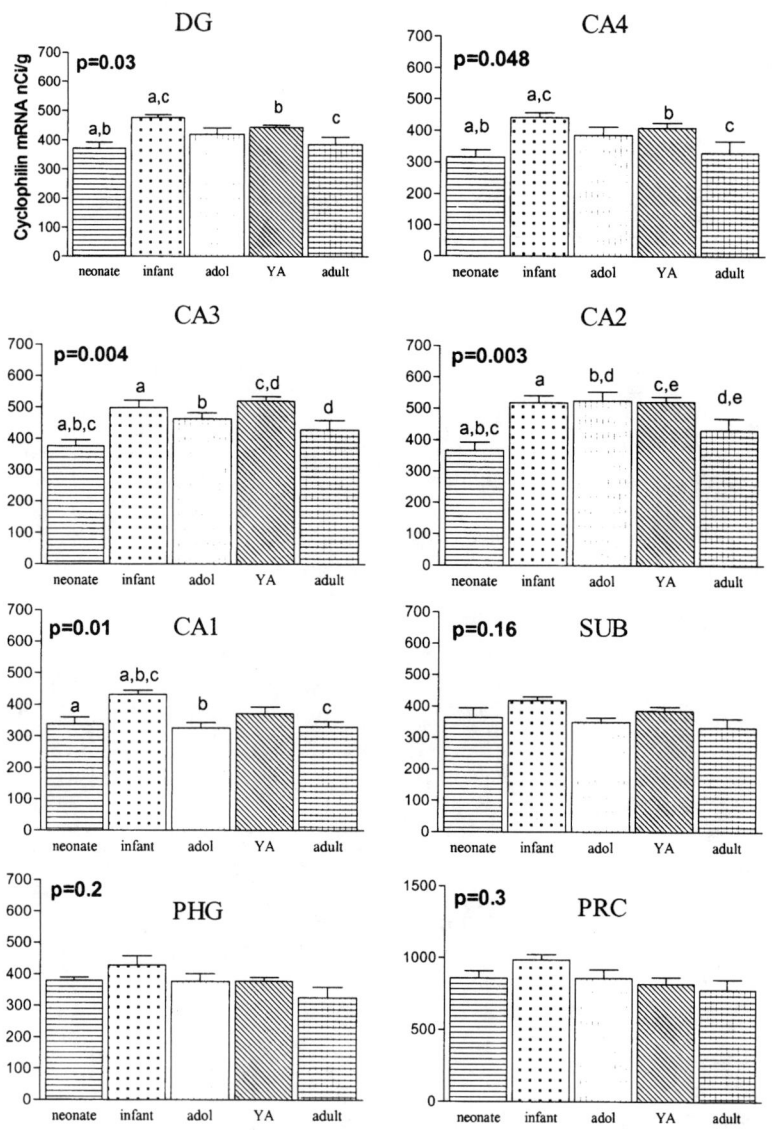

FIGURE 1. Expression of cyclophilin mRNA in the human hippocampus during development. *P* values represent overall ANOVA. Letters (a,b,c,d) depict groups significantly different from each other ($P < .05$). Abbreviations: DG, dentate gyrus; SUB, subiculum; PHG, parahippocampal gyrus; PRC, perirhinal cortex.

DEVELOPMENTAL PROFILE OF HUMAN HIPPOCAMPAL NMDAR SUBUNIT mRNAs

TABLE 2 shows the expression of the three NMDAR subunit mRNAs in each age group. NR1 mRNA was lower in the neonates compared to all other age groups in CA4, CA3, and CA2. NR2B mRNA levels were higher in the neonate compared to the older age groups in the dentate gyrus, subiculum, and parahippocampal gyrus. Conversely, NR2A mRNA levels remained constant throughout the developmental time points surveyed. These phenomena lead to an age-related increase in the NR2A/2B transcript ratio, as previously seen in the rat.[3] Postmortem interval, brain pH, and brain weight did not correlate with transcript abundance (not shown).

DEVELOPMENTAL PROFILE OF HUMAN HIPPOCAMPAL CYCLOPHILIN mRNA

A complex pattern of age-related changes in the abundance of cyclophilin mRNA was observed (FIG. 1). Less cyclophilin mRNA was found in the neonate compared to the infant, adolescent, and young adults groups, suggesting that the weaker NR1 mRNA signal in the neonates compared to older age groups is a nonspecific finding. However, when each subject's NR1 mRNA was normalized to his or her cyclophilin mRNA, the finding of lower neonatal NR1 expression remained. Furthermore, other aspects of the data emphasize the differential changes of NR subunit mRNAs. For example, in contrast to NR1 mRNA, levels of cyclophilin mRNA were similar in the neonate and the adult groups and, notably, NR2B mRNA was highest in the neonate group, in contrast to cyclophilin mRNA.

CONCLUSION

NMDAR subunit mRNAs undergo age-related changes during human hippocampal development. As in the rat, there is a postnatal increase in NR1 mRNA and decrease in NR2B mRNA; unlike the rat, there is no increase in NR2A mRNA.[2,3] These findings are of interest with regard to the role of NMDARs during human brain development and to the pathophysiology of neurodevelopmental disorders, including schizophrenia. The different maturational profile of cyclophilin mRNA may be indicative of developmental shifts in overall levels of gene expression.

ACKNOWLEDGMENTS

This work was supported by a Wellcome Trust Biomedical Collaboration grant and the Stanley Medical Research Institute.

REFERENCES

1. CULL-CANDY, S., S. BRICKLEY & M. FARRANT. 2001. NMDA receptor subunits: diversity, development and disease. Curr. Opin. Neurobiol. **11:** 327–335.

2. MONYER, H., N. BURNASHEV, D.J. LAURIE, *et al.* 1994. Developmental and regional expression in the rat brain and functional properties of four NMDA receptors. Neuron **12:** 529–540.
3. ZHONG, J., D.P. CARROZZA, K. WILLIAMS, *et al.* 1995. Expression of mRNAs encoding subunits of the NMDA receptor in developing rat brain. J. Neurochem. **64:** 531–539.

Blockade of the GlyT1 Glycine Transporter Prolongs Response to VTA Stimulation in Nucleus Accumbens Neurons

BARBARA LEWIS AND PATRICIO O'DONNELL

Center for Neuropharmacology and Neuroscience, Albany Medical College, Albany, New York 12208, USA

KEYWORDS: GlyT1; schizophrenia; NMDA receptors

INTRODUCTION

Extensive evidence indicates that a glutamatergic deficit may be a significant component of the pathophysiology of schizophrenia.[1] NMDA receptors may play a crucial role in this dysfunction as NMDA receptor antagonists produce cognitive and behavioral deficits that mimic those of schizophrenia.[2] NMDA receptor function is controlled by both the glutamate binding site on the NR2 subunit and glycine binding sites of the NR1 subunit, and both are required for activation.[3] It has been suggested that the glycine site does not remain saturated *in vivo*.[4] Therefore increasing glycine by blocking its uptake into astrocytes may facilitate NMDA function. This uptake is mediated by the high-affinity glycine transporter GlyT1. The use of the potent inhibitor JRF/NPS 1000 (Janssen and NPS/Allelix) has been shown to increase extracellular glycine levels.[4] Here, we investigated the effects of increasing glycine levels by blockade of the Gly T1 glycine transporter on an NMDA-dependent response in the nucleus accumbens *in vivo*.

Simultaneous intracellular and extracellular recordings have revealed that extracellular local field potentials in the nucleus accumbens correlate with glutamate-driven "up" states.[5] Up states in the nucleus accumbens are sensitive to administration of NMDA antagonists;[6] thus, NMDA is an important component in these depolarizations, which can also be detected with field potentials. Ventral tegmental area (VTA) stimulation evokes local field potentials resembling up states, which are reduced but not eliminated by combined administration of D_1 and D_2 antagonists,[5] suggesting that a glutamatergic component is responsible for the initial depolarization and dopamine contributes to sustain the depolarization in the nucleus accumbens. Thus, in order to test the effects of GlyT1 blockade on NMDA responses *in vivo*, VTA–evoked local field potentials were tested before and after applying the GlyT1 blocker.

Address for correspondence: Patricio O'Donnell, Center for Neuropharmacology and Neuroscience, Albany Medical College, Albany, NY 12208. Voice: 518-262-5904; fax: 518-262-0687.
odonnep@mail.amc.edu

METHODS

Field potential recordings were conducted in 25 Sprague Dawley adult male rats (300–440 g). Animals were anesthetized with chloral hydrate (400 mg/kg, i.p.) and placed in a stereotaxic apparatus. Anesthesia was maintained during recording with continuous delivery through an intraperitoneal cannula. Concentric 0.5-mm bipolar stimulating electrodes (David Kopf) were placed in the VTA (5.8 mm caudal to bregma; 0.5 mm lateral; 8.3 mm ventral to brain surface). Recording electrodes were made with fine tungsten wire (125 µm; A-M Systems) and lowered into the nucleus accumbens (bregma +1.6, lateral: 1.5 mm, −7.0 to −7.2 mm from brain surface) inside a glass capillary tube with 7 mm or less of the wire exposed. Signals were acquired unfiltered and sampled at 10 KHz. Custom-made analysis software

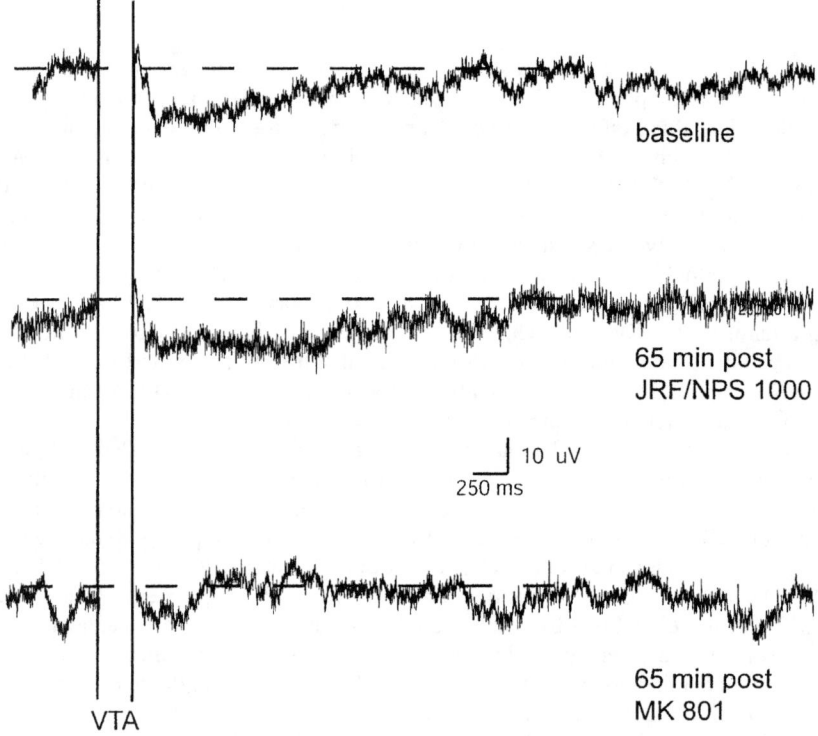

FIGURE 1. JRF/NPS 1000 (5 mg/kg, i.p.) prolonged VTA stimulation-evoked local field potentials in accumbens neurons. This effect was reversed by systemic administration of MK-801. (*Top*) Field potential response evoked by VTA stimulation. The vertical lines over all three traces indicate onset and end of VTA trains of pulses (stimulus artifacts were eliminated). The *horizontal dashed lines* indicate the zero potential value. (*Middle*) Field potential obtained in the same site following GlyT1 blocker administration, showing a prolonged duration of the negative deflection. (*Bottom*) Field potential evoked after administration of the NMDA antagonist MK-801.

(Neuroscope) was used to filter the waveform (0.1–10 KHz) for analysis. Recordings were performed for 5–10 minutes to establish baseline activity prior to VTA stimulation. Electrical stimulation was delivered as a train of 5 pulses at 20 Hz to mimic dopamine cell burst firing. Baseline recordings and VTA stimulation were repeated following systemic administration of the Gly T1 inhibitor NPS 1000 (5 mg/kg) via an intraperitoneal cannula. In some cases, the NMDA antagonist MK-801 was delivered systemically (0.1 mg/kg, i.p.) followed by baseline and stimulation recordings. Saline and vehicle injections were used for control. Brains were removed and postfixed in formalin and processed using standard histological techniques for verification of electrode placement.

RESULTS

Spontaneous negative shifts in local field potentials were recorded in the nucleus accumbens, as previously observed.[5] The frequency of these field potential shifts was 1.06 ± 0.13 Hz ($N = 8$). Following systemic administration of the GlyT1 inhibitor JRF/NPS 1000 (5 mg/kg, i.p.), the frequency of these events increased to 1.26 ± 0.17 Hz ($N = ; P < .01$, paired t test). This effect was seen 30 minutes following drug administration and persisted for at least 2 hours.

The VTA was stimulated with trains of 5 pulses at 20 Hz to mimic dopamine cell burst firing. This procedure had been shown to elicit persistent negative deflections in field potentials that matched persistent intracellular depolarizations resembling up states.[5] Following systemic administration of JRF/NPS 1000, VTA–evoked field potentials were significantly increased in duration, from 654.3 ± 283.7 ms to 1049.4 ± 512.6 ms ($P < .001$, paired t test), and this effect persisted for up to 90 minutes in three cases. Subsequent systemic administration of the NMDA antagonist MK-801 (0.1 mg/kg) reduced the duration of the response to VTA stimulation to 805 ± 465 ms 30 minutes following delivery, and to 732 ± 518 ms at 60 minutes following injection ($N = 7$). In some experiments, vehicle was administered systemically instead of the GlyT1 blocker. The responses to VTA stimulation remained constant. Electrode placements outside of the VTA failed to evoke a consistent field potential.

DISCUSSION

Blocking glycine uptake enhances spontaneous and VTA-evoked field potentials in the nucleus accumbens, an effect reversed by the NMDA antagonist MK-801. This indicates that increasing glycine levels by blocking its uptake enhances NMDA responses *in vivo*. The negative shifts in accumbens local field potentials evoked by VTA stimulation are glutamate-dependent events sustained by dopamine. They reflect membrane potential depolarizations (up states) in a population of neurons.[5] Indeed, up states in the nucleus accumbens can be eliminated with systemic administration of phencyclidine,[6] indicating they are glutamatergic in nature and exhibit a prominent NMDA component. The blockade of GlyT1 causing an increase in spontaneous negative field potentials is consistent with this concept. Thus, increasing glycine levels could produce an upregulation of NMDA activity in the nucleus accumbens; this action opens a promising new tool in the pharmacology of schizophrenia. It is possible that increasing NMDA function could contribute to improvements in the cognitive deficits in this disorder.

ACKNOWLEDGMENTS

This study was funded by U.S. Public Health Service Grants MH60131 and by Allelix/NPS and Janssen Pharmaceuticals.

REFERENCES

1. CARLSSON, M. & A. CARLSSON. 1990. Interactions between glutamatergic and monoaminergic systems within the basal ganglia—implications for schizophrenia and Parkinson's disease. Trends Neurosci. **13:** 272–276.
2. HERESCO-LEVY, U. & D.C. JAVITT. 1998. The role of N-methyl-D-aspartate (NMDA) receptor-mediated neurotransmission in the pathophysiology and therapeutics of psychiatric syndromes. Eur. Neuropsychopharmacol. **8:** 141–152.
3. KLECKNER, N.W. & R. DINGLEDINE. 1988. Requirement for glycine in activation of NMDA-receptors expressed in *Xenopus* oocytes. Science **241:** 835–837.
4. ATKINSON, B.N., S.C. BELL, M. DE VIVO, et al. 2001. ALX 5407: a potent, selective inhibitor of the hGlyT1 glycine transporter. Mol. Pharmacol. **60:** 1414–1420.
5. GOTO, Y & P. O'DONNELL. 2001. Network synchrony in the nucleus accumbens *in vivo*. J. Neurosci. **21:** 4498–4504.
6. O'DONNELL, P. & A.A. GRACE. 1998. Phencyclidine interferes with the hippocampal gating of nucleus accumbens neuronal activity *in vivo*. Neuroscience **87:** 823–830.

Modulation of Inhibitory Transmission in the Rat Globus Pallidus by Activation of mGluR4

MICHAEL J. MARINO, ORNELLA VALENTI, JULIE A. O'BRIEN,
DAVID L. WILLIAMS JR., AND P. JEFFREY CONN

Neuroscience Department, Merck Research Laboratories,
Merck & Company Inc., West Point, Pennsylvania 19486, USA

KEYWORDS: mGluR4; rat globus pallidus; Parkinson's disease

The striatopallidal synapse is a critical point in the cortico-basal ganglia-thalamic-loops that may have significance for a number of neurological and neuropsychiatric disorders, including schizophrenia and Parkinson's disease (PD). In the motor basal ganglia, the inhibitory GABAergic synapse between the striatal medium spiny neurons and the globus pallidus (GP) GABAergic output neurons represents the first synapse in the indirect pathway and is thought to be involved in the pathophysiology of Parkinson's disease. Recordings from Parkinsonian nonhuman primates reveal a marked increase in rhythmic oscillatory spike discharge in the external GP.[1,2] Consistent with this, recordings from human Parkinson's patients reveals similar abnormal firing patterns that appear to correlate with symptom severity and drug treatment.[3-5] In addition, several studies in rodent models lend support to the hypothesis that increased GABAergic input may underlie alterations in GP firing patterns. Increasing GABA concentrations in the GP has been demonstrated to have an akinetic effect.[6] Furthermore, the GABA A antagonist bicuculline produces an antiparkinsonian effect when injected into the GP.[7] Therefore, any manipulation that decreases striatopallidal transmission may act to restore a balance to the basal ganglia circuit and provide palliative benefit for the treatment of PD. This potential therapeutic benefit is further underscored by the recent studies suggesting antiparkinsonian actions of A2a adenosine receptor antagonists in both animal models,[8-11] and in human clinical trials.[12,13] A2a antagonists act, at least in part, by decreasing transmission at the striatopallidal synapse.[14] Therefore, directly decreasing transmission at this synapse, possibly through the activation of a presynaptic G-protein coupled receptor, may provide a novel approach for the treatment of PD.

In addition to the importance of the dorsal striatopallidal circuit in the control of motor behavior, a large body of literature implicates the ventral striatopallidal sys-

---

Address for correspondence: Michael J. Marino, Merck Research Laboratories, Merck & Co., Inc., 770 Sumneytown Pike, P. O. Box 4, WP 46-200, West Point, PA 19486-0004. Voice: 215-652-0852; fax: 215-652-3811.
  michael_marino@merck.com

tem in the pathophysiology of schizophrenia (for review see O'Donnell and Grace[15]). In particular, alterations in the dopamine system and subsequent effects in information processing by the ventral striatum have played a key role in the development of models of this disorder. The current hypothesis of ventral striatopallidal function states that the accumbens nucleus of the ventral striatum acts as an integrator of input from the prefrontal cortex, hippocampus, and amygdala, all under the neuromodulatory influence of the dopamine system. The processed information is next relayed by the GABAergic projection neurons of the nucleus accumbens (NAc) to the ventral pallidum (VP). The VP provides inhibitory projections to the thalamus, in particular to the reticular thalamic nucleus, which regulates thalamocortical activity. This cortical-ventral striatopallidal-thalamic loop provides a mechanism by which alterations in subcortical processes can produce widespread disruptions in cortical function. This is consistent with the global deficits in cognition and information processing observed in schizophrenic patients. While this model views the VP as a relay nucleus, it is now becoming clear that substantial information processing also takes place within this structure (for review see Napier and Metrovic[16]). Glutamatergic projections from the prefrontal cortex and the basolateral amygdala provide the main excitatory drive to the VP. This glutamatergic input is particularly interesting because in addition to providing a direct excitation by actions on ionotropic glutamate receptors, glutamate can modulate the VP by actions on metabotropic glutamate receptors (mGluRs).

The metabotropic glutamate receptors (mGluRs) are a family of eight G-coupled receptors divided into three groups according to pharmacology, coupling to signal transduction systems, and sequence homology. Our previous studies have found that mGluR4 is highly localized to striatal terminals within the globus pallidus.[17] This finding suggests that activation of mGluR4 could reduce inhibitory striatal transmission in this nucleus and therefore play a possible role in the processing of information in this pathway. We have performed whole-cell patch clamp recordings from neurons in the GP and measured striatopallidal GABA A–mediated inhibitory postsynaptic currents (IPSCs) evoked by stimulation of the striatum. Activation of group III mGluRs by the selective group III mGluR agonist L-AP4 produces a marked inhibition of transmission at the striatopallidal synapse. A biophysical analysis of this effect suggests that it is presynaptically mediated. This finding is consistent with the previous anatomical localization of mGluR4 at this synapse.[17] The group III mGluR-mediated inhibition of striatopallidal transmission has a pharmacology consistent with L-AP4 activation of mGluR4 and is absent in mGluR4 knockout mice. In an effort to develop more selective tools for the study of mGluR4 function, we have found that N-phenyl-7-(hydroxyimino)cyclopropa[b]chromen-1a-carboxamide, a novel group I mGluR antagonist,[18] acts as a positive allosteric modulator of mGluR4 and potentiates the effects of low concentrations of group III mGluR-selective agonist at inhibiting striatopallidal transmission. Taken together, these findings suggest that mGluR4 mediates an inhibition of transmission at the striatopallidal synapse and that native mGluR4 is amenable to modulation by compounds acting at allosteric sites. This raises the exciting possibility that allosteric modulators of mGluR4 may provide novel antiparkinsonian agents and could potentially provide tools to help determine how information is processed in both the dorsal and ventral striatopallidal circuits.

REFERENCES

1. NINI, A., A. FEINGOLD, H. SLOVIN & H. BERGMAN. 1995. Neurons in the globus pallidus do not show correlated activity in the normal monkey, but phase-locked oscillations appear in the MPTP model of parkinsonism. J. Neurophysiol. **74:** 1800–1805.
2. BERGMAN, H. et al. 1998. Physiological aspects of information processing in the basal ganglia of normal and parkinsonian primates. Trends Neurosci. **21:** 32–38.
3. LOZANO, A. et al. 1996. Methods for microelectrode-guided posteroventral pallidotomy. J. Neurosurg. **84:** 194–202.
4. MAGNIN, M., A. MOREL & D. JEANMONOD. 2000. Single-unit analysis of the pallidum, thalamus and subthalamic nucleus in parkinsonian patients. Neuroscience **96:** 549–564.
5. BROWN, P. et al. 2001. Dopamine dependency of oscillations between subthalamic nucleus and pallidum in Parkinson's disease. J. Neurosci. **21:** 1033–1038.
6. PYCOCK, C., R.W. HORTON & C.D. MARSDEN. 1976. The behavioural effects of manipulating GABA function in the globus pallidus. Brain Res. **116:** 353–359.
7. MANEUF, Y.P., I.J. MITCHELL, A.R. CROSSMAN & J.M. BROTCHIE. 1994. On the role of enkephalin cotransmission in the GABAergic striatal efferents to the globus pallidus. Exp. Neurol. **125:** 65–71.
8. GRONDIN, R. et al. 1999. Antiparkinsonian effect of a new selective adenosine A2A receptor antagonist in MPTP-treated monkeys. Neurology **52:** 1673–1677.
9. KANDA, T. et al. 2000. Combined use of the adenosine A(2A) antagonist KW-6002 with L-DOPA or with selective D1 or D2 dopamine agonists increases antiparkinsonian activity but not dyskinesia in MPTP-treated monkeys. Exp. Neurol. **162:** 321–327.
10. KOGA, K., M. KUROKAWA, M. OCHI, et al. 2000. Adenosine A (2A) receptor antagonists KF17837 and KW-6002 potentiate rotation induced by dopaminergic drugs in hemi-Parkinsonian rats. Eur. J. Pharmacol. **408:** 249–255.
11. SHIOZAKI, S. et al. 1999. Actions of adenosine A2A receptor antagonist KW-6002 on drug-induced catalepsy and hypokinesia caused by reserpine or MPTP. Psychopharmacology (Berl.) **147:** 90–95.
12. SHERZAI, A. et al. 2002. Adenosine A2a antagonist treatment of Parkinson's disease. American Academy of Neurology Abstracts P06.104.
13. HUBBLE, J.P. & R. HAUSER. 2002. A novel adenosine antagonist (KW-6002) as a treatment for advanced Parkinson's disease with motor complications. American Academy of Neurology Abstracts, S21.001.
14. SHINDOU, T., A. MORI, H. KASE & M. ICHIMURA. 2001. Adenosine A(2A) receptor enhances GABA(A)-mediated IPSCs in the rat globus pallidus. J. Physiol. **532:** 423–434.
15. O'DONNELL, P. & A.A. GRACE. 1998. Dysfunctions in multiple interrelated systems as the neurobiological bases of schizophrenic symptom clusters. Schizophr. Bull. **24:** 267–283.
16. NAPIER, T.C. & I. MITROVIC. 1999. Opioid modulation of ventral pallidal inputs. Ann. N.Y. Acad. Sci. **877:** 176–201.
17. BRADLEY, S.R. et al. 1999. Immunohistochemical localization of subtype 4a metabotropic glutamate receptors in the rat and mouse basal ganglia. J. Comp. Neurol. **407:** 33–46.
18. ANNOURA, H., A. FUKUNAGA, M. UESUGI, et al. 1996. A novel class of antagonists for metabotropic glutamate receptors, 7-(hydroxyimino)cyclopropchromen-1a-carboxylates. Bioorg. Med. Chem. Lett. **6:** 763–766.

Expression of Transcripts for the Vesicular Glutamate Transporters in the Human Medial Temporal Lobe

ROBERT E. McCULLUMSMITH AND JAMES H. MEADOR-WOODRUFF

Department of Psychiatry and Mental Health Research Institute, University of Michigan Medical School, Ann Arbor, Michigan 48109-0720, USA

KEYWORDS: vesicular glutamate transporters; medial temporal lobe; *in situ* hybridization

Originally characterized as plasma membrane inorganic phosphate transporters, the vesicular glutamate transporters (vGluts) facilitate proton-dependent uptake of glutamate into presynaptic vesicles.[1–3] In contrast to the plasma membrane-localized excitatory amino acid transporters, vGlut1 (BNPi) and vGlut2 (DNPi) do not transport aspartate or rely on a Na^+ gradient for glutamate transport.[1] vGlut2, but not vGlut1, transcript expression is widespread in subcortical structures, while both are expressed in the cortex and striatum.[2–4] Interestingly, a third vesicular transporter, vGlut3, was recently characterized and reportedly expressed in serotonergic and cholinergic neurons.[5–7] Differential patterns of distribution of vGlut mRNAs likely define subtypes of glutamatergic neurons, potentially yielding novel information regarding glutamatergic circuitry. For this study, we utilized *in situ* hybridization to characterize transcript expression of vGlut 1, 2, and 3 in medial temporal lobe structures to determine if subpopulations of glutamatergic neurons exist in this brain region.

Human brain sections, provided by the Stanley Foundation Neuropathology Consortium, were obtained at autopsy and 14 μm sections were prepared as previously described.[8] To generate subclones we amplified unique regions of vGlut1 (NCBI Genebank ascension number AB032436, nucleotide coding region 953-1481), vGlut2 (AB032435, 1337-1843), and vGlut3 (NM_139319, 826-1673) from a human cDNA brain library using PCR. Amplified cDNA segments were extracted, subcloned and confirmed by nucleotide sequencing. Riboprobes were synthesized from linearized plasmid DNA and *in situ* hybridization was performed as previously described.[9,10] Radiolabeled sections were dipped in NTB-2 emulsion, incubated for 2–12 weeks, developed, and counterstained as previously described.[11]

Address for correspondence: Robert E. McCullumsmith, M.D., Ph.D., Mental Health Research Institute and Department of Psychiatry, University of Michigan, 205 Zina Pitcher Place, Ann Arbor, MI 48109-0720. Voice: 734-936-2061; fax: 734-647-4130.
smithrob@umich.edu

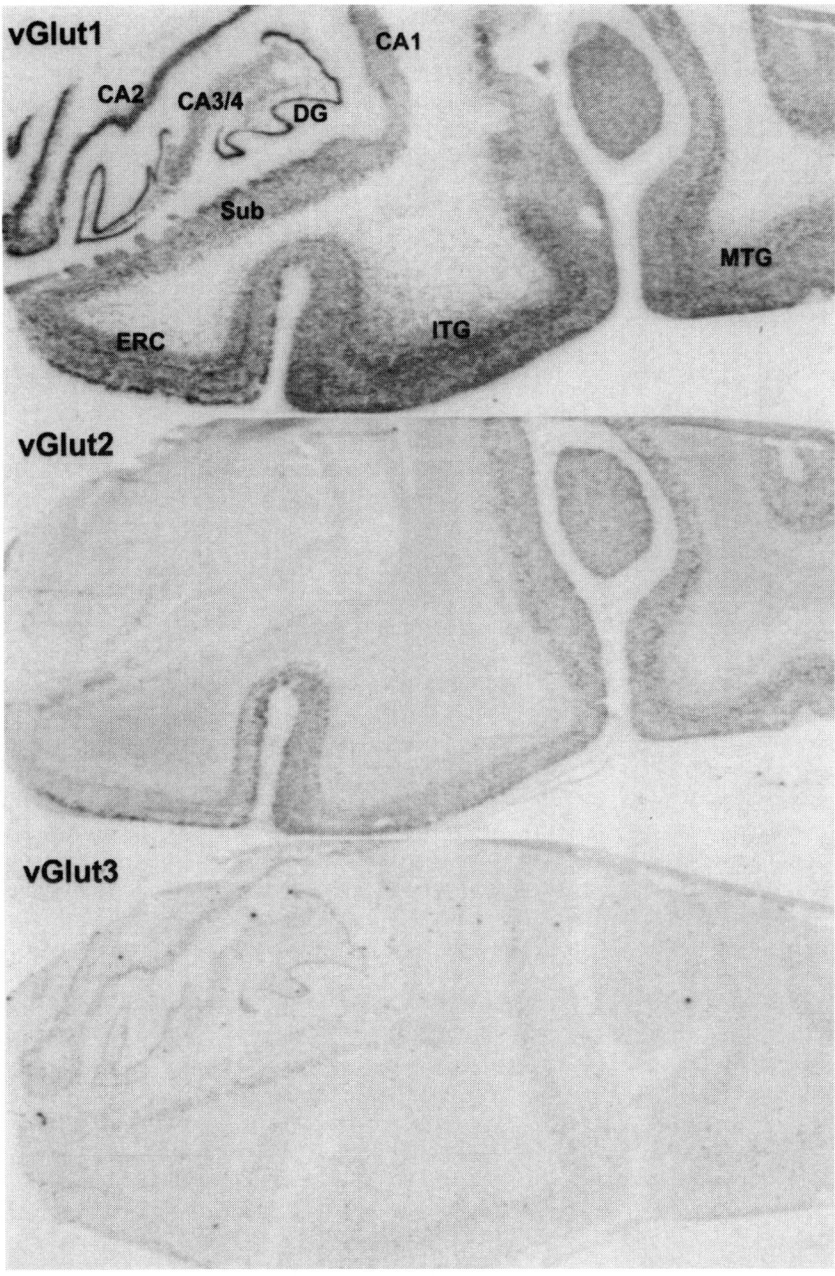

FIGURE 1. *In situ* hybridization with [^{35}S]-riboprobes demonstrating vGlut 1, 2, and 3 transcript expression in the human medial temporal lobe. Abbreviations: vesicular glutamate transporter (vGlut), dentate gyrus (DG), subiculum (Sub), entorhinal cortex (ERC), inferior temporal gyrus (ITG), middle temporal gyrus (MTG).

vGlut1 transcript expression was apparent after a 3-day exposure of [^{35}S]-radiolabeled sections exposed to film, versus a 6-week exposure for vGlut2 and vGlut3. We detected vGlut1 and vGlut3 transcript expression in the hippocampal subfields CA1, CA2, CA3, and CA4, the dentate gyrus (DG), the subiculum (Sub), the entorhinal cortex (ERC), the inferior temporal gyrus (ITG), and the middle temporal gyrus (MTG) (FIG. 1). vGlut2 transcripts were detected in DG, Sub, ERC, ITG,

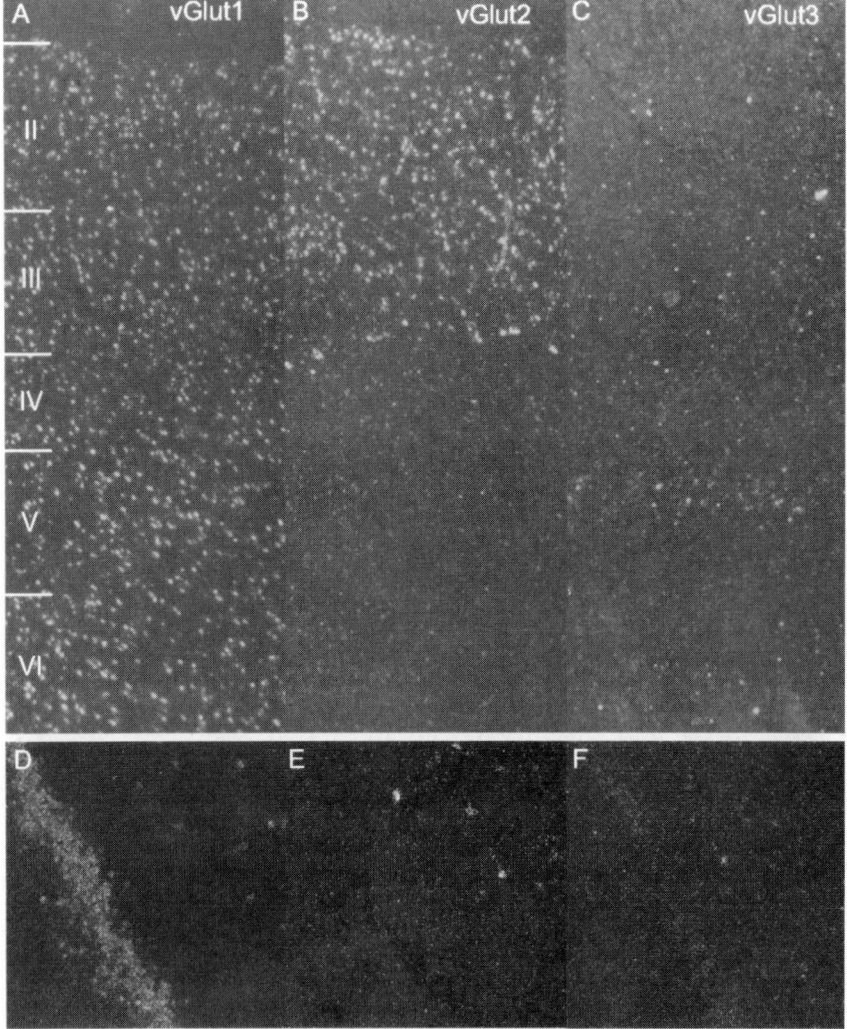

FIGURE 2. Darkfield microscopy of [^{35}S]-radiolabeled vGlut1 (**A, D**), vGlut2 (**B, E**), and vGlut3 (**C, F**) transcripts in emulsion dipped slides. Cortical layers of the inferior temporal gyrus (**A–C**) and identification of the dentate gyrus (**D–F**) were confirmed by Nissl counterstaining.

MTG and also in scattered neurons of CA4 (FIGS. 1 and 2). Cortical vGlut1 transcripts were differentially expressed in layers II-VI of the ERC, ITG, and MTG, while vGlut2 expression was primarily restricted to the superficial pyramidal cell layers (II and III) in these same cortical regions (FIGS. 1 and 2). Cortical vGlut3 expression was primarily limited to layer V in the ITG and MTG, and layer II in the ERC (FIGS. 1 and 2). The level of expression of vGlut 1, 2, and 3 mRNAs in the DG exemplifies the differential distribution of transcripts for these genes in the medial temporal lobe (FIG. 2, D–F).

We have detected unique patterns of expression of vGlut 1, 2, and 3 transcripts in the medial temporal lobe, consistent with previous reports demonstrating expression of vGlut1 in the cortex and hippocampus, and vGlut2 in the cortex.[2–4] Interestingly, vGlut2 transcript expression was restricted to superficial pyramidal cell layers in the cortex, suggesting an association with glutamatergic neurons in layers II and III that primarily form cortico–cortical connections. In contrast, vGlut1 was expressed in both superficial and deeper cortical layers. Evaluation of emulsion-dipped sections revealed that expression of vGlut3 mRNA in the cortex is primarily in large pyramidal neurons and is restricted in large part to layer V of the ITG and MTG, suggesting an association with glutamatergic neurons that project to subcortical structures. These findings support the hypothesis that the vGluts probably define subpopulations of glutamatergic neurons and are likely associated with specific efferent pathways and circuits.

ACKNOWLEDGMENTS

This work was supported by a Pfizer Postdoctoral Fellowship (R.E.M.) and MH53327 (J.M.W.).

REFERENCES

1. BELLOCCHIO, E.E., et al. 2000. Uptake of glutamate into synaptic vesicles by an inorganic phosphate transporter. Science **289:** 957–960.
2. FREMEAU, R.T., JR., et al. 2001. The expression of vesicular glutamate transporters defines two classes of excitatory synapse. Neuron **31:** 247–260.
3. FUJIYAMA, F., T. FURUTA & T. KANEKO. 2001. Immunocytochemical localization of candidates for vesicular glutamate transporters in the rat cerebral cortex. J. Comp. Neurol. **435:** 379–387.
4. BELLOCCHIO, E.E., et al. 1998. The localization of the brain-specific inorganic phosphate transporter suggests a specific presynaptic role in glutamatergic transmission. J. Neurosci. **18:** 8648–8659.
5. FREMEAU, R.T., JR., et al. 2002. The identification of vesicular glutamate transporter 3 suggests novel modes of signaling by glutamate. Proc. Natl. Acad. Sci. USA **99:** 14488–14493.
6. TAKAMORI, S. et al. 2002. Molecular cloning and functional characterization of human vesicular glutamate transporter 3. EMBO Rep. **3:** 798–803.
7. GRAS, C. et al. 2002. A third vesicular glutamate transporter expressed by cholinergic and serotoninergic neurons. J. Neurosci. **22:** 5442–5451.
8. TORREY, E.F. et al. 2000. The Stanley Foundation brain collection and neuropathology consortium. Schizophr. Res. **44:** 151–155.
9. SMITH, R.E. et al. 2001. Vesicular glutamate transporter transcript expression in the thalamus in schizophrenia. Neuroreport **12:** 2885–2887.

10. MCCULLUMSMITH, R.E. & J.H. MEADOR-WOODRUFF. 2002. Striatal excitatory amino acid transporter transcript expression in schizophrenia, bipolar disorder, and major depressive disorder. Neuropsychopharmacology **26:** 368–375.
11. MEADOR-WOODRUFF, J.H. *et al.* 1989. Distribution of D2 dopamine receptor mRNA in rat brain. Proc. Natl Acad Sci. USA **86:** 7625–7628.

Metabotropic Glutamate Receptor Regulation of Extracellular Glutamate Levels in the Prefrontal Cortex

ROBERTO I. MELENDEZ AND PETER W. KALIVAS

Neuroscience Institute, Medical University of South Carolina, Charleston, South Carolina 29425, USA

KEYWORDS: mGluRs; nucleus accumbens; prefrontal cortex; extracellular glutamate

Studies employing *in vivo* microdialysis have been successfully used to determine the involvement of group I[1,2] and group II[3] metabotropic glutamate receptors (mGluRs) in modulating the extracellular levels of glutamate in the nucleus accumbens. To date, the *in vivo* demonstration of direct modulation of extracellular glutamate by mGluR in the prefrontal cortex (PFC) remains unknown. The role of mGluR regulation of extracellular glutamate in the PFC may provide insights on the neurobiological mechanisms of cognition and reward. The objective of the present study was to characterize the modulation of extracellular glutamate by direct perfusion of group I and group II agonists and antagonists into the PFC.

While under anesthesia, male Sprague-Dawley rats (250–275 g) were implanted with guide cannulae aimed above the PFC (+2.7 mm anterior, +1.1 mm mediolateral, −2.0 mm ventral using a 6° angle). Concentric dialysis probes (with 2 mm active membrane) were inserted into the PFC the evening before microdialysis testing. Dialysis buffer (5 mM KCl, 140 mM NaCl, 1.4 mM $CaCl_2$, 1.2 mM $MgCl_2$, 5 mM glucose, pH of 7.4) was advanced through the probe at a flow rate of 2 µL/min via syringe pump. Baseline samples were collected at 20-min intervals. Thereafter, multiple doses of each mGluR agonist or antagonist were administered alone or in combination. The concentration of glutamate in the dialysis samples was determined using HPLC with flourometric detection as described previously.[3] A one-way ANOVA with repeated measures over dose was used to determine the effects of individual drugs on extracellular glutamate levels in the PFC.

Data analysis of the group I mGluR data indicated that local perfusion of DHPG (0.5 – 50 µM; $N = 6$), an agonist for mGluR 1/5, dose-dependently increased the extracellular levels of glutamate [$F(9,54) = 2.37$; $P < .05$]. Post hoc comparisons

Address for correspondence: Dr. Roberto I. Melendez, Department of Physiology and Neuroscience, Medical University of South Carolina, 173 Ashley Avenue, Suite 409 BSB, Charleston, SC 29425. Voice: 843-792-4272; fax: 843-792-4423.
melendez@musc.edu

revealed that 5 and 50 µM DHPG significantly elevated the extracellular levels of glutamate (maximal increase: 164% of baseline, $P < .05$). Conversely, perfusion of CPG (50–500 µM; $N = 7$), an antagonist for mGluR 1/5, failed to significantly alter the extracellular levels of glutamate (maximal decrease: 78% of baseline). Analysis of the group II mGluR data showed that local perfusion of APDC (0.5–500 µM; $N = 7$), an agonist for mGluR 2/3, failed to significantly alter the extracellular levels of glutamate (maximal decrease: 90% of baseline). However, perfusion of the mGluR 2/3 antagonist LY 341495 (0.1–10 nM), produced a dose-dependent increase in the extracellular levels of glutamate $[F(9,54) = 3.64; P < .005]$. Post hoc comparisons revealed that 1 and 10 nM LY 341495 significantly elevated the extracellular levels of glutamate (maximal increase: 164% of baseline, $P < .05$). Furthermore, the LY 341495-induced increase in extracellular glutamate levels was blocked by co-perfusion of 500 µM APDC.

These data provide the first *in vivo* evidence that direct pharmacological activation of group I mGluRs, as well as, blockade of group II mGluRs, significantly elevate the extracellular levels of glutamate in the PFC. These results are in agreement with previous microdialysis studies performed in the nucleus accumbens[2,3] and are consistent with an action on group I and group II presynaptic glutamate terminals. Our findings demonstrating that pharmacological blockade of group I mGluRs or the activation of group II mGluRs does not significantly decrease the extracellular levels of glutamate in the PFC, suggest maximal tone by endogenous glutamate on mGluRs 1/5 and mGluRs 2/3. Other possibilities include high affinity transport of extracellular glutamate and/or complex feedback regulatory mechanisms. Future studies will be necessary to further explore these issues.

REFERENCES

1. MORONI, F., A. COZZI, G. LOMBARDI, *et al.* 1998. Presynaptic mGlu1 type receptors potentiate transmitter output in the rat cortex. Eur. J. Pharmacol. **347:** 189–195.
2. SWANSON, C.J., D.A. BAKER, D.S. CARSON, *et al.* 2002. Repeated cocaine administration attenuates group I metabotropic glutamate receptor-mediated glutamate release and behavioral activation: a potential role for homer. J. Neurosci. **21:** 9043–9052.
3. XI, Z.X., D.A. BAKER, H. SHEN, *et al.* 2002. Group II metabotropic glutamate receptors modulate extracellular levels of glutamate in the nucleus accumbens. J. Pharmacol. Exp. Ther. **300:** 162–171.

Cystine/Glutamate Antiporter Regulation of Vesicular Glutamate Release

MEGAN M. MORAN, ROBERTO MELENDEZ, DAVID BAKER,[a] PETER W. KALIVAS, AND JEREMY K. SEAMANS

Department of Physiology and Neuroscience, Medical University of South Carolina, Charleston, South Carolina 29425, USA

[a]*Department of Biomedical Sciences Marquette University, Milwaukee, Wisconsin, USA*

KEYWORDS: antiporter regulation; vesicular glutamate release

Changes in extracellular glutamate levels have been linked with a variety of pathological neural conditions. For example, withdrawal from chronic cocaine administration causes a significant reduction in basal levels of extracellular glutamate within the nucleus accumbens (NAc).[1] Blunted glutamate neurotransmission within the prefrontal cortex (PFC), is thought to play a critical role in schizophrenia.[2] Thus, understanding the processes involved in glutamate regulation may provide insight into various neurological disorders. Moreover, brain regional differences in receptor expression, cell type, and morphology may play a role in the regulation of glutamate release.

Although the cystine/glutamate antiporter is found throughout the body, in the brain it is located primarily on the plasma membrane of glial cells.[3,4] The antiporter exchanges extracellular cystine for intracellular glutamate, driving cystine into the cell and glutamate out. Previous studies have shown that the antiporter plays a significant role in the regulation of extracellular glutamate levels within the NAc.[5] In the NAc, glutamate released by the antiporter provides tone to group II metabotropic glutamate receptors (mGluRs).[6] These receptors have been shown to regulate vesicular release of glutamate. Thus, manipulation of the antiporter by bath application of cystine, should increase extracellular glutamate, activate group II metabotropic receptors, and subsequently decrease synaptic activity, as measured by changes in spontaneous glutamate release. The objectives of the study were (1) to examine the effect of cystine/glutamate exchanger activity on synaptic activity and (2) to examine whether there are regional differences in the effects of the antiporter on synaptic activity in the PFC versus the Nac.

All experiments were conducted using acute brain slices containing either the PFC or the NAc. Rats (P14–P20), were anesthetized by intraperitoneal injection of chloral hydrate (400 mg/kg) and decapitated. The brain was removed, and coronal

---

Address for correspondence: Megan M. Moran, 173 Ashley Avenue, BSB 408, Charleston, SC 29425. Voice: 843-792-5444; fax: 843-792-4423.
moranm@musc.edu

slices (300–600 μm) were collected and incubated in oxygenated artificial cerebrospinal fluid (aCSF) at room temperature, while measurements were performed at 35°C.

Microdialysis was used to determine the effect of bath application of cystine on glutamate levels in the slices. Concentric microdialysis probes (2 mm of active membrane) were inserted into 600 μm PFC slices ($N=5$) and perfused with aCSF through the probe at a flow rate of 3 μL/min via a syringe pump. Following a wash-out period, baseline dialysis samples were collected at 5-min intervals for 30 minutes. After collecting baseline samples, cystine (0.1 μM) was perfused into the bath. Dialysis samples were collected at 5-min intervals for 30 minutes. The concentration of glutamate in the dialysis samples was determined using HPLC with fluorimetric detection. Mean data were compared as a percent of baseline. Although cystine administration did not result in a significant elevation in glutamate levels, a trend towards increasing levels was noted.

Whole-cell patch-clamp electrophysiology was used to examine synaptic activity. Electrodes were filled with a cesium chloride–based solution. Mini EPSCs (mESPCs) were recorded in the presence of tetrodotoxin (TTX, 1 μM) and bicuculline (10 μM) to assess solely changes in glutamate release probability independent of action potentials. After a baseline period, cystine, the cystine/glutamate antiporter blocker (S)-4-carboxyphenylglycine (CPG), and/or the group II metabotropic receptor antagonist (2S)-2-Amino-2-[(1S,2S)-2-carboxycycloprop-1-yl]-3-(xanth-9-yl)propanoic acid (LY341495) were perfused via the bath. Data were evaluated as a percent change in mEPSC frequency from baseline ± standard error. Significant differences were noted by a $P<.05$

In the mPFC, cystine (0.1 μM) caused a 42±5.5% reduction in the frequency of mEPSCs. However, when the cystine/glutamate antiporter was blocked by CPG (5 μM), there was no significant difference in mEPSC frequency between baseline and cystine. In addition, when group II metabotropic receptors were blocked by the antagonist LY341495 (0.3 μM), cystine had no effect on mEPSC frequency. Thus, it appears that in the PFC when action potentials are blocked by TTX, antiporter activity has a significant effect on synaptic transmission. However, when action potentials were present (in the absence of TTX), cystine had no effect on synaptic activity, as the mean frequency of spontaneous EPSCs (sEPSCs) was similar during baseline and cystine application. In contrast, when action potentials were present in the NAc, cystine administration caused a 42±4.6% reduction in the frequency of sEPSCs.

These data indicate that the cystine glutamate antiporter plays a significant role in regulation of synaptic activity through group II metabotropic receptors, within both the mPFC and the NAc. However, the ability of the antiporter to regulate synaptic activity is brain region specific. In PFC, the antiporter appears to only regulate action potential–independent release of glutamate, whereas in the NAc it plays a more general role. These regional variations may be a function of differences in the properties of the antiporter or in differences in mGluR receptor tone, number, or kinetics.

REFERENCES

1. PIERCE, R.C., K. BELL, P. DUFFY & P.W. KALIVAS. 1996. Repeated cocaine augments excitatory amino acid transmission in the nucleus accumbens only in rats having developed behavioral sensitization. J. Neurosci. **16:** 1550–1560.

2. CHEN, L. & C.R. YANG. 2002. Interaction of dopamine D1 and NMDA receptors mediates acute clozapine potentiation of glutamate EPSPs in rat prefrontal cortex. J. Neurophysiol. **87:** 2324–2336.
3. CHO, Y. & S. BANNAI. 1990. Uptake of glutamate and cysteine in C-6 glioma cells and in cultured astrocytes. J. Neurochem. **55:** 2091–2097.
4. POW, D.V. 2001. Visualising the activity of the cystine-glutamate antiporter in glial cells using antibodies to aminoadipic acid, a selectively transported substrate. Glia **34:** 27–38.
5. BAKER, D.A., Z.X. XI, H. SHEN, *et al.* 2002. The origin and neuronal function of in vivo nonsynaptic glutamate. J. Neurosci. **22:** 9134–9141.
6. XI, Z.X., D.A. BAKER, H. SHEN, *et al.* 2002. Group II metabotropic glutamate receptors modulate extracellular glutamate in the nucleus accumbens. J. Pharmacol. Exp. Ther. **300:** 162–171.

Expression of the NR3A Subunit of the NMDA Receptor in Human Fetal Brain

HELENA T. MUELLER AND JAMES H. MEADOR-WOODRUFF

Department of Psychiatry and Mental Health Research Institute,
University of Michigan Medical School, Ann Arbor, Michigan 48109-0720, USA

KEYWORDS: NMDA receptor; schizophrenia; NR3A subunit; human fetal brain

NMDA receptors (NMDARs), a specific class of glutamate receptors, play an important role in many normal and pathological processes in the brain. NMDARs are ligand gated ion channels comprised of heteromeric assemblies of subunits encoded by seven different genes (NR1, NR2A-2D, and NR3A-3B). Distinct pharmacological, physiological, and signal-transducing properties are associated with each subunit, indicating that fundamental properties of NMDARs are in part determined by subunit composition.[1] Changes in subunit composition may in turn underlie pathophysiological alterations linked to NMDAR dysfunction.

There is growing evidence that glutamatergic dysfunction may be involved in the pathophysiology of schizophrenia; in particular, it has been hypothesized that the activity of the NMDAR is decreased in this illness.[2] Several studies examining the expression of the NR1 and NR2 subunits in schizophrenia have found brain region-specific changes in transcript levels for these subunits in schizophrenia.[3] Expression of the NR3A subunit has not yet been investigated in schizophrenia.

Previous studies, however, do provide evidence that the NR3A subunit may act as a unique and important modulator of NMDAR activity and function. In rodents, incorporation of the NR3A subunit into NR1/NR2-containing receptors decreases NMDAR activity.[4] NR3A can also indirectly modulate NMDAR activity by altering the phosphorylation state of the obligatory NR1 subunit,[5] and when coassembled with NR1, generates a functional excitatory glycine receptor.[6]

Anatomical data in rodents also indicate that the NR3A subunit is highly expressed in brain regions implicated in schizophrenia, including prefrontal cortex, thalamus, hippocampus, and amygdala. Many of the abnormalities observed in these regions in schizophrenia are thought to be associated with abnormal developmental processes occurring during the second trimester of pregnancy.[7] Collectively, these data suggest that alterations in NR3A expression may also contribute to NMDAR dysfunction implicated in schizophrenia.

Address for correspondence: Helena T. Mueller, Mental Health Research Institute and Department of Psychiatry, University of Michigan, 205 Zina Pitcher Place, Ann Arbor, MI 48109-0720. Voice: 734-936-2061; fax: 734-647-4130.
  hmueller@umich.edu

Ann. N.Y. Acad. Sci. 1003: 448–451 (2003). © 2003 New York Academy of Sciences.
doi: 10.1196/annals.1300.049

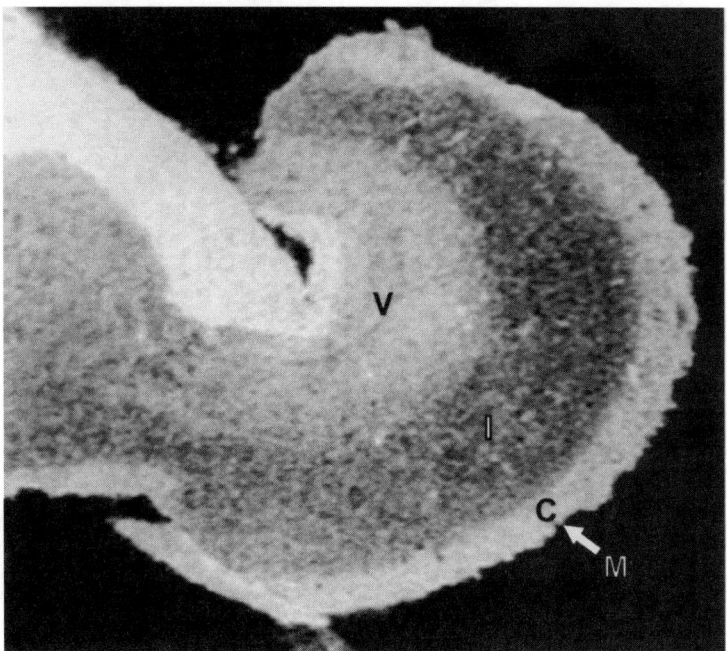

FIGURE 1. NR3A mRNA expression in the human fetal cortex. *In situ* hybridization was performed using a [^{35}S]-labeled riboprobe containing a subclone for the rat NR3A subunit (NCBI GeneBank accession number AF073379, nucleotide coding region 1139-1676). Transcript for the NR3A subunit of the NMDA receptor was detected in all four cortical layers in the developing human neocortex. Abbreviations: M= marginal zone, C=cortical plate, I= intermediate zone, V=ventricular zone.

To begin to characterize NR3A expression in human brain and determine if the NR3A subunit is expressed at an appropriate time in human brain development to play a role in the etiology of schizophrenia, we examined NR3A transcript levels by *in situ* hybridization in human fetal cortex. Human fetal tissue was obtained from Dr. Alan Unis, University of Washington, and has previously been described in an earlier report.[8] The riboprobe used for this study was synthesized from a linearized plasmid containing an NR3A subclone (NCBI GeneBank accession number AF073379, nucleotide coding region 1139-1676) and labeled with [^{35}S]. NR3A mRNA was detected in all layers of the developing cortex (FIG. 1) at different gestational weeks within the second trimester (FIG. 2). NR3A mRNA was especially prominent in the marginal and ventricular zones during this time, which corresponds to a period of cortical migration in human cortical development,[9] suggesting that NR3A may play a role in cortical migration. Interestingly, NMDARs are known to be involved in cortical migration[10] and there is evidence of abnormal cortical neuronal migration in schizophrenia.[11,12] Another developmental abnormality observed in schizophrenia is decreased dendritic spine density in the cortex and mice lacking NR3A show increased

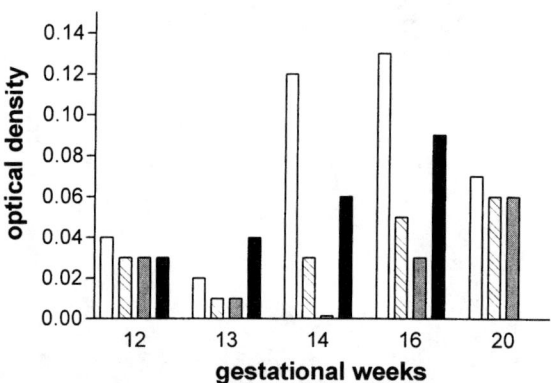

FIGURE 2. Time course of NR3A mRNA expression in the developing human cortex. NR3A transcript levels were measured in the developing human cortex during gestational weeks 12–20. Transient increases were seen in all cortical regions during this period. NR3A expression was particularly high in the marginal and ventricular zones in the second trimester. *Open bars,* marginal zone; *striped bars,* cortical plate; *gray bars,* intermediate zone; *black bars,* ventricular zone.

cortical spine density suggesting that NR3A may play a role in dendritic growth during development.[13]

Taken together, these data on the NR3A subunit of the NMDAR suggest that it is highly expressed in the developing human neocortex, may be associated with neuronal migration and dendritic spine density, and may play a role in developmental abnormalities associated with the pathophysiology of schizophrenia.

ACKNOWLEDGMENTS

This work was supported by Grants MH650101 (H.T.M.) and MH53327 (J.M.W.) from the National Institute of Mental Health.

REFERENCES

1. HOLLMANN, M. & S. HEINEMANN. 1994. Cloned glutamate receptors. Annu. Rev Neurosci. **17:** 31–108.
2. OLNEY, J.W., J.W. NEWCOMER & N.B. FARBER. 1999. NMDA receptor hypofunction model of schizophrenia. J. Psychiatr Res. **33:** 523–533.
3. MEADOR-WOODRUFF, J.H. & D.J. HEALY. 2000. Glutamate receptor expression in schizophrenic brain. Brain Res. Brain Res Rev. **31:** 288–294.
4. SUCHER, N.J. *et al.* 1995. Developmental and regional expression pattern of a novel NMDA receptor-like subunit (NMDAR-L) in the rodent brain. J. Neurosci. **15:** 6509–6520.

5. CHAN, S.F. & N.J. SUCHER. 2001. An NMDA receptor signaling complex with protein phosphatase 2A. J. Neurosci. **21:** 7985–7992.
6. CHATTERTON, J.E. *et al.* 2002. Excitatory glycine receptors containing the NR3 family of NMDA receptor subunits. Nature **415:** 793–798.
7. MARENCO, S. & D.R. WEINBERGER. 2000. The neurodevelopmental hypothesis of schizophrenia: following a trail of evidence from cradle to grave. Dev. Psychopathol. **12:** 501–527.
8. RITTER, L.M., A.S. UNIS & J.H. MEADOR-WOODRUFF. 2001. Ontogeny of ionotropic glutamate receptor expression in human fetal brain. Brain Res. Dev Brain Res. **127:** 123–133.
9. RAKIC, P. 1972. Mode of cell migration to the superficial layers of fetal monkey neocortex. J. Comp. Neurol. **145:** 61–83.
10. BEHAR, T.N. *et al.* 1999. Glutamate acting at NMDA receptors stimulates embryonic cortical neuronal migration. J. Neurosci. **19:** 4449–4461.
11. AKBARIAN, S. *et al.* 1993. Altered distribution of nicotinamide-adenine dinucleotide phosphate-diaphorase cells in frontal lobe of schizophrenics implies disturbances of cortical development. Arch. Gen. Psychiatry **50:** 169–177.
12. AKBARIAN, S. *et al.* 1996. Selective alterations in gene expression for NMDA receptor subunits in prefrontal cortex of schizophrenics. J. Neurosci. **16:** 19–30.
13. DAS, S. *et al.* 1998. Increased NMDA current and spine density in mice lacking the NMDA receptor subunit NR3A. Nature **393:** 377–381.

Coupling of Glutamatergic Neurotransmission and Neuronal Glucose Oxidation over the Entire Range of Cerebral Cortex Activity

ANANT B. PATEL,[a] ROBIN A. DE GRAAF,[b] GRAEME F. MASON,[a]
DOUGLAS L. ROTHMAN,[b] ROBERT G. SHULMAN,[c] AND KEVIN L. BEHAR[a]

*Departments of [a]Psychiatry, [b]Diagnostic Radiology, [c]Molecular Biophysics and Biochemistry, and Magnetic Resonance Research Center,
Yale University School of Medicine, New Haven, Connecticut 06520, USA*

KEYWORDS: glutamate/glutamine cycle; neuronal glucose oxidation; neuron–astroglia interaction; NMR spectroscopy; seizures

INTRODUCTION

In vivo ^{13}C NMR experiments have shown that the glutamate/glutamine cycle flux ($V_{cycle(Glu/Gln)}$) is comparable in magnitude to the rate of neuronal glucose oxidation ($CMR_{glc(ox)N}$) in the cortices of resting humans and of anesthetized rats.[1–3] In anesthetized rats, $V_{cycle(Glu/Gln)}$ and $CMR_{glc(ox)N}$ change proportionately (~1:1 ratio) over a substantial range of brain activity.[2] This proportionality has been explained theoretically on the basis of the calculated energetic contribution of specific ion flows associated with glutamatergic neurotransmission.[4] The proportionality between $V_{cycle(Glu/Gln)}$ and $CMR_{glc(ox)N}$ supports a proposed coupling mechanism where glucose uptake is stoichiometrically coupled to the glutamate/glutamine cycle.[5] Positron emission tomography study in human subjects during visual stimulation show a greater consumption of glucose than oxygen, leading to the suggestion that the energy required for neuronal activity during stimulation arises primarily from nonoxidative glucose metabolism.[6] In this study we show that the proportionality seen between $\Delta V_{cycle(Glu/Gln)}$ and $\Delta CMR_{glc(ox)N}$ in nonstimulated anesthetized cortex holds during neuronal stimulation and that oxidative glucose metabolism supports neuronal activity over the entire physiological range.

METHODS

Two groups of overnight fasted, Male Wistar rats were anesthetized (1.5% halothane) and ventilated (70% N_2O, 30% O_2). ^1H-observed ^{13}C-edited NMR spectra were recorded continuously at 7 tesla (Bruker Avance Spectrometer) from a localized

Address for correspondence: Dr. Anant Patel or Dr. Kevin Behar, MRRC, P.O. Box 208043, 300 Cedar Street, New Haven, CT 06520. Voice: 203-785-6199; fax: 203-785-6643.
anant.patel@yale.edu or kevin.behan@yale.edu

volume ($7\times4\times7$ mm^3) centered in the frontoparietal cortex during the infusion of [1,6-$^{13}C_2$]glucose.[7] Seizures were induced by injection of bicuculline (1 mg/kg, i.v.) 4 min after beginning the ^{13}C-glucose infusion. Spectra were acquired serially (3 min/spectrum) for 85 min (control) and 55 min (seizures). At the end of the experiments, metabolites were extracted from the frozen cortical tissues and analyzed for their concentration and percent ^{13}C enrichment.[8] The time courses of glutamate-C4 and glutamine-C4 ^{13}C enrichments were fitted to a three-compartment metabolic model to obtain the fluxes of neuronal glucose oxidation and glutamate/glutamine cycling.[3]

RESULTS AND DISCUSSION

The ^{13}C labeling of glutamate-C4 and glutamine-C4 from [1,6-^{13}C]glucose was faster during seizures than in the control animals indicating that metabolic fluxes had increased. Lactate increased rapidly and leveled off at ~12.0 μmol/g. Total glucose utilization during seizures increased 455%, whereas $CMR_{glc(ox)N}$ increased 232%, which was comparable to the increase in $V_{cycle(Glu/Gln)}$ (221%). The ratio, $\Delta CMR_{glc(ox)N}/\Delta V_{cycle(Glu/Gln)}$ during seizures was 1.09, similar to the value reported previously (1.04) in the nonactivated cortex of rats subjected to varying depths of anesthesia.[2] Our findings indicate that neuronal glucose oxidation, rather than total glucose consumption, is proportional to glutamatergic neurotransmission. Nonoxidative glucose consumption, as reflected by the excess lactate produced during the intense neuronal activation, may be required to support rapid neuronal firing.[9] In this mechanism the lactate generated from the excess glucose uptake is transported out of the brain to the blood.

REFERENCES

1. SHEN, J., K.F. PETERSEN, et al. 1999. Determination of the rate of the glutamate-glutamine cycle in the human brain by in vivo ^{13}C NMR. Proc. Natl. Acad. Sci. USA **96:** 8235–8240.
2. SIBSON, N.R., A. DHANKHAR, et al. 1998. Stoichiometric coupling of brain glucose metabolism and glutamatergic neuronal activity. Proc. Natl. Acad. Sci. USA **95:** 316–321.
3. SIBSON, N.R., G.F. MASON, et al. 2001. In vivo ^{13}C NMR measurement of neurotransmitter glutamate cycling, anaplerosis and TCA cycle flux in rat brain during [2-^{13}C]glucose infusion. J. Neurochem. **76:** 975–989.
4. ATTWELL, D. & S.B. LAUGHLIN. 2001. An energy budget for signaling in the gray matter of the brain. J. Cereb. Blood Flow Metab. **21:** 1133–1145.
5. MAGISTRETTI, P., L. PELLERIN, et al. 1999. Perspective: Neuroscience "Energy on Demand." Science **283:** 496–497.
6. FOX, P.T., M.E. RAICHLE, et al. 1988. Nonoxidative glucose consumption during focal physiologic neural activity. Science **241:** 462–464.
7. DE GRAAF, R.A., P.B. BROWN et al. 2003. Detection of [1,6-$^{13}C_2$]-glucose metabolism in rat brain by in vivo ^1H-[^{13}C]-NMR spectroscopy. Magn. Reson. Med. **49:** 37–46.
8. PATEL, A.B., D.L. ROTHMAN, et al.2001. Glutamine is the major precursor for GABA synthesis in rat neocortex in vivo following acute GABA-transaminase inhibition. Brain Res. **919:** 207–220.
9. SHULMAN, R.G., F. HYDER & D.L. ROTHMAN. 2001. Cerebral energetics and the glycogen shunt: neurochemical basis of functional imaging. Proc. Natl. Acad. Sci. USA **98:** 6417–6422.

Real Time *in Vivo* Measures of L-Glutamate in the Rat Central Nervous System Using Ceramic-Based Multisite Microelectrode Arrays

F. POMERLEAU, B. K. DAY, P. HUETTL, J. J. BURMEISTER, AND G. A. GERHARDT

Center for Sensor Technology, University of Kentucky, Lexington, Kentucky 40536, USA

KEYWORDS: microelectrode arrays; L-glutamate; rat central nervous system

L-Glutamate (Glu), the predominant excitatory neurotransmitter in the CNS, is associated with a wide variety of functions including motor behavior, cognition, memory and sensory perception. Studies support that complex mechanisms involving glial cell metabolism, uptake reversal, and neuronal vesicular release contribute to the extracellular pool of Glu.[1] Microdialysis studies have lacked sufficient spatial and temporal resolution to study the fast dynamics of Glu, although major advances in temporal resolution have been made when this technique is coupled to capillary electrophoresis.[2] However, there has been the need for a Glu sensor that is less invasive, which has a low detection limit, fast response time, small recording area, and is free from interferents.

Multisite (four Pt sites, 15 μm × 333 μm) ceramic-based microelectrode arrays configured for enzyme-based (Glu-oxidase) detection of Glu with subsecond time resolution (800 msec) and low limits of detection (~0.2 μM) were developed, characterized, and used to directly study Glu levels and regulation in CNS of anesthetized and freely moving rats. The Pt sites were first coated with Nafion™ to repel anionic interferents, namely ascorbic acid and DOPAC. Second, a layer of 2% glutamate oxidase (GluOx) solution (BSA 1%, glutaraldehyde 0.125%) was applied. The GluOx in the presence of Glu/FAD generates H_2O_2, which is oxidized on the Pt sites and creates a current that is amplified and digitized.[3,4] Amperometry (+0.7 V vs. Ag/AgCl) was used to measure basal Glu in two areas of the rat brain (striatum and prefrontal cortex) and to monitor Glu levels in the somatosensory cortex of unrestrained awake animals.

Address for correspondence: F. Pomerleau, Center for Sensor Technology, University of Kentucky, 800 Rose Street, 306 Davis-Mills Bldg., Lexington, KY 40536. Voice: 859-873-4531; fax 859-257-5310.
Francois.Pomerleau@uky.edu

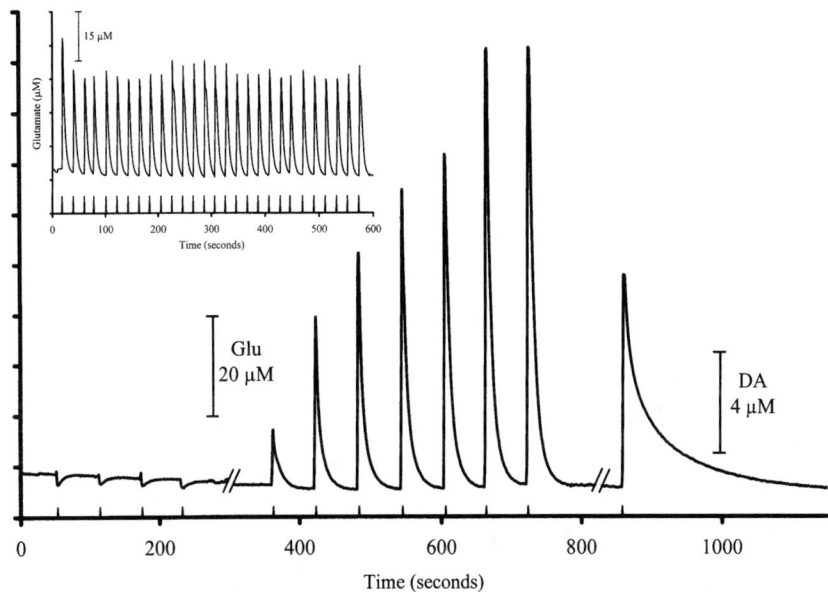

FIGURE 1. Microelectrode responses *in vivo*.

Characterization of the signals we measured is critical when using electrochemical methods. *In vitro* calibration has shown that the enzyme-based microelectrodes are insensitive to application of ascorbic acid (AA) and DOPAC because of the anionic repellent properties of Nafion™. Furthermore, it did not respond to additions of aspartate (a close relative of Glu) at concentrations up to 200 μM. Addition of Glu produced signals that were proportional (linear) to the amount of Glu present. FIGURE 1 demonstrates that the implanted microelectrode arrays do not respond to local applications of saline but respond in a dose-dependant manner to increasing volumes (~100 nL to 800 nL) of Glu (5 mM). The inset demonstrates that the Glu-evoked release by ejection of high potassium solution (70 mM, ~100 nL, every 20 s) is constant and robust. Furthermore, when compared to the signals obtained with local application of dopamine (DA) (which can be an inteferent or a molecule of interest), the kinetics (especially the rate of uptake) are quite different and allow us to discriminate between the two signals. These results, combined with the fact that the Glu oxidase enzyme is nearly exclusively selective for Glu as a substrate to form peroxide, support that we are measuring Glu.

Variation of the applied voltage can also be used to further identify the Glu signals. FIGURE 2 demonstrates that dropping the applied potential from +0.7 V to +0.2 V (vs. Ag/AgCl) nearly abolishes the peroxide (H_2O_2) and the Glu signals, while the responses seen from DA are largely unchanged. These results demonstrate that the detected Glu signals coming from the generation of peroxide by the GluOx are not oxidized at a lower voltage, supporting that our amperometric recording approach to measure Glu is selective. Furthermore, since DA can be detected at a lower applied potential, this approach can be used to identify Glu released *in vivo*.

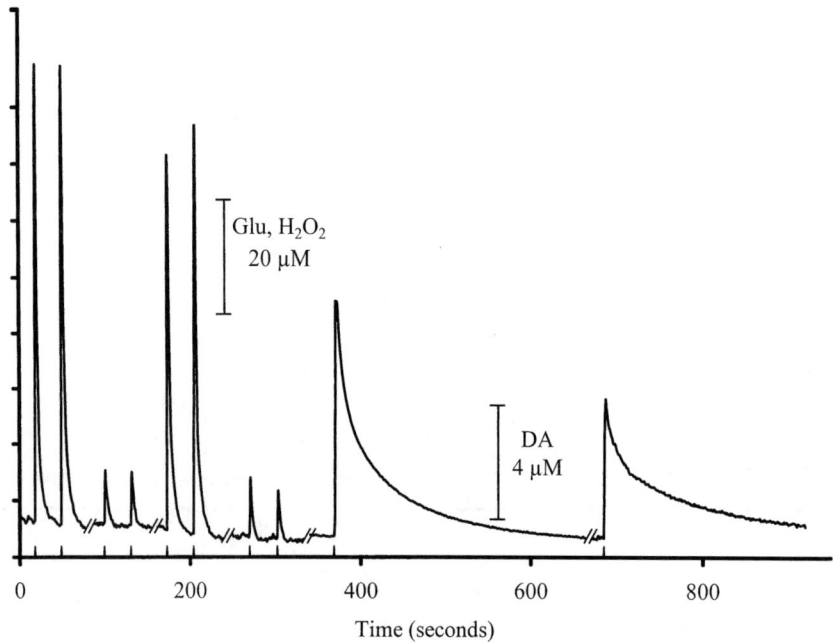

FIGURE 2. Response of the microelectrode at various applied potentials.

Pharmacological manipulations of basal Glu were also performed using the reversible Na^+-channel blocker tetrodotoxin (TTX) and the Glu uptake inhibitor DL-threo-β-benzyloxyaspartate (TBOA). Our results showed that local application of TTX (0.1 mM, ~100 nL), caused a significant decrease (1–3 μM) in basal Glu in both brain regions supporting that part of the basal Glu signal we are measuring is neuronal in origin. Local application of the Glu uptake inhibitor TBOA (0.1 mM, ~100 nL), induced a rapid increase in basal Glu (1–3 μM) supporting that by decreasing Glu uptake, increases in extracellular concentrations of Glu can be measured.

More recent studies in freely moving rats (FIG. 3) have demonstrated that stimulations of the contralateral vibrissae induced significant Glu release (4–6 μM) lasting the duration of stimulation (10–60 s), while ipsilateral vibrissae stimulations had no effect (not shown). These results support that Glu is involved in whisker movement perception in the somatosensory cortex of rats and our recordings can be performed in awake animals.

Taken together, our data support that the multisite microelectrode arrays can be used for fast, sensitive, and reliable measures of basal Glu in various brain regions in anesthetized and freely moving rats.

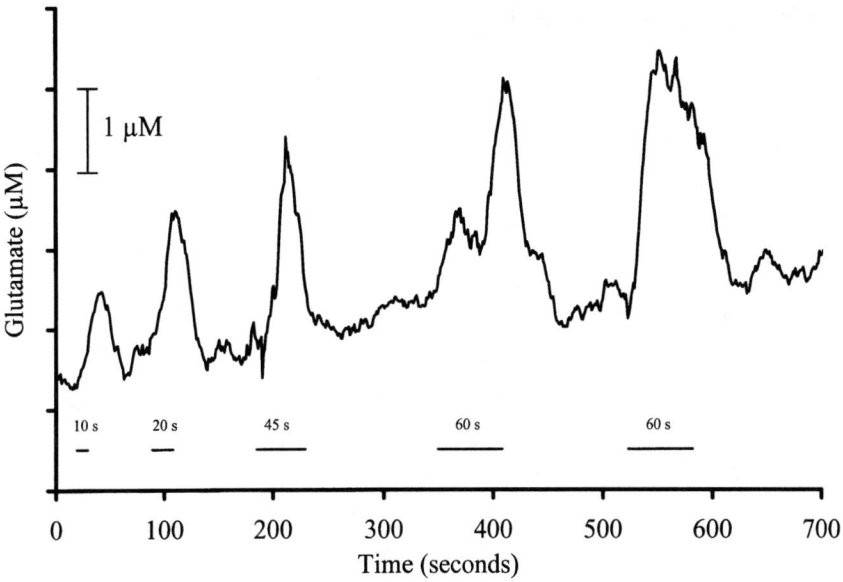

FIGURE 3. Recordings of Glu signals in freely moving rats.

ACKNOWLEDGMENTS

This work was supported by National Science Foundation Grant DBI-9987807 and U.S. Public Health Service grants AG01245 and DA14944.

REFERENCES

1. DANBOLT, N.C. 2001. Glutamate uptake. Prog. Neurobiol. **65:** 1–105.
2. KENNEDY, R.T. et al. 2002. In vivo neurochemical monitoring by microdialysis and capillary separations. Curr. Opin. Chem. Biol. **6:** 659–665.
3. BURMEISTER, J.J. & G.A. GERHARDT. 2001. Self-referencing ceramic-based multisite microelectrodes for the detection and elimination of interferences from the measurement of L-glutamate and other analytes. Anal. Chem. **73:** 1037–1042.
4. BURMEISTER, J.J. et al. 2002. Improved ceramic-based multisite microelectrode for rapid measurements of L-glutamate in the CNS. J. Neurosci. Methods **119:** 163–171.

TRH and Related Peptides

Homeostatic Regulators of Glutamate Transmission?

A. SATTIN,[a,b,d] A. E. PEKARY,[a,c] R. L. LLOYD,[e] M. PAULSON,[e]
J. A. MEYERHOFF,[f] P. M. HINKLE,[g] AND K. FAULL[b,d]

[a]*Veterans Administration Greater Los Angeles Healthcare System, Los Angeles, California 90073, USA*

Departments of [b]Psychiatry, [c]Medicine, the [d]Brain Research Institute, University of California at Los Angeles, Los Angeles, California, USA

[e]*Department of Psychology, University of Minnesota, Duluth, Minnesota, USA*

[f]*Walter Reed Army Institute of Research, Silver Spring, Maryland, USA*

[g]*Department of Pharmacology, University of Rochester, Rochester, New York, USA*

KEYWORDS: TRH; glutamate transmission; colocalization

Thyrotropin-releasing hormone (TRH) immunohistochemistry has shown colocalization with glutamate (Glu) throughout the neuraxis. In rat hippocampus (HC), TRH is seen only in the CA and granule neurons and fimbria-fornix fibers[1] and is released from HC slices by high K^+.[2] Blockade of Glu receptors by TRH indicates negative feedback regulation.[3] Clinical observations of antidepressant effects of intrathecal TRH and neuropathological findings in prefrontal cortex led us to propose a pathophysiology of major depression involving hyperactivation of prefrontal-to-limbic Glu circuits in which TRH regulation of Glu has decompensated.[4] It is likely that augmented synthesis of TRH within Glu neurons during therapeutic seizures (ECT) contributes to the clinical antidepressant effects.

More recently, we have discovered an endogenous family of TRH-like peptides in CNS with the structure pGlu-X-Pro-NH_2 where X may be any amino acid (for TRH, X=His; –T=TRH-like peptide). By combining reverse-phase HPLC with RIA of eluates from standards and brain extracts using antisera to this tripeptide that are blind to X, we have matched retention times to X=Glu, Phe, Tyr, Leu, Val, Tyr + unidentified peaks. Each TRH-like peptide exhibits a unique pharmacobehavioral profile in rats, including (collectively) ± antidepressant (in preparation), ± analeptic, and ± wet-dog shakes after intraperitoneal injection.[5] Glu-T has a proven somatic function as sperm capacitation factor.[6] Tyr-T has now also been structurally proven

---

Address for correspondence: A. Sattin, Veterans Administration Greater Los Angeles Healthcare System, P.O. Box 84122, Los Angeles, CA 90073. Voice: 310-268-4440; fax: 310-478-6259.
asattin@ucla.edu

by gel chromatography mass spectrometry (Faull, in preparation). The affinities of eight of the TRH-like peptides to TRH1 and TRH2 receptors range from two to five orders of magnitude lower than TRH, making it unikely that those receptors directly mediate their effects.[5]

Given the antidepressant effects of TRH, our recent finding that chronic therapeutic blood levels of Li^+ significantly increase [^{3}H]3Me-TRH binding in accumbens and amygdala of Wistar and WKY rats is relevant to the known augmentation by Li^+ of antidepressant treatment.[7] This may also be relevant to the recent evidence-based clinical finding that the addition of Li^+ to antidepressant drug treatment significantly improves relapse prevention after completion of a course of ECT.[8] Thus, ECT induces increased synthesis of TRH within the (putatively hyperactivated) Glu circuits,[4] reducing Glu activation of AMPA, NMDA and voltage-regulated postsynaptic effects.[3] Subsequent treatment with Li^+ might complement the increased TRH by increasing TRH receptor number, which might contribute to the clinically proven relapse prevention.

We have also demonstrated that hormonal alterations and drug treatments given acutely, chronically, and in acute withdrawal can profoundly alter the regional content of TRH and many of the measurable TRH-like peptides. These effects were specific to the agent, the treatment condition, and the brain region from which the peptides were extracted. For example, both thyroid hormone depletion with PTU and castration of male rats significantly increased the content of TRH, Glu-T, and PS4 (a cotranscript of pro-TRH) in the accumbens.[9] In the accumbens, acute Li^+ increased Glu-T 2.6-fold, while the contents of TRH and Leu-T were both reduced to <50%.[7] Acute cocaine induced a 4.1-fold increase in Val-T in the medulla.[10] The drug-induced acute changes in these peptides is meaningful in that the rats were sacrificed 2 h postinjection, which is too short an interval to allow synthesis to play a role. Therefore reduction in content indicates release and increased content indicates inhibition of release. It is likely that, like TRH itself, the TRH-like peptides are colocalized in Glu neurons throughout the neuraxis, because most of the antisera that have been used used for immunohistochemistry identify pGlu-X-Pro-NH_2, not specifically TRH. Definitive localization will have to await the identification of the precursor proteins of each member of this new class of neuropeptides.

ACKNOWLEDGMENT

This work was supported by Veterans Administration Research Service.

REFERENCES

1. Low, W.C, S.D. Farber, T.G. Hill, et al. 1989. Evidence for extrinsic and intrinsic TRH in hippocampus by RIA and immunocytochemistry. Ann. N.Y. Acad. Sci. **553:** 574–578.
2. Knoblach, S.M. & M.J. Kubek. 1994. TRH release is enhanced in hippocampal slices after electroconvulsive shock. J. Neurochem. **62:** 119–125.
3. Koenig, M.L., D.L. Yourick & J.L. Meyerhoff. 1996. TRH attenuates glutamate-stimulated increases in calcium in primary neuronal cultures. Brain Res. **730:** 143–149.
4. Sattin, A. 1999. The role of TRH and related peptides in the mechanism of action of ECT. J. ECT **15:** 76–92.

5. HINKLE, P.M., A.E. PEKARY, S. SENANAYAKI & A. SATTIN. 2002. Role of TRH receptors as possible mediators of analeptic actions of TRH-like peptides. Brain Res. **935:** 59–64.
6. GREEN, C.M., S.M. COCKLE, P.F. WATSON & L.R. FRASER. 1996. A possible mechanism of action for fertilization promoting peptide, a TRH-related peptide that promotes capacitation and fertilizing ability in mammalian spermatozoa. Mol. Reprod. Dev. **45:** 244–252.
7. SATTIN, A., S. SENANAYAKE & A.E. PEKARY. 2002. Lithium modulates expression of TRH receptors and TRH-related peptides in rat brain. Neuroscience **115:** 263–273.
8. SACKEIM, H.A., R.F. HASKETT, B.H. MULSANT, *et al.* 2001. Continuation pharmacotherapy in the prevention of relapse following electroconvulsive therapy: a randomized controlled trial. J. Am. Med. Assoc. **285:** 1299–1307.
9. PEKARY, A.E. & A. SATTIN. 2001. Regulation of TRH and TRH-related peptides by thyroid and steroid hormones. Peptides **22:** 1161–1173.
10. PEKARY, A.E., S. SENANAYAKE & A. SATTIN. 2002. Cocaine regulates TRH-related peptides in rat brain. Neurochem. Int. **41:** 415–428.

L-Homocysteine Sulfinic Acid and L-Homocysteic Acid Stimulate Phosphoinositide Hydrolysis in Rat Cortical Neurons

QI SHI, SANDRA J. HUFEISEN, JARDA T. WROBLEWSKI,[a] JOSEPH H. NADEAU, AND BRYAN L. ROTH

Departments of Biochemistry and Genetics,
Case Western Reserve University, Cleveland, Ohio 44106, USA

[a]*Department of Pharmacology,*
Georgetown University Medical Center, Washington, DC 20007, USA

KEYWORDS: L-homocysteine sulfinic acid; L-homocysteic acid; phosphoinositide hydrolysis; metabotropic GluR

Moderate hyperhomocysteinemia is characterized by a moderate elevation of total plasma homocysteine level (16–31 µM)[1] and is statistically associated with coronary artery disease,[2,3] stroke,[2,3] Alzheimer's disease,[4] schizophrenia,[5] and other illnesses. The molecular mechanisms for the pathologic actions of homocysteine are unknown. We have recently demonstrated that several acidic oxidized derivatives of homocysteine, including L-homocysteine sulfinic acid (L-HCSA), L-homocysteic acid (L-HCA), L-cysteine sulfinic acid (L-CSA), and L-cysteic acid (L-CA), are potent and selective agonists at metabotropic glutamate receptors (mGluRs) based on an unbiased screening of 60 different cloned cell-surface receptors, ion channels, and transporters.[6] To extend our findings from the non-neuronal expression system, we examined the effect of L-HCSA and L-HCA on phosphoinositide (PI) hydrolysis in cultured cortical neurons. The cortical neuron cultures were prepared from frontal cortices of Sprague-Dawley rat fetuses (E17.5) and cultured in supplemented Neurobasal media (Invitrogen) on 24-well cell culture plates coated with poly-D-lysine in 5% CO_2 incubator at 37°C for 7 days.[7] On day 8, the neuronal culture media were switched to a modified Krebs bicarbonate buffer (1 mL) containing 0.5 µCi/mL myo-[^{3}H]-inositol. The cultures were subsequently incubated in 5% CO_2 incubator at 37°C for 1 hour. Thereafter, agonists in increasing concentration (10 µL) were administered to the cultures and followed by incubation for 30 minutes. PI hydrolysis

---

Address for correspondence: Qi Shi, M.D., Ph.D., Departments of Biochemistry and Genetics, Case Western Reserve University School of Medicine, BRB, Room 609B, 2109 Adelbert Rd., Cleveland, OH 44106. Voice: 216-368-0626; fax: 216-368-3432.
    qxs7@po.cwru.edu

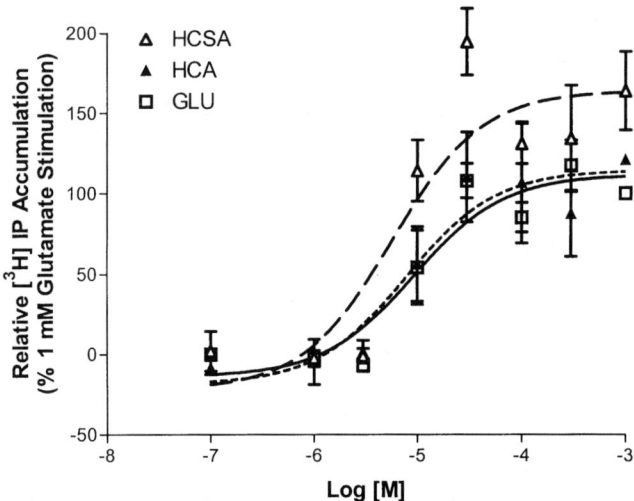

FIGURE 1. Stimulatory effect of L-homocysteine sulfinic acid (HCSA) and L-homocysteic acid (HCA) on phosphoinositide hydrolysis in rat embryonic cortical neurons. After preincubation with myo-[^{3}H]-inositol for 1 hour, neurons were treated with HCSA (*open triangle with dashed line*), HCA (*filled triangle with dotted line*), and L-glutamate (GLU, *open square with solid line*) in the indicated concentrations for 30 min at 37°C. [^{3}H]-Inositol phosphates were determined by anion exchange chromatography. Readings of dpm were normalized to the effect induced by 1 mM L-glutamate and presented in percentage (mean±SEM). The dose-response curves were established based on three independent replicate experiments ($P > .05$).

was determined in duplicate by measurement of [^{3}H]-inositol phosphate as previously described.[6,8,9] The normalized data were used to generate dose-response curves by non-linear regression curve fit method. The EC_{50} and E_{max} values were determined by best-fitting of the normalized data using GraphPad Prism 3.02.

Our data have demonstrated that: (1) L-HCSA and L-HCA stimulate PI hydrolysis in rat frontal cortical neurons in a dose-dependent manner (FIG. 1); (2) the EC_{50} values for L-HCSA, L-HCA, and L-glutamate are 5.75±1.77, 7.76±1.92, and 9.22 ±1.58 µM, respectively ($P > .05$); and (3) the Emax for L-HCSA, L-HCA, and L-glutamate are 163.9±15.13, 114.5±13.16, and 111.6±8.93% of the stimulation of 1 mM glutamate, respectively ($P > .05$), indicating that L-HCSA and L-HCA stimulate PI hydrolysis in the neurons as potently and efficaciously as L-glutamate, a well-characterized endogenous neurotransmitter for mGluRs.

It is well established that Group I mGluRs involve activation of phospholipase C resulting in intracellular PI hydrolysis, whereas Group II and III mGluRs participate in inhibition of adenylyl cyclase. A recent immunocytochemistry study of Group I mGluRs during cortical development has reported that mGluR1 and mGluR5 are expressed in rat cerebral cortex as early as embryonic day 18 and they are continuously expressed in neocortical neurons of postnatal rats (P0–P10).[10] Therefore, we hypothesize that L-HCSA and L-HCA activate Group I mGluRs on the frontal cortical neurons, which in turn accelerate PI hydrolysis in the neurons. Interestingly, several previous studies of murine models suggested that Group I mGluRs were involved in neuronal apoptosis during stroke,[11] neuronal degeneration and intraneuronal amyloid peptides production,[12] and loss of prepulse inhibition of acoustic startle response (a measurement of schizophrenia).[13] The ultimate significance of our findings is contingent upon the demonstration that sufficient elevations in the local concentrations of L-HCSA and L-HCA are present *in vivo*, particularly in diseases associated with moderate hyperhomocysteinemia. Future experiments will be needed to determine whether the local concentrations of acidic homocysteine derivatives are elevated in animal models of moderate hyperhomocysteinemia.

REFERENCES

1. KANG, S. S., P.W. WONG & M.R. MALINOW. 1992. Annu. Rev. Nutr. **12:** 279–298.
2. BRATTSTROM, L. & D.E. WILCKEN. 2000. Am. J. Clin. Nutr. **72:** 315–323.
3. UELAND, P.M., H. REFSUM, S.A. BERESFORD & S.E. VOLLSET. 2000. Am. J. Clin. Nutr. **72:** 324–332.
4. MILLER, J.W. 2000. Nutrition **16:** 675–677.
5. LEVINE, J. *et al.* 2002. Am. J. Psychiatry **159:** 1790–1792.
6. SHI, Q. *et al.* 2003. J. Pharmacol. Exp. Ther. **305:** 131–142.
7. BREWER, G.J. 1997. J. Neurosci. Methods **71:** 143–155.
8. RAUSER, L., J.E. SAVAGE, H.Y. MELTZER & B.L. ROTH. 2001. J. Pharmacol. Exp. Ther. **299:** 83–89.
9. SHAPIRO, D.A., K. KRISTIANSEN, W.K. KROEZE & B.L. ROTH. 2000. Mol. Pharmacol. **58:** 877–886.
10. LOPEZ-BENDITO, G., R. SHIGEMOTO, A. FAIREN & R. LUJAN. 2002. Cereb. Cortex **12:** 625–638.
11. BRUNO, V. *et al.* 2000. Neuropharmacology **39:** 2223–2230.
12. STEPHENSON, D.T. & J.A. CLEMENS. 1998. Neurochem. Int. **33:** 83–93.
13. GRAUER, S.M. & K.L. MARQUIS. 1999. Psychopharmacology (Berl.) **141:** 405–412.

Distinct Contributions of Glutamate Receptor Subtypes to Cognitive Set-Shifting Abilities in the Rat

MARK R. STEFANI AND BITA MOGHADDAM

Department of Psychiatry, Yale University School of Medicine,
VA Medical Center, West Haven, Connecticut 06516, USA

KEYWORDS: set-shift ability; schizophrenia; glutamatergic neurotransmission; NMDA receptor

Impaired cognitive function is a fundamental aspect of the psychopathology of schizophrenia.[1] One aspect of cognition impaired in schizophrenia is "behavioral flexibility," which describes the ability of an animal to redirect, or shift, ongoing behavior in response to changes in internal goals or environmental stimuli.[2] Also referred to as "set-shifting" ability, behavioral flexibility is a complex construct, requiring attention, working memory, response selection, and response inhibition.

Behavioral flexibility is commonly assessed in patients with schizophrenia using the Wisconsin Card Sort Task (WCST). Performance of this task requires subjects to sort cards according to periodically shifting rules based on stimulus categories that include color, shape, and number.[2,3] The impaired WCST performance of schizophrenics is characterized by increased perseverative behavior: although schizophrenics learn the first sorting rule at the same rate as healthy individuals, they are unable to suppress responding according to the initial rule once the first category shift has been made and the initial sorting rule is no longer correct.[4] The neural mechanisms responsible for the WCST deficits observed in schizophrenia are not known.

Aberrant NMDA receptor–mediated glutamatergic neurotransmission has been suggested as an underlying cause of the cognitive deficits associated with schizophrenia (for review, see Tamminga[5]). NMDA receptor antagonists, such as phencyclidine, produce in humans symptoms closely resembling those of schizophrenia, including cognitive deficits. In rodents, NMDA receptor antagonists also influence cognitive abilities, impairing spatial memory, reversal learning, and response inhibition.[6-8] Abnormal prefrontal cortical function is strongly implicated in schizophrenia-associated cognitive deficits, as well as impaired behavioral flexibility in

---

Address for correspondence: Mark R. Stefani, Department of Neuroscience, University of Pittsburgh, 446 Crawford Hall, Pittsburgh, PA 15260. Voice: 412-624-9309; fax: 412-624-9198.
stefani@bns.pitt.edu

general. Prefrontal lesions or inactivation impairs set-shifting ability in both humans[3] and rodents.[9,10]

We examined the effects of glutamate receptor blockade in the rat medial prefrontal cortex (mPFC) on performance of a maze-based task, analogous to the WCST, that measures behavioral flexibility in rats (described in detail by Stefani and colleagues[8]). The effects of antagonism of both AMPA and NMDA glutamate receptor subtypes were studied.

The set-shift task was conducted in a four-arm maze whose arms varied along two stimulus dimensions, brightness and texture. Rats were trained in a single session on Set 1, in which they learned a simple discrimination based on either brightness (dark vs. light arms) or texture (rough vs. smooth arms) to receive a food reward. Rats were trained to a criterion performance level of eight consecutive correct arm choices on Set 1. On the following day, rats were shifted to Set 2, that is, they were trained to discriminate maze arms on the basis of the alternate stimulus dimension, e.g., from light to rough, and trained for 80 trials. Bilateral intra-cortical injections into the mPFC of a vehicle solution, the AMPA receptor antagonist LY293558 (1 μg per hemisphere), or the NMDA receptor antagonist MK-801 (1 or 3 μg per hemisphere) were made 20 min prior to the start of Set 2 training. An interval of 15 s separated trials during Sets 1 and 2.

Rats required 51.8±2.2 trials and 31.1±1.6 min (means ± SEM) to acquire the Set 1 rule to criterion level. Glutamate receptor blockade during Set 2 training impaired acquisition of the Set 2 stimulus-reward association rule. Both LY293558 and the 3 μg per hemisphere dose of MK-801 significantly increased the number of trials required to reach the criterion performance level. There were no significant differences between the effects of the 1 μg per hemisphere dose of MK-801 and vehicle on any measure. The number of trials required to reach criterion for the treatment groups were (mean±sem): Vehicle, 54.0±4.7; 1 μg MK-801, 49.1±5.1; 3 μg MK-801, 72.5±3.3; and 1 μg LY293558, 76.5±2.5. Glutamate receptor blockade significantly decreased the rate of learning across trial blocks, relative to vehicle-injected controls (FIG. 1, A). There were no significant treatment-dependent differences in the amount of time required to complete the 80 trials of Set 2 (data not shown), suggesting that the set-shifting deficit observed was not due to impaired motor abilities or decreased motivation.

Further analysis indicated that AMPA and NMDA receptor blockade differed qualitatively in their effects on set-shifting ability. Perseverative responding on Set 2 was assessed by evaluating performance from each of two start arm types: Perseveration Arms (PA) and Reinforcement Arms (RA). For a given rat, Perseveration Arms were the start arms from which, during Set 2, responding according to the correct stimulus-reward contingency from Set 1 produced an incorrect response. Reinforcement Arms were the start arms from which, during Set 2, responding according to the correct stimulus-reward contingency from Set 1 produced a (spuriously) correct response. LY293558 impaired performance nonspecifically, decreasing performance to near-chance levels from both PA and RA starts. MK-801 impaired set-shifting abilities specifically by increasing the number of perseverative responses across trial blocks (FIG. 1, B and C).

Our data demonstrate that glutamatergic neurotransmission within the mPFC is required for normal set-shifting performance in rats. Furthermore, we observed a dissociation between the effects of AMPA and NMDA receptor blockade. AMPA

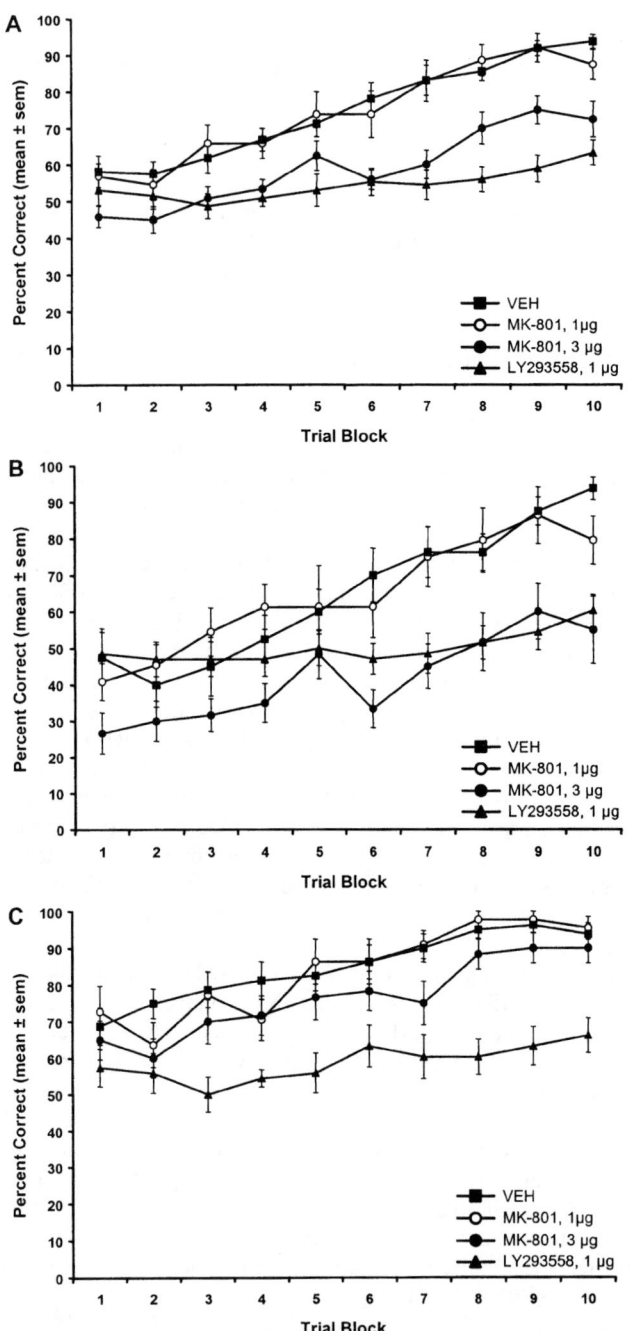

FIGURE 1. *See following page for legend.*

receptor blockade within the mPFC, which would be expected to decrease overall prefrontal glutamatergic transmission, produced a general deficit in learning and memory. That is, rats were neither able to learn the new, Set 2, strategy, nor did they demonstrate significant memory for, or ability to use, the strategy learned during previous training on Set 1. Blockade of NMDA receptors instead impaired only the shift to the new strategy; rats continued to respond according to the previously learned Set 1 response rule. The data suggest that dysfunctional prefrontal NMDA receptor-mediated neurotransmission may underlie the deficits in behavioral flexibility associated with schizophrenia.

REFERENCES

1. GREEN, M.F. 1996. What are the functional consequences of neurocognitive deficits in schizophrenia? Am. J. Psychiatry **153:** 321–330.
2. GREEN, M.F., P. SATZ, S. GANZELL & J.F. VACLAV. 1992. Wisconsin card sorting test performance in schizophrenia: Remediation of a stubborn deficit. Am. J. Psychiatry **149:** 62–67.
3. MILNER, B. 1963. Effects of different brain lesions on card-sorting, the role of the frontal lobes. Arch. Neurol. **9:** 90–100.
4. GOLDBERG, T.E., D.R. WEINBERGER, K.F. BERMAN, et al. 1987. Further evidence for dementia of the prefrontal type in schizophrenia? A controlled study of teaching the Wisconsin Card Sorting Test. Arch. Gen. Psychiatry **44:** 1008–1014.
5. TAMMINGA, C.A. 1998. Schizophrenia and glutamatergic transmission. Crit. Rev. Neurobiol. **12:** 21–36.
6. AULTMAN, J.M. & B. MOGHADDAM. 2001. Distinct contributions of glutamate and dopamine receptors to temporal aspects of rodent working memory using a clinically relevant task. Psychopharmacology **153:** 353–364.
7. JENTSCH, J.D. & J.R. TAYLOR. 2001. Impaired inhibition of conditioned responses produced by subchronic administration of phencyclidine to rats. Neuropsychopharmacology **24:** 66–74.
8. STEFANI, M.R., K. GROTH & B. MOGHADDAM. 2003. Glutamate receptors in the rat medial prefrontal cortex regulate set-shifting ability. Behav. Neurosci. **117:** 728–737.
9. BIRRELL, J.M. & V.J. BROWN. 2000. Medial frontal cortex mediates perceptual attentional set shifting in the rat. J. Neurosci. **20:** 4320–4324.
10. RAGOZZINO, M.E., S. DETRICK & R.P. KESNER. 1999. Involvement of the prelimbic-infralimbic areas of the rodent prefrontal cortex in behavioral flexibility for place and response learning. J. Neurosci. **19:** 4585–4594.

FIGURE 1. (**A**) Overall performance. LY293558 and MK-801 impaired acquisition of the stimulus discrimination task across trial blocks. At Block 10, rats receiving either LY293558 or 3 µg/hemisphere MK-801 performed significantly worse than did those receiving vehicle (VEH) ($P<.05$ vs. VEH). There were no significant performance differences between the VEH and the 1 µg/hemisphere MK-801 group. (**B**) Perseveration arm (PA) performance. Rats in the LY293558 and 3 µg/hemisphere MK-801 groups performed significantly worse on the Set 2 discrimination from PA starts than did rats in the VEH or 1 µg/hemisphere MK-801 groups ($P<.05$). (**C**) Reinforcement arm (RA) performance. Only rats treated with LY293558 failed to significantly improve their performance across trial blocks from RA starts. At Trial Block 10, the LY293558 group performed significantly worse than those in the VEH, 1 or 3 µg/hemisphere MK-801 groups ($P<.05$).

Evidence for a Relationship between Group 1 mGluR Hypofunction and Increased Cocaine and Ethanol Sensitivity in Homer2 Null Mutant Mice

KAREN K. SZUMLINSKI,[a] SHIGENOBU TODA,[a] LAWRENCE D. MIDDAUGH,[a,b] PAUL F. WORLEY,[c] AND PETER W. KALIVAS[a]

[a]*Department of Physiology and Neuroscience and* [b]*Department of Psychiatry and Behavioral Sciences, Medical University of South Carolina, Charleston, South Carolina 29425, USA*

[c]*Department of Neurosciences, The Johns Hopkins University School of Medicine, Baltimore, Maryland, USA*

KEYWORDS: Homer proteins; mGluR hypofunction; cocaine; ethanol; Homer2 null mutant mice

Homer proteins are encoded by three genes (*Homer1-3*), providing both constitutive forms of Homer (Homer1b/c, Homer2 and Homer3) and an immediate-early gene product, Homer1a.[1,2] Constitutively expressed Homer proteins interact, via an EVH1 binding domain, with the C-terminus of Group 1 metabotropic receptors (mGluRs), the inositol triphosphate receptor, and the postsynaptic scaffolding protein Shank.[1,3] Earlier reports revealed that, when compared to wild-type (WT) mice, Homer2 protein null mutant mice (*Homer2* knock-out; KO) exhibit increased sensitivity to the psychomotor-activating and conditioned rewarding effects of cocaine.[4] Consistent with a role for Homer proteins in the regulation of glutamate transmission,[1–3] the "presensitized" behavioral phenotype of *Homer2* KO mice was associated with a 50% reduction in basal extracellular levels of glutamate and an enhanced capacity of acute cocaine to elevate glutamate in the ventral striatum.[4] These alterations in ventral striatal glutamate transmission in *Homer2* KO mice are similar those reported in cocaine-sensitized rats,[5] thus implicating the constitutive expression of Homer proteins and presumably their regulation of Group 1 mGluR function in the neural plasticity underlying the development of addiction-related behaviors.

Ethanol is a drug of abuse that is well-documented to influence glutamate receptor function[6–9] and to affect glutamate transmission in the ventral striatum.[10,11]

Address for correspondence: Karen K. Szumlinski, Ph.D., Department of Physiology and Neuroscience, Basic Science Building, Suite 403, Medical University of South Carolina, 173 Ashley Avenue, Charleston, SC 29425. Voice: 843-792-1838; fax: 843-792-4423.
szumlink@musc.edu

Ann. N.Y. Acad. Sci. 1003: 468–471 (2003). © 2003 New York Academy of Sciences.
doi: 10.1196/annals.1300.055

Because Homer proteins regulate glutamate receptor function, basal extracellular levels of glutamate, and drug-induced changes in glutamate in the ventral striatum, WT and *Homer2* KO mice were compared with respect to ethanol-induced neural and behavioral plasticity. As increased sensitivity to certain behavioural effects of both cocaine and ethanol has been related to drug-induced alterations in Group 1 mGluR function,[6,7,9] genotypic differences in the expression of the Group 1 mGluR subtypes mGluR1a and mGluR5 as well as the capacity of an intraventral striatum infusion of the nonselective Group 1 mGluR agonist DHPG to elicit an increase in motor hyperactivity and extracellular glutamate were assessed. The results of all the behavioral and neurochemical assays conducted for this report are summarized in TABLE 1.

Homer2 deletion blunted ethanol reward, an effect associated with impaired development of ethanol-induced neural plasticity. Eight pairings of ethanol (0, 0.3, 1.0, and 3.0 g/kg, i.p., vol=0.02 mL/kg) with the nonpreferred compartment of a biased place conditioning apparatus elicited dose-dependently the expression of place conditioning in WT mice. In contrast, no place conditioning was apparent in KO animals. When allowed 24-h home cage access to two sipper tubes, one containing water and the other containing increasing concentrations of ethanol (0, 3, 6, and 12% vol/vol), *Homer2* KO mice exhibited aversion to 12% ethanol and a corresponding reduction in ethanol consumption. The reduction in ethanol reward may be attributed to an increased sensitivity to the motor-impairing/sedative effects of ethanol as KO mice exhibited longer sedation in response to an acute injection of 5 g/kg ethanol. The acute sedative effects of 3 g/kg ethanol were associated with an ethanol-induced reduction in extracellular glutamate in the ventral striatum. Upon repeated administration of 3 g/kg ethanol (8 injections, i.p.) an increase in ethanol-induced locomotion was observed across injections in WT mice, which was associated with an increase in the capacity of an ethanol challenge to elevate ventral striatal levels of both dopamine and glutamate. Consistent with their increased sensitivity to ethanol-induced motor-impairment/sedation, no such increase in the locomotor or the neurochemical effects of repeated ethanol administration was observed in KO mice. Collectively, these data demonstrate that the constitutive expression of Homer2 is necessary for the development of ethanol-induced neural plasticity that contributes to the psychomotor-activating and incentive motivational effects of ethanol.

Homer2 deletion blunted ventral striatal Group1 mGluR function and reduced mGluR1a protein content. The capacity of an intraventral striatal infusion of the Group1 mGluR agonist DHPG to elicit an increase in both motor activity (10 nmol/side) and extracellular glutamate (0, 3, 30, and 300 µmol/side) was blunted markedly in KO versus WT subjects. The reduction in ventral striatal Group1 mGluR function observed for *Homer2* KO mice was associated with a reduction in the total protein content for the dimer form of mGluR1a, but not mGluR5. These data provide *in vivo* evidence for the functional regulation of Group1 mGluRs by constitutively expressed Homer2 protein.[1,2] Furthermore, as Group1 mGluR antagonists are reported to decrease ethanol consumption/preference and to increase the sedative effects of ethanol,[12,13] the observed genotypic differences in both cocaine and ethanol reward and motor activity implicate the regulation of mGluRs by Homer2 in the determination of acute drug sensitivity, drug-induced neural plasticity, and addiction vulnerability.

TABLE 1. Summary of the observed genotypic differences in drug-induced changes in behavior and neurochemistry between *Homer2* KO and WT mice and their relationship to basal genotypic differences in ventral striatal glutamate transmission

Drug	Assay	Drug Regimen	Observed Effect	Statistics
Cocaine	Place conditioning	4 × 0–50 mg/kg	KO > WT	Genotype effect: $P=.008$ ($N\geq6$ per dose)[4]
	Locomotion	1 × 0–50 mg/kg	KO > WT	Genotype effect: $P=.002$ ($N\geq6$ per dose)[4]
	Glutamate	1 × 15 mg/kg	KO > WT	Genotype×Time: $P=.006$ ($N\geq6$)[4]
	Dopamine	1 × 15 mg/kg	KO = WT	Genotype×Time: $P=.94$ ($N\geq6$)[4]
Ethanol	Place conditioning	8 × 0–3 g/kg	KO < WT	Genotype×Dose: $P=.004$ ($N\geq6$)
	Preference	0–12% free access	KO < WT	Genotype×Dose: $P=.004$ (KO: $N=17$; WT: $N=22$)
	Consumption	0–12% free access	KO < WT	Genotype×Dose: $P=.05$ (KO: $N=17$; WT: $N=22$)
	Righting reflex	1 × 5 g/kg	KO > WT	Genotype effect: $P=.04$ ($N=10$)
	Locomotion	8 × 0–3 g/kg	KO < WT	Genotype×Injection: $P=.05$ ($N\geq6$)
	Glutamate	8 × 3 g/kg	KO < WT	Genotype×Injection×Time: $P=.08$ ($N\geq5$)
	Dopamine	8 × 3 g/kg	KO < WT	Genotype×Injection×Time: $P=.007$ ($N\geq5$)
	Basal glutamate	drug-naïve	KO < WT	Genotype effect: $P=.0008$ ($N\geq5$)[4]
	mGluR1a dimer	drug-naïve	KO < WT	t test: $P=.05$ ($N\geq7$)[4]
	mGluR5 dimer	drug-naïve	KO = WT	t test: $P=.43$ ($N\geq7$)[4]
DHPG	Locomotion	10 nmol/side (IC)	KO = WT	Genotype effect: $P>.05$ ($N\geq7$)[4]
	Rearing	10 nmol/side (IC)	KO < WT	Genotype effect: $P=.05$ ($N\geq7$)[4]
	Glutamate	0–300 µM (via probe)	KO < WT	Genotype×Dose: $P<.0001$ ($N\geq6$)[4]

NOTE: IC refers to an intra-ventral striatal injection.

ACKNOWLEDGMENTS

The authors would like to thank Kelly A. Frys, Jennifer K. Walker, and Ashley R. Mason for their assistance with the behavioral studies, as well as Marlin H. Dehoff, Shin H. Kang, and Glory Harris (Johns Hopkins) for generation of the KO mouse and genotyping. This work was supported by National Institute of Drug Abuse Grants DA-03906 (P.W.K.) and DA-11742 (P.R.W.), National Institute on Alcohol Abuse and Alcoholism Grant P50-AA1076, Project #2 (L.D.M.), and National Institute of Mental Health Grant MH-40817 (P.W.K.). K.K.S. is supported by a postdoctoral fellowship from the Canadian Institutes of Health Research.

REFERENCES

1. BRAKEMAN, P.R., A.A. LANAHAN, R. O'BRIEN, *et al.* 1997. Homer: a protein that selectively binds metabotropic glutamate receptors. Nature **386:** 284–288.
2. XIAO, B., J.C. TU & P.F. WORLEY. 2000. Homer: a link between neural activity and glutamate receptor function. Curr. Opin. Neurobiol. **10:** 370–374.
3. KATO, A., F. OZAWA, Y. SAITOH, *et al.* 1998. Novel members of the Vesl/Homer family of PDZ proteins that bind metabotropic glutamate receptors. J. Biol. Chem. **273:** 23969–23975.
4. SZUMLINSKI, K.K., M.H. DEHOFF, S.H. KANG, *et al.* 2003. *Homer2* deletion pre-sensitizes cocaine reward and motor activity: role of accumbens glutamate. Proc. Natl. Acad. Sci. Submitted for publication.
5. SWANSON, C.J., D.A. BAKER, D. CARSON, *et al.* 2001. Repeated cocaine administration attenuates group I metabotropic glutamate receptor-mediated glutamate release and behavioral activation: a potential role for Homer. J. Neurosci. **21:** 9043–9052.
6. CHRISTIAN, M., G.Y. SUN & A. SIMONYI. 2002. Chronic ethanol-induced decrease in the expression of metabotropic glutamate receptors in rat hippocampus. Soc. Neurosci. Abstr. no. 339.8.
7. GRUOL, D.L., K.L. PARSONS & N. DIJULIO. 1997. Acute ethanol alters calcium signals elicited by glutamate receptor agonists and K^+ depolarization in cultured cerebellar Purkinje neurons. Brain Res. **773:** 82–89.
8. LOVINGER, D.M., G. WHITE & F.F. WEIGHT. 1989. Ethanol inhibits NMDA-activated ion current in hippocampal neurons. Science **243:** 1721–1724.
9. MINAMI, K., R.W. GEREAU, IV, M. MINAMI, *et al.* 1998. Effects of ethanol and anesthetics on type 1 and 5 metabotropic glutamate receptors expressed in Xenopus oocytes. Mol. Pharmacol. **53:** 148–156.
10. MOGHADDAM, B. & M.L. BOLINAO. 1994. Biphasic effect of ethanol on extracellular accumulation of glutamate in the hippocampus and the nucleus accumbens. Neurosci. Lett. **178:** 99–102.
11. DAHCHOUR, A., A. HOFFMAN, R. DEITRICH & P. DE WITTE. 2000. Effects of ethanol on extracellular amino acid levels in high- and low-alcohol sensitive rats: a microdialysis study. Alcohol Alcohol. **35:** 548–553.
12. MCMILLAN, B.A., M.S. CRAWFORD, C.M. KULERS & H.L. WILLIAMS. 2002. Effect of a mGlu5 metabotropic glutamate receptor antagonist on ethanol consumption by genetic drinking rats. Soc. Neurosci. Abstr. no. 607.8.
13. SHARKO, A.C., C.W. HODGE, K. ILLER, *et al.* 2002. Involvement of metabotropic glutamate receptor subtype 5 (mGlu5) in alcohol self-administration and sedation. Soc. Neurosci. Abstr. no. 783.1.

Bidirectional Modulation of Cystine/Glutamate Exchanger Activity in Cultured Cortical Astrocytes

XING-CHUN TANG AND PETER W. KALIVAS

Department of Physiology and Neuroscience, Medical University of South Carolina, Charleston, South Carolina 29425, USA

KEYWORDS: cystine/glutamate exchanger; astrocyte; cystine; glutamate; group II metabotropic glutamate receptor; protein kinase A; protein kinase C

Transport of cystine across the cell membrane is essential for the synthesis of the major cellular antioxidant glutathione and the maintenance of extracellular basal glutamate levels.[1] Cystine uptake in the brain occurs by both the Na^+-independent cystine/glutamate exchanger (system x_c^-), which mediates the entry of cystine into cells coupled to the efflux of glutamate and the X_{AG-} family of high-affinity, Na^+-dependent glutamate transporter. System x_c^- is found in the brain at a density near to that of the Na^+-dependent glutamate transporter. Control of extracellular glutamate concentration in the CNS by these membrane transporters is important for terminating synaptic transmission and for keeping glutamatergic tone at an ideal level.[2] An earlier study in this lab has confirmed a close relationship between system x_c^- and extracellular level of glutamate *in vivo*.[3] Since there are consensus phosphorylation sites within system x_c^-,[4] it is possible that the phosphorylation state may play an important role in regulating the activity of system x_c^-. This study was directed towards investigating the modulation of system x_c^- by kinase activity in primary-subculture cortical glia cultures.

Primary cultures of astrocytes were prepared from cerebral cortices of 1- to 2-day-old rat pups.[5] The tissue was dissociated and plated in 150 cm² flasks at a density of 2.5×10^5 cells/mL. When the primary culture reached confluence, the monolayer of cells were trypsinized and plated into 12-well plates. Cells cultured for 21–35 days were used in the experiments. Uptake studies were performed in triplicate in bicarbonate buffered Krebs-Ringer solution (pH 7.4).[6] To study the specific cystine/glutamate antiporter, asparate and acivicin were added into each well to block system X_{AG}-and γ-glutamyl transpeptidase.[7] Cystine uptake experiments were initi-

---

Address for correspondence: Xing-Chun Tang, Ph.D., Department of Physiology and Neuroscience, Basic Science Building, Suite 403, Medical University of South Carolina, 173 Ashley Avenue, Charleston, SC 29425. Voice: 843-792-1838; fax: 843-792-4423.
tangxc@musc.edu

Ann. N.Y. Acad. Sci. 1003: 472–475 (2003). © 2003 New York Academy of Sciences.
doi: 10.1196/annals.1300.056

TABLE 1. Modulation of sodium-independent L-[^{35}S]-cystine uptake by metabotropic glutamate 2,3 receptor agonist and antagonist in primary-subcultured cortical astrocytes

Drugs	Concentration (μM)	Cystine uptake (% of control)
Control		100 ± 2.5
LY341495	0.01	99 ± 3.8
	0.1	92 ± 4.5
	1.0	66 ± 4.6**
	10	94 ± 4.0
APDC	0.1	101 ± 3.4
	1.0	135 ± 14*
	10	141 ± 5.4*
	100	104 ± 12
	1000	66 ± 5.3*
	10 (plus 1 μM of LY341495)	109 ± 3.8
	10 (plus 1,000 μM of APICA)	132 ± 2.6**
	1,000 (plus 1 μM of LY341495)	65 ± 2.5**
	1,000(plus 1 μM of APICA)	73 ± 6.4**

Cultured astrocytes were incubated with 0.5 μCi/mL of L-[^{35}S]-cystine for 15 min at room temperature in the presence of different drugs at the concentration indicated. The results are means ± SEM of at least three independent observations, measured in triplicate. **$P<.01$, *$P<.05$ compared to control.

ated by adding 0.5 μCi/mL of L-^{35}S-cystine and incubate the culture for 15 min at room temperature. The uptake was terminated by rapidly washing the cells three times with ice-cold media. Cells were solubilized in 0.5 mL of RIPA buffer. Radioactivity and protein content in the cell lysate were measured.

(2R,4R)-APDC, a metabotropic glutamate receptor group II (mGluR2,3) agonist (1,000 μM) greatly decreased the uptake of cystine (30~40%), which could not be blocked by the group II (mGluR2,3) antagonists (R,S)-APICA or LY341495. Interestingly, 1 and 10 μM of APDC significantly increased the uptake of cystine (20~40%), which could be abolished by LY341495 (1 μM). Astrocytes also showed ~35% reduction in cytine uptake when in the presence of LY341495 (1 μM). The results of effects of mGluR2,3 agonist and antagonist on cystine uptake are summarized in TABLE 1. Similar to the effect of APDC, H89, a PKA inhibitor, also produced a bell-shaped dose-response curve with stimulation of system x_c^- from 10 nM to 1 μM of H89 and inhibition of system x_c^- at 1 μM of H89. The PKA activator forskolin showed an inhibitory effect on the antiporter activity. Furthermore, elimination of the activity of PKC by application of 0.1 or 1 μM chelerythrine chloride potentiated the cystine uptake and activation of PKC resulted in the inhibition of cystine uptake. The modulation of system x_c^- activity by activators and inhibitors of PKA and PKC are summarized in TABLE 2. Finally, the activity of system x_c^- was stimulated by blockade of calcium/calmodulin-dependent kinase II with KN93. Taken together,

TABLE 2. Effects of inhibitors, activators of different kinases on sodium-independent L-[^{35}S]-cystine uptake in primary-subcultured cortical astrocytes

Drugs	Concentration (μM)	Cystine uptake (% of control)
Control		100 ± 4.1
Forskolin	1.0	107 ± 6.2
	10	82 ± 1.6**
	100	102 ± 4.0
H-89	0.001	104 ± 5.2
	0.01	142 ± 1.6**
	0.1	139 ± 3.8**
	1.0	132 ± 3.9**
	10	99 ± 11
	100	43 ± 6.5**
	0.01 (plus 10 μM of forskolin)	97 ± 7.8
	0.1 (plus 10 μM of forskolin)	105 ± 3.4
	1.0 (plus 10 μM of forskolin)	102 ± 2.2
PMA	0.0001	104 ± 2.7
	0.001	81 ± 5.5**
	0.01	79 ± 6.0**
	0.1	98 ± 6.9
Chelerythrine chloride	0.01	103 ± 6.7
	0.1	132 ± 7.5**
	1.0	138 ± 8.6**
	10	77 ± 1.3
	1.0 (plus 0.001 μM of PMA)	105 ± 5.8
	1.0 (plus 0.01 μM of PMA)	99 ± 9.3
	10 (plus 0.001 μM of PMA)	64 ± 6.1**
	10 (plus 0.01 μM of PMA)	60 ± 3.5**

**$P<.01$, *$P<.05$ compared to control.

these data suggest that group II metabotropic glutamate receptor and PKA signaling system can bidirectionally modulate the activity of system x_c^-.[8] Our data also illustrate that PKC and calcium/calmodulin protein kinase II pathways are also involved in the modulation of the activity of system x_c^-.

ACKNOWLEDGMENTS

The authors would like to thank Judson Chandler and Ezekiel Carpenter-Hyland for their assistance with the astrocyte culture. This work was supported by National Institute of Drug Abuse Grant DA-03906 (P.W.K.) and National Institute of Mental Health Grant MH-40817 (P.W.K.).

REFERENCES

1. WARR, O., M. TAKAHASHI & D. ATTWELL. 1999. Modulation of extracellular glutamate concentration in rat brain slices by cystine-glutamate exchange. J. Physiol. **514:** 783–793.
2. MCBEAN, G.J. 2002. Cerebral cystine uptake: a tale of two transporters. Trends Pharmacol. Sci. **23:** 299–302.
3. BAKER, D.A., Z.X. XI, H. SHEN, et al. 2002. The origin and neuronal function of in vivo nonsynaptic glutamate. J. Neurosci. **22:** 9134–9141.
4. BRIDGES, C.C., R. KEKUDA, H. WANG, et al. 2001. Structure, function, and regulation of human cystine/glutamate transporter in retinal pigment epithelial cells. Invest. Ophthalmol. Vis. Sci. **42:** 47–54.
5. DUCIS, I., L.O. NORENBERG & M.D. NORENBERG. 1990. The benzodiazepine receptor in cultured astrocytes from genetically epilepsy-prone rats. Brain Res. **531:** 318–321.
6. BENDER, A.S., W. REICHELT & M.D. NORENBERG. 2000. Characterization of cystine uptake in cultured astrocytes. Neurochem. Intern. **37:** 269–276.
7. COTGREAVE, I.A. & I. SCHUPPE-KOISTINEN. 1994. A role for gamma-glutamyl transpeptidase in the transport of cystine into human endothelial cells: relationship to intracellular glutathione. Biochim. Biophys. Acta **1222:** 375–382.
8. GOCHENAUER, G.E. & M.B. ROBINSON. 2001. Dibutyryl-cAMP (dbcAMP) up-regulates astrocytic chloride-dependent L-[^{3}H]glutamate transport and expression of both system xc(–) subunits. J. Neurochem. **78:** 276–286.

Dopamine–Glutamate Interactions in the Control of Cell Excitability in Medial Prefrontal Cortical Pyramidal Neurons from Adult Rats

KUEI-YUAN TSENG AND PATRICIO O'DONNELL

*Center for Neuropharmacology and Neuroscience,
Albany Medical College, Albany, New York 12208, USA*

KEYWORDS: dopamine–glutamate interactions; prefrontal cortex; electrophysiology

Understanding the cellular basis of dopamine (DA) actions on prefrontal cortical (PFC) activity has implications for cognitive functions and the pathophysiology of neuropsychiatric disorders. Important elements in this regard are the interactions between DA and glutamate (GLU). DA can affect GLU function depending on the receptor subtypes involved. In the striatum, D1 enhances NMDA currents, whereas D2 depresses AMPA-mediated responses.[1,2] In the PFC, D1 enhancement of NMDA increase of pyramidal neuron excitability has been observed in slices from P24- to 28-day-old rats.[3] However, the role of D2 DA receptor on NMDA/AMPA-mediated responses remains unexplored. Given that neocortical pyramidal neurons undergo several changes in both morphological and physiological properties during postnatal development until P42,[4] it is possible that the effects observed in previous *in vitro* studies (<P42) may not necessarily reflect the function in adulthood. As an attempt to address this issue, whole-cell patch clamp recordings in layer V medial PFC pyramidal neurons were performed in adult (P42–65) rat brain slices, measuring changes in cell excitability induced by combinations of DA and GLU agonists.

Medial prefrontal cortical slices were obtained from adult male Sprague-Dawley rats (P42–65). Rats were anesthetized with chloral hydrate (400 mg/kg, i.p.) before being decapitated. Brains were rapidly removed into ice-cold artificial CSF (aCSF: in mM; 125 NaCl, 25 $NaHCO_3$, 10 glucose, 3.5 KCl, 1.25 NaH_2PO_4, 0.5 $CaCl_2$, 3 $MgCl_2$; pH 7.45; 295±5 mOsm). Coronal slices (300 µm) containing the medial PFC were cut on a Vibratome in ice-cold aCSF, transferred and incubated in warm (±35°C) aCSF solution constantly oxygenated with 95% O_2–5% CO_2 for at least 60 min before recording. Whole-cell current clamp recordings were performed in pyra-

---

Address for correspondence: Kuei-Yuan Tseng, Center for Neuropharmacology and Neuroscience, Albany Medical College (MC-136), 47 New Scotland Avenue, Albany, NY 12208. Voice: 518-262-0688; fax: 518-262-5799.
tsengky@mail.amc.edu

midal neurons located in layer V of the medial PFC identified under visual guidance using infrared-differential interference contrast (IR-DIC) video microscopy with a 40× water-immersion objective. The image was detected with an IR-sensitive CCD camera and displayed on a monitor. All experiments were conducted at 33–35°C. In the recording aCSF (perfused at 2 mL/min), $CaCl_2$ was increased to 2 mM and $MgCl_2$ was decreased to 1 mM. Patch pipettes (5–8 MΩ) were filled with (in mM): 115 K-gluconate, 10 HEPES, 2 $MgCl_2$, 20 KCl, 2 MgATP, 2 Na_2-ATP, 0.3 GTP (pH=7.3, 280±5 mOsm). In each cell, membrane potential, number of spikes and latency to the first spike evoked by a 500 ms duration depolarizing current pulse, as well as input resistance, were analyzed before (baseline) and after drug treatment. All drugs were mixed into oxygenated aCSF and applied in the recording solution in known concentrations. Both control and drug-containing aCSF were continuously oxygenated throughout the experiments.

Both D1 and D2 agonists affected NMDA-mediated responses. Bath application of NMDA induced concentration-dependent (from 1 to 8 μM) excitability increases. This effect was enhanced in presence of a marginally effective dose of SKF38393 (2 μM), and reduced by quinpirole (0.4 μM). The D1-NMDA synergism was blocked by addition of the D1 antagonist (SCH23390 10 mM) to the bath, and requires both an intracellular Ca^{2+} increase and PKA activation. The D2-NMDA interaction was blocked in presence of the D2 DA receptor antagonist eticlopride (20 μM), or the $GABA_A$ antagonist bicuculline (10 μM). This interaction is not mediated by postsynaptic intracellular Ca^{2+} increases and PKA activity. Thus, in pyramidal cells from adult animals, D1 increase and D2 decrease postsynaptic NMDA responses.

AMPA-induced excitability increase was affected by D2, not D1 agonists. The D1 agonist SKF38393 failed to modify AMPA-mediated responses. However, AMPA-mediated concentration-dependent (from 0.05 to 0.4 μM) excitability increase was strongly reduced by activation of D2 DA receptors. In contrast to NMDA-mediated responses, the D2-AMPA interaction was not mediated by $GABA_A$ receptor activation, but instead involved both an intracellular Ca^{2+} increase (mediated by the PLC-IP_3 cascade) and PKA inhibition. These results indicate that activation of D2 DA receptors strongly reduces both NMDA and AMPA-mediated excitability increase through different cellular mechanisms. Together, this divergent role of D1 and D2 DA receptors on PFC GLU responses could play a crucial role in shaping synaptic plasticity processes with a strong impact on PFC functions depending on the receptors involved.

ACKNOWLEDGMENT

This work was supported by MH57683, MH60131, DA14020, and a National Alliance for Research on Schizophrenia and Depression Independent Investigator Award (P.O'D.).

REFERENCES

1. CEPEDA, C., N.A. BUCHWALD & M.S. LEVINE. 1993. Neuromodulatory actions of dopamine in the neostriatum are dependent upon the excitatory amino acid receptor subtypes activated. Proc. Natl. Acad. Sci. USA **90**: 9576–9580.

2. CEPEDA, C. & M.S. LEVINE. 1998. Dopamine and N-methyl-D-aspartate receptor interactions in the neostriatum. Dev. Neurosci. **20:** 1–18.
3. WANG, J. & P. O'DONNELL. 2001. D1 dopamine receptors potentiate NMDA-mediated excitability increase in layer V prefrontal cortical pyramidal neurons. Cereb. Cortex **11:** 452–462.
4. ZHU, J.J. 2000. Maturation of layer 5 neocortical pyramidal neurons: amplifying salient layer 1 and layer 4 inputs by Ca^{2+} action potentials in adult rat tuft dendrites. J. Physiol. **526:** 571–587.

Modulation of Excitatory Transmission onto Midbrain Dopaminergic Neurons of the Rat by Activation of Group III Metabotropic Glutamate Receptors

ORNELLA VALENTI, MICHAEL J. MARINO, AND P. JEFFREY CONN

Merck & Co. Inc., Neuroscience Department, West Point, Pennsylvania 19486, USA

KEYWORDS: neuromodulatory control; dopamine; mGluRs

Excitatory inputs onto dopamine neurons have been implicated in a number of disorders, including Parkinson's disease and schizophrenia. A great deal of both anatomical and functional data points to the importance of interactions between the dopaminergic and glutamatergic systems,[1] and both systems have been implicated in schizophrenia.[2] Midbrain dopaminergic neurons receive excitatory glutamatergic inputs from the cortex and subthalamic nucleus, as well as a mixed glutamatergic and cholinergic excitatory input from the pedunculopontine nucleus. Based on the central role of the dopamine system in multiple neurodegenerative and neuropsychiatric disorders, an understanding of the neuromodulatory control of synaptic inputs to the dopamine neurons may provide critical insight on potentially novel therapeutic mechanisms. To this end, we have been interested in the roles that metabotropic glutamate receptors (mGluRs) may play in the regulation of excitatory inputs onto midbrain dopamine neurons.

Eight mGluR subtypes have been cloned (designated mGluR1–mGluR8) from mammalian brain. These mGluRs are classified into three major groups based on sequence homologies, coupling to second messenger systems, and selectivities for various agonists.[3] Group I mGluRs, which include mGluR1 and mGluR5, couple primarily to Gq-type G-proteins and associated effectors including phospholipase C. Group II mGluRs (mGluR2 and mGluR3) and group III mGluRs (mGluR4, 6, 7, and 8) couple to Gi/Go-type G-proteins and associated effectors, such as adenylyl cyclase. The mGluRs are widely distributed throughout the central nervous system and play important roles in regulating cell excitability and synaptic transmission.[3,4]

A previous report demonstrated that agonists selective for each of the three groups of mGluRs induced an inhibition of excitatory transmission in the midbrain

Address for correspondence: Ornella Valenti, Merck Research Laboratories, Merck & Co., Inc., 770 Sumneytown Pike, P.O. Box 4, WP 46-300, West Point, PA 19486-0004. Voice: 215-652-5156; fax: 215-652-3811.

ornella_valenti@merck.com

Ann. N.Y. Acad. Sci. 1003: 479–480 (2003). © 2003 New York Academy of Sciences.
doi: 10.1196/annals.1300.058

dopamine neurons.[5] With the development of more selective pharmacological tools, it should now be possible to determine which mGluR subtypes mediate these responses. To this end, we have begun to characterize the group III mGluR–mediated decrease in excitatory transmission in the substantia nigra pars compacta (SNc) using whole-cell patch clamp recording. Low (1–10 µM) concentrations of the group III mGluR–selective agonists L-AP4 produced a robust and reversible decrease in excitatory postsynaptic current (EPSC) amplitude. This high potency suggests that L-AP4 is not acting at mGluR7, which normally is not activated by L-AP4 concentrations under 100 µM. Interestingly, the mGluR8-selective agonist S-(3,4)-DCPG (300 nM) failed to mimic the actions of L-AP4. This, combined with the fact that mGluR6 is not broadly expressed in the central nervous system, suggests that this effect may therefore be mediated by mGluR4. A biophysical analysis of the L-AP4-induced inhibition of excitatory transmission in midbrain dopamine neurons revealed that the effect is mediated through a presynaptic mechanism of action. Taken together, these studies suggest that a group III metabotropic glutamate receptor, likely mGluR4, mediates a presynaptic inhibition of excitatory transmission in midbrain dopamine neurons.

The finding that mGluR4 is the most likely group III mGluR mediating the inhibition of excitatory transmission in midbrain dopamine neurons was surprising. While previous anatomical studies have demonstrated a low level of mGluR4 immunoreactivity in the substantia nigra,[6,7] this immunostaining appears to be restricted to the pars reticulata. However, it must be kept in mind that pars compacta dopamine neurons extend dendrites to layers deep within the pars reticulata. Therefore, the excitatory synapses that are modulated by group III mGluR activation may lie outside of the dopaminergic cell body region. Further studies of the distribution of mGluR4 that explore the synaptic and subsynaptic localization of mGluR4 should provide information on the likelihood of this hypothesis. In addition, ongoing studies with mGluR4 knockout mice should allow for a definitive statement as to the receptor subtype that mediates the inhibition of transmission at this synapse.

REFERENCES

1. Tzschentke, T.M. 2001. Pharmacology and behavioral pharmacology of the mesocortical dopamine system. Prog. Neurobiol. **63:** 241–320.
2. Marino, M.J. & P.J. Conn. 2002. Direct and indirect modulation of the N-methyl D-aspartate receptor: Potential for the development of novel antipsychotic therapies. Curr. Drug Targets CNS Neurol. Disorders **1:** 1–16.
3. Conn, P.J. & J.P. Pin. 1997. Pharmacology and functions of metabotropic glutamate receptors. Annu. Rev. Pharmacol. Toxicol. **37:** 205–237.
4. Anwy, L.R. 1999. Metabotropic glutamate receptors: electrophysiological properties and role in plasticity. Brain Res. Rev. **29:** 83–120.
5. Wigmore, M.A. & M.G. Lacey. 1998. Metabotropic glutamate receptors depress glutamate-mediated synaptic input to rat midbrain dopamine neurones in vitro. Br. J. Pharmacol. **123:** 667–674.
6. Corti, C., L. Aldegheri, P. Somogyi & F. Ferraguti. 2002. Distribution and synaptic localisation of the metabotropic glutamate receptor 4 (mGluR4) in the rodent CNS. Neuroscience **110:** 403–420.
7. Bradley, S.R. *et al.* 1999. Immunohistochemical localization of subtype 4a metabotropic glutamate receptors in the rat and mouse basal ganglia. J. Comp. Neurol. **407:** 33–46.

Difference in mGluR5 Interaction between Positive Allosteric Modulators from Two Structural Classes

DAVID L. WILLIAMS JR., JULIE A. O'BRIEN, WEI LEMAIRE, TSING-BAU CHEN, RAYMOND S. L. CHANG, MARLENE A. JACOBSON, SOOKHEE N. HA,[b] DAVID D. WISNOSKI,[a] CRAIG W. LINDSLEY,[a] CYRILLE SUR, MARK E. DUGGAN,[a] DOUGLAS J. PETTIBONE, AND P. JEFFREY CONN

Departments of Neuroscience-WP, [a]Medicinal Chemistry and [b]Molecular Systems, Merck Research Laboratories, West Point, Pennsylvania and Rahway, New Jersey

KEYWORDS: mGluR; allosteric modulators

A number of selective and nonselective mGluR agonists and antagonists have been developed from structural analogues of glutamate, quisqualate, or phenylglycine[1] that act through binding at or near the agonist site in the amino terminal domain. With the advent of high capacity functional assays, it has been possible to expand the search for compounds that modulate GPCRs to include molecules that act through allosteric sites in addition to those acting through the normal orthosteric site.[2] We are interested in developing additional pharmacological tools to explore the function of mGluRs and have used this novel approach to search for compounds that act through allosteric mechanisms. Such compounds may afford very high subtype selectivity, since while there is considerable evolutionary pressure to maintain the structure of the natural agonist binding site, this seems unlikely to be the case for allosteric sites.

We have identified specific and selective positive allosteric modulators of the metabotropic glutamate receptor subtype 5 (mGluR5) from two different structural classes. 3,3'-Difluorobenzaldazine (DFB, FIG. 1) has no agonist activity, but acts as a selective positive allosteric modulator of the Ca^{2+} response to activation of human and rat mGluR5 expressed in Chinese hamster ovary (CHO) cells. DFB potentiates threshold responses to glutamate, quisqualate, and 3,5-dihydroxyphenylglycine 3- to 6-fold, with EC_{50} values in the 2–5 µM range, and at 10–100 µM shifts mGluR5 agonist concentration response curves approximately twofold to the left. N-{4-chloro-2-[(1,3-dioxo-1,3-dihydro-2H-isoindol-2-yl)methyl]phenyl}-2-hydroxybenzamide (CPPHA, FIG. 1), a compound from a different structural class, has no agonist activity, and acts as a selective positive allosteric modulator of human and rat mGluR5. CPPHA potentiates threshold Ca^{2+} responses to glutamate 6- to 13-fold with EC_{50}

Address for correspondence: David L. Williams, Jr., Department of Neuroscience, WP46-300, Merck Research Laboratories, West Point, PA 19486. Voice: 215-652-6798; fax: 215-652-3811.
david_williams1@merck.com

Ann. N.Y. Acad. Sci. 1003: 481–484 (2003). © 2003 New York Academy of Sciences.
doi: 10.1196/annals.1300.059

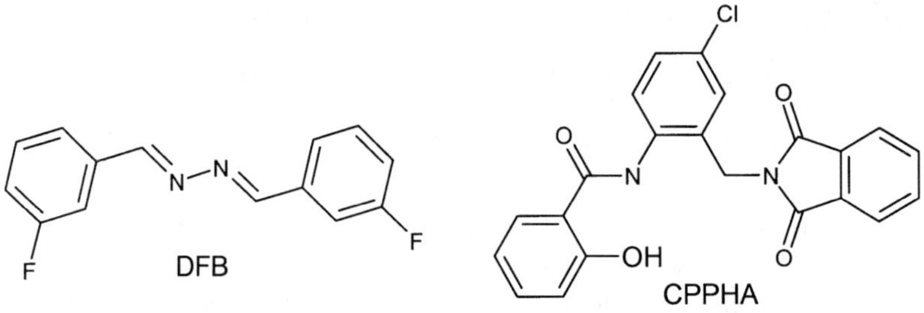

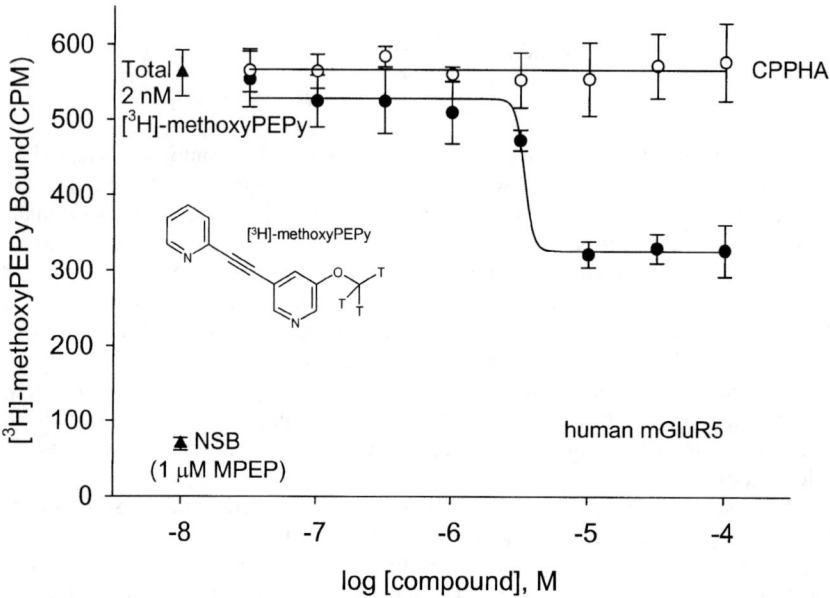

FIGURE 1.

FIGURE 2.

values in the 150–400 nM range, and at 10 µM shifts mGluR5 agonist concentration response curves to glutamate, quisqualate, and 3,5-dihydroxyphenylglycine 4- to 9-fold to the left.

At 100 µM, neither of these compounds affects [^{3}H]-quisqualate binding to mGluR5. However, DFB partially competes for binding of the 2-methyl-6-(phenylethynyl)-pyridine (MPEP) analogue [^{3}H]-3-methoxy-5-(2-pyridinylethynyl) pyridine (methoxyPEPy,[3] FIG. 2), while CPPHA has no effect on the binding of MPEP analogues to mGluR5.

This difference on the effect of these two classes of compounds on the binding site for the negative allosteric modulator MPEP indicates differences in interaction with mGluR5 and suggests the possibility of different binding sites and mechanisms of action. The binding sites of these compounds will be characterized in detail through the use of chimeric and mutant mGluRs.

REFERENCES

1. SCHOEPP, D.D., D.E. JANE & J.A. MONN. 1999. Pharmacological agents acting at subtypes of metabotropic glutamate receptors. Neuropharmacol. **38:** 1431–1476
2. CHRISTOPOULOS, A. 2002. Allosteric binding sites on cell surface receptors: novel targets for drug discovery. Nature Rev. Drug Disc. **1:** 198–210.
3. COSFORD, N.D.P., J. ROPPE, L. TEHRANI, et al. 2003. [^{3}H]-methoxymethyl-MTEP and [^{3}H]-methoxy-PEPy: potent and selective radioligands for the metabotropic glutamate subtype 5 (mGlu5) receptor. Bioorg. Med. Chem. Lett. **13:** 345–348.

Index of Contributors

Abi-Dargham, A., 138–158
Alonso, G., 212–225
Andrzejewski, M.E., 159–168
Atzori, M., 346–348

Baker, D., 169–175, 349–351, 445–447
Baldwin, A.E., 159–168
Behar, K.L., 452–453
Beneyto, M., 75–93, 352–355
Boutrel, B., 415–418
Bowers, M.S., 169–175, 356–357, 419–421
Brady, A.M., 358–363
Bressan, R.A., 364–367
Buccafusco, J.J., 381–385, 386–390
Burmeister, J.J., 454–457
Bymaster, F., 412–414

Carl, G.F., 381–385, 386–390
Carlezon, W.A., Jr., 368–371
Carr, D.B., 36–52
Chang, R.S.L., 481–483
Charara, A., 53–74
Charney, D.S., 273–291
Chartoff, E.H., 368–371
Chen, T.-B., 481–483
Choi, K.-H., 372–374
Clemens-Smith, A., 412–414
Clinton, S.M., 75–93
Conn, P.J., 12–21, 435–437, 479–480, 481–483
Coyle, J.T., 318–327
Cunningham, V.J., 364–367

D'Souza, D.C., 176–184
Dackis, C., 328–345
Davidkova, G., 375–377
Davis, R., 412–414
Day, B.K., 454–457
de Graaf, R.A., 452–453
de Lecea, L., 196–211
Denicoff, K.D., 273–291
Dong, J-Y., 419–421

Du, J., 273–291, 378–380, 402–404
Duggan, M.E., 481–483

Eastwood, S.L., 94–101
Edwards, G.L., 381–385, 386–390
Edwards, S., 372–374
Elkins, R.L., 381–385, 386–390
Ell, P.J., 364–367
Erlandsson, K., 364–367
Everitt, B.J., 410–411

Fagen, Z.M., 185–195
Faleiro, L.J., 391–394
Falke, C.S., 378–380, 402–404
Farber, N.B., 119–130
Faull, K., 458–460
Fei, Y.J., 381–385, 386–390
Floresco, S.B., 53–74
Flores-Hernandez, J., 346–348

Gao, X.-M., 113–118
Gasparini, F., 415–418
Gerhardt, G.A., 454–457
Ghasemzadeh, M.B., 395–397
Glantz, L.A., 102–112
Glick, S.D., 358–363
Goff, D., 318–327
Goldman, D., 22–35
Goto, Y., 398–401
Grace, A.A., 53–74
Gray, N.A., 273–291, 378–380, 402–404
Gregory, M.L., 405–409
Gunn, R.N., 364–367

Ha, S.N., 481–483
Hall, S., 372–374
Harrison, A.A., 415–418
Harrison, P.J., 94–101, 426–430
Herman, M.M., 426–430
Hernandez, P.J., 159–168
Hinkle, P.M., 458–460
Hobbs, S.H., 381–385, 386–390

Hoffman, B.J., 412–414
Holcomb, H.H., 113–118
Huettl, P., 454–457
Hufeisen, S.J., 461–463
Hutcheson, D.M., 410–411

Jackson, M.E., 131–137
Jacobson, M.A., 481–483
Johnson, K.W., 412–414
Jones, S., 391–394

Kalivas, P.W., 169–175, 349–351, 356–357, 395–397, 405–409, 419–421, 443–444, 445–447, 468–471, 472–475
Kanold, P., 346–348
Katner, J., 412–414
Kauer, J.A., 391–394
Keath, J.R., 185–195
Kegeles, L.S., 138–158
Kelley, A.E., 159–168
Kenny, P.J., 415–418
Kleinman, J.E., 426–430
Konradi, C., 368–371
Koob, G.F., 415–418
Krupitsky, E., 176–184
Krystal, J.H., 176–184, 292–308

Lahti, A.C., 113–118
Lake, R.W., 349–351, 356–357, 395–397, 419–421
Lanier, S.M., 356–357
Lapish, C.C., 356–357
Laruelle, M., 138–158
Lavin, A., 422–425
Law, A.J., 94–101, 426–430
Lemaire, W., 481–483
Lewis, B., 431–434
Lewis, D.A., 102–112
Lindsley, C.W., 481–483
Lipsky, R.H., 22–35
Lloyd, R.L., 458–460
Love, P., 412–414
Lovinger, D.M., 226–240

Madamba, S., 196–211
Malenka, R.C., 1–11

Mangiavacchi, S., 241–249
Manji, H.K., 273–291, 378–380, 402–404
Mansvelder, H.D., 185–195
Manzoni, O.J., 212–225
Marino, M.J., 435–437, 479–480
Markou, A., 415–418
Martin, G., 196–211
Mason, G.F., 292–308, 452–453
McCullumsmith, R.E., 75–93, 375–377, 438–442
McFarland, K., 169–175, 349–351, 356–357
McGehee, D.S., 185–195
Meador-Woodruff, J.H., 75–93, 352–355, 375–377, 438–442, 448–451
Medoff, D.R., 113–118
Melendez, R., 443–444, 445–447
Meyerhoff, J.A., 458–460
Middaugh, L.D., 468–471
Moghaddam, B., xiii–xiv, 131–137, 464–467
Moran, M.M., 445–447
Mueller, H.T., 448–451
Mulligan, R.S., 364–367

Nadeau, J.H., 461–463
Neve, R.L., 372–374
Nie, Z., 196–211
Nomikos, G., 412–414

O'Brien, C., 328–345
O'Brien, J.A., 435–437, 481–483
O'Donnell, P., 358–363, 398–401, 431–434, 476–478
Omelchenko, N., 36–52
Orr, T.E., 381–385, 386–390
Owens, J., 364–367
Owens, R.W., 405–409

Papadopoulou, M., 368–371
Partridge, J.G., 226–240
Patel, A.B., 452–453
Paterson, N.E., 415–418
Paul, I.A., 250–272
Paulson, M., 458–460
Pekary, A.E., 458–460

INDEX OF CONTRIBUTORS

Permenter, L.K., 395–397
Perry, K., 412–414
Peterson, Y.K., 356–357
Petrakis, I.L., 176–184
Pettibone, D.J., 481–483
Phebus, L., 412–414
Pierri, J.N., 102–112
Pilowsky, L.S., 364–367
Pineda, J.C., 346–348
Pinto, A., 36–52
Pomerleau, F., 454–457
Pratt, W.E., 159–168

Quiroz, J., 273–291

Rahman, Z., 372–374
Rausch, J.L., 381–385, 386–390
Robbe, D, 212–225
Roberto, M., 196–211
Rosenkranz, J.A., 53–74
Roth, B.L., 461–463
Rothman, D.L., 292–308, 452–453
Rubinchik, S., 419–421

Sanacora, G., 292–308
Sattin, A., 458–460
Schoepp, D.D., 309–317
Schütz, C., 176–184
Seamans, J.K., 445–447
Self, D.W., 372–374
Semenova, S., 415–418
Sesack, S.R., 36–52
Shannon, H., 412–414
Shen, H., 349–351
Shi, Q., 461–463
Shulman, R.G., 452–453
Siggins, G.R., 196–211
Singh, J., 273–291
Skolnick, P., 250–272

Skoubis, P.D., 415–418
Stech, N.E., 405–409
Stefani, M.R., 464–467
Sun, X., 241–249
Sur, C., 481–483
Swanson, C.J., 309–317
Sweet, R.A., 102–112
Szabo, S., 378–380
Szumlinski, K., 169–175, 468–471

Tamminga, C.A., 113–118
Tang, K-C., 226–240
Tang, X.-C., 472–475
Toda, S., 349–351, 468–471
Trevisan, L., 176–184
Tsai, G., 318–327
Tseng, K.-Y., 476–478

Valenti, O., 435–437, 479–480

Webster, M.J., 426–430
Weickert, C.S., 426–430
West, A.R., 53–74
Williams, D.L., Jr., 435–437, 481–483
Wisnoski, D.D., 481–483
Wolf, M.E., xiii–xiv, 241–249
Worley, P.F., 468–471
Wroblewski, J.T., 461–463

Xi, Z-X., 169–175

Yu, H., 412–414
Yuan, P., 378–380, 402–404

Zarate, C.A., Jr., 273–291